Lecture Notes in Computer Science 11229

Commenced Publication in 1973
Founding and Former Series Editors:
Gerhard Goos, Juris Hartmanis, and Jan van Leeuwen

More information about this series at http://www.springer.com/series/7408

Hervé Panetto · Christophe Debruyne
Henderik A. Proper · Claudio Agostino Ardagna
Dumitru Roman · Robert Meersman (Eds.)

On the Move to Meaningful Internet Systems

OTM 2018 Conferences

Confederated International Conferences:
CoopIS, C&TC, and ODBASE 2018
Valletta, Malta, October 22–26, 2018
Proceedings, Part I

 Springer

Editors
Hervé Panetto
CNRS
University of Lorraine
Vandoeuvre-les-Nancy, France

Christophe Debruyne
Trinity College Dublin
Dublin, Ireland

Henderik A. Proper
Luxembourg Institute of Science
and Technology
Esch-sur-Alzette, Luxembourg

Claudio Agostino Ardagna
Università degli Studi di Milano
Crema, Italy

Dumitru Roman
SINTEF and University of Oslo
Oslo, Norway

Robert Meersman
TU Graz
Graz, Austria

ISSN 0302-9743   ISSN 1611-3349 (electronic)
Lecture Notes in Computer Science
ISBN 978-3-030-02609-7   ISBN 978-3-030-02610-3 (eBook)
https://doi.org/10.1007/978-3-030-02610-3

Library of Congress Control Number: 2018957822

LNCS Sublibrary: SL2 – Programming and Software Engineering

This Springer imprint is published by the registered company Springer Nature Switzerland AG
The registered company address is: Gewerbestrasse 11, 6330 Cham, Switzerland

General Co-chairs and Editors' Message for OnTheMove 2018

The OnTheMove 2018 event held October 22–26 in Valletta, Malta, further consolidated the importance of the series of annual conferences that was started in 2002 in Irvine, California. It then moved to Catania, Sicily in 2003, to Cyprus in 2004 and 2005, Montpellier in 2006, Vilamoura in 2007 and 2009, in 2008 to Monterrey, Mexico, to Heraklion, Crete in 2010 and 2011, Rome in 2012, Graz in 2013, Amantea, Italy, in 2014, and lastly to Rhodes in 2015, 2016, and 2017.

This prime event continues to attract a diverse and relevant selection of today's research worldwide on the scientific concepts underlying new computing paradigms, which of necessity must be distributed, heterogeneous, and supporting an environment of resources that are autonomous yet must meaningfully cooperate. Indeed, as such large, complex, and networked intelligent information systems become the focus and norm for computing, there continues to be an acute and increasing need to address the software, system, and enterprise issues involved and discuss them face to face in an integrated forum that covers methodological, semantic, theoretical, and application issues. As we all realize, e-mail, the Internet, and even video conferences are not by themselves optimal or even sufficient for effective and efficient scientific exchange.

The OnTheMove (OTM) International Federated Conference series has been created precisely to cover the scientific exchange needs of the communities that work in the broad yet closely connected fundamental technological spectrum of Web-based distributed computing. The OTM program every year covers data and Web semantics, distributed objects, Web services, databases, information systems, enterprise workflow and collaboration, ubiquity, interoperability, mobility, as well as grid and high-performance computing.

OnTheMove is proud to give meaning to the "federated" aspect in its full title1: it aspires to be a primary scientific meeting place where all aspects of research and development of internet- and intranet-based systems in organizations and for e-business are discussed in a scientifically motivated way, in a forum of interconnected workshops and conferences. This year's 15th edition of the OTM Federated Conferences event therefore once more provided an opportunity for researchers and practitioners to understand, discuss, and publish these developments within the broader context of distributed, ubiquitous computing. To further promote synergy and coherence, the main conferences of OTM 2018 were conceived against a background of their three interlocking global themes:

- Trusted Cloud Computing Infrastructures Emphasizing Security and Privacy
- Technology and Methodology for Data and Knowledge Resources on the (Semantic) Web
- Deployment of Collaborative and Social Computing for and in an Enterprise Context.

Originally the federative structure of OTM was formed by the co-location of three related, complementary, and successful main conference series: DOA (Distributed Objects and Applications, held since 1999), covering the relevant infrastructure-enabling technologies, ODBASE (Ontologies, DataBases and Applications of Semantics, since 2002) covering Web semantics, XML databases and ontologies, and of course CoopIS (Cooperative Information Systems, held since 1993), which studies the application of these technologies in an enterprise context through, e.g., workflow systems and knowledge management. In the 2011 edition, security aspects, originally started as topics of the IS workshop in OTM 2006, became the focus of DOA as secure virtual infrastructures, further broadened to cover aspects of trust and privacy in so-called Cloud-based systems. As this latter aspect came to dominate agendas in this and overlapping research communities, we decided in 2014 to rename the event to the "Cloud and Trusted Computing (C&TC) Conference," and it was originally launched in a workshop format.

These three main conferences specifically seek high-quality contributions of a more mature nature and encourage researchers to treat their respective topics within a framework that simultaneously incorporates (a) theory, (b) conceptual design and development, (c) methodology and pragmatics, and (d) applications in particular case studies and industrial solutions.

As in previous years, we again solicited and selected additional quality workshop proposals to complement the more mature and "archival" nature of the main conferences. Our workshops are intended to serve as "incubators" for emergent research results in selected areas related, or becoming related, to the general domain of Web-based distributed computing. We were very glad to see that our earlier successful workshops (EI2N, META4eS, FBM) re-appeared in 2018. The Fact-Based Modeling (FBM) workshop in 2015 succeeded and expanded the scope of the successful earlier ORM workshop. The Industry Case Studies Program, started in 2011, under the leadership of Hervé Panetto, Wided Guédria, and Gash Bhullar, further gained momentum and visibility in its seventh edition this year.

The OTM registration format ("one workshop and/or conference buys all workshops and/or conferences") actively intends to promote synergy between related areas in the field of distributed computing and to stimulate workshop audiences to productively mingle with each other and, optionally, with those of the main conferences. In particular EI2N continues to so create and exploit a visible cross-pollination with CoopIS.

We were very happy to see that in 2018 the number of quality submissions for the OnTheMove Academy (OTMA) noticeably increased. OTMA implements our unique, actively coached and therefore very time- and effort-intensive formula to bring PhD students together, and aims to carry our "vision for the future" in research in the areas covered by OTM. Its 2018 edition was organized and managed by a dedicated team of collaborators and faculty, Peter Spyns, Maria-Esther Vidal, inspired as always by the OTMA Dean, Erich Neuhold.

In the OTM Academy, PhD research proposals are submitted by students for peer review; selected submissions and their approaches are presented by the students in front of a wider audience at the conference, and are independently and extensively analyzed and discussed in front of this audience by a panel of senior professors. One may readily appreciate the time, effort, and funds invested in this by OnTheMove and especially by the OTMA Faculty.

As the three main conferences and the associated workshops all share the distributed aspects of modern computing systems, they experience the application pull created by the Internet and by the so-called Semantic Web, in particular developments of big data, increased importance of security issues, and the globalization of mobile-based technologies.

The three conferences seek exclusively original submissions that cover scientific aspects of fundamental theories, methodologies, architectures, and emergent technologies, as well as their adoption and application in enterprises and their impact on societally relevant IT issues.

- CoopIS 2018, Cooperative Information Systems, our flagship event in its 26th edition since its inception in 1993, invited fundamental contributions on principles and applications of distributed and collaborative computing in the broadest scientific sense in workflows of networked organizations, enterprises, governments, or just communities
- C&TC 2018 (Cloud and Trusted Computing 2018) the successor of DOA (Distributed Object Applications), focused on critical aspects of virtual infrastructure for cloud computing, specifically spanning issues of trust, reputation, and security
- ODBASE 2018, Ontologies, Databases, and Applications of Semantics covered the fundamental study of structured and semi-structured data, including linked (open) data and big data, and the meaning of such data as is needed for today's databases; as well as the role of data and semantics in design methodologies and new applications of databases

As with the earlier OnTheMove editions, the organizers wanted to stimulate this cross-pollination by a program of engaging keynote speakers from academia and industry and shared by all OTM component events. We are quite proud to list for this year:

- Martin Hepp, Universität der Bundeswehr Munich/Hepp Research GmbH
- Pieter De Leenheer, Collibra
- Richard Mark Soley, Object Management Group, Inc. (OMG)
- Tom Raftery, SAP/Instituto Internacional San Telmo

The general downturn in submissions observed in recent years for almost all conferences in computer science and IT has also affected OnTheMove, but this year the harvest again stabilized at a total of 173 submissions for the three main conferences and over 50 submissions in total for the workshops. Not only may we indeed again claim success in attracting a representative volume of scientific papers, many from the USA and Asia, but these numbers of course allow the respective Program Committees (PCs) to again compose a high-quality cross-section of current research in the areas covered by OTM. Acceptance rates vary but the aim was to stay consistently at about 1 accepted full paper for 3 submitted, yet as always these rates are subordinated to professional peer assessment of proper scientific quality.

As usual we separated the proceedings into two volumes with their own titles, one for the main conferences and one for the workshops and posters. But in a different approach to previous years, we decided the latter should appear post-event and so allow workshop authors to eventually improve their peer-reviewed papers based on critiques by the PCs and on live interaction at OTM. The resulting additional complexity and effort of editing the proceedings were professionally shouldered by our leading editor, Christophe Debruyne, with the general chairs for the conference volume, and with Hervé Panetto for the workshop volume. We are again most grateful to the Springer LNCS team in Heidelberg for their professional support, suggestions, and meticulous collaboration in producing the files and indexes ready for downloading on the USB sticks. It is a pleasure to work with staff that so deeply understands the scientific context at large, and the specific logistics of conference proceedings publication.

The reviewing process by the respective OTM PCs was performed to professional quality standards: Each paper review in the main conferences was assigned to at least three referees, with arbitrated e-mail discussions in the case of strongly diverging evaluations. It may be worthwhile to emphasize once more that it is an explicit OnTheMove policy that all conference PCs and chairs make their selections in a completely sovereign manner, autonomous and independent from any OTM organizational considerations. As in recent years, proceedings in paper form are now only available to be ordered separately.

The general chairs are once more especially grateful to the many people directly or indirectly involved in the setup of these federated conferences. Not everyone realizes the large number of qualified persons that need to be involved, and the huge amount of work, commitment, and the financial risk in the uncertain economic and funding climate of 2018, that is entailed in the organization of an event like OTM. Apart from the persons in their roles mentioned above, we therefore wish to thank in particular explicitly our main conference PC chairs:

- CoopIS 2018: Henderik A. Proper, Markus Stumptner and Samir Tata
- ODBASE 2018: Dumitru Roman, Elena Simperl, Ahmet Soylu, and Marko Grobelnik
- C&TC 2018: Claudio A. Ardagna, Adrian Belmonte, and Mauro Conti

And similarly we thank the PC (co-)chairs of the 2018 ICSP, OTMA, and Workshops (in their order of appearance on the website): Wided Guédria, Hervé Panetto, Markus Stumptner, Georg Weichhart, Peter Bollen, Stijn Hoppenbrouwers, Robert Meersman, Maurice Nijssen, Gash Bhullar, Ioana Ciuciu, Anna Fensel, Peter Spyns, and Maria-Esther Vidal.

Together with their many PC members, they performed a superb and professional job in managing the difficult yet vital process of peer review and selection of the best papers from the harvest of submissions. We all also owe a serious debt of gratitude to our supremely competent and experienced conference secretariat and technical admin staff in Guadalajara and Dublin, respectively, Daniel Meersman and Christophe Debruyne.

The general conference and workshop co-chairs also thankfully acknowledge the academic freedom, logistic support, and facilities they enjoy from their respective institutions: Technical University of Graz, Austria; Université de Lorraine, Nancy, France; Latrobe University, Melbourne, Australia—without which such a project quite simply would not be feasible. Reader, we do hope that the results of this federated scientific enterprise contribute to your research and your place in the scientific network, and we hope to welcome you at next year's event!

September 2018

Robert Meersman
Tharam Dillon
Hervé Panetto
Ernesto Damiani

Organization

OTM (On The Move) is a federated event involving a series of major international conferences and workshops. These proceedings contain the papers presented at the OTM 2018 Federated conferences, consisting of CoopIS 2018 (Cooperative Information Systems), C&TC 2018 (Cloud and Trusted Computing), and ODBASE 2018 (Ontologies, Databases, and Applications of Semantics).

Executive Committee

OTM Conferences and Workshops General Chairs

Robert Meersman	TU Graz, Austria
Tharam Dillon	La Trobe University, Melbourne, Australia
Hervé Panetto	University of Lorraine, France
Ernesto Damiani	Politecnico di Milano, Italy

OnTheMove Academy Dean

Erich Neuhold	University of Vienna, Austria

Industry Case Studies Program Chairs

Hervé Panetto	University of Lorraine, France
Wided Guédria	LIST, Luxembourg
Gash Bhullar	Control 2K Limited, UK

CoopIS 2018 PC Co-chairs

Henderik A. Proper	LIST, Luxembourg
	University of Luxembourg, Luxembourg
	Radboud University, The Netherlands
Markus Stumptner	University of South Australia, Australia
Samir Tata	IBM Reasearch, USA

C&TC 2018 PC Co-chairs

Claudio A. Ardagna	Università degli Studi di Milano, Italy
Adrian Belmonte	ENISA, Greece
Mauro Conti	University of Padua, Italy

ODBASE 2018 PC Chair and Vice-chairs

Dumitru Roman	SINTEF and University of Oslo, Norway
Elena Simperl	University of Southampton, UK
Ahmet Soylu	NTNU and DNV GL, Norway
Marko Grobelnik	Jozef Stefan Institute, Slovenia

Publication Chair

Christophe Debruyne   Trinity College Dublin, Ireland

Logistics Team

Daniel Meersman

CoopIS 2018 Program Committee

Agnes Nakakawa
Alex Norta
Amal Elgammal
Amel Bouzeghoub
Amel Mammar
Andreas Opdahl
Athman Bouguettaya
Baazaoui Hajer Baazaoui
Barbara Pernici
Barbara Weber
Bas van Gils
Beatrice Finance
Bruno Defude
Carlo Combi
Chengzheng Sun
Chirine Ghedira
Daniel Florian
Daniela Grigori David Aveiro
Djamal Benslimane
Doing Hai Eduard Babkin
Elisa Yumi Nakagawa
Epaminondas Kapetanios
Ernesto Damiani
Ernesto Exposito
Eva Kühn
Faouzi Ben Charrada
Francisca Peréz
Francisco Moo Mena
Francois Charoy
George Feuerlicht
Gerald Oster
Gil Regev
Hamid Motahari Nezhad
Huemer Christian
Jan Mendling

Janusz Szpytko
Jean-Sebastien Sottet
Jian Yang
Joao Paulo Almeida
Jolita Ralyte
Joonsoo Bae
Josephine Nabukenya
Joyce El Haddad
Juan Manuel
Julius Köpke
Kais Klai
Khalid Belhajjame
Khalil Drira
Laura Margarita Rodríguez Peralta
Liang Zhang
Lijie Wen
Lucinia Heloisa Thom
Luis Garrido Jose
Manfred Reichert
Marcelo Fantinato
Marco Aiello
Marco Comuzzi
Markus Stumptner
Marlon Dumas
Massimo Mecella
Matulevicius Raimundas
Mehdi Ahmed-Nacer
Messai Nizar
Michael Mrissa
Michael Rosemann
Michael Sheng
Michele Missikoff
Mohamed Graiet
Mohamed Mohamed
Mohamed Sellami

Mohand-Said Hacid
Narjes Bellamine-Ben Saoud
Nour Assy
Noura Faci
Oana Balan
Olga Nabuco
Olivier Perrin
Oscar Pastor
Pnina Soffer
Raibulet Claudia
Richard Chbeir
Rik Eshuis
Romero David
Rüdiger Pryss
Salima Benbernou
Sami Bhiri
Sami Yangui
Sanjay K. Madria

Saul Pomares
Schahram Dustdar
Sebastian Steinau
Selmin Nurcan
Shazia Sadiq
Sietse Overbeek
Stefan Jablonski
Stefan Schönig
Stephan Aier
Stijn Hoppenbrouwers
Sybren de Kinderen
Valérie Issarny
Walid Gaaloul
Xavier Blanc
Yemna Sayeb
Zhangbing Zhou
Zohra Bellahsene

CoopIS 2018 Additional Reviewers

Amartya Sen
Azadeh Ghari Neiat
Bryden Da Yang Cho
Carlos Azevedo
Carlos Habekost dos Santos
Chahrazed Labba
Chaima Ghribi
Chamseddine Hamdeni
Diego Toralles Avila
Duarte Gouveia
Elio Mansour
Emna Hachicha Belghith
Fadoua Ouamani
Georg Grossmann
Geri Joskowicz
Guillaume Rosinosky
Haithem Mezni
Joaquin Garcia Alfaro
Josephine Nabukenya
Julio Cesar Nardi
Karamjit Kaur
Kevin Andrews
Lara Kallab

Laura Rodríguez
Lil Rodríguez Henríquez
Maha Riad
Marc Schickler
Marco Franceschetti
Martina Sengstschmid
Matt Selway
Michael Zimoch
Mohamed Ramzi Haddad
Olga Nabuco
Pavel Malyzhenkov
Quentin Laporte-Chabasse
Robin Kraft
Sarra Slimani
Sebastian Steinau
Slim Kallel
Stefan Crass
Victorio Carvalho
Wei Steve Wang
Weiliang Zhao
Weiwei Cai
Yemna Sayeb

C&TC 2018 Program Committee

Alberto Compagno
Belmonte Adrian
Chia-Mu Yu
Christos Xenakis
Claudio A. Ardagna
Claus Pahl
Conti Mauro
Daniele Sgandurra
David Chadwick
Ernesto Damiani
Eugenia Nikolouzou
Francesco Di Cerbo
George Karabatis
Gwanggil Jeon
Joerg Schwenk

Julian Schutte
Luca Vigano
Luis Vega
Marco Anisetti
Marit Hansen
Meiko Jensen
Michele Bezzi
Miguel Vargas Martin
Nabil El Ioini
Patrick Hung
Pierluigi Gallo
Rasool Asal
Scharam Dustdar
Stefan Schulte
Stefanos Gritzalis

CTC 2018 Additional Reviewers

Cedric Hebert
Christos Kalloniatis
Vaios Bolgouras

ODBASE 2018 Program Committee

Ademar Crotti Junior
Ahmet Soylu
Alessandra Mileo
Alfredo Maldonado
Andreas Harth
Anna Fensel
Annika Hinze
Antonis Bikakis
Axel Ngonga
Carlos A. Iglesias
Christoph Bussler
Christoph Benzmuller
Christophe Debruyne
Costin Badica
Cristina Feier
Csaba Veres
Dieter Fensel
Dietrich Rebholz
Dimitris Plexousakis

Divna Djordjevic
Dumitru Roman
Elena Simperl
Evgenij Thorstensen
Fabrizio Orlandi
Flavio De Paoli
Georg Rehm
George Vouros
George Konstantinidis
Giorgos Stoilos
Giorgos Stamou
Grigoris Antoniou
Guido Governatori
Harald Sack
Harry Halpin
Ioan Toma
Irene Celino
Irini Fundulaki
Jacek Kopecky

James Hodson
Jan Jurjens
Juan Miguel Gomez Berbis
Judie Attard
Kai-Uwe Sattler
Luis Ibanez Gonzalez
Manolis Koubarakis
Marko Tadic
Marko Grobelnik
Markus Stumptner
Markus Luczak-Roesch
Martin Hepp
Matteo Palmonari
Mihhail Matskin
Nick Bassiliades
Nikolay Nikolov

Oscar Corcho
Paul Fodor
Ruben Verborgh
Simon Scerri
Simon Krek
Soeren Auer
Stefano Pacifico
Stefano Modafieri
Steffen Lamparter
Sung-Kook Han
Till C. Lech
Tomi Kauppinen
Uli Sattler
Vadim Ermolayev
Vladimir Alexiev
Witold Abramowicz

ODBASE 2018 Additional Reviewers

Blerina Spahiu
Federico Bianchi
Mohammed Nadjib Mami
Volker Hoffmann
Yuchen Zhao

OnTheMove 2018 Keynotes

Web Ontologies: Lessons Learned from Conceptual Modeling at Scale

Martin Hepp

Universität der Bundeswehr Munich/Hepp Research GmbH, Germany

Short Bio

Martin Hepp is a professor of E-business and General Management at the Universität der Bundeswehr Munich and the CEO and Chief Scientist of Hepp Research GmbH. He holds a master's degree in business management and business information systems and a PhD in business information systems from the University of Würzburg (Germany). His key research interests are shared data structures at Web scale, for example Web ontology engineering, both at the technical, social, and economical levels, conceptual modeling in general, and data quality management. As part of his research, he developed the GoodRelations vocabulary, an OWL DL ontology for data interoperability for e-commerce at Web Scale. Since 11/2012, GoodRelations is the e-commerce core of schema.org, the official data markup standard of major search engines, namely Google, Yahoo, Bing, and Yandex. Martin authored more than 80 academic publications and was the organizer of more than fifteen workshops and conference tracks on conceptual modeling, Semantic Web topics, and information systems, and a member of more than sixty conference and workshop program committees, including ECIS, EKAW, ESWC, IEEE CEC/EEE, ISWC, and WWW.

Talk

Ever since the introduction of the term "ontology" to Computer Science, the challenges for information exchange, processing, and intelligent behavior on the World Wide Web, with its vast body of content, huge user base, linguistic and representational heterogeneity, and so forth, have been taken as a justification for ontology-related research. However, despite two decades of work on ontologies in this context, very few ontologies have emerged that are used at Web scale in a way compliant with the original proposals by a diverse, open audience.

In this talk, I will analyze the differences between the original idea of ontologies in computer science, and Web ontologies, and analyze the specific economic, social, and technical challenges of building, maintaining, and using socially agreed, global data structures that are suited for the Web at large, also with respect to the skills, expectations, and particular needs of owners of Web sites and potential consumers of Web data.

Data Governance: The New Imperative to Democratize Data Science

Pieter De Leenheer

Collibra, USA

Short Bio

Pieter De Leenheer is a cofounder of Collibra and leads the company's Research & Education group, including the Collibra University, which offers a range of self-paced learning and certification courses to help data governance professionals and data citizens gain new skills and expertise. Prior to co-founding the company, Pieter was a professor at VU University of Amsterdam. Today he still serves as adjunct professor at Columbia University in the City of New York and as visiting scholar at several universities across the globe including UC San Diego and Stanford.

Talk

We live in the age of abundant data. Through technology, more data is available, and the processing of that data easier and cheaper than ever before. Data science emerged from an unparalleled fascination to empirically understand and predict societies businesses and markets. Yet there is an understated risk inherent to democratizing data science such as data spills, cost of data exploration, and blind trust in unregulated, incontestable and oblique models. In their journey to unlock competitive advantage and maximize value from the application of big data, it is vital that data leaders find the right balance between value creation and risk exposure. To realize the true value of this wealth of data, data leaders must not act impulsively, but rethink assumptions, processes, and approaches to managing, governing, and stewarding that data. And to succeed, they must deliver credible, coherent, and trustworthy data and data access clearing mechanisms for everyone who can use it. As data becomes the most valuable resource, data governance delivers a imperative certification for any business to trust one another, but also increasingly sets a precondition for any citizen to engage in a trustworthy and endurable relationship with a company or government.

Learning to Implement the Industrial Internet

Richard Mark Soley

Object Management Group, Inc. (OMG), USA

Short Bio

Dr. Richard Mark Soley is Chairman and CEO of the Object Management Group (r), also leading the Cloud Standards Customer Council (tm) the Industrial Internet Consortium (r). Previously cofounder and former Chairman/CEO of A. I. Architects, he worked for technology companies and venture firms like TI, Gold Hill, Honeywell & IBM. Dr. Soley has SB, SM and PhD degrees in Computer Science and Engineering from MIT.

A longer bio is available here: http://www.omg.org/soley/

Talk

- The Industrial Internet Consortium and its members develop testbeds to learn more about Industrial Internet implementation: hiring, ecosystem development, standards requirements
- The Object Management Group takes real-world standards requirements and develops standards to maximize interoperability and portability
- The first insights into implementation and the first requirements for standards are underway
- Dr. Soley will give an overview of both processes and talk about the first insights from the projects

The Future of Digital: What the Next 10 Years Have in Store

Tom Raftery

SAP/Instituto Internacional San Telmo, Spain

Short Bio

Tom Raftery is a Global Vice President for multinational software corporation SAP, an adjunct professor at the Instituto Internacional San Telmo, and a board advisor for a number of start-ups.

Before joining SAP Tom worked as an independent industry analyst focusing on the Internet of Things, Energy and CleanTech and as a Futurist for Gerd Leonhardt's Futures Agency.

Tom has a very strong background in technology and social media having worked in the industry since 1991. He is the co-founder of an Irish software development company, a social media consultancy, and is co-founder and director of hyper energy-efficient data center Cork Internet eXchange – the data centre with the lowest latency connection between Europe and North America.

Tom also worked as an Analyst for industry analyst firm RedMonk, leading their GreenMonk practice for over 7 years. Tom serves on the Advisory Boards of Smart-Cities World and RetailEverywhere.com.

Talk

Digital Transformation, the Internet of Things and associated technologies (block-chain, machine learning, edge computing, etc.) are the latest buzz words in technology. Organisations are scrambling to get up to speed on them before their competitors, or some young start-up gets there first and completely disrupts them.

Right now, these digital innovation systems are, roughly speaking at the same level of maturity as the web was in 1995. So where are these new digital technologies taking us? What is coming down the line, and how will these changes affect my organisation, my wallet, and the planet?

Join Tom Raftery for our OnTheMove keynote as he unpacks what the Future of Digitisation is going to bring us.

Contents – Part I

Contents – Part II

ODBASE Experience Papers

ODBASE Short Papers

International Conference on Cooperative Information Systems (CoopIS) 2018

CoopIS 2018 PC Co-chairs' Message

CoopIS 2018 (the 26th CoopIS) takes place in Valetta on the beautiful island of Malta. The CoopIS series of conferences can look back upon a quarter of a century of scientific excellence and industrial success. The longevity of CoopIS demonstrates its continuing success in addressing the key challenges in the engineering of Cooperative Information Systems, a topic which is as relevant as it ever was.

Over the past quarter of a century, the CoopIS conference series has witnessed a plethora of IT trends come and go. Above, and beyond these trends, the CoopIS conference remains committed to the deeper challenge of achieving seamless cooperation of the socio-cyber-physical mix of actors that make up the ecosystem involved in the execution, development, and maintenance of modern day information systems.

As in previous years, this year we also were able to produce a high quality conference programme, thanks to the many high-quality submissions. A total of 123 papers were submitted, of which 37 were accepted as regular papers, and 12 as research in progress papers. The authors of the accepted papers originate from many countries and cultures around the world, in keeping with the tradition of the international nature of CoopIS.

We would like to thank everyone who contributed to the success of CoopIS 2018. In particular We thank both the authors who contributed their papers on their research to CoopIS 2018, and the PC members and additional reviewers who have reviewed the submissions in a timely manner while providing valuable and constructive feedback to the authors.

September 2018

Henderik A. Proper
Samir Tata
Markus Stumptner

Digitization of Government Services: Digitization Process Mapping

Heloise Acco Tives Leão(✉)(iD), Edna Dias Canedo(✉)(iD),
and João Carlos Felix Souza(✉)

Computer Science Department - Professional Masters in Applied Computing,
University of Brasília (UnB), P.O. Box 4466, Brasília-DF 70910-900, Brazil
heloise.acco@gmail.com, {ednacanedo,jocafs}@unb.br

Abstract. The search for improvement and standardization of the digitization of government services has led governments around the world to focus on solutions that seek satisfaction, engagement and involvement of society in general. In addition, governmental systems are seeking constantly to renew the digital governance environment through good planning, use of best practices, and offering greater opportunities to establish collaborative and participatory relationships among all stakeholders (Government and Society). This paper presents a systematic literature review of the digitization of services contributing to the knowledge of the processes and methodologies adopted by these governments to provide their services to the citizen. The main contribution of this work is the proposal of a process mapping model that can be adopted during the stages of providing digital services by interested agencies in offering services focused on the needs of the citizens. The proposed model can be used by any government agency or private company interested in updating its processes, tools and methods of digitization and services automation according to their necessities.

Keywords: Service digitization · Government services
Systematic literature review · Digitization process mapping

1 Introduction

The digitization of services has emerged as a way to provide services with greater efficiency, efficiency and quality [13] and with less bureaucracy existing in the current processes [12]. Governments around the world have a renewed focus on citizens' perception and engagement, and a well-planned digital government environment offers greater opportunities for building collaborative and participatory relationships among all relevant stakeholders.

In the case of public services, access must be universal, in other words, available to all citizens regardless of income, level of education, geographical location or conditions of access to technological resources [13]. Providing services under such conditions is a challenge for governments around the world. The major

© Springer Nature Switzerland AG 2018
H. Panetto et al. (Eds.): OTM 2018 Conferences, LNCS 11229, pp. 3–20, 2018.
https://doi.org/10.1007/978-3-030-02610-3_1

trends fueling the public sector are driven by rising citizen expectations, which in turn drives four key aspects: Pressure to deliver for more consumer-like citizen services; Need to refocus resources in areas that boost government program delivery and make it visible to citizens; Drive to improve citizen outcomes and install a government culture of service excellence and accountability; Necessity to diversify the economy, attract and nurture new businesses utilizing new business models under the umbrella of government as a facilitator.

Driven by these pressures, government objectives cannot be limited to just the introduction of digital technologies and process automation within departments. It goes far beyond that, requiring a focused effort on digitally engaging citizens to modernize the public sector as a whole. A key measure of success for modern countries is the level of engagement that its people undertake with their government. As the OECD (Organization for Economic Cooperation and Development) states, good decision-making requires the knowledge, experiences, views, and values of the public, and unless citizens themselves understand and are engaged in the decision-making, trust is easily lost.

There are several benefits to citizens driving public policy reform and modernization as part of a digital government transformation. Citizen engagement drives the success of e-government, or digital government, by increasing the acceptance and uptake with the government through digital channels. This helps departments scale up services while reducing cost without compromising sustainability. It improves governance and creates a more informed government. This marks the shift in viewing citizens as customers of the government rather than subjects, which dictates a higher degree of interaction and engagement.

Engaged citizens can make important contributions to policies and programs related to every aspect of city life and government services. It reinforces government success by introducing a critical and honest feedback mechanism, building public trust in their leadership [24]. With regard to Brazil, the government has sought to encourage its Agencies to transform their digital services for access, sharing and monitoring of information, registration of demands and requests for official documents. The main objective of the Brazilian Government is to have a Digital Government From Citizen-Centric To Citizen-Driven, modernizing all public services. Since 2016, important decrees have been published in this sense, defining a Digital Governance Policy [21] and the Digital Citizenship Platform [8] in the scope of the Federal Public Administration.

The Digital Citizenship Platform [8] aims to broaden and simplify the access of Brazilian citizens to digital public services, including through mobile devices. The Federal Government Service Portal should be a single integrated channel for the provision of information, electronic request and monitoring of public services, whose objective is, in addition to providing practicality and agility for citizens and entrepreneurs, for digital services to reduce the cost to government.

The actions of the Platform are aligned with the Digital Governance Strategy (DGS) [22, 27] that will guide the actions of Information and Communications Technology (ICT) of the Brazilian Government until 2019. It is necessary to identify tools that measure the reach of the services not scanned in Brazil, to assist in the implementation and expansion of the services provision. These tools can help increase citizen participation in the use of current services and even in the development of new services. Tools for pricing the costs of all these operations, still need to be studied and/or even developed.

This paper presents a study about the digitization of Government services. Its main contribution is to identify in the literature how the governments of other countries are promoting the automation and digitization of their public services, as well as to present the best practices adopted for the process of government automation and digitization. In addition, it presents the registry of the technological solutions adopted in the processes used by success cases.

The remainder of this paper is organized as follows. Section 2 presents the research methodology adopted, the protocol and the result of the systematic literature review. In the Sect. 3 the results and the discussions are presented. Section 4 presents the proposed model of Digitization Process Mapping. The conclusions and future work are presented in Sect. 5.

2 Systematic Literature Review

Systematic Literature Review (SLR) is a way of identifying, analyzing and interpreting available evidence related to a particular research question, area or phenomena of interest [15,18]. During the SLR, the Planning, Conducting and Reporting phases of the results were followed [6,9,18].

The tool StArt (State of the Art through Systematic Review) [14], assisting in the planning and conduction stages of the Systematic Review Literature. The SLR was carried out with the objective of identifying and presenting the best practices and technologies currently adopted in the automation and digitization processes of services, as well as presenting guidelines on how to include automated and digitized processes in Brazilian federal public services.

The search strategy involved the use of Automatic Search [31], which consists of searching through the Search String in the electronic databases, followed by the Manual Search [31], through which searches were performed for works in conferences, newspapers or magazines. The Automatic Search was performed in the following databases: Digital library ACM; Digital Library IEEE Xplore; DBLP-Computer Science Bibliography. The Manual Search activity was performed in the Annals of the Conferences and Periodicals specific to the e-government area.

2.1 Selection of Primary Studies

We started the automatic selection process of the primary studies by executing the search string in the digital databases. The automatic search in the 3 bases defined resulted in a total of **727 articles**, being **354 or 49% of articles** from Digital library ACM and **255 or 35% of articles** came from Digital library IEEE Xplore and **118 or 16% of articles** came from

Table 1. Evolution of the work selection strategy

Adopted strategy	Stage 1	Stage 2	Stage 3	Stage 4
Automatic search	727	118	65	19
Manual search	56	40	31	7
Selected primary studies	**783**	**158**	**96**	**26**

Digital library DBLP. It is important to note that only **6 papers** were identified as duplicates, and the occurrence of duplicity occurred on the basis of ACM and DBLP. The Table 1 presents the evolution of the steps adopted in the selection strategy, which were applied in the papers identified during the automatic search.

The manual search performed in the Annals of Conferences and Periodicals was performed through various combinations of the search string defined in the protocol. This variation of the String was necessary because the bases of the Conferences and Periodicals have a smaller volume of publications and with the complete String often no work was identified. The manual search resulted in a total of **56 articles**, which followed the stages defined in the adopted protocol, adding **783**. The evolution of the selection of these in the systematic literature review (SLR) is presented on the Table 1. After the final application of all stages of the work selection strategy, a total of **26 articles** were identified to be used in data extraction.

3   Results Systematic Literature Review

This **research identified 26 primary studies**. The data extraction occurred in all 26 articles selected in the last step of the strategy defined in the protocol of this research. From the complete reading of these articles it was possible to answer the research questions.

RQ.1. How to promote the automation and digitization of federal public services in Brazil?

The deployment of government-digitized services requires more than technological sophistication, requires a shift in the mindset of public administration to citizen-centered service delivery [29]. The work presented by [32], highlights 5 key methods for the digitization of services by governments, being:

1. All citizens should be taken into account by promoting reliable, innovative and easily accessible services for all;
2. Efficiency and effectiveness must be a reality in the services provided, contributing to the high satisfaction, transparency and responsibility of users, relieving management and providing quality gains and resource savings;
3. The services implemented must start with the essential and high impact for citizens and companies. In order to identify such services, citizens and society must be included in the process of definition of scope and design;

4. Enabling elements should be added to services, enabling citizens and businesses to benefit from convenient, secure and interoperable forms of access;
5. Participation in developing ideas and choosing priorities should be democratic, using tools for effective public debate and empowering citizens and society in decision-making.

These key methods can be followed by the Brazilian government in order to allow the expansion and improvement of the digitized services provided. Other factors that may contribute to this process are: 1. Analysis, measurement and quality assurance of the ways of making these services available. For this, the user's perceptions about technology, satisfaction and trust must be taken into account [1]; 2. Providing services tailored to the needs of each citizen, respecting their profile (which may be related to age group, educational level, economic situation and others) and from this determine the amount of information and the level of detail to be provided [11].

The choice of service delivery channels is another issue that should be widely evaluated by the Brazilian government to support the digitization of services. Communication technologies have been evolving over the years, starting with traditional and personal communications, including options such as telephone and mail, reaching the use of the internet, social media and mobile phones with a wide variety of applications [26]. Artificial Intelligence techniques are also being inserted in this process, for example with the use of social and conversational robots that interact more and more with the citizens [26]. Currently, the most widely used means for delivering services digitized by governments and companies is the use of website portals, which deliver online services 24 h a day in any part of the world, provided there is connection with internet [10].

Other channels for providing services to citizens are: 1. Mobile devices to access portals or service applications [10]; 2. Self-service kiosks that have the option of delivering official documents [2]; 3. SMS (Short Messaging Service) for the delivery of public services, mainly warnings and information, in order to maintain proximity, connectivity, interactivity and continuous communication with all citizens [30]; 4. Social Networks that has a fast and wide reach of the population when it is necessary to disclose urgent information. This channel can also be used to identify citizens' needs not formally expressed [33]; 5. Chat used as a way of asking the citizen's doubts about various issues [33].

RQ.2. How to include in the automation process and the digitization of public services the citizens and agencies that provide the services?

Deciding which services are to be deployed, identifying which services need improvement and finding out why some services do not have the expected volume of access are tasks that can only be performed if they have the joint involvement of citizens, society at large, public servants and decision makers of the provider of this service. The inclusion of stakeholders in digitization is a good practice for the success of the process, since treating the citizen as a partner in government activities provides a number of contributions, such as: saving time, experiencing real need and great interest in achievement of positive results [19, 24]. In addition, citizen participation enhances the transparency, confidence, acceptability and legitimacy of decisions taken by policy makers [7]. Some initiatives that seek for this greater participation of the citizen and of the own agencies in the process of automation and digitization of services are:

- In addition to the application of technologies, it should be taken into account in the provision of digitized services 3 main categories: (1) the ability of decision-makers to communicate constantly with implementers in order to deliver as well as obtain the right information when needed; (2) competence of decision-makers to assign responsibilities to implementers in order to reduce bureaucracy and allow greater agility between processes; and (3) the ability to define clear rules in the provision and use of services, which should be widely disseminated to all stakeholders [23];
- The use of open data may be conducive to increasing the transparency of processes and can also be used to identify new demands for services to be made available. Some examples to achieve this increase in value for citizens can be: 1. Through the analysis of the questions answered by the citizens during the use of the services [25]; 2. Identification of the most accessed information [25]; 3. The mining of data to identify information expressed by citizens in non-official media, such as the use of social networks [25] and [7];
- The gamification techniques can be used to involve citizens in the process of ideas of new types of services, helping in the elaboration of new concepts of digital services or improvement of existing services [17].

In addition to involving the citizen in the participation of the development of services, it is also important to identify ways in which citizens are more interested in using these services. Some examples are:

- Keep the information and forms of access centralized in a single point, in addition to automating and simplifying the processes in order to make the citizen more independent to meet their needs [12];
- Information portals, booklets or other means of disseminating knowledge should not be static, rather, they should be able to deliver personalized information to citizens with the volume of data and details appropriate to each profile [11];
- Realization of more investments in marketing, advertising and promotion of services, being through different channels and with various forms of access. This seeks to raise the level of awareness and knowledge of citizens regarding the services provided by the Government [7];
- Demographic and socioeconomic conditions such as gender, age, formal education, economic income and political attitudes are factors that must be raised and understood so that the services are adequate to the different profiles existing for citizens [20]. Systems and services should be prepared to have user-friendly interfaces, adapting whenever possible to the profiles that have been identified [7];
- Providing public infrastructures, accessible and prepared to support the services offered, in order to guarantee availability and avoid access problems [7];

- Provide complete and high quality systems and processes, solving the needs of citizens in their entirety [7];
- Provide security and privacy of individuals and their data that must be kept intact and confidential [7];
- Gamification parts of the systems and services to thus involve and motivate the users to adopt the new processes with greater interest [3].

RQ.3. Which are the best practices adopted for the automation process and digitization of public services?

In the provision of digitized services, it is not enough to provide countless services without understanding the real needs of citizens and the factors that influence the use of this type of service. Among the factors that are pointed out as essential for the interaction of citizens with the digitized services, one can cite **quality, agility, privacy and security**, all those present at each stage of interaction between the interested parts [1]. It is worth emphasizing that quality must be present in different aspects, such as: **quality of the system, quality of service, quality of information, quality of content and product quality** [1]. The work presented by Bertot et. al [5] complements the list presented by Akram and Malik [1], of the essential elements to be evaluated in the implementation of digitized services, which are:

1. **Infrastructure:** Digital infrastructure is a necessary prerequisite, including robust digital technology infrastructure within governments, between citizens and industry. Without connectivity, access to systems and service applications is not possible;
2. **Capacity:** Different capacities, including organizational, human, regulatory, collaborative and other, must be present in all governments, industry, communities and citizens. These capabilities are needed to leverage the digital technology infrastructure and broadcast digital innovations;
3. **Ecosystems:** Innovative services, empowered by governments, should be part of a broader social innovation ecosystem, facilitating cultural change to adopt a positive attitude towards risk and product acceptance;
4. **Partnerships:** While governments may face challenges with their ability to innovate, they can take advantage of the innovative capabilities and resources of partners. Developing the capacity to partner with the private and non-profit sectors and engaging citizens in defining new services are important mechanisms for delivering innovative public services;
5. **Inclusion:** If innovative services must be ubiquitous and benefit all, they need to be available and usable by everyone. Implemented innovations should ensure that all actors have the ability to use and benefit from these services;
6. **Value:** Innovations must offer public value and be valued;
7. **Delivery Channels**: Many factors, including age, preferences, digital literacy, infrastructure, among others, affect the acceptance of digital services and opportunities for citizens to get involved. Therefore, several service delivery channels are required for engagement as well as multichannel delivery strategies to decide the most appropriate channels for each service;

8. **Security:** Digital service innovations can not be deployed without ensuring the security of interactions and stored content;

9. **Privacy:** Security focuses on content protection, while privacy belongs to citizens' ability to opt out of digital public services. Innovations can not be mandatory, but citizens must retain the right to select the services they wish to receive, use or wish to engage with. For this to happen, privacy must be ensured;

10. **Authentication:** Secure and verifiable authentication is required, but we also need appropriate authentication measures to ensure that the recipients of the service are indeed recipients. This requires layers of security and authentication across all services.

RQ.4. What technological solutions are adopted in the automation processes and digitization of public services?

The use of existing Information and Communication Technologies (ICT) is vital in the process of accelerating the digitization of services, providing services increasingly adapted to the individual needs of citizens, providing greater satisfaction and use of these services. This process is important for citizens, who will have access to a more convenient, preferable and economical form of interaction with public agencies. For governments, this approach is important because it allows for cross-departmental synchronization, decreasing queues, reducing response time and financial expenses [2].

New technologies arise at all times and are analyzed in order to identify benefits of their use in existing processes. One of the major concerns of governments and citizens' demands [10] is with respect to the security of the transactions and the data used in these operations. Some examples of innovative practices with respect to electronic government security are: 1. SecureGov is a mechanism that has been implemented in Korea's Public Information Sharing Center (PISC) [10]; 2. A prototype is being tested by the government of India in that the proposal is an integrated digital signature approach based on cloud computing to enable electronic authentication and data security in the transaction phase of digitized services [16]; 3. The Government of India also uses another mechanism involving the digital signature to perform authentication of the data through the temporary proxy signature, where the owner of the signature transfers the power of its use to a signatory authority during a specific period of time and any misuse of the resources is prevented through the signature key generation procedure [4]; 4. The government of Georgia has an exclusive department to deal with the cyber security of its digitized services, among its practices is the monitoring of the use of services and the volume of transactions carried out; 5. The government of Greece uses an online system for requesting services by citizens where the processing of the orders made takes place almost completely in an automated way, to reduce response time and minimize the need for human intervention, with the aim of avoiding fraud at any stage of the request process [13]; 6. The use of SmartGates, which are kiosks for citizens' recognition through the evaluation of biometric data and even the face which are common at airports and customs for validation of documents such as passports, is already being tested by several governments, such as Russia, which in self-service kiosks allows the citizen to make requests for documents and makes their delivery using this technology [2].

Continued innovation in the provision of public services is essential to meet the diverse social needs, raising social aspirations, economic pressure and unequal conditions for the provision of public services within and between countries. The results of this systematic literature review allowed us to identify several Government initiatives with this intention and to establish a panorama for the needs of the Brazilian Government, considering the best practices that were identified. In addition, a proposal for a model for the digitization of Brazilian public services was constructed.

4   Digitization Process Model for Government Services

With the accomplishment of the literature review it was possible to answer the research questions that were elaborated to conduct this work and to propose a service digitization model. The proposed model can be implemented by any government agency that wishes to offer digitized services to the citizen. Figure 1 shows the step flow of the model, which is composed of six phases, namely: Question, Customize, Innovate, Facilitate, Integrate and Communicate [28].

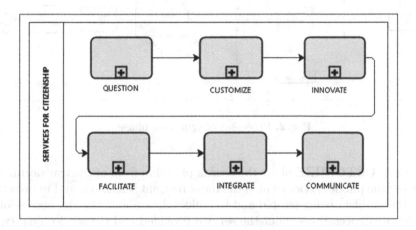

Fig. 1. Proposed model for digitization of public services.

During the **QUESTION** phase it is expected that the Federal Public Administration Body (FPA Body) will be able to identify its main services and the degree of maturity in the management of the services offered to the citizen. In this phase the following activities/tasks are indicated:

1. **Decide on the digitization of services:** the decision on the digitization of services should be strategic and involve all possible areas of the agency.

2. **Map services:** the complete mapping of the provided services situation must be carried out by surveying the existing systems, identifying the non-digitized provided services and verifying the requests that have not yet been met;
3. **Evaluate system interfaces:** the services already provided must provide a minimum set of information and have a friendly standard in their interfaces, which should be evaluated;
4. **Integrate data:** should be performed the integration of data or identify mechanisms to standardize its formats;
5. **Identify bottlenecks:** delays in processes, exaggerated bureaucracy or rework must be verified and recorded for improvements to be implemented.

The activity flow to be performed in the **QUESTION** phase of the proposed model is presented in Fig. 2.

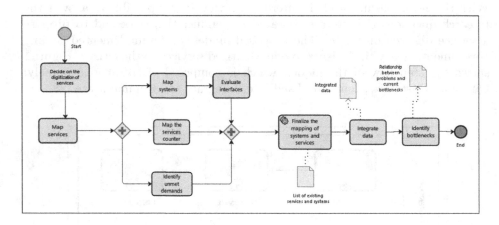

Fig. 2. Processes of Question phase

In the **CUSTOMIZE** phase the user is placed in front of the transformation process of the public service to map the most relevant sensations and impressions about the problem being treated and to collect data about the use, access form, satisfaction, expectations about the service provided and the service quality. In this phase the following activities/tasks are indicated:

1. **Understand the Actor/Citizen:** draw the general picture and the description of the different individuals, groups and organizations that are inter-related, directly or indirectly, with the service between them. The form of execution of this activity will be further detailed in Fig. 4;
2. **Identify communication channels:** check which communication methods will be used. They may be formally structured: surveys, interviews; formally unstructured: email and chat; informal: social networks;
3. **Collect needs:** map and classify quantitatively the priorities of services and/or functionalities to be implemented;

4. **Apply techniques to identify the profile of citizens:** profiles related to age group, educational level, economic situation, etc., must be identified to deliver the appropriate volume of information. Profiles related to difficulty of access or disability (auditory or visual) should be identified to fit the service delivery form - user type study (model according to the citizen/user profile);
5. **Provide ways to evaluate the services offered:** services must be continuously evaluated in order to identify changes or improvements to be developed, such as below-expected access, as well as to perceive infrastructure or security problems;
6. **Identify Changes and Improvements:** from the already performed services mapping, the modifications and improvements to be implemented should be listed and prioritized;
7. **Apply Improvements in provided services:** improvements or changes must be implemented to fit the perceptions of citizens;
8. **Implement mechanisms to maintain and/or encourage the use of services:** citizens should be encouraged to use services as if they were part of their daily lives.

The flow of activities proposed to be performed in the **CUSTOMIZE** phase is shown in Fig. 3.

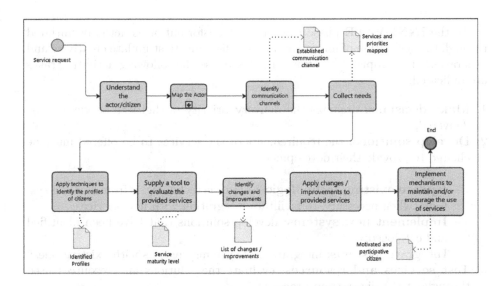

Fig. 3. Processes of Customize phase

The Actor Mapping process from the **CUSTOMIZE** phase consists of the identification and application of techniques to detect the citizens who use or can use the digitized services, as shown in Fig. 4. The techniques to be applied to identify the actors/citizens may include from conducting direct interviews with interested citizens or already users of the services, the definition of focus groups to delineate profiles of citizens, as well as through surveys and application of questionnaires with the objective of obtaining a greater range in the identification of the actors.

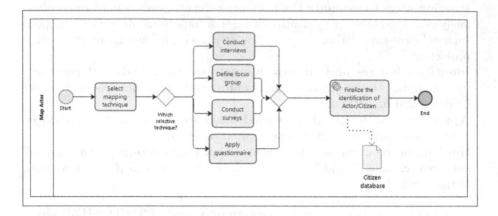

Fig. 4. Processes of Actor Mapping phase

In the **INNOVATE** phase the service transformation is actually initiated through the generation of innovative ideas, reflections to stimulate creativity and generation of appropriate solutions. In this phase the following activities/tasks are indicated:

1. **Make decisions:** organize the delivery priority of digitized services to be delivered;
2. **Develop solutions:** the requirements of the services to be offered must be elicited to provide their development;

   – **Update existing technologies:** upgrade technologies from existing systems, when necessary, to facilitate integration with new services;
   – **Implement new systems:** develop solutions that have been identified and cataloged;
   – **Integrate systems:** integrate new systems/services with existing ones;
3. **Test services and resources:** evaluate the solutions and resources used throughout the digitization process;
4. **Ensure security:** implement security mechanisms in the services provided.

The activity flow to be executed during the **INNOVATE** phase is shown in Fig. 5.

In the **FACILITATE** phase, resources and tools are provided to simplify and digitize services through the identification of supporting tools and/or technologies. In this phase the following activities/tasks are indicated:

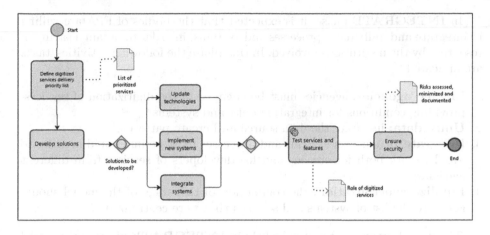

Fig. 5. Processes of Innovate phase

1. **Identify ways to facilitate access:** a broad mapping should be done to identify mechanisms to expand the access of digitized services;
2. **Expand service channels:** several channels for the provision of services should be implemented, including mechanisms for printing and/or delivery of official documents;
3. **De-bureaucratizing processes:** for services to be digitized, the existing bureaucracy must be minimized without losing the necessary controls;
4. **Automate services:** automate services in order to minimize dependencies and increase the autonomy of the citizen;
5. **Validate digitized services:** the services developed must be validated before being available for use.

The activity flow to be performed in the **FACILITATE** phase of the proposed model is presented in Fig. 6.

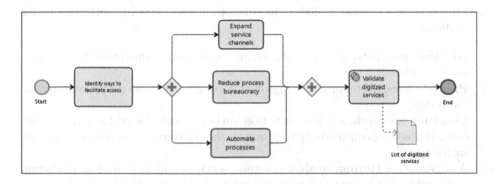

Fig. 6. Processes of Facilitate phase

In **INTEGRATE** phase, it is expected that the bodies of FPA are willing to integrate and unify data, processes and systems, in order to obtain a saving of resources by the institutions involved. In this phase the following activities/tasks are indicated:

1. **Engage agencies:** agencies must be engaged in the digitization of services, providing conditions for integrating data and systems;
2. **Unify data:** integrate the data source and create patterns;
3. **Integrate processes:** systems and processes must be integrated in order to avoid rework both for citizens and for developers of services from different agencies;
4. **Finalize centralization:** the completion of this stage of the model should generate the list of systems and services that were centralized.

The flow of activities stated to be held in **INTEGRATE** phase of the model is shown in Fig. 7.

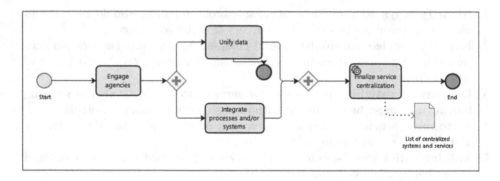

Fig. 7. Processes of Integrate phase

In the **COMMUNICATE** phase citizens are informed about the changes and improvements made available. In this phase the following activities/tasks are indicated:

1. **Disclose services:** invest in marketing, advertising and promotion of the services offered;
2. **Provide access to services:** digitized services should be made available to the public;
3. **Conduct research:** service evaluation surveys should be performed periodically, if possible, continuously to promote the continuous improvement of the entire cycle;
4. **Analyze the timing:** analyze whether services and information are being delivered at the correct and expected speed and timing by the citizen;
5. **Evaluate service availability:** implement mechanisms to evaluate and guarantee the availability of services;

6. **Determine scope of services:** check that all citizens who need access to the service are having it. Otherwise, improvements must be implemented;
7. **Make identified adjustments:** implement mechanisms to support the continuous improvement of digitized services;
8. **Ensure continuity:** implement mechanisms to keep the service active.

The activity flow indicated to be carried out in the **COMMUNICATE** phase of the proposed model is presented in Fig. 8.

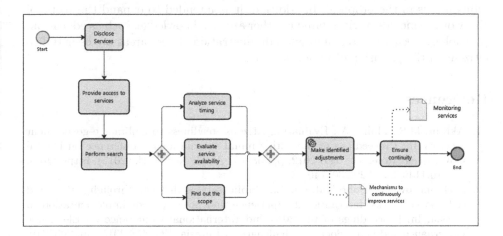

Fig. 8. Processes of Communicate phase

We intend to carry out a practical case study in a Federal Public Administration Body in order to validate the proposed model and verify that the processes are adequate and satisfactorily serve the process of implementation and maintenance of digitized services in the provision of public services by government to the citizen.

5   Conclusion

It is expected that the results found in the systematic review of the literature contribute scientifically in the area of digital government and digitization of public or private services, through the identification of mechanisms to promote the automation and digitization of federal public services; verification of models to include the citizen in the process of automation and digitization of services; survey of good practices adopted for the process of automation and digitization of services. As well as a synthesis of the technological solutions adopted in the processes of automation and digitization of services by any government that wishes to use the proposed process.

Based on this research, it was possible to propose a digitization model of citizen centered services in order to assist government service providers in reaching their goals with the offer of digitized services. This model aims to provide a direction of excellence and conditions necessary to engage citizens in the consumption of these services, thus achieving greater satisfaction and adherence by those involved in the process as a whole.

As a future work, it is intended to apply the processes of the proposed model in a Brazilian Federal Public Administration Body in order to collect information related to the proposed model and, if necessary, to implement improvements and adjustments in the proposal. In addition, it is intended to expand the scope of this work to include works related to other areas of knowledge, such as education, psychology, security, economics and administration. These areas are essential to citizens in the provision of public services.

References

1. Akram, M.S., Malik, A.: Evaluating citizens' readiness to embrace e-government services. In: Proceedings of the 13th Annual International Conference on Digital Government Research, dg.o 2012, pp. 58–67. ACM, New York (2012). https://doi.org/10.1145/2307729.2307740
2. Aleksandrov, O., Dobrolyubova, E.: Public service delivery through automated self-service kiosks: international experience and prospects for implementation in Russia. In: Proceedings of the 2015 2nd International Conference on Electronic Governance and Open Society: Challenges in Eurasia, EGOSE 2015, pp. 205–210, ACM, New York (2015). https://doi.org/10.1145/2846012.2846048
3. Alloghani, M., Hussain, A., Al-Jumeily, D., Aljaaf, A.J., Mustafina, J.: Gamification in e-governance: development of an online gamified system to enhance government entities services delivery and promote public's awareness. In: Proceedings of the 5th International Conference on Information and Education Technology, ICIET 2017, pp. 176–181. ACM, New York (2017). https://doi.org/10.1145/3029387.3029388
4. Bannore, A., Devane, S.R.: Use of proxy signature in e-governance. In: Proceedings of the Second International Conference on Information and Communication Technology for Competitive Strategies, ICTCS 2016, pp. 82:1–82:6. ACM, New York (2016). https://doi.org/10.1145/2905055.2905141
5. Bertot, J.C., Jaeger, P.T., Gorham, U., Greene, N.N., Lincoln, R.: Delivering e-government services through innovative partnerships: public libraries, government agencies, and community organizations. In: DG.O, pp. 126–134. ACM (2012)
6. Biolchini, J., Mian, P.G., Natali, A.C.C., Travassos, G.H.: Systematic review in software engineering. System Engineering and Computer Science Department COPPE/UFRJ, Technical report ES, vol. 679, no. 05, p. 45 (2005)
7. Boudjelida, A., Mellouli, S.: A multidimensional analysis approach for electronic citizens participation. In: Proceedings of the 17th International Digital Government Research Conference on Digital Government Research, dg.o 2016, pp. 49–57. ACM, New York (2016). https://doi.org/10.1145/2912160.2912195
8. da Republica do Brasil, P.: Plataforma de cidadania digital - decreto n. 8.936 de 19 de dezembro de 2016 (2016). http://www.planalto.gov.br/ccivil_03/_ato2015-2018/2016/decreto/D8936.htm

9. Brereton, P., Kitchenham, B.A., Budgen, D., Turner, M., Khalil, M.: Lessons from applying the systematic literature review process within the software engineering domain. J. Syst. Softw. **80**(4), 571–583 (2007)
10. Choi, J.J.U., Ae Chun, S., Kim, D.H., Keromytis, A.: SecureGov: secure data sharing for government services. In: Proceedings of the 14th Annual International Conference on Digital Government Research, dg.o 2013, pp. 127–135. ACM, New York (2013). https://doi.org/10.1145/2479724.2479745
11. Colineau, N., Paris, C., Linden, K.V.: Automatically generating citizen-focused brochures for public administration. In: DG.O, ACM International Conference Proceeding Series, pp. 10–19. Digital Government Research Center (2011)
12. Cordella, A., Tempini, N.: E-government and organizational change: reappraising the role of ICT and bureaucracy in public service delivery. Gov. Inf. Q. **32**(3), 279–286 (2015)
13. Drigas, A., Koukianakis, L.: Government online: an e-government platform to improve public administration operations and services delivery to the citizen. In: Lytras, M.D., et al. (eds.) WSKS 2009. LNCS (LNAI), vol. 5736, pp. 523–532. Springer, Heidelberg (2009). https://doi.org/10.1007/978-3-642-04754-1_53
14. Fabbri, S., Silva, C., Hernandes, E., Octaviano, F., Di Thommazo, A., Belgamo, A.: Improvements in the start tool to better support the systematic review process. In: Proceedings of the 20th International Conference on Evaluation and Assessment in Software Engineering, p. 21. ACM (2016)
15. Felizardo, K.R., Nakagawa, E.Y., Fabbri, S.C.P.F., Ferrari, F.C.: Revisão Sistemática da Literatura em Engenharia de Software: Teoria e Prática. Elsevier, Brasil (2017)
16. Jain, V., Kumar, R., Saquib, Z.: An approach towards digital signatures for e-governance in India. In: Proceedings of the 2015 2nd International Conference on Electronic Governance and Open Society: Challenges in Eurasia, EGOSE 2015, pp. 82–88. ACM, New York (2015). https://doi.org/10.1145/2846012.2846014
17. Kauppinen, S., Luojus, S., Lahti, J.: Involving citizens in open innovation process by means of gamification: the case of welive. In: Proceedings of the 9th Nordic Conference on Human-Computer Interaction, NordiCHI 2016, pp. 23:1–23:4. ACM, New York (2016). https://doi.org/10.1145/2971485.2971526
18. Kitchenham, B.: Procedures for performing systematic reviews. Keele University, Keele, UK, vol. 33, no. 2004, pp. 1–26 (2004)
19. Linders, D.: We-government: an anatomy of citizen coproduction in the information age. In: Proceedings of the 12th Annual International Digital Government Research Conference: Digital Government Innovation in Challenging Times, dg.o 2011, pp. 167–176. ACM, New York (2011). https://doi.org/10.1145/2037556.2037581
20. Ma, L., Zheng, Y.: Good wine needs bush: a multilevel analysis of national e-government performance and citizen use across European countries. In: Proceedings of the 17th International Digital Government Research Conference on Digital Government Research, dg.o 2016, pp. 184–193. ACM, New York (2016). https://doi.org/10.1145/2912160.2912166
21. Moura, M.A.: Política de governança digital brasileira: em pauta a participação social e a transparência ativa. Revista Ágora: políticas públicas, comunicação e governança informacional **1**(1), 121–125 (2016)
22. Musafir, V.E.N.: Brazilian e-government policy and implementation. In: Alcaide Muñoz, L., Rodríguez Bolívar, M.P. (eds.) International E-Government Development, pp. 155–186. Springer, Cham (2018). https://doi.org/10.1007/978-3-319-63284-1_7

23. Omar, A., Weerakkody, V., Millard, J.: Digital-enabled service transformation in public sector: institutionalization as a product of interplay between actors and structures during organisational change. In: Proceedings of the 9th International Conference on Theory and Practice of Electronic Governance, ICEGOV 2015–2016, pp. 305–312. ACM, New York (2016). https://doi.org/10.1145/2910019.2910080
24. Patón-Romero, J.D., Baldassarre, M.T., Piattini, M., García Rodríguez de Guzmán, I.: A governance and management framework for green it. Sustainability 9(10), 1761 (2017)
25. Pereira, G.V., Macadar, M.A., Luciano, E.M., Testa, M.G.: Delivering public value through open government data initiatives in a smart city context. Inf. Syst. Front. 19(2), 213–229 (2017)
26. Pieterson, W., Ebbers, W., Madsen, C.Ø.: New channels, new possibilities: a typology and classification of social robots and their role in multi-channel public service delivery. In: Janssen, M., et al. (eds.) EGOV 2017. LNCS, vol. 10428, pp. 47–59. Springer, Cham (2017). https://doi.org/10.1007/978-3-319-64677-0_5
27. Ministerio do Planejamento, G.e.D.: Estrategia de governanca digital (egd) (2016). http://www.planejamento.gov.br/EGD
28. Ministerio do Planejamento, G.e.D.: Kit de transformacao de serviços publicos (2018). http://www.planejamento.gov.br/antigocidadaniadigital/transformacao/arquivos/transformacao-de-servicos-guia-referencial.pdf
29. Qian, H.: Global perspectives on e-governance: from government-driven to citizen-centric public service delivery. In: ICEGOV, ACM International Conference Proceeding Series, vol. 444, pp. 1–8. ACM (2010)
30. Shareef, M.A., Dwivedi, Y.K., Kumar, V., Kumar, U.: Reformation of public service to meet citizens' needs as customers: evaluating SMS as an alternative service delivery channel. Comput. Hum. Behav. 61, 255–270 (2016)
31. Silva, F.S., et al.: Using CMMI together with agile software development: a systematic review. Inf. Softw. Technol. 58, 20–43 (2015)
32. Vaezi, S.K.: Measurement and evaluating frameworks in electronic government quality management. In: Proceedings of the 2nd International Conference on Theory and Practice of Electronic Governance, ICEGOV 2008, pp. 160–165. ACM, New York (2008). https://doi.org/10.1145/1509096.1509128
33. Valle-Cruz, D.: Dynamic interaction between emerging technologies and organizational factors in government agencies. In: Proceedings of the 9th International Conference on Theory and Practice of Electronic Governance, ICEGOV 2015–2016, pp. 428–431. ACM, New York (2016). https://doi.org/10.1145/2910019.2910096

Modeling Process Interactions with Coordination Processes

Sebastian Steinau(✉), Kevin Andrews, and Manfred Reichert

Institute of Databases and Information Systems, Ulm University, Ulm, Germany
{sebastian.steinau,kevin.andrews,manfred.reichert}@uni-ulm.de

Abstract. With the rise of data-centric process management paradigms, small and interdependent processes, such as artifacts or object lifecycles, form a business process by interacting with each other. To arrive at a meaningful overall business process, these process interactions must be coordinated. One challenge is the proper consideration of one-to-many and many-to-many relations between interacting processes. Other challenges arise from the flexible, concurrent execution of the processes. Relational process structures and semantic relationships have been proposed for tackling these individual challenges. This paper introduces *coordination processes*, which bring together both relational process structures and semantic relationships, leveraging their features to enable proper coordination support for interdependent, concurrently running processes. Coordination processes contribute an abstracted and concise model for coordinating the highly complex interactions of inter-related processes.

Keywords: Process interactions · Semantic relationships
Many-to-many relationships · Relational process structure
Coordination process

1 Introduction

In enterprises, different entities need to collaborate to reach business objectives. The processes used to reach these objectives are not entirely executed in isolation, but have relations and are therefore interdependent. In particular, processes may depend on the execution status of several other processes, i.e., process dependencies may involve one-to-many or many-to-many-relationships. Corresponding interdependencies must be taken into account for the proper coordination of these concurrently executed processes. The proper coordination includes the challenge of coordinating multiple process instances, whose exact quantity is unknown at design-time and which may have different kinds of complex relationships with other process instances. Furthermore, the concurrent execution of processes may be asynchronous, i.e., a process depending on another process may only be synchronized at certain points in time. Finally, any coordination mechanism should impact the execution of the involved process instances as little as possible.

© Springer Nature Switzerland AG 2018
H. Panetto et al. (Eds.): OTM 2018 Conferences, LNCS 11229, pp. 21–39, 2018.
https://doi.org/10.1007/978-3-030-02610-3_2

For dealing with the interdependencies between processes in one-to-many relationships, basic coordination patterns have been identified [15]. These patterns are denoted as *semantic relationships* and may be used to describe complex coordination constraints among multiple process instances. As a prerequisite, semantic relationships require precise knowledge about which process instances are related to which other process instances at run-time. Furthermore, dynamic changes to the relations of process instances, i.e., the creation or deletion of process instances, must be tracked. A solution is the *relational process structure* [14]. While semantic relationships and the relational process structure each solve a part of the problem of process coordination, an overall concept bringing together both parts is still missing. Such a concept requires the specification of semantic relationships at design-time as well as the consideration of dynamic changes to process relations and the concurrent execution of process instances at run-time.

This paper presents *coordination processes*, which leverage both semantic relationships and the advantages of the relational process structure to provide a comprehensive coordination of interrelated process instances. Coordination processes not only support the concise specification of semantic relationships, but additionally allow for the appropriate semantic relationship to be automatically derived based on the relational process structure. Semantic relationships may be combined to express more complex constraints for process coordination. Furthermore, coordination processes take asynchronous execution of the coordinated processes into account by design. A coordination process interferes only when necessary, at certain points during the execution of a process instance, impacting its execution as little as possible. The concept of coordination processes originates from the object-aware approach to process management, where the coordination of the lifecycle processes of objects constitutes an integral part [9]. This paper contributes the support of *many-to-many relationships* in process coordination, which until now has been an open research challenge [6]. Further, the paper contributes a concise model and the ability to express sophisticated coordination constraints, allowing for the proper coordination of vast structures of interdependent processes in a comprehensive fashion.

The remainder of the paper is organized as follows: Sect. 2 recaps semantic relationships and relational process structures and characterizes their basic features. Section 3 introduces the concept of coordination processes. Additionally, it discusses the combination of different semantic relationships using ports and the customization of semantic relationships with expressions. Section 5 covers related work and discusses other approaches to process coordination. Finally, Sect. 6 concludes the paper with a summary and an outlook.

2   Semantic Relationships and Process Structures

Semantic Relationships and the relational process structure provide the fundamental concepts that enable the definition of coordination processes. This section provides a recap of relational process structures and semantic relationships. A running example from the human resource domain is used throughout the paper (cf. Example 1) to illustrate the concepts.

Example 1 (Recruitment Business Process). In the context of recruitment, applicants may apply for job offers. The overall process goal for a company is to determine who of the many applicants is best suited for the job. Applicants must write their application for a specific job offer and send it to the company. The company employees then evaluate each application by performing reviews. To reject an application or proceed with the application, a sufficient number of reviews need to be performed, e.g., the majority of reviews determines whether or not an application is rejected. If the majority of reviews are in favor of the application, the applicant is invited for one or more interviews, after which he may be hired or ultimately rejected. In the meantime, more applications may have been sent in, for which reviews are required, i.e., the evaluation of different applications may be handled concurrently, as well as the conduction of interviews. In particular, when an applicant is hired for the job offer, all other applicants are rejected.

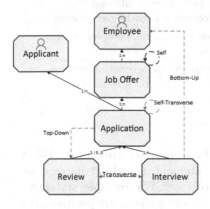

Example 1 describes many individual processes that are related to each other. At design-time, a *relational process structure* captures these processes and their relations [14]. A relation between processes indicates a dependency; on one hand, this explicit capturing of relations allows using these relations for various other purposes, such as specifying message exchanges between two related processes. On the other, it enables the detailed monitoring of process relations at run-time. The detailed knowledge about which process instances are related to which other process instances is crucial for proper process coordination. Figure 1 shows a relational process structure at design-time the processes of Example 1. In detail, the processes are *Job Offer*, *Application*, *Review*, and *Interviews*, whereas *Employee* and *Applicant* represent users, indicating responsibilities for creating and executing other processes.

Fig. 1. Relational process structure and examples of semantic relationships

Example 2. A *Job Offer* may be related to one or more *Applications*, which may have one or more relations to *Interviews*. In case of *Reviews*, the relation is restricted to at minimum three and at most five *Reviews* per *Application*. This cardinality restriction on a process relation is also enforced at run-time by the relational process structure. Furthermore, *Interview* and *Job Offer* are not directly related, but *transitively* via a path of relations. Allowing transitive relations allows for more expressiveness in the coordination of processes. At runtime, the relational process structure tracks the creation and deletion of process instances and their relations. Consequently, a relational process structure is able to give always up-to-date information about which process instances are related with each other. Semantic relationships leverage this capability of relational process structures to specify dependencies between processes and enforce them at run-time.

Table 1. Overview over semantic relationships

Name	Description of the semantic relationship
Top-Down	The execution of one or more lower-level processes depends on the execution status of one common higher-level process
Bottom-Up	The execution of one higher-level process depends on the execution status of one or more lower-level processes of the same type
Transverse	The execution of one or more processes is dependent on the execution status of one or more processes of different type. Both types of processes have a common higher-level process
Self	The execution of a process depends upon the completion of a previous step of the same process
Self-Transverse	The execution of a process depends on the execution process of other processes of the same type. All processes have a common higher-level process

Semantic relationships may be used to model *coordination constraints* [15]. A coordination constraint is a formal or informal statement describing one or more conditions or dependencies that exist between processes. For example, statement "An application may only be accepted if three or more reviews are positive" is a coordination constraint. A coordination constraint must be expressed in terms of semantic relationships for the use in a coordination process. For a proper representation of coordination constraints, the combination of different semantic relationships might be necessary. A semantic relationship describes a recurring semantic pattern inherent in the coordination of processes in a one-to-many or many-to-many relationship (cf. Table 1). As one example of a pattern, several process instances may depend on the execution of one other process instance. A semantic relationship may only be established between processes if a path of relations in the relational process structure, i.e., a dependency, exists between these processes. Figure 1 shows examples of semantic relationships between different processes. In this context, the terms *lower-level* and *higher-level* refer to the fact that the relations are directed (cf. Fig. 1). Process A is denoted as higher-level process in respect to a reference process B if there is a directed relation from B to A. Analogously, there may be many source processes C_i denoted as lower-level processes in respect to a reference process D. This terminology applies with transitive relations as well.

The execution status referred to in Table 1 is represented by a state-based view of the process [15]. Thereby, the process to be coordinated is abstracted and partitioned into different states that provide significant meaning for process coordination. Furthermore, as the state-based view abstracts from the underlying process language, any language might be used to model the process. For example, an *Application* has states *Sent* and *Checked*, which are important milestones for its coordination. An *active state* represents the current execution status. At run-time, based on the execution status, semantic relationships possess a logical value that indicates whether the represented condition is currently satisfied and, therefore, whether the execution of processes may progress or must halt. For example, a *Job Offer* has active state *Published*, and a top-down semantic relationship has value *true* to indicate that now *Applications* may be created for the *Job Offer*. Apart from the state-based view, processes may provide access to data attributes for use in a coordination process.

However, a method to *effectively specify semantic relationships is still missing*. Furthermore, coordination constraints often need several semantic relationships to be expressed. As semantic relationships have a logical value to indicate whether or not they are satisfied, boolean operators are required to express more complicated coordination logic. Coordination processes combine the effective specification of semantic relationships with a graphical representation of boolean logic. A coordination process leverages the relational process structure and properties of semantic relationships to automatically derive the appropriate semantic relationship between two processes at design-time.

3   Coordination Processes

Coordination processes are a generic concept for coordinating processes by expressing coordination constraints with the help of semantic relationships, which are then enforced at run-time. The concept allows specifying sophisticated coordination constraints for vast structures of interrelated process instances with an expressive, high-level graphical notation using a minimum amount of modeling elements. The modeling elements are the *coordination step*, the *coordination transition,* and the *port*. Coordination processes follow a type-instance schema, where types (denoted ^{T}) represent design-time entities and instances (^{I}) run-time entities. Consequently, an instance is created by instantiating a type. The dot (.) represents the access operator.

Definition 1 (Coordination Process Type). *A coordination process type c^T has the form $(\omega^T, B^T, \Delta^T)$ where*

- ω^T *is the process type to which the coordination process type c^T belongs.*
- B^T *is a set of coordination step types β^T (cf. Definition 3).*
- Δ^T *is a set of coordination transition types δ^T (cf. Definition 4).*

The *coordinating process type ω^T* determines the overall context of the coordination process, e.g., it determines the start and end coordination steps of the coordination process and which processes may be coordinated.

Definition 2 (Process Type). *A process type ω^T has the form (n, Σ^T, c^T) where*

- *n is the name of the object type.*
- *Σ^T is a set of state types σ^T, representing the state-based view of ω^T.*
- *c^T is an optional coordination process type (cf. Definition 1). Default is $\bot$.*

For handling the complexities of dozens or hundreds of interrelated processes at design-time, abstraction in a coordination process is crucial. As a part of this effort, process types are represented with a state-based view, which abstracts from process details and only exposes properties which are useful for process coordination.

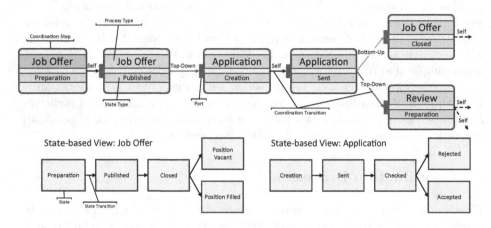

Fig. 2. Coordination process model and state-based views, Part I

3.1   Coordination Steps and Coordination Transitions

Coordination processes are represented as a directed graph that consists of *coordination steps*, *coordination transitions* and *ports*. Figure 2 shows a part of the coordination process for Example 1 with *Job Offer* as the coordinating process type. Coordination steps are the vertices of the graph referring to a process type as well as to one of its states, e.g. *Job Offer* and state *Published*. For the sake of convenience, a coordination step is addressed with referenced process type and state in the form of *ProcessType:State*, e.g. *Job Offer:Published*. A formal definition for coordination steps is presented in Definition 3.

Definition 3 (Coordination Step Type). *A coordination step type β^T has the form $(c^T, \omega^T, \sigma^T, \Delta_{out}^T, H^T)$ where*

- *c^T is the coordination process type (cf. Definition 1).*
- *ω^T is a reference to a process type.*
- *σ^T is a reference to a state type belonging to ω^T, i.e., $\sigma^T \in \omega^T.\Sigma^T$.*

- Δ^T_{out} is a set of outgoing coordination transition types δ^T (cf. Definition 4).
- H^T is a set of port types η^T (cf. Definition 5).

A coordination step type represents a collection of process instances of type ω^T at run-time. As such, a coordination step type provides a succinct and abstract way to represent *multiple process instances at run-time*, thus constraining much of the complexity of interdependent multiple process instances to the run-time instead of the design-time.

A coordination transition is a directed edge that connects a *source coordination step* with a *target coordination step* (cf. Fig. 2). Both source and target coordination step reference a process type of the relational process structure. By creating a coordination transition between source and target step, a semantic relationship is created as well. Conceptually, a semantic relationship is attached to a coordination transition. With the relations from the relational process structure and the definitions of semantic relationships (cf. Table 1), it can be automatically derived which semantic relationship is established between the process types referenced by the two coordination steps.

Example 3 (Top-Down and Bottom-Up Semantic Relationships). Connecting *Job Offer:Published* with *Application:Sent* constitutes a top-down relationship (cf. Fig. 2). The sequence in which the steps occur is important for determining the type of semantic relationship. Connecting *Application:Sent* with *Job Offer:Closed*, a bottom-up semantic relationship is established instead, as *Application* is a lower-level process type of *Job Offer*.

A formal definition for coordination transitions can be found in Definition 4.

Definition 4 (Coordination Transition Type). *A coordination transition type δ^T has the form $(\beta^T_{src}, \eta^T_{tar}, s^T)$ where*

- β^T_{source} is the source coordination step type (cf. Definition 3).
- η^T_{target} is the target port type (cf. Definition 5).
- s^T is a semantic relationship between $\beta^T_{src}.\omega^T$ and $\eta^T_{tar}.\beta^T.\omega^T$.

For the sake of convenience, the terminology of source or target coordination step of a coordination transition applies for the corresponding semantic relationships as well. The strict distinction between coordination transition and semantic relationship is crucial at run-time and is therefore reflected in the design-time model. For establishing a semantic relationship between two processes, the state σ^T of any coordination step is not relevant, only the process types are relevant. However, states becomes crucial for the actual representation and enforcement of coordination constraints at run-time. Depending on the activation of states at run-time, semantic relationships become enabled or disabled.

Example 4 (Top-down Semantics). Figure 2 depicts coordination steps *Job Offer:Published* and *Application:Creation*. The top-down semantic relationship between these coordination steps enforces that a *Job Offer* must reach state *Published* before any application may be created (i.e., *Creation* is the start state of an *Application*). Once a particular *Job Offer* reaches state *Published* and the state becomes active, the top-down semantic relationship becomes enabled and subsequently allows creating any number of *Applications* for the *Job Offer*, at different points in time.

Coordination processes only permit or prohibit the activation of states. The actual activation is determined by the process itself, i.e., by its progress. It is therefore possible that a coordination process allows activating a state long before actually reaching this state of the process. On the other side, a coordination process may halt process execution until the specified coordination constraints are fulfilled, i.e., the semantic relationships become enabled.

Enabling a semantic relationship requires that all predecessor semantic relationships in the coordination process have been enabled as well, e.g., enabling the bottom-up semantic relationship of *Application*:*Sent* with *Job Offer*:*Closed* requires that the top-down semantic relationship *Job Offer*:*Published* with *Applica-tion*:*Sent* has already been enabled (cf. Fig. 2). As a consequence, a coordination process graph must be acyclic and connected. If a coordination process contained a cycle, it would result in an immediate deadlock once a coordination step of the cycle is reached. Due to the cycle, its incoming semantic relationships never become enabled. Furthermore, the start and end coordination steps of a coordination process must reference the coordinating process type, i.e., $\beta^T.\omega^T = c^T.\omega^T$, ensuring its proper start and completion.

Semantic relationships are based on one-to-many relationships. This includes transitive relations, e.g., a semantic relationship may be established between *Job Offer* and *Interview*. If the processes are in a many-to-many ($m{:}n$) relationship and are in a (w.l.o.g.) top-down semantic relationship, a coordination process replicates the top-down semantic relationship m times at run-time, depending on the number m of higher-level processes. It is thereby established that each of the n lower-level processes depends on each of the m higher-level processes. In consequence, many-to-many-relations may be elegantly coordinated.

So far, just based on coordination steps and coordination transitions, only simple coordination constraints may be expressed, i.e., constraints that may be represented by a single semantic relationship. However, coordination constraints may require multiple semantic relationships to be properly represented. Additionally, the states of a process may be involved in several coordination constraints, requiring all of them to be fulfilled in order to become enabled. Coordination processes therefore incorporate the concept of *ports*, which allow combining multiple semantic relationships for a state to become active.

3.2   Ports

Coordination transitions do not target a coordination step directly, but instead target a port attached to a coordination step. Any coordination step must have one or more ports (with the exception of the start coordination step).

Definition 5 (Port Type). *A port type η^T has the form (β^T, Δ_{in}^T) where*

- *β^T is the coordination step type to which this port type belongs.*
- *Δ_{in}^T is the set of all incoming coordination transitions δ^T (cf. Definition 4).*

Ports allow realizing different semantics for combining semantic relationships. Connecting multiple transitions to the same port corresponds to AND-semantics, i.e., all semantic relationships attached to the incoming transitions must be enabled for the port to become enabled as well. Enabling a port also enables the coordination step, allowing the state of the coordination step to become active. Generally, at least one port of a coordination step must be enabled for the coordination step to become enabled as well. Consequently, connecting transitions to different ports corresponds to OR-semantics.

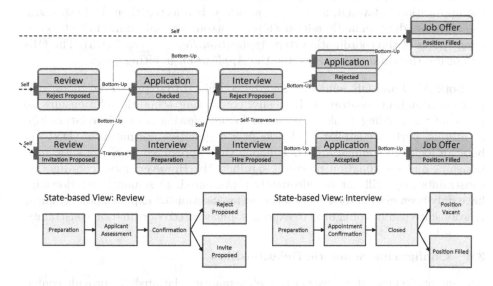

Fig. 3. Coordination process model and state-based views, Part 2

Example 5 (Port AND Semantics). A coordination constraint may state that for an application to be accepted, a sufficient number of interviews must propose a hire. Implicitly, a job may not be given to two different applicants; therefore, no other application must have already been accepted for the same job offer. To model this coordination constraint, coordination step *Application:Accepted* (cf. Fig. 3) has one port with two incoming coordination transitions. Therefore, for an *Application* to be accepted, both conditions represented by the semantic relationships need to be fulfilled. The bottom-up semantic relationship outgoing from coordination step *Interview:Hire Proposed* requires a sufficient number of *Interviews* to have reached state *Hire Proposed* before an *Application* may be accepted. The process type and state combination required to enable the semantic relationship is determined by its source coordination step. The exact number of *Interviews* required to be in state *Hire Proposed* is a design choice.

The implicit condition is represented by a *self-transverse* semantic relationship, i.e., an *Application* depends on the execution status of other *Applications* in context of the same *Job Offer*. In this case, only exactly one *Application* may reach state *Accepted*, i.e., once an *Application* has been accepted, other *Applications* are prevented from reaching state *Accepted*. The AND-semantics of the target port require that both conditions are *true* at the same time so the *Application* may be accepted.

Example 6 (Port OR Semantics). Rejecting an *Application* may be achieved in two different ways. First, the *Reviews* corresponding to the *Application* favor an immediate rejection. Second, during one or more *Interviews*, a rejection of the application is favored, and the *Application* is rejected then. In both cases, the corresponding semantic relationship is bottom-up, but connects to two different ports of the coordination step *Application:Rejected* (cf. Fig. 3). The OR-semantics, therefore, allows rejecting the *Application* in either case.

Both AND and OR semantics may be combined to express even more complex coordination constraints. Basically, connecting semantic relationships to ports allows building boolean formulas. When viewing semantic relationships as literals, ports are similar to clauses in a disjunctive normal form (DNF) of boolean logic. In summary, ports enhance the expressiveness of semantic relationships and coordination processes significantly. However, most coordination constraints may still not be adequately represented, as semantic relationships have only been used in their basic forms so far. Section 3.3 explores how semantic relationships can be configured to express sophisticated coordination constraints.

3.3   Configuring Semantic Relationships

All semantic relationships, except the self semantic relationships, provide configuration options to the process modeler [15]. This allows for fine-grained control over the basic semantics of the semantic relationship, increasing the degree to which complex coordination constraints may be expressed. A top-down semantic relationship must specify when it is no longer enabled, due to progression in the higher-level process (cf. Table 1). For example, a *Job Offer* may no longer accept new *Applications* after state *Closed* has become active. However, the exact means to achieve this are not specified by the top-down semantic relationship. Coordination processes rectify the situation by introducing a *state set*. A top-down semantic relationship becomes enabled once the state of the source coordination step, denoted as the *base state* of the top-down semantic relationship, becomes active. For example, reaching base state *Published* of a *Job Offer* enables the outgoing top-down semantic relationship (cf. Fig. 2).

Consequently, the base state must automatically be part of the state set. As long as the currently active state of the higher-level process belongs to this state set, the top-down semantic relationship remains enabled. Successor states of the base state may also be added to the state set by the process modeler. This keeps the top-down semantic relationships enabled while the higher-level process progresses, as long as its active state belongs to the state set.

Example 7 (Top-Down Configuration). Suppose that state *Closed*, a successor state of state *Published*, is added to the state set of the top-down semantic relationship. Then, new *Applications* may still be added even when the *Job Offer* is closed, i.e., is in state *Closed*. The top-down semantic relationship becomes disabled as soon as the *Job Offer* reaches either state *Position Filled* or state *Position Vacant*, i.e., *Applications* may no longer establish new relations to the *Job Offer*.

As opposed to top-down semantic relationships, bottom-up, transverse, and self-transverse semantic relationships can be configured by using expressions. Such an expression is denoted as a *coordination expression* and represents more specialized conditions, in addition to the basic semantics of the respective semantic relationship. For example, the bottom-up relationship between *Interview:Hire Proposed* and *Application:Accepted* requires a sufficient number of *Interviews* in state *Hire Proposed*. With an expression, this condition can be specified formally and with a concrete number.

Example 8 (Expressions). A process modeler may specify that at least two *Interviews* are necessary for a hire. This may be represented as *Count(Interview, Hire Proposed)* ≥ 2, where *Count* is a function. In principle, the required expressions may be of arbitrary complexity, allowing for the full range of boolean and arithmetic functions, constants, and variables based on process data. In particular, negating the semantics of semantic relationships is possible.

In Example 8, the notation of the expression does not incorporate the given context for which this expression must be evaluated at run-time. In fact, *Count* must not be evaluated on a global level, i.e., counting every *Interview* of every *Applicant* for every *Job Offer*, but must be evaluated in context of a single *Application*. Otherwise, this would have undesired and even absurd side effects, such as that two positive interviews (for any two applicants) would allow additional applicants to be accepted as well. Therefore, it is essential that the context is reflected in the expression framework, while keeping the expressions simple. Often, it is desired that an expression framework shows high expressiveness to solve the particular problem at hand. However, high expressiveness usually comes with a number of drawbacks. Among these drawbacks, two may be considered as the most severe. First, high expressiveness is generally correlated with high complexity. This causes difficulties when specifying expressions, as substantial knowledge of the expression framework is necessary for modeling. Second, high complexity poses problems in the implementation of the expressions, requiring considerable efforts to support all possible expressions in all possible combinations. Thus, less used or more complex expressions are often not implemented due to time and resource constraints, limiting the use of the expression framework in practice. With the clear focus of expressions in semantic relationships, it becomes possible to reduce the complexity of the expressions.

Several models[1] that involve coordination processes have shown that a high percentage of the expressions used for configuring semantic relationships require the *counting* of process instances. The instances to be counted are represented by the source and target coordination steps. Furthermore, counting depends on the state of the process, e.g., it is important how many processes are in a particular active state. In other cases, it is important whether a particular state has been active, has not yet been active, or has been skipped due to the selection of alternative execution paths in the process. Due to the state-based views of the involved processes, the status of states is of particular concern to the coordination of processes at run-time. Therefore, it is beneficial to define specialized counting functions as part of the expression framework.

Definition 6 (Coordination Expression Counting Functions). *Let s^T be a semantic relationship and δ^T be a coordination transition. Let Ω^I_{source} be the process instances of type $\delta^T.\beta^T_{source}.\omega^T$ coordinated by s^T. Let Σ^T be the state type set of $\delta^T.\beta^T_{source}.\omega^T$. Then:*

- *$\#AllSource : \Omega^I_{source} \to \mathbb{N}_0$*
  Determines the total number of process instances for s^T.
- *$\#InSource : \Omega^I_{source} \times \Sigma^T \to \mathbb{N}_0$*
  Determines the number of process instances of s^T where state σ^I of type $\sigma^T \in \Sigma^T$ is currently active.
- *$\#BeforeSource : \Omega^I_{source} \times \Sigma^T \to \mathbb{N}_0$*
  Determines the number of process instances of s^T where state σ^I of type $\sigma^T \in \Sigma^T$ has not yet been active, i.e., a predecessor state is active.
- *$\#AfterSource : \Omega^I_{source} \times \Sigma^T \to \mathbb{N}_0$*
  Determines the number of process instances of s^T where state σ^I of type $\sigma^T \in \Sigma^T$ has been active in the past, i.e., a successor state is active.
- *$\#SkippedSource : \Omega^I_{source} \times \Sigma^T \to \mathbb{N}_0$*
  Determines the number of process instances of s^T where state σ^I of type $\sigma^T \in \Sigma^T$ is not on the execution path of the process instances, i.e., a mutual exclusive state to σ^I is active.

Analogously, functions can be defined that count process instances of Ω^I_{target} with type $\delta^T.\eta^T_{target}.\beta^T.\omega^T$.

With these functions, expression *Count(Interview, Hire Proposed)* ≥ 2 can be redefined, explicitly taking context, i.e., the respective semantic relationship, into account. Figure 4 shows an excerpt from the coordination process from Figs. 2 and 3. The semantic relationships have been annotated with their respective coordination expressions. For the bottom-up semantic relationship, counting function *#InSource* has been used, as the source coordination step is

---

[1] A selection has been approved for publication, available at https://bit.ly/2yo6GTe.

Interview:Hire Proposed. Using the counting function *#InSource*, in conjunction with the source or target coordination step and the respective semantic relationship, therefore clearly defines the context for evaluating the expression.

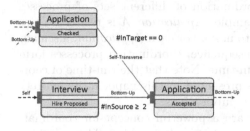

Fig. 4. Counting functions example

At run-time, as soon as two *Interviews* reach state *Hire Proposed*, the semantic relationship becomes enabled. For the self-transverse semantic relationship, the counterpart counting function *#InTarget* has been used, counting *Applications*. In line with the semantics of a self-transverse semantic relationship (cf. Table 1), an *Application* may only be accepted if no other *Application* has already been accepted. As at run-time initially no *Application* is accepted and thus, the coordination expression is *true*, which means the self-transverse semantic relationship becomes enabled once an *Application* reaches state *Checked*. If an *Application* becomes accepted, the self-transverse semantic relationship is disabled due to the coordination expression *#InTarget == 0* no longer evaluating to *true*. Therefore, no more *Applications* may reach state *Accepted*.

In case a modeler has not specified a coordination expression, bottom-up, transverse, and self-transverse semantic relationships default to the expression *#InSource = #AllSource*, meaning the referenced state must be active in all process instances. Note that these functions are intended to facilitate frequently encountered use cases when specifying semantic relationships, the expression framework is not limited to using these functions. In previous work [9], expression were limited to counting source processes, severely limiting the expressiveness of semantic relationships. With the addition of the target coordination expressions (e.g., *#InTarget*) and other types of expressions, which are not replicated here for the sake of brevity, a wider range of coordination constraint can be realized.

3.4 Operational Semantics of Coordination Processes

The concept of coordination processes not only comprises the modeling of process interactions, but includes *operational semantics* as well. The operational semantics defines the run-time behavior of coordination processes. The highly dynamic nature of the relational process structure at run-time and the frequent state changes of processes create a unique set of challenges for a coordination process at run-time. This requires a high flexibility to tackle these challenges on part of the coordination process. For example, as processes may execute concurrently, semantic relationships must cope with different processes reaching particular states at different points in time and in different order. Furthermore, creating and deleting processes or changing relations of interconnected processes all affect a coordination process, i.e., coordination constraints become fulfilled or are no longer fulfilled. The operational semantics must account for all eventualities to ensure a correct process execution of all involved process instances.

For this purpose, a coordination step type may be instantiated multiple times at run-time, each representing one process instance of the relational process structure. Consequently, semantic relationships are also instantiated multiple times at run-time, allowing for the coordination of different sets of processes independently and contextually. For example, *Application A* is related to two *Interviews* and *Application B* is related to none. For each *Application*, a semantic relationship is instantiated. As a consequence, coordination processes form complex, interconnected structures at run-time. Note that the run-time of coordination processes is too complex to be presented in this paper in its entirety and must therefore be reserved for future publications.

In summary, coordination processes are a powerful concept for coordinating processes in one-to-many and many-to-many relationships. They combine the capabilities of relational process structures and the semantic relationships to model coordination constraints and enforce these constraints at run-time. With ports and coordination expressions, coordination processes may be customized to fit individual needs while retaining the basic semantics of the individual semantic relationships. Modeling is facilitated by the comparatively low number of modeling elements and the fact that semantic relationships may be derived automatically when connecting coordination steps with a coordination transition, which is possible due to the underlying relational process structure. With this, managing the challenges of multiple interrelated processes at run-time is possible. Furthermore, semantic relationships, as the cornerstone of coordination processes, allow reacting correctly to the changes during lifecycle execution. The relational process structure ensures that a coordination process has always up-to-date information on every process instance and its relations. Finally, a coordination process model is designed so that it is immediately executable upon instantiation, i.e., there is no distinction between functional and technical model.

4   Proof-of-Concept: Demonstrating the Feasibility of Coordination Processes

Object-aware process management [9] has been centered around the idea that objects with lifecycles and their interactions constitute a business process. As a data-centric paradigm, objects acquire data according to their lifecycle processes, i.e., the change in progress is data-driven. In particular, many process instances of a type may exist, which have interdependencies to other process instances. Consequently, coordination processes have been developed to steer these different interacting lifecycles in order to reach a meaningful overall business process. Objects, their lifecycles and coordination processes constitute the core of a business process management system prototype. This prototype is based on the object-aware approach and has been developed in the PHILharmonicFlows[2] project at Ulm University.

---

[2] For more details on the prototype visit https://bit.ly/2KYvyT9.

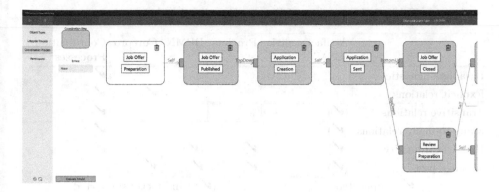

Fig. 5. PHILharmonicFlows modeling tool showing a coordination process

The prototype comprises a tool for modeling data models with objects together with their lifecycle processes and relations to other objects, i.e., a relational process structure. The tool also supports the modeling of coordination processes as described in this paper. Figure 5 shows the coordination process of the running example (cf. Figs. 2 and 3) modeled with the tool. The prototype additionally comprises a run-time environment, which is able to asynchronously execute both lifecycle and coordination processes with the required flexibility. The prototype uses a micro service architecture for high scalability and parallel, asynchronous execution of processes, as objects and their attached processes are uniquely suited to be distributed among such micro services. This raises further challenges for coordination processes when employed in large-scale, distributed relational process structures.

Furthermore, coordination processes were successfully used to model various processes, both real-world and exploratory examples. Some have been modeled in cooperation with industrial partners. All models comprise dozens of object types and multiple coordination processes[3]. The models showed that, in general, coordination processes are able to represent coordination constraints adequately. While modeling of coordination processes requires extensive knowledge of several concepts, e.g., semantic relationships, it is by far compensated by the built-in executability of the models.

5 Related Work

Coordination processes support various features rarely seen in other process coordination approaches, most notably the support of many-to-many relationships. This gives coordination processes a unique advantage. Table 2 shows a comparison between coordination processes and selected related work. Note that Table 2 compares approaches according to specific features and therefore does not represent an overall quality assessment of the individual approaches.

---

[3] A selection has been approved for publication, available at https://bit.ly/2yo6GTe.

Table 2. Comparison of Process Coordination Approaches

	Artifact-centric (GSM)	Proclets	BPMN	Corepro	Coordination Processes
Paradigm-agnostic					✔
Explicit relations			✔		✔
Transitive relations	(✔)				✔
Many-to-many relations	(✔)				✔
Process cardinality	(✔)	✔	(✔)		✔
Message-based	(✔)	✔	✔	✔	

✔ : Supported   (✔) : Indirectly supported

Artifact-centric process management [12] uses the Guard-Stage-Milestone (GSM) meta-model [7,8] for process modeling. Central to this approach is the *artifact*, which holds all process-relevant information. It may further interact with other artifacts. However, GSM does not provide dedicated coordination mechanisms or explicit artifact relations, and therefore does not support any criterion of Table 2. Instead, GSM incorporates an arbitrary information model and a sophisticated expression framework that, in principle, allow fulfilling the comparison criteria with expressions and custom data. As a drawback, expressions might become very complex and explicitly need to be integrated into the process model. Therefore, model verification [1,2,4] constitutes an important aspect of artifact-centric process management. Further, [6] recognizes the need for supporting many-to-many relationships in artifact-centric choreographies.

For artifact-centric process models based on Finite-State-Machines (FSMs), [16] developed a message-based declarative artifact-centric choreography. This approach proposes the use of exactly one master artifact to coordinate all artifacts in a correlation graph. The approach explicitly considers the run-time presence of multiple instances. While the approach shows some similarities to coordination processes, the message-based coordination mechanism neither provides the run-time flexibility of semantic relationships nor the succinct model of a coordination process. Moreover, it is unclear if and how the findings translate from FSM-based to GSM-based artifacts.

Proclets [17] are lightweight processes with focus on process interactions. Proclets interact via messages called *Performatives*. Proclets allow specifying the cardinality for a message multicast, i.e., the number of Proclets that receive a performative. Proclets are capable of asynchronous and concurrent execution. However, relations between different Proclets are not considered. Proclets are defined using Petri nets, which are extended with ports that send and receive performatives. The concept of ports in the Proclet approach is fundamentally different than ports in coordination processes.

The coordination of large process structures with focus on the engineering domain is considered in [10,11]. The COREPRO approach explicitly considers process relations with one-to-many cardinality and dynamic changes at run-time, but transitive relations are not considered. In comparison to COREPRO, semantic relationships correspond, in principle, to external state transitions of a Lifecycle Coordination Model. However, the external state transitions do not take the semantics of the respective process interaction into account.

Regarding the activity-centric process modeling paradigm, several approaches enable a specific kind of coordination. For activity-centric processes, workflow patterns have been identified [18]. Several workflow patterns describe interactions between processes, which may be used for coordinating processes. The business process architecture approach [5] also identifies generic patterns to describe a coordination between processes. iBPM [3] enhances BPMN to support coordination of processes by modeling process interactions.

The BPMN standard [13] provided the choreography diagram explicitly dedicated to model the interactions between processes. Similar to coordination processes, choreography diagrams abstract from the coordinated processes and only display interactions themselves. The coordinated process types can be annotated with single-instance and multi-instance markers, showing very limited support in restricting process cardinality. Similar to coordination processes, they are displayed on separate diagrams and posses few modeling elements.

Common to all these approaches, with the exception of artifacts, is the use of messages as a mechanism for coordination. While the exchange of messages allows for a detailed process coordination, all message flows have to be identified, the contents of the messages defined, and the recipients determined. This constitutes an enormous complexity when facing numerous processes that need to be coordinated, and in many cases, it impairs the flexible execution of the involved processes. Except Proclets, the modeling of coordination aspects is integrated into the actual process models, increasing the complexity to the process models. Coordination processes allow expressing complex interdependencies concisely using semantic relationships.

6   Summary and Outlook

A coordination process is an advanced concept for coordinating a collection of individual processes. It provides the superstructure to effectively employ relational process structures and semantic relationships. A coordination process itself is specified in a concise and comprehensive manner using coordination steps, coordination transitions and ports, abstracting from the complexity of coordinating a multitude of interrelated processes. Coordination processes allow for the automatic derivation of semantic relationships from connecting two coordination steps with a coordination transition. Complex coordination constraints are expressed by combining multiple semantic relationships using ports, and are configured using a comprehensive context-aware expression framework.

For future work, the operational semantics of coordination processes are the main focus, as coordinating multiple concurrently running instances poses unique challenges. In particular, the multi-instance nature of the run-time requires that semantic relationships are instantiated multiple times, once for each context. This leads to a highly complex instance representation of a coordination process at run-time, which must be kept synchronized with each process instance in the relational process structure and the execution status of each process instance. Both operational semantics and large process structure coordination will be investigated in future work. Additionally, a thorough empirical investigation of the coordination process modeling concept shall demonstrate its applicability in practice.

Acknowledgments. This work is part of the ZAFH Intralogistik, funded by the European Regional Development Fund and the Ministry of Science, Research and the Arts of Baden-Württemberg, Germany (F.No. 32-7545.24-17/3/1).

References

1. Belardinelli, F., Lomuscio, A., Patrizi, F.: Verification of GSM-based artifact-centric systems through finite abstraction. In: Liu, C., Ludwig, H., Toumani, F., Yu, Q. (eds.) ICSOC 2012. LNCS, vol. 7636, pp. 17–31. Springer, Heidelberg (2012). https://doi.org/10.1007/978-3-642-34321-6_2
2. Damaggio, E., Hull, R., Vaculín, R.: On the equivalence of incremental and fixpoint semantics for business artifacts with guard-stage-milestone lifecycles. Inf. Syst. **38**(4), 561–584 (2013)
3. Decker, G., Weske, M.: Interaction-centric modeling of process choreographies. Inf. Syst. **36**(2), 292–312 (2011)
4. Deutsch, A., Li, Y., Vianu, V.: Verification of hierarchical artifact systems. ArXiv e-prints (2016)
5. Eid-Sabbagh, R.-H., Dijkman, R., Weske, M.: Business process architecture: use and correctness. In: Barros, A., Gal, A., Kindler, E. (eds.) BPM 2012. LNCS, vol. 7481, pp. 65–81. Springer, Heidelberg (2012). https://doi.org/10.1007/978-3-642-32885-5_5
6. Fahland, D., de Leoni, M., van Dongen, B.F., van der Aalst, W.M.P.: Many-to-many: some observations on interactions in artifact choreographies. In: 3rd Central-European Workshop on Services and their Composition (ZEUS), 2011. CEUR Workshop Proceedings, vol. 705, pp. 9–15. CEUR-WS.org (2011)
7. Hull, R., et al.: business artifacts with guard-stage-milestone lifecycles: managing artifact interactions with conditions and events. In: 5th ACM International Conference on Distributed Event-based System (DEBS) 2011, pp. 51–62. ACM (2011)
8. Hull, R., et al.: Introducing the guard-stage-milestone approach for specifying business entity lifecycles. In: Bravetti, M., Bultan, T. (eds.) WS-FM 2010. LNCS, vol. 6551, pp. 1–24. Springer, Heidelberg (2011). https://doi.org/10.1007/978-3-642-19589-1_1
9. Künzle, V., Reichert, M.: PHILharmonicFlows: towards a framework for object-aware process management. J. Softw. Maint. Evol.: Res. Pract. **23**(4), 205–244 (2011)
10. Müller, D., Reichert, M., Herbst, J.: Data-driven modeling and coordination of large process structures. In: Meersman, R., Tari, Z. (eds.) OTM 2007. LNCS, vol. 4803, pp. 131–149. Springer, Heidelberg (2007). https://doi.org/10.1007/978-3-540-76848-7_10

11. Müller, D., Reichert, M., Herbst, J.: A new paradigm for the enactment and dynamic adaptation of data-driven process structures. In: Bellahsène, Z., Léonard, M. (eds.) CAiSE 2008. LNCS, vol. 5074, pp. 48–63. Springer, Heidelberg (2008). https://doi.org/10.1007/978-3-540-69534-9_4
12. Nigam, A., Caswell, N.S.: Business artifacts: an approach to operational specification. IBM Syst. J. **42**(3), 428–445 (2003)
13. Object Management Group: Business Process Model and Notation (BPMN), Version 2.0 (2011)
14. Steinau, S., Andrews, K., Reichert, M.: The relational process structure. In: Krogstie, J., Reijers, H.A. (eds.) CAiSE 2018. LNCS, vol. 10816, pp. 53–67. Springer, Cham (2018). https://doi.org/10.1007/978-3-319-91563-0_4
15. Steinau, S., Künzle, V., Andrews, K., Reichert, M.: Coordinating business processes using semantic relationships. In: 19th IEEE Conference on Business Informatics (CBI), pp. 33–43. IEEE Computer Society Press (2017)
16. Sun, Y., Xu, W., Su, J.: Declarative choreographies for artifacts. In: Liu, C., Ludwig, H., Toumani, F., Yu, Q. (eds.) ICSOC 2012. LNCS, vol. 7636, pp. 420–434. Springer, Heidelberg (2012). https://doi.org/10.1007/978-3-642-34321-6_28
17. van der Aalst, W.M.P., Barthelmess, P., Ellis, C.A., Wainer, J.: Proclets: a framework for lightweight interacting workflow processes. Int. J. Coop. Inf. Syst. **10**(04), 443–481 (2001)
18. van der Aalst, W.M.P., ter Hofstede, A.H.M., Kiepuszewski, B., Barros, A.: Workflow patterns. Distrib. Parallel Databases **14**(1), 5–51 (2003)

Blockchain-Based Collaborative Development of Application Deployment Models

Ghareeb Falazi[✉], Uwe Breitenbücher, Michael Falkenthal,
Lukas Harzenetter, Frank Leymann, and Vladimir Yussupov

Institute of Architecture of Application Systems, University of Stuttgart,
Stuttgart, Germany
{falazi,breitenbuecher,falkenthal,harzenetter,leymann,
yussupov}@iaas.uni-stuttgart.de

Abstract. The automation of application deployment is vital today as the manual alternative is too slow and error-prone. For this reason, many technologies for deploying applications automatically based on deployment models have been developed. However, in many scenarios, these models have to be created in collaborative processes involving multiple participants that belong to independent organizations. However, the potential competing interests of these organizations hinder the degree of trust they have in each other. Thus, without a guarantee of accountability, iterative collaborative deployment modeling is not possible in such domains. In this paper, we propose a decentralized deployment modeling approach that achieves accountability by utilizing public blockchains and decentralized storage systems to store intermediate states of the collaborative deployment model. The approach guarantees integrity of deployment models and allows obtaining the history of changes they went through while ensuring participants' authenticity.

Keywords: Declarative deployment models · Blockchains
Accountability

1 Introduction

The automation of application deployment is vital in many modern scenarios because manual deployment is typically too slow and error-prone—especially in cloud-based environments that are completely designed for automation. For this reason, the idea of deployment models was introduced [16]. These models describe the application to be deployed which includes all the required components as well as their relationships. Furthermore, many deployment automation technologies, such as Chef [2] and Kubernetes [3], have been developed to automate the entire deployment process based on the aforementioned models.

However, the creation of such models is becoming more difficult as many scenarios nowadays require different companies and experts to participate in th

© Springer Nature Switzerland AG 2018
H. Panetto et al. (Eds.): OTM 2018 Conferences, LNCS 11229, pp. 40–60, 2018.
https://doi.org/10.1007/978-3-030-02610-3_3

process. For example, the creation of a deployment model for a data analytics application would likely involve multiple participants that belong to independent organizations. Moreover, such a collaborative development process is likely iterative which makes it fairly complex, and even worse, the various participants of such a distributed scenario may have competing interests in the market, with a low level of mutual trust. This raises the importance of enforcing accountability in the collaborative deployment modeling process, so that the changes made to the deployment model over the course of the creation process are immutably documented and linked with their authors. However, existing version control technologies such as Git and Subversion (SVN) are not enough to support the desired process as their main aim is aiding collaboration rather than accountability. For example, both approaches do not prevent that the history of log changes on the server are altered secretly by a malicious party, which destroys accountability.

In this paper, we tackle these issues by presenting a decentralized collaboration approach that allows developing deployment models in business-critical scenarios where accountability is of high importance. The approach we present in this paper is based on blockchains which we use as an immutable store for the various states a deployment model goes through. Participants use a modeling tool, like Winery [22], to enhance a shared deployment model, and before forwarding it to other participants, they register the version they have produced in the blockchain. Through the application of the approach, we achieve process integrity which means that we guarantee having an irrefutable proof of who is responsible for which changes. Furthermore, we also guarantee maintaining the provenance of the model and its sub-components in the form of an immutable history of changes that allows us to inspect how the model evolved, and who is responsible for potential malicious acts. We prove the technical feasibility of the approach based on a prototype integrated into Winery.

The remainder of this paper is structured as follows: in Sect. 2, we introduce the basics of deployment modeling, and motivate our work. In Sect. 3, we provide a formal description of the problem, whereas in Sect. 4 we present our blockchain-based approach. We prove the feasibility of the approach in Sect. 5, and evaluate the prototype in Sect. 6. Finally, we discuss the related work, and provide concluding remarks with a glance at the future work.

2 Motivation and Background

In this section, we provide some basic domain knowledge and emphasize the problem we aim to solve in our work, as well as framing its scope.

2.1 Deployment Automation Technologies and Models

Cloud computing environments nowadays support a high degree of service provisioning automation which allows customers to acquire cloud resources required

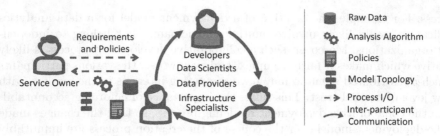

Fig. 1. Collaborative development of application deployment models.

for deploying their applications in a dynamic manner. Such automatic provisioning reduces the number of potential errors and accelerates the process. A provisioning engine is usually responsible for deployment automation, and supports describing the deployment configuration in the form of a *deployment model* [16]. A *declarative* deployment model describes the structure of the application to be deployed including all components it consists of as well as their configuration. For example, a declarative deployment model may specify that a certain java application has to be deployed on a Tomcat webserver listening on port 80, which shall be installed on a virtual machine running on Amazon EC. Thus, it describes only the desired goal, not how to execute the actual deployment technically. In contrast, *imperative* models explicitly describe *how* provisioning should take place in the form of a process describing each task to be executed. In this work, we focus on declarative deployment models, as they are less intrusive to developers, and because they are more widely accepted by most deployment automation technologies [11]. Developing such deployment models is often done by a single company as part of a regular DevOps process without external participants; however, in the following, we show how the development of specific kinds of applications requires collaborating with external parties.

2.2 Motivation: Business-Critical Deployments

The development of business-critical applications often requires the collaboration of multiple companies and experts. For example, a data-analytics application requires critical data, which a specialized data-owning company stores, to be accessed and processed by algorithms provided by data analysts. However, for such algorithm developments that require a significant level of expertise, often external companies are engaged [8,18], and the final results of the analysis could be consumed by a different company, such as an advertisement agency. If we want to deploy such an application to a cloud environment using the aforementioned declarative deployment models, we need a collaborative process in which each of the relevant companies participates by providing certain portions of the model.

Figure 1 shows an exemplary scenario to develop the deployment model of a cloud-based data-analytics application. In this scenario a company, which could be, e.g., an advertisement agency, is interested in the analysis of data from a

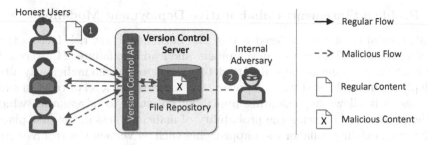

Fig. 2. Honest interaction with a centralized VCS (1) vs. malicious interaction performed by an internal adversary (2).

certain domain and requests the creation of a deployment model for a cloud application that fulfills its needs. The company, which we denote as a *Service Owner*, outsources the development process and associates its request with the set of requirements and policies it wants to enforce. The creation of a deployment model for this application would likely involve multiple participants that belong to independent organizations: a (i) *Data Scientist* provides the respective algorithms, and indicates the requirements they have for execution, such as the appropriate runtimes, the used libraries, etc. Furthermore, a (ii) *Data-owning Company* supplies the possibly business-critical data on which the algorithms operate, and specifies privacy and security requirements for the deployment, which could be, for example, that the deployed algorithms are only allowed to access specific databases of the company, and that these algorithms must not run on a shared IT resource. However, certain security requirements, e.g., that the deployed analysis algorithm must not access the Internet as otherwise business-critical data may leave the company without permission, have the potential of interfering with needs of the algorithm. Thus, a (iii) *Infrastructure Specialist* is required that refines the deployment model based on (1) the data scientists' technical requirements and the (2) data-owning company's security and privacy policies.

Moreover, such a collaborative development process can easily become iterative, e.g., if the final analysis results are not satisfying and changes need to be made to the provided algorithms by tuning their parameters, or to the set of accessible databases by allowing access to more of them. This makes the whole process fairly complex, and to make things worse, the various participants of such a distributed scenario may have competing interest in the market, and thus the degree of trust they have in each other is limited. For example, if the *Infrastructure Specialist* was malicious, they could alter the analysis algorithms to produce misleading results and then the blame could be falsely pointed at the *Data Scientist*, or they can alter the security policies provided by the *Data-owning Company* to allow illegal exportation of business-critical data.

This means that accountability plays a vital role in facilitating collaborative development of business-critical deployment models as it enables tracking the various changes made throughout the process and associating them with their authors. This allows the blame for malicious acts to be correctly directed.

2.3 Problem Statement: Collaborative Deployment Modeling

The absence of trust is considered one of the most important risk factors that can affect collaborative software development and lead it to failure [24]. Accountability, which we define as the ability to detect the actor(s) who maliciously alter the deployment model under development, can increase the level of trust in such processes as it allows implementing punishing measures to adversaries, which has the potential of reducing the probability of malicious acts in the first place.

However, existing collaboration approaches such as Version Control Systems (VCS) do not facilitate accountability. Centralized VCSs such as Github or SVN constitute a single-point of failure for collaborative processes. For example, a large DDoS attack stalled Github for five days in 2015 [20]. Furthermore, participants of collaborative processes need to trust the impartiality of parties operating these systems, and that they are well-protected against adversaries. Figure 2 shows that, although regular users interact with such centralized systems through their exposed APIs only, an adversary who has access to the internals of the server can directly alter the contents of the repository holding the shared files without having to use the API. These malicious edits would later propagate to the honest participants with the next pull from the server without them noticing the attack.

On the other hand, the Git protocol, which is the basis of Github and similar services, allows working in a peer-to-peer style without the need for a centralized origin server. However, this usage is discouraged as it increases the potential of confusing who is doing what, and it requires peers to be always online to provide access to their local repositories [15, p. 101]. Furthermore, both centralized and decentralized VCSs allow changing various aspects of the recorded history such as the author or the message of a commit [5,29], and although Git, as an example, allows using cryptographic signatures to associate commits to their authors, it does not enforce it [15, pp. 233–237], and even if a set of rules that enforces signing all commits were introduced on top of Git, this would not prevent unilaterally removing commits completely from the history log by one or more partners which cannot be proven by others, especially if a trusted intermediary is not used.

In this paper, we propose a blockchain-based approach that enforces accountability in the collaborative development process of deployment models while avoiding the need for trust in third-parties. Here, we do not aim at replacing the role of VCSs, but finding a way to ensure accountability regardless of the underlying technologies used to exchange and version files. So our approach may be combined with VCSs in a complementary way. For example, each participating organization can have a local VCS to organize the local development process.

3 Assumptions and Formal Definitions

The sensitive nature of certain collaborative processes makes it crucial to have the ability of tracking changes of the evolving deployment model even when the level of trust among the participants performing these changes is low. In this regard, we can identify two general aspects of change traceability that should be considered in order to achieve the desired accountability: (i) the **integrity** of the collarborative deployment modeling process, and (ii) the **provenance** of all artifacts contained in the deployment model. Before defining these two terms, we explain the assumptions we build upon.

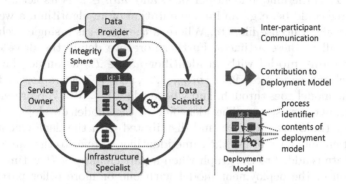

Fig. 3. Maintaining accountability in the collaborative deployment model development.

3.1 Identity Establishment Assumptions

The basic assumption we have regarding the deployment model and its sub-resources, such as algorithm implementations, data files, policies, etc., is that they can be uniquely identified during the collaboration process. Figure 3 shows an exemplary process in which multiple participants operate on a deployment model depicted as a single entity consisting of sub-resources which get added or modified in an iterative manner. The entity, shown as a square containing multiple artifacts, is associated with an identity seen by all participants. To achieve this, we assume that the participant who initiated the development processes creates an initial version of the deployment model associated with a unique identifier. Then, all other participants operate on this initial model while always referring to it using the same identifier. We refer to the set of tasks that allow participants to have a common view on the deployment model as the *integrity sphere*. Finally, we assume that all involved participants can also be uniquely identified. Later in this paper, we demonstrate how our approach fulfills these assumptions.

3.2 Formal Problem Definition

Now, we formally describe a data structure that captures the development of a deployment model and show why all participants should have access to it in order to ensure integrity and provenance, the two pillars of accountability.

Assuming the following notations: (i) $\mathcal{M}$ refers to all potential deployment models, (ii) $\mathcal{I}_m$ refers to all potential identifiers of deployment models, (iii) $\mathcal{S}$ refers to all potential sub-resources, (iv) $\mathcal{I}_s$ refers to all potential identifiers of sub-resources, (v) $\mathcal{D}$ refers to all potential contents of sub-resources, and (vi) $\mathcal{P}$ refers to all potential participants of the process, we define the **state of the**

deployment model at the point in time $t \in \mathcal{T}$ as a tuple $m_t = (author, S) \in \mathcal{M}$ where: (i) $author \in \mathcal{P}$ is the participant who authored this specific version, and (ii) $S \subseteq \mathcal{S}$ is the set of all sub-resources contained within the main resource at time t. Moreover, we define each **sub-resource** $s \in \mathcal{S}$ as a tuple $s = (id_s, data)$ where $id_s \in \mathcal{I}_s$ is the unique identifier of s, and $data \in \mathcal{D}$ is its actual content. A sub-resource could be, e. g., an implementation of an algorithm, a web server configuration file, a license file, etc. Whereas the model is a single archive that encapsulates all of these artifacts. Furthermore, we define the **development of a deployment model** with an identifier $id_m \in \mathcal{I}_m$ as an acyclic digraph $G_{id_m} = (V, E)$ whose vertices $V = \{m_{start}, \ldots, m_{end}\} \subseteq \mathcal{M}$ represent the various states the model goes through, and whose edges $E \subseteq V \times V$ represent causal dependencies between states. The identifier of the model development graph, id_m, is set by the *Service Owner*, and distributed with the deployment model. Further participants have to use the same identifier to refer to the same process.

A new state is added to the graph when a participant $p \in \mathcal{P}$ at time t shares a new version of the deployment model with one or more other participants, thus, our notion of time is discrete, and new time points are added only when new states emerge in the model development graph. Our approach, as explained in Sect. 4.1, guarantees that no two distinct deployment model states can share the exact same point in time of creation. Furthermore, we define $\mathcal{A} : \mathcal{I}_m \to \mathcal{P}(\mathcal{P}_{bc})$ as a mapping that, given a deployment model identifier, returns the set of participants authorized to take part in the process. For convenience and to enhance readability, we define the projection $\pi_e(tuple)$, which returns the element labeled e of a given $tuple := (.., e, ..)$. Based on these notations, we define the notion of integrity:

Definition 1 (Integrity of Collaboration Process). *Integrity is the fact that the author of any deployment model state of a given development graph G_{id_m} is an authorized participant:* $\forall m_t \in \pi_V(G_{id_m}) : \pi_{author}(m_t) \in \mathcal{A}(id_m)$ □

This means that if the deployment model transitions to a new state which we cannot associate to any of the authorized participants of the process, we should consider this state as invalid, and refuse to operate on it. Furthermore, integrity guarantees the non-repudiation of participants' actions meaning that a participant causing the state of the deployment model to change will not be able to deny their responsibility for this action.

On the other hand, provenance refers to the ability to identify all states a sub-resource goes through, which includes knowing the set of participants who operated on it, and when state changes occurred:

Definition 2 (Provenance of a Sub-resource). *Given a model development graph G_{id_m}, provenance is a mapping:* $p : \mathcal{S} \to \mathcal{P}(\mathcal{T} \times \mathcal{P} \times \mathcal{D})$ *which is defined as:*

$$p(s) := \{(t, \pi_{author}(m_t), \pi_{data}(\sigma)) \in \mathcal{T} \times \mathcal{P} \times \mathcal{D} \mid$$
$$m_t \in \pi_V(G_{id_m}) \wedge \sigma \in \pi_S(m_t) \wedge \pi_{id_s}(\sigma) = \pi_{id_s}(s)\}$$

□

We notice that sub-resources are always transmitted as part of the deployment model, so their evolution is coupled with its evolution.

4   Blockchain-Based Approach to Ensuring Accountability

As seen in the previous section, guaranteeing accountability requires having the development graph of the deployment model accessible to all participants so they can add to it when producing a new version, or read from it when validating the integrity of a given one, and even discover how it evolved over time. Our approach, which aims at storing and managing this graph so that accountability is achieved, works by combining three complementing aspects: (i) using blockchains to create an immutable trace of all state changes of the deployment model, (ii) storing the detailed differences between resource states in a decentralized storage system to facilitate a fine level of accountability, and finally (iii) building a tree-of-trust that allows authorizing participants and maintaining their identities. Before giving more details about the approach, we first summarize why we have chosen blockchains as the core technology to implement it.

4.1   Suitability of Public Blockchains

Maintaining the globally accessible model development graph is straight forward in a centralized setup. However, such an approach requires the involvement of a trusted third-party, which is sometimes an unrealistic assumption [32]. Consequently, we aim to find a decentralized approach that does not have such a requirement by delegating the responsibility of managing the model development graph to the set of network peers collectively. However, the task is more challenging in a decentralized setup as we do not trust any specific peer to behave according to the protocol we set, nor do we trust the transportation channels among peers. This problem is known as the Byzantine Fault Tolerance (BFT) problem [23], and public blockchains provide a probabilistic approach to solve it [25].

Blockchain systems are usually used in situations were several parties need to share a common state regardless of the fact that they do not trust each other nor do they trust a third-party. They started as the backbone of crypto-currencies such as Bitcoin [25], but later involved use-cases in other fields, such as finance [28], and supply chains [13]. Moreover, looking at the high-level properties of the public blockchains, we can judge their suitability to our use-case. Public blockchain consensus protocols, such as Proof-of-Work [25] and Proof-of-Stake [21], periodically produce blocks of transactions which are broadcast to the peers of the network. These transactions are ordered within the blocks, and each of them alters the world state represented by the blockchain. For example, in the case of Bitcoin, the state is the set of balances of all accounts, and transactions represent currency transfer operations. However, if we apply this concept to our use-case, the state would be the deployment model itself, and each transaction would represent the creation of a new version of it. The aforementioned consensus

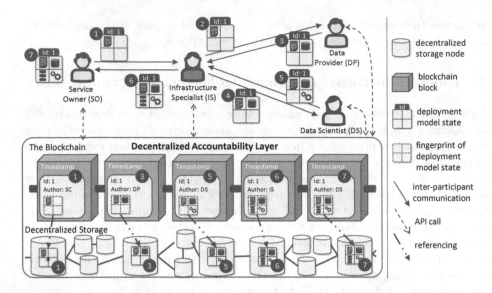

Fig. 4. Blockchain-based accountability approach applied to an exemplary use-case

protocols also ensure that altering an existing portion of the blockchain is practically infeasible without controlling either a large portion of the network's computing power (in the case of Proof-of-Work), or a large portion of the network's
stakes (in the cased of Proof-of-Stake). This ensures that the blockchain provides
not only the current state of the shared resource, but also an immutable history
of it.

This means that each participant will be able to locally construct the model
development graph G_{id_m} by traversing the blockchain and applying all relevant transactions in order. What makes blockchains even more suitable for the
use case at hand is that they use public-key cryptography to sign all submitted
transactions, which ensures their authenticity, i.e., that the author of each transaction is actually who they claim to be. Being a decentralized and an immutable
ledger of discrete events that ensures authenticity of its records, makes the public
blockchain suitable for our use-case.

4.2   Maintaining an Immutable History of State Changes

In this section, we explain the first angle of the approach. The basic idea here
is demonstrated in Fig. 4 which shows its application to the motivational scenario we presented earlier in Sect. 2.2. The collaboration process starts when the
Service Owner creates an initial version of the deployment model, which contains a set of policies, and gives it its unique identifier. Before forwarding the
resource to the next participant (the *Infrastructure Specialist* in this scenario),
the he registers it in the blockchain by issuing a transaction which contains
enough information about this specific version of the model and the associated

sub-resources to allow their unique identification. We call this information a fingerprint, and in Sect. 4.3 we explain it in more details. After making sure that the transaction containing the fingerprint is durably persisted in the blockchain, the *Service Owner* sends the model to the *Infrastructure Specialist*. It is irrelevant to our approach how the model is sent to the next participant as long as it was registered in the blockchain beforehand. When the model is received, the *Infrastructure Specialist* first checks the blockchain for stored versions of the same model and compares them to the received one. If the model is found to be registered beforehand in the blockchain, then its integrity is guaranteed; otherwise, the ownership of the received model cannot be proven, which means that it lacks integrity, and in such a case she should refuse working on it. We call this check integrity verification, and if it is successful, the *Infrastructure Specialist* can operate on the model. In this case, she needs the inputs of the *Data Provider*, and the *Data Scientist* before she can decide on the appropriate infrastructure, so she just forwards it to them without modifying it, and since no new version of the model is generated, she does not need to store anything in the blockchain. Of course this is just an exemplary scenario; in other cases a common default infrastructure may be specified before the others enhance it with their artifacts. The process continues similarly: whenever participants receive a new version of the deployment model they: (i) verify its integrity by checking its stored versions in the blockchain, and if the check passes, they (ii) optionally operate on it by, e. g., adding new sub-resources to it, or altering the content of existing ones. If such changes were performed, (iii) they store a fingerprint of the new version in the blockchain, and then (iv) they further forward the model.

In Fig. 4, the solid arrows represent sending versions of the deployment model directly from one participant to the other, whereas the dashed ones represent communication between participants and the decentralized accountability layer. Furthermore, the numbered circles represent the order in which events happen in this specific scenario. The figure also shows how the blockchain gives total order to its transactions, and consequently to the stored deployment model states. This allows the reconstruction of the provenance of any given sub-resource by traversing the transactions in order, and collecting the partial fingerprints belonging to the sub-resource along with the author and the timestamp of each version. But how can we find out details about the exact changes that happened to each sub-resource? We answer this question in the following section.

4.3   Storage of Deployment Model States

The storage of arbitrary data in public blockchains is generally very expensive because all peers of the protocol need to store the entire blockchain locally. This led Ethtereum, for example, to set high fees for data storage. The general approach to address this issue is putting only a digest of the actual content in the blockchain rather than storing it entirely there. However, our scenario requires identifying the exact changes made by each participant, but storing merely the hash of the deployment model only helps detecting that something has changed from one state to another, but not exactly what. We present two approaches that ensure capturing enough details about the deployment model state changes to facilitate accountability while not storing the entire model in the blockchain.

In the first approach we try to identify the data items of the deployment model that correspond to changes exclusive to one participant. For example, in a certain scenario we could expect that a configuration file of an application server is expected to be created and edited by a single participant only, so storing its hash in the blockchain allows us to identify if some other participant breaks this rule. However, in other cases we could expect that such a configuration file can be altered by multiple participants, but that each entry of it, e. g., the listening port of the server, can be modified by a single participant only. In this case we store a map of the hashes of these entries in the blockchain instead. When we identify the correct level of data items that corresponds to changes exclusive to a single participant for each of the sub-resource types we have, i.e., algorithm implementations, configuration files, etc., we create a collection of the hashes of these data items and store it in the blockchain for every state change we go through. We call this collection the **fingerprint** of the state. This fingerprint allows us to detect malicious participants that edit data items that are not expected to be altered by them. However, if the level of exclusive changes is too low, e. g., at the level of single characters, storing the fingerprint in the blockchain becomes even more expensive than storing the whole state there.

In the second approach we store the whole deployment model in a content-addressable decentralized file storage accessible by all participants, such as Swarm [6], or Inter Planetary File System (IPFS) [9]. A decentralized storage provides at least the following two functions for storing and retrieving arbitrary files: $store : \mathcal{D} \rightarrow \mathcal{H}$ and $retrieve : \mathcal{H} \rightarrow \mathcal{D}$ where $\mathcal{H}$ is the range of some hash function h. At the same time, we store only the hash in the blockchain, which serves both as the address of the content in the decentralized storage and as a guarantee that it has not been altered there. However, in this approach encryption needs to be utilized, or else the privacy of stored content is lost.

In our work here, we use a hybrid of these two approaches: in the blockchain, we store a fingerprint listing all sub-resources identifiers and their hashes, whereas in the decentralized storage, we store the actual artifacts. In order to generate the fingerprint of a deployment model state m_t, we use the following function: $fingerprint(m_t) := \{(id_s, h(data))|s = (id_s, data) \in \pi_S(m_t)\}$, then the blockchain transaction itself would be: $tx(m_t) := (id_m, author, fingerprint(m_t))$. To reconstruct the actual sub-resources of m_t, the fingerprint can be used to locate content in the decentralized storage as demonstrated in Fig. 4.

4.4   Establishing Identity and Authenticity

Public blockchains allow public access to their content; anyone can run a peer node and submit new transactions to it or read existing transactions. Consequently, anyone can pretend to be a participant in the development process, and inject fake or altered fingerprints of the deployment model. Thus, we need an additional mechanism to ensure the authenticity of participants without imposing the requirement of knowing all participants at the beginning of the process which is unrealistic in large-scale collaborations. Moreover, to identify their users, public blockchains use pseudonyms, which are hashes of the public key of each participant [25,34], thus guaranteed to be unique while hiding the true identity. However, regarding accountability, it might be beneficial to expose actual identities.

A potential way to approach these issues is the usage of permissioned blockchains like Hyperledger Fabric [12], which establishes the identity of all participants and implements access-control to its content. However, the usage of such an architecture increases the centralization of the system. For example, a Membership Service Provider (MSP) is needed for the purposes of certificate issuing, validation, and user authentication, and its configuration is stored in the genesis block of the ledger [4]. MSPs are centralized services whose setup and operation need to be agreed upon by the participants beforehand. This has the same drawbacks as needing a trusted third-party, and further constitutes a single point of failure for the system. We propose a simple alternative approach to solving these problems which also builds upon public blockchains.

The basic idea is building a tree of trusted participants rooted at the *Service Owner*, which can be augmented with the Real-World Identity (RWI) of each one of them if needed by the use case. This tree is built with the assumption that if a participant A forwards the deployment model to another participant B, then A trusts B and knows their RWI. Based on this assumption, the approach simply requires A to insert a new node in the tree including this trust information before actually forwarding the model to B (providing that B is not already part of the tree). If all participants follow this rule, then we build a tree of trust rooted at the *Service Owner*, who, by transitivity, would trust all included participants.

Formally, based on the previous notations, and providing that $\mathcal{P}_{bc}$ refers to all potential blockchain-based participant identities (pseudonyms), referred to earlier as $\mathcal{P}$, and that $\mathcal{P}_{rwi}$ refers to all potential RWIs of participants, e. g., their names or email addresses, the tree of trust for a given deployment model id_m is identified as a tuple $T_{id_m} := (ownerId, V_T, E_T)$ where: (i) $ownerId$ is the blockchain-identity of the *Service Owner*, (ii) $V_T \subseteq \mathcal{P}_{bc} \times \mathcal{P}_{rwi}$ is the set of tree nodes each of which, $v := (bcid, rwid)$, represents the blockchain-identity and, optionally, the RWI of a trusted participant, and (iii) $E_T \subseteq V_T \times V_T$ is the set of tree edges that represent the inclusion of new participants by an existing one. In order to build this tree, we also use a public blockchain; when A wants to include B in the tree they submit a blockchain transaction $tx := (id_m, authorizer, authorized, rwi)$ where: (i) id_m: the identifier of the deployment model, (ii) $authorizer, authorized$: the blockchain identities of A and B, and (iii) rwi: the RWI of B. Reconstructing the tree of trust based on the set of all relevant transactions R_{id}, which can be obtained by traversing the blockchain, happens as follows:

$$ownerId := \pi_{authorizer}(tx) : \nexists tx_1 \in R_{id} \ s.t. \ tx_1 < tx$$

$$V_T := \{(ownerId, \_)\} \bigcup \{(\pi_{authorized}(tx), \pi_{rwi}(tx)) | \ tx \in R_{id}\}$$

$$E_T := \{(v_1, v_2) \in V_T \times V_T | \ tx \in R_{id} \ s.t.$$

$$\pi_{authorizer}(tx) = \pi_{bcid}(v_1) \ \wedge \ \pi_{authorized}(tx) = \pi_{bcid}(v_2)\}$$

where $<$ uses the total order of transactions guaranteed by the blockchain, and ($\_$) refers to an empty value (as the RWI of the Service Owner is not important). Now, we can show how we implement the mapping $\mathcal{A}$ that describes the authorized participants using the tree-of-trust (cf. Sect. 3.2).

Definition 3 (Authorized Participants Mapping). *Given a tree-of-trust of a collaboration process T_{id_m}, the mapping $\mathcal{A} : \mathcal{I}_m \rightarrow \wp(\mathcal{P}_{bc})$ is defined as:*

$$\mathcal{A}(id_m) := \{p \in \mathcal{P}_{bc} | \ \exists path = \langle v_{start}, ..., v_{end} \rangle \ in \ T_{id_m} \ s.t.$$

$$\pi_{bcid}(v_{start}) = ownerId \wedge \pi_{bcid}(v_{end}) = p\}$$

$\square$

Here, we simply find the participants reachable form the root with a path.

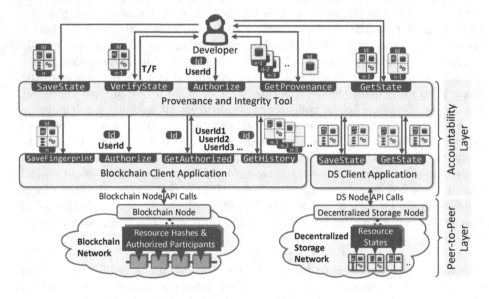

Fig. 5. The system architecture

5   Proof of Concept

The goal of this chapter is demonstrating how we realized the proposed approach by introducing a system architecture that supports it and by providing a prototypical implementation of this architecture that proves its feasibility.

5.1   System Architecture

Figure 5 depicts the suggested system architecture. The architecture is divided into two layers: the first layer, i.e., the **Peer-to-peer Layer** is located at the bottom of the architecture, and its components are not part of the client application itself, but rather allow it to access both the blockchain network, and the decentralized storage network. The layer represents the point of view in which a peer sees these networks: in order for peers to communicate with a public blockchain network, which is collectively responsible for storing the blockchain data, they need to run a local node implementing the corresponding blockchain protocol, and exposing an API that allows utilizing it. The same applies to the decentralized storage network; a local node is also needed here.

On the other hand, our approach relies on smart contracts, which are decentralized applications that provide an easy way to store and manage arbitrary data on the blockchain, to create two necessary repositories: (i) a repository for storing and managing the various states of the deployment model, and (ii) a repository for storing and managing the tree-of-trust for the collaboration process. The tasks of reading from these repositories and writing to them are part of the node's functionality, and are exposed via its API. The second layer, i.e., the **Accountability Layer** is responsible for implementing the logic of the proposed approach. It is divided into two sub-layers: The one at the bottom consists of two client applications that communicate with the local peer-to-peer nodes on one hand, and simplify interacting with them on the other hand by exposing easy-to-use domain-specific operations to the sub-layer above it. The operations `SaveState`, and `SaveFingerprint` are responsible respectively for storing the state of a deployment model in the decentralized storage, and for storing its fingerprint in the blockchain. Besides its original purpose as a means to establish integrity of the deployment model, the fingerprint stored in the blockchain allows us to retrieve the state from the content-addressable decentralized storage using the `GetState` operation. On the other hand, the operation `GetHistory` gets the history of fingerprints for a given collaboration process, whereas `Authorize`, and `GetAuthorized` are used to authorize a new participant, and to read the list of

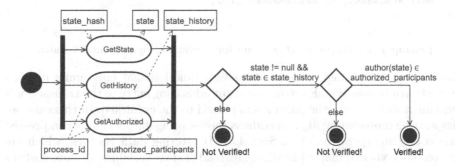

Fig. 6. UML activity diagram showing the state verification process

all authorized participants respectively. The role of this sub-layer is similar to
the role of the Data Source Layer of the 3-layer architecture [19].

The sub-layer at the top has the role of providing the "business-logic", and it
divides its functionality into a set of operations: GetState, and Authorize simply expose the corresponding operations of the sub-layer below it. However, the
other three operations are more sophisticated; GetProvenance gets the provenance of a specific sub-resource of the whole deployment model. To this end,
it utilizes the GetHistory operation and uses its results to retrieve a complete
history using the GetState operation, and finally analyzes the results. Moreover,
SaveState is called before sending a deployment model to the next participant
and it stores the current state of the resource in the decentralized storage using
the SaveState operation at the lower level, and it also stores a fingerprint of
the state in the blockchain by invoking the SaveFingerprint operation. On the
other hand, VerifyState is invoked when receiving a new version of the deployment model in order to determine its integrity. The execution of this operation
is depicted in Fig. 6 which shows that it uses the GetHistory, GetState and
GetAuthorized operations of the sub-layer below to achieve its tasks.

5.2   Prototype

To validate the feasibility of the architecture, we have implemented a prototype that realizes it which is publicly accessible via Github[1]. To implement
the bottommost layer of the architecture, we have chosen Ethereum [34] as the
underlying blockchain protocol since it is the most mature public blockchain
that supports smart contracts and because it can be easily integrated with other
peer-to-peer protocols from the Ethereum Foundation [17]. Smart contracts are
programmed using Solidity, a specialized Turing-complete language. Listing 1
shows the smart contract responsible for managing the repository of the deployment model states:

```
contract Provenance {
  event NewState(string indexed _id, address indexed _author, bytes _fp);
  function addState(string _id, bytes _fp) public {
    emit NewState(_id, msg.sender, _fp);
  }
}
```

Listing 1. Ethereum smart contract for storing deployment model states

The contract has a single event and a single function that only emits it when
invoked via a transaction. Emitting an event adds a log entry to the transaction's
permanent storage with the parameters passed to the event [34]. In this case, we
pass a string representing id_m, an address representing $author$, and a compressed
version of $fingerprint(m_t)$ (c.f. Sect. 4.3). By storing data via logs, which are
only readable via external applications, instead of state variables, we save costs as
the price for storing them is cheaper [34]. Moreover, adding the keyword *indexed*

---

[1] https://github.com/OpenTOSCA/winery/releases/tag/paper%2Fgf-accountability.

to an event parameter speeds up queries that retrieves log entries with a filter based on this parameter. Finally, the smart contract for managing the tree-of-trust is formulated similarly. We connect our local node to a testnet instead of the mainnet as it allows us to run experiments free of charge.

On the other hand, we have chosen Swarm [6] as the implementation of the decentralized storage as it is meant to be self-sustainable in the sense that network peers are motivated to keep copies of the shared resources on their local nodes because they get paid for this with Ethereum tokens (Ethers), which allows peers to operate in an upload and disconnect mode [31].

Furthermore, we have integrated the *Accountability Layer* as a reusable sub-module into Winery [22], a web-based environment that allows the graphical modeling of TOSCA [1] topologies. Winery sub-modules are written in Java, and packaged using Maven. The prototype supports Cloud Service ARchives (CSARs) as deployment models. These archives are part of the TOSCA standard, and they contain the topology, management plans, and the various other software artifacts required for the correct provisioning, and operation of a cloud application. An important part of the CSAR is the *TOSCA Meta File* which allows to interpret its various components properly. This file is divided into blocks that describe each of the artifacts contained in the CSAR.

We use the TOSCA meta file as a fingerprint of the whole CSAR, so instead of storing the actual archive in the blockchain, we store the meta file (cf. Sect. 4.3). To this end, we augment each block with the digest of the corresponding artifact (sub-resource), so we make sure we capture any changes made to it over the stored state versions. When we store these fingerprints in the blockchain, they become immutable, and providing that all blockchain transactions are signed by their creators, we can identify the author of each version of the CSAR, and we can detect whenever a contained artifact is changed and by whom. Furthermore, we compress the contents of the meta file before storing it due to cost reasons. Finally, to allow showing detailed difference between the various states of a given sub-resources, when we store the fingerprint in the blockchain, we also store the changed artifacts in the Swarm. These artifacts can be referenced using their hashes which are parts of the fingerprint stored in the blockchain. This allows us to dynamically retrieve them, e.g., when the user wants to visualize the provenance of a specific sub-resource. In a previous work [35], we have shown how to address the issue of securing parts of a CSARs in the context of collaborations. This can also be applied in the context of the files we store in the Swarm to guarantee sensitive data is not leaked.

6   Evaluation

In this section we show the applicability of the aforementioned prototypical implementation to real-world use cases by evaluating the costs and additional execution times incurred when using it.

6.1 Cost

The costs incurred by the approach are divided into: (i) infrastructure costs, (ii) Ethereum transaction costs, and (iii) Swarm storage costs.

To allow the prototype to access the Ethereum and Swarm networks, participants are advised to run and maintain local nodes. These nodes need to be in sync with other nodes in their networks before they are usable, so it is suggested that they are always online. An alternative is connecting to publicly accessible nodes instead, but that increases security risks. The costs incurred here are due to maintaining these nodes and providing them with Internet connection. On the other hand, each Ethereum transaction incurs certain costs [34]. A transaction that invokes a smart contract has some fixed base cost in addition to costs related to the complexity of the code executed and the size of data newly stored in the blockchain due to this execution. These costs are introduced to prevent malicious or buggy code from running indefinitely or for too long on Ethereum nodes and they are calculated in terms of *Gas*. The author of a transaction pays for the gas it consumes, and the node mining it receives this payment as a fee. The price of the unit of gas is determined by the author, and it affects how quickly the transaction is processed by the network since miners are motivated to pick transactions with higher fees first when formulating new blocks. In the following, we show an estimation of the costs of the transactions issued by the prototype assuming a price of 22×10^{-9} Ethers per unit of gas which usually results in processing the transaction in the next 1–2 blocks. Furthermore, we assume an Ether to Euro conversion rate of 237.49 which is valid at the time of writing. The prototype issues two types of transactions: (i) *authorization transactions* to add entries to the tree-of-trust (c.f. Sect. 4.4). These transactions incur a minor cost due to the low amount of data stored. On average, a transaction of this type consumes 31000 gas units which corresponds to ~0.0007 Ethers or ~**0.17 Euros**. (ii) *provenance transactions* that contribute to building the graph G_{id_m} (c.f. Sect. 3.2). Most of the gas consumed here is due to the TOSCA meta file which we store compressed in them as a model fingerprint (c.f. Sect. 5.2), thus the more complex the model is, the higher the cost it incurs in this context since the complexity of the model is reflected in a larger TOSCA meta file. Table 1 summarizes the costs incurred by sample CSAR files of various complexities. Here we measure the complexity by the number of associated sub-resources (files).

Finally, as mentioned earlier, a decentralized storage system requires a mechanism to incentivize peers to store content of other peers. The amount of the

Table 1. Costs incurred by storing TOSCA metadata files of various sizes in Ethereum

CSAR complexity	Metadata file size	Gas consumed	Price (Ether/Euro)
Low (73 files)	9958 B	461106	(0.010144332/**2.41**)
Medium (111 files)	13786 B	639123	(0.014066706/**3.34**)
High (133 files)	17234 B	795659	(0.017504498/**4.16**)

incentive would depend on the size of files stored and the degree of replication required. However, Swarm has not implemented such a mechanism yet (planned for 2019 [30]). Thus for now no additional costs are incurred by using Swarm apart from operating a local node.

6.2 Execution Time

The usage of the prototype in Winery increases the execution times of both exporting a CSAR and importing it. When exporting a CSAR, additional time is required to (i) hash the contents of the sub-resources in order to formulate the fingerprint, (ii) store the contents of the CSAR in a local Swarm node, and (iii) issue a blockchain transaction that contains the fingerprint. Whereas, when importing a CSAR additional time is required to validate its integrity by querying the local Ethereum node for relevant information (c.f. Fig. 6). Ethereum blocks are generated by the network at almost a constant rate of 15 s [34]. Thus, depending on the traffic and the gas price, time factor (iii) is usually 15–30 s pertaining to a delay of 1 to 2 blocks. On the other hand Table 2 shows the effect of the remaining factors on the import and export times of 5 exemplary CSAR files with different sizes and sub-resource counts. These measurements were performed on a computer running Windows 10 64-bits with a (Intel(R) Core(TM) I7-4710MQ CPU @ 2.5 GHz) processor and 16 GB of RAM.

Table 2. Increase in CSAR import/export times when using the prototype

CSAR	Export time (mm:ss)		Import time (mm:ss)	
	Original	Increase	Original	Increase
Model 1 (73 files, 19 MB)	02:12	00:09	00:02	00:04
Model 2 (106 files, 27.8 MB)	02:37	00:08	00:03	00:04
Model 3 (111 files, 20 MB)	02:48	00:08	00:02	00:05
Model 4 (121 files, 11 MB)	02:49	00:06	00:03	00:05
Model 5 (133 files, 27 MB)	03:01	00:11	00:04	00:06

7 Related Work

Web-of-Trust (WoT) (Philip [26]) provides a distributed alternative to the centralized Public-Key Infrastructure (PKI). Our identity and authenticity approach shares common features with WoT; the act of inserting another participant's address in the blockchain is comparable to signing another person's certificate in the WoT because all transactions in the blockchain are signed by their authors, and a participant's address is derived from their public key. On the other hand, our approach is simple and suitable specifically for collaborations initiated by a Service Owner, whereas, WoT is a general-purpose scheme which addresses further aspects irrelevant to the use case at hand.

Furthermore, several centralized approaches were suggested to support collaborative development processes in various domains [14,33]. However, these approaches share common properties such as centralized access-control mechanisms or strong assumptions about the underlying data structures which make them not suitable in the context of business-critical collaborative deployment modeling. Furthermore, they do not focuses on maintaining model provenance.

On the other hand, few decentralized approaches exist for supporting collaboration in software development scenarios. These approaches focus on providing a decentralized implementation of the Git protocol; Rashkovskii [27] proposes a Git implementation in which files are distributed on a Bitcoin-like proof-of-work-based blockchain network, which would ensure immutability of commit histories, and thus promote accountability of participants' actions. However, the approach requires a new proof-of-work blockchain, and since the security of proof-of-work depends mainly on the hash-rate of the network, many participants need to adopt it before it becomes usable, which reduces the trust in the approach altogether. Furthermore, Ball [7] shows how the file storage for Git can be implemented using the BitTorrent P2P protocol. The approach is completely decentralized, and uses Bitcoin as a name directory for repository addresses. However, the history of commits is not immutably persisted, and thus accountability is not achieved. Finally, Beregszaszi [10], also suggests a decentralized storage of Git objects, but with the help of IPFS [9] which is a content-addressed P2P filesystem. The approach further depends on Ethereum smart contracts to provide access-control and to manage pointers of the latest repository revisions. However, the history of commits is not stored in the blockchain, and thus resource provenance and process accountability cannot be guaranteed.

8   Concluding Remarks and Future Work

In this paper we presented a novel approach to enable accountability in the collaborative development processes of declarative deployment models for business-critical applications. Our approach is decentralized and based on blockchains. This allowed us to ensure integrity and provenance, the two aspects of accountability, without the need for a trusted third-party which could be difficult to agree upon, and would constitute a single point of failure. Our approach also provided a decentralized mechanism to maintain the identity and authenticity of all participants as part of the process integrity. Finally, as a future work, we plan to enhance the security of the approach via tackling the issue of an authorized participant losing their private key by making the tree-of-trust dynamic based on participants' behavior, or by adding a blacklisting mechanism to it.

Acknowledgments. This research was funded by the Ministry of Science of Baden-Württemberg for the doctoral program "Services Computing", and by the project SePiA.Pro (01MD16013F) of the BMWi program Smart Service World.

References

1. Topology and orchestration specification for cloud applications, November 2012. http://docs.oasis-open.org/tosca/TOSCA/v1.0/os/TOSCA-v1.0-os.html
2. Chef, May 2018. https://www.chef.io/
3. Kubernetes, May 2018. https://kubernetes.io
4. Membership service providers (MSP), July 2018. http://hyperledger-fabric.readthedocs.io/en/release-1.1/msp.html
5. Rewriting history – Git commit -amend and other methods of rewriting history (2018). https://www.atlassian.com/git/tutorials/rewriting-history
6. Swarm, July 2018. https://github.com/ethersphere/swarm
7. Ball, C.: Announcing GitTorrent: a decentralized GitHub, May 2015. https://blog.printf.net/articles/2015/05/29/announcing-gittorrent-a-decentralized-github/
8. Baumann, F.W., Breitenbücher, U., Falkenthal, M., Grünert, G., Hudert, S.: Industrial data sharing with data access policy. In: Luo, Y. (ed.) CDVE 2017. LNCS, vol. 10451, pp. 215–219. Springer, Cham (2017). https://doi.org/10.1007/978-3-319-66805-5_27
9. Benet, J.: IPFS-content addressed, versioned, P2P file system. arXiv preprint arXiv:1407.3561 (2014)
10. Beregszaszi, A.: Mango, July 2016. https://medium.com/@alexberegszaszi/mango-git-completely-decentralised-7aef8bcbcfe6
11. Bergmayr, A.: A systematic review of cloud modeling languages. ACM Comput. Surv. (CSUR) 51(1), 22 (2018)
12. Cachin, C.: Architecture of the hyperledger blockchain fabric. Technical report, IBM Research - Zurich (2016)
13. Cecere, L.: Seven use cases for hyperledger in supply chain, January 2017. http://www.supplychainshaman.com/big-data-supply-chains-2/10-use-cases-in-supply-chain-for-hyperledger/
14. Cera, C.D.: Role-based viewing envelopes for information protection in collaborative modeling. Comput.-Aided Des. 36(9), 873–886 (2004)
15. Chacon, S., Straub, B.: Pro Git. Apress, Berkeley (2014)
16. Endres, C., Breitenbücher, U., Falkenthal, M., Kopp, O., Leymann, F., Wettinger, J.: Declarative vs. imperative: two modeling patterns for the automated deployment of applications. In: Proceedings of the 9th International Conference on Pervasive Patterns and Applications, pp. 22–27. Xpert Publishing Services (XPS) (2017)
17. Ethereum Foundation: Web3 base layer services, August 2018. http://ethdocs.org/en/latest/contracts-and-transactions/web3-base-layer-services.html
18. Falkenthal, M., et al.: Towards function and data shipping in manufacturing environments: how cloud technologies leverage the 4th industrial revolution. In: Proceedings of the 10th Advanced Summer School on Service Oriented Computing, pp. 16–25. IBM Research Division (2016)
19. Fowler, M.: Patterns of Enterprise Application Architecture. Addison-Wesley Longman Publishing Co., Inc., Boston (2002)
20. Goodin, D.: Massive denial-of-service attack on GitHub tied to Chinese government, March 2015. https://arstechnica.com/information-technology/2015/03/massive-denial-of-service-attack-on-github-tied-to-chinese-government/
21. King, S., Nadal, S.: PPCoin: peer-to-peer crypto-currency with proof-of-stake, August 2012. https://peercoin.net/assets/paper/peercoin-paper.pdf
22. Kopp, O., Binz, T., Breitenbücher, U., Leymann, F.: Winery – a modeling tool for TOSCA-based cloud applications. In: Basu, S., Pautasso, C., Zhang, L., Fu, X. (eds.) ICSOC 2013. LNCS, vol. 8274, pp. 700–704. Springer, Heidelberg (2013). https://doi.org/10.1007/978-3-642-45005-1_64

23. Lamport, L., Shostak, R., Pease, M.: The Byzantine generals problem. ACM Trans. Program. Lang. Syst. **4**(3), 382–401 (1982)
24. Mohtashami, M., Marlowe, T., Kirova, V., Deek, F.P.: Risk management for collaborative software development. Inf. Syst. Manag. **23**(4), 20–30 (2006). https://doi.org/10.1201/1078.10580530/46352.23.4.20060901/95109.3
25. Nakamoto, S.: Bitcoin: a peer-to-peer electronic cash system (2008)
26. Philip, Z.: PGP user's guide, volume I: essential topics. Phil's Pretty Good Software, version 2.6.2, October 1994. https://web.pa.msu.edu/reference/pgpdoc1.html
27. Rashkovskii, Y.: Gitchain, September 2014. https://www.kickstarter.com/projects/612530753/gitchain/description
28. Schwartz, D., Youngs, N., Britto, A., et al.: The Ripple protocol consensus algorithm (2014). https://ripple.com/files/ripple_consensus_whitepaper.pdf
29. Steinbeis, G.: Change author of SVN commit, June 2011. https://blog.tinned-software.net/change-author-of-last-svn-commit/
30. Trón, V.: Announcing Swarm proof-of-concept release 3, June 2018. https://blog.ethereum.org/2018/06/21/announcing-swarm-proof-of-concept-release-3/
31. Trón, V., Fischer, A., Nagy, D.A., Felföldi, Z., Johnson, N.: Swap, swear and swindle: incentive system for swarm, May 2016. https://swarm-gateways.net/bzz:/theswarm.eth/ethersphere/orange-papers/1/sw
32. Viriyasitavat, W., Martin, A.: In the relation of workflow and trust characteristics, and requirements in service workflows. In: Abd Manaf, A., Zeki, A., Zamani, M., Chuprat, S., El-Qawasmeh, E. (eds.) ICIEIS 2011. CCIS, vol. 251, pp. 492–506. Springer, Heidelberg (2011). https://doi.org/10.1007/978-3-642-25327-0_42
33. Wang, Y.: Intellectual property protection in collaborative design through lean information modeling and sharing. J. Comput. Inf. Sci. Eng. **6**(2), 149–159 (2006)
34. Wood, G.: Ethereum: a secure decentralised generalised transaction ledger - Byzantium version (2018). https://ethereum.github.io/yellowpaper/paper.pdf
35. Yussupov, V., Falkenthal, M., Kopp, O., Leymann, F., Zimmermann, M.: Secure collaborative development of cloud application deployment models. In: Proceedings of the 12th International Conference on Emerging Security Information, Systems and Technologies (SECURWARE) (2018)

Evaluating Multi-tenant Live Migrations Effects on Performance

Guillaume Rosinosky[1]([✉]) [iD], Chahrazed Labba[1] [iD], Vincenzo Ferme[2,3],
Samir Youcef[1] [iD], François Charoy[1] [iD], and Cesare Pautasso[3]

[1] Université de Lorraine, CNRS, Inria, LORIA, 54000 Nancy, France
{guillaume.rosinosky,chahrazed.labba,samir.youcef,
francois.charoy}@loria.fr
[2] Reliable Software Systems, University of Stuttgart, Stuttgart, Germany
[3] Software Institute, University of Lugano, Lugano, Switzerland
{vincenzo.ferme,cesare.pautasso}@usi.ch

Abstract. Multitenancy is an important feature for all Everything as
a Service providers like Business Process Management as a Service. It
allows to reduce the cost of the infrastructure since multiple tenants
share the same service instances. However, tenants have dynamic workloads. The resource they share may not be sufficient at some point in
time. It may require Cloud resource (re-)configurations to ensure a given
Quality of Service. Tenants should be migrated without stopping the service from a configuration to another to meet their needs while minimizing operational costs on the provider side. Live migrations reveal many
challenges: service interruption must be minimized and the impact on
co-tenants should be minimal. In this paper, we investigate live tenants
migrations duration and its effects on the migrated tenants as well as
the co-located ones. To do so, we propose a generic approach to measure
these effects for multi-tenant Software as a Service. Further, we propose
a testing framework to simulate workloads, and observe the impact of
live migrations on Business Process Management Systems. The experimental results highlight the efficiency of our approach and show that
migration time depends on the size of data that have to be transferred
and that the effects on co-located tenants should not be neglected.

Keywords: Live migration · Multitenancy · BPMS · Performance

1 Introduction

A software service provider that wants to provide its service in the Cloud needs
to adjust the Cloud resources it consumes to the needs of his customers. A Web
based business application requires the deployment of a software stack including
application servers and databases. In order to accommodate multiple customers,
the provider can rely on a multi-tenant architecture that allows managing several
customers on the same server. Thus, to accommodate all customers with the minimal set of resources, it might be necessary to «move» tenants from one installation to another one. During peak hours, many resources might be required, while

© Springer Nature Switzerland AG 2018
H. Panetto et al. (Eds.): OTM 2018 Conferences, LNCS 11229, pp. 61–77, 2018.
https://doi.org/10.1007/978-3-030-02610-3_4

at night a single resource could respond to all the requests from all customers. *Live migrations* can be used to ensure such kind of optimization. It can lead to significant gains in the operational costs. Migrating a tenant from a ressource to another has consequences. It takes time during which the service is not available for the customer. It consumes resources as CPU, networking, and disk IO that may affect the Quality of Service (QoS) for other co-located customers. In this work, we investigate the effects of multi-tenant Web applications live migrations. To the best of our knowledge, our work is the first to focus on evaluating the impact of live migrations on both service interruption time and co-located tenants performances for a service including both application and database servers. To investigate the effects of tenants live migration, we propose an approach to evaluate: **(i)** the duration of service interruption when a tenant is migrated; **(ii)** the effects of a live migration on the migrated tenant; and **(iii)** the effects on the performances of other tenants hosted on both origin and destination resources during the migration. The proposed approach is generic and can be applied for all multi-tenant Web applications sharing a set of resources (computing and/or storage). We apply it to a multi-tenant Business Process Management System (BPMS), representing a classical transactional application. To do so, we propose a performance test framework to investigate the efficiency of our approach in determining the effects of migrations for multi-tenant BPMS. The main design goal of the proposed framework is to create a dynamic testing environment that scales in terms of resources and tenants numbers.

To summarize, the main contributions of this paper are as follows:

- A generic approach to measure the impact of the live migrations in terms of service interruption time and performances (execution time) for multi-tenant Web applications. (Section 3)
- A performance test framework for multi-tenant BPMS. Indeed a detailed description of the framework architecture is presented. (Section 4)
- An example on how the framework can be used for measuring the effects of live migrations on the Performance edition of the BPMS Bonita 7.4.3[1]. (Section 5)

The rest of the paper is organized as follows. Section 2 describes a motivating example that we use throughout this paper to explain the different steps of the proposed method. Our proposed approach for evaluating migrations effects is introduced in Sect. 3. The test framework for multi-tenant BPMS is presented in Sect. 4. Experiments and empirical results are presented in Sects. 5 and 6. Related work, Threats to Validity and Conclusion and Future Work are presented respectively in Sects. 7, 8, and 9.

2   Motivating Example

Live migrations of tenants is important for the service provider. Let's consider a Business Process Management as a Service setting consisting of a BPMS

---

[1] http://www.bonitasoft.com.

deployed on a Cloud provider infrastructure. The Cloud provider offers a set of computing resources. For performance reasons, a BPMS installation is deployed on a resource that consists of at least two compute instances for both a database and an application server. Each resource is characterized by its price per unit of time as well as a minimum and a maximum tasks throughput per second. Each active BPMS instance is used by customers named tenants. Each tenant has a QoS requirement expressed in terms of BPM task throughput. In Fig. 1, the tenant $T4$ has an initial BPM task throughput of (40 task/second) at time $(t = t0)$. It goes up to (60 tasks/second) at time $(t = t1)$. The widths of the resources and the tenants correspond respectively to the capacity and the requirements in terms of task throughput.

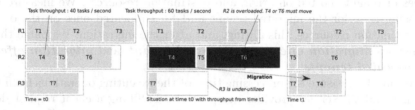

Fig. 1. Tenants distribution at time $t = t_0$ and $t = t_1$

At time $(t = t0)$, the tenants are placed on the different resources. For example, in Fig. 1, the tenants $T1$, $T2$ and $T3$ share the first installation deployed on the resource $R1$. At time $(t = t1)$, the tasks throughput of the different tenants (presented in black) changes, which leads to over and under utilization, respectively of $R2$ and $R3$. We reorganize the distribution of tenants on the different resources. An appropriate solution consists in migrating $T4$ to $R3$. Migrations may induce a service disruption on the tenant side and is not instantaneous. The interruption time for $T4$, includes its deactivation on the origin installation $R2$, its migration to $R3$, and its activation on $R3$. Further, it may have an impact on the performances of the co-located tenants. Thus, we are interested in providing a way to investigate the effects of the migrations on the service interruption time and the performances of the co-located tenants. The next section presents the proposed approach used to solve the aforementioned challenge.

3   Measuring the Impact of Migration

In this section we present our approach to evaluate the impacts of tenants live migrations on both service time interruption and tenants performances. First, we present the metrics we use to detect these impacts. After this, we describe the assumptions we make related to the effects of live migrations. Then, we explain the generic setup we propose to do the measurements in a way that can be replicated for Web applications including application and database servers.

To the best of our knowledge, there is no zero-downtime technique to perform tenants migrations from an origin to a destination resource. Further, with regular relational databases, there exist no way to perform live migrations from one resource to another, with a zero-downtime for the tenant. During a migration, the tenant may undergo a limited availability of the services provided by the Web application, causing QoS breaks. We characterize the impact of tenants migrations on the overall software operations, using three metrics.

The **migration duration** or service interruption time is the time that can be measured between the beginning of the migration on origin resource, when it stops accepting requests and the time when it starts accepting requests on the destination installation.

The second and third metric measures the impact of migrations on performances of tenants hosted on origin and destination resources. We measure the processing time for the migrated tenant and for co-located ones on the origin and destination resources, this during, and after the migration. We name these two metrics *migration effects on migrated tenant*, and *migration effects on co-located tenants*.

We want to measure the processing time of the executing operations from the customer and the service provider perspectives. Fulfilling a client request does not imply the termination of the associated operations on the service provider side. A BPMS engine may take additional time to terminate the operation after satisfying the client request as there could be long executing processes running after the answer to the client.

We have several assumptions about the negative effects of migrations on the service interruption time and the performances, that we want to verify and measure using our approach:

- *Migrations duration* is predictable and increases with the size of the data. Migrations consist of copying Web application data from the origin resource to the destination resource. More data means more time to copy.
- *Migration effects on migrated tenant* exist and are provoked by the creation of a new tenant on the destination installation. This could have various causes such as negative cache hit or asynchronous libraries initialization.
- *Migration effects on co-located tenants* exist and must be taken into account. The live migration will have effects on both the origin and destination resource of the migration, and their tenants. As tenants are activated, deactivated, and their data is being read, copied across the network and written on the resources, it may induce performance degradation on both resources executing co-located tenants operations.

In order to measure these metrics, we simulate users interactions with the studied Web application. For each measurement, we observe variations such as the nature of the workload and the quantity of operations initiated in the system. We store every duration of each launched operation on the Web application as well as of each migration.

We propose an experiment where we first execute a defined workload for a tenant, hosted on the origin resource. We then migrate this tenant from the origin resource to the destination resource, and observe the corresponding migration duration. For each experiment, we store each corresponding time stamp, including the relevant internal steps (such as the beginning and end of the movement of data, the eventual phases of deactivation and reactivation of tenants, etc.).

In order to evaluate the *Migration effects on migrated tenant*, we measure the performances of the tenant before migration on the origin resource, and after migration on the destination resource. We assume we can retrieve Web application response times, if possible. This is usually stored inside the database of the Web application. It should also be possible to retrieve these metrics from the load testing tool point of view. All the duration of the HTTP queries and corresponding processing durations are timestamped and fully identified so we can compare them with the point of view of the application.

We want to compare the effects of a migration on the other tenants currently hosted on the origin and destination resources. In this case, we propose to initiate workloads on tenants on both origin and destination resource. We also initiate a shorter workload for a tenant on the origin resource who will be migrated after to the destination resource, as in previous experiments. A workload is then executed on the migrated tenant, after the migration.

The *Migration effects on co-located tenants* are retrieved for the co-located tenants hosted on both origin and destination resources. In order to do this one should retrieve response times of the HTTP queries and corresponding processing durations for tenants. We consider as "query concerned by migration" every query running between the beginning of the migration and the end of the migration, i.e., where the timestamp of the beginning of the query is before the end of the migration, and the end timestamp of the query is after the beginning of the migration. We then compare to other processing durations, when there is the same number of tenants, with the same load and on the same resources. The differences in processing durations will show the effects of the migration on the co-located tenants.

4   Experiment Framework

In order to validate our approach, we developed a testing framework to study the impact of live migrations for multi-tenant BPMS. According to the type of the performance metric to be investigated, the framework takes as input a set of parameters and provides as output a set of measurements. The Fig. 2 depicts the architecture of the proposed framework. We distinguish the following elements:

- The migration method used to move the tenants from one installation to another (out of the scope of this paper).
- The load testing components used to generate variable workloads to the System Under Test (SUT).
- The SUT represents the BPMS to be studied.

In the rest of this section, we present a detailed description of each component as well as the orchestration of the framework to perform the measurements.

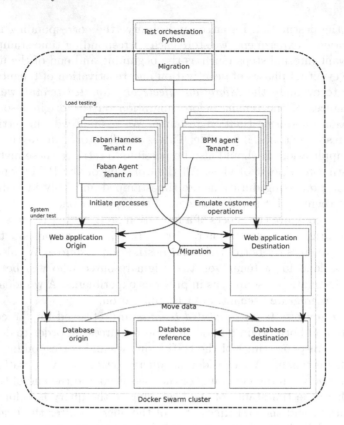

Fig. 2. The framework architecture for studying the impact of live migrations on multi-tenant BPMS

The load testing components (Fig. 2) emulates the behaviour of users within tenants interacting with the BPMS.

In the case of a BPMS, we used two different categories of load testing components: a load tester for tenant and processes initialization and BPM agents for BPM tasks retrieval and processing.

The goal of the load tester is the following:

- Initialize all the tenants as well as the users within each one.
- Deploy the BPM processes needed for the experiments, and add the authorizations and roles needed for the proper functioning of the BPMS.
- Start process instances according to the specific test requirements, by initializing calls to the HTTP API of the BPMS.

The load tester is implemented relying on the Faban framework[2] used for the BenchFlow framework [3]. It enables the specification of complex workload, as well as the number of simulated users that are interacting with the SUT, and the duration of the interaction. The Faban framework then takes care of executing

---

[2] http://faban.org/about.html.

the specified load test, and ensure a correct and verified execution of the defined load.

A BPMS tenant has an organizational structure that consists of different groups of users. According to their profiles and roles, users carry out different business activities within the organization. We use a multi agent system (MAS) to model and simulate the users involved in the execution of business activities within each tenant. A MAS consists of a number of **agents** in a common environment (real/virtual) where they can act and cooperate to achieve system objectives [10]. In our work, we focus on the mapping of the organizational settings within tenants to agents. Using agents on behalf of tenants users provides a suitable means of reproducing human behaviour. Due to their distinct capabilities in terms of autonomy, flexibility, and adaptability agents present effective solutions in modeling and simulating behaviours of human resources. We use agents to imitate human resources in order to: (i) work with process instances including various proportions of human tasks, which are similar to those deployed within real organizations; and (ii) maintain a given threshold of active tasks during tenants migrations to investigate its impact on the migration duration as well as on the co-localized tenants on both origin and target installations.

Each agent represents an active entity that performs specific tasks on behalf of a user within a BPMS tenant. The agent user behaviour consists of the following actions: the agent first connects to the BPMS platform. Then, it starts retrieving the available tasks. If the number of ready tasks is under a given threshold, which represents the sought tasks number before tenant migration, the agent waits until more process instances are started. Else, the agent executes the task after assigning it to itself.

In order to fulfill reproducibility, our framework uses Docker containers. Docker permits easy operating-system-level virtualization. We use it to the launch of the SUT, and for the injection and migration scripts and tools. We used Docker Swarm[3] for our cluster management, and Docker Compose for the description of the distributed system. Swarm is a functionality of Docker to manage containers on several nodes. Docker Compose allows to describe a complete stack, which can be composed by several containers having dependencies between them, as the use of the database by the Web application.

We also used Faban and BPM agent Docker Compose files in order to launch them easily, with all the required parameters such as the names of the users, tenants, and processes.

To use the framework, we must prepare a test descriptor containing the various parameters of the experiments. The parameters can be customized in order to evaluate the behaviour depending on the BPM schema, the BPMS, the duration of the BPM processes injection, etc.

Once a test is launched, the following steps are triggered:

1. Test initialization: creation of unique identifier for the test and experiment directory.

---

[3] https://docs.docker.com/engine/swarm/.

2. Deploy initialization: reset and initialize Docker Swarm on the resources and needed files for the experiment.
3. Deploy BPMS: deploy the BPMS and corresponding databases containers on the resources.
4. Launch test: deploy the Faban load testers, the BPM agents on the concerned resources, and execute the tests for the concerned tenants.
5. Copy results: launches queries on BPMS database in order to get processing time, retrieve results and store them.

In step 4 we launch multiple loads on a set of tenants for a defined duration, named "background tenants" while a loaded tenant is migrated. The tests for the background tenants are launched in an asynchronous manner. A launch of the Faban load tester coupled to the BPM agent is triggered on the studied tenant ("migrated tenant") for a short duration. Once this first step is finished, the framework waits for a defined duration to let all processes finish. Then the migration is triggered. After this migration, a second process injection is launched on the destination resource. Results of the tests can be then retrieved from the BPMS, the BPM agents and the Faban agents.

In step 5, we retrieve the measures. The results consist in the duration of the migration, the response time and the duration of processing of each process and tasks launched during the test. These results have a timestamp stored in CSV file with the identifier of the BPM task and its corresponding process, and their durations.

5   Experimentation

We explain in this section how we have adapted our framework[4] for our use case. As shown in the Fig. 2, the SUT is composed of five software elements, including the origin and destination stacks consisting of one BPMS engine and its corresponding database, and a reference database whose goal is to host the archive and system data.

For our tests, we use BonitaBPM 7.4.3[5] in its commercial Performance edition[6].

We conducted these tests with different parameters for the origin and target resources, and for multiple processes quantities. We needed a method to migrate tenants, a realistic workload, test scenarios, and test infrastructure. All the experiments characteristics are described in the rest of this section.

Live migration of tenants requires a series of steps. BonitaBPM allows stopping and restarting the tenant operations. When a tenant or his underlying operations are in a stopped state, new operations cannot be launched, and current operations are put on hold.

---

[4] https://github.com/guillaumerosinosky/migration_bpms.
[5] http://www.bonitasoft.com.
[6] Multi-tenancy is only available in this commercial licence.

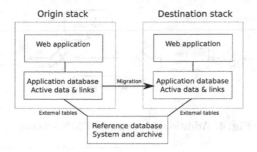

Fig. 3. Proposed database architecture. Software uses a relational database. A "reference" database for system-related and archive data is used. External tables are used for the links between the reference database and the database used by the Web application.

We have used specific Docker Compose file referencing two Bonita Docker containers, two Postgresql 10 Docker containers for the application databases, and one for the reference database. Figure 3 shows the corresponding five containers, which will be generated for each experiment.

We use three different business processes models to investigate the impact of the different workflow structures on the migration duration. The processes are modeled using the graphical standard language BPMN[7]. The first model consists of a single human task. The other models represent a combination of human and automatic tasks as well as parallel and exclusive gateways. These models were initially defined and used in [4]. The second model, in Fig. 4, consists of a set of automatic tasks implemented as script tasks and a single human task. The process starts by initializing an integer number ($x = 0$), which is incremented by the script task situated after the first exclusive gateway. With respect to the value of the variable (x), the upper and lower branches are followed. The model, shown in Fig. 5, has a more complex structure compared to the processes models above. It consists of two human tasks executed in parallel. The process starts by initializing an integer number ($x = 0$), which will be later incremented within the script task situated after the gateway "endAsk". With respect to the value of the variable (x), the upper and lower branches are followed. To have a deterministic behaviour, the upper path, which contains the last exclusive gateway is executed till ($x == 4$) before the ending of the process.

In order to evaluate the three metrics presented in Sect. 3, we have conducted two sets of experiments:

- The first experiment evaluates the *Migration duration*. In this case, we have launched the framework for only one tenant. We executed the load on the origin resource for iteratively 0, 5, 10, 30, 60, 120, 300, 600 s for the three types of BPM schemas, and then we performed three migrations of the processes from the origin resource to the destination resource, and vice versa. We observed the total duration of the migration, including its different steps: deactivation of tenant, migration of the data and activation of the tenant.

---

[7] Business Process Modeling Notation.

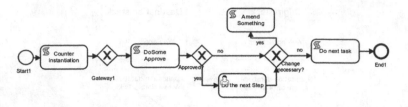

Fig. 4. AdditionalApproval BPMN schema

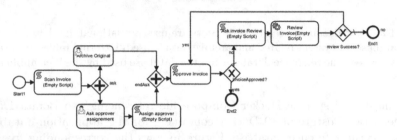

Fig. 5. M3Process BPMN schema

– The second experiment observes *Migration effects on migrated tenant* and *Migration effects on co-located tenants* tests. In this case, we have three tenants: one background tenant running on the origin resource (*tenant1*), one background tenant running on the destination resource (*tenant3*). Each one has one Faban agent injecting processes for 20 min, while 100 BPM agents close their tasks. A third tenant, tenant2, is launched first for 2 min on the origin resource. The framework waits for 5 min in order to let every process finish, and to have comparison for Test 3. Then a migration from the origin resource to the destination resource is triggered, and a 2 min load is launched for this tenant on the destination resource.

We have conducted the tests on Azure Public Cloud. We used the following instances types:

– Databases: Standard E2s v3 (2 vcpus, 16 GB memory) - 3 instances
– BPMS: Standard F4s (4 vcpus, 8 GB memory) - 2 instances
– Faban load tester (Harness and Agent): Standard F1s (1 vcpus, 2 GB memory) - 1 or 3 instances respectively for the first and second experiment
– BPM Agents: Standard F2s (2 vcpus, 4 GB memory) - 0 or 3 instances respectively for the first and second experiment
– Orchestrator: Standard B2ms (2 vcpus, 8 GB memory) - 1 instance

E-series are memory-optimized instances[8]. We used it for the database instances. F-series[9] are computing-optimized instances. We have chosen it for

---

[8] Standard memory optimized instance: https://docs.microsoft.com/en-us/azure/virtual-machines/linux/sizes-memory.
[9] Standard computing optimized instance: https://docs.microsoft.com/en-us/azure/virtual-machines/linux/sizes-compute.

the BPMS, the BPM agent and the Faban load tester. The instance we have used for the orchestration and data collection is a small size burstable instance.

6   Empirical Results and Analysis

In this section, we present the results we obtained following our experimentation. Due to space limitations, detailed results are shared here[10].

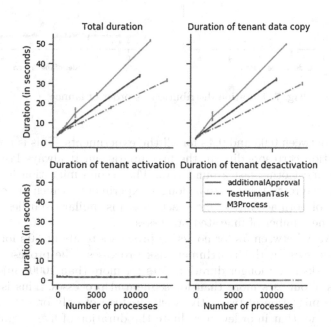

Fig. 6. Mean duration vs number of processes

Figure 6 shows the duration for: the deactivation of the tenant on the origin resource, the migration of the data, the activation on the destination resource, and the total duration of the migration.

The deactivation of tenants is very fast (less than 1 s) and stable regardless the number of active processes or the BPM schema. The activation of tenants is less stable, and last for a few seconds. The copy of the data seems very linear. It is the fastest for the smaller schema, TestHumanTask (about 20 s for 10000 processes), then longer for the schema addtionalApproval (more than 30 s), and even longer for the M3Process schema. All durations stay the same for 0 processes, and there is some variability, which may be linked to the uncertainties of Cloud behaviour.

Figure 7 provides details on the duration of the deactivation and activation of tenants compared to the number of invocation. The duration is stable for the

---

[10] http://dx.doi.org/10.5281/zenodo.1402632.

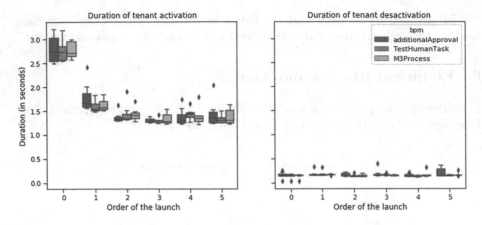

Fig. 7. Duration distribution vs order of launch

deactivation, between 0.05 and 0.4 s for all the experiments. This is not the case for the activation corresponding to the first migration. It always last for more than 3 s when most other last around 1.5 s. The second migration has a higher duration than the rest. Apart of some longer experiments, which are occasional, the behaviour of both activation and deactivation is similar regardless the BPM schema, and the number of migrated processes.

Duration vary between 5 s for no active processes to about 30 s for about 10 thousands processes for the TestHumanTask processes. The process comprised of a parallel tasks has a longer duration: 50 s for more than 10000 injected processes, this is about 20 s more than for sequential processes. This is probably caused by the migration of two active tasks instead of only one for other processes. This shows that in order to evaluate the duration of live migrations, one must consider the BPM schema structure. The duration of the migration of the schema M3process is also a bit longer than the duration of the HumanTask. The presence of numerous automated tasks and gateways could provoke the storage of additional data for this process.

We have seen in the last part that the duration of the activation part of the migrations is longer for the first migrations. The reader must remember that in this experimentation we have launched iteratively 6 migrations, 3 from the origin to the destination and 3 for the destination to the origin, using the same tenant. The probable explanation here is that the first activation of the tenant in a resource needs some initialization in the tenant's metadata (such as library, cache initialization, filesystem based resources, etc.), which do not exist yet in the filesystem or memory of the target resource's installation when it is migrated for the first time. This behaviour happens only one time, and the duration of the activation part stays stable after this first.

Now, we present the results for processes durations on the origin resource, and on the destination resource of a migrated tenant. The Table 1 shows the general statistics depending on the observed schema, and when the results have been retrieved in the experiment(before -pre- or after -post- the migration). The mean duration of TestHumanTask process is about 1300 ms shorter after the migration compared to before, but 1500 ms longer for the M3Process schema and 2700 ms for the AdditionalApproval process. The standard deviations are similar except for the TestHumanTask process.

Table 1. Duration of tasks grouped by BPM schema and moment compared to the migration.

BPM schema	When	Count	Mean	Std	Min	25%	50%	75%	Max
M3Process	Post	10292.0	38449.2	16217.7	13657.0	26046.0	34401.0	44808.0	76321.0
M3Process	Pre	14630.0	36950.8	15909.0	4398.0	24837.0	33933.5	47169.0	87303.0
TestHumanTask	Post	19529.0	12579.3	9378.1	581.0	3081.0	9588.0	15921.0	32269.0
TestHumanTask	Pre	19688.0	13892.8	15594.2	466.0	2969.5	12282.0	19553.5	168560.0
AdditionalApproval	Post	17710.0	113370.0	42819.4	22915.0	78601.0	116355.0	146674.0	199938.0
AdditionalApproval	Pre	25749.0	110685.8	42710.7	12907.0	83799.0	114749.0	141261.0	199856.0

We obtain slower mean durations for M3Process and additionalApproval processes after the live migration, and faster mean duration for the TestHumanTask process. At the time of writing, we don't have rational explanation for that.

We compare in Fig. 8 the duration of processes from the BPM provider perspective (left), and task duration from the BPM agent perspective (right). On both figures, the blue (left) boxplot for each number of active processes is the duration when no migration is executed, and the orange (right) boxplot is the duration during the migration.

There are a lot of variation probably due to the Cloud behaviour, and to some non deterministic effects after the live migrations. Some processes last for more than 200 s, up to 1000 s. Even with these effects, the duration of both processes and tasks are positively correlated to the number of active processes. Durations are a bit longer and less stable during migrations.

In order to have a more homogeneous comparison, we removed from the data the processes having a duration superior to 200 s and tasks superior to 15 s. The Table 2 shows the corresponding results for the BPM tasks response time, aggregated by tenant. The table shows that the duration is always longer during migrations, both for the tenant on the origin resource (tenant1) and on the destination resource (tenant3). The difference in duration is between 26 and 650 ms. We cannot find correlation with the BPM schema. The performances of tenant1 and tenant3 are similar, sometimes faster, and sometimes slower.

As in the previous experiments, additional process is much longer than M3Process, which is longer than human task. The same behaviour occurs for the composing tasks at a much smaller scale.

The effects of the live migrations on the co-located tenants are not important but they exist. We observe a few hundred more milliseconds for each type of BPM schema. It is higher for the more complex BPM schemas than for the simpler. This is the same when we compare the effects on the tenant of the origin

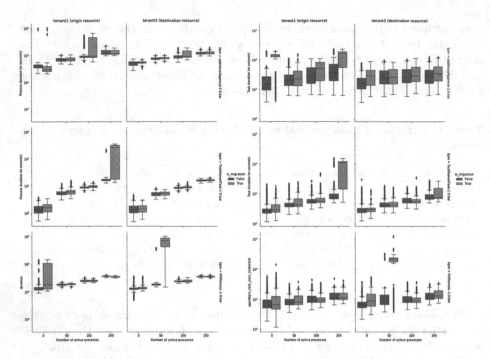

Fig. 8. Process duration -BPMS point of view- distribution (left) and task duration -BPM agent point of view- distribution (right) vs process schema and number of active processes. (Color figure online)

Table 2. Duration of tasks during migration compared to before grouped by BPM schema and tenant

BPM schema	Tenant	Migration running	Count	Mean	Std	Min	25%	50%	75%	Max
M3Process	Tenant1	False	11745.0	1018.9	511.4	183.0	668.0	884.0	1157.00	5819.0
M3Process	Tenant1	True	4602.0	1126.4	695.8	115.0	738.0	962.0	1302.75	13926.0
M3Process	Tenant3	False	11351.0	1003.2	475.9	208.0	674.0	874.0	1174.00	3612.0
M3Process	Tenant3	True	3995.0	1208.9	600.1	297.0	795.0	1025.0	1533.50	5270.0
TestHumanTask	Tenant1	False	17375.0	718.0	1171.1	114.0	330.0	482.0	687.00	11671.0
TestHumanTask	Tenant1	True	4107.0	734.1	464.6	124.0	439.0	617.0	791.50	4665.0
TestHumanTask	Tenant3	False	18857.0	609.5	393.7	122.0	345.0	505.0	704.00	2959.0
TestHumanTask	Tenant3	True	4667.0	738.6	491.7	145.0	426.5	603.0	791.00	3444.0
AdditionalApproval	Tenant1	False	15131.0	2790.5	2055.7	360.0	1281.0	2100.0	3404.00	12241.0
AdditionalApproval	Tenant1	True	5405.0	2883.6	2359.1	398.0	1313.0	2152.0	3484.00	14949.0
AdditionalApproval	Tenant3	False	19530.0	3050.3	2109.8	354.0	1458.0	2255.5	4110.00	12673.0
AdditionalApproval	Tenant3	True	8299.0	3702.0	2349.8	558.0	1751.0	2935.0	5521.00	12882.0

resource to those on the destination resource. Migrations have more effects on the AdditionalApproval process destination resource than on the origin resource. This will require more investigations. The results we obtain are similar for the process duration.

7 Related Work

Performance is a major concern for multi-tenant service providers. In the literature, performance evaluation for multi-tenant environments is tackled from different perspectives including evaluation of tenant placement [11,12], live migrations [2,5] and the suitability of the sharing approach [6,8]. The target applications are mainly multi-tenant databases, whereas in our work, we focus on evaluating the tenants migration effects for a Web application including both application and database servers. Initially, the live migration is used for Virtual Machines [1]. The approach consists in moving a virtual machine and its dependencies from one hardware resource to another without being shutdown. The aim is to decommission the hosting of resources, which are not needed anymore, while minimizing the downtime and the negative effects. In [6], the authors extend the TCP application for benchmark to support multi-tenant platforms. The performance evaluations focus on determining the maximum throughput as well as the tenants number supported by the platform, which can be used later to provide insights about the suitable sharing approach. Although, in our work, the number of the tenants as well as their requirements in terms of throughput are known in advance, our main focus is on investigating tenants live migrations impacts on duration and response time, which is not supported in [6]. In [12], the authors present the STeP framework for scheduling multi-tenant databases on the suitable resources. SteP provides a new set of tenants packing algorithms to optimize both static and dynamic resource allocation. Further, it uses a set of metrics including performance objective violation, the operations cost and the monetary penalty to determine the efficiency of the placement approach. In our work, we focus on the impact of the dynamic placement and its effects on the tenants where the authors consider only scheduling. In [9], the authors present a profit-driven tenant placement strategy for multi-tenant databases. Similar to [12], a set of placement algorithms are proposed, which were evaluated based on operations cost and the SLA penalty costs. Although the authors discuss the suitable deployment of multiple tenants, the migration of tenants and its impacts are kept as a future extension of their work. In [7], the authors propose a framework that furnishes policies for hardware provisioning and tenants scheduling according to the tenants classes and their performance SLOs. While in our work, we assume we know in advance the required resources to deal with the evolving requirements of tenants and we provide an approach to measure the impact of the live migration in terms of duration and response time.

8 Threats to Validity

The proposed approach has different limits, that we tried to mitigate when possible:

- We have tested the effects of co-located tenants only for small numbers of active processes (0 to 200). We need to test with more active processes.

- Using Public Cloud resources induces variability problems. As we have seen in the results, even when executing multiple tests we still have many outliers, sometimes on a whole experiment. The only solution is to launch even more experiments. We plan to make a more intensive analysis of migration for a next iteration, and the framework will help us in this task.
- We have not studied the effects of multitenancy on the performance. In this case we have compared results for a fixed number of tenants. Indeed, there could be additional operations related to the management of tenants, but non related to the workload. Response time could not be totally proportional to the injection if there are more tenants. We plan to study this in a future work.
- The results concern only one BPMS and one live migration method. We plan to test with more BPMS from different vendors.
- In its current implementation, the framework scales in the number of Faban/BPM agents. However, when we tried to scale up the number of Faban load tester instances, we had issues with the parameters. The SUT should to be tuned in order to take advantage of this (resource-wise, and parameter-wise). In this experiment, we have tuned the SUT in the same way for all the experiments. We plan to execute tests with different configurations of the parameters and test different resource configurations in future work.

9   Conclusion and Future Work

In this paper, we presented two contributions: (i) a generic approach to measure the impact of live migrations in terms of service interruption and performances evaluation for Web applications; and (ii) a performance test framework for multi-tenant BPMS. We present an example on how the framework can be used for measuring the effects of live migrations using a well known BPMS. The analysis of the experimental results showed that the duration of a migration depends linearly on the size of the active data and that the effects on the co-located tenants cannot be neglected. Indeed, live migrations may last for long and providers should take them into account and study their effects before on their applications if they want to guarantee the same QoS. This fact could included in the migration decision of elasticity algorithms. Further, our framework can be easily used for other Web applications or other live migration methods with the appropriate adaptation mentioned in the paper.

As a future work, we plan to perform deeper analysis on the effects of live migrations on co-located tenants, for various number of tenants, live migration methods, and Web applications in order to have better view on their QoS impacts. We also intend to enhance our framework, by improving the BPM agent component through the modeling and simulation of more advanced human behaviours. Our work allows to have a better view on live-migrations effects and pinpoint important criteria Software as a Service (SaaS) providers should not underestimate. It can be used to evaluate deployment decision and to parameterize elasticity algorithms when QoS constraints are very strong.

Acknowledgments. This work has been partly supported by the German Research Foundation (HO 5721/1-1, DECLARE), and by the Swiss National Science Foundation (project no. 178653). This work has been supported by Azure Research Grant. We thank heartfully Bonitasoft without whom this analysis could not have been done.

References

1. Clark, C., et al.: Live migration of virtual machines. In: Proceedings of the 2nd Conference on Symposium on Networked Systems Design and Implementation, vol. 2, pp. 273–286. USENIX Association (2005)
2. Das, S., Nishimura, S., Agrawal, D., El Abbadi, A.: Albatross: lightweight elasticity in shared storage databases for the cloud using live data migration. Proc. VLDB Endow. **4**(8), 494–505 (2011)
3. Ferme, V., Ivanchikj, A., Pautasso, C.: A framework for benchmarking BPMN 2.0 workflow management systems. In: Motahari-Nezhad, H.R., Recker, J., Weidlich, M. (eds.) BPM 2015. LNCS, vol. 9253, pp. 251–259. Springer, Cham (2015). https://doi.org/10.1007/978-3-319-23063-4_18
4. Ferme, V., Ivanchikj, A., Pautasso, C.: Estimating the cost for executing business processes in the cloud. In: La Rosa, M., Loos, P., Pastor, O. (eds.) BPM 2016. LNBIP, vol. 260, pp. 72–88. Springer, Cham (2016). https://doi.org/10.1007/978-3-319-45468-9_5
5. Elmore, A.J., Das, S., Agrawal, D., El Abbadi, A.: Zephyr: live migration in shared nothing databases for elastic cloud platforms. In: 2011 ACM SIGMOD. ACM, June 2011
6. Krebs, R., Wert, A., Kounev, S.: Multi-tenancy performance benchmark for web application platforms. In: Daniel, F., Dolog, P., Li, Q. (eds.) ICWE 2013. LNCS, vol. 7977, pp. 424–438. Springer, Heidelberg (2013). https://doi.org/10.1007/978-3-642-39200-9_36
7. Lang, W., Shankar, S., Patel, J.M., Kalhan, A.: Towards multi-tenant performance SLOs. In: 2012 IEEE 28th International Conference on Data Engineering, pp. 702–713, April 2012
8. Liu, R., Aboulnaga, A., Salem, K.: DAX: a widely distributed multitenant storage service for DBMS hosting. Proc. VLDB Endow. **6**(4), 253–264 (2013)
9. Liu, Z., Hacigümüs, H., Moon, H.J., Chi, Y., Hsiung, W.P.: PMAX: tenant placement in multitenant databases for profit maximization. In: EDBT (2013)
10. Russell, S.J., Norvig, P.: Artificial Intelligence: A Modern Approach, 2nd edn. Pearson Education, Upper Saddle River (2003)
11. Schaffner, J., et al.: RTP: robust tenant placement for elastic in-memory database clusters. In: SIGMOD Conference (2013)
12. Taft, R., Lang, W., Duggan, J., Elmore, A.J., Stonebraker, M., DeWitt, D.: STeP: scalable tenant placement for managing database-as-a-service deployments. In: Proceedings of the Seventh ACM Symposium on Cloud Computing, pp. 388–400. SoCC 2016. ACM, New York (2016)

Probability Based Heuristic for Predictive Business Process Monitoring

Kristof Böhmer[(✉)] and Stefanie Rinderle-Ma

Faculty of Computer Science, University of Vienna, Vienna, Austria
{kristof.boehmer,stefanie.rinderle-ma}@univie.ac.at

Abstract. Predictive business process monitoring concerns the unfolding of ongoing process instance executions. Recent work in this area frequently applies "blackbox" like methods which, despite delivering high quality prediction results, fail to implement a transparent and understandable prediction generation process, likely, limiting the trust users put into the results. This work tackles this limitation by basing prediction and the related prediction models on well known probability based histogram like approaches. Those enable to quickly grasp, and potentially visualise the prediction results, various alternative futures, and the overall prediction process. Furthermore, the proposed heuristic prediction approach outperforms state-of-the-art approaches with respect to prediction accuracy. This conclusion is drawn based on a publicly available prototypical implementation, real life logs from multiple sources and domains, along with a comparison with multiple alternative approaches.

Keywords: Business process · Predictive monitoring · Probability

1 Introduction

Predictive process monitoring enables to predict the *unfolding* of ongoing process executions based on behaviour extracted from historic executions. This includes the prediction of, e.g., the activity to be executed next. Hereby, the planning and prioritisation of instances and their resource utilisation can be supported, e.g., to *prevent* the *violation* of Service-Level Agreements (SLAs), cf. [24,27].

We found that predictive monitoring work, especially if it strives to predict upcoming activity executions and their timestamps, can largely be categorised into two main groups based on the applied approach, cf. [18]: At first, probability based works which transform historic logged execution behaviour into probability based prediction models to predict, e.g., the most probable future execution behaviour (e.g., the next activity), cf. [26]. We found those models to be small, easy to understand, follow, and interpret. Secondly, neural networks are gaining interest, cf. [18]. Especially, as it was found that they *outperform* probability based approaches, for example, with regards to the prediction accuracy, cf. [27].

Unfortunately, the latter *struggle* with regards to prediction result transparency and understandability – as neural networks are generally assumed as

© Springer Nature Switzerland AG 2018
H. Panetto et al. (Eds.): OTM 2018 Conferences, LNCS 11229, pp. 78–96, 2018.
https://doi.org/10.1007/978-3-030-02610-3_5

"black boxes", cf. [3]. For example, fully understanding and grasping the relation and inner organisation between hundreds of neurons, which today's increasingly *larger* and *complex* neural networks (prediction models, resp.) are composed of, cf. [27], is extremely challenging. Further, given today's network complexity, it is hardly possible to fully grasp why/how such a complex neural network generated a specific outcome or how changing the network/neurons would affect it, cf. [21]. So, novel techniques are required which: (*a*) combine the advantages of probability based techniques (model/result traceability and understandability); and (*b*) neural network based techniques (high prediction result accuracy).

Hence, inspired by this observation this paper proposes a probability based prediction technique to predict the next execution event (activity, timestamp). Formally: Let p be an execution trace of process P for which the next execution event should be predicted. Further let L hold all historic execution traces t of P. The key idea is to: first, identify the *relevant* traces in L for this task, and secondly to create a probability based prediction model M from them. Here, trace relevance is measured based on the similarity of the execution events in p and t. Finally, the most probable next event for p is determined based on M.

So, instead of complex almost unfathomable neural network based prediction models this work generates and applies simplistic histogram based probability distributions. Its feasibility is analysed by comparing a prototypical implementation of the proposed approach with state-of-the-art neural network and probability based approaches using real-life execution logs from multiple domains.

This paper is organised as follows: Prerequisites and the proposed approach are introduced in Sect. 2. The proposed prediction approach (i.e., prediction model generation and its application) is, in detail, described in Sects. 2 and 3. Related work is discussed in Sect. 5 while Sect. 4 holds the evaluation. Finally, conclusions, discussions, and future work is given in Sect. 6.

2 Prerequisites and General Approach

The presented approach enables to predict, based on a given (sub) trace p, the next execution event (i.e., activity and timestamp). For this a prediction model M is generated from historic traces L (log, resp.), those are: (*a*) automatically generated by process engines; (*b*) representing real behaviour (including noise and ad hoc changes); and (*c*) independent from outdated documentation, cf. [20]. This section, for the sake of understanding, focuses on next event activity prediction to outline the general proposed prediction approach. The more complex prediction of next event timestamps builds on and extends this approach in Sect. 3.

Definition 1 (Execution log). *Let L be a set of execution traces $t \in L$; $t := \langle e_1, \cdots, e_n \rangle$ holds a non-empty ordered list of execution events $e_i :=$ (ea, et); e_i represents the execution of activity $e_i.ea$ at timestamp $e_i.et \in \mathbb{R}_{>0}$; t's order is given by $e_i.et$, i.e., the events' timestamp, cf. [5]. Based on a given event e_i and trace, $\bullet e_i$ determines its preceding event, i.e., $\bullet e_i = e_{i-1}$ if $i > 1$.*

This notion represents information provided by process execution log formats, such as, the eXtensible Event Stream[1], but also holds the necessary information (activities and timestamps) for the prediction of execution events. Accordingly, the first event e_1 for trace t_1 of the running example, cf. Table 1, is $t_1.e_1 = (A, 23)$. *Auxiliary* functions: $\langle \cdot \rangle_i$ and $\langle \cdot \rangle_{[f,k]}$ retrieve the element with index i (former) or a range of indexes (latter) where $f \leq i \leq k$, while $\langle \cdot \rangle^l$ retrieves the last element. $|T|$ determines the length and T^0 retrieves a random element from a list/set T.

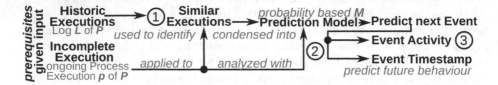

Fig. 1. Proposed probability based predictive monitoring approach – overview

Figure 1 gives an overview on the proposed three staged prediction heuristic: Sects. 2 (activity) and 3 (timestamp). The core component is a probability based prediction model M which is generated based on a selection of given historic execution traces $t \in L$ and an incomplete executions trace $p \notin L$ – for which the next event should be predicted. Both, p and L are assumed as given input (prerequisites). Not each trace $t \in L$ is relevant for the prediction model M.

This is because some traces $t \in L$ are too dissimilar from $p \notin L$ to provide a glimpse on p's future behaviour ①. For example, because t and p follow dissimilar control flow execution paths in P so that behaviour in t does not allow to draw reliable conclusions for p's upcoming events. Compare, for example, trace t_4 and t_1 in Table 1, both represent an execution of P_1 with vastly different activity orders and occurrences which could stem, e.g., from different control flow decision node evaluations. Accordingly, we propose to utilise the dissimilarity/distance between the given traces when deciding which traces in L are used to build M.

Table 1. Realistic running example log L, cf. Helpdesk-Logs in Sect. 4

Process P	Trace t	Event $e_i := (ea, et)$ where $ea =$ activity, $et =$ timestamp					
		e_1	e_2	e_3	e_4	e_5	e_6
P_1	t_1	(A,23) $\rightarrow$	(E,32) $\rightarrow$	(E,37) $\rightarrow$	(F,40) $\rightarrow$	(E,47) $\rightarrow$	(D,53)
P_1	t_2	(A,49) $\rightarrow$	(E,54) $\rightarrow$	(F,61) $\rightarrow$	(E,68) $\rightarrow$	(B,69) $\rightarrow$	(D,78)
P_1	t_3	(A,40) $\rightarrow$	(F,45) $\rightarrow$	(E,49) $\rightarrow$	(F,51) $\rightarrow$	(E,57) $\rightarrow$	(D,63)
P_1	t_4	(C,17) $\rightarrow$	(A,21) $\rightarrow$	(A,22) $\rightarrow$	(A,25) $\rightarrow$	(F,30) $\rightarrow$	(E,37)

---

[1] http://www.xes-standard.org – IEEE 1849-2016 XES Standard.

The applied distance measurement is inspired by the *Damerau-Levenstein distance* (DL for short) [10] – a common algorithm to measure the dissimilarity between two sequences (traces, resp.). Here, this metric was chosen, over other approaches, such as, the Levenstein distance [23], which is frequently applied by existing prediction work, such as, [12]. This is because DL explicitly supports the transposition of a sequences' elements, cf. Definition 2 – enabling to support parallel executions with varying activity orders but still comparable behaviour, cf. [27].

Definition 2 (Damerau-Levenstein activity based trace dissimilarity).
Let t and t' be two traces. Their dissimilarity is measured by determining the most `cost efficient sequence` *of* `insert`, `delete`, `substitution`, *and* `transpositions` `operations` *required so that the order of event activity labels (given by e.ea) in t, t' becomes* `equal`. *Accordingly, each edit operation gets assigned an individual cost:* $ins, del, sub, tran \in \mathbb{N}_{>0}$. *Finally, the dissimilarity of t and t' is* `recursively` *calculated by applying $\Delta(t, t') \mapsto \mathbb{N}$, i.e., comparable to [10]:*

$$
\Delta(t, t') := \begin{cases} \max(|t|, |t'|) & \text{if } \min(|t|, |t'|) = 0 \\ \min \begin{cases} \Delta(t_{[1,|t|-1]}, t') + ins \\ \Delta(t, t'_{[1,|t'|-1]}) + del \\ \Delta(t_{[1,|t|-1]}, t'_{[1,|t'|-1]}) + \begin{cases} 0 & \text{if } t^l.ea = t'^l.ea \\ sub & \text{otherwise} \end{cases} \\ \Delta(t_{[1,|t|-1]}, t'_{[1,|t'|-1]}) + tran \\ \quad \text{if } t^l.ea = t'_{|t|-1}.ea \wedge t_{|t|-1}.ea = t'^l.ea \end{cases} \end{cases}
$$

In Sect. 3 this definition is extended to analyse activity and temporal behaviour at once, cf. Definition 7. Auxiliary functions $\max(a, b)$ and $\min(a, b)$ determine and return the maximum (minimum, resp.) value in $\{a, b\}$.

Assume that for the event $e_3 = (E, 37)$ in trace t_1 (i.e., $t_1.e_3$) the directly successive next event should be predicted, cf. Table 1. For this task we assume, e.g., that the subtraces $t_{2[1,3]}$ and $t_{3[1,3]}$ are relevant information sources while $t_{4[1,3]}$ is not. This is because we assume that $t_{4[1,3]}$'s execution behaviour (e.g., given by the order and occurrence of the respective event activities) is too different from $t_{1[1,3]}$ to draw, based on $t_{4[1,3]}$, conclusions on $t_1.e_3$'s next event. The DL distance reflects this assumption, i.e., the DL distance of $t_{1[1,3]}$ and $t_{2[1,3]} = 1$, $t_{3[1,3]} = 1$, and $t_{4[1,3]} = 3$; assuming a general edit cost of one, cf. Definition 2 and [10].

Further, the DL can naturally be applied on discrete values, such as, event activity labels, cf. Definition 2 and [12]. However, this work also takes values into account which origin from a continuous data range, such as, the timestamp $et \in \mathbb{R}_{>0}$ of each event. For this the original DL approach is extended into a two step approach. Hereby, the first step follows the original idea of exact equality between event activity labels (i.e., $e.ea$) while the second step factors in the partial similarity of an events' temporal behaviour, cf. Definition 7. Without this extension very similar traces could be classified as completely dissimilar, solely because of minimal temporal fluctuations, which we found to be likely, cf. [5].

Subsequently, the $ms \in \mathbb{N}_{\geq 1}$ *most similar* traces $MS \subseteq L$ are transformed into a probability based prediction model M ②. Hereby, the key idea is that the most probable behaviour, based on the most relevant traces, should become the predicted behaviour. To implement this key idea the proposed approach:

(1) searches $\forall t \in MS$ the events $er \in t$ which are most representative for the last known (i.e., most recently occurred) event in p (i.e., p^l), cf. Definition 2; to

(2) extract the directly successive event of each event er as a potential representation of probable successive behaviour for p^l, cf. Definition 3; and

(3) stores extracted information in M, which is inspired by weighted histograms, cf. Definition 4. The weights represent the relevance of extracted behaviour based on the similarity between p and the traces the behaviour was extracted from.

Definition 3 (Prediction behaviour extraction). *Let $p \notin L$ be a trace for which the **next event** should be predicted. Let further $MS \subseteq L$ hold historic traces ($t \in MS$, resp.) which were found to be **similar** to p. Finally, for each trace t the events with an index in the range of $[|p| - |p| \cdot s, |p| + |p| \cdot s]$ are analysed. Hereby, $s \in \mathbb{R}$ controls which indexes "**around**" $|p|$ are taken into consideration. Behaviour extraction function $ext(p, MS, s) \mapsto L$ extracts (sub) traces by:*

$$ext(p, MS, s) := \{t_{[1, i+1]} \in MS | i \geq (|p| - |p| \cdot s) \wedge i \leq (|p| + |p| \cdot s) \wedge t_i.ea = p^l.ea\}$$

Taking the indexes into account enables to represent our assumption that there is a correlation between the position of an event in a trace and its successive events. For example, it was observed that for a given process typically a correlation between the number of already traversed loop iterations and the likelihood that another iteration occurs (or not) can be found. Accordingly, the number of iterations (roughly represented by the event index) also has an impact on the to be predicted successive events. We factor this observation in by focusing on events which have a similar *index* than the last event in p (i.e., $|p|$).

Assume that for $t_1.e_3 = (E, 37)$, i.e., $p = t_{1[1,3]}$, cf. Table 1, the successive activity should be predicted while $MS = \{t_2, t_3\}$. For this, a set of all possible subtraces $PS := ext(p, MS, s)$ is extracted from the traces in MS for which the same activity, as given in $t_1.e_3.ea = E$, occurs roughly at the *same index* as the last event in p ($p^l.ea = E$) so $MS = \{\langle (A, 49), (E, 54), (F, 61) \rangle, \langle (A, 49), (E, 54),$ $(F, 61), (E, 68), (B, 69) \rangle, \langle (A, 40), (F, 45), (E, 49), (F, 51) \rangle\}$ when $s = 0.\dot{3}$, cf. Definition 3.

From the subtraces given in PS the prediction model M is formed, cf. Definition 4. For this, the last two events (t^l and $t_{|t|-1}$) of each (sub) trace $t \in PS$ are extracted and its **weight** is determined by its reciprocal DL distance to p, so:

Definition 4 (Weighted prediction model). *Let PS hold subtraces from L which were identified as **relevant behaviour** sources because of their similarity to p to form the prediction model $M := \{(t^l, t_{|t|-1}, 1/(\Delta(t, p) + 1)) | t \in PS\}$ Hereby, for each $m \in M$; $m := (e_1, e_2, w)$ holds **two events** e_1 and e_2 and a* *weight* $w \in \mathbb{R}_{>0}$ *representing the subtrace similarity based* **relevance** *of* m *for the current prediction task at hand, such as, activity or timestamp prediction.*

Subsequently, M is used to predict event activities, cf. Definition 5, and timestamps, cf. Definition 8 ③. For the sake of understanding solely the prediction of activities is described here while the timestamp prediction is given in Sect. 3.4.

Definition 5 (Predicting activities). *Let* M *be extracted relevant behaviour (i.e., the prediction model), cf. Definition 4. Prediction function* $pa(M)$ *predicts the most probable activity to be executed* **next** *for* p *(after* p^l *resp.) by* $pa(M) \mapsto ea$:

$$pa(M) := \{v | (v, \cdot, \cdot) \in M, \forall (v', \cdot, \cdot) \in M; sa(M, v) \geq sa(M, v')\}^0.ea$$

hereby $sa(M, v) := \sum_{m \in M} m.w$ *where* $m.e_1.ea = v.ea$, *i.e.,* $sa(M, v)$ *sums up the weights in* M *for a given event* v *based on the events' activity* $v.ea$. *This enables the identification of the* **most probable** *activity to be executed next.*

For example, when predicting the successive event for $t_1.e_3 = (\mathsf{E}, 37)$ then $M = \{(\mathsf{F}, \cdot, 0.5), (\mathsf{B}, \cdot, 0.\dot{3}), (\mathsf{F}, \cdot, 0.\dot{3})\}$; this prediction model is visualised in Fig. 2. Based on that model M, $pa(M) = \mathsf{F}$ as the summed up weight for F is $0.8\dot{3}$. In comparison the second most probable activity B only achieves a summed up weight of $0.\dot{3}$ – cf. running example in Table 1. Here, in Sect. 2, we have outlined the proposed event activity prediction approach; in Sect. 3 it is extended to predict temporal behaviour (event timestamps).

3 Probability Based Predictive Temporal Monitoring

This section gives additional details on the approach set out in Fig. 1. It focuses on the prediction of *temporal behaviour* (i.e., p's next event timestamp). Note, that the prediction of the next events' activity was already outlined in Sect. 2.

3.1 Applying Intervals to Analyse Continuous Variables

The similarity calculation and prediction approach proposed in Sect. 2 can naturally be applied to *discrete values*, such as, activities (labels, resp.). However, to apply them to values which originate from a *continuous data* range, such as, timestamps or timespans, they must be extended to prevent the generation of largely incorrect prediction results: similar temporal behaviour would be recognised as dissimilar due to minor temporal *fluctuations*. For this, we propose to represent *continuous values* as *intervals*, cf. [4]. This increases the flexibility as slightly varying temporal business process execution behaviour is still recognised as similar as, for example, $t_1.e_1.et = 1, t_2.e_1.et = 3$ both fit in the interval $[0, 4]$.

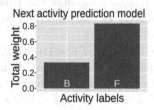

Fig. 2. Exemplary, weighted histogram based visualisation of the prediction model M

In the following the temporal process execution behaviour in L (to define intervals and perform predictions) is represented as *timespans*. Here, such timespans refer to the time which has passed between two directly successive execution event observations. This enables to predict the most probable timespan between the last known event (i.e. p^l) and the time of execution of the to be predicted next/successive process execution event activity – which can subsequently be mapped on p's next event execution timestamp while not being affected by fluctuations in the specific event execution times. For example, the timespan between $t_1.e_5 \to t_1.e_6$ and $t_3.e_5 \to t_3.e_6$ is equal (i.e., 6) while the individual event timestamps are different (e.g., $t_1.e_5.et = 47$, $t_3.e_5.et = 57$), cf. Table 1.

Definition 6 (Timespan extraction). *Let L be a set of execution traces and a, a' two activities for which all timespans should be extracted from L. Extraction function $ate(a, a', L) \mapsto \{d_1, \cdots, d_n\}$ extracts $\forall t \in L$ the timespans ($d \in \mathbb{R}_{\geq 0}$) between directly successive executions of the activities a, a' as a multiset:*

$$ate(a, a', L) := \{|e.et - e'.et||t \in L; e, e' \in t; e.ea = a \wedge e'.ea = a' \wedge e = \bullet e'\}$$

The timespan extraction starts by selecting a pair of activities (i.e., a and a'). Subsequently all traces in L are searched for directly successive events (i.e., e_i, e'_{i+1}) where $e.ea = a$ and $e'.ea = a'$. Finally the timespan d between e, e' is calculated by executing $d = |e.et - e'.et|$, cf. Definition 6. Accordingly, for the running example in Table 1: $ate(\mathsf{A}, \mathsf{E}, L) = \{6, 5\}$ (from t_1 and t_2) when $a = \mathsf{A}$ and $a' = \mathsf{E}$.

To determine the size and amount of *intervals* required to represent the extracted timespans the *Freedman-Diaconis rule* [4] is utilised. It determines, for a given list of timespans X, i.e., $X = ate(a, a', L)$, a suitable interval size: $int(X) := 2 \cdot (IQR(X)/\sqrt[3]{|X|})$ where $IQR(X)$ is the interquartile range for X.

Based on the prerequisites given in Definition 6 and the Freedman-Diaconis rule multiple auxiliary functions for temporal behaviour are defined. These functions are applied, in the following, when predicting the most probable timespan which has to pass after the timestamp $p^l.et$ till the next execution event can be observed, cf. Sect. 3.3 (i.e., enabling the prediction of the next events' timestamp).

First, $bc(a, a', L) := \lceil (max(X) - min(X))/int(X) \rceil$ where $X := ate(a, a', L)$. It determines the number of intervals the timespans between two successive activities a, a' can be divided in – based on the behaviour in L. Secondly, $bi(e, e', L) := \{i | i = 0, \cdots, bc(e.ea, e'.ea, L); mi(X, i) < d; mx(X, i) \geq d\}^0$ where $d = |e.et - e'.et|$, $X := ate(e.ea, e'.ea, L)$, $mi(X, i) = min(X) + int(X) \cdot i$ and $mx(X, i) = mi(X, i) + int(X)$. It determines how many intervals i must be summed up to cover the timespan d between the directly successive events e, e'.

Finally, $bt(e, e', L) := \lceil min(X) + int(X) \cdot bi(e, e', L) + int(X)/2 \rceil$ maps timespans/intervals which are related to the events e and e' (based on L) on a single value based on the interval size identified by $int(X)$. This is utilised in Sect. 3.3, for example, to determine if given pairs of events have equal/similar temporal behaviour. Further, $bt(e, e', L)$ is applied when determining the most probable timespan between p^l an the to be predicted next execution event occurrence.

Given the example traces in Table 1 the auxiliary functions would behave as follows: $bc(\mathsf{F}, \mathsf{E}, L) = 2$, i.e., the Freedman-Diaconis rule determines that two intervals (here: $[4, 6.4]$ and $[6.4, 8.8]$) are required to cover the timespan between the activities $\mathsf{F} \to \mathsf{E}$ for L's traces. Hereby, $X = \{7, 7, 4, 6, 7\}$ and $int(X)$ becomes 2.4. Accordingly, the first interval always starts at $min(X)$ and has a size of $int(X)$. Subsequent intervals always start at the end of the previous one. Additional intervals, if necessary, are generated till all timespans in X are covered.

In this example $bi(t_1.e_4, t_1.e_5, L) = 1$, i.e., the timespan between $t_1.e_4 \to t_1.e_5$ is covered by the second interval (which is $[6.4, 8.8]$) as the timespan between both events is 7. Finally, $bt(t_1.e_4, t_1.e_5, L) = \lceil 4 + 2.4 \cdot 1 + 2.4/2 \rceil = \lceil 7.6 \rceil$. Based on these auxiliary functions the Damerau-Levenshtein distance metric (DL), cf. Definition 2, is extended to incorporate temporal process execution behaviour.

3.2   Temporal Behaviour Based Trace Similarity

Section 2 argues that the relevance of historical execution traces $t \in L$, for the prediction of future events, for a given incomplete trace p, is related to the similarity between p and L's traces. So, Sect. 2 applies the DL distance to measure activity focused trace similarities. However, the *unaltered* DL algorithm is too *sensible* to be applied on continuous data, such as, timestamps or timespans. This is because temporal behaviour is frequently fluctuating, e.g., the timespan between two activity executions is sometimes a bit shorter or longer. Such fluctuations would result in determining similar execution behaviour (traces) as completely dissimilar. So, we propose to extend the DL algorithm to address this limitation.

Definition 7 (Extended Damerau-Levenstein operation cost). *Let e, e' be two events in L's traces. Further, let $c \in \mathbb{N}_{>0}$ be the cost assigned to a chosen DL* edit operation, *such as,* ins. *The individual operation cost, taking the temporal differences into account for e, e', is calculated by $tc(c, e, e') \mapsto \mathbb{R}$:*

$$tc(c, e, e') := \begin{cases} c & \text{if } eqa(e, e') = \textit{false} \\ co(c, e, e', L) & \text{if } eqa(e, e') = \textit{true} \land eqt(e, e', L) = \textit{false} \\ 0 & \text{if } eqa(e, e') = \textit{true} \land eqt(e, e', L) = \textit{true} \end{cases}$$

where $eqa(e, e') := e.ea = e'.ea \land \bullet e.ea = \bullet e'.ea$ and $eqt(e, e', L) := bi(e, \bullet e, L) = bi(e', \bullet e', L)$ determine if the activity ($eqa(e, e')$) or timespan interval ($eqt(e, e', L)$) of e, e' and their directly preceding events are equal. Further $co(c, e, e', L) := c \cdot (1 - (|bi(e, \bullet e, L) - bi(e', \bullet e', L)|)/bc(e.ea, e'.ea, L))$ calculates the relative edit cost if $eqa(e, e') = \textit{true}$ and $eqt(e, e', L) = \textit{false}$. The latter is the case if the activities represented by both events are equal but the temporal behaviour is not.

The proposed extension of the DL algorithm, cf. Definitions 2 and 7, replaces the cost variables ins, del, sub, tran with a cost function $tc(c, e, e')$; where

$c \in \{\texttt{ins}, \texttt{del}, \texttt{sub}, \texttt{tran}\}$ represent the configured costs and e, e' represent two events to be compared. Hereby three scenarios can emerge: (1) unequal event activity: full cost (i.e., c); (2) equal activity, dissimilar temporal behaviour: fraction of c, relative to the temporal dissimilarity; (3) equal activity and timespan interval: cost $= 0$. The following examples are based on the running example given in Table 1 and cover all three scenarios given above with exemplary event pairs:

(1) **Unequal event activity**: e.g., $t_1.e_1 = (\texttt{A}, 23)$ and $t_2.e_2 = (\texttt{E}, 54)$, cost $= c$;
(2) **Equal activity, dissimilar temporal behaviour**: $t_1.e_5 = (\texttt{E}, 47)$ and $t_3.e_3 = (\texttt{E}, 49)$ both cover the same activity (i.e., equal activity label, $t_1.e_5.ea = t_3.e_3.ea$). Accordingly, the timespan d between these two events ($t_1.e_5$ and $t_3.e_3$) and their relative directly preceding event (i.e., $t_1.e_4$ and $t_3.e_2$) is analysed to take temporal differences into account: for the activity transition $F \to E$ five timespans can be extracted from L, such that, $ate(\texttt{F}, \texttt{E}, L) = \{7, 7, 4, 6, 7\} = X$. For this $int(X) = 2.4$, such that, two timespan intervals become relevant, first, $[4, 6.4]$ representing $t_3.e_3$ and, secondly, $[6.4, 8.8]$ representing $t_1.e_5$. So both events are represented by different adjacent intervals, such that, the cost becomes $c \cdot (1 - (|0 - 1|)/2) = c \cdot 0.5$; and finally
(3) **Equal activity and temporal behaviour**: e.g., $t_1.e_6 = (\texttt{D}, 53)$ and $t_3.e_6 = (\texttt{D}, 63)$. For these events the activity and the transition timespan (i.e., 6) interval from the preceding events is equal, i.e., the dissimilarity/edit cost$=0$.

3.3   Temporal Behaviour: Predicting Event Timestamps

Predicting *timestamps* of upcoming/next events follows the *same* key idea as the prediction of upcoming event *activities*, cf. Definition 5. However, instead of directly predicting the most probable next event execution timestamp an indirect approach is applied. So, the proposed approach predicts the most probable timespan between the last known activity execution event in p (i.e., p^l) and the most probable occurrence of the "to be predicted" next event. For this, initially, comparable to the activity prediction, the model M, which holds the most relevant behaviour for the current prediction task, is formed, cf. Sect. 2 and Definition 8.

Definition 8 (Predicting timespans). *Let M be extracted relevant behaviour in L where $m \in M$; $m := (e_1, e_2, w)$, cf. Definition 5. Here the weight $w \in \mathbb{R}_{>0}$ is calculated by using the DL algorithm given in Definition 2 and its extension given in Definition 7, i.e., the proposed cost function; which enables to take temporal dissimilarity into account. Further, let $mwp \in \mathbb{R}_{>0}$ control the minimum relevant relative weight to handle highly fluctuating temporal behaviour. Finally, function $pt(M, mwp) \mapsto \mathbb{R}_{>0}$ predicts the most probable timespan, which is expected to pass after $p^l.et$, till p's next execution event (activity execution) will be observed:*

$$pt(M, mwp) := \left\lceil \frac{\sum_{i=0, (at, \cdot, o) \in SF}^{o} at}{\sum_{(\cdot, \cdot, o) \in SF} o} \right\rceil$$

assuming $MF := \{(e_1, e_2, w) \in M | e_2.ea = pa(M)\}$ filters M based on the activity prediced by $pa(M)$ and $gt(m) = bt(m.e_1, m.e_2, L)$ determines an average timespan for m; $MFF(at) := \{m \in MF | gt(m) = at\}$ filters MF based on the average transition timespan. Further, $S := \{(at, g, o) | m \in MF; at := gt(m), o := |MFF(at)|, g := \sum_{m \in MFF(at)} m.w\}$ maps triplets in MF onto average interval driven timespans along with the relevant metadata. Finally, $minW(S, mwp) := \{s.g | s \in S; \forall s' \in S; s.g \geq s.g'\}^0 \cdot mwp$ determines the minimal relevant weight and $SF := (s \in S | s.g \geq minW(S, mwp))$ filters S accordingly.

We found that, the behaviour hold by M can further be focused/filtered. For this the next activity $a = pa(M)$, cf. Definition 5, can be predicted and utilised to remove all triples from $m \in M$ where $m.e_2.ea \neq a$ (such that, M becomes MF). Hereby, the data is further condensed to only hold the most relevant behaviour which is related to the most probable next activity execution (optional step).

Subsequently, all $m \in MF$ are condensed to form triplets $(at, g, o) \in S$. Hereby a single triplet in S can represent one or more entries in M. In each triplet at identifies the relevant average timespan given by $bt(m.e_1, m.e_2, L)$, $g \in \mathbb{R}_{>0}$ represents the summed up weights (cf., $m.w$), and $o \in \mathbb{N}_{>0}$ represents the number of entries $m \in MF$ which were condensed to form this triplet in S. Note, all entries in $m \in M$ which result in the same $at := bt(m.e_1, m.e_2, L)$ are mapped on the same triplet in S. Finally, the triplets in S are utilised to predict the timespan till p's next execution event will most probably be observed.

For this, all $s \in S$ are filtered to identify the ones which have a $s.g \geq minW(S, mwp)$ where $mwp \in \mathbb{R}_{>0}$. Here, the user chosen mwp enables to handle situations were multiple timespans have an equal or close probability. For example, it was found that when timespans are heavily fluctuating (e.g., from minutes to days) between successive events in L then the Freedman-Diaconis rule does not choose the interval sizes perfectly. In such cases a large amount of observations is assigned, for example, in two adjacent intervals which would, if only the most probable timespan is determined, result in large prediction errors.

The use of mwp enables to take this into account. So, *not* the single most probable timespan is used but the average timespan of the mwp "most probable" ones. For this, the average timespans (i.e., at) of the filtered ($SF := (s \in S | s.g \geq minW(S, mwp))$) triplets $(at, g, o) \in SF$ are multiplied by o and summed up. Finally, the resulting summed up timespan is divided by the summed up number of condensed entries (cf. o) of all relevant triples to determine the most suitable average timespan between the last known and the, to be predicted, next event.

Assume that, based on the running example in Table 1, the timespan between $t_1.e_4$ and $t_1.e_5$ (F → E) should be predicted (i.e., $p = t_{1[1,4]}$). Assuming that $ms = 4$; all traces t_2, t_3, t_4 are relevant. Accordingly, $M = \{((F, 61), (E, 68), 0.289), ((F, 45), (E, 49), 0.227), ((F, 51), (E, 57), 0.217), ((F, 30), (E, 37), 0.2)\}$ when $s = 1$ and a general DL edit cost of 1 is assumed (following related raw DL distances were calculated: $2.4\dot{6}, 3.4, 3.59, 4$). Hereby, the first triplet, is motivated by t_2, the last triplet from t_4 and the remaining ones represent behaviour from t_3.

In this example, $MF = M$, i.e., as $pa(M) = \mathbb{E}$ filtering M has no effect. Subsequently, $S = \{(5.2595, 0.444, 2), (7.7785, 0.489, 2)\}$ as two intervals ($[4, 6.519]$ and $[6.519, 9.038]$) need to be created when mapping the triplets from MF; $int(\{7, 4, 6, 7\}) = 2.519$. When assuming $mwp = 0.9$ then $minW(S, mwp) = 0.4401$ (i.e., $0.489 \cdot 0.9$) so that both triplets in S are identified as being relevant, such that $S = SF$. Accordingly $pt(M, mwp) = \lceil (5.2595 + 5.2595 + 7.7785 + 7.7785)/(2 + 2) \rceil = 7$ which matches to the observed timespan for $t_1.e_4 \rightarrow t_1.e_5$.

3.4 Predicting Future Execution Events

Predicting the next execution event $en := (ea, et)$, cf. Definition 1, for an incomplete execution trace p requires to predict the events' activity $en.ea$ and timestamp $en.et$. [27] found that both are interdependent. Accordingly, this work initially predicted, based on Definition 5, the next activity $en.ea$. Subsequently $en.ea$ can be applied when prediction the timespan $d \in \mathbb{R}_{\geq 0}$ between the last known event p^l and $en.et$. Hence, for predicting $en.et$ the most probable timespan d between $p^l.et$ and $en.et$ is predicted, cf. Definition 8, and added to $p^l.et$ to get $en.et := p^l.et + d$.

3.5 Fostering Understandability and Trust

Recent predictive monitoring approaches are commonly applying *blackbox* like prediction techniques, such as, neural networks, see Sect. 5. For those approaches typically an outstanding prediction performance (e.g., a high activity prediction accuracy) is reported. However, simultaneously they fall short when required to:

(1) explain *alternative* futures which were not classified as being most probable;
(2) explain the aspects which *motivated* the specific prediction results; or
(3) understand how and why a prediction result would most probably *change* when adapting the prediction logic, configuration, or model in certain ways.

We assume, that these drawbacks limit the acceptance and applicability in the respective predictive monitoring target group (e.g., management or production planning staff). For example, when only providing the most probable futures but not providing any alternatives expert knowledge can hardly be incorporated into the prediction. However, based on expert knowledge a user can decide, e.g., that not the most probable, but the second or third most probable activity will most probably be observed next, potentially, because of some external factors which cannot (or not yet) be grasped by the prediction algorithms, cf. [19].

Further, providing information also about less probable futures enables to perform random walks, cf. [14]. Those enable, for example, to depict and clarify how multiple potential futures will unfold – enabling users to build trust into the results and quickly grasp the related uncertainty and potential developments. For this, we assume, that the low number of simple configuration parameters, applied here, is advantageous, i.e., it gives the users a simple "knob" to play with and explore different configurations, predictions, and futures quickly, cf. [19].

The proposed approach can support users during result and prediction process interpretation based on weighted histogram like visualisations which can directly be derived from the generated and utilised weighted prediction models M and MF, cf. Fig. 2. We assume that this enables users to grasp, but also learn, for example, how a (potential) configuration change has affected the performed prediction steps and results – fostering trust and understandability, cf. [28]. However, the question remains if the listed limitations of blackbox like prediction work may not be acceptable given the overall prediction performance of such neural network based techniques. To address this question, the following section will evaluate the feasibility of the proposed approach and compare it with multiple state-of-the-art predictive monitoring approaches.

4   Evaluation

The evaluation utilises real life process execution logs from multiple domains in order to assess the prediction quality and feasibility of the proposed approach, namely: BPI Challenge 2012[2] (BPIC) and Helpdesk[3]. Both logs are also utilised by existing state-of-the-art approaches, such as, [27] – enabling to compare the proposed approach with multiple alternative approaches: [1,6,16,27]. Hereby, we especially focus on incorporating neural network based prediction approaches as those are generally assumed as outperforming probability based ones, such as, the one proposed in this work, cf. [27]. Overall the evaluation, execution logs, and how the evaluation is carried out is *similar* to [27] to ensure comparability.

BPIC 2012 Log: The BPIC 2012 log is provided by the Business Process Intelligence Challenge (BPIC) 2012 in conjunction with a large financial institution. It contains traces generated by the execution of a finance product application process. This process consists out of one manually and two automatically executed subprocesses: (1) application state tracking; (2) handling of application related *work items*; and (3) offer state tracking. The comparison approaches, such as, [27] are only interested in the prediction of manually performed events. Accordingly, this and the comparison work narrow down the events to the work items subprocess to ensure comparability. The same motivation resulted in filtering this log to only contain events whose type is defined as *complete*. Overall 9,657 traces with 72,410 execution events were retained for the work items subprocess.

Helpdesk Log: This log is provided by an Italian software company and contains execution traces generated by a support-ticket management process. Each trace starts by generating a novel ticket and ends with closing the ticket when the related issues were resolved. Overall the utilised process consists of 9 activities while the log holds 3,804 traces which consist of 13,710 execution events. We assume the helpdesk log as being more *challenging* than the BPIC log. This is because the temporal fluctuation is larger and the number of activities is higher while the amount of traces, from which behaviour can be learned, is lower.

---

[2] DOI: 10.4121/uuid:3926db30-f712-4394-aebc-75976070e91f.

[3] DOI: 10.17632/39bp3vv62t.1.

Comparison Approaches: The proposed approach is compared with seven alternative prediction approaches described in [1,6,16,27]. Four of those can be categorised as probability based techniques which apply finite state automata or transition systems. The latter can further be subdivided into set, bag, or sequence abstraction – depending on the applied transition system building and abstraction techniques. Three approaches apply neural networks. Either in the form of Recurrent Neural Networks (RNN) – which incorporate feedback channels between the neurons a network is composed of – or Long Short-Term Memory (LSTM) based neural networks. The latter (LSTM) were found to deliver consistent high quality results in a wide range of domains by coupling neural networks with the capabilities to "remember" previous internal states, cf. [27].

4.1 Metrics and Evaluation

This section analyses the feasibility of the presented prediction approach. For this, two metrics are applied to determine and compare the prediction result quality of multiple prediction approaches. First, *Mean Absolute Error* (MAE) is utilised to analyse the prediction quality for temporal behaviour, i.e., the difference between the observed real event timestamps in the given log traces and the predicted event timestamps. Hereby, MAE was chosen as it is less affected by outliers, where the timespan between two events is unusually large, cf. [27], than other approaches, such as, the Root Mean Square Error. Secondly, for activity prediction the *accuracy* is measured. For this, the percentage of predicted events for which the predicted and the observed event activities are equal is determined.

While performing the evaluation the first 2/3 of the chronologically ordered traces hold by the logs are utilised as training data, e.g., to form prediction models. The remaining 1/3 of the traces are utilised to evaluate the activity and event timestamp prediction performance of the proposed and comparison approaches (testing data). For this, basically, all possible subtraces with $t_{[1,n]}$ where $2 \leq n < |t|$ are generated from the testing data and the $n + 1$ event (activity and timestamp) is predicted. Note, that only subtraces with a size of ≥ 2 are used so that sufficient behaviour is known to base the prediction on, cf. [27]. Further, the evaluation results were only calculated once as the proposed approach contains no random aspects, i.e., deterministic results were observed.

4.2 Evaluation Results

The results were generated based on the BPIC 2012 and Helpdesk process execution logs. The implementation was found to execute the required preparatory (e.g., similar trace extraction) and next event prediction steps fast enough to output the predicted events almost instantaneously (i.e., below one second).

It was found that this performance could only be achieved because the proposed approach filters and separates the given traces into traces which are relevant or non-relevant for the prediction task at hand. This significantly reduces the amount of data which must be processed during the main prediction steps.

Further, the initial similarity based relevance calculations were found to be executed quickly. Computationally intense temporal behaviour focused calculations and their results can be stored and reused for multiple predictions in a row. This suggests an applicability onto even larger process repositories and execution logs.

Primary tests were applied to identify appropriate configuration values for each log and prediction task – which are summarised in Table 2. This is $ms \in \mathbb{N}_{>0}$, i.e., the amount of most similar traces hold by MS, $s \in \mathbb{R}$, i.e., the index spreading control variable for the prediction behaviour extraction, cf. Definition 3, and, $mwp \in \mathbb{R}$ which enables to configure how less probable temporal behaviour is incorporated to compensate less than ideal interval definitions which can stem from heavily fluctuating temporal execution behaviour, Definition 8. Finally, $ps \in \mathbb{N}_{>0}$ controls the maximum number of events which are taken into account during the similarity calculation by creating and using a subtrace with the length of at most ps events (e.g., $t_{[|p|-ps,|p|]}$) for the similarity based trace relevance analysis.

Different configurations were used for different prediction tasks (activity vs. timestamp) and logs to reflect the unique characteristics of each task and log. Given the low amount of simple numeric configuration values those can, likely, also be automatically optimised and defined based on computer algorithms. In, general, it was observed that choosing an overly high value for the configuration variables ms, s, and mwp could result in being affected by irrelevant behaviour and noise while too low values could result in not extracting sufficient behaviour for the prediction task at hand. In comparison, the value ps seems to mainly affect the amount of computational effort which must be invested (higher=more).

Finally, following edit costs are utilised for the original and the extended DL algorithm: `del` = 2, `sub` = 3, `tran` = 2, `ins` = 1. Hereby, each cost was chosen based on our assumption how strongly the related edit operation (e.g., to delete an event) would affect the effective trace behaviour. For example, sub was defined as three as it combines a delete (cost 2) and insert (cost 1) operation at once. In comparison `tran` "solely" moves events towards a new trace index.

Table 2. Evaluation configuration for each execution log and prediction task

Configuration	ms	s	mwp	ps
BPIC event activity prediction	200	0.2	0.05	10
Helpdesk event activity prediction	10	0.1	0.05	10
BPIC event timestamp prediction	50	0.2	0.2	20
Helpdesk event timestamp prediction	200	0.2	0.2	6

The achieved evaluation results are summarised in Table 3 (event timestamp prediction) and 4 (event activity prediction) for all compared approaches (bold = best). As can be seen, the proposed probability based approach consistently outperforms the alternative probability *and* neural network based comparison approaches for all logs. Overall, an average improvement of 5% for the event timestamp and 4% for the event activity prediction can be observed over the best performing comparison approach given in [27]. In general, we found that the observed advantage, of the proposed approach, over the compared approaches is even increasing when the analysed process execution behaviour along with the prediction task becomes more challenging. This indicates that the observed advantage would further increase at more challenging prediction tasks/behaviour.

Further it was found that the time, but also computational effort which is required to prepare and execute the predictions is substantially lower for the proposed approach than for the compared neural network based approaches. For example, the authors in [27] utilise an expensive professional high end NVidia Tesla k80 GPU and still need between *"15 and 90 seconds per training iteration"* [27, p. 483] – of which, typically, tens of thousands are required for a single neural network to achieve reasonable results. Moreover, multiple prediction models (neural networks) must be prepared for each process and sub-trace length, i.e., for $t_{[1,2]}, t_{[1,3]}, t_{[1,4]} \cdots t_{[1,n]}$ where $n \in \mathbb{N}_{>0}$ is the longest expected trace length: if traces become longer than n, approaches, such as, [27] are no longer applicable.

In comparison cheap general purpose processors used in today's office PCs are sufficient for the proposed approach to quickly perform predictions. This enables users to dynamically and quickly explore the impact of different configuration values and the related futures – which we assume as helpful when in need to understand the unfolding of complex process executions. Finally, given the computational performance of the proposed approach it is not necessary to execute lengthy prediction model preparations for individual sub-trace lengths, i.e., there is no predetermined upper limit on the trace lengths which are supported.

Table 3. Evaluation results: execution event timestamp prediction MAE

	Mean Absolute Error (MAE) in days					
	Proposed Probability	Set abstraction Probability [1]	Bag abstraction Probability [1]	Sequence abstraction Probability [1]	LSTM Neural Network [27]	Recurring Neural Network [27]
Helpdesk	**3.54**	5.83	5.74	5.67	3.75	3.98
BPIC 2012	**1.49**	1.97	1.97	1.91	1.56	N.A.[a]

[a]Results denoted as "N.A." are not available as the compared/related work does not cover the respective log or prediction task during its respective evaluation.

Table 4. Evaluation results: execution event activity prediction accuracy

	Activity prediction accuracy				
	Proposed Probability	LSTM Neural Network [16]	Finite automaton Probability [6]	LSTM Neural Network [27]	Recurring Neural Network [27]
Helpdesk	**0.77**	N.A.[a]	N.A.[a]	0.71	0.66
BPIC 2012	**0.77**	0.62	0.72	0.76	N.A.[a]

[a]Results denoted as "N.A." are not available as the compared/related work does not cover the respective log or prediction task during its respective evaluation.

This evaluation shows the feasibility of the proposed approach. However, due to space restrictions the user focused application benefits, compared to related neural network based approaches, are only discussed, but not yet evaluated, cf. Fig. 2 – which will be done, based on user studies, in future work, cf. Sect. 6.

5   Related Work

Overall, it was found that existing work mainly addresses four areas: (a) predicting the next event [27]; (b) estimating remaining execution times [27]; (c) classifying and predicting instance outcomes [9]; and (d) predicting risks which could hinder successful instance completions [2]. Here, we assume (a) as most relevant.

In general it was observed that early related work was mainly focusing on *probability* based approaches using transition networks, state automata, (Hidden) Markov Models, frequent (sub) sequences, and fuzzy logic, cf. [7,17,18,25]. Later work, mainly starting in 2015, heavily focused on *neural networks*, initially, starting with recurring neural networks (RNN) and later extending RNNs with Long Short Term Memory (LSTM) capabilities, cf. [13,15,18,27]. Extending neural networks with LSTM capabilities enables the neurons, a neural network is composed of, to remember historic internal "states" over arbitrary time intervals, cf. [27], resulting in an improved prediction quality. Overall, recent work, such as, [27] has indicated that neural network based approaches significantly outperform alternative approaches, e.g., the probability based ones. However, this should be reconsidered as the probability based approach given in this work was found to outperform both, i.e., RNN and LTSM based neural networks, cf. Sect. 4. Alternative machine learning techniques, such as, Support Vector Regression [8] or (regression) trees [11], seem to be rarely applied – in comparison.

We assume that the reported advantages, cf. [27], of RNN and LTSM based approaches over alternative approaches origin from the *memory* capabilities of RNN (limited capabilities) and LTSM (extensive capabilities). Hence, LTSM based approaches can factor in a wide range of instant dependent historic states and observations throughout the prediction. In comparison alternative approaches, such as, Markov Chains, heavily focus on the most recently observed event (p^l) during the prediction phase without taking previously observed instant behaviour (e.g., the number of loop iterations) sufficiently into account. Further, existing probability based work was found to apply a *global* prediction approach, namely, incorporating all known historic trace behaviour during each prediction task at hand – even if a majority of the historic traces are unsuitable for this task as they are significantly dissimilar from the instance p which is predicted upon, cf. [22]. Accordingly, the behaviour representation of *previous* alternative probability based approaches is assumed by us as *overly generic and abstract* – resulting in the observed *unsatisfying* prediction performance and quality.

In comparison the proposed approach builds, on the fly, individual prediction models which are tailored specifically for each incoming novel and unique prediction task – exploiting that it requires only a low amount of computational effort to perform the prediction model generation and to execute the required prediction steps. Accordingly, "optimal" prediction models are generated which factoring in the most similar and so most relevant behaviour – enabling so called "local" prediction, cf. [22]. Section 4 shows that this enables to *outperform* existing work in the area of activity and timestamp based predictive monitoring.

The dynamic and fast prediction model generation also enables to react quickly on changes, such as, *concept drift*, as time intense training phases, required by alternative machine learning approaches, such as, neural networks, are no longer necessary. Further, this also enables to apply the proposed approach in dynamic flexible *online* prediction scenarios where the timespan between (*a*) when the most recent execution event becomes available; and (*b*) the prediction result becomes obsolete (as the process progressed) is small. This becomes relevant, for example, when execution events are constantly streamed by a process execution engine at a steady high pace, e.g., as production facilities and their *ever changing* processes cannot be "halted" till a prediction algorithm has drawn a conclusion.

6   Discussion and Outlook

This paper focuses on two main challenges (*a*) to outperform existing state-of-the-art predictive monitoring approaches (next event prediction); while (*b*) striving to implement a transparent prediction approach to foster the trust users put into the results. We conclude that this paper was able to meet the first challenge as the conducted evaluation shows that we outperform the second best compared prediction approach (based on LTSM neural networks) by 4 to 5 percent.

With regards to the second challenge we assume that the results generated by predictive monitoring approaches can have a significant impact on an organisation. For example, a prediction result could trigger a reorganisation of staff/ project assignments resulting in the fulfilment or, given incorrect results, the violation of SLA agreements. Accordingly, we argue that prediction processes should be transparent and understandable. One the one hand to foster the trust in the results but also to enable experts to draw informed decisions while factoring in their unique domain knowledge. Accordingly, the proposed approach strives to provide a transparent prediction process which enables users to grasp the prediction model and the possible different prediction results. For this the prediction models are deliberately simple, contain all relevant futures along with their probability, and could, as we assume, be visualised as weighted histograms.

Future work will concentrate on two aspects: (*a*) further improving the prediction result quality; and (*b*) evaluating if the proposed approach is capable of fostering the trust in the predicting results, e.g., by providing a transparent and understandable prediction result generation process. The first aspect will be tackled by further extending the proposed approach, for example, by combining it with alternative techniques to incorporate their unique advantages. For the latter we will concentrate on expanding and evaluating the proposed approaches' result and prediction process understandability/transparency. Accordingly, visualisation, cf. Fig. 2, and management tools will be created that enable to handle the provided information (e.g., which potential futures are probable) in an interactive manner. Further, user studies will be performed to assess the benefits of the proposed approach on predictive monitoring driven management decisions.

Acknowledgment. This work has been funded by the Vienna Science and Technology Fund (WWTF) through project ICT15-072.

References

1. Van der Aalst, W.M., Schonenberg, M.H., Song, M.: Time prediction based on process mining. Inf. Syst. **36**(2), 450–475 (2011)
2. van Beest, N.R.T.P., Weber, I.: Behavioral classification of business process executions at runtime. In: Dumas, M., Fantinato, M. (eds.) BPM 2016. LNBIP, vol. 281, pp. 339–353. Springer, Cham (2017). https://doi.org/10.1007/978-3-319-58457-7_25
3. Benítez, J.M., Castro, J.L., Requena, I.: Are artificial neural networks black boxes? Trans. Neural Netw. **8**(5), 1156–1164 (1997)
4. Birgé, L., Rozenholc, Y.: How many bins should be put in a regular histogram. ESAIM: Probab. Stat. **10**, 24–45 (2006)
5. Böhmer, K., Rinderle-Ma, S.: Multi instance anomaly detection in business process executions. In: Carmona, J., Engels, G., Kumar, A. (eds.) BPM 2017. LNCS, vol. 10445, pp. 77–93. Springer, Cham (2017). https://doi.org/10.1007/978-3-319-65000-5_5
6. Breuker, D., Matzner, M., Delfmann, P., Becker, J.: Comprehensible predictive models for business processes. MIS Q. **40**(4), 1009–1034 (2016)
7. Ceci, M., Lanotte, P.F., Fumarola, F., Cavallo, D.P., Malerba, D.: Completion time and next activity prediction of processes using sequential pattern mining. In: Džeroski, S., Panov, P., Kocev, D., Todorovski, L. (eds.) DS 2014. LNCS (LNAI), vol. 8777, pp. 49–61. Springer, Cham (2014). https://doi.org/10.1007/978-3-319-11812-3_5
8. Cesario, E., Folino, F., Guarascio, M., Pontieri, L.: A cloud-based prediction framework for analyzing business process performances. In: Buccafurri, F., Holzinger, A., Kieseberg, P., Tjoa, A.M., Weippl, E. (eds.) CD-ARES 2016. LNCS, vol. 9817, pp. 63–80. Springer, Cham (2016). https://doi.org/10.1007/978-3-319-45507-5_5
9. Conforti, R., Fink, S., Manderscheid, J., Röglinger, M.: PRISM – a predictive risk monitoring approach for business processes. In: La Rosa, M., Loos, P., Pastor, O. (eds.) BPM 2016. LNCS, vol. 9850, pp. 383–400. Springer, Cham (2016). https://doi.org/10.1007/978-3-319-45348-4_22
10. Damerau, F.J.: A technique for computer detection and correction of spelling errors. Commun. ACM **7**(3), 171–176 (1964)
11. de Leoni, M., van der Aalst, W.M.P., Dees, M.: A general framework for correlating business process characteristics. In: Sadiq, S., Soffer, P., Völzer, H. (eds.) BPM 2014. LNCS, vol. 8659, pp. 250–266. Springer, Cham (2014). https://doi.org/10.1007/978-3-319-10172-9_16
12. Di Francescomarino, C., et al.: Clustering-based predictive process monitoring. IEEE Trans. Serv. Comput. (2016). https://ieeexplore.ieee.org/document/7797472

13. Di Francescomarino, C., Ghidini, C., Maggi, F.M., Petrucci, G., Yeshchenko, A.: An eye into the future: leveraging a-priori knowledge in predictive business process monitoring. In: Carmona, J., Engels, G., Kumar, A. (eds.) BPM 2017. LNCS, vol. 10445, pp. 252–268. Springer, Cham (2017). https://doi.org/10.1007/978-3-319-65000-5_15

14. Durrett, R.: Probability: Theory and Examples. Cambridge University Press, Cambridge (2010)

15. Evermann, J., Rehse, J.R., Fettke, P.: Predicting process behaviour using deep learning. Decis. Support Syst. **100**, 129–140 (2017)

16. Evermann, J., Rehse, J.-R., Fettke, P.: A deep learning approach for predicting process behaviour at runtime. In: Dumas, M., Fantinato, M. (eds.) BPM 2016. LNBIP, vol. 281, pp. 327–338. Springer, Cham (2017). https://doi.org/10.1007/978-3-319-58457-7_24

17. Ferilli, S., Esposito, F., Redavid, D., Angelastro, S.: Extended process models for activity prediction. In: Kryszkiewicz, M., Appice, A., Ślęzak, D., Rybinski, H., Skowron, A., Raś, Z.W. (eds.) ISMIS 2017. LNCS (LNAI), vol. 10352, pp. 368–377. Springer, Cham (2017). https://doi.org/10.1007/978-3-319-60438-1_36

18. Di Francescomarino, C., Ghidini, C., Maggi, F.M., Milani, F.: Predictive process monitoring methods: Which one suits me best? In: Weske, M., Montali, M., Weber, I., vom Brocke, J. (eds.) BPM 2018. LNCS, vol. 11080, pp. 462–479. Springer, Cham (2018). https://doi.org/10.1007/978-3-319-98648-7_27

19. Gleicher, M.: Explainers: expert explorations with crafted projections. Vis. Comput. Graph. **19**(12), 2042–2051 (2013)

20. Greco, G., Guzzo, A., Pontieri, L.: Mining taxonomies of process models. Data Knowl. Eng. **67**(1), 74–102 (2008)

21. Idri, A., Khoshgoftaar, T.M., Abran, A.: Can neural networks be easily interpreted in software cost estimation? In: Fuzzy Systems, vol. 2, pp. 1162–1167. IEEE (2002)

22. Klinkmüller, C., van Beest, N.R.T.P., Weber, I.: Towards reliable predictive process monitoring. In: Mendling, J., Mouratidis, H. (eds.) CAiSE 2018. LNBIP, vol. 317, pp. 163–181. Springer, Cham (2018). https://doi.org/10.1007/978-3-319-92901-9_15

23. Levenshtein, V.I.: Binary codes capable of correcting deletions, insertions, and reversals. In: Soviet physics doklady, vol. 10, pp. 707–710 (1966)

24. Mehdiyev, N., et al.: A multi-stage deep learning approach for business process event prediction. In: Business Informatics, vol. 1, pp. 119–128. IEEE (2017)

25. Pandey, S., Nepal, S., Chen, S.: A test-bed for the evaluation of business process prediction techniques. In: Collaborative Computing, pp. 382–391. IEEE (2011)

26. Rogge-Solti, A., Weske, M.: Prediction of business process durations using non-Markovian stochastic Petri nets. Inf. Syst. **54**, 1–14 (2015)

27. Tax, N., Verenich, I., La Rosa, M., Dumas, M.: Predictive business process monitoring with LSTM neural networks. In: Dubois, E., Pohl, K. (eds.) CAiSE 2017. LNCS, vol. 10253, pp. 477–492. Springer, Cham (2017). https://doi.org/10.1007/978-3-319-59536-8_30

28. Verenich, I., Nguyen, H., La Rosa, M., Dumas, M.: White-box prediction of process performance indicators via flow analysis. In: Proceedings of the 2017 International Conference on Software and System Process, pp. 85–94. ACM (2017)

Indulpet Miner: Combining Discovery Algorithms

Sander J. J. Leemans[1(✉)], Niek Tax[2], and Arthur H. M. ter Hofstede[1]

[1] Queensland University of Technology, Brisbane, Australia
{s.leemans,a.terhofstede}@qut.edu.au
[2] Eindhoven University of Technology, Eindhoven, The Netherlands
n.tax@tue.nl

Abstract. In this work, we explore an approach to process discovery that is based on combining several existing process discovery algorithms. We focus on algorithms that generate process models in the process tree notation, which are sound by design. The main components of our proposed process discovery approach are the Inductive Miner, the Evolutionary Tree Miner, the Local Process Model Miner and a new bottom-up recursive technique. We conjecture that the combination of these process discovery algorithms can mitigate some of the weaknesses of the individual algorithms. In cases where the Inductive Miner results in overgeneralizing process models, the Evolutionary Tree Miner can often mine much more precise models. At the other hand, while the Evolutionary Tree Miner is computationally expensive, running it only on parts of the log that the Inductive Miner is not able to represent with a precise model fragment can considerably limit the search space size of the Evolutionary Tree Miner. Local Process Models and bottom-up recursion aid the Evolutionary Tree Miner further by instantiating it with frequent process model fragments. We evaluate our approaches on a collection of real-life event logs and find that it does combine the advantages of the miners and in some cases surpasses other discovery techniques.

Keywords: Process mining · Process discovery · Boosting
Process trees · Bottom-up recursion

1 Introduction

Process Mining [1] is a scientific discipline that bridges the gap between process analytics and data analysis, and focuses on the analysis of event data logged during the execution of a business process. Events contain information on what was done, by whom, for whom, where, when, etc. Such event data is often readily available from information systems such as Enterprise Resource Planning (ERP), Customer Relationship Management (CRM) or Business Process Management (BPM) systems. *Process discovery*, which plays a prominent role in process mining, is the task of automatically generating a process model that

© Springer Nature Switzerland AG 2018
H. Panetto et al. (Eds.): OTM 2018 Conferences, LNCS 11229, pp. 97–115, 2018.
https://doi.org/10.1007/978-3-030-02610-3_6

accurately describes a business process based on such event data. Many process discovery techniques have been developed over the last decade (e.g. [3,5–7,12,13,16,17,31]), producing process models in various forms, such as Petri nets [25], process trees [6] and Business Process Model and Notation (BPMN) models [26].

In the research field of Machine Learning, it has long been studied how to combine multiple predictive models into a single combined model. This so-called *ensemble learning* gained traction when Schapire [27] showed that a strong classifier could be generated by combining a collection of weak classifiers through a procedure he called *boosting*. In later years, many different approaches have been developed to combine several predictors into a single more accurate predictor, including *bagging* [4], *stacking* [30] and *Bayesian model averaging* [14].

Dahari et al. [7] recently explored combining multiple process discovery approaches to obtain a single process model. The model that they obtain is based on multiple process models that originate from the Inductive Miner infrequent (IMi) [17] with different parameter settings of the algorithm. In this work, we take this idea one step further by exploring the combination of multiple process discovery algorithms to jointly discover a single process model, thereby bringing some of the ideas of ensemble learning to the field of process mining.

We focus on process discovery algorithms that generate process models in one consistent process representation, namely the *process tree* [6], in order to enable combining the results of discovery approaches. Three existing process discovery algorithms that generate process models in process tree notation are the Inductive Miner (IM) [16,17], the Evolutionary Tree Miner (ETM) [6] and the Local Process Model (LPM) Miner [28]. The Inductive Miner algorithm is a computationally very fast algorithm, however, the process models that it generates often allow for too much behaviour (i.e., they are imprecise) when it is applied to event logs that originate from highly variable or unstructured processes. The ETM uses a genetic algorithm to find a process tree that optimises multiple quality criteria for process models and it can, therefore, find more precise models from logs of unstructured processes. However, finding a high-quality process model with the ETM can be time-consuming when the process or the event log is large or complex. Our combination of process discovery algorithms, Indulpet Miner (IN), aims to combine the strengths of four process discovery techniques: Inductive Miner, Local Process Models, the Evolutionary Tree Miner and a new bottom-up recursive technique (BUR). First, we apply IM on the parts that it can describe precisely. Second, we use the LPM Miner and BUR to mine local patterns of process behaviour that we use as a starting point for the ETM, which prevents that the genetic search of the ETM has to start from scratch. Third, we only apply the ETM locally for the remaining parts.

The remainder of this paper is structured as follows: in Sect. 2 we discuss related research. In Sect. 3 we introduce basic concepts and notation that we use throughout the later sections of the paper. In Sect. 4 we introduce the Indulpet Miner, our novel mining approach. Section 5 evaluates the Indulpet Miner and compares it to existing techniques. Finally, Sect. 6 concludes the work.

2   Related Work

Several process discovery techniques have been proposed before. We briefly discuss the ones most relevant for this paper; for a more elaborate overview, please refer to [1]. Evoluationary Tree Miner, Inductive Miner and the Local Process Models technique will be described in Sect. 4.

The Split Miner [3] is a process discovery algorithm that extracts a set of directly follows relations from the event log and, from these relations, mines a process model using several heuristics. The Split Miner is often able to discover more precise process models than the Inductive Miner. Split Miner guarantees that the discovered process models are free of deadlocks, however, unlike soundness-guaranteeing algorithms, Split Miner does not guarantee that the final state of the model can be reached (no *weak soundness*).

Dahari et al. [7] recently developed the Fusion Miner, which is related to the Indulpet Miner in the sense that it mines several process trees and combines them into a single process model. However, the process trees that are combined by the Fusion Miner are restricted to those that are generated by the Inductive Miner infrequent (IMf) [17], while the Indulpet Miner combines process trees that originate from multiple algorithms, thereby making use of the strengths of different algorithms.

Mannhardt et al. [23] proposed a combined approach based on Local Process Models (LPMs) and the Inductive Miner (IM), thereby closely linking to the Indulpet Miner. This approach first uses LPMs to abstract the event log to a different event log where the events are on a higher level of granularity, then applies the IM to this higher level log, and replaces the high-level activities in the discovered model with the LPM patterns to obtain a model on the granularity level of the original log. In this work, we propose to start with IM, thereby reducing the application of the more computationally expensive LPM miner to fragments of the log where the IM fails to find a satisfactory result, while in the solution proposed by Mannhardt et al. [23] the LPM miner always needs to process the full log. Furthermore, this work incorporates the Evolutionary Tree Miner (ETM) and a novel bottom-up recursion (BUR) strategy.

3   Preliminaries

Given an alphabet of activities Σ containing all the basic process steps, Σ^* denotes the set of all sequences over Σ and $\sigma = \langle a_1, a_2, \ldots, a_n \rangle$ denotes a sequence of length n, with $|\sigma| = n$. $\langle \rangle$ denotes the empty sequence and $\sigma_1 \cdot \sigma_2$ is the concatenation of sequences σ_1 and σ_2. A multiset (or bag) over X is a function $B : X \to \mathbb{N}$ which we write as $[a_1^{w_1}, a_2^{w_2}, \ldots, a_n^{w_n}]$, where for $1 \leq i \leq n$ we have $a_i \in \Sigma$ and $w_i \in \mathbb{N}^+$. The set of all multisets over X is denoted with $\mathcal{B}(X)$.

An *event log* is a multiset of traces that denote process executions. For instance, the log $[\langle a, b, c \rangle, \langle b, d \rangle^2]$ consists of one trace that consists of activity a followed by b and c, plus two traces of b followed by d.

A *directly follows graph* is an abstraction of an event log or a language. The nodes are the activities of the log or language, while the directed edges denote whether in the log or language, an activity may be directly followed by another activity. For instance, the directly follows graph of our example log is

$$a \longrightarrow b \overset{\frown}{\longrightarrow} c \quad \nearrow d .$$

A frequently used process-model notation is the Petri-nets notation [25]. A Petri net is a directed bipartite graph consisting of places (depicted as circles) and transitions (depicted as rectangles), connected by arcs. Transitions represent activities, while places represent the enabling conditions of transitions. A special label τ is used to represent invisible transitions (depicted as narrow rectangles), which are only used for routing purposes and not recorded in the execution log.

In a *labelled Petri net*, labels are assigned to transitions to indicate the type of activity that they model.

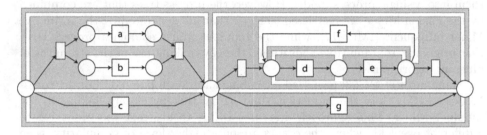

Fig. 1. An example of a labeled Petri net. This labeled Petri net has a block structure which is denoted by the filled regions. (Obtained from [18]).

A state of a Petri net is defined by its *marking*, and it is often useful to consider a Petri net in combination with an initial marking and a set of possible final markings. This allows us to define the language accepted by the Petri net as a set of sequences of activities ($\mathfrak{L}$). We refer to a Petri net with an initial and a set of final states as an *accepting* Petri net.

A *process tree* [6] is an abstract representation of a block-structured hierarchical process model, in which the leaves represent the activities and the operators describe how their children are to be combined. τ denotes the activity (leaf) whose execution is not visible in the event log. We consider four operators: $\times$, $\rightarrow$, $\wedge$ and $\circlearrowright$ ($\oplus$ denotes any process tree operator). $\times$ describes the exclusive choice between its children, $\rightarrow$ the sequential composition and $\wedge$ the parallel composition. The first child of a loop $\circlearrowright$ is the body of the loop; all other children are redo children. First, the body must be executed, followed by zero or more iterations of a redo child and the body; after each iteration, execution can stop. Figure 1 shows the Petri net corresponding to the process tree $\rightarrow (\times(\wedge(a,b),c), \times(\circlearrowright (\rightarrow (d,e),f),g))$. Process trees can be straightforwardly translated to Petri nets, and these translated nets are inherently sound.

Notation-wise, let t be a trace and let A be a set of activities, then $t|_A$ refers to a *projected* trace containing only the events of t of which the activities are in A. Notice that this projected trace may be empty. Similarly, for a log L, $L|_A$ denotes a log consisting of the traces of L projected on A.

In an *alignment* procedure, an event log and a process model, which can be a normative or a descriptive, are compared. That is, for each trace of the log, a matching execution path through the model is searched for. This matching path might deviate from both the log, by skipping events, and the model, by skipping activities. A matching with the lowest number of skips is referred to as an *optimal* alignment. For instance, an optimal alignment of the trace $\langle a, b, c \rangle$ and the process tree $\rightarrow (\times(\wedge(a, b), c), \times(\circlearrowright (\rightarrow (d, e), f), g))$ is: $\begin{array}{|c|c c c|} \hline \text{trace} & a & b & c & - \\ \hline \text{model} & a & b & - & g \\ \hline \end{array}$.

4   Indulpet Miner

Indulpet Miner (IN) aims to combine the strengths of four process discovery techniques: Inductive Miner (IM), Local Process Models (LPM), the Evolutionary Tree Miner (ETM) and a new bottom-up recursive technique (BUR). We illustrate IN using Fig. 2: starting with a full event log Fig. 2(a), as a first step, Inductive Miner is applied, which tries to discover some structure in the event log and splits it accordingly into sub-logs, until it is unable to find structure in the sub-logs Fig. 2(b). We start with the Inductive Miner since it is the only process tree algorithm that guarantees fitness, and hence, in situations where this miner manages to find a precise model, no further work is needed. This step increases the number of event logs but decreases their complexity.

Second, the new bottom-up recursive technique (BUR) is applied Fig. 2(c). Where IM aims to find the highest-level structure in the log (starting with the root of the process tree), BUR aims to find lowest-level structure in the log (starting with individual activities and combining these into subtrees). The combination of IM and BUR is applied exhaustively until the event logs cannot be reduced in complexity anymore.

Third, LPM is applied to gather candidate local process models, which serve as starting seeds for ETM. That is, rather than starting from an arbitrary population of models, ETM begins its iterations using the local process models as its population. Initializing the ETM with an initial population of LPMs reduces the search space of the ETM, thereby improving computational efficiency. Finally, the ETM is applied to obtain a complete process tree Fig. 2(d).

Using these steps, the low-hanging fruit of well-structured behaviour is captured by Inductive Miner and the bottom-up recursive method, thereby minimizing the heavy lifting that has to be performed by the Evolutionary Tree Miner.

In this section, we describe each of the steps of Indulpet Miner. We first describe the three existing techniques Inductive Miner, Evolutionary Tree Miner and Local Process Models in more detail. Second, we introduce the new bottom-up recursive technique. We finish with the pseudo code of IN and a description of its implementation.

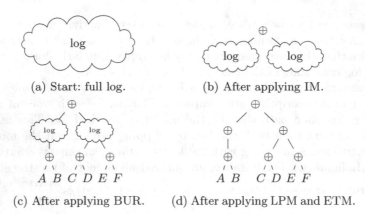

(a) Start: full log.   (b) After applying IM.

(c) After applying BUR.   (d) After applying LPM and ETM.

Fig. 2. An illustration of Indulpet Miner applying several steps to obtain a process tree from an event log.

4.1 Inductive Miner

Inductive Miner (IM) applies a recursive divide-and-conquer approach to process discovery, using four distinct steps [16]. That is, first the "most important" behaviour in the event log is identified, consisting of a process tree operator ($\times$, $\to$, $\wedge$, $\circlearrowleft$) and a proper division of the activities in the event log (a *cut*, denoted as $(\oplus, S_1, \ldots S_n)$ for operator $\oplus$ and sets of activities $S_1 \ldots S_n$). The operator is recorded as the root of the resulting process tree. Second, the cut found is used to split the event log into smaller sub-logs. Third, IM recurses on each sub-log until a *base case* is encountered, for instance, a log containing only a single activity, which is returned as a process tree leaf. Finally, if IM cannot find a cut, a *fall-through* is returned that overestimates the behaviour of the event log to be able to continue the recursion. For instance, if a particular activity occurs precisely once in each trace, then this activity is filtered out of the event log, the recursion continues, and the activity is put in parallel with the resulting process tree. As a last resort, IM will return a flower model that represents all behaviour over the activities in the event log, thereby guaranteeing fitness. In practice, on event logs of loosely structured processes, this last-resort fall-through of IM may cause a decrease in precision. With the Indulpet Miner, we aim to improve the fall-through of the Inductive Miner, by adding a more involved approach that combines several other process discovery techniques to the original fall-through strategies.

    Due to its flexibility, several variants of IM have been proposed, in particular to handle noise and infrequent behaviour (IMf) [17] and to handle lifecycle information (IMlc) [19], and even their combination (IMflc) [19], which we use in Indulpet Miner[1].

---

[1] Lifecycle information handling capabilities are necessary for Indulpet as the bottom-up recursion might insert such information in the event log, even if it was not present in the input event log.

Indulpet Miner uses all four steps of IMflc, and the cut selection, recursion and base case steps are used without change. If no cut can be found, before trying the default fall-throughs of IM and possibly overgeneralising, Indulpet first attempts to apply bottom-up recursion and, if that is successful, recurse further using IM. Second, if bottom-up recursion is not successful, Indulpet will apply LPM mining and finally the ETM. Should these not return a result, for instance when running out of a user-chosen time limit, the original fall-throughs of IM are applied to ensure that a process tree is returned at all times.

4.2   Evolutionary Tree Miner

The Evolutionary Tree Miner (ETM) [6] is a technique that applies a genetic search approach to process discovery: it starts with a population of process models and repeatedly evaluates, selects and mutates the models in the population to optimise it towards a set of chosen quality criteria. In each iteration, the models in the population are evaluated, and only the best-performing models are selected and considered further. The mutation steps of ETM are mutation and crossover. Crossover combines the well-performing parts of multiple models, while mutation replaces the ill-performing parts with random variations. To guarantee that all models can, eventually, be considered, a certain degree of randomness is inserted into the selection and mutation steps.

As ETM limits itself to process tree models, soundness is guaranteed. Furthermore, the Evolutionary Tree Miner can optimise for any log- or model-based quality criterion imaginable, and any process-tree construct, including duplicate activities, can be discovered. However, ETM is computationally expensive, which can be addressed by providing an initial population of models to give ETM a head start. In Indulpet Miner, we use the results of LPM mining for this purpose. Furthermore, to decrease the amount of work to be performed by ETM, Indulpet Miner reduces the size and complexity of the event log by a bottom-up recursion technique before calling ETM.

4.3   Local Process Models

Local Process Models (LPMs) [28] are process models that describe frequent but partial behaviour. That is, they model only a subset of the activities that were seen in the event log. An iterative expansion procedure is used to generate a ranked collection of LPMs. The iterative expansion procedure of LPM is often bounded to a maximum number of expansion steps (in practice often to 4 steps), as the expansion procedure is a combinatorial problem of which the size depends on the number of activities in the event log as well as the maximum number of activities in the LPMs that are mined.

LPM keeps a set of process trees, LPM, starting by considering a single activity from Σ, for instance $LPM_1 = a$. In each expansion step, the process trees in the set that occur often enough in the log, that is, their *support* exceeds some threshold, are expanded into larger trees. That is, an arbitrary activity is replaced with a sub-process tree containing that activity. For instance, one of the possible expansions of a is $\rightarrow (a, b)$, thereby creating $LPM_2 =\rightarrow (a, b)$ as an expansion of LPM_1. This procedure is repeated for every possible expansion using the operators $\times$, $\rightarrow$, $\wedge$ and $\circlearrowleft$. For instance, $\rightarrow (a, b)$ could be expanded into $LPM_3 =\rightarrow (a, \wedge(b, c))$, which is an expansion of LPM_2. The expansion procedure is guided by several heuristics and monotonicity properties [28], and ends at a certain user-chosen number of activities (for instance, 4).

All trees in the set that meet the given support threshold are returned. In Indulpet Miner, these returned trees serve as inputs to the ETM step.

4.4   Bottom-Up Recursion

Inductive Miner recurses in a top-down fashion, that is, it looks for the 'largest' behaviour in an event log and uses this behaviour to split the log into multiple smaller sublogs. In this section, we propose a novel approach, bottom-up recursion (BUR), that looks for the 'smallest' behaviour in an event log and uses this behaviour to reduce the complexity of the event log. BUR applies four steps: first a *partial cut* $(\oplus, A, B)$ is identified, consisting of a process tree operator $\oplus$ and two sets of activities A, B^2. Intuitively, a partial cut $(\oplus, A, B)$ denotes that $\oplus(M_A, M_B)$ has been identified as being a subtree of the resulting process tree, where M_A and M_B are process trees representing the choices between the activities of A and B respectively: $M_A = \times(a_1, \ldots a_n), M_B = \times(b_1, \ldots b_m)$. Second, this partial cut is used to *collapse* the event log: the activities of the partial cut are replaced by a dummy activity in each trace of the log. Third, BUR recurses and a process tree is returned. Fourth, in the resulting process tree, the dummy activity is replaced with a subtree corresponding to the partial cut.

For instance, let $L = [\langle a, b, d \rangle, \langle a, d, b \rangle]$ be an event log. BUR could identify the partial cut $(\rightarrow, \{a\}, \{b\})$. Next, L is collapsed by replacing a and b with a dummy activity y: $L' = [\langle y_s, y_c, d \rangle, \langle y_s, d, y_c \rangle]$. Then, L' is recursed on and, as y and d overlap in time, a process tree $\wedge(y, d)$ results. As a final step, BUR replaces y with a subtree representing the partial cut, and the final model becomes $\wedge(\rightarrow (a, b), d)$.

In the remainder of this section, we first define partial cuts formally and show how they can be identified. Second, we explain the log collapsing.

Partial Cuts and Detecting Them. A *partial cut* $(\oplus, A, B)$ consists of an operator $\oplus$ and two sets of activities A and B:

Definition 1 (Partial cut). *Let Σ be an alphabet, $\oplus \neq \times$ be a process tree operator and A, B be sets of activities such that $A \cup B \subseteq \Sigma$. Then, $(\oplus, A, B)$ is a partial cut of Σ.*

Notice that every binary cut of Inductive Miner is also a partial cut. Intuitively, a partial cut $(\oplus, \{A_1, \ldots A_m\}, \{B_1, \ldots B_n\})$ means that

---

^{2} For simplicity, we use only binary partial cuts in this paper. The definitions extend to n-ary partial cuts and this does not change the expressivity of the method [18].

$\oplus(\times(A_1,\ldots A_m),\times(B_1,\ldots B_n))$is a subtree of the to-be discovered model. That is, $\oplus$ defines the relation between the set of activities A and B, while the relation between the activities within sets A and B is defined by $\times$. Therefore, partial cuts do not need to consider $\times$-operators: we limit ourselves to $\rightarrow$, $\wedge$ and $\circlearrowleft$.

We formalise this intuition in *fitting* partial cuts: a partial cut is fitting if and only if, in each trace, the tree corresponding to the partial cut is executed zero or more times, that is, it is never violated:

Definition 2 (Fitting partial cut). *Let Σ be an alphabet, let L be an event log over Σ and let $C = (\oplus, A, B) = (\oplus, \{A_1,\ldots A_m\}, \{B_1,\ldots B_n\})$ be a partial cut. Then, C is a* fitting *partial cut if each trace in the log follows the semantics of the partial cut:*

$$L|_{A \cup B} \subseteq \mathcal{L}(\qquad\qquad\qquad)$$

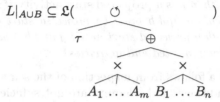

Notice that this definition gives another reason not to consider $\times$-operators in partial cuts: a partial cut $(\times, \{a\}, \{b\})$ would always fit and thus not express any new information.

In Indulpet Miner, BUR exhaustively considers all partial cuts to find one that is fitting. As soon as a fitting partial cut is found, BUR continues as described before.

It is rather expensive to test whether a given partial cut is fitting, as this requires a pass over the entire event log. To limit the number of times that this time-consuming step has to be performed, we identified some necessary, though not sufficient, conditions that a partial cut has to satisfy in order to be fitting. Thus, BUR uses these conditions to prune the search space of partial cuts.

For these conditions, and for the remainder of this paper, we assume that the partial cut is disjoint and non-empty, i.e. $A \cap B = \emptyset$, $A \neq \emptyset$ and $B \neq \emptyset$.

Lemma 3 (Necessary conditions for fitting partial cuts). *Let Σ be an alphabet, L be an event log over Σ and $C = (\oplus, A, B)$ be a partial cut such that C fits L, $A \cap B = \emptyset$ and $A, B \neq \emptyset$. Then:*

1. *Within the sets of activities, there are no connections in the directly follows graph:*
   $$\forall_{X \in \{A,B\}} \forall_{a,b \in X \wedge a \neq b} a \not\rightarrow b$$
2. *Between the sets of activities, the directly follows graph exhibits an "approved" pattern, depending on the operator $\oplus$. Let $a \in A$ and $b \in B$:*
   $\oplus = \rightarrow$ *There is a directly follows relation between a and b: $a \rightarrow b$.*
   $\oplus = \wedge$ *A directly follows relation is present in both directions: $a \twoheadrightarrow b \wedge b \twoheadrightarrow a$.*
   $\oplus = \circlearrowleft$ *A directly follows relation is present in both directions: $a \twoheadrightarrow b \wedge b \twoheadrightarrow a$.*
3. *For each pair of activities in A, B in relation to each other activity, the directly follows graph exhibits an "approved" pattern, depending on the operator $\oplus$. Let $a \in A$, $b \in B$ and $c \in \Sigma \setminus (A \cup B)$:*

$\oplus = \rightarrow$ $a \longrightarrow b$ $a \longrightarrow b$ $a \longrightarrow b$ $a \longrightarrow b$ $a \longrightarrow b$

c c c c c

$\oplus = \wedge$ $a \rightleftarrows b$ $a \rightleftarrows b$ $a \rightleftarrows b$ $a \rightleftarrows b$

c c c c

$\oplus = \circlearrowright$ $a \rightleftarrows b$ $a \rightleftarrows b$ $a \rightleftarrows b$ $a \rightleftarrows b$ $a \rightleftarrows b$

c c c c c

4. *Let x be an activity, then x is a projected start activity if $\exists_{t \in L|_{A \cup B}} t = \langle x, \ldots \rangle$. Then, for $\oplus = \wedge$, both a and b must be projected start activities. Then, for $\oplus = \circlearrowright$, a must be both a projected start activity and b must not. (a symmetric requirement holds for projected end activities)*

The proof of this lemma follows from inspection of the semantics of partial cuts. We show that the conditions of the lemma are not sufficient to conclude that a partial cut is fitting by means of a counterexample: consider the event log $L_1 = [\langle a, b, c, b \rangle, \langle c, a, c \rangle]$ and the partial cut $C_1 = (\rightarrow, a, b)$. The directly follows graph of L_1 is $a \rightleftarrows c \rightleftarrows b$. The partial cut C_1 satisfies all conditions of Lemma 3, but does not fit L_1, as, for instance, the second trace of L_1 projected on a and b yields $\langle a \rangle$, which violates the partial cut, as this cut indicates that after each a there should be a b.

To handle noise, the fitness requirement of Definition 2 can be relaxed. That is, BUR searches for the partial cut with the highest fitness, as long as the fitness (measured as the fraction of traces that adhere to the partial cut) reaches a certain user-chosen threshold.

Log Collapsing. As the second step, given a partial cut, BUR collapses the event log, depending on the operator $\oplus$ of the cut.

In LPM [23], a similar step is performed ("log abstraction"). However, the LPM collapsing procedure is computationally much more expensive due to the need to use alignments in order to obtain fitness and precision measures, which BUR does not need in this step.

In BUR, any activity that is not part of the partial cut is ignored. Let y be a fresh activity that does not appear in the log. Then, every execution instance of the partial cut is replaced with an execution instance of y. That is, the first event of each instance is replaced with y_{start}, the last event with $y_{complete}$, and all other events of the activities in the partial cut in between are removed. In case the partial log is not fitting, non-fitting events are removed.

For instance for $(\rightarrow, \{a\}, \{b\})$: $\langle a, b, c, a, b, a, b \rangle$ is collapsed into $\langle y_{start}, y_{complete}, c, y_{start}, y_{complete}, y_{start}, y_{complete} \rangle$. As another example, for $(\circlearrowright, \{a\}, \{b\})$: $\langle a, b, c, a, c, a, b, a, b, a \rangle$ is collapsed into $\langle y_{start}, c, y_{complete}, c, y_{start}, y_{complete} \rangle$.

Example of BUR. Let L_1 be $\{\langle a,b \rangle, \langle b,a \rangle, \langle a,b,c,d,a,b \rangle, \langle b,a,c,d,b,a \rangle\}$. Its directly follows graph is $a \rightleftarrows b$.

$$a \rightleftarrows b$$
$$\uparrow \times \downarrow$$
$$d \longleftarrow c$$

- The partial cut $(\wedge, \{a\}, \{c\})$ does not preserve fitness, with as a counterexample the trace $\langle a,b \rangle$, as this trace contains a but not c;
- The partial cut $(\circlearrowright, \{a\}, \{c\})$ preserves fitness, but the directly follows graph does not contain the required edge $c \twoheadrightarrow a$;
- The partial cut $(\wedge, \{a\}, \{b\})$ preserves fitness and has an approved directly follows pattern;
- The partial cut $(\rightarrow, \{c\}, \{d\})$ preserves fitness and has an approved directly follows pattern;

Arbitrarily choose the partial cut $(\wedge, \{a\}, \{b\})$ and merge L_1 using this cut and the fresh activity e: $L_2 = \{\langle e_s, e_c \rangle, \langle e_s, e_c \rangle, \langle e_s, e_c, c, d, e_s, e_c \rangle, \langle e_s, e_c, c, d, e_s, e_c \rangle\}$.

On L_2, the algorithm recurses. Its directly follows graph is $e \twoheadleftarrow d \rightleftarrows c$.

- The partial cut $(\wedge, \{e\}, \{c\})$ does not preserve fitness, with as a counterexample the trace $\langle e_s, e_c \rangle$, as e appears but c does not.
- The partial cut $(\circlearrowright, \{e\}, \{c\})$ preserves fitness, but the directly follows graph does not contain the edge $c \twoheadrightarrow e$, so an approved pattern is not present;
- The partial cut $(\rightarrow, \{c\}, \{d\})$ preserves fitness and has an approved directly follows pattern;

Choose the partial cut $(\rightarrow, \{c\}, \{d\})$ and merge L_2 using this cut and the fresh activity f: $L_3 = \{\langle e_s, e_c \rangle, \langle e_s, e_c \rangle, \langle e_s, e_c, f_s, f_c, e_s, e_c \rangle, \langle e_s, e_c, f_s, f_c, e_s, e_c \rangle\}$. On L_3, BUR recurses and obtains the directly follows graph $e \rightleftarrows f$.

- The partial cut $(\wedge, \{e\}, \{f\})$ does not preserve fitness, with as a counterexample the trace $\langle e_s, e_c \rangle$;
- The partial cut $(\circlearrowright, \{f\}, \{e\})$ does not preserve fitness, with as a counterexample the trace $\langle e_s, e_c \rangle$;
- The partial cut $(\circlearrowright, \{e\}, \{f\})$ preserves fitness and has an approved directly follows pattern;

Thus, choose the partial cut $(\circlearrowright, \{e\}, \{f\})$ and construct a process tree top-down by replacing the introduced fresh activities with process trees corresponding to their partial cuts: $\circlearrowright (e, f)$ to $\circlearrowright (\wedge(a,b), f)$ and as the final result $\circlearrowright (\wedge(a,b), \rightarrow (c,d))$.

Complexity and Rediscoverability. All possible partial cuts are explored, and the first one that is encountered that satisfies Definition 2 is returned. The number of possible partial cuts is $O(2^{2^{|\Sigma|}})$, all of which need to be considered. The properties of Lemma 3 are applied for each partial cut to minimise the time spent per partial cut. If a property fails, the partial cut is discarded. If all properties of Lemma 3 hold, the fitness of the partial cut with respect to the event log is measured to find the best-fitting partial cut. In practice, this last step is rarely called.

Acknowledging these run-time considerations, BUR could be used as a stand-alone process discovery algorithm for smaller logs. Such an algorithm would be able to distinguish all process trees consisting of the four operators $\times$, $\rightarrow$, $\wedge$ and $\circlearrowleft$ (but excluding τ leaves). In particular, so-called short loops can be handled, which have been shown to pose difficulties for discovery algorithms such as the IM and the α-algorithm [1]. For instance, $\wedge(\circlearrowleft (a, b), \circlearrowleft (c, d))$ can be distinguished from $\circlearrowleft (\wedge(a, c), \wedge(b, d))$, even though these two trees have the same directly follows graph.

4.5   Indulpet: Algorithm and Implementation

To summarise, Indulpet Miner applies the following steps:

function INDULPET(log L)
    if IMflc finds a base case b in L **then return** b **end if**
    if IMflc finds a cut c of operator $\oplus$ in L **then**
        $L_1 \ldots L_n \leftarrow$ split L using c
        return $\oplus(\text{INDULPET}(L_1), \ldots \text{INDULPET}(L_n))$
    end if
    if BUR finds a partial cut $c = (\oplus, \{Q_1, \ldots Q_q\}, \{R_1, \ldots R_r\})$ in L **then**
        $L' \leftarrow$ collapse L using c and a fresh activity a'
        $T' \leftarrow$ INDULPET(L')
        return T with all a''s replaced by $\oplus(\times(Q_1, \ldots Q_q), \times(R_1, \ldots R_r))$
    end if
    $X \leftarrow$ LPM(L)
    if $M \leftarrow$ ETM(L, X) **then return** M **end if**
    return the first fallthrough of IMflc that applies
end function

Please note that Indulpet adheres to the IM framework, thus all guarantees and proofs provided for the IM framework apply [15]. The proof obligations for fitness do not hold due to the ETM step, thus fitness is not guaranteed. However, rediscoverability holds for Indulpet, that is, the ability to rediscover the language of a system-model underlying the event log, while making some assumptions on the class of the system-model and the event log [15, Theorem 6.43].

Indulpet Miner has been implemented as a plug-in of the ProM framework [11] and is distributed in the package manager of ProM 6.8. Given the high complexity of bottom-up recursion (there are $O(2^{2^n})$ possible partial cuts that all need to be considered), a time limit of 10 min is applied for this step. Furthermore, the ETM step is time-limited to 10 min as well. Both time limits can be overridden by a user. Please note that both steps can be called many times, thus Indulpet might take longer than the set time limit.

5   Evaluation

Several process discovery techniques have been proposed, and in this section, we compare Indulpet Miner to existing discovery techniques. In particular, we aim to answer two questions: (1) does Indulpet Miner strike a new balance in log-quality criteria compared to existing techniques, and (2) does Indulpet Miner combine the advantages of Inductive Miner, Local Process Models, bottom-up recursion and the Evolutionary Tree Miner?

To perform this evaluation, we applied the discovery algorithms to several real-life event logs. The real-life event logs that were included are described in Table 1. We perform our measures using 3-fold cross-validation, in which each log is split into three buckets. That is, each trace is put in one of the three buckets randomly. Each combination of two buckets is used for discovery, while one bucket is used for evaluation.

To address the randomness of some algorithms (Indulpet, Evolutionary Tree Miner), we repeat the 3-fold validation ten times. That is, for each real-life log, each algorithm is applied 30 times in total, yielding 30 models.

All models are translated to accepting Petri nets (to reduce the impact of this translation on the results, models were structurally reduced after the translation), and fitness is measured using alignments [2]. For most event logs, precision is measured using ETC [24] (as the procedure from process tree via accepting Petri net to alignment and ETC is not deterministic, it was applied 5 times for each model). For the bpic15 events logs, computing the alignments proved too time- and memory consuming and we evaluated fitness and precision using Projected Conformance Checking (PCC) [20]. As we are interested in the trade-offs between quality criteria, we do not report on the f-score. Furthermore, the number of places, transitions, and arcs in the accepting Petri nets is reported as simplicity and we verified the boundedness of models using the LoLa tool [29].

All existing discovery techniques that guarantee soundness and of which an implementation is publicly available were included: Evolutionary Tree Miner (ETM) [6] and Inductive Miner - infrequent (IMf) [17]. Furthermore, we added Split Miner (SM) [3], which guarantees models without deadlocks, though neither weak soundness nor boundedness. Finally, a baseline flower model (F) was included, which is a model that supports all behaviour over all activities seen in the event log. It would be interesting to include the Fusion Miner [7] and the approach of [23] as well, however both techniques lack an implementation that is compatible with the ProM framework, therefore disallowing the application of evaluation measures provided by ProM such as ETC, PCC, simplicity measures, and soundness checking.

Reproducibility. All event logs are publicly available, and the source code we used to run these experiments is available at https://svn.win.tue.nl/repos/prom/Packages/SanderLeemans/Trunk/.

Table 1. Event logs used in the evaluation.

		Events	Traces	Activities
bpic12	BPI Challenge 2012 [8]: a mortgage application-to-approval process in a Dutch financial institute	262,200	13,087	36
bpic15	BPI Challenge 2015 [9]: a building-permit approval process in five Dutch municipalities	52,217	1199	395
		44,354	832	410
		59,681	1409	383
		47,293	1053	356
		59,083	1156	389
bpic18	BPI Challenge 2018 [10]: an agricultural grant-application process of the European Union. There are eight logs in this data set, each representing a sub-process for a particular document	161,296	43,808	7
		46,669	29,297	6
		293,245	15,260	20
		569,209	29,059	16
		197,717	5,485	15
		132,963	14,750	10
		984,613	43,809	24
		128,554	43,802	6
rtf	Road Traffic Fines [21]: a process of collecting road traffic fines by an Italian government	561,470	150,370	11
sps	Sepsis [22]: a sepsis-treating process in a hospital	15,214	1,050	16

5.1   Results and Discussion

The results of the evaluation are shown in Table 2. Some results could not be obtained and have been denoted with exclamation marks: ETM ran out of memory or ran for multiple days, while SM returned unbounded models that could not be measured by either alignment-based or PCC conformance checking measures.

Run Time. Table 2 shows indicative run times, given as $\log_{10}$, such that 1 is one second, 2 is hundred seconds, etc. These results suggest that IN is faster than ETM, while IMf is faster than IN, both by several orders of magnitude. Comparing IN with SM yields mixed results: on some logs (e.g. bpic12, bpic15_1), SM is faster, while on others (e.g. bpic18_1, bpic18_2) IN uses its top-down recursion more and is faster, although the logs on which IN is faster tend to be the smaller logs, where IN behaves as IMf. It is clear that IMf and SM are preferred choices if a process model is to be obtained fast. Nevertheless, to put

Table 2. Results of the evaluation. Fitness: f, precision: p, simplicity: s, time in $\log_{10}$ seconds: t. The numbers are given as average $\pm$ sample standard deviation.

	bpic12				bpic15-1				bpic15-2			
	f	p	s	t	f	p	s	t	f	p	s	t
IMf	0.97±0.01	0.59±0.03	186.57±9.59	0	1.00±0.00	0.66±0.04	1286.07±149.54	1	0.99±0.00	0.76±0.04	1048.10±167.84	1
IN	0.36±0.07	0.88±0.05	116.30±30.52	3	0.78±0.01	0.97±0.01	91.50±54.38	3	0.74±0.01	0.96±0.01	114.10±23.71	3
ETM			!	!			!	!			!	!
SM	0.96±0.00	0.68±0.00	239.00±0.00	1		!	4715.20±8.88	2		!	5077.27±10.28	
F	1.00±0.00	0.11±0.00	85.00±0.00	-1	1.00±0.00	0.64±0.01	1145.30±20.02	-1	1.00±0.00	0.64±0.01	1172.70±24.44	-1

	bpic15-3				bpic15-4				bpic15-5			
	f	p	s	t	f	p	s	t	f	p	s	t
IMf	1.00±0.00	0.71±0.03	1201.63±133.03	1	0.99±0.00	0.75±0.03	1082.47±98.26	0	0.99±0.00	0.79±0.05	1108.60±141.55	1
IN	0.79±0.01	0.97±0.01	118.37±80.59	3	0.76±0.02	0.93±0.03	272.17±63.98	4	0.75±0.02	0.96±0.02	244.17±120.50	3
ETM			!	!			!	!			!	!
SM		!	3995.47±8.96			!	3882.73±5.00	2		!	4730.20±11.38	2
F	1.00±0.00	0.66±0.01	1114.50±16.26	-1	1.00±0.00	0.65±0.02	1024.30±27.16	-1	1.00±0.00	0.65±0.02	1116.40±26.41	-1

	bpic18-1				bpic18-2				bpic18-3			
	f	p	s	t	f	p	s	t	f	p	s	t
IMf	1.00±0.00	0.95±0.02	54.87±0.73	-0	0.96±0.00	0.96±0.05	43.70±6.73	-1	0.93±0.00	0.53±0.01	74.77±1.28	1
IN	1.00±0.00	0.95±0.02	54.87±0.73	-0	0.96±0.00	0.96±0.05	43.70±6.73	-1	0.79±0.04	0.95±0.04	55.60±17.73	3
ETM	0.97±0.04	0.95±0.12	77.50±86.35	3	0.99±0.00	0.76±0.21	183.93±91.24	4			!	!
SM	1.00±0.00	0.97±0.03	59.00±0.00	1	1.00±0.00	0.95±0.02	90.00±0.00	1	1.00±0.00	0.67±0.01	291.00±0.00	1
F	1.00±0.00	0.32±0.00	30.00±0.00	-0	1.00±0.00	0.51±0.00	27.00±0.00	-1	1.00±0.00	0.14±0.00	68.90±0.55	-0

	bpic18-4				bpic18-5				bpic18-6			
	f	p	s	t	f	p	s	t	f	p	s	t
IMf	0.86±0.04	0.35±0.03	96.57±14.44	1	0.80±0.01	0.67±0.01	129.10±15.84	-0	0.97±0.00	0.55±0.02	73.83±2.74	-1
IN	0.61±0.19	0.83±0.27	71.97±50.49	4	0.78±0.01	0.69±0.05	164.00±43.89	2	0.97±0.00	0.55±0.02	73.83±2.74	-0
ETM			!	!			!	!			!	!
SM	0.99±0.00	0.55±0.00	247.00±0.00	1	0.88±0.00	0.74±0.00	131.00±0.00	1	1.00±0.00	0.72±0.01	147.00±0.00	1
F	1.00±0.00	0.18±0.00	57.00±0.00	-0	1.00±0.00	0.15±0.00	54.00±0.00	-1	1.00±0.00	0.34±0.01	39.00±0.00	-1

	bpic18-7				bpic18-8				rtf			
	f	p	s	t	f	p	s	t	f	p	s	t
IMf	0.93±0.02	0.83±0.01	164.67±17.54	2	1.00±0.00	0.89±0.06	54.87±3.17	-0	0.98±0.00	0.79±0.04	97.97±4.78	1
IN	0.89±0.03	0.84±0.02	220.43±70.34	3	1.00±0.00	0.89±0.05	54.87±3.17	-0	0.98±0.00	0.79±0.02	96.50±7.00	2
ETM			!	!			!	!	0.89±0.08	0.68±0.27	117.93±85.12	3
SM	0.02±0.00	0.96±0.00	333.00±0.00	2	1.00±0.00	0.97±0.03	66.00±0.00	1	1.00±0.00	0.96±0.00	82.00±0.00	1
F	1.00±0.00	0.64±0.01	81.00±0.00	-0	1.00±0.00	0.41±0.01	27.00±0.00	-0	1.00±0.00	0.38±0.01	46.00±0.00	-0

	sps			
	f	p	s	t
IMf	0.91±0.02	0.41±0.06	140.23±13.85	-1
IN	0.91±0.02	0.43±0.05	122.43±8.66	2
ETM			!	!
SM	0.76±0.00	0.73±0.01	138.00±0.00	0
F	1.00±0.00	0.21±0.00	61.00±0.00	-2

things in perspective, the maximum run time of IN observed in this experiment was 5 hours, which suggests that IN is still feasible, even for large and complex logs.

Quality. For bpic12, IN achieves the simplest model (except baseline F) and the highest precision, but fitness is significantly lower than the other algorithms. On all bpic15 logs, IN consistently achieves the best simplicity (even surpassing baseline F) and precision, with a lower fitness. On these logs, SM discovers very complex unbounded models, which can be measured using neither alignments nor the PCC framework.

The bpic18 logs differ much in complexity (6 to 24 activities) and show mixed results: on bpic18_1 and bpic18_2, IN discovers the same models as IMf, with their measures differing marginally from SM, though always being Pareto optimal. On bpic18_3 and bpic18_4, all tested algorithms perform Pareto optimal. To illustrate these models, four have been included in Fig. 3: in the model by IMf, some concurrent activities can be arbitrarily repeated, leading to a rather low precision. The model by SM does not contain any concurrency and has a high fitness. However, it is much less simple and precise than the other models. The model by IN has a lower fitness, but a much higher precision and is very simple. To provide some intuition: in the vast majority of traces in this log, the activities "begin editing" and "finish editing" are alternating according to intuition: for instance, only in 34 of the 29,059 traces, the trace ends after "begin editing" rather than "finish editing", and only 15,325 of the 295,621 executions of both activities are repeated. The model by IMf only requires that after each "finish editing" there must be a "begin editing", the model by SM requires "begin editing" to be executed first, but afterwards no further constraint is posed on these two activities, and the model by F does not express this constraint at all. Only IN discovers this constraint correctly by duplicating the "begin editing" activity.

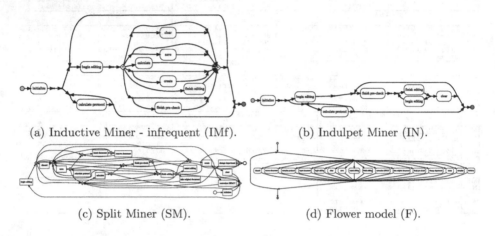

(a) Inductive Miner - infrequent (IMf).   (b) Indulpet Miner (IN).

(c) Split Miner (SM).   (d) Flower model (F).

Fig. 3. Results for bpic18-4 ("Geo parcel document").

On bpic18_5, SM is the clear winner on all dimensions. Remarkably, on this log IMf has a large standard deviation on simplicity, which, as IMf is deterministic, indicates that the 3-fold procedure withheld useful information from discovery. As the standard deviation of SM is much smaller, SM uses less information from the event logs. On bpic18_6, IMf and IN discover the same models, but SM achieves the highest fitness and precision, and F the best simplicity (all models are Pareto optimal). On bpic18_7, IMf achieves the highest fitness, while IN discovers models with slightly higher precision and lower fitness. The models by SM have the highest precision, but a very low fitness, as the log contains many activities that are repeated (so-called short loops), which SM does not discover. On bpic18_8, all algorithms achieve perfect fitness, IN and IMf achieve the best simplicity, and SM achieves the highest precision.

On rtf, SM discovers a model that is better on all measures than IMf, IN and ETM. Finally, on sps, IN is Pareto-optimal over IMf, and SM achieves the highest precision, at the cost of fitness and simplicity.

The results show that IN in many cases combines the advantages of its subalgorithms: the speed, feasibility and fitness of IMf with the precision of ETM. However, it is difficult to appoint a clear winner: algorithms strike different balances of log-quality criteria, and all routinely achieve Pareto optimality. Nevertheless, we can conclude that Indulpet Miner can achieve a different balance in log-quality criteria than the other tested techniques and might be a useful discovery algorithm in the toolbox of analysts, depending on the use case at hand.

6   Conclusion and Future Work

We have presented a novel process discovery approach which combines several existing process discovery algorithms that are based on process trees: the Inductive Miner (IM) [16], the Evolutionary Tree Miner (ETM) [6] and the Local Process Model (LPM) miner [28]. Additionally, we have proposed a novel process tree mining approach that is based on bottom-up recursion, which is too computationally complex to process a full event log but is useful as an element in the Indulpet miner to find local structures of process behaviour. The Indulpet miner is guaranteed to find sound process models, and we have shown on a collection of real-life event logs that the Indulpet miner often provides a Pareto optimal, yet different, trade-off between fitness, precision and simplicity compared to the Inductive Miner and the Split Miner algorithms.

An interesting direction of future work would be to explore heuristic search or other intelligent approaches to search the space of all possible cuts of the proposed bottom-up recursion technique. This would enable the use of bottom-up recursion as a process discovery technique by itself, rather than an element in a hybrid mining approach, which is currently rendered infeasible by the computational complexity of exploring the full space of bottom-up recursion cuts. Another interesting direction for future work would be to make the ETM and LPM approaches able to deal with lifecycle transitions, which would make the Indulpet Miner fully compatible with logs that have lifecycle transitions. In the case of IM, there already exists a version that is able to handle lifecycles [19]. Making the Indulpet Miner compatible with lifecycles amongst others makes it possible to calculate more accurate process performance statistics on the process model.

Acknowledgement. We thank Joos Buijs for his help in integrating the Evolutionary Tree Miner (ETM) with the Indulpet Miner, and Eric Verbeek for coming up with its name.

References

1. van der Aalst, W.M.P.: Process Mining: Data Science in Action. Springer, Heidelberg (2016). https://doi.org/10.1007/978-3-662-49851-4
2. van der Aalst, W.M.P., Adriansyah, A., van Dongen, B.F.: Replaying history on process models for conformance checking and performance analysis. Wiley Interdiscip. Rew: Data Min. Knowl. Discov. **2**(2), 182–192 (2012). https://doi.org/10.1002/widm.1045
3. Augusto, A., Conforti, R., Dumas, M., La Rosa, M., Polyvyanyy, A.: Split miner: automated discovery of accurate and simple business process models from event logs. Knowl. Inf. Syst., 1–34 (2018). https://dblp.org/rec/bibtex/conf/icdm/AugustoCDR17

4. Breiman, L.: Bagging predictors. Mach. Learn. **24**(2), 123–140 (1996)
5. vanden Broucke, S.K.L.M., De Weerdt, J.: Fodina: a robust and flexible heuristic process discovery technique. Decis. Support Syst. **100**, 109–118 (2017)
6. Buijs, J.C.A.M., van Dongen, B.F., van der Aalst, W.M.P.: A genetic algorithm for discovering process trees. In: IEEE Congress on Evolutionary Computation, pp. 1–8. IEEE (2012)
7. Dahari, Y., Gal, A., Senderovich, A., Weidlich, M.: Fusion-based process discovery. In: Krogstie, J., Reijers, H.A. (eds.) CAiSE 2018. LNCS, vol. 10816, pp. 291–307. Springer, Cham (2018). https://doi.org/10.1007/978-3-319-91563-0_18
8. van Dongen, B.: BPI challenge 2012 dataset (2012). https://doi.org/10.4121/uuid: 3926db30-f712-4394-aebc-75976070e91f
9. van Dongen, B.: BPI challenge 2015 dataset (2015). https://doi.org/10.4121/uuid: 31a308ef-c844-48da-948c-305d167a0ec1
10. van Dongen, B., Borchert, F.: BPI challenge 2018 dataset (2018). https://doi.org/ 10.4121/uuid:3301445f-95e8-4ff0-98a4-901f1f204972
11. van Dongen, B.F., de Medeiros, A.K.A., Verbeek, H.M.W., Weijters, A.J.M.M., van der Aalst, W.M.P.: The ProM framework: a new era in process mining tool support. In: Ciardo, G., Darondeau, P. (eds.) ICATPN 2005. LNCS, vol. 3536, pp. 444–454. Springer, Heidelberg (2005). https://doi.org/10.1007/11494744_25
12. Goedertier, S., Martens, D., Vanthienen, J., Baesens, B.: Robust process discovery with artificial negative events. J. Mach. Learn. Res. **10**(Jun), 1305–1340 (2009)
13. Günther, C.W., van der Aalst, W.M.P.: Fuzzy mining – adaptive process simplification based on multi-perspective metrics. In: Alonso, G., Dadam, P., Rosemann, M. (eds.) BPM 2007. LNCS, vol. 4714, pp. 328–343. Springer, Heidelberg (2007). https://doi.org/10.1007/978-3-540-75183-0_24
14. Hoeting, J.A., Madigan, D., Raftery, A.E., Volinsky, C.T.: Bayesian model averaging: a tutorial. Stat. Sci. **14**(4), 382–401 (1999)
15. Leemans, S.J.J.: Robust process mining with guarantees. Ph.D. thesis, Eindhoven University of Technology (2017)
16. Leemans, S.J.J., Fahland, D., van der Aalst, W.M.P.: Discovering block-structured process models from event logs - a constructive approach. In: Colom, J.-M., Desel, J. (eds.) PETRI NETS 2013. LNCS, vol. 7927, pp. 311–329. Springer, Heidelberg (2013). https://doi.org/10.1007/978-3-642-38697-8_17
17. Leemans, S.J.J., Fahland, D., van der Aalst, W.M.P.: Discovering block-structured process models from event logs containing infrequent behaviour. In: Lohmann, N., Song, M., Wohed, P. (eds.) BPM 2013. LNBIP, vol. 171, pp. 66–78. Springer, Cham (2014). https://doi.org/10.1007/978-3-319-06257-0_6
18. Leemans, S.J.J., Fahland, D., van der Aalst, W.M.P.: Discovering block-structured process models from incomplete event logs. In: Ciardo, G., Kindler, E. (eds.) PETRI NETS 2014. LNCS, vol. 8489, pp. 91–110. Springer, Cham (2014). https:// doi.org/10.1007/978-3-319-07734-5_6
19. Leemans, S.J.J., Fahland, D., van der Aalst, W.M.P.: Using life cycle information in process discovery. In: Reichert, M., Reijers, H.A. (eds.) BPM 2015. LNBIP, vol. 256, pp. 204–217. Springer, Cham (2016). https://doi.org/10.1007/978-3-319-42887-1_17
20. Leemans, S.J.J., Fahland, D., van der Aalst, W.M.P.: Scalable process discovery and conformance checking. Softw. Syst. Model. **17**(2), 599–631 (2018). https://doi.org/10.1007/s10270-016-0545-x
21. de Leoni, M., Mannhardt, F.: Road traffic fine management process (2015). https:// doi.org/10.1007/s00607-015-0441-1

22. Mannhardt, F.: Sepsis cases - event log (2018). https://doi.org/10.4121/uuid:915d2bfb-7e84-49ad-a286-dc35f063a460

23. Mannhardt, F., Tax, N.: Unsupervised event abstraction using pattern abstraction and local process models. In: Working Conference on Enabling Business Transformation by Business Process Modeling, Development, and Support, pp. 55–63. CEUR-WS.org (2017)

24. Munoz-Gama, J.: Conformance Checking and Diagnosis in Process Mining - Comparing Observed and Modeled Processes. LNBIP, vol. 270. Springer, Cham (2016). https://doi.org/10.1007/978-3-319-49451-7

25. Murata, T.: Petri nets: properties, analysis and applications. Proc. IEEE **77**(4), 541–580 (1989)

26. Object Management Group: Notation (BPMN) version 2.0. OMG Specification (2011)

27. Schapire, R.E.: The strength of weak learnability. Mach. Learn. **5**(2), 197–227 (1990)

28. Tax, N., Sidorova, N., Haakma, R., van der Aalst, W.M.P.: Mining local process models. J. Innov. Digit. Ecosyst. **3**(2), 183–196 (2016)

29. Tredup, R., Rosenke, C., Wolf, K.: Elementary net synthesis remains np-complete even for extremely simple inputs. In: Khomenko, V., Roux, O.H. (eds.) PETRI NETS 2018. LNCS, vol. 10877, pp. 40–59. Springer, Cham (2018). https://doi.org/10.1007/978-3-319-91268-4_3

30. Wolpert, D.H.: Stacked generalization. Neural Netw. **5**(2), 241–259 (1992)

31. van Zelst, S.J., van Dongen, B.F., van der Aalst, W.M.P.: Avoiding over-fitting in ILP-based process discovery. In: Motahari-Nezhad, H.R., Recker, J., Weidlich, M. (eds.) BPM 2015. LNCS, vol. 9253, pp. 163–171. Springer, Cham (2015). https://doi.org/10.1007/978-3-319-23063-4_10

Towards Event Log Querying for Data Quality

Let's Start with Detecting Log Imperfections

Robert Andrews[✉], Suriadi Suriadi, Chun Ouyang, and Erik Poppe

Queensland University of Technology, Brisbane, Australia
{r.andrews,s.suriadi,c.ouyang,e.poppe}@qut.edu.au

Abstract. Process mining is, by now, a well-established discipline focussing on process-oriented data analysis. As with other forms of data analysis, the quality and reliability of insights derived through analysis is directly related to the quality of the input (*garbage in - garbage out*). In the case of process mining, the input is an event log comprised of event data captured (in information systems) during the execution of the process. It is crucial then that the event log be treated as a first-class citizen. While data quality is an easily understood concept little effort has been directed towards systematically detecting data quality issues in event logs. Analysts still spend a large proportion of any project in 'data cleaning', often involving manual and *ad hoc* tasks, and requiring more than one tool. While there are existing tools and languages that query event logs, the problem of different approaches for different log imperfections remains. In this paper we take the first steps to developing QUELI (Querying Event Log for Imperfections) a log query language that provides direct support for detecting log imperfections. We develop an approach that identifies capabilities required of QUELI and illustrate the approach by applying it to 5 of the 11 event log imperfection patterns described in [29]. We view this as a first step towards operationalising systematic, automated support for log cleaning.

Keywords: Process mining · Event log query language
Data quality · Event log imperfection patterns

1   Introduction

Process mining is, by now, a well-established discipline focussing on process-oriented data analysis. As with other forms of data analysis, the quality and reliability of insights derived through analysis is directly related to the quality of the input (*garbage in - garbage out*). In the case of process mining, the input is an event log comprised of event data captured (in information systems) during the execution of the process. It is crucial then that the event log be treated as a first-class citizen.

© Springer Nature Switzerland AG 2018
H. Panetto et al. (Eds.): OTM 2018 Conferences, LNCS 11229, pp. 116–134, 2018.
https://doi.org/10.1007/978-3-030-02610-3_7

From it's inception to the present day, process mining has focused on three key areas; *discovery*, (taking an event log and generating a process model), *conformance* (comparing an existing process model with an event log drawn from the same process), and *enhancement* (extending an existing process model using information recorded in an event log) [3]. Process mining then is a model-based discipline with the event log being seen as the enabler for model development and subsequent process analysis. The Process Mining Manifesto [2] provides a star-rating for event logs in terms of their readiness for use in a process mining analysis. Existing data quality frameworks for event logs that have been proposed [9,21] are useful in classifying/labelling identified quality issues, but are not useful in detecting specific examples of quality issues.

Historically, analysts have devoted much time and effort to cleaning data (i.e. ensuring the data is 'fit for purpose') prior to analysis. A recent survey revealed data scientists spend more than half their time collecting, labeling, cleaning and organising data [11] and, while some cleaning tasks can be automated, "far too much handcrafted work...is still required"[18] with (compatibility) issues arising through the use of multiple tools [30] in collecting and preparing data.

Quality issues commonly found in event logs have been described [9,29] and, although solutions in the form of filtering, detection and repair algorithms [13, 15,19,31] have been proposed, the problem (different approaches to deal with different log imperfections), persists. A common feature of all these approaches however is an underlying capability to, in some way, query the log for the presence (or absence) of certain characteristics. Hence, our long-term goal is to develop an event log query language that can directly detect log imperfection issues. We call this Querying Event Log for Imperfections (QUELI). We take into consideration the view expressed by Behashti et al. [6] that "querying methods need to enable users to express their data analysis and querying needs using process-aware abstractions rather than other lower level abstractions". This simply means that more than being **somehow possible**, it should be **actually convenient** to encode event log and process constructs in a query language.

The key questions we consider are (i) what are the capabilities required to achieve QUELI?, and (ii) how do we identify these capabilities? In Sect. 3 we address these questions as a three stage approach involving (i) considering *how* log imperfections manifest in a log (i.e. what to look for in detecting log imperfections), (ii) stepwise refinement (narrative to rigorous algorithmic) of detection strategies, and (iii) consolidation of the algorithms to abstract 'building blocks'.

The major contributions of this paper are:

- definition of algorithms to detect 5 out of 11 log imperfection patterns described in [29];
- a preliminary (and by no means complete) consolidation of building blocks needed by QUELI; and
- a systematic approach to identifying remaining/further building blocks needed by QUELI.

The remainder of this paper is structured as follows: Sect. 2 describes the background to the project with a focus on (event log) data quality and discusses related work in the areas of detecting log quality issues and event log querying. In Sects. 3 and 4 we introduce our approach to developing detection strategies and algorithms for the selected log imperfection patterns and in Sect. 5 we consolidate the pattern-at-a-time requirements into a set of QUELI constructs and briefly assess a range of existing log query languages against their ability to support the QUELI constructs. Section 6 concludes the paper with a brief discussion and some thoughts on future work.

2   Background and Related Work

Data quality is an easily understood concept, at least to the extent that "high" quality data is generally regarded as being desirable and "poor" quality data undesirable as input for any analysis. However, the numerous published attempts to objectively describe the characteristics of high/poor quality data (see [4, 16, 34] for various surveys of data quality research), testify to the fact that the 'devil is in the detail' as far as a universal understanding of data quality goes. High quality data has been defined as "data that is fit for use by data consumers" [28] with data quality considered as "the degree to which the characteristics of data satisfy stated and implied needs when used under specified conditions" [1]. Data quality is frequently described as being a multi-dimensional concept [33] with dimensions (i.e. *measurable* data quality properties) such as accuracy/correctness, completeness, unambiguity/understandability and timeliness/currentness [1, 5, 33] being frequently mentioned.

Table 1. Manifestation of quality issues in event log entities [9]

| Quality issues | Event log entities | | | |
	Missing data	Incorrect data	Imprecise data	Irrelevant data
Case	I1	I10		I26
Event	I2	I11		I27
Relationship	I3	I12	I19	
Case attrs.	I4	I13	I20	
Position	I5	I14	I21	
Activity name	I6	I15	I22	
Timestamp	I7	I16	I23	
Resource	I8	I17	I24	
Event attrs.	I9	I18	I25	

The issue of data quality for event logs was first considered in [2] with event log quality frameworks being proposed in [9, 21, 32]. Mans et al. [21] describes event log quality as a two-dimensional spectrum with the first dimension concerned about the *level of abstraction* of the events and the second one concerned with the *accuracy* of the timestamp (in terms of its (i) *granularity*, (ii) *directness of registration* (i.e. the currency of the timestamp recording) and (iii) *correctness*. Bose et al. [9] identify four broad categories of issues affecting process mining event log quality: Missing, Incorrect, Imprecise and Irrelevant data. The authors then show where each of these issues may manifest themselves in the various entities of an event log resulting in 27 separate data quality issues (see Table 1). Such frameworks are useful in **classifying** quality issues, however, they do not provide guidance as to how to **discover** quality issues in a log, nor do they provide a mechanism to quantify the extent to which a log is affected by any identified quality issue. For instance, how many cases are affected by missing events?

In Suriadi et al. [29] the authors describe a set of 11 *log imperfection patterns*[1] that capture some data quality issues commonly found in event logs (see Table 2 for relationship between patterns and quality issues in Table 1). In this paper we use 5 of the 11 log imperfection patterns described in [29] as the basis for understanding the requirements of an event log query language.

2.1 Related Work

Table 2. Relationship between individual patterns and quality framework.

Pattern	Quality issue/s
Form-based event capture	I16, I27
Inadvertent time travel	I16
Unanchored event	I23, I16
Scattered event	I2
Elusive case	I3
Scattered case	I12
Collateral events	I27
Polluted label	I15, I17
Distorted label	I15
Synonymous labels	I15
Homonymous label	I22

Suriadi et al. [29] adopt a patterns-based approach to describing event log quality issues and provide *indicative rules* for detecting the presence of each described log imperfection pattern. As the name suggests, an indicative rule describes conditions that make it likely that the related imperfection pattern is present in the log. The indicative rules are not however, at a low enough level to be immediately operationalised to provide direct support for pattern detection. Lu et al. [19] apply so-called *behaviour patterns* ("*partial orders* of activities where direct and indirect succession of activities are specified") to detect and visualise areas of complexity in an event log. While not specifically designed to detect log quality issues, the approach can be used to visualise event concurrency and hence provide an indication of the possibility of the existence of Form-based Event Capture and Collateral Events patterns [29]. Unsupervised event log pattern detection approaches such as [8,17] use statistical methods to detect frequently occurring behaviours in logs. These approaches suffer from "pattern explosion" (return many patterns) and require the user to sift through returned patterns to decide which are interesting. As these approaches are targeted at frequently occurring behaviours, infrequent behaviours (which may represent data quality issues) are harder to detect. Mannhardt et al. [20] by contrast, describe an approach involving manual specification of behavioural *activity patterns* which encode assumptions about how high-level activities manifest in lower-level events in a log. This approach is neither quality focused nor domain agnostic (requires domain knowledge), and relies on the expertise of the user in specifying patterns with the risk that the user may miss important patterns. Research in the area of record-linkage, data matching and ontology matching exists [10,27], but generally deals with less complex issues, e.g. only matching labels based on similarity measures.

Log repair, by definition, involves rectifying some identified quality issue. Log repair approaches necessarily require a filtering or querying step to identify log elements that are the subject of repair actions. As an example, in [13] the authors discuss three indicators of event order-related quality issues in event logs (mixed

---

[1] http://www.workflowpatterns.com/patterns/logimperfection/.

granularity timestamps, unusual or low-frequency directly followed relations or statistical anomalies in timestamp values) and describe techniques by which the log may be queried to detect each anomaly.

Unlike approaches that implement only one (or a limited range of) pattern types, event log query languages can potentially be used to encode multiple types of patterns (depending on the language constructs and the formulation of the event log). For instance, where the event log is represented as a table, SQL may be applied to log querying. In [12] the authors point out that formulating conceptually simple, but nevertheless fundamental, process related questions such as 'retrieve all directly follows relations between events' is inefficient and difficult to phrase in standard SQL (requiring joining the events table to itself and a NOT EXISTS nested sub-query) and has performance implications. Dijkman et al. [12] then propose (but do not implement) a relational algebaraic operator to extract the 'directly follows' relation from a log. In [25] the authors exploit *RelationalXES* [26] a relational database architecture for storing event log data and show how conventional SQL can be used to encode declarative constraints to extract process knowledge from event logs stored in relational tables. The approach, implemented as SQLMiner, has at least the following limitations: (i) data-perspective constraints have not yet been implemented, (ii) intermediate results are not available for follow-on queries, and (iii) native SQL is not process-aware, therefore the encoded declarative constraints are complex (and possibly beyond the ability of all but expert level SQL users).

FQSPARQL [7] is a process event query language and graph-based querying process engine derived from SPARQL [24]. The FPSPARQL approach models an event log as a graph of typed nodes and edges. In [22] de Murillas, Reijers and van der Aalst describe DAPOQ, the Data-Aware Process Oriented Query Language for querying event data. DAPOQ was purpose built for process querying and so has the advantages of improved query development time and readability of queries. The language supports the traditional process view (events, instances and processes), with the data perspective (data models, objects and object versions). Process Instance Query Language (PIQL) [23] describes a query language specifically designed to report on various Process Performance Indicators (PPIs). PPI queries are formulated in PIQL to return the number of process instances or tasks that are (i) (not) finalized, (ii) (not) cancelled, (iii) executed by {name}, (iv) start before {date}, etc. The fact that the language is specifically designed to report on a PPIs limits its generality. XSLT (Extensible Stylesheet Language Transformations) is a declarative data transformation language developed by the W3C and used for transforming XML documents. If the event log is encoded as a tree structure (a log contains several cases and each case contains one or more events) in XES/XML, XSLT can be used to filter, transform and query the event log. Durand et al. [14] leverage this prevalence of XML-based standards for encoding event logs to build an XML vocabulary and execution language on top of XSLT, that can be used to query and analyse event sequences.

3   Approach

Our approach for identifying log query constructs, which are key building blocks for a query language that is suitable for detecting imperfections in event logs, is illustrated in Fig. 1. By 'constructs', we mean a collection of functionality that is required by a language (e.g. a log query language) to serve its purpose (e.g. to detect log imperfections).

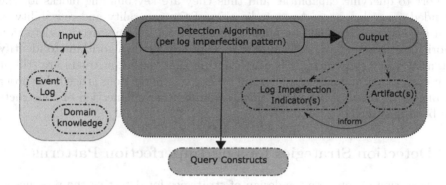

Fig. 1. Approach for identifying query constructs for log imperfection detection

Inputs. The inputs to our approach are an event log and, if available, domain knowledge (e.g. the valid ranges of values for the timestamps of certain activities). An event log is a collection of events. An event can be defined as a tuple consisting of various attributes. An attribute may be mandatory or optional. For example, there are three mandatory event attributes, known as *case identifier*, *activity label*, and *timestamp*, required for process mining.

Definition 1 (Event).   Let *Case* be the set of case identifiers, *Act* the set of activity labels, and *Time* the set of timestamps. $\mathscr{E} \subseteq Case \times Act \times Time$ is the set of events. For any event $e \in \mathscr{E}$, $case(e) \in Case$, $act(e) \in Act$, and $time(e) \in Time$, represent the case identifier, activity label, and timestamp of event e.   □

Definition 2 (Event Log).   An event log $\mathcal{L} \subseteq \mathscr{E}$ is a set of events.   □

Detection Algorithms. In order to identify the constructs needed for querying an event log for the existence of various imperfections, we start by developing detection strategies to address certain log imperfections. These detection strategies should be *generic*, i.e. independent of any specific language and system/tool, and *rigorously* defined in the form of an algorithm (for each strategy per log imperfection pattern) so that they can be implemented in a precise and unambiguous way. In the next section of this paper, we elaborate on the detection strategies for five event log imperfection patterns among those defined in [29].

Outputs. Each algorithm will produce some artifacts (such as statistical summary of certain characteristics in an event log, or the derivation of sub-logs that meet certain criteria) that will then be used to reason about the existence, or lack thereof, of certain log imperfections.

Query Constructs. Having defined the algorithms for detecting various log imperfections, we consolidate them to identify which part(s) of the algorithms can potentially be used as *query constructs*. These constructs provide direct support to querying capabilities and thus they are key building blocks for the intended query language. As an indication of the generality and reusability of such a construct, it should be applicable to several log imperfection detection algorithms. In Sect. 5, we explain how we consolidate the algorithms to identify potential query constructs. It is worth noting that these constructs are still at their early stage, and further analysis of other log imperfections is needed in order for one to be reasonably assured that a comprehensive collection of constructs has been identified.

4   Detection Strategies for Log Imperfection Patterns

In this section we propose the design of strategies for detecting the presence of log imperfection patterns. For each pattern, we start with describing how the pattern manifests in an event log, what could be the main reason that has led to the existence of the pattern in the log, and how the pattern exemplifies a data quality issue that is defined at a more abstract level as discussed in Sect. 2. From there, we continue to present our detection strategy for the pattern which consists of an outline of the underlying detection mechanism in natural language and then a conceptual design of the strategy specified in the form of an algorithm.

4.1   Form-Based Event Capture

The pattern manifests in an event log as multiple occurrences of groups of events characterised by a common set of activity labels with each event in the group having the same or very nearly the same timestamp (allowing for physical logging by the system) in the same case. An example of the pattern is shown in Fig. 2. A main reason that may cause this pattern to arise is that when process-related data is entered into fields on a computerised form, updates to form field values are logged as *separate* events (e.g. one event per field) in response to a user action (e.g. clicking 'Save' on the form). In this case, the

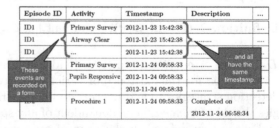

Fig. 2. Example of *form-based event capture*

activity labels are usually informed directly by the corresponding form field names. It can be observed, taking the example in Fig. 2, that the presence of

---

Algorithm 1. DETECTFORM-BASEDEVENTCAPTURE

---

input : event log $\mathcal{L}$, timestamp difference Δt
output: $\mathcal{F_G}$
begin
    $\mathcal{G} \leftarrow$ findSimultaneousEvents$(\mathcal{L}, \Delta t)$ /* *Step-1*: $\mathcal{G}$ is a set of event groups */
    $\mathcal{A} \leftarrow \varnothing$ /* *Step-2.a*: To compute a set of groups of activity labels $\mathcal{A}$ */
    foreach $G_i \in \mathcal{G}$ **do**
        $\mathcal{A} \leftarrow \mathcal{A} \cup \{$getActLabels$(G_i)\}$
    $\mathcal{D} \leftarrow \mathcal{A}$ /* *Step-2.b*: To compute a set of distinct groups of activity labels $\mathcal{D}$ */
    do
        $\mathcal{D}_t \leftarrow \mathcal{D}$
        $D_0 \leftarrow$ getOneSetElement$(\mathcal{D})$
        $\mathcal{D} \leftarrow \mathcal{D} \setminus \{D_0\}$
        foreach $D_i \in \mathcal{D}$ **do**
            if $D_0 \cap D_i \neq \varnothing$ **then**
                $D_0 \leftarrow D_i \cup D_0$
                $\mathcal{D} \leftarrow \mathcal{D} \setminus \{D_i\}$
        $\mathcal{D} \leftarrow \mathcal{D} \cup \{D_0\}$
    while $\mathcal{D} \neq \mathcal{D}_t$;
    if $|\mathcal{D}| = 1$ **then**
        $\mathcal{F_G} \leftarrow \{\mathcal{D}, |\mathcal{L}|\}$ /* Detection ends if only *one* distinct group of activity labels */
    else
        $\mathcal{F_G} \leftarrow \varnothing$ /* *Step-3*: To compute the likelihood of form existence $\mathcal{F_G}$ */
        foreach $D \in \mathcal{D}$ **do**
            count$_D \leftarrow 0$
            foreach $G \in \mathcal{G}$ **do**
                if getActLabels$(G) \subseteq D$ **then**
                    count$_D$++
            $\mathcal{F_G} \leftarrow \mathcal{F_G} \cup (D, \text{count}_D)$

---

this pattern in an event log leads to the recording of incorrect timestamps of affected events representing the time the form was saved, not the time each of affected events actually occurred. As such, this pattern presents a more concrete example of data quality issue *I16 - Incorrect data: timestamp*.

Detection Strategy: The main idea is to discover the existence of a form by identifying the group of activity labels that are likely informed by the corresponding fields on that form. An outline of our detection strategy follows.

Step-1. Due to the fact that these activities are logged as events that have the same or very nearly the same timestamps in the same case, the first step of detection is to find groups of such *simultaneous events*.

Step-2. It is important to realise that updates may be applied to different fields on a form in different cases. For example, as shown in Fig. 2, in one case (Episode ID1) 'Primary Survey' and 'Airway Clear' are among those updated, and in another case (Episode ID2) update occurred to 'Primary Survey' and 'Pupils

Responsive' on the form. Hence, it is necessary to find the group of all possible fields on each form (given the scope of the data available in a log).

This can be achieved by (a) extracting the groups of activity labels from the groups of simultaneous events and (b) traversing the activity labels group by group and merging the groups that have at least one overlapping label between them. As such, the second step of detection will yield *distinct* groups of activity labels meaning that each activity label belongs to only one of the groups.

However, it is possible that only *one* distinct group of activity labels is identified and the detection process will then end with an output of mainly this group of activity labels. In this case, a conclusion likely to be drawn by end users is that either all the events in the log are recorded from a single form or no form exists.

Step-3. This is to help understand how likely each group of activity labels may inform the existence of a form. Certain quantitative measures can be computed to provide reasonable indication for the likelihood of form existence. An example of such a measure is to count how often each group of activity labels, including its sub-groups, have appeared in the event log. This can help end users to make their decision given necessary domain knowledge. E.g., if the above count of a specific group of activity labels is larger than a certain threshold value, a conclusion can be drawn that there exists a form that contains the corresponding field names.

4.2   Collateral Events

This pattern manifests as an event log containing groups of activities with timestamps that are very close to each other (e.g. within seconds). The problem may be introduced into the log through incorrect or too fine grained logging of event data. As illustrated by the example in Fig. 3, the snippet of an event log contains a list of micro-steps of a process activity, whereas it is the activity, but not its micro-steps, that is of interest to process analysis. As such, this pattern presents a more concrete example of data quality issue *I27 - Irrelevant data: event.*

caseID	Activity	Timestamp
1234567	Adjust recovery cost	19/06/2014 12:15:18
1234567	Adjust recovery cost	19/06/2014 12:16:53
1234567	Email	19/06/2014 12:19:25
....	....	....
1234567	Pay assessor fee	19/06/2
1234567	Adjust admin cost	19/06/2014 12:22:48

All events refer to single process step 'Pay Insurance Claim Assessor'

Fig. 3. Example of *collateral events*

Detection Strategy: The key objective is to discover a high-level (parent) activity by identifying the list of micro-steps (instead of the activity) that are possibly recorded in the event log. Hence, the detection strategy is similar to that of form-based event capture. The differences are: (1) the central objects for detection of collateral events are parent process activities and their micro-steps (instead of forms and their fields in the case of form-based event capture); and (2) the input time difference Δt for detection of collateral events has a greater value than that of form-based event capture. Algorithm 1 can be re-used for detection of collateral events, because it is generic (e.g. being independent of specific objects to identify, such as a form and its fields vs. a process activity and its micro-steps) and configurable (e.g. Δt being a user input variable).

4.3   Inadvertent Time Travel

This pattern manifests as a number of cases in the log where the temporal ordering of events deviates significantly from the majority of cases in the log or from a mandatory temporal ordering property. For example, Fig. 4 shows a snippet of a log with this imperfection pattern: the activity Arrival first hospital (henceforth referred to as activity A) was recorded to take place on 2011-09-08 00:30:00; however, the time of the Injury activity (henceforth referred to as activity B), which triggered the patient being sent to the hospital, was recorded to take place more than 23 h *later* (at 2011-09-08 23:47:01). The cause of this problem is the 'midnight problem' whereby a hospital staff recorded the correct 'time' of patient arrival but failed to change the 'date' portion of the timestamp (it should have been 2011-09-09). The occurrence of this pattern is often associated with manual recording of timestamp data and results from the 'proximity' between correct value and the recorded, incorrect value. Proximity errors occur through a user pressing an incorrect key on a keyboard, or as in our example above, a user failing to recognise the recently-crossed date/time boundary (such as

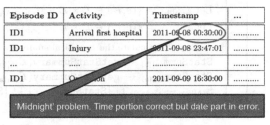

Fig. 4. Example of *inadvertent time travel*

the mentioned midnight problem, or a new year). This pattern negatively impacts the attribute accuracy of the log in that the temporal ordering of the events no longer reflects the actual ordering of events. Therefore, this pattern is a manifestation of the data quality issue *I16 - Incorrect data: timestamp.*

Detection Strategy: The main idea is to discover the existence of pairs of activities, within the same case, with 'unusual' temporal ordering, i.e., it either deviates from the majority of the cases or violates some mandatory ordering. Once such pairs of activities are discovered, we then extract statistical summary information (such as the proportion of cases with the deviant temporal ordering) to be presented to users to determine if the unusual temporal ordering is indeed a data quality issue. An outline of our detection strategy is as follows.

Step-1. Using the example from Fig. 4, an unusual temporal ordering of two events is seen when in one or more cases, the activity B *succeeds* A, while in the majority of cases B (the injury event) *precedes* A. In other words, we say A and B occurred in any order (in parallel). The first detection step is thus to identify *all pairs of activity names* that can occur in any order.

Step-2. Next, for each pair of activity names that can occur in any order, we extract the corresponding *pairs of events*. Using our example above, the idea here is to obtain two sets of pairs of events: the first set consists of all pairs of events where A was followed by B, and the second group consists of all event pairs where B was followed by A.

Step-3. Finally, we obtain some statistical summary of those two groups. The intended statistical summary includes information such as the proportion of cases of usual vs. unusual temporal ordering and the frequency of each pair of events.

This statistical summary information is then presented to the user to determine if it is an acceptable deviance or if it is an event log quality issue.

Algorithm 2. DetectInadvertentTimeTravel

---

input : event log $\mathcal{L}$
output: $\mathbb{S}_{(A||B)}$
begin

  /* *Step-1*: $\mathcal{A}_{||}$ is a set of activity names that can occur in any order */
  $\mathcal{A}_{||} \leftarrow$ **findParallelEventPairs**$(\mathcal{L})$.

  $\mathbb{S}_{(A||B)} \leftarrow \emptyset$ /* Initialise the return value */

  foreach $(a, b) \in \mathcal{A}_{||}$ **do**
    /* *Step-2*: $\mathcal{L}_{(a||b)}$ and $\mathcal{L}_{(b||a)}$ are the corresponding sets of pairs of events with activities that can occur in any order */
    Let $\mathcal{L}_{(a||b)} = \{(e, e') \in \mathcal{L} \times \mathcal{L} \mid \exists_{(a,b) \in \mathcal{A}_{||}}: act(e) = a \wedge act(e') = b\}$
    Let $\mathcal{L}_{(b||a)} = \{(e, e') \in \mathcal{L} \times \mathcal{L} \mid \exists_{(a,b) \in \mathcal{A}_{||}}: act(e) = b \wedge act(e') = a\}$

    /* *Step-3*: Calculate the statistical summary */
    StatSumm$_{(a||b)} \leftarrow$ getStatSummary$(\mathcal{L}_{(a||b)})$
    StatSumm$_{(b||a)} \leftarrow$ getStatSummary$(\mathcal{L}_{(b||a)})$
    $\mathbb{S}_{(A||B)} \leftarrow \mathbb{S}_{(A||B)} \cup \{$StatSumm$_{(a||b)}\} \cup \{$StatSumm$_{(b||a)}\}$

4.4 Synonymous Labels

This pattern manifests as the existence of multiple values of a particular attribute that seem to share a similar meaning but are nevertheless, distinct. For example, Fig. 5 shows a snippet of a log with this imperfection pattern: the activities Medical Assign and DrSeen refer to the activity of consulting a medical doctor. However, the labels (or names) given to the activity are different.

This log imperfection pattern may arise when an event log is constructed from multiple source logs, each of which represents the same process, but uses a different label to represent essentially the same process step. The existence of multiple names for the same attribute creates ambiguity in an event log. As such, this imperfection pattern is a manifestation of the *I22 - Imprecise data: event attributes* quality issue.

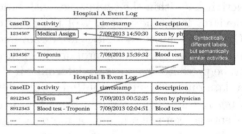

Fig. 5. Example of *synonymous labels*

Detection Strategy: The main idea is to discover the existence of pairs of activities that never occur together within the same case. Using the example above, the underlying assumption is that some cases were recorded in one particular system using the activity name Medical Assign while other cases were recorded in another system using the activity name DrSeen. Then, we examine

---

Algorithm 3. DETECTSYNONYMOUSLABELS

Input : event log $\mathcal{L}$
Output: $\{\mathcal{A}_{\text{synonymous}}\}$
begin

/* *Step-1*: $\mathcal{A}_{\#}$ is a set of activity names that are in conflict */
$\mathcal{A}_{\#} \leftarrow$ findConflictPairs($\mathcal{L}$) is a set consisting of pairs of events with conflict relation.

$\mathcal{A}_{\text{synonymous}} \leftarrow$ NULL /* Initialise the return value */

foreach $(a, b) \in \mathcal{A}_{\#}$ **do**

/* *Step-2*: Obtain the corresponding events for each pair of activity names that are in conflict */
Let $\mathcal{L}_{\#a} = \{e \in \mathcal{L} \mid \exists_{(a,b) \in \mathcal{A}_{\#}} : act(e) = a \}$
Let $\mathcal{L}_{\#b} = \{e \in \mathcal{L} \mid \exists_{(a,b) \in \mathcal{A}_{\#}} : act(e) = b \}$

/* *Step-3*: Obtain the context variable and check for similarity of the context variables */
$\mathcal{C}_{\text{context}(a)} =$ getContextVariables($\mathcal{L}_{\#a}, \mathcal{L}$)
$\mathcal{C}_{\text{context}(b)} =$ getContextVariables($\mathcal{L}_{\#b}, \mathcal{L}$)
if $\mathcal{C}_{context(a)} \approx \mathcal{C}_{context(b)}$ **then**
$\quad \mathcal{A}_{\text{synonymous}} \leftarrow \mathcal{A}_{\text{synonymous}} \cup (a, b)$

---

the contextual variables surrounding this pair of activity names. Contextual variables include the surrounding activity names preceding, succeeding, or running in parallel with the activity name being examined. If the contextual variables are similar, then this pair of activity names may be candidate for synonymous label. An outline of our detection strategy is as follows.

Step-1. The first step in our detection strategy is to identify those activity labels that never occur together within the same case, across all cases seen in the event log. When two activity labels never occur together within the same case, we call these activity labels to be *in conflict*. By examining the whole event log, we will get a list of pairs of activity names that are in conflict.

Step-2. Next, for each pair of activity names that are in conflict, we extract the corresponding contextual variables as explained above. To do so, we need to extract two groups of events: each group consists of all events whose activity names is the same as one of those activity names in conflict. Using our example above, the first group of events will be those events whose activity names are equal to Medical Assign, while the other group consists of all events with activity name DrSeen.

Step-3. For each group of events extracted in the previous step, we obtain the contextual variables and then compare them to see if they are similar. Using the example above, if the contextual variables between the activity names DrSeen and Medical Assign are similar enough, we store this pair of activity names into a list to be presented to users for determination.

4.5   Homonymous Labels

This pattern manifests as the existence of an activity name being repeated multiple times within a case (i.e. the same activity name applied to each occurrence of the activity), but the interpretation of the activity, from a process perspective, differs across the various occurrences. For example, in Fig. 6, the activity name Triage Assessment occurred multiple times within the same case. The second and third occurrences happened roughly 7 days after the first. From a process perspective, the first occurrence referred to an actual triage activity of a patient, while the second and third occurrences referred to a doctor reviewing the triaging decision made earlier (it is impossible to be triaged again after a patient has been discharged from the hospital). This log imperfection pattern may arise when the original logging or the subsequent event log extraction does not record the context information necessary to distinguish between the different occurrences of the activity. For instance, in our example, the first triage activity activity should have

caseID	activity	timestamp	Description
1234567	Triage Assessment	06/09/2013 12:33:17	....
1234567	Progress Note	06/09/2013 13:10:23	
1234567	Discharged	06/09/2013 13:15:00	....
1234567	Triage Assessment	13/09/2013 07:24:36	....
1234567	Triage Assessment	13/09/2013 07:28:51	....

Fig. 6. Example of *homonymous labels*

been further qualified by using the name Triage - Initial, while the second and third should be further qualified by using the name Triage - Review. The occurrence of this log imperfection pattern makes certain activity names too coarse to reflect the different connotations associated with the recorded events. As such, this is a manifestation of the *I12 - Imprecise data: activity name* data quality issue.

Detection Strategy: A homonymous label pattern manifests itself as the occurrence of an activity name at rather 'odd' places within a case, e.g. the occurrence of a triage activity after the patient was discharged (in the example above). In other words, similar to the *Inadvertent Time Travel* pattern, the first detection step is thus to discover the existence of pairs of activities, within the same case, with 'unusual' temporal ordering. Next, from those pairs of activities with unusual temporal ordering, we identify those activities that are repeated at least twice within the same case (a homonymous label pattern will only be seen if the activity is repeated). Finally, for each activity that is repeated, we obtain the contextual variable to identify if we see one or more distinct contextual variables. We then present the contextual variables along with the activity names to users to determine if there is a homonymous label pattern in the log. An outline of our detection strategy is as follows.

Step-1. As explained above, the first step in our detection strategy is to identify those pairs of activity names with unusual temporal ordering. As explained in the detection of *Inadvertent Time Travel* pattern, we can identify such pairs of activity names by extracting those pairs of activities that happened in any order.

Step-2. The logic behind this step is similar to *Step-2* of the *Inadvertent Time Travel* pattern: for each pair of activity names that can occur in any order, we extract the corresponding *pairs of events*.

Step-3. Next, for each pair of activities that can occur in any order, we extract those duplicated activity names within a case. This step is needed to narrow down the list of potential homonymous activity labels for further examination. As explained above, homonymous label pattern exists for an activity name that occurs at least twice within a case.

Step-4. Finally, for the set of duplicated activity names extracted from *Step-3*, we obtain the contextual variables for each occurrence of this activity name. We then return the set of contextual variables for each duplicated activity name from *Step-3* to users to determine the existence of the homonymous label pattern.

5   Towards Required Capabilities for QUELI

By presenting detection strategies for a number of event log imperfection patterns we have shown that interacting with event log data in a way that enables the detection of data quality issues is often non-trivial. Rather than expecting users to manually apply these strategies to their data sets it would be useful to provide them with "building blocks" that they can use to apply these strategies

---

Algorithm 4. DETECTHOMONYMOUSLABELS

Input   : event log $\mathcal{L}$
Output: $N_{context}$
begin

/* *Step-1*: $\mathcal{A}_{||}$ is a set of activity names that can occur in any order */
$\mathcal{A}_{||} \leftarrow \texttt{findParallelEventPairs}(\mathcal{L})$.

/* *Step-2*: $\mathcal{L}_{(a||b)}$ and $\mathcal{L}_{(b||a)}$ are the corresponding sets of pairs of events for all activities that can occur in any order */
$\mathcal{L}_{(a||b)} \leftarrow \emptyset$
$\mathcal{L}_{(b||a)} \leftarrow \emptyset$
foreach $(a, b) \in \mathcal{A}_{||}$ **do**

Let $L_{(a||b)} = \{(e, e') \in \mathcal{L} \times \mathcal{L} \mid \exists_{(a,b) \in \mathcal{A}_{||}}: act(e) = a \wedge act(e') = b\}$
Let $L_{(b||a)} = \{(e, e') \in \mathcal{L} \times \mathcal{L} \mid \exists_{(a,b) \in \mathcal{A}_{||}}: act(e) = b \wedge act(e') = a\}$
$\mathcal{L}_{(a||b)} \leftarrow \mathcal{L}_{(a||b)} \cup L_{(a||b)}$
$\mathcal{L}_{(a||b)} \leftarrow \mathcal{L}_{(a||b)} \cup L_{(b||a)}$

/* *Step-3*: Identify duplicated activity names in a case. The set $\mathcal{A}_{candidates}$ consists of all activity names that can be duplicated within a case. */
$\mathcal{A}_{candidates} = \texttt{getDuplicateNames}(\mathcal{L}_{(a||b)}, \mathcal{L}_{(b||a)})$

/* Initialise the return value */
$N_{(context)} \leftarrow \emptyset$

foreach $a \in \mathcal{A}_{candidates}$ **do**

/* *Step-4*: Extract the contextual variables. */
$\mathcal{L}_{candidates} = \{e \in \mathcal{L} \mid e.act = a\}$
$C_{context(a)} \leftarrow \texttt{getContextVariables}(\mathcal{L}_{candidates}, \mathcal{L})$
$N_{context} \leftarrow N_{context} \cup (a, N_{context(a)})$

through querying. Our detection strategies already show reoccurring operations and data structures used in the detection. In Table 3 we aggregate the primitives used in the presented detection strategies, generalise them, and show which primitives are relevant to detecting more than one pattern in order to identify some potential "building blocks".

Table 3. Aggregated primitives to detect log imperfection patterns [29]

Primitive: `findSimultaneousEvents(`$\mathcal{L}$`, `Δt`)`
Relates to: Form-based Event Capture, Collateral Events
Primitive: `getOneSetElement(`$\mathcal{D}$`)`
Relates to: Form-based Event Capture, Collateral Events
Primitive: `findRelationshipPairs(`$\mathcal{L}$`,`$[||, \#, <, >]$`)`
Relates to: Inadvertent Time Travel, Synonymous Label, Homonymous Label
Generalisation: We combine `findParallelEventPairs(`$\mathcal{L}$`)` and `findConflictPairs(`$\mathcal{L}$`)` and anticipate the need to extract direct-follow and direct-precede relations
Primitive: `getActLabels(`$\mathcal{L}$`,`$A \subseteq \mathcal{A}$`, `$\delta(\mathcal{L})$`)`
Relates to: Form-based Event Capture, Collateral Events
Primitive: `getStatSummary(`$\mathcal{L}$`, `$\mathcal{L}$`)`
Relates to: Inadvertent Time Travel, Homonymous Label
Primitive: `getContextVariables(`$a \in \mathcal{A}$`,`$\mathcal{L}$`)`
Relates to: Synonomous Label, Homonymous Label
Primitive: `getDuplicateNames(`$\mathcal{L}$`, `$\mathcal{L}$`)`
Relates to: Homonymous Label

On a higher level, we can summarise that, to apply the presented detection strategies, a mixture of high-level language features are required. These are (i) support for selection/projection of data, (ii) support for aggregation of results, (iii) support for set-operations, (iv) support for loops, and (v) support for event-relations (e.g. parallel, conflict) (see Table 4). In the following we show that the required high-level language features are not supported by any single query language. As at least one of the referenced languages is Turing complete and can therefore theoretically perform any operation that can be defined as an algorithm, we more specifically check for direct support. We define *direct support* for a primitive as the availability of parameterised function calls, so that the query can be performed without writing procedural code. An example of such a parameterised operator for filtering events in a log is the WHERE clause in an SQL statement.

The getActLabel() primitive returns activity labels associated with events. Hence, this primitive requires selection (to identify events) and projection (to return the activity label attribute values) functionality. All query languages support this functionality. All our proposed detection algorithms make use of set operations. As a set-based language, SQL provides support for set operations. PIQL supports querying the numbers of process or task instances and is therefore not able to provide the actual sets of event (tasks) as required for our primitives. The getStatSummary() primitive returns aggregates of low-level data. SQL provides support for aggregation through the GROUP BY clause and built-in functions such as MIN(), MAX(), AVG(). XSLT is Turing complete and should therefore in theory be able to provide support for all our primitives, however, aggregations are, in fact, not well supported and are practically infeasible. All our detection algorithms involve some form of repetition, generally count-controlled, with two algorithms (Form-based Event Capture and Collateral Events) requiring condition-controlled iteration. Only XSLT provides direct support for both forms of repetition, DAPOQ supports only count-controlled repetition, while the other languages do not support either form of repetition. The findRelationshipPairs() primitive requires identifying relationships between pairs of events. While this is possible in a language such as SQL, the query is complex and, using only standard SQL features, requires manual specification of pairs of events and relationship type. Hence, it is not reasonable to conclude that SQL provides direct support for this primitive. Only FPSPARQL, through its notion of paths, provides direct support for this primitive.

Table 4. High-level features required for application of detection strategies

Features	SQL	FPSPARQL	DAPOQ	PIQL	XSLT
Support for selection/projection of data *Relates to*: getActLabels()	Y	Y	Y	Y	Y
Support for set-operations *Relates to*: All strategies	Y	Y	Y	N	N
Support for aggregation of results *Relates to*: getStatSummary()	Y	Y	N	Y	N
Support for repetition *Relates to*: Form-based Event Capture, Collateral Events	N	N	N	N	Y
Support for event-relations *Relates to*: findRelationshipPairs()	N	Y	N	N	N

6 Conclusion

This work was motivated by our experiences in data preparation for multiple process mining case studies. For each study, the objective was to construct event logs with the highest possible data quality. Often, the starting point was source log(s) (i) drawn from non-process aware information systems, (ii) all of which exhibited a mixture of the issues described in [9,21,29] and (iii) required the use of multiple off-the-shelf tools and, sometimes, custom-developed software to identify and rectify quality issues (with the associated save, open, save-as different format, import operations to move from one environment to the next). In this paper we have outlined an approach that identified a small set of function primitives for detecting a range of data quality issues commonly found in event logs, *viz.* log imperfection patterns. We note that, unsurprisingly, none of the tools or query languages we considered provide **direct** support for **all** of the functional requirements we derived. The detection strategies and algorithms provided in this paper meet our aim of providing guidance to process analysts in detecting the log imperfection patterns and form the basis of future work in implementing the primitives in QUELI.

Acknowledgement. The contributions to this paper of Robert Andrews and Chun Ouyang were supported through ARC Discovery Grant DP150103356.

References

1. ISO/IEC 25010:2011: Systems and software engineering - Systems and software product Quality Requirements and Evaluation (SQuaRE) - System and software quality models (2011)
2. van der Aalst, W., et al.: Process mining manifesto. In: Daniel, F., Barkaoui, K., Dustdar, S. (eds.) BPM 2011. LNBIP, vol. 99, pp. 169–194. Springer, Heidelberg (2012). https://doi.org/10.1007/978-3-642-28108-2_19
3. van der Aalst, W.: Process Mining: Discovery Conformance and Enhancement of Business Processes. Springer, Heidelberg (2011). https://doi.org/10.1007/978-3-642-19345-3
4. Batini, C., Palmonari, M., Viscusi, G.: Opening the closed world: a survey of information quality research in the wild. In: Floridi, L., Illari, P. (eds.) The Philosophy of Information Quality. SL, vol. 358, pp. 43–73. Springer, Cham (2014). https://doi.org/10.1007/978-3-319-07121-3_4
5. Batini, C., Scannapieco, M.: Data Quality: Concepts, Methodologies and Techniques. Springer, Heidelberg (2006). https://doi.org/10.1007/3-540-33173-5
6. Beheshti, S.-M.-R., Benatallah, B., Motahari-Nezhad, H.R.: Scalable graph-based OLAP analytics over process execution data. Distrib. Parallel Datab. **34**(3), 379–423 (2016)
7. Beheshti, S.-M.-R., Benatallah, B., Motahari-Nezhad, H.R., Sakr, S.: A query language for analyzing business processes execution. In: Rinderle-Ma, S., Toumani, F., Wolf, K. (eds.) BPM 2011. LNCS, vol. 6896, pp. 281–297. Springer, Heidelberg (2011). https://doi.org/10.1007/978-3-642-23059-2_22
8. Jagadeesh Chandra Bose, R.P., van der Aalst, W.M.P.: Abstractions in process mining: a taxonomy of patterns. In: Dayal, U., Eder, J., Koehler, J., Reijers, H.A. (eds.) BPM 2009. LNCS, vol. 5701, pp. 159–175. Springer, Heidelberg (2009). https://doi.org/10.1007/978-3-642-03848-8_12
9. Jagadeesh Chandra Bose, R.P., Mans, R.S., van der Aalst, W.M.: Wanna improve process mining results? CIDM **2013**, 127–134 (2013)
10. Christen, P.: Data Matching: Concepts and Techniques for Record Linkage, Entity Resolution, and Duplicate Detection. Springer, Heidelberg (2012). https://doi.org/10.1007/978-3-642-31164-2

11. CrowdFlower: 2017 Data Scientist Report (2017). https://visit.crowdflower.com. Accessed 25 July 2018
12. Dijkman, R., Gao, J., Grefen, P., ter Hofstede, A.: Relational algebra for in-database process mining. arXiv preprint arXiv:1706.08259 (2017)
13. Dixit, P.M., et al.: Detection and interactive repair of event ordering imperfection in process logs. In: Krogstie, J., Reijers, H.A. (eds.) CAiSE 2018. LNCS, vol. 10816, pp. 274–290. Springer, Cham (2018). https://doi.org/10.1007/978-3-319-91563-0_17
14. Durand, J., Cho, H., Moberg, D., Woo, J.: XTemp: event-driven testing and monitoring of business processes. In: Proceedings of Balisage, The Markup Conference 2011, vol. 7. Balisage Series on Markup Technologies (2011)
15. Günther, C.W., Rozinat, A.: Disco: discover your processes. BPM (Demos) **940**, 40–44 (2012)
16. Laranjeiro, N., Soydemir, S.N., Bernardino, J.: A survey on data quality: classifying poor data. In: PRDC 2015, pp. 179–188. IEEE (2015)
17. Leemans, M., van der Aalst, W.M.P.: Discovery of frequent episodes in event logs. In: Ceravolo, P., Russo, B., Accorsi, R. (eds.) SIMPDA 2014. LNBIP, vol. 237, pp. 1–31. Springer, Cham (2015). https://doi.org/10.1007/978-3-319-27243-6_1
18. Lohr, S.: For big-data scientists, 'janitor work' is key hurdle to insights. New York Times, 17 August 2014
19. Lu, X., et al.: Semi-supervised log pattern detection and exploration using event concurrence and contextual information. In: Panetto, H., et al. (eds.) OTM On the Move to Meaningful Internet Systems, pp. 154–174. Springer, Cham (2017). https://doi.org/10.1007/978-3-319-69462-7_11
20. Mannhardt, F., de Leoni, M., Reijers, H.A., van der Aalst, W.M.P., Toussaint, P.J.: From low-level events to activities - a pattern-based approach. In: La Rosa, M., Loos, P., Pastor, O. (eds.) BPM 2016. LNCS, vol. 9850, pp. 125–141. Springer, Cham (2016). https://doi.org/10.1007/978-3-319-45348-4_8
21. Mans, R.S., van der Aalst, W.M., Vanwersch, R., Moleman, A.: Process Support and Knowledge Representation in Health Care. LNCS, vol. 7738, pp. 140–153. Springer, Heidelberg (2013). https://doi.org/10.1007/978-3-642-36438-9
22. González López de Murillas, E., Reijers, H.A., van der Aalst, W.M.P.: Everything you always wanted to know about your process, but did not know how to ask. In: Dumas, M., Fantinato, M. (eds.) BPM 2016. LNBIP, vol. 281, pp. 296–309. Springer, Cham (2017). https://doi.org/10.1007/978-3-319-58457-7_22
23. Perez-Alvarez, J.M., Gomez-Lopez, M.T., Parody, L., Gasca, R.M.: Process instance query language to include process performance indicators in DMN. In: EDOCW 2016, pp. 1–8. IEEE (2016)
24. Prud'hommeaux, E., Seaborne, A.: SPARQL query language for RDF. W3C recommendation, January 2008 (2008)
25. Schönig, S., Rogge-Solti, A., Cabanillas, C., Jablonski, S., Mendling, J.: Efficient and customisable declarative process mining with SQL. In: Nurcan, S., Soffer, P., Bajec, M., Eder, J. (eds.) CAiSE 2016. LNCS, vol. 9694, pp. 290–305. Springer, Cham (2016). https://doi.org/10.1007/978-3-319-39696-5_18
26. Shabani, S., et al.: Relational XES: data management for process mining. In: CAiSE 2015. CEUR-WS. org (2015)
27. Shvaiko, P., Euzenat, J.: Ontology matching: state of the art and future challenges. IEEE Trans. Knowl. Data Eng. **25**(1), 158–176 (2013)
28. Strong, D.M., Lee, Y.W., Wang, R.Y.: Data quality in context. Commun. ACM **40**(5), 103–110 (1997)

29. Suriadi, S., Andrews, R., ter Hofstede, A., Wynn, M.: Event log imperfection patterns for process mining: towards a systematic approach to cleaning event logs. Inf. Syst. **64**, 132–150 (2017)
30. Suriadi, S., Wynn, M.T., Ouyang, C., ter Hofstede, A.H.M., van Dijk, N.J.: Understanding process behaviours in a large insurance company in australia: a case study. In: Salinesi, C., Norrie, M.C., Pastor, Ó. (eds.) CAiSE 2013. LNCS, vol. 7908, pp. 449–464. Springer, Heidelberg (2013). https://doi.org/10.1007/978-3-642-38709-8_29
31. Vázquez-Barreiros, B., Mucientes, M., Lama, M.: Mining duplicate tasks from discovered processes. In: ATAED@ Petri Nets/ACSD, pp. 78–82 (2015)
32. Verhulst, R.: Evaluating quality of event data within event logs: an extensible framework. Ph.D. thesis, Technische Universiteit Eindhoven (2016)
33. Wand, Y., Wang, R.Y.: Anchoring data quality dimensions in ontological foundations. Commun. ACM **39**(11), 86–95 (1996)
34. Wang, R.Y., Storey, V., Firth, C.: A framework for analysis of data quality research. IEEE Trans. Knowl. Data Eng. **7**(4), 623–640 (1995)

Towards a Collective Awareness Platform for Privacy Concerns and Expectations

Giorgos Flouris[1]([✉]), Theodore Patkos[1], Ioannis Chrysakis[1],
Ioulia Konstantinou[2], Nikolay Nikolov[3], Panagiotis Papadakos[1],
Jeremy Pitt[4], Dumitru Roman[3], Alexandru Stan[5],
and Chrysostomos Zeginis[1]

[1] ICS-FORTH, N. Plastira 100, P.O. Box 1385, 70013 Heraklion, Greece
{fgeo,patkos,hrysakis,papadako}@ics.forth.gr
[2] Vrije Universiteit Brussel (VUB), Pleinlaan 2, 1050 Brussels, Belgium
ioulia.konstantinou@vub.ac.be
[3] SINTEF, Forskningsveien 1a, 0373 Oslo, Norway
{nikolay.nikolov,dumitru.roman}@sintef.no
[4] Imperial College London, South Kensington Campus, London SW7 2AZ, UK
j.pitt@imperial.ac.uk
[5] IN2 Digital Innovations GmbH, Auf dem Hasenbank 23a, Lindau, Germany
as@in-two.com

Abstract. In an increasingly instrumented and inter-connected digital world, citizens generate vast amounts of data, much of it being valuable and a significant part of it being personal. However, controlling who can collect it, limiting what they can do with it, and determining how best to protect it, remain deeply undecided issues. This paper proposes *CAPrice*, a socio-technical solution based on collective awareness and informed consent, whereby data collection and use by digital products are driven by the expectations and needs of the consumers themselves, through a collaborative participatory process and the configuration of collective privacy norms. The proposed solution relies on a new innovation model that complements existing top-down approaches to data protection, which mainly rely on technical or legal provisions. Ultimately, the CAPrice ecosystem will strengthen the trust bond between service developers and users, encouraging innovation and empowering the individuals to promote their privacy expectations as a quantifiable, community-generated request.

Keywords: Collective Awareness Platforms · Collaborative platforms
Collaborative design · Privacy · Digital social innovation · Crowdsourcing
Terms of Service · Privacy expectations

1 Introduction

Privacy and anonymity in the digital world are becoming increasingly difficult to achieve. While we recognize the dramatic progress brought about by Information and Communication Technology (ICT) in almost every aspect of our everyday life, we realize that, in the process, we handed over privacy management to businesses and corporations that are primarily driven by a profit motive, making our personal data

© Springer Nature Switzerland AG 2018
H. Panetto et al. (Eds.): OTM 2018 Conferences, LNCS 11229, pp. 135–152, 2018.
https://doi.org/10.1007/978-3-030-02610-3_8

vulnerable to exploitation in ways that are harmful to us. As society in general acknowledges that privacy preservation is essential in human relations, democracy, independence and reputation, nowadays it is openly stated that businesses often offer digital products and services that are inconsistent with consumer values[1]. Yet, for a variety of reasons, the more pronounced being limited awareness of the involved risks, we tolerate and give our consent to untrustworthy software to collect, store and process our data, having limited or no evidence as to how this sensitive information will be protected, who has access to it, or even what the intended purpose is.

The need to forge sound laws to regulate business policies for data protection is judged necessary by many stakeholders in the digital market. Europe, in particular, is pioneering such efforts by recently enacting a new, reformed data protection regulation[2] and by constantly updating its e-Privacy-related directives[3].

Legal frameworks alone are not always effective, as exemplified by the many digital products caught not only breaching national or European laws, but even violating their own privacy policies. The Norwegian Consumer Council (NCC), for example, has been revealing a multitude of such cases[4], having filed a series of complaints for apps that violated both Norwegian and European laws[5]. Similar stories about digital products that have clear discrepancies between their terms and what actually happens when consumers use them reach frequently the press, even regarding products whose main task is to offer a trusted and safer online experience[6].

At the same time, the ease with which we often give our consent to the processing of our data not only hinders the efficacy of legal regulations, but also makes it difficult for technical countermeasures to achieve a broad, society-wide impact to consumers privacy protection. The industry seems to lack incentives to adopt a more privacy-respecting attitude; the much debated Do Not Track[7] policy proposal is a characteristic example: despite its adoption by all main browsers, most web sites ignore it, having no significant reason to do otherwise [3].

Our limited understanding of the potency of digital services and the low degree of awareness on the privacy risks involved help accentuate the problem. The situation is sustained, and implicitly supported, by the current scheme with General Terms and Conditions, Terms of Service, Privacy Policy or End-User License Agreement documents (collectively referred to as *ToS* in this paper), which represent the most direct

---

[1] http://webfoundation.org/2017/03/web-turns-28-letter/.

[2] Regulation 2016/679 of the European Parliament and of the Council of 27 April 2016 on the protection of natural persons with regard to the processing of personal data and on the free movement of such data, and repealing Directive 95/46/EC (General Data Protection Regulation - GDPR), L 119/14.5.2016.

[3] https://ec.europa.eu/digital-single-market/en/proposal-eprivacy-regulation.

[4] https://www.forbrukerradet.no/appfail-en/.

[5] http://www.forbrukerradet.no/side/norwegian-consumer-council-files-complaint-against-tinder-for-breaching-european-law, http://www.forbrukerradet.no/side/happn-shares-user-data-in-violation-of-its-own-terms/.

[6] https://www.ndr.de/nachrichten/netzwelt/Nackt-im-Netz-Millionen-Nutzer-ausgespaeht,nacktimnetz100.html,

[7] http://donottrack.us/.

means a consumer has to understand how his/her personal data are handled. A recent study by NCC showed that just reading the ToS for apps on a typical smartphone would take more than 24 h[8]. Considering their scope, length and complexity, it comes as no surprise that the average consumer is not investing sufficient time to study ToS before agreeing to them, thus unintentionally granting permission to apps to access and process a wealth of personal information.

With the number of privacy violations growing though, it is becoming obvious that the contrasting views between what consumers want and what firms offer can hurt the industry in the long run. As privacy concerns crystallize in public perception, small businesses will be the first to experience the consequences of consumers turning their back on privacy-suspicious products[9]. Furthermore, recent studies provide evidence that privacy policy is interlinked with innovation policy and consequently has impact for innovation and economic growth [4]. A collaborative scheme, built on trust relations, can benefit all involved parties (consumers, app developers, service providers). Within such a scheme, *data protection and privacy will not be seen as barriers to business growth, but as a competitive advantage and an innovation opportunity*. The ensuing competition will provide opportunities for start-ups to enter the market, as well as for established firms to improve their market share by appropriately adjusting the privacy-related characteristics of their digital products/services, all for the benefit of the end-user (consumer) of these products/services.

The main thesis motivating this paper is that technical solutions and solid legal regulations are necessary but not fully sufficient for accomplishing a paradigm shift towards a new data economy. In addition, we firmly believe that *data protection can be powered by the society itself*. By mobilizing consumers to become active players in digital marketplaces and by developing socio-technical tools to harness our collective power, the adoption of the technical and regulatory frameworks will become more effective and ubiquitous, and the market will act with responsiveness. As stated in [14], to protect privacy adequately, society needs awareness, but also consensus about privacy protecting measures and processes that generate social norms, with which service providers will voluntarily comply because it is profit maximizing.

This paper proposes *CAPrice, a suite of mechanisms to facilitate community interaction, enabling the explicit declaration of consumers' privacy expectations of the various digital products*. Through a combination of socio-technical methods, such as community-generated design contractualism, crowd sourcing and a knowledge commons approach to privacy policy, the outcome is a new innovation model that will allow consumers to collectively and collaboratively express their concerns, and developers to adopt more privacy-friendly practices and respond to the needs of consumers with novel products and services. To support this aim, a community is being formed that wishes to support actions towards the vision discussed in this paper (details

---

[8] http://www.forbrukerradet.no/side/the-consumer-council-and-friends-read-app-terms-for-32-hours/
http://www.forbrukerradet.no/side/250000-words-of-app-terms-and-conditions/.

[9] http://www.bloomberg.com/bw/articles/2013-03-05/why-mobile-apps-privacy-policies-are-so-important
https://www.cognizant.com/whitepapers/the-business-value-of-trust-codex1951.pdf.

are given below). The current paper describes the long-term vision of the CAPrice idea, as well as the current results of applying this vision in practice.

In Sect. 2 we describe the theoretical framework upon which or work is based, whereas Sect. 3 describes the complete vision associated with CAPrice. The current progress of CAPrice is described in Sect. 4. We conclude in Sect. 5. An earlier version of this idea appears in [10].

2 Theoretical Framework

Against the current landscape in the digital world, the protection of privacy is not just the result of applying legal and technical requirements. It seems to be also connected with the idea of the personal privacy expectations of each individual, an expectation that also depends strongly on the context in which the user is interacting (e.g., media sharing sites, social networks, apps), the social status of the user (gender, marital status, age, employment, etc.), and, of course, his/her personality and privacy sensitivity. Digital awareness has become a key issue and, consequently, citizens are another link in the chain of protecting their own privacy. In this regard, the improvement of the individual's empowerment may be the missing link in the implementation of a comprehensive and effective global strategy for the protection of privacy in the digital age. This empowerment, achieved through collaboration, crowd sourcing and collaborative open innovation management, is the main focal point of CAPrice. Before describing the software tools that will enable and facilitate this collaboration, we analyze here the main theoretical principles associated with CAPrice.

Collective Awareness Platforms for Sustainability and Social Innovation (CAPS) is a research initiative endorsed and supported by the European Commission, aiming to explore new solutions at the confluence of social networks, knowledge networks and networks of things [1]. Officially, CAPS is an initiative that "pioneers new models to create awareness of emerging sustainability challenges and of the role that each and every one of us can play to ease them through collective action"[10]. It aims at designing online platforms for creating social awareness and for allowing collective solutions to emerge through the interaction among participants, exploiting the hyper-connectivity characteristic of the digital society. Several projects associated with this initiative have been funded[11] and have already produced (or will produce soon) important results showing how collective action can support and enhance many different facets of human activity.

CAPrice leverages this idea towards *creating a community centered around privacy that will both contribute to, and benefit from, the improved, community-wide awareness on privacy.* More precisely, CAPrice complements top-down efforts by creating a community including consumers, industrial stakeholders, decision-makers and the general public. This community will engage in a *multi-directional communication*, aided by software tools that will help promote awareness and cooperation among

---

[10] https://ec.europa.eu/digital-single-market/en/collective-awareness.

[11] https://capssi.eu/caps-projects/.

different stakeholders, towards the mutual benefit of everyone. Unlike other initiatives, in which a group of experts tries to inform other users on the privacy-related dangers of certain actions or products, we try to break this asymmetry: every person in the CAPrice community can, potentially, play the role of both the "teacher" and the "student", or both the "informer" and the "informed".

In fact, mechanisms for specifying the intended use of information have been suggested in the past (e.g., P3P [7]), but never achieved wide acceptance. The bottom-up participatory innovation paradigm of CAPS offers the means to achieve a more substantial impact, but society-wide participation and engagement are key aspects for its success. The most important difficulty that most "young" CAPS face is how to reach a critical size above which payoff for the platform (however defined) becomes positive. In order to overcome this initial threshold effect [2], a multi-dimensional strategy is needed to promote user engagement and foster social innovation.

Design contractualism is the idea that developers make legal, moral or ethical decisions and then (a) embed these decisions in the code itself and (b) make those decisions manifest to the user. Part (a) is achieved by encoding rules of order for appropriate behavior in computational logic as above, so the second critical innovation is to make those rules manifest to the different actors in the system. Since we are dealing with a knowledge commons, one approach is to extend an idea from the Creative Commons[12]. For example, Creative Commons supports six different licenses in three layers, each of which constitutes a norm, as it serves to coordinate expectations. However, one can imagine a user group operating under one license, but reaching a point where they would prefer to operate under a different license: the question is how to agree changes in licensing arrangements. CAPrice proposes a similar approach through the annotation of ToS documents.

Privacy protection and management, as well as information sensitivity, are inherently user-centered, thus it cannot be claimed that a given set of norms for a given app is suitable for all users and contexts. In the CAPrice model, we encourage debates for norm generation that will allow the identification of groups of people sharing common opinions. Once this happens, a separate debate per group can specify the corresponding fit-for-purpose norms.

These guidelines can be adapted to enhance the privacy policies of diverse digital services. Apps for mobile devices, for instance, specify the groups of capabilities or information (permission groups - PGs) that they need access to. Many platforms operate on a take-it-or-leave-it style, leaving a lot of aspects inadequately supported; in particular, developers are not required to explain why they need access to the requested PGs and what they plan to do with the respective data. CAPrice expands the current scheme with support for explanation generation and justification modeling: for each PG that some app requests access to, the justification can comprise a set of aspects denoting why the app developer needs this PG and a set of aspects denoting the user benefits. Our proposed solutions intend to facilitate discussion about the privacy scope of apps with regards to data access, and enable users become aware and understand how their data is manipulated, as well as to express their privacy expectations.

---

[12] https://creativecommons.org/.

3 Methodology

Our limited understanding as consumers of the capabilities of digital technologies in collecting and processing our personal data, and our inability to easily request guarantees for data protection or to prevent collection and sharing, lead us to adopt behaviors in our digital interactions that would seem unreasonable in the physical world.[13] For the time being, as the current data economy has obvious benefits for both firms and individuals, it comes as no surprise that we seem to feel comfortable with, or at least tolerate, the existing situation. Nevertheless, the protection of privacy in the digital world is becoming a vital societal problem and many stakeholders worldwide ring the bell for appropriate action. Inevitably, as privacy concerns solidify in public perception, the implications of consumers' suspicions towards digital products will eventually hurt the industry, especially the smaller players.

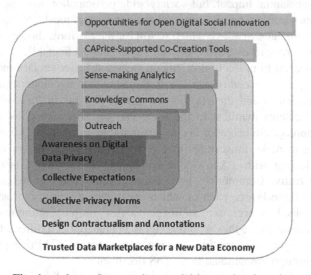

Fig. 1. A layered approach to social innovation for privacy

Unfortunately, the protection of personal data is not "a few clicks away" for the average consumer; changing application configurations, installing technological countermeasures, even reading the privacy policies and understanding the risks, is a needlessly difficult task, especially for consumers who have grown accustomed to quick-and-easy interactions with technology, or for those with a low level of technological competence. *Our intention is to offer solutions that will make privacy-friendly digital interactions for the consumer as easy to accomplish as it currently is to neglect caring about privacy protection.*

Accomplishing this goal requires a paradigm shift in the way we understand and experience technology, which cannot occur overnight, but needs a methodical approach that will steadily empower passive consumers of digital products to understand the value of their data and take control of it. In this effort, policy makers and ICT tools will offer indispensable leverage; yet, a key step for achieving effective impact will be to convince developers that they have many benefits to reap in the new trusted data economy, by seeing privacy and data protection as a competitive advantage, rather than a barrier. CAPrice is a solution that will enable consumers to express their privacy expectations and desires about digital products, while offering innovation opportunities

---

[13] The following video is instructive: https://www.youtube.com/watch?v=xYZtHIPktQg.

for developers who are willing to listen and respond to their needs. Our approach for contributing towards this paradigm shift will happen along the socio-technical actions and innovations shown in Fig. 1 and explained next.

3.1 Awareness on Digital Data Privacy

The first vital step is to approach individuals from different social and demographic groups who share similar values regarding privacy, and make them aware of the privacy risks that are hidden in the careless use of digital technology. Although digital privacy protection is included in the agenda of many organizations and institutions, in order to achieve a society-wide paradigm shift, it is important to create a global community of citizens that not only subsumes the already established groups, but expands to consumers who never before considered the protection of privacy a key concern of their daily interaction with technology.

Towards this end, we initiated an attempt to create a grassroots community of privacy-aware consumers. Securing participation in virtual online communities is not trivial, and simply bringing together individuals who share similar goals or purposes is not sufficient. To successfully foster and sustain engagement in the CAPrice virtual community, we followed the well-known 3-stages process [8], described below.

First, we need to identify and understand the needs of community members that create the intrinsic motivations for participation. As the numbers from our social channels indicate (see Sect. 4.1), real stories about smart toys, baby monitors, mHealth apps, even about future autonomous cars can have dramatic effect in driving awareness of diverse audiences, compared to other material.

Second, member participation must be promoted, by highlighting the value of collective actions, by creating enjoyable experiences or by encouraging content creation, among others. In fact, similar community creation attempts in other domains showed that any grassroots community is prone to lose interest, unless a vibrant, self-motivated group of users exists in its core to make it sustainable and to help establish self-definition[14]. In our case, this group is called the *CAPrice Privacy Ambassadors*, a group of individuals with specific technical and social skills, who have taken over the task of engaging citizens in this effort (see Sect. 4.2).

Finally, the third stage is to sustain member engagement by motivating cooperation, enabling members not only to meet specific needs, but also to co-create value for themselves and the community. We have designed a number of ICT tools to foster cooperation among ordinary consumers, researchers, privacy-enthusiasts, hackers, as well as general-purpose digital-product developers. On top of these tools, a rewards program will incentivize participation, driving user engagement and supporting reputation mechanisms to assure members that their contributions are recognized.

It is important to repeat here that this is different from top-down efforts, where awareness is achieved through a group of experts. Our aim is to complement such efforts by creating a community including consumers, everyday people, industrial stakeholders and decision-makers, who will engage in a multi-directional communication that will

---

[14] http://www.scp-centre.org/wp-content/uploads/2016/05/Final_Report_CATALYST.compressed-2.pdf.

help promote awareness and cooperation among different stakeholders, towards the mutual benefit of everyone. Also, CAPrice differs from technical solutions to privacy (such as, e.g., the PlusPrivacy tool[15]), whose aim are to ensure that digital interactions respect the privacy preferences of the user, by imposing such preferences at a technical (software or hardware) level.

Awareness corresponds to the first, innermost layer shown in Fig. 1. So far, we have been quite successful in growing our community; details regarding community creation and sustainability can be found in Sects. 4.1, 4.2.

3.2 Collective Expectations

The second step (Fig. 1) is to *capture consumers' expectations* regarding the privacy policies of the different digital products they ordinarily use. This is achieved by allowing consumers to explicitly state their own expectations and treating these expectations as a common-pool resource. By enabling users to specify which access permissions they find reasonable for products of a given category and which they consider too intrusive, we aim at generating shared content that will be directly exploited by many stakeholders, ranging from simple consumers and developers, to policy makers, even to social scientists that will attempt to interpret the dynamics of the community and their stance towards privacy. Towards this, we are in the process of creating a global repository of human-readable and machine-processable privacy-related content (consumers' expectations, annotated ToS, application ratings, and others) in the form of a semantic privacy wiki (see Sects. 4.4 and 4.5).

With the generation of citizens' collective intelligence about privacy expectations in the form of measurable data, the accent is not only on the peer pressure that can be used to drive more privacy-respecting practices by developers, but also on the realization by consumers that expressing privacy needs and requesting solutions is not exclusively a top-down process, but can also be accomplished by each individual user uniting her or his voice to that of other members of the community.

3.3 Collective Privacy Norms

The basic position of CAPrice is based on the acknowledgement that, when it comes to privacy, one solution that can serve all needs is not feasible. Within the privacy protection boundaries set by legal regulations, one should listen to the plurality of opinions issued by consumers regarding the level of privacy space they wish to have, which leads to different privacy needs and expectations. Identifying these differences is of course beneficial for innovative developers who can design flexible services that adapt to the various needs. But this is even more critical for building a society that respects and supports the different trends, and where policy makers can recognize and act upon the dynamics behind the contradicting mindsets of citizens.

One of the key points of our approach is related to the *identification of collective privacy norms* (3rd layer in Fig. 1). In contrast to legal regulations, which apply

---

[15] https://plusprivacy.com/.

ubiquitously, social norms are more flexible: they can be contradicting, as different attitudes may be considered "ordinary" by different people; they are more dynamic, being easily adapted to societal trends; and they have no geographical restrictions. On the other hand, law and policy making require a thorough understanding of a situation before being issued to guarantee just treatment; however, this reduces their adaptability and makes them unable to confront the astonishing speed with which ICT progresses. And there is always the risk that the country our data go to does not have the desired level of protection (although this problem is being mitigated, at least in Europe, by the introduction of European regulations such as the GDPR). We argue that collective privacy norms that exist inside the boundaries of regulations, despite being less stringent and reliable than legal regulations, can be equally powerful to control market dynamics if appropriately supported.

The aggregation and analysis of consumers' expectations into collective norms that will conceptualize the stance of citizens towards privacy products introduces certain challenges. First, the result should be measurable, to enable developers to weigh their profit-loss trade-off, but also semantically rich, to allow for meaningful interpretations of the data. Otherwise, the industry will find no incentive to adopt a different attitude towards privacy protection, as has happened many times in the past already.

In addition, the privacy principles underlying these norms need to be simple and comprehensive, in order to clearly capture the intuition of consumers and to secure society-wide coverage. We consider for our approach the experience of other initiatives that try to model users' preferences about privacy settings, relying on principles such as transparency and minimization of use.

Simplicity is key for users' comprehension, so we base our approach on a what/who/why/how/how-long scheme, i.e.: what data are being collected and processed; who is collecting or has access to the data (data controller/processor); why are the data collected or processed; how are they published; and for how long are they stored and processed. This is in close compliance with Opinion 02/2013 of the EU Article 29 Working Party on apps for smart devices[16] that provides, among others, the smallest set of recommendations that developers should follow in their privacy policies.

3.4   Design Contractualism and Annotation

It is well-understood that the contribution of users in isolation towards a common goal and the aggregation of their data is only half-way towards achieving the collective intelligence needed to address a societal problem. What is also imperative is the *participation of users in co-creation processes* that will empower them to collectively generate new ideas and decide collective actions. This

Fig. 2. Visual cues for terms of service documents (Taken from https://disconnect.me/icons)

---

[16] http://ec.europa.eu/justice/data-protection/article-29/documentation/opinion-recommendation/files/2013/wp202_en.pdf.

co-creation process, which also fosters group awareness and understanding of the problems at hand, requires well-structured deliberation and discussion tools that can support goal-driven exchange of opinions, and where conclusion making is equally important to the identification of the different trends in the dialogue.

In CAPrice, we reuse and extend tools with proven impact, focusing on generating bottom-up solutions on privacy, and incorporating for the first time the consumer's point of view, following the ideas of design contractualism (4th layer in Fig. 1). In fact, our advanced notion of design contractualism goes two steps further.

Firstly, because instead of making legal or ethical decisions, designers and developers construct a legal or ethical decision space, and enable the point in that space to be selected by the users. This is the basis of algorithmic self-governance [9], whereby those affected by a set of rules (of an embedded, socio-technical, data-processing system) also participate in the selection, modification and application of those rules.

Secondly, we advance design contractualism by not just encoding this decision space in the software, but crucially also in the interface. This user-centric approach to governance modeling entails the use of visualizations to ensure that the commonly agreed privacy principles are manifested by visually identifiable and interpretable means. Using visual cues, such as the ones shown in Fig. 2, CAPrice intends to employ crowdsourcing techniques that will augment privacy policy documents with annotations easy for consumers to check and understand. Appropriate ICT means and personalization algorithms will hide the complexity of the task for users who decide to offer annotation services to the community, and, implicitly, to the general public.

3.5 Trusted Data Marketplaces for a New Data Economy

The ultimate objective of this stepwise approach (outermost layer in Fig. 1) is to contribute towards *a new marketplace*, where the interactions between consumers and developers are based on trust relations. By associating consumers with their privacy expectations, while providing the technological means for developers to exploit this information for undertaking novel, more privacy-friendly and respectful to consumers practices, we aim towards creating the substrate for developing new ICT tools and services. This will allow the provision of added-value services on top of the open architecture of CAPrice, and will lead to new and innovative privacy-enhancing applications. The engagement of consumers will overcome the problems faced by purely legal or purely technical solutions, creating a novel data economy for developers.

Of course, the legal and technical aspects are also necessary to ensure trust among all involved parties. Existing data marketplaces are essentially centralized systems, where participants (data providers and consumers) have to trust a third party, the data marketplace provider/operator, with managing their data. Typically, access to data on a marketplace is governed by a set of privacy policies, often rather vague, unclear, and difficult to understand, leaving data providers with little control over their data. The guarantees that current data marketplace players receive give them little confidence that data recipients will treat the received data in the way they promise.

In order to ensure trust, a data market-place must be transparent with all stake-holders. Transparency is a fundamental principle in data protection and highlighted in the GDPR. This means that the participants in a data exchange should have knowledge about what data are shared and what operations are done over the data, and be in agreement that the data can be used for that purpose.

In CAPrice, we make steps towards offering a starting point for developers to adopt more privacy-respecting practices. In particular, we leverage emerging technological concepts, such as smart contracts and blockchains, and incorporate them into a

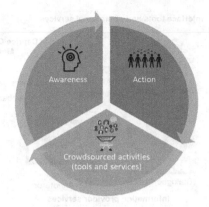

Fig. 3. The Best Practice Lifecycle of CAPrice

trusted data marketplace, thereby enabling the processing of data with "by-design" trust and transparency. Smart contracts are self-executing contractual states, stored on the blockchain, and represent computer programs that can automatically execute the terms of a contract. Blockchain, as a decentralized technology, provides security, anonymity and data integrity. An example of a reference architecture for trusted data marketplaces was proposed in [11] where more details are provided on how such emerging technologies can be combined to achieve more trust and transparency in data sharing.

A trusted data marketplace caters to the interests of both application providers and their end-users. Application providers will have the opportunity to develop applications which technically guarantee their end-users' privacy, thus making them more attractive and competitive. End-users, on the other hand, will benefit from the fact that any system based on the trusted marketplace will provide transparency and unbreakable assurances that the promises of the data consumer will be kept.

3.6 The Best Practice Lifecycle of CAPrice

To summarize, the CAPrice Best Practice Lifecycle (Fig. 3) aims at maximum impact through three conceptual phases. The first phase is *awareness*: only through awareness can people understand the problem and start considering solutions. The second step is *action*: in the context of CAPrice, action consists in participating in the collaborative process of annotating ToS documents, stating privacy concerns, creating and configuring collective privacy norms, and participating in the co-creation process. The third step is the exploitation of the acquired knowledge through *crowd sourced activities*. In this respect, a series of tools will allow the users to better implement the second step (action), and also other relevant stakeholders (policy makers, developers, legislators) understand better the needs of the public in order to contribute towards making digital products and services more transparent and privacy friendly.

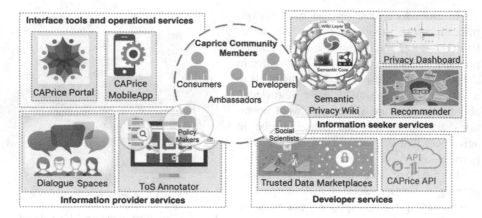

Fig. 4. The CAPrice ecosystem

4   CAPrice: Architecture and State of Development

The high-level overview of the CAPrice ecosystem is depicted in Fig. 4. According to the purpose of use, the members of the CAPrice community will be offered different groups of services, from user interfaces and services for information seekers to services for developers and information providers. These are briefly described below.

Harnessing the power of crowdsourcing tools and methodologies to collect, organize, annotate and simplify this knowledge can achieve immediate results and produce valuable content. At the heart of the CAPrice ecosystem lies the *CAPrice Semantic Privacy Wiki*, an open repository containing, among others, privacy-related information regarding digital products. The repository combines the benefits of semantic technologies with the collaborative editing capabilities associated with wikis, offering a set of functionalities that go beyond simple wiki-style catalogue for ToS: it enables the user to express privacy preferences about each product or category of products, it permits developers to explain their policies and automatically access the underlying data, it offers a public place for experts to post findings about products, and others.

The Semantic Privacy Wiki will be populated with information from the *ToS annotator* (see Sect. 4.4) and the *Dialogue Spaces* which will facilitate structure discussions and the creation of privacy norms.

The content of the Semantic Privacy Wiki is leveraged by all the other CAPrice services. In particular, the *Information Seeker Services* enables the user to understand better the privacy policies of popular applications, and provides application recommendations that can satisfy the needs of the user while being as compatible as possible with the user's privacy expectations. The functionalities are exposed through appropriate UIs, accessible through the Web (*CAPrice portal*), while being also mobile-friendly (through the *CAPrice Mobile App*). Last but not least, we provide a set of *Developer Services* to allow external developers to improve or enhance the CAPrice functionality by providing new services or by improving existing ones.

As already mentioned, the current maturity level of these tools varies, as the development of the CAPrice platform is work in progress. As a result, some of these tools are still at the planning stage (e.g., Dialogue Spaces, CAPrice API), whereas others have progressed to the implementation phase with varying levels of progress (e.g., Semantic Privacy Wiki, Recommender, Privacy Dashboard, CAPrice Portal, ToS Annotator). In the rest of this section, we give details on the most important activities currently undertaken towards developing the CAPrice platform, including a short presentation of the most mature tools and the results of applying these tools in practice (where available).

4.1 Communication Channels Towards a Grassroots Community

CAPrice is a holistic solution towards improved privacy awareness. Even though we have not yet implemented the CAPrice solution to its full extent, some early efforts have led to the implementation of a series of *communication channels* (consisting of a frequently-updated website and social media accounts for improving our engagement and penetration potential), and to the creation of the initial core of the CAPrice ecosystem, including an *Ambassadors' Group* and the *CAPrice Community*.

The CAPrice website[17] offers multiple ways to users to provide feedback, mention their personal stories, and express their opinion. We have been active in continuously providing information regarding the latest policies on privacy of digital apps and services. In addition, the CAPrice website acts as a digital privacy portal presenting privacy leaks, potential solutions and multimedia content regarding digital privacy, focusing on privacy concerns and data protection issues that arise daily. To achieve this, we follow relevant sites, scientific reports/papers, news by privacy experts and hackers, and we publish relevant articles to the CAPrice website. Moreover, there are a lot of short videos/animations in the privacy related section that can be used by teachers and parents to inform kids in a visual and more entertaining way about privacy issues.

The relevant content is also shared through the CAPrice social media accounts (Facebook[18], Twitter[19], Youtube[20]) and is used to gather feedback or interact with users upon relevant posts or issues for our active online community. The content we share is not technical and is intended to the general public. Special focus is given in news concerning toys for kids, student apps and other subjects that, from experience, seem to attract the most attention, in order to ensure that the interest level of CAPrice community members remains high. The use of social media accounts is a key tool towards maximizing the community outreach and achieving optimal results. Social media are very popular in children and teenagers, which are critical age groups for achieving real, time-enduring change in privacy-related practices.

The aforementioned communication channels have contributed to the creation of the CAPrice community. Although the CAPrice community is by no means a sizable

---

[17] https://www.caprice-community.net.

[18] https://www.facebook.com/CapriceCommunity/.

[19] https://twitter.com/CapriceSociety.

[20] https://www.youtube.com/watch?v=4L8gOfU9MXg.

virtual community (yet), the initial statistics show not only prominent indications that the critical mass needed to make the community self-sustainable can indeed be reached, but also that the topic of digital privacy has become a key concern for the average consumer, despite the fact that the current scheme of interacting with digital technology shows otherwise. At the time of writing, the website had 114 unique visitors per day on average and 171 email subscribers, while the Facebook page had 569 likes, the Twitter account 232 followers, and the Youtube video had been viewed 1483 times. An indicator for the impact of this effort is the fact that our tweets overall have earned around 14000 impressions over the last 3 months (May 29, 2018 to August 26, 2018) while the pinned tweet earned 4946 impressions with 63 engagements. Furthermore, the latest 30 posts that have been published in our Facebook page during the same period have earned 9253 reaches and 381 reactions.

4.2   The CAPrice Privacy Ambassadors

Perhaps the most challenging part when transferring a socio-technical solution from paper to practice is to achieve the right balance between communities and technology. This is one of the most emphasized lessons learned by almost all past collective intelligence initiatives. Indeed, practice shows that for any established community to grow or for any new community to obtain substance, a group of highly committed and internally motivated individuals needs to be at its core. These individuals support and energize the whole community and maintain the social processes within; they initiate action, generate ideas, and motivate others. Members of the core, which is usually only a small fraction of the community, are characterized by both specific psychological traits (engagement, motivation and charisma), as well as specific structural positions in the social network [13].

Within the CAPrice ecosystem, these members are the *CAPrice Privacy Ambassadors*[21]. The group of Ambassadors is an evolving entity that has a specific role in the entire lifecycle of our initiative. Our intention is to exert only minor control over this group's dynamics, fuelling it with the proper means to help it obtain self-definition, but still leaving the necessary flexibility required to grow in size and adapt to the community's evolving needs.

The key role of the Ambassadors in our effort led us to start contacting and securing the support of the first Ambassadors as one of our first tasks. Currently, the CAPrice Privacy Ambassadors group is a core group of high-profile privacy enthusiasts from Europe and around the world. The founding members were carefully selected to combine three profile characteristics: privacy consciousness, more than average knowledge about digital technology, and confirmed desire to motivate society into adopting a more privacy-aware behavior. Currently, CAPrice has employed 20 ambassadors with various characteristics and expertise, ranging from academics to lawyers, developers and entrepreneurs.

---

[21] https://www.caprice-community.net/privacy-community/.

4.3 Improving Engagement Through the CAPrice Game

To keep our community active, and also to help them become more aware of privacy-related issues, we created the *CAPrice Game*[22], a simple, interactive mobile quiz game that tests the knowledge of kids, parents and teachers regarding the privacy of popular digital apps. The CAPrice game is available through the Kahoot platform[23] and requires only network access and a teacher/manager to control the whole game. This game contains a lot of fun features (music and sound effects, scoreboard to show the current top-scoring players, extra points for correct sequential answers and awards for the top-3 players) in order to increase motivation and engagement. Furthermore, it offers a single-player and a multiplayer mode. The game is highly configurable and scalable to include more questions or request relevant feedback from the players. The results of the game could be saved and exported in various formats in order to gain more knowledge by drawing conclusions about users' privacy expectations and by paying attention to the correct answers.

The CAPrice Game can be easily modified to include more questions or request relevant feedback from the players and can be played in the English or Greek language. We have already tested it to high schools that have visited the Institute of Computer Science at FORTH, and it was also demonstrated at the TEDxUniversityOfCrete conference[24].

4.4 Annotating Terms of Service Documents (ToS Annotator)

To cope with the complexity of ToS documents, there are efforts along the following two directions: (a) formal privacy policy languages readable by machines that can be used by both the users and the services for describing their privacy expectations, concerns and policies, and (b) through annotating the ToS with privacy related information.

A lot of work is currently conducted along the direction of enriching and annotating privacy policies with privacy related information (e.g., specifically designed tags embracing different privacy concerns like data collection, data retainment, etc.). Such tags can be pinned in ToS either by privacy experts or through machine learning algorithms. Unfortunately, although experts are able to provide accurate annotations, the task of annotating the available ToS in the huge and dynamic Internet/Web environment is possibly a Sisyphean one for the limited number of privacy experts. On the other hand, the current machine learning approaches are only able to annotate ToS segments with the correct but general privacy concern categories, while they are not able to identify more fine-grained information related with the specific values for this category [6, 12].

In CAPrice, we put forward another alternative for annotating ToS that revolves around the wisdom of the crowds. Since the problem of privacy awareness is a social issue, we believe that users should be active producers and reviewers of privacy related

---

[22] https://www.caprice-community.net/game.

[23] https://www.kahoot.com.

[24] http://tedxuniversityofcrete.com/.

content, and not just consumers. Towards this, we have designed and developed *a crowd sourced platform for engaging users in the annotation of privacy policies* [5]. Our aim is to provide to the CAPrice community and all interested users a reference open-source and public platform for the creation, review and evaluation of privacy policy annotations. We already implemented a first pilot version to test various interaction modes for non-expert users and to verify that the content created can be of high quality. Our initial comparative results conducted over the only available expert based OPP-115 ToS privacy annotated collection[25] from the Usable Privacy project, show that the crowd-sourced privacy policy annotations, cooperatively created and reviewed in our platform, are of high importance and quality, comparable in most cases to the annotations created by the expert users [5].

4.5   Interacting with CAPrice Data

We are implementing the first release of an *open semantic repository* that constitutes the core of the CAPrice ecosystem and will store a multitude of privacy-related information. Through this tool, all visitors will be able to find information about digital products, such as the requested access policies or the related ToS documents. Facilities are being developed to assist exploration on various axes, e.g., by categorizing products based on their type (smartphone apps, smart products), their purpose (entertainment, weather, travel), the community rating (highly trusted, suspicious), etc.

The system also allows CAPrice members to specify their own expectations and views regarding the privacy policy of each product, e.g., how comfortable they feel about the privacy requests of a particular product, under which conditions they would grant access, and others. We are designing a set of visual cues to help users in expressing their expectations, without overwhelming them with questionnaires and textboxes (see, e.g., Figure 5).

Finally, developers will also be able to add input, specifying their access policies and justifying them as appropriate. Note that this latter input is not necessary to ensure a smooth operation of our platform; due to the collaborative nature of CAPrice, simple or expert users can provide relevant information, although of course the active involvement of developers will also be encouraged and supported, in order to help them build a more privacy-sensitive profile.

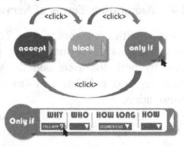

Fig. 5. Multi-button for expressing privacy expectations on a specific data access request

Our current implementation of the semantic repository stores the aforementioned data in RDF format, the standard Semantic Web language for semantically enriched content. This format allows posing expressive queries that enable more sophisticated forms of automated information seeking and analysis, while also permitting the

---

interconnection of the content with other datasets, following the Linked Open Data paradigm and enhancing the interoperability of the ecosystem. Our current version of the repository uses the open source Blazegraph triple store and currently contains around 2.8 M triples concerning information about 241 K applications on 55 categories that were automatically extracted from the Android Play Store, our starting point for the first release of the platform.[26]

On top of the repository, we are currently developing a graphical user interface that will be the frontend of the CAPrice portal, along with the first version of the Recommender and the Dashboard. The Recommender uses SPARQL, a standard query language for RDF data, that suggests -among others- similar smart products of comparable quality (based on the Android scoring system), but with fewer (or more compatible with the user's preferences) permission requests, or higher privacy-related rating by CAPrice users. The Dashboard, on the other hand, will aggregate data, in order to extract norms and trends with respect to the CAPrice users' expectations and will visualize analytics in various forms.

There are other implementation tasks that are pending in order to materialize the CAPrice ecosystem shown in Fig. 4, but of higher priority is the creation of the engagement and reputation mechanism discussed earlier, that will motivate and reward community members in the generation of new content, while helping them iron out contributions of limited value.

5 Conclusions

In this paper, we presented CAPrice, a *socio-technical solution based on collective awareness and informed consent that allows better engagement and awareness of the average consumer towards (digital) privacy.* Our approach aims to make the gains of adopting a more privacy-respecting attitude obvious and measurable, both for consumers, and for service/software providers, while also allowing decision makers and social scientists understand better the consumer needs. This way, the collective pressure of citizens, combined with market forces, will lead to synergies, healthy competition and attitude change for all involved stakeholders.

Acknowledgements. The authors thank the following individuals for contributions in earlier versions of this work: G. Baroutas, A. Dimitriadis, K. Doerr, G. Ioannidis, Y. Marketakis, N. Minadakis, G.M. Moen, F. Myrstad, A.K. Ravna, Y. Rousakis, M. Titorencu. The work of N. Nikolov and D. Roman was partly funded by the H2020 projects euBusinessGraph (#732003), EW-Shopp (#732590), and TheyBuyForYou (#780247).

---

[26] The endpoint can be accessed from here (using "caprice" as the namespace and "http//caprice/" as the named graph): http://bit.ly/2z3k9jt. The Blazegraph rest API can be found here: https://wiki. blazegraph.com/wiki/index.php/REST_API.

References

1. Arniani, M., et al.: Collective Awareness Platform for Sustainability and Social Innovation: An Introduction, Brussels, EC, CAPS (2014)
2. Bagnoli, F., Guazzini, A., Pacini, G., Stavrakakis, I., Kokolaki, E., Theodorakopoulos, G.: Cognitive structure of collective awareness platforms. In: IEEE 8th International Conference on Self-Adaptive and Self-Organizing Systems Workshops (2014)
3. Carrascosa, J.M., Mikians, J., Cuevas, R., Erramilli, V., Laoutaris, N.: I always feel like somebody's watching me: measuring online behavioural advertising. In: 11th International Conference on Emerging Networking Experiments and Technologies (2015)
4. Goldfarb, A., Tucker, C.: Privacy and innovation. Innov. Policy Econ. **12**(1), 65–90 (2012)
5. Hompis, G.: CAPPA: a collective awareness platform for privacy policy annotations. M.Sc. thesis, University of Crete (2018)
6. Liu, F., Ramanath, R., Sadeh, N., Smith, N.A.: A step towards usable privacy policy: automatic alignment of privacy statements. In: 25th International Conference on Computational Linguistics (2014)
7. Olurin, M., Adams, C., Logrippo, L.: Platform for privacy preferences (P3P): current status and future directions. In: 10th Conference on Privacy, Security and Trust (2012)
8. Porter, C.E., Donthu, N., MacElroy, W.H., Wydra, D.: How to foster and sustain engagement in virtual communities. Calif. Manag. Rev. **53**(4), 80–110 (2011)
9. Pitt, J., Diaconescu, A.: Interactive self-governance and value-sensitive design for self-organising socio-technical systems. In: 1st International Workshop on Foundations and Applications of Self* Systems (2016)
10. Patkos, T., et al.: Privacy-by-norms privacy expectations in online interactions. In: 9th International Conference on Self-Adaptive and Self-Organizing Systems (2015)
11. Roman, D., Gatti, S.: Towards a reference architecture for trusted data marketplaces: the credit scoring perspective. In: 2nd International Conference on Open and Big Data (2016)
12. Sathyendra, K.M., Wilson, S., Schaub, F., Zimmeck, S., Sadeh, N.: Identifying the provision of choices in privacy policy text. In: Empirical Methods in Natural Language Processing (2017)
13. Schroer, J., Hertel, G.: Voluntary engagement in an open web-based encyclopedia: Wikipedians and why they do it. Media Psychol. **12**(1), 96–120 (2009)
14. Sloan, R., Warner, R.: Unauthorized Access: The Crisis in Online Privacy and Security, 1st edn. CRC Press Inc., Boca Raton (2013)

Shadow Testing for Business Process Improvement

Suhrid Satyal[1,2]([envelope]), Ingo Weber[1,2], Hye-young Paik[1,2], Claudio Di Ciccio[3],
and Jan Mendling[3]

[1] Data61, CSIRO, Sydney, Australia
{suhrid.satyal,ingo.weber}@data61.csiro.au
[2] University of New South Wales, Sydney, Australia
hpaik@cse.unsw.edu.au
[3] Vienna University of Economics and Business, Vienna, Austria
{claudio.di.ciccio,jan.mendling}@wu.ac.at

Abstract. A fundamental assumption of improvement in Business Process Management (BPM) is that redesigns deliver refined and improved versions of business processes. These improvements can be validated online through sequential experiment techniques like AB Testing, as we have shown in earlier work. Such approaches have the inherent risk of exposing customers to an inferior process version during the early stages of the test. This risk can be managed by offline techniques like simulation. However, offline techniques do not validate the improvements because there is no user interaction with the new versions. In this paper, we propose a middle ground through *shadow testing*, which avoids the downsides of simulation and direct execution. In this approach, a new version is deployed and executed alongside the current version, but in such a way that the new version is hidden from the customers and process workers. Copies of user requests are partially simulated and partially executed by the new version as if it were running in the production. We present an architecture, algorithm, and implementation of the approach, which isolates new versions from production, facilitates fair comparison, and manages the overhead of running shadow tests. We demonstrate the efficacy of our technique by evaluating the executions of synthetic and realistic process redesigns.

Keywords: Shadow testing · Business process management
DevOps · Live testing

1 Introduction

Business process improvement ideas often do *not* lead to actual improvements. Works on business improvement ideas found that only a third of the ideas observed had a positive impact [11,13,14]. If improvements can only be achieved in a fraction of the cases, there is a need to rapidly validate the assumed benefits.

© Springer Nature Switzerland AG 2018
H. Panetto et al. (Eds.): OTM 2018 Conferences, LNCS 11229, pp. 153–171, 2018.
https://doi.org/10.1007/978-3-030-02610-3_9

Simulation and AB testing techniques for business processes provide incremental validation support [18, 20]. A new process version can be simulated using historical data from the old version. If the performance projections from the simulation are satisfactory, the two versions can be deployed simultaneously on their production system such that each version receives a portion of customer requests. This method of simultaneous live-testing in production, called AB testing, is a method from DevOps [2], and compares two versions (A and B) in a fair manner. The speculative projections from the simulation are validated through performance data from the production system. This approach treats off-line simulation as a sanity-check before deployment of the new version, which reduces the risk of deploying versions with problems that can be anticipated beforehand. Nevertheless, the risk of exposing a significant number of customers to a bad version during the early stages of AB testing still remains. In addition, with these techniques, the performance of the two versions can only be compared on a collective level. There is no one-to-one mapping between process instances of two versions: each instance is executed on either version A or B.

In this paper, we propose a middle ground between simulation and AB testing with the idea of shadow testing. A new version of a process is deployed alongside the current version in the production system, but it is hidden from the customers and the process workers. A copy of user requests on the current version is forwarded to this hidden *shadow* version. When a process instance of the current version runs to completion, a corresponding shadow version in instantiated in *shadow mode*. This shadow process instance is partially executed as if it were the current version, and partially simulated with the execution information obtained from the completed process instance. This approach allows us to take a particular customer request as a reference and observe the performance and behaviour differences between the corresponding instances of the current version and the new version. We implemented and evaluated the approach.

The remainder of this paper starts with a discussion of the requirements and related work in Sect. 2. Section 3 describes our approach to the requirements and the design trade-offs. In Sect. 4, we discuss the implementation architecture and the details of shadow test execution. We evaluate our approach in Sect. 5, discuss the strengths and weaknesses in Sect. 6, and draw conclusions in Sect. 7.

2 Background

There are essentially two broad approaches to process improvement in business process management. First, *business process re-engineering* (BPR) offers a methodology for redesigning processes from a clean slate [5, 10]. Second, approaches to *business process improvement* take a more cautious and incremental method [12]. The BPM lifecycle integrates process improvement into a continuous management approach [7]. This lifecycle puts a strong emphasis on modelling and analysis before engaging with redesign, and also assumes that processes models are incrementally improved. Using these approaches, the old version of a process is replaced by a new version in the production system.

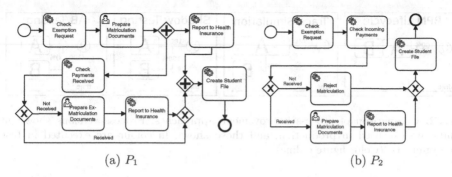

(a) P_1 (b) P_2

Fig. 1. Two versions of student matriculation process model: P_1 (as-is) and P_2 (to-be). Adapted from [8].

The AB-BPM [20] methodology provides an extended BPM lifecycle that facilitates rapid validation of improvement assumptions without needing to replace the old version. To date, this is the only approach that supports the idea of using the principles of DevOps [2] to address this need for business processes. Rapid validation builds on the existence of a newly redesigned process versions, which are typically modelled using the BPMN notation. Figure 1 shows as example of an *as-is* version of process and a redesigned *to-be* version of a student matriculation process originating from a case study in the literature [8]. In AB-BPM, such versions are progressively validated by simulation and AB testing.

In the following, we describe the problems with the AB-BPM methodology, derive key requirements for an improved solution, and describe the related work.

2.1 Problem Description and Requirements

Customer preferences are difficult to anticipate before deployment and there is a need to carefully test improvement hypotheses in practice [14]. If an improvement hypothesis has to be validated through user interaction, the new versions have to be exposed to the users. However, exposing inferior versions to the user can have an adverse effect on the business. We call this problem *risk of exposure*.

In AB testing, the risk of exposure can be reduced by gradually allocating more user requests to a better version during the tests [18–20]. However, this risk is not completely eliminated.

A better solution should eliminate this risk completely while still providing support for validating process improvements. Based on the above analysis, such a system should satisfy the following requirements:

R1 Fair Comparison: The system should compare execution of process versions under similar circumstances and reduce bias that may result from the execution environment and the behaviour of process workers.
R2 Minimal Impact from Overhead: The process version under test should have minimal impact on the performance of the production version.

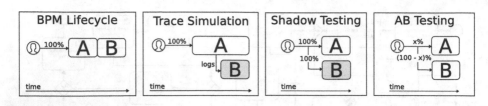

Fig. 2. Comparison of process improvement approaches. Process versions shown in white are executed in production, and those shown in colour are executed in test environments. (Color figure online)

R3 Isolation: The process version under test should not be visible to users, and should not change the process in the production system.

One way to eliminate this risk of exposure is by a form of live testing known as *shadow testing*. This approach is commonly used to observe functional issues that were undetected in earlier testing phases, and non-functional properties that can be known only after deployment. A copy of a request from the production system is forwarded to a *shadow version*, which runs in the background.

Our research adopts this idea of shadow testing for validating the improvement assumption inherent in new versions of a business process. Following the discussion of related work below, we devise dedicated architecture and execution techniques that addresses requirements R1-R3.

2.2 Related Work

We start this section with discussion on existing works on business process simulation, AB testing, and shadow testing in general. Then, we compare these techniques from a lifecyle perspective. Finally, we describe the gaps in shadow testing literature and how we address that gap.

Process simulation techniques can be useful for analyzing process redesigns. Advanced simulation techniques can extract knowledge from the historical logs of a process and predict performance of new versions [1,16,17,20]. These off-line techniques make many assumptions about process execution, and do not execute the code associated with activities. Therefore, the results of such techniques are not always reliable.

AB testing is a live testing approach that overcomes this limitation of simulation. Typically, a pool of user are selected for test, and their requests are evenly allocated to the two deployed versions (*A* and *B*) [2,4,14]. The *risk of exposure* in AB testing can be reduced by gradually allocating more users to the version that is estimated to best serve them as the test progresses [19].

Another form of live testing is *shadow* (or *dark*) *testing*. A copy of a request from the production system is forwarded to a hidden *shadow version*. Such tests are typically executed for a shorter duration than AB tests because these tests are not dependent on getting statistically significant user behaviour data [21].

Shadow testing can also be found in the area of aircraft operation control where a shadow version collects real time data and performs simulations [6].

Referring to the as-is version of a process as A and the to-be version as B, Fig. 2 illustrates how user requests for process instantiation are allocated to the two versions. In the traditional approach, all requests are allocated to version A and then to version B when the version A is phased out. In simulation, all requests are allocated to version A. After the execution of some process instances, the logs of version A are used offline to estimate the performance of version B. In shadow testing, a single user request instantiates both the version A, which serves the user, and version B, which runs in the shadow. Finally, in AB testing, requests are dynamically routed to the two versions which run in production.

The practice and effects of shadow testing for non-functional testing of web applications are documented in surveys and industry reports [9,21]. However, there is a lack of scientific literature on dedicated architectures and techniques. Furthermore, shadow testing has not been adopted in business process management. This paper adds to this body of knowledge by providing an architecture design and execution mechanism for shadow tests for business processes.

3   Solution Approach

In this section, we analyse how two process versions can differ and classify these differences. We then discuss how to address the requirements, and the encountered challenges and trade-offs.

3.1   Classification of Processes, Changes, and Activities

We need two process versions to execute shadow tests: the customer facing *production version* and a new *shadow version*. In order to understand the scope of shadow tests, we classify the types of changes between the versions as follows: *(i)* process model, *(ii)* back-end implementation, and *(iii)* user interface. Validation of user interface changes requires customer exposure. So, shadow tests are targeted towards the other two types of changes.

Processes models are composed of elements such as activities, gateways, and sequence flows. We classify these activities in three categories: *(i)* automatic tasks, *(ii)* user tasks, and *(iii)* manual tasks. They respectively represent fully automated, semi-automated, and unautomated activities. Service, script, send, receive, and business-rule tasks from the BPMN standard [7] belong to the first category. The difference between user and manual task is that manual tasks are performed without the aid of the process engine, according to definitions in the BPMN standard[1] . Therefore, we limit our scope to user tasks. Tasks from the shadow version can be further classified into two types: tasks that are shared between the two process models, henceforth *common* tasks, and *new* tasks introduced in the shadow version.

---

[1] BPMN 2.0 Specification, http://www.omg.org/spec/BPMN/2.0/PDF/, Retrieved 25-07-2018.

3.2   Conceptual Design

To facilitate fair comparison (R1), it is desirable to gather data from the full execution of process instances of the new version. However, this is at odds with our requirement for minimizing overhead (R2). If we instantiate the shadow version for each instance of the production version using the concept of shadow testing, we double the total work because the activities are enacted twice. Scheduled tasks, especially user tasks, require the involvement of valuable resources (machines and workers). This inevitably affects the performance of the production version. Furthermore, executing the same task twice for the same inputs can confuse the process workers.

As a compromise, we can instantiate the shadow version for every instance of the production version, but only partially execute the shadow instances. The executable parts of shadow version should run in isolation (R3), separate from the production version. We can execute *new* tasks in the shadow version, and estimate *common* tasks by copying information from the corresponding instance from production. This eliminates the problem of double execution but introduces a degree of speculation.

Partial execution can be ineffective for reducing overhead in scenarios where shadow versions have many new tasks. Execution of these new tasks inevitably affects the performance of the the production version. In such cases, we can reduce the amount of work related to the shadow version. One way to do this is to instantiate the shadow version when the production workload is low. However, this approach inhibits fair comparison because the production and the shadow instances may not execute under similar circumstances. Another approach is to instantiate the shadow version only when the load is expected to be low. Creating less instances of the shadow version in this way reduces the need to execute new tasks. However, this deviates from the concept of shadow testing where one shadow instance is created for every production instance thus establishing a one-to-one mapping between them.

To find a balance between fair comparison through one-to-one mapping and overhead reduction, we create a coupling between the rate of shadow version instantiation with the performance degradation. When the observed performance degradation is below a user defined threshold, we instantiate the shadow version for every instance of the production version. We reduce the instantiation rate when the degradation is above the given threshold, and increase the instantiation rate when the degradation is below the threshold.

One can consider the shadow process instance executions as experiments. Knowing that a certain task is being executed solely for the experiment may prompt the process workers to perform their work differently to influence the outcome. Therefore, another way in which we strive for fairness (R1) is by hiding the experiment. We propose a unified user interface that aggregates scheduled tasks from both versions and hides the internals of the experimentation platform. The limitation of this design choice is that user interface changes cannot be evaluated in the shadow mode. Such changes could be evaluated through separate AB tests for the user interface changes.

Operations on the test environment should be safe, i.e., tasks in shadow process instances should not override data in the production system and external services. The design choice of avoiding double scheduling puts isolation (R3) at odds with fairness (R1) and overhead management (R2). Consider the example of using a global budget stored in the application database, where a new budget drawing task in the shadow version is followed by a common task. In this example, the common task and every other task that follows afterwards have to be executed so that the effect of this new budget drawing task is reflected on the execution of the instance of the shadow version. This is only possible if we execute the common tasks, which introduces the problem of double execution.

We devise a solution that lets the shadow version independently interact with a separate test database. We can manage test database in three ways: by synchronizing with production database, by keeping it un-synchronized, or by replacing it with snapshots of production during the tests. If we synchronize or overwrite with snapshots, we cannot see the impact of shadow version at the data level. If we don't synchronize, we cannot see the true impact because common tasks are not executed. Therefore, we chose the option of not synchronizing.

4   Conceptual Solution and Architecture

In this section, we present the conceptual solution by first describing the modular architecture we adopted. Then we discuss how the modules operate internally and how they interact.

4.1   Architecture

We propose a layered architecture which uses a single process engine to execute the production and shadow versions. The architecture is divided into four layers: the User Interface, the BPMS Layer, the Service Layer, and the Persistence Layer. Figure 3 shows the modular architecture diagram. The components shown in colour facilitate shadow testing.

The BPMS Layer is responsible for instantiating, executing, and estimating process instances. The shadow version and the production version are deployed alongside each other. The Process Engine enacts process instances and manages their state. The Execution Controller determines when the shadow version can be instantiated, and how tasks can be executed. User tasks are allocated to process workers. Automatic tasks are delegated to the Service Layer. If a task in the shadow version can be estimated, the Execution Controller delegates the task to the Estimator instead. The Estimator mimics the execution of the task and estimates its performance.

The Service Layer is responsible for executing automatic tasks. These tasks are implemented as services. Common Services and New Services are implementations of common and new tasks respectively. External Services represent third party services. Test Doubles are services that replace other services for testing purposes. These services are exposed to the BPMS Layer through a Service API.

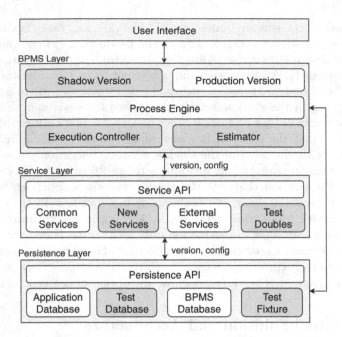

Fig. 3. Modular architecture for shadow testing. Components shown in colour facilitate shadow testing. (Color figure online)

The Persistence Layer handles database operations. The Application Database hosts the application data. The BPMS Database hosts the configuration, the process instance mapping, and the execution data for both versions. The Test Database is a copy of the Application Database. The Test Fixture is a database loaded with a known set of data used specifically to make tests repeatable. These databases are exposed through a Persistence API.

A unified user interface hides the presence of two process versions. Process version and configuration information are passed down to the Service Layer and the Persistence Layer during the execution of process instances. Based on this information, these layers decide which services and data operations should be invoked. Test Doubles, Test Database, and Test Fixture are optional. However, either a Test Database or a Test Fixture is required for the execution of shadow versions that contain new automatic tasks.

4.2 Process Instance Execution

The shadow version is instantiated after the completion of an instance of the production version. A one-to-one mapping is created between these two instances so that they can be executed similarly and then compared after the experiment.

We can execute an instance of shadow version by progressively copying over the execution data of its corresponding instance of production version. We capture this data in the form of *snapshots*. A snapshot is composed of the task name, the start and end timestamps of the task, other metrics such as cost, and the values of all process variables at the time of completion of the task. A snapshot is created when a task in a process instance of the production version runs to completion. These snapshots are stored in the BPMS database, and retrieved during execution of corresponding shadow instances for estimation.

Consider the case of tasks reordering shown in Fig. 4 for versions V_1 to V_2. Execution sequences for V_1 begin with $Old = \langle A, B, C \rangle$. Snapshots are created at the completion of tasks A, B, and C. The shadow version begins with $New = \langle B, A, C \rangle$. We can estimate B by copying over snapshot data from Old even though the trace does not conform with V_2. Since the same tasks are in different order, we cannot copy over raw timestamps from the snapshot. Instead, we aggregate the timestamps into duration and waiting time metric and make an estimate of execution times of these tasks.

Since we avoid executing common tasks and copy over aggregated information from snapshots, shadow process instances do not need to be executed concurrently with their corresponding process instances.

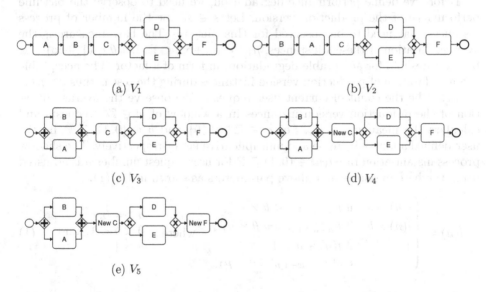

(a) V_1

(b) V_2

(c) V_3

(d) V_4

(e) V_5

Fig. 4. Original process model version V_1 and redesigned versions $V_2 - V_5$.

Adjusting Task Estimates Using Redesign Information. When the shadow version is deployed in production, common tasks may run differently than those in the old production version. These differences cannot be captured by copying over snapshots. Therefore, we allow analysts to adjust estimates available in the snapshots through user defined functions. These user defined functions are executed by the Estimator.

Information on redesigns can be used in making better estimates of duration and waiting times of tasks. Consider some redesigns chosen from known best practices [7,15] illustrated in Fig. 4. In the example of parallelisation, from V_1 to V_4, the redesign may make a process instance faster, but it can consume more resources in the given time-frame. Analysts can make explicit assumptions about how duration and waiting times of the parallelised tasks are affected and provide a function that adjusts the values retrieved from snapshots.

Managing Performance Overhead. In the presence of numerous new tasks, the execution of shadow process instances can slow down the production instances, because those new tasks have to be enacted. We approach this challenge by reducing the rate at which the shadow versions are instantiated. We instantiate shadow versions only when there is evidence that doing so does not degrade the quality of instances in production. Our method to adjust the rate of instantiation was inspired by the congestion control algorithm of TCP/IP [3]: its Additive Increase Multiplicative Decrease (AIMD) algorithm exponentially decreases the instantiation rate when performance degradation is observed, and linearly increases the instantiation rate when performance is acceptable.

Before we define performance degradation, we need to observe the baseline performance of the production version. Let $s \in \mathbb{Z}^+$ be the number of process instances that need to be observed for this purpose. The base measure is the average duration of these s instances, $\mu_{1,s} \in (0, +\infty) \subset \mathbb{R}$. Let $\theta \in (1, +\infty) \subset \mathbb{R}$ be the threshold for acceptable degradation, in form of a factor. The acceptable average duration of production version instances during the test is thus $\theta \times \mu_{1,s}$.

Let n be the count of current user requests. We observe the average duration of the production version instances in a window of $w \in \mathbb{Z}^+$ requests and calculate the mean $\mu_{n-w+1,n}$. Let $a \in \mathbb{Z}^+ \cup \{0\}$ and $b \in (0,1) \subseteq \mathbb{R}$ be the user-defined additive increase and multiplicative factor respectively. The shadow process instantiation rate $i(n) \in (0,1) \subseteq \mathbb{R}$ for user request number n is adjusted using the base measures and above parameters as shown in Eq. (1).

$$
i(n) = \begin{cases} i(n) + a & \text{if } \mu_{n-w+1,n} < \theta \times \mu_{1,s} \\ i(n) \times b & \text{if } \mu_{n-w+1,n} >= \theta \times \mu_{1,s} \\ 1 & \text{if } i(n) + a > 1 \\ & \text{and } \mu_{n-w+1,n} < \theta \times \mu_{1,s} \end{cases} \quad \text{where} \quad \begin{array}{l} a >= 0, \\ 0 < b < 1, \text{ and} \\ \theta > 1 \end{array} \qquad (1)
$$

Task Execution. The layered architecture and configuration information dictate how and where a task is executed. Figure 5 summarizes how these layers decide the way of handling tasks.

Tasks created by a production process instance are executed as per the norm. Automatic tasks are invoked; user tasks are scheduled. Execution and application data are written to the corresponding databases. Tasks created by instances of a shadow version can only be executed if either the Test Fixtures or the Test Database is available. This is essential for providing isolation (R3) because test data can be read from but should not be written to the production database. Test Doubles can be executed in place of automatic tasks if they are available. New user tasks are always scheduled.

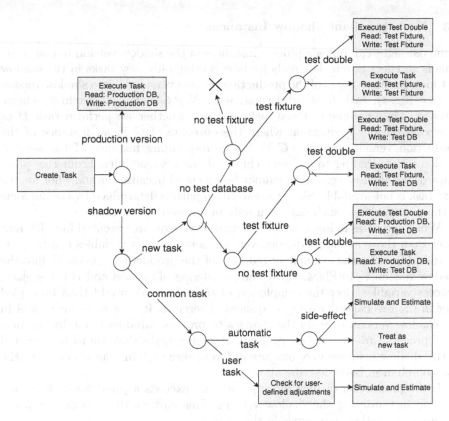

Fig. 5. Decision tree for task execution

Common user tasks can be executed as if they were new tasks, if they do not have a *side-effect* to the business. The tasks have to be marked as such, and it is up-to the analyst to determine what constitutes a side-effect. For example, a task that requires workers to make phone calls to a third party may be said to have a side-effect. Common automatic tasks are treated as new tasks and executed by default. However, these tasks can also be configured to be estimated.

Consider the example of student matriculation process versions shown in Fig. 1 where we execute P_1 in production and P_2 in shadow. Assume that we provide a Test Database and say that user tasks do not produce side-effects. Using the decision tree in Fig. 5, we see that all tasks in P_1 will be executed in the same way as if P_1 was the only version that was deployed. For P_2, only the tasks with new labels (e.g. "Reject Matriculation") will be executed. The implementation of these tasks will write data to the configured Test Database. Tasks that are available in P_1 but not in P_2 (e.g. "Prepare Ex-Matriculation Documents") do not need to be executed in the shadow version.

4.3   Non-compliant Shadow Instances

Some redesigns can impede process instances of the shadow version to complete. This is an effect of our design choice of executing only new tasks in the shadow instance. In the redesign of the production process version V_3 into shadow version V_4 (see Fig. 4), task C is substituted with *New C* before an exclusive-choice gateway. The outcome of these tasks dictates whether we perform task D or E next. Consider a situation where the outcome of C in an instance of the production version and *New C* in the corresponding instance of the shadow version differ, leading to different choice of the next activity. From this point onward, the shadow version V_4 cannot be executed because the snapshot for the next task is not available. Since the execution path is determined by the outcome of the previous task, such cases can only be resolved at runtime.

When activities in instances of the shadow version are executed in a different order than those in the production version, some process variables can be over-written. Consider the example of redesign of the production version V_1 into the shadow version V_2 in Fig. 4, where the re-ordering of tasks A and B takes place. Process variables after the completion of tasks from V_1 would then be copied over in the reverse order using snapshots. Therefore, it is possible for task A in the shadow version to *undo* the changes to process variables made by the task B. If process variables are overwritten, then new tasks that are to be executed in the shadow instance may produce different results. This can also lead to the non-compliance. issue describe above.

If the proportion of such incomplete cases exceeds a pre-defined threshold, we stop instantiating the shadow version. This ensures that we do not waste resource on fruitless executions in the future.

5   Evaluation

In this section, we evaluate our approach using two sets of redesigned process models: the five versions created by using redesign strategies, as shown in Fig. 4, and the two versions of the realistic process models in Fig. 1. Using synthetic data, we first investigate how the instantiation rate of a shadow version changes in response to varying production load. Then we test if shadow tests can accurately estimate the performance of new versions in terms of execution time, by comparing the estimated execution time from shadow testing to the execution time of the same version when it is deployed in production. For this purpose, we use both the synthetic and realistic models (Sects. 5.2 and 5.3 respectively). Finally, we compare simulation and shadow tests qualitatively.

Experiment Setup. We have designed our shadow testing experiments in such a way that there is one (emulated) process worker who can only work on one task at a given time. Upon creation of a user task, the task is placed on a First-In-First-Out (FIFO) task queue. The process worker completes items in the queue. We implemented the shadow testing architecture and emulated the process worker on the Java Virtual Machine with Activiti BPMS and PostgreSQL database. We executed the experiments on a Ubuntu 16.04 machine with 8 GB RAM and Intel Core i7-6700 CPU at 3.4 GHz.

For the synthetic process models, we execute the new version in shadow mode with V_1 in production and estimate the execution time of the instances of shadow version. We assume that $V_2, \ldots, V_5$ are the new versions designed to replace V_1. Table 1 shows the execution and waiting times for each task in the five versions. For the realistic process models, we execute the as-is version P_1 in production, estimate the execution time of instances of the to-be version P_2, and then compare the results with the execution time of P_2 as deployed as production. The shadow tests are performed without resorting on user defined functions for waiting-time and execution-time adjustment.

Table 1. Waiting and execution times of activities in $V_1 - V_5$

Activity	Min. wait time	Execution time
A	1 s	1 s
B	1 s	2 s
C	3 s	2 s
D	5 s	1 s
E	1 s	5 s
F	2 s	1 s
New C	1 s	3 s
New F	1 s	2 s

5.1   Adaptation to Production Load

In this section, we describe how we observed the impact of the shadow process instances on the performance of the execution environment running process instances in production, thus assessing the effectiveness of the overhead management algorithm. In particular we execute V_1 in production and V_5 in shadow mode with one process worker operating with both versions. Since V_5 requires the process worker to complete new tasks, executing a shadow instance slows down the other concurrently active process instances.

We denote with l the number of process instances that are serving a request. We start with the assumption that the workload on production is $l = 1$. We set 5 requests as the bootstrap period during which only the production version is instantiated. The average duration of these 5 instances is taken as the baseline. We set up the overhead management strategy shown in Eq. (1) with arbitrarily chosen parameters $a = 0.1, b = 0.5, w = 5$, and $\theta = 1.1$. If the most recent 5 production instances slow down to 1.1 times the baseline, the shadow version instantiation rate is halved. Otherwise, the rate is increased by 0.1. With this setup, we execute the shadow tests, and intermittently increase the load to test how an increase in the load affects the shadow version instantiation rate $i(n)$.

Figure 6 shows the performance of the production instances under various workloads. It also shows the number of shadow processes instantiated at diverse workloads, and how instantiation rate is adjusted in response to performance degradation. We observe that the shadow version is instantiated less often when the load is high. This is reflected in how the cumulative number of shadow instances change over time. For instance, when the workload increases to $l = 2$ after 10 requests, the execution time of instances surpasses the threshold. As a response to this degradation, the instantiation rate drops by half to 0.5. As a result, the shadow version is instantiated probabilistically and the cumulative shadow instances does not increase linearly as before. We choose the average execution time in window $w = 5$ for detecting overhead. Therefore, the adjustment in instantiation rate lags the first observation of overhead.

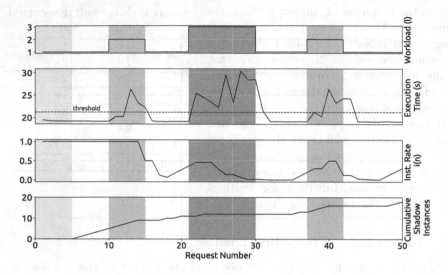

Fig. 6. Comparison of performance of production version V_1 and instantiation rate of shadow version V_5.

5.2   Accuracy of Estimated Execution Time – Synthetic Processes

In this section, we describe our observation on the accuracy of estimates of execution time from shadow tests. Figure 7 shows the differences between the performance of V_1 in production and the respective shadow version. We observe the average performance of for 50 instances of each version. In the case of V_1 vs V_2, the result of the shadow test is accurate because the waiting time and the execution time of the re-ordered tasks are the same. In the case of V_1 vs. V_3, V_3 is faster than estimated because the waiting time between the parallelised tasks is reduced. Since all tasks can be estimated, the waiting time and the execution time are copied over using snapshots. We do not use user defined functions for adjustment of waiting times. This effect of reduction of waiting time can also be seen in V_4 and V_5.

V_4 and V_5 replace task C with *New C*. We have implemented these tasks in such a way that task C determines that the branches through D and E are traversed equally at random, whereas *New C* determines D to be enacted in 60% of the cases and E in 40%. Because of these implementation differences, the corresponding instances between production and shadow versions do not always take the same path. Figure 7 shows the estimates of the shadow instances

Table 2. Waiting time execution time of activities in P_1 and P_2

Activity	Min. wait time	Execution time
All automated tasks	0	1 min
Check Payments Received	3 days	2 min
Check Incoming Payments	3 days	2 min
Prepare Matriculation Documents	$\mathcal{N}(\mu = 1\ day, \sigma^2 = 1^2)$	$\mathcal{N}(\mu = 60\ \text{min}, \sigma^2 = 5^2)$
Prepare Ex-Matriculation Documents	$\mathcal{N}(\mu = 1\ day, \sigma^2 = 1^2)$	$\mathcal{N}(\mu = 30\ \text{min}, \sigma^2 = 2^2)$

that *matched* the same path as their corresponding production instances. It also shows how much of the estimate was comprised of the actual execution and how much was based on the estimates from snapshots.

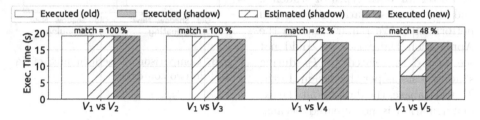

Fig. 7. Performance of old version V_1 in production, and new versions $V_2 - V_5$ in production and shadow mode

5.3   Accuracy of Estimated Execution Time – Realistic Processes

We perform shadow tests on a student matriculation process from a German university. This process requires eligible students to confirm their enrolment by paying tuition fees. In this experiment, we aim to mitigate unintended bias from using simpler synthetic versions, and to mimic realistic scenario by adding more complexity. We add more complexity by treating processing times as distributions and introducing request context.

We have adopted the *as-is* and the *to-be* versions of these processes, depicted in Fig. 1, based on a case study reported in the business process improvement literature [8]. The goal of this redesign was to reduce the cycle time. In this experiment, we use shadow testing to check if this goal can be achieved.

Since the execution details are not available, we implement the tasks such that they follow the waiting and execution times shown in Table 2. We assume that there are two types of requests: those from graduating students and those from students who wish to continue their studies. For the purpose of illustration, we arbitrarily set the success rate of receiving payments from graduating students to 75%, and from continuing students to 80%. We assume that each request includes a student id. Requests raised during the process execution also include the process generated id sent to the health insurance for report, the matriculation document, the ex-matriculation document, and the resulting matriculation status. In our implementation, we generate unique identifiers and random text to represent this information. We also use a process variable to test whether payment was received. With this implementation, we can compare the performance and resulting data from the production environment with those in the shadow mode.

Figure 8 shows the results upon the execution of 250 process instances of P_1 in production (old), P_2 in shadow mode during the shadow test (shadow), and P_2 in production as a stand-alone version (new). It can be observed that P_2 is estimated to be better for both types of students, and that this estimation reflects real performance. We remark that in the Application Database, we could not see any data related to P_2 during shadow tests as per the design. In the Test Database, the data about the ex-matriculation is not created because this activity is not available in P_2.

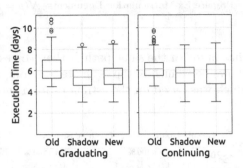

Fig. 8. Comparison of performance of P_1 and P_2 in two contexts

5.4   Qualitative Comparison with Simulation – Realistic Processes

In this scenario, the simulation approach from AB-BPM [20] would fail to produce accurate results. To simulate P_2, trace simulation techniques requires the following inputs from the analyst: (i) historical event logs of P_1, (ii) the default value for the process variable that indicates whether payment was received in P_2, and (iii) estimates for duration of new activities, e.g., "Reject Matriculation". In shadow testing we do not require these inputs because the implementation of $P2$ is available. In trace simulation, we cannot rely on the event log of P_1 after the completion of the "Check Incoming Payments" activity in P_2. Therefore, the simulation uses a predefined user-provided value dictating whether the payment was received or not, to choose the next activity after the exclusive-or gateway.

The trace simulation approach of AB-BPM is unaware of request contexts and implementation details. For every context (graduating or continuing students), a separate simulation using filtered logs should be run. During the simulation, potential implementation bugs regarding data storage cannot be discovered because the new version is not actually executed. For instance, if the implementation of the "Create Student File" task in P_2 depends on an ex-matriculation document created for failed payments, the shadow instances would raise an exception (notice that the dedicated activity is missing in P_2).

6   Discussion

We instantiate the shadow version only when the load is manageable. By design, this approach cannot be used to evaluate shadow versions under high load. If new versions are designed to handle high load, AB testing can be used instead. Our approach instantiates a shadow version after the completion of an instance in the production. This simplifies the execution because we do not have to synchronize the two instances. For fairness, we have to ensure that the corresponding instances execute under similar external conditions.

In this paper, our focus was on architecture design and test execution. We assumed a simplistic approach of resolving data dependencies. Data consistency for general cases was out of scope for this work, and requires further research. In our evaluation, we adopted process models from the redesign literature and implemented them. We cannot use real world log data because of their complexity and lack of implemented code behind the activities. To mitigate the unintended bias from using synthetic redesigns, we also evaluated our approach using a redesign from a case study. Since we propose a new architecture and execution mechanism, we could not find industry implementations for evaluation.

7   Conclusion

Business process improvement ideas do not necessarily yield actual improvements. Simulation and AB testing techniques can be utilized for validating process improvements. However, the results of simulation are speculative and AB tests introduce risk of exposure. In this paper, we propose a shadow testing approach that provides a middle ground between these techniques.

A new version of a process is deployed alongside the current version in such a way that the new version is hidden from the customers and the process workers. User requests that instantiate the current version are also forwarded to the new version. If the performance overhead falls under a user defined threshold, the new version is instantiated. Process instances of the new version are partially executed and partially estimated by copying over *snapshots* from executing the current version. This way of shadow testing facilitates fair comparison between the two versions, isolates the new version, and limits the overhead of running tests. Using synthetic and realistic process redesigns, we demonstrate that shadow testing provides accurate performance estimates of new versions.

In future work, we plan to include shadow testing as a step in the AB-BPM methodology, and run case studies where we compare the costs and benefits of the methodology as a whole.

Acknowledgements. The work of Claudio Di Ciccio and Jan Mendling has received funding from the EU H2020 programme under MSCA-RISE agreement 645751 (RISE_BPM).

References

1. van der Aalst, W.M.P.: Business process simulation survival guide. In: vom Brocke, J., Rosemann, M. (eds.) Handbook on Business Process Management 1. IHIS, pp. 337–370. Springer, Heidelberg (2015). https://doi.org/10.1007/978-3-642-45100-3_15
2. Bass, L., Weber, I., Zhu, L.: DevOps - A Software Architect's Perspective. SEI Series in Software Engineering. Addison-Wesley, Boston (2015)
3. Chiu, D., Jain, R.: Analysis of the increase and decrease algorithms for congestion avoidance in computer networks. Comput. Netw. **17**, 1–14 (1989)
4. Crook, T., Frasca, B., Kohavi, R., Longbotham, R.: Seven pitfalls to avoid when running controlled experiments on the web. In: KDD, pp. 1105–1114 (2009)
5. Davenport, T.H.: Process Innovation: Reengineering Work Through Information Technology. Harvard Business Press, Boston (1993)
6. Denery, D.G., Erzberger, H.: The center-TRACON automation system: simulation and field testing (1995)
7. Dumas, M., La Rosa, M., Mendling, J., Reijers, H.A.: Fundamentals of Business Process Management, 2nd edn. Springer, Heidelberg (2018). https://doi.org/10.1007/978-3-662-56509-4
8. Falk, T., Griesberger, P., Leist, S.: Patterns as an artifact for business process improvement - insights from a case study. In: vom Brocke, J., Hekkala, R., Ram, S., Rossi, M. (eds.) DESRIST 2013. LNCS, vol. 7939, pp. 88–104. Springer, Heidelberg (2013). https://doi.org/10.1007/978-3-642-38827-9_7
9. Feitelson, D.G., Frachtenberg, E., Beck, K.L.: Development and deployment at Facebook. IEEE Internet Comput. **17**(4), 8–17 (2013)
10. Hammer, M., Champy, J.: Reengineering the Corporation: A Manifesto for Business Revolution. HarperCollins, New York (1993)
11. Holland, C.W.: Breakthrough Business Results With MVT: A Fast, Cost-Free "Secret Weapon" for Boosting Sales, Cutting Expenses, and Improving Any Business Process. Wiley, Hoboken (2005)
12. Kettinger, W.J., Teng, J.T.C., Guha, S.: Business process change: a study of methodologies, techniques, and tools. MIS Q. **21**(1), 55–98 (1997)
13. Kevic, K., Murphy, B., Williams, L.A., Beckmann, J.: Characterizing experimentation in continuous deployment: a case study on bing. In: ICSE-SEIP, pp. 123–132. IEEE Press (2017)
14. Kohavi, R., Longbotham, R., Sommerfield, D., Henne, R.M.: Controlled experiments on the web: survey and practical guide. Data Min. Knowl. Discov. **18**(1), 140–181 (2009)
15. Reijers, H.A., Mansar, S.L.: Best practices in business process redesign: an overview and qualitative evaluation of successful redesign heuristics. Omega **33**(4), 283–306 (2005)
16. Rogge-Solti, A., Weske, M.: Prediction of business process durations using non-Markovian stochastic Petri nets. Inf. Syst. **54**, 1–14 (2015)
17. Rozinat, A., Wynn, M.T., van der Aalst, W.M.P., ter Hofstede, A.H.M., Fidge, C.J.: Workflow simulation for operational decision support. Data Knowl. Eng. **68**(9), 834–850 (2009)
18. Satyal, S., Weber, I., Paik, H., Di Ciccio, C., Mendling, J.: AB-BPM: performance-driven instance routing for business process improvement. In: Carmona, J., Engels, G., Kumar, A. (eds.) BPM 2017. LNCS, vol. 10445, pp. 113–129. Springer, Cham (2017). https://doi.org/10.1007/978-3-319-65000-5_7

19. Satyal, S., Weber, I., Paik, H., Di Ciccio, C., Mendling, J.: AB testing for process versions with contextual multi-armed bandit algorithms. In: Krogstie, J., Reijers, H.A. (eds.) CAiSE 2018. LNCS, vol. 10816, pp. 19–34. Springer, Cham (2018). https://doi.org/10.1007/978-3-319-91563-0_2
20. Satyal, S., Weber, I., Paik, H., Di Ciccio, C., Mendling, J.: Business process improvement with the AB-BPM methodology. Inf. Syst. (2018)
21. Schermann, G., Cito, J., Leitner, P., Zdun, U., Gall, H.C.: We're doing it live: a multi-method empirical study on continuous experimentation. Inf. Softw. Technol. (2018)

A DevOps Implementation Framework for Large Agile-Based Financial Organizations

Anitha Devi Nagarajan and Sietse J. Overbeek[(✉)]

Department of Information and Computing Sciences, Utrecht University,
Utrecht, The Netherlands
{A.D.Nagarajan2,S.J.Overbeek}@uu.nl

Abstract. Modern large-scale financial organizations show an interest in embracing a DevOps way of working in addition to Agile adoption. Implementing DevOps next to Agile enhances certain Agile practices while extending other practices. Although there are quite some DevOps maturity models available in the literature, they are either not specific to large-scale financial organizations or do not include the Agile aspects within the desired scope. This study has been performed to identify why such organizations are interested in implementing DevOps and how this implementation can be guided by a conceptual framework. As a result, a list of drivers, a generic DevOps implementation framework and driver-dependent variations are presented. The development of these artifacts has been realized through a design science research method and they have been validated by practitioners from financial organizations in the Netherlands. The practitioners have identified the developed artifacts as useful, mainly to educate people within their organizations. Moreover, the artifacts have been applied to real organizational goals to demonstrate how they can be of help to identify the useful measurement units, which in turn can help to measure and achieve their DevOps transformation goals. Thus, the developed artifacts are not only serving as a baseline for future research but are also useful for existing financial organizations to commence and get ahead with their DevOps implementations.

Keywords: Agile · DevOps implementation framework
DevOps drivers · Large financial organizations
Transformation measurement

1 Introduction

Agile methodologies have gained a widespread acceptance due to advantages like faster software development with improved quality, and the ability to welcome changes throughout the project leading to improved customer satisfaction in comparison with traditional software development approaches such as the waterfall method [2] and the incremental method [34]. However, the structural

© Springer Nature Switzerland AG 2018
H. Panetto et al. (Eds.): OTM 2018 Conferences, LNCS 11229, pp. 172–188, 2018.
https://doi.org/10.1007/978-3-030-02610-3_10

division between the functional departments such as development and operations remained, and it leads to delays in the deployment of the developed software in the actual production environment [9, 35]. As a solution to the latter, the DevOps movement has emerged with the purpose of closing the gap between development and operations [38, 40].

Although DevOps has emerged from cloud-based product organizations, the DevOps paradigm is not exclusive to organizations that are surrounded with cloud computing [38, 40]. Due to the DevOps adoption benefits reported by several other organizations such as better quality assurance and enhanced collaboration and communication [31], large financial organizations are also willing to embrace this new way of working. However, the objectives behind a DevOps implementation may differ considerably among different types of organizations and so does the corresponding means to measure their success [10].

DevOps is often coalesced with Agile principles and practices. Those studies that compared DevOps with other software development methodologies have identified that both Agile and DevOps have similar goals and values, but their scope varies. When DevOps is 'laid over' an Agile implementation, it enhances several Agile practices while extending others outside development activities [19, 20, 23]. There are several studies that list the factors required for a successful Agile transformation from various perspectives [4, 5]. The existing DevOps implementation models have focused only on DevOps and so little attention has been paid on its relationship with Agile principles and practices [24].

Motivated by these concerns and encouraged by the current needs of participating organizations, it was intended to develop a conceptual framework, which depicts the various aspects involved in the DevOps implementation of Agile-based financial organizations. We wanted to connect it to their drivers and the measurement units to make it more complete since the progress towards the DevOps transformation goals are measured by measurement units, which in turn help to get closer to the goals. This paper is organized as follows: Sect. 2 describes the research approach and the involved methods and techniques. Section 3 elaborates the developed artifacts and subsequently Sect. 4 discusses the evaluation of those artifacts. Finally, Sects. 5 and 6 reviews and concludes the study respectively.

2   Research Approach: The Design Cycle

This study design is inspired from the design science methodology by Wieringa [39] and so a brief overview of this study's approach is shown in Fig. 1. Each of the phases mentioned in the figure is explained further.

Problem Investigation and Data Collection. First, the research objectives and research questions were formulated by understanding the interests and needs of the involved stakeholders, one of the participating financial organizations in the Netherlands and also by identifying a gap in the available literature in this context. Next to that, the literature study was carried out to review other relevant scientific studies to lay a stable theoretical foundation and to gather data

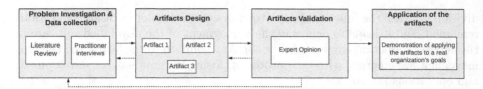

Fig. 1. Research approach inspired by a design science framework [39]

from them. Finally, semi-structured interviews were conducted with practitioners from different organizations in order to collect industry-specific data. Furthermore, the requirements about the prospective artifacts were also collected from the stakeholders.

Artifacts Design. In the second phase, the following results were established: (1) Drivers for Agile and DevOps implementation at the large financial organizations, (2) Generic DevOps implementation framework, and (3) Driver-dependent framework variations based on the relationship identified between the developed framework and the identified drivers. These results were realized by performing the directed content analysis on the data collected from the literature and practitioner interviews. In order to perform the content analysis, we used the tool named Nvivo. The resulted artifacts from this phase are elaborated in Sect. 3.

Artifacts Validation. After identifying the mentioned results and documenting them as artifacts, they were iteratively validated with industrial experts by means of the expert opinion method [39]. The validation session results are detailed in Sect. 5. Based on the validation session outcome, the artifacts were refined and their updated versions were used for further validation sessions. These refinements included a missing driver and an incomplete focus area.

Application of the Artifacts. It is demonstrated how the developed artifacts can be applied to an organization's goals to identify the possible and useful measurement units, which in turn can help to measure their progress towards their DevOps implementation goals. For this, we have considered certain goals of a large Dutch bank that participated in the research and performed the demonstration. The purpose of this demonstration is to show how the developed framework can uncover the various possible obstacles towards the organizational goals and how to select the suitable measurement units based on that.

3   Solution Design : Development of the DevOps Implementation Framework

This chapter is presenting the results of the data analysis performed on the data collected from both practitioners and the available literature. Based on the results, the basic components namely, drivers, perspectives and focus areas are identified. Later by grouping the discerned perspectives and focus areas, a DevOps implementation framework is developed. Thereafter the relationship between the developed framework and the identified drivers are revealed.

3.1 Agile and DevOps Implementation Drivers at Financial Organizations

The list of drivers provide the reasons why large financial organizations are interested in implementing Agile and DevOps. There are six such drivers identified namely, (1) agility and customer-centricity, (2) efficient value delivery to customers, (3) cooperative culture, (4) empowered people, (5) focus on continuous improvement and, (6) process and stakeholder alignment as shown in Fig. 2. These drivers have been identified based on the data collected from the practitioners and their importance has been legitimized by cross-checking with the literature.

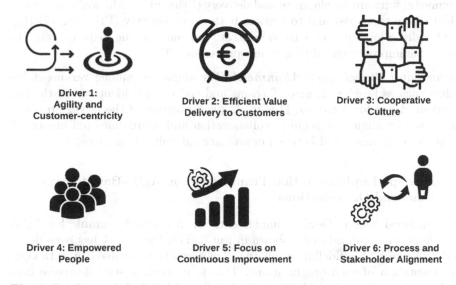

**Driver 1:
Agility and
Customer-centricity**

**Driver 2: Efficient Value
Delivery to Customers**

**Driver 3: Cooperative
Culture**

**Driver 4: Empowered
People**

**Driver 5: Focus on
Continuous Improvement**

**Driver 6: Process and
Stakeholder Alignment**

Fig. 2. DevOps and Agile implementation drivers for large financial organizations

Agility and Customer-Centricity. Agility is an ability of organizations to respond faster to changes as such from customer and market [36]. Customer-centricity is the ability of the organizations to develop systems according to customer preferences and needs [21]. The need for these abilities are found to be the main drivers for large financial organizations to embrace DevOps and Agile.

Efficient Value Delivery to Customers. Agile has proven to be increasing the speed of the upstream processes of software development such as, identifying business needs and developing software accordingly. On the other hand, the speed of downstream processes like verification, validation and delivery of the software can be improved with the support of DevOps [15].

Cooperative Culture. Thanks to the Agile software development process, the wall between the customers and development team was brought down as a consequence of the frequent communication possibilities and smaller iterations [27].

DevOps has enhanced this scenario further by allowing all possible roles within system development, operations and maintenance to work closely with each other, which has largely enriched the communication among the involved IT stakeholders [20,31].

Empowered People. Organizations want their people to be more empowered by taking ownership on their tasks and be capable of doing more than what is described in their job descriptions. People are expected to focus on achieving the group goal instead of focusing only on their individual achievements.

Focus on Continuous Improvement. Traditionally, organizations were interested mostly in the improvement of their delivery. However, Agile promoted the incremental software development and delivery (Principle 3) which allowed teams to learn from their past and to become better progressively (Principle 12) [13]. DevOps drives organizations to concentrate not only on the delivery part but also on the improvement of people and processes [19].

Process and Stakeholder Alignment. The alignment among various stakeholders such as business teams, IT teams and end users is identified as the last important driver for Agile and DevOps implementations of the target organizations. Such an alignment requires collaboration and communication among IT teams and customers, and between organizational units themselves.

3.2 DevOps Implementation Framework for Agile-Based Large Financial Organizations

The developed generic DevOps implementation framework suitable for Agile-based financial organizations is shown in Fig. 3. This framework has been developed to serve as a guideline to all types of employees involved in a DevOps implementation of such organizations. This framework is suitable to be used for those that have already implemented Agile or that are interested in implementing Agile along with a DevOps implementation. The framework has been developed with two levels of constructs namely, perspectives and focus areas.

Perspectives are the dimensions that the corresponding implementation factors belong to and there are four of them namely, (1) organizational perspective, (2) people perspective, (3) process perspective and (4) technology perspective. These perspectives are identified from the agile software development literature and they are maintained here since the interviewees agreed that these perspectives are appropriate to this context.

The focus areas are the principal sections that require attention within every perspective regarding the Agile and DevOps adoption. These focus areas are patterns identified mainly from the interviews and they are further explained with the corresponding literature studies. Every focus area is related to one of the given perspectives.

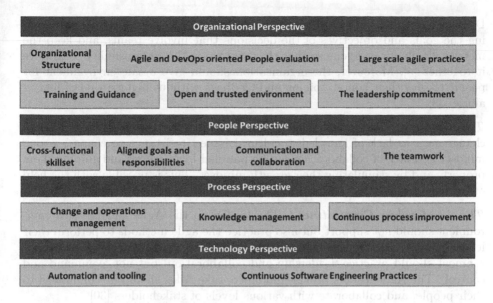

Fig. 3. DevOps implementation framework for Agile-based large financial organizations

Organizational Perspective. The organizational perspective includes all the focus areas that the management of a typical non-DevOps organization should consider and facilitate in order to constitute the landscape within their organization to foster a DevOps mentality. The following focus areas are identified as the significant ones within this perspective.

(Sub-)Organizational Structure. The first focus area is about the structure of the organization and the sub-organizations. DevOps leads to teams that bring together experts such as software development professionals and operations professionals enabling them to share their skills and experiences [19]. The team structure should allow for live and peer-to-peer communication within the team but not via other means such as through management or tickets [32,33].

Agile and DevOps Oriented People Evaluation. The next one is regarding the people evaluation and performance reviews that are commonly conducted within an organization. It is imperative for the organization to make sure that the method of evaluation is team-based, encourages collaboration over competition among team members and teams, and is not conflicting with the behavioral needs of Agile and DevOps [6].

Large-Scale Agile Practices. This focus area emphasizes on tailoring the agile practices specific to the organization and following them throughout the organization, not only at the team levels but also at the project and portfolio levels across the enterprise [17].

Open and Trusted Environment. Having an open and transparent environment is an important characteristic not only for the teams but for the entire

DevOps enterprise. Therefore, the management should be clearly communicating the goals and objectives of the decisions that involve teams, and keep the metrics visible for everyone so that they can share the responsibility to achieve it together [6]. Moreover, the organization should give a safe environment for people to give their honest opinion and feedback without being afraid of fear or abuse [33].

Training and Guidance. The human impediments towards organizational change such as lack of knowledge, cultural issues, resistance to change, wrong mindset and lack of collaboration can be handled by coaching and guiding them properly and by stimulating their growth mindset. This is possible with the help of training and human facilitators such as coaches and champions [14,25].

The Leadership Commitment. The leaders in the Agile and DevOps environment should not support but also practice the agile methods to perform their leadership activities wherever applicable. Moreover, the managers in such environment should practice 'leadership and collaboration' but not 'command and control'. The Agile leaders provide guidance, take risks, should be committed to their people, and collaborate with various levels of stakeholders [30].

People Perspective. This perspective identifies the most important people characteristics required for the effective working at Agile and DevOps based organizational environments.

Cross-Functional Skillset. The cross-functional teams are an important component of DevOps environment which in general is formed initially by involving experts from various functional domains such as programmers, functional testers, performance testers and operations personnel. Ideally, this can lead to a situation in which these experts communicate and collaborate to become cross-functional team members who are multi-skilled and flexible [1,7,19].

Aligned Goals and Responsibilities. Hutterman defines a team as a group of people working together to achieve a shared group goal [19]. Within a DevOps organization, the team goals should not conflict with each other but focus on achieving a common goal that is beneficial to an user group. Based on our understanding of the collected data, we say that the sub-organizational goal should be aligned to the main goals of the enterprise and in the same way, an (agile) team's goals should be in line with the corresponding sub-organizational goals. Thus the people goals and their responsibilities should be driven by the shared team goals which are associated with them. Figure 4 depicts this.

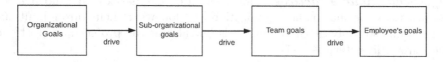

Fig. 4. Emphasizing aligned goals within DevOps environment

Communication and Collaboration. The next focus area accentuates the need for effective communication and intense collaboration among the team members, IT management and business. With the act of communication, people do exchange knowledge, influence each other, recognize each other's work and build a community. By working collaboratively within the community, people build trust and empathy for each other [6].

The Teamwork. The next focus area draws attention to the teamwork aspect of people working in Agile and DevOps organizations. Teamwork boosts not only the performance of the team but also the individual performance and together it contributes to the overall performance of the organization [22].

Process Perspective. This perspective includes the important process areas that the agile organizations need to consider within the context of DevOps.

Change and Operations Management. This focus area insists on developing a change and operations management plan and integrating it with the project management method. DevOps practices intend to reduce the time between the code commits of a change in the development system and placing the change in the production system [31]. Involving the operations group in the Change Advisory Board and by coordinating with the operations maintenance personnel will help to make sure current operations will not be negatively impacted [26].

Knowledge Management. Thanks to Agile and also DevOps, the functional groups of people are disseminated and restructured into cross-functional DevOps teams, which are formed around value streams. This brings in a clear need for effective knowledge management processes and activities so that continuous learning and coordination can happen within the enterprise. The organization and the people need to identify suitable knowledge management processes that work for them and support them with relevant tools and infrastructure. People should be aware of the advantages and importance of knowledge management practices and so are encouraged to share knowledge with each other.

Continuous Process Improvement. The Agile and DevOps adoption by large complex organizations require experimentation and adaptation of the methods and processes to the organization's structure, culture, product/service strategy, human resource management policies, customer interfaces, project roles and governance structures, including program and project portfolio management [17].

Technology Perspective. This perspective identifies and describes the focus areas which require attention from the technological standpoint within the DevOps implementation.

Automation and Tooling. According to several studies, automation is found to be the technological enabler of DevOps [19,31,40]. Being aware of both the benefits and possible pitfalls, organizations should perform effective automation so that it can be an advantage but not an impediment.

Continuous Software Engineering Practices. Continuous Software Engineering practices help to eliminate waste in the context of lean software development. Some examples of waste are, (a) delays, due to lack of communication and understanding; (b) unnecessary additional work that does not yield expected business value; (c) defects due to poor execution of tasks; (d) partial completion of work. These wastes can possibly be removed or reduced with the implementation of continuous software engineering practices such as continuous integration, continuous testing, continuous monitoring, continuous delivery and other such practices [12]. Because of their contribution to the faster value delivery, they are here involved in the context of DevOps implementation.

3.3   Relationship Matrix Between Drivers and Focus Areas of the Developed Framework

The final research outcome is developed to understand which among the fifteen focus areas should be first aimed at, based on what has driven an organization to go with a DevOps implementation. Based on these relationships, we have developed the variations of the presented framework for every driver, which highlights the related focus areas from Table 1. However, they are not shown here due to the space restrictions. Also the rationale behind the given relationships are not described here for the same reason but, those justifications are either based on literature or from the collected interview data, or both. For example, the second focus area 'Agile and DevOps oriented people evaluation' can influence the cooperative work culture (driver 3) if the reward structure is utilized appropriately [3]; the empowered people (driver 4), since people's self-development is encouraged and self-confidence is improved when the feedback is constructive [19,29]; the focus on continuous improvement (driver 5), because rewarding the whole team can encourage them to achieve more together [37] and giving them regular feedback help them to refine themselves progressively [19]. However, this focus area's direct influence on other drivers were not found from the collected data sources.

From Table 1, we infer that the focus areas namely, the leadership commitment from organizational perspective, aligned goals and responsibilities, team work from the people perspective and the continuous software engineering practices from the technology perspective are significant for all identified DevOps drivers. In addition to that, from Table 1 it can be deduced that the people perspective is the most contributing one in the case of a DevOps transformation as this is influencing most of the drivers and the DevOps implementation goals.

Table 1. Relationship matrix between the identified drivers and the focus areas

Focus Areas Drivers	Driver 1: Agility and customer-centricity	Driver 2: Efficient value delivery to customers	Driver 3: Cooperative culture	Driver 4: Empowered people	Driver 5: Focus on continuous improvement	Driver 6: Process and stakeholder alignment
Focus area 1: (Sub-)organizational structure	X	X	X	X		X
Focus area 2: Agile and DevOps oriented people evaluation			X	X	X	
Focus area 3: Large-scale agile practices	X	X	X			X
Focus area 4: Open and trusted environment			X	X		
Focus area 5: Training and guidance			X	X	X	
Focus area 6: The leadership commitment	X	X	X	X	X	X
Focus area 7: Cross-functional skillset	X	X	X	X		X
Focus area 8: Aligned goals and responsibilities	X	X	X	X	X	X
Focus area 9: Communication and collaboration	X	X	X		X	X
Focus area 10: Teamwork	X	X	X	X	X	X
Focus area 11: Change and operations management	X	X				X
Focus area 12: Knowledge management		X	X		X	X
Focus area 13: Continuous process improvement			X	X	X	X
Focus area 14: Automation and tooling	X	X			X	X
Focus area 15: Continuous software engineering practics	X	X	X	X	X	X

4 Application of Drivers and Framework to Identify Measurement Units

This section demonstrates how the developed framework and identified drivers can be used to achieve the DevOps transformation goals of an organization with the help of an example. For this, we have considered one of the goals of a Dutch financial organization, which is a multinational banking and financial services company and it is one of the largest banks in the Netherlands. Their goal is about expediting their delivery so that their customers can enjoy their service and products earlier than before. Since this goal is suitable to be analyzed from the perspectives of both people and process, we have chosen to present it here.

As mentioned above, in order to explain the derivation of suitable measurement units, the goal has been analyzed in two ways: (a) from the people and organizational perspective and (b) from the process and technology perspective as shown in Fig. 5. In order to improve the time taken for delivery, it is important to first know how much time it currently takes for any requirement including new feature related requirements and change requests. Thus it is relevant to measure the (1) **time passing between the initiation and the actual delivery of those requirements**. However, it might not really be enough to keep looking at the overall time that is being taken for the life-cycle of a requirement when the time stays indifferent. In that case, we can have a deeper look into the time by checking it in two different ways, which makes the given Fig. 5 to get separate branches into people and process perspectives.

From the people perspective, the requirements can be checked to see (2) **the time period that a requirement stayed with different roles such as tester or operations**. It is useful to link the organizational perspective with the people perspective here. Based on the identified time taken by different roles to handle requirements, the following questions may arise:

1. Why does a specific role keep these requirements longer?
2. What is the average time spent by that role on other requirements?
3. What can be done to reduce the time spent by that role?
4. Is this a common scenario with anyone taking that role or is it something specific to the person who took that role?

The above questions are formulated to get a deeper understanding on the source of the problem (i.e., longer processing time of requirements) based on the identified perspectives. For example, the above questions may help to reveal the existing issues such as communication issues among roles, specific role's inability in taking up other role's tasks, management interference or less commitment of the involved people. One or more of these issues may be identified as the obstacle towards achieving the goal and so they need to be paid attention to.

Similarly, from the process and technology perspective, the requirements can be checked to see (3) **the time periods that a requirement stayed with involved processes such as development or functional testing**. This helps to identify which process takes longer which in turn initiates an analysis, such as:

1. Why that specific process takes longer than others for a requirement?
2. What is the average time spent on that process for other requirements?
3. How can that process be improved to reduce time?
4. Is the improvement required on the identified process or any other dependent process?
5. Is it really the process that needs improvement or the people who are involved in it?

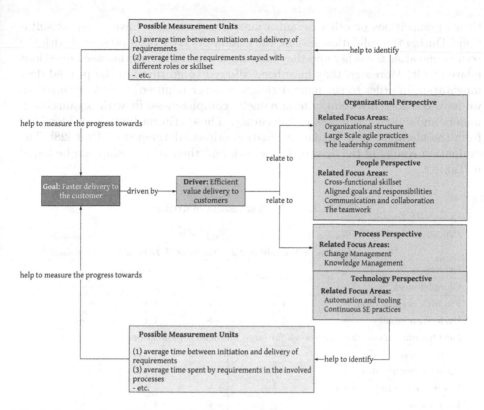

Fig. 5. Application of drivers and the developed framework on an organizational goal

As explained, asking such relevant questions helps to identify where the obstacle is and how that obstacle can be removed. The more important note is that these analyses should always lead to the identification of metrics that provoke the discussion of improvement points in terms of people, process or technology but not blaming each other.

Knowing the relevant focus areas which are related to the corresponding driver helps to ask the relevant questions. Moreover, they can be of help to go on and check the next relevant focus area from different perspectives so that the obstacles indirectly related to the goal can be identified and so the measurement units can be adjusted to measure the right focal point that needs attention.

5   Validation of the Artifacts

The expert opinion sessions were conducted with five experts who have various levels of experience working at financial organizations to evaluate the mentioned artifacts. These experts fulfilled the roles of DevOps engineer, DevOps consultant, DevOps architect and Delivery Manager in their respective organizations. All the experts have been part of one or more DevOps implementations within their organizations or other organizations for which they have given consultations. During the validation session, the evaluators have been presented with the results one after the other and they have been asked the criteria-based questions related to it. Moreover, they have been allowed to go through the printed documentation in order to get more details whenever required. For the evaluation, we have considered several criteria namely, completeness, fit with organization, understandability, usefulness and accuracy. These criteria have been identified from the hierarchy of IS artifact evaluation criteria developed by Prat [28]. The evaluation results of the drivers, framework and their relationship can be found in Table 2.

Table 2. Evaluation results of artifacts

Criterion	Session 1		Session 2		Session 3
	Evaluator 1	Evaluator 2	Evaluator 3	Evaluator 4	Evaluator 5
Drivers					
Completeness	++	++	++	++	++
Fit with organization	+ -	+ -	++	++	++
Understandability	++	++	++	++	++
DevOps implementation framework for large financial organizations					
Completeness	++	++	++	+ -	++
Fit with organization	++	++	++	+ -	++
Understandability	++	++	++	++	++
Usefulness	++	++	++	+ -	++
Relationship drawn between artifacts 1 and 2					
Accuracy	N/A	N/A	++	++	++
Usefulness	+ -	++	++	+ -	++

++ Fully Agreed +- Partially Agreed - - Rejected N/A Not Assessed

According to the validation results, the evaluators agreed that the identified list of drivers is complete and the developed DevOps implementation framework includes all the required perspectives and focus areas. They confirmed that the results are easy to understand and most of them agreed that they are suitable to their organization. The evaluators agreed that the developed framework is suitable for the organizations who are yet to implement DevOps or those who are at the initial and immature stages of DevOps implementation. On the other hand, they mentioned that the framework may not help for those who are already mature with their DevOps implementations.

As it can be noted with the given results in Table 2, none of our results have been completely rejected by the evaluators. It could be because of one of the limitations of the study i.e. both data collection and evaluation was performed by companies based in one country (Netherlands) and the number of participated companies is limited to three.

6 Discussion

The current study has developed a high-level framework that encompasses the various perspectives and the focus areas that are relevant for the successful DevOps implementation of a large-scale organization. To maximize the usefulness of the framework and to make it specific for financial organizations, from where the practitioners are selected to participate in this study, we have also identified the drivers and we have demonstrated the derivation of measurement units based on the DevOps implementation goals.

Implications. In this research, several factors in terms of perspectives and focus areas as part of a DevOps implementation have been introduced. As can be seen in Table 1, the identified focus areas have many-to-many relationships with the collected drivers. Overall, a DevOps implementation is a collective effort of people working at different levels within an organization. For a successful implementation, an organization should discover which areas of the organization need what kind of changes and how to proceed from there. The developed conceptual framework is beneficial to realize such an implementation. In short, an exemplary DevOps organization underlines the need for people development and for process improvement. Moreover, it has a culture in which competition is of less importance compared to the importance of learning.

Comparison with Related Studies. There are several studies which identified the success factors of an Agile implementation from different perspectives [4,5,11]. The current study is also inspired on those studies and followed the list of perspectives taken from those studies. However, this one is different from them since the other mentioned studies concentrated on Agile whereas, the current one has concentrated on a DevOps implementation where Agile is also followed. This study expects the involved organization to already follow Agile or to implement Agile together with a DevOps implementation. Next to that, there are some DevOps maturity models available in the literature [8,24]. This study is different from those studies in the following aspects: the current study focused on large organizations and is specific for the finance industry; the current study has the possibility to be expanded to an 'organization specific maturity model' in which every focus area is defined with the list of capabilities that the involving organization wants to progressively reach. This can be achieved by analyzing the organization's situation and identifying the specific capabilities based on where they are and what areas they want to reach with DevOps. This process of developing an organization-specific maturity model using the developed conceptual framework is comparable to Situational Method Engineering [16]. On the contrary, the other mentioned studies focused on developing a generic maturity model which may not be suited to every organization and also they may not be suitable for tailoring.

Limitations. The current study considered several sources of data which came from both practitioners and other scientific studies. Although the participated practitioners are originally from various geographical areas, they all currently belong to a few financial organizations in the Netherlands. Moreover, the evaluation part is performed by means of expert opinion, which focused on evaluating the artifacts against the given criteria. These evaluation results are not enough to quantitatively prove the usefulness of the developed artifacts in a real DevOps implementation scenario.

7   Conclusions and Future Research

As like with any other industry, DevOps is becoming popular among financial software organizations. Because of the advantages observed with Agile software development methods such as faster development time, improved quality and high customer-satisfaction, several large-scale financial organizations prefer Agile methods over traditional software development methods like waterfall software development method. However, before the start of this study it was still not clear why they are interested in implementing DevOps along with or on-top of an Agile implementation. Thus, this paper concentrated on identifying the drivers for large-scale financial organizations to 'go for' DevOps along with an Agile software development method. Nevertheless, with a DevOps adoption, several existing factors get affected and many other new factors need attention. Thus, the current study brings up a framework based on high-level factors from different perspectives that are required for the DevOps implementation in such organizations and develops the variations of the framework based on its relationship with the identified drivers. To justify the usage of the developed DevOps implementation framework along with identified drivers, an application scenario with a real financial organization's goal has been presented.

This study provides quite some future research opportunities. Since the data for the current study were collected mostly from banks in the Netherlands, the future studies can concentrate on performing similar research by taking other financial institutions into account, such as insurance companies and possibly organizations from various geographical locations. Subsequently, comparing the current study with studies in those different but comparable domains may even bring interesting results. As suggested by one of the evaluators, an useful note for similar research is to develop more specific focus areas that reduce the overlap between the driver-dependent variations of the developed framework. Furthermore, the current study has established the relationship between drivers and the focus areas and it was mostly based on the theoretical data found from the available literature. Every driver - focus area pair can be empirically researched to identify the concrete relationships between them. Since the developed conceptual framework has not been applied to a full fledged DevOps implementation, the framework itself can be revised and improved after being utilized in its entirety.

References

1. Abidin, Z.F.A., Jawawi, D., Ghani, I.: Agile transition model based on human factors. Int. J. Innov. Comput. **7**(1), 23–32 (2017)
2. Bhadoriya, N., Mishra, N., Malviya, A.: Agile software development methods, comparison with traditional methods and implementation in software firm. Int. J. Eng. Res. Technol. **3**(7) (2014)

3. Chatman, J.A., Barsade, S.G.: Personality, organizational culture, and cooperation: evidence from a business simulation. Adm. Sci. Q. **40**, 423–443 (1995)
4. Chow, T., Cao, D.B.: A survey study of critical success factors in agile software projects. J. Syst. Softw. **81**(6), 961–971 (2008)
5. Darwish, N.R., Rizk, N.M. : Multi-dimensional success factors of agile software development projects. Int. J. Comput. Appl. **118**(15) (2015)
6. Davis, J., Daniels, R.: Effective DevOps: building a culture of collaboration, affinity, and tooling at scale. O'Reilly Media Inc., Sebastopol (2016)
7. Demirkan, H., Spohrer, J.: T-shaped innovators: identifying the right talent to support service innovation. Res.-Technol. Manag. **58**(5), 12–15 (2015)
8. de Feijter, R., Overbeek, S., van Vliet, R., Jagroep, E., Brinkkemper, S.: DevOps competences and maturity for software producing organizations. In: Gulden, J., Reinhartz-Berger, I., Schmidt, R., Guerreiro, S., Guédria, W., Bera, P. (eds.) BPMDS/EMMSAD -2018. LNBIP, vol. 318, pp. 244–259. Springer, Cham (2018). https://doi.org/10.1007/978-3-319-91704-7_16
9. de FranSa, B.B.N., Jeronimo Jr., H., Travassos, G.H.: Characterizing DevOps by hearing multiple voices. In: Proceedings of the 30th Brazilian Symposium on Software Engineering, pp. 53–62 (2016)
10. Elberzhager, F., Arif, T., Naab, M., Süß, I., Koban, S.: From agile development to DevOps: going towards faster releases at high quality – experiences from an industrial context. In: Winkler, D., Biffl, S., Bergsmann, J. (eds.) SWQD 2017. LNBIP, vol. 269, pp. 33–44. Springer, Cham (2017). https://doi.org/10.1007/978-3-319-49421-0_3
11. El Hameed, T.A., Latif, M.A.E., Kholief, S.: Identify and classify critical success factor of agile software development methodology using mind map. Int. J. Adv. Comput. Sci. Appl. **7**(5), 85–92 (2016)
12. Farid, A.B., Helmy, Y.M., Bahloul, M.M.: Enhancing lean software development by using DevOps practices. Int. J. Adv. Comput. Sci. Appl. **8**(7), 267–277 (2017)
13. Fowler, M., Highsmith, J.: The agile manifesto. Softw. Dev. **9**(8), 28–35 (2001)
14. Gandomani, T.J., Zulzalil, H., Ghani, A.A.A., Sultan, A.B.M., Nafchi, M.Z.: Obstacles in moving to agile software development methods; at a glance. J. Comput. Sci. **9**(5), 620–625 (2013)
15. Geurts, W.J.: Faster is better and cheaper. In: INCOSE International Symposium, vol. 26, no. 1, pp. 1002–1015 (2016)
16. Harmsen, A.F., Brinkkemper, J.N., Oei, J.H.: Situational method engineering for information system project approaches, pp. 169–194, Department of Computer Science, University of Twente, pp. 1–32 (1994)
17. Hobbs, B., Petit, Y.: Agile methods on large projects in large organizations. Proj. Manag. J. **48**(3), 3–19 (2017)
18. Horney, N., Pasmore, B., O'Shea, T.: Leadership agility: a business imperative for a VUCA world. Hum. Resour. Plan. **33**(4), 32–38 (2010)
19. Huttermann, M.: DevOps for Developers. Apress, New York City (2012)
20. Jabbari, R., bin Ali, N., Petersen, K., Tanveer, B.: What is DevOps?: A systematic mapping study on definitions and practices. In: Proceedings of the Scientific Workshop Proceedings of XP2016, p. 12. ACM (2016)
21. Liang, T.P., Tanniru, M.: Special section: customer-centric information systems. J. Manag. Inf. Syst. **23**(3), 9–15 (2006)
22. Lindsjørn, Y., Sjøberg, D.I., Dingsøyr, T., Bergersen, G.R., Dybå, T.: Teamwork quality and project success in software development: a survey of agile development teams. J. Syst. Softw. **122**, 274–286 (2016)

23. Lwakatare, L.E., Kuvaja, P., Oivo, M.: Relationship of DevOps to agile, lean and continuous deployment. In: Abrahamsson, P., Jedlitschka, A., Nguyen Duc, A., Felderer, M., Amasaki, S., Mikkonen, T. (eds.) PROFES 2016. LNCS, vol. 10027, pp. 399–415. Springer, Cham (2016). https://doi.org/10.1007/978-3-319-49094-6_27
24. Mohamed, S.: DevOps maturity calculator DOMC-value oriented approach. Int. J. Eng. Sci. Res. **2**(2), 25–35 (2016)
25. Parizi, R.M., Gandomani, T.J., Nafchi, M.Z.: Hidden facilitators of agile transition: agile coaches and agile champions. In: 8th Malaysian Software Engineering Conference (MySEC 2014), pp. 246–250. IEEE (2014)
26. Phifer, B.: Next-generation process integration: CMMI and ITIL do devops. Cut. IT J. **24**(8), 28 (2011)
27. Pikkarainen, M., Haikara, J., Salo, O.: The impact of agile practices on communication in software development. Empir. Softw. Eng **13**(3), 303–337 (2008)
28. Prat, N., Comyn-Wattiau, I., Akoka, J.: Artifact evaluation in information systems design-science research-a holistic view. In: PACIS, pp. 23–39 (2014)
29. Quinn, R.E., Spreitzer, G.M.: Seven questions every leader should consider. Organ. Dyn. **26**(2), 37–49 (1997)
30. Rigby, D.K., Sutherland, J., Takeuchi, H.: Embracing agile. Harv. Bus. Rev. **94**(5), 40–50 (2016)
31. Riungu-Kalliosaari, L., Mäkinen, S., Lwakatare, L.E., Tiihonen, J., Männistö, T.: DevOps adoption benefits and challenges in practice: a case study. In: Abrahamsson, P., Jedlitschka, A., Nguyen Duc, A., Felderer, M., Amasaki, S., Mikkonen, T. (eds.) PROFES 2016. LNCS, vol. 10027, pp. 590–597. Springer, Cham (2016). https://doi.org/10.1007/978-3-319-49094-6_44
32. Sharma, S.: The DevOps Adoption Playbook: A Guide to Adopting DevOps in a Multi-Speed IT Enterprise. Wiley, Hoboken (2017)
33. Swartout, P.: Continuous Delivery and DevOps: A Quickstart Guide. Packt Publishing Ltd., Birmingham (2014)
34. Tarhan, A., Yilmaz, S.G.: Systematic analyses and comparison of development performance and product quality of Incremental Process and Agile Process. Inf. Softw. Technol. **56**(5), 477–494 (2014)
35. Tessem, B., Iden, J.: Cooperation between developers and operations in software engineering projects. In: Proceedings of the 2008 International Workshop on Cooperative and Human Aspects of Software Engineering, pp. 105–108 (2008)
36. Tseng, Y.H., Lin, C.T.: Enhancing enterprise agility by deploying agile drivers, capabilities and providers. Inf. Sci. **181**(17), 3693–3708 (2011)
37. Walls, M.: Building a DevOps Culture, 1st edn. O'Reilly Media Inc., Sebastopol (2013)
38. Wettinger, J., Breitenbcher, U., Leymann, F.: Standards-based DevOps automation and integration using TOSCA. In: Proceedings of the 2014 IEEE/ACM 7th International Conference on Utility and Cloud Computing, pp. 59–68 (2014)
39. Wieringa, R.J.: Design Science Methodology for Information Systems and Software Engineering. Springer, Heidelberg (2014). https://doi.org/10.1007/978-3-662-43839-8
40. Zhu, L., Bass, L., Champlin-Scharff, G.: Devops and its practices. IEEE Softw. **33**(3), 32–34 (2016)

Utilizing Twitter Data for Identifying and Resolving Runtime Business Process Disruptions

Alia Ayoub[✉] and Amal Elgammal

Faculty of Computers and Information, Cairo University, Cairo, Egypt
{a.magdy, a.elgammal}@fci-cu.edu.eg

Abstract. The advent of web 2.0 technologies represents a paradigm shift in how individuals collaborate in their businesses and daily lives. Web 2.0 opens new opportunities for businesses to reconsider their strategies and operating models by taking a customer-centric approach, which creates a competitive advantage. Business Process Management (BPM) is taking advantage from this phenomenon (aka social business processes or business processes 2.0), embracing 'social' and embed it through different stages of the BP lifecycle. This paper contributes by a novel framework for the real-time monitoring and improvement of business processes by analyzing the huge amounts of social data, providing visibility and control, which leads to informed decision making and immediate corrective actions. Thus, the proposed framework bridges in the gap between the social and business worlds. The applicability, efficiency and utility of the proposed approach is validated through its application on a real-life case study of a leading telecommunication company.

Keywords: Web 2.0 · Business Process Management (BPM)
Social business processes · Social data · Process improvement
Customer-centric · Human empowerment · Sentiment analysis
Tweets analysis · Clustering · Classification

1 Introduction

In contrast to web 1.0 that was limited to the passive viewing of content to users in a static way, the emergence of web 2.0 technologies allow users to communicate and collaborate [1] through using social media, which comes in many different forms, including blogs, forums, business networks, photo-sharing platforms, social gaming, microblogs, chat apps, and social networks. Social networks such as Facebook, Twitter, Wikis, etc., result in a massive amounts of data, however, data alone does not create competitive advantage. Only when companies analyze and act on data when competitive advantage and potential economic growth can be achieved.

On the other hand, business processes explicitly capture the set of activities participating in the accomplishment of a specific organizational goal, and their control flow [1]. Business Process Management (BPM) is the discipline that combines knowledge from information technology and knowledge from management sciences and applies this to operational business processes to enable their efficient design,

© Springer Nature Switzerland AG 2018
H. Panetto et al. (Eds.): OTM 2018 Conferences, LNCS 11229, pp. 189–206, 2018.
https://doi.org/10.1007/978-3-030-02610-3_11

execution, control, measurement and optimization [1]. Recently BPM has gained much interest from the industrial and academic communities due to the promise it brings for increasing productivity and significant cost reduction. Therefore, business processes form the foundation for all organizations and subsequently business entities are striving for the utilization of information and communication technologies for their continuous improvement.

Social BPM (also known as business processes 2.0) represents a paradigm shift and a gateway to enhanced process efficiency [2]. Organizations taking social BPM initiative have recognized that its processes, supportive tools and technologies is the focal point of this shift, and that a customer-centric approach should be adopted that represents a collaborative effort between process designers and customers to improve the entire process. This creates a closed feedback loop from customers and other stakeholders for continuous BP improvement throughout the various stages of the BP lifecycle. Business process areas that are most prone to improvement include [2]: (i) *Collaborative process improvisation and implementation*: where feedback from social media is continuously collected and used to enhance process designs, as well as aiding the implementation through constructing the interplay between unstructured social data and BP implementation, (ii) *Process discovery and analysis*: this creates a communication loop involving not only process engineers, but customers as a key stakeholder, realizing a customer-centric approach. (iii) *Real-time monitoring*: the massive amount of social data is continuously monitored by tracking key people and events in real-time, which leads to informed decision-making and immediate (semi-automated) corrective actions, and (iv) *Spontaneous Status Updates and Feedback*: this ensures timely and effective process improvisations and enhancements during various BP stages, from design to implementation.

However, the gap between the social and the business worlds is still non-tackled. In essence, the majority of the proposed solutions are taking a marketing or business perspective and lacking a structured approach with concrete IT implementation. The main contribution of this article is a novel runtime monitoring framework that incorporates and integrates the social and business realms, and utilizes social unstructured data for the identification and resolution of BP disruptions/disturbances taking a customer-centric approach. We define a BP disruption (disturbance) as any event that hinders a customer's satisfaction in the delivery and/or operation of a specific service offered by a service provider. BP disruptions are domain-specific that need to be identified and analyzed for the considered domain; for instance, if we consider mobile services offered by a telecommunication company, customers might be complaining (that's BP disruptions) of a network disconnect at a specific time period, slow internet connection, payment error, etc.

To achieve this, the framework entails: a formal approach with associated supportive tools that continuously collect social data; filter and analyze it on the basis of utilizing and integrating data mining and machine learning techniques; relate it to the business realm; identify and detect possible online BP disruptions; automatically propose a recovery plan; and visualize the results in a user-friendly dashboard by accommodating various stakeholders' perspectives. The proposed framework addresses the "real-time monitoring" and "spontaneous status updates and feedback" challenges discussed above. To keep the discussion focused, the paper concerns itself with presenting in detail the peculiarities of the 'analysis' component of the framework.

To validate the applicability, efficiency and utility of the proposed framework, a prototypical implementation has been developed by considering Twitter as the social medium source of data, and applied on a case study of the Customer Relationship Management (CRM) BP of a leading telecommunication company. Our empirical results show that pairing the analysis of the social side with its equivalent business side provides visibility and control throughout the execution phase of the BP/lifecycle, which enables informed decision-making and immediate corrective action(s) taking. These collectively lead to enhanced customer relationship, continuous BP improvements, and ultimately, a competitive advantage creation.

The rest of the paper is organized as follows: Sect. 2 summarizes related work efforts. Section 3 discusses the proposed framework. In Sect. 4, we discuss in detail the analysis approach proposed as a vital component of realizing the proposed framework. Section 5 demonstrates the application of the analysis approach on a real-life case study. This is followed by Sect. 6, which presents the conclusions and highlights future work directions.

2 Related Work

With the growth of blogs, social networks and opinion mining, social data analysis becomes a field of interest for many researches and practitioners. The majority of proposals in the literature consider Twitter as the target social media because Twitter is a widely used social media site for posting comments through short statuses called tweets [3]. Each tweet was of 140 characters and now it is expanded to 280 characters however, it still has a size limit that means it is easier to be analyzed. Moreover, one can keep track of tweets talking about a specific topic through using the hashtag symbol (e.g., #topic). The millions of tweets received every year could be subjected to sentiment analysis and many other types of analytics. However, handling such a huge amount of unstructured data is a tedious task to take up. In a parallel context, there are some studies that consider other social media sites, such as Facebook [4–8].

This article considers Twitter for the previously cited reasons, and therefore the next discussion will focus on summarizing related work efforts in this direction. One of the prominent areas of Twitter analysis is the indication/prediction of the level of satisfaction of the customer with respect to a specific service or product, which is widely known as sentiment analysis [9]. Sentiment analysis is a type of data mining that measures the inclination of people's opinions through Natural Language Processing (NLP), computational linguistics and text analysis, which are used to extract and analyze subjective information from the Web, mostly social media and similar sources. The analyzed data quantifies the public's sentiments or reactions toward certain products, people or ideas and reveal the contextual polarity of the information[1]. It is also called opinion mining.

---

[1] https://www.techopedia.com/definition/29695/sentiment-analysis.

Conversely, few studies exist in the literature that attempt to embrace 'social' and embed it with BPM, which is known as social business processes or business processes 2.0. In the following, prominent work efforts in Twitter analysis and the few attempts towards social business processes are discussed and appraised against the approach presented in this paper. The main contribution of this article as compared to related work efforts is the establishment of a formal framework that bridges in the gap between the social and the BP worlds, embracing 'social' to BPM for disturbances/deficiencies analysis, identification and their proactive correction.

2.1   Twitter Data Analysis

Sentiment Analysis for Market Research using Text Mining. Text mining is a technology that attempts to extract meaningful information from unstructured textual data. The study in [7] describes a case study that applies text mining to analyze unstructured text content on Facebook and Twitter sites of three largest pizza chains. The study revealed that Domino's Pizza got higher level of commitment and consumer engagement than the other two pizza chains through the number of posts and user comments on social media.

Analogously, in [10] the authors proposes a sentiment analysis method based on N-gram classification approach to measure the reputation of a given company by using particularly tweets of Twitter. A given tweet has either negative or positive impact on the company's reputation or product. Similar approaches are also proposed in [12–14]. While the proposals in this category aims at getting insights about the weak points of a specific business by utilizing sentiment analysis, they do not link these insights to the business process realm. This is tackled in our approach in the 'analysis' component/phase of the framework through first clustering the data to identify the classes/clusters of disturbances and any unforeseen/unexpected patterns.

Improving Sentiment Score Results by Using Sentiment Analysis. Another track in the same area of research is analyzing the social side for improving the sentiment score or the sentiment results by using domain ontologies [15–18]. The authors in [14] conducted sentiment analysis based on a domain ontology to produce more accurate results than any other sentiment analysis classification. The domain ontology has been developed using a semi-automated ontology learning technique that deploys text-mining techniques via user-friendly interface that reduces development time and complexity called OntoGen. Ontologies enable the sharing of a common understanding of the domain of interest among people and software tools; enable the reuse and extension of the domain knowledge; make assumptions regarding the domain explicit; separate domain knowledge from the operational knowledge; and enable the analysis of the knowledge leading to improved decision making. We regard the integration of ontologies to our proposed framework as future work direction.

Prediction Using Sentiment Analysis. This track combines sentiment analysis with machine learning techniques to predict something about a product or service [19–21]. The study in [20] analyzed how machine learning techniques and twitter sentiment analysis can be used to predict stock market fluctuations. The authors applied various machine-learning models such as Linear Regression, Support Vector Machines (SVM) and Neural Networks and tuned them up in order to maximize the efficiency. The authors worked on historical data of stocks taken from Yahoo Finance website and built a classifier based on the Movies Reviews dataset, which we consider as an incompatible issue. The study concluded that stock markets are heavily sentiment driven. Similarly, the study in [21] uses sentiment information mined from current movie tweets for predicting movie's performance. The authors developed a prototype, which may be useful to marketers in the of course correcting marketing campaigns to garner positive sentiments before the release of the movie. Similarly, studies in [8, 22, 23] take the same direction.

The approach proposed in this paper is tightly related to this category, however, we can distinguish ourselves by: (i) a comprehensive framework for disruptions monitoring, analysis, planning of corrective/proactive plans and their execution; the article focuses on presenting the details of the analysis component, (ii) the proposed "analysis" component integrates clustering and classification techniques to get insights and visibility over unexpected/unforeseen patterns, while all related work efforts directly conduct classification, (iii) the clustering activity has the main objective of linking the classification/prediction results to the BP world.

Social Business Process Management. Some research efforts in this direction view social BPM as designing and implementing business processes socially using any enterprise collaboration platforms, such as Yammer, and Chatter, or by employing hybrid Wikis. Prominent work efforts in this direction are: [25, 26]. Another stream of research in social BPM utilizes social BP. At diagnosis time, a BP execution component is implemented to discover and build the networks of social relations between the business process components (task, machine, person) based on process execution logs as well as the BP model. This is mainly reported in [26], where a model is being introduced (called SUPER standing for Social based bUsiness Process managEment fRamework) that leverages social computing principles for the design and development of social business processes). SUPER identifies task (t), person (p), and machine (m) as the core components of a business process. The authors defined all social relation states of t, p, & m. At diagnosis time, authors implemented a social analysis component to discover and build the networks of social relations between the business process components (t, p, m) based on process execution logs as well as the BP model. Every time a task is suspended, its resource is checked to identify the reason of being idle.

The work in [27] built a platform that bridges in the gap between the social and the business world through meet-in-the-middle platform, just for integration purposes without any analysis and/or improvement mechanisms.

The work in this paper proposes a novel approach for the analysis and resolution of BP disruptions through the utilization of social data and by integrating clustering and classification techniques. The clustering technique aids us to identify disturbances patterns and unforeseen/unexpected patterns and link them to the BP world, and then the classification approach aims at predicting future disturbances. Therefore, we consider our approach to fall under the categories of "Social BPM" and "Prediction Using Sentiment analysis". To the best of our knowledge, such an integration does not exist in the literature.

3 Proposed Social BPM Monitoring Framework

Figure 1 presents a schematic view of the proposed framework for Social BP monitoring and improvement.

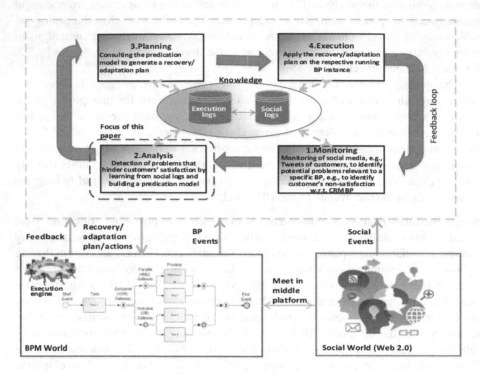

Fig. 1. Proposed framework

The framework is presented as an instantiation of the well-recognized IBM MAPE-K adaptation loop [29, 30], which is an efficient and novel approach for self-adaptation in autonomic computing. Autonomic computing is a computing environment with the ability to manage itself and dynamically adapt to changes in accordance with business policies and objectives [30]. As discussed in [31], self-adaptiveness in the general level inhibits a number of self-* properties in the major level, including self-configuring, self-healing, self-optimizing and self-protecting. We consider the approach presented in this article to fall under the *self-healing* category. Self-healing is the capability of the system (BPM in our case) of discovering, diagnosing and reacting to disruptions [30].

Self-healing can be classified into self-diagnosing and self-repairing, where the former concerns itself with diagnosing errors, faults and failures, and the latter focuses on recovering from detected disturbances. While the focus of the proposed framework is to provide an integrated approach for self-diagnosis and self-repairing by exploiting social media for BP improvement, due to space limitations and to keep the next discussion focused, this article focuses only on presenting the details of the proposed *self-diagnosis* approach. The other components will be presented in future publications by referring to the framework.

The upper part of Fig. 1 represents the five MAPE-K self-adaptation loop components corresponding to its acronyms; that's: K: Knowledge, M: Monitoring, A: Analysis, P: Planning, E: Execution and K: Knowledge.

The *knowledge* component in our framework constitutes the interlink between: (i) execution log(s), which maintains and relates business process execution logs, and (ii) social logs. This includes a structured representation of social data, e.g., tweets, in addition to predicted & extracted features that entails more value to the business (details are presented in Sect. 4). The knowledge component is the backbone of the four MAPE activities defined next.

The *monitoring* component constitutes monitoring running BP instances, which has been continuously acknowledged in the literature as key to ensure the successful completion of running BP instances. With the advent of web 2.0 technologies and their growing adaptation in business organizations, social artefacts and social events (see the bottom-right of Fig. 1) bring together the key parties and events, which can naturally be used to track key people and events in runtime. This enables quick decision making and inferring (semi-) automated corrective/prevention actions. Complex Event Processing (CEP) [32] is adopted by the proposed framework for realizing this component by applying our previous work in this area as reported in [34, 35]. CEP technology mainly combines data from multiple sources to infer events or patterns that suggest more complicated circumstances. For example, if we consider that the BP model under consideration is the Customer Relationship Management (CRM) BP of a telecommunication company as introduced in Sect. 5, then the events of interest are the problems/disruptions that hinder customers' satisfaction, e.g., a customer complaining about extra charges added to her mobile phone bill, or a customer suffering from no network coverage in his/her area etc.

Based on the monitoring results, the *analysis* component is responsible for performing complex data analysis and reasoning, by the continuous interaction with the knowledge component. Particularly, the analysis component carries out processing, correlation, and analysis of event streams to detect the occurrence of disturbances. To realize this component, our analysis approach exploits and integrates data mining and machine learning techniques, i.e., clustering and classification techniques. The next discussion focuses on presenting in detail the concrete analysis approach that realizes this component, which represents the main focus of this article.

Based on the results of the analysis component, the *planning* component is responsible for establishing a preventive/corrective action plan to avoid/minimize the impact of the problems detected by the analysis component. The action plan will be mainly constructed semi-automatically by a planning agent that infers from the knowledge base -based on the history of recovery plans- the most appropriate recovery plan(s), ranked based on a number of selected features. Then, the BP expert can make the final decision through the intuitive interaction with the dashboard.

Finally, the *execution* component involves the automated application of the self-healing plan produced from the planning component to the respective running BP instance. For example, if the analysis component detects a network degradation issue, and the planning component proposes the recovery plan as including an ad-hoc BP activity that assigns extra minutes to the affected customer, then the execution component actually sends signals to the BP execution engine to execute this ad-hoc BP instance in the new planned sequence. The work to realize the concrete approach of the related planning and execution components is ongoing and will be considered for future publications.

4 The Analysis Approach: Integrating Data Mining and Machine Learning Techniques

Figure 2 depicts the major activities of the analysis approach as described in the framework presented in Sect. 3. Starting from the left hand-side of the figure, users of Twitter typically tweet by complaints/compliments/inquires of the service provided by a specific service provider, for example, services of a telecommunication company as used as the running scenario in this article. Once the tweets are received, they are stored in its raw format, so that they can be eventually used for learning and analysis purposes. When sufficient amount of raw tweets data is available in the raw tweets log, the flow then goes to the *"Filtration"* activity. This activity is required to remove tweets of positive sentiment since we are only concerned with negative tweets representing complaints/problems the customer is facing.

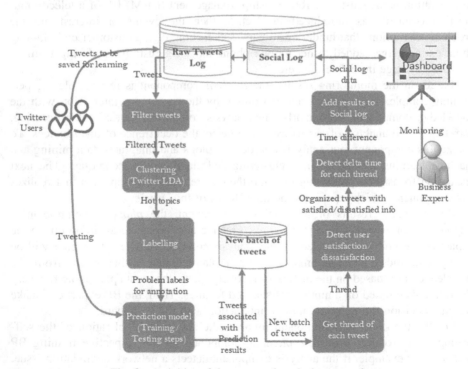

Fig. 2. Activities of the proposed analysis approach

Then the process goes to *"clustering tweets"* activity. The main objective of clustering here is to identify the labels/topics of problems faced by the customers, and to detect any unforeseen/unexpected patterns. Clustering is unsupervised machine learning problem and is performed in our approach by using Twitter LDA [35] (Latent Dirichlet Allocation). Twitter LDA is a text data mining clustering algorithm to extract hot topics from text. It includes a two-step approach to twitter data analysis. The first is to generate a topic model and the second to cluster tweets into topic-based categories. Twitter LDA clusters the tweets by automatically inferring/identifying hot topics (customer's problems in our case), and then generating a number of tweets clusters based on these topics.

Then, the flow goes to the *"Labelling"* activity, which is responsible for annotating each tweet by one or more problem topic(s), identified by the prior "Clustering" activity. The "Labelling" activity operates semi-automatically, such that it refers to a dictionary we have built representing the words that are related to each problem topic, and requires an expert's validation. If a tweet contains words that are not defined in the dictionary, the human expert will be prompted to manually decide on the label. These annotation labels that we call problem topic labels are required for the next classification activity that builds the prediction model.

The *"Creation of the Prediction Model"* activity applies the Sequential Minimal optimization (SMO) classification technique [36]. Our selection of this technique is based on an analytical evaluation of the accuracy of several evaluation techniques, which includes: Random Forest [37], Naïve Bayes [38], Naïve Bayes Multinomial [39], Sequential Minimal optimization (SMO) [36] classification algorithms. The evaluation accuracy of SMO was 76.7% in our controlled experiment. We have used WEKA for this purpose, which implements a large collection of machine learning algorithms for data mining tasks[2]. The goal of this activity is to build a prediction model that once given a new tweet, it predicts whether it represents a problem and then identifies its problem topic. Then, afterwards, this tweet associated with its predicted topic label will be stored in a transient data store (called New batch of tweets in the diagram) that receives batches of new tweets outputted from the prediction model for applying further analysis steps on them (The batch size could be adjusted by size or on daily basis according to the density of the tweets received).

Following the two-fold classical machine learning methodology, first, the model is *trained*; second, the model is *tested* to determine its accuracy. Typically, classification of texts usually follows a mathematical approach by representing words as vectors called *wordToVector* representation, then the selected classification algorithm works on classifying these vectors and building the prediction model.

After building the prediction model and being deployed for the analysis of new streamed tweets, *"Getting the thread of each tweet"* activity is responsible for organizing each tweet with its replies sequentially to be viewed as a dialogue between the customer and the service provider in an easy to visualize way. This component is implemented in Python programming language.

---

[2] https://www.cs.waikato.ac.nz/ml/weka/.

The tweet thread is then given as an input to *"Detecting the customer satisfaction/dissatisfaction"* activity, which indicates whether the customer of this tweet thread is satisfied, dissatisfied, neutral, or incomplete thread (incomplete occurs when the customer does not reply). In other words, this activity concerns itself with computing the sentiment score of the customer's tweet thread, and then amending it with the result of the sentiment score.

"Detecting the delta time for each thread" activity computes the time between the user tweet and the company's reply (i.e., retweet). In many organizations, there is a violation if the company took longer time than allowed to reply to the customer. This metric/feature will be added to detect time violations, which can be tuned according to the company's policies.

Afterwards, these extracted measures are added to the *social log* (see Fig. 4 for a snap shot of the social log) and then presented on the dashboard for business experts, by highlighting/alerting events that need attention. This is considered as the output of the analysis component in our proposed framework demonstrated in Sect. 3. The social log is formed from 8-tuple as follows:

$$< id, t, twt, Pid, username, Pb, Sent, Res > \tag{1}$$

Where:

- *id*: *tweet id.*
- *t*: *timestamp of the tweet.*
- *twt*: *tweet text.*
- *Pid*: *parent id of the tweet that detects whether the tweet is a retweet or it's the first tweet.*
- *Username*: *name of the user who sent the tweet.*
- *Pb*: *problem topic predicted by the classifier.*
- *Sent*: *sentiment of the tweet whether it is positive, negative, or neutral.*
- *Res*: *result of the issue between the client & the service provider. It can be satisfied, unsatisfied, not complete case, or neutral.*
- *time measure is a calculated measure not added in the social log.*

The next *Planning* component as presented in the framework in Sect. 3 also follows a machine learning technique, which assumes that the business expert first (semi-automatically) decides on how to respond to sufficient number of detected disruptions/problems by the aid of the clusters inferred in the analysis phase (the results of the "Clustering" activity in Fig. 2). These clusters will point to relevant BP activities/fragments that the business expert needs to interrogate to resolve the detected disruptions. The resolution plan/actions decided by the business expert for each disruption is/are also stored in the repository, along with an indication of how much the customer was satisfied with this reconciliation. This data is then used to build a new classification model, which is capable of automatically inferring the possible recovery plan/actions to resolve future detected violations. Given the fact that humans/experts should always be in the loop, during the *Execution* phase, the business expert has to

check the automatically generated recovery plan and can make modifications, if needed. The details of our integrated *"Planning & Execution"* approach is left for a next
publication.

5 Application of the Analysis Approach on a Case Study

In order to validate the applicability, efficiency and utility of the proposed analysis
approach described in Sect. 4, we have considered a real-life case study of a leading
telecommunication company that provides mobile communication services, and
applied the analysis approach on. For this purpose, we have collected tweets using Java
twitter API [3], by filtering them based on #TheCompanyName. The attributes of interest
include *tweet ID*, *created* (which specifies the date and time of the tweet), *text of tweet*,
user name of the sender, parent ID, and *sentiment score* ('0' means neutral, '+ve'
integer means satisfied, and '-ve' integer means unsatisfied). Figure 3 shows excerpt of
the collected tweets.

1	id	created	text	username	parent_or	sentiment
2	##	2017-10-2	Morning I'm looking to join your network but as a pay as you go customer and was wandering if the bundles expire after 30 days?	Rianna_Ar senal		0
3	##	2017-10-2	worst customer service provided by ------- on webchat. Doesn't understand my needs and handling	Atiksh Singh	######	-99
4	##	2017-10-2	What about if your network is extremely over priced and has no option for unlimited data?	Alec Houston	######	0
5	##	2017-10-2	hi, I'd like a call back from one of your team. I've been lied to by one of your staff and my bill is more than it should be.	Tom		-60
6	##	2017-10-2	every email requesting help answered by the same automated bot which tells me information I already know..........	Tim James		-60
7	##	2017-10-2	Hi Kirsty, I've still no 3/4G despite putting my phone in aeroplane mode for 15 mins.	Nicola	######	-10
8	##	2017-10-2	There is no signal in my phone since yesterday	Adams		-50

Fig. 3. Example of raw tweets

As shown in Fig. 3, the raw tweets cannot give clues about anything if left without
further analysis. We will apply the analysis approach discussed in Sect. 4 on these
unstructured/raw tweets.

First, the *Filtration activity* is applied to remove the tweets of positive sentiment
since we are only concerned with unsatisfied customers/tweets. This activity resulted in
5,191 tweets.

Then the flow goes to the *Clustering* activity, where we run *Twitter LDA* [35] on
the filtered dataset of tweets, and nine clusters were generated corresponding to nine
topics of problems (Problem topic labels): (1) Internet (2) Customer-service (3) Mobile-

---

[3] https://dev.twitter.com/overview/api.

data/cellular (4) Network (5) Signal (6) Financial (7) Phone-services (8) Sales (9) Other[4].

Then *Creation of the prediction model* is done using classification on WEKA. The previously mentioned topics of problems represent the annotation problem topic labels for the training step in the classification model. Then, 80% of the tweets are used as a training set and 20% as a testing set. Accordingly, we have 4,152 tweet as a training set and 1,039 tweets for testing purpose, which have been semi-automatically annotated with one or more of the nine identified topics above (as the result of the prior clustering activity). The annotations have been conducted by developing a python program to automatically detect the words of interest and suggest the annotations. For example, if annotation program finds in the tweet key-words like 'credit', 'bill', 'billing', 'charge', 'overcharge' and 'credit card' then it will annotate this tweet with 'Financial' as the problem topic. Having the data sets annotated, the classification model is then built.

Id	T	Twt	Pid	Username	Pb	Sent	Res
12	10/25/2017 9:45	I have a problem in my router, It doesn't work from 2 days!!		@George	Internet	-13	
13	10/25/2017 10:40	What is the kind of your router @George?	12	@TheCompan		0	
14	10/25/2017 12:05	The router is Huawei 6000.What should I do?	12	@George	Internet	0	
15	10/25/2017 4:00	A representative employee will call you to tell you the resolving steps	12	@TheCompan yName		0	
16	10/25/2017 5:00	Thank you	12	@George		5	Satisfied
17	10/26/2017 5:29	Why does my internet and phone signal keep dropping?		@Tom	signal	-15	
18	10/26/2017 5:29	Hi Tom, please provide the postcode of your area so we can check this for you	17	@TheCompan yName		0	Incomplete
19	10/26/2017 6:29	Can't remember last time I had 4G with your company.		@Ashly Hamilton	Mobile data	-12	
20	10/26/2017 7:30	Hi Ashly, this is not good to hear. Is anyone else in the area effected by the issue or just yourself?	19	@TheCompan yName		0	Incomplete
21	10/26/2017 8:30	I have paid my bill and the app even says I have but I have had a text saying my phone is being cut off as I haven't paid??		@kim	Financial	-18	
22	10/26/2017 10:30	Hi Kim, our Billing team on Live Chat can look into this. Please contact them here: https://t.co/VpH12jMJZ Kirsty	21	@TheCompan yName		0	
23	10/27/2017 8:01	This is so frustrating, the issue is not yet solved & I still receive the same message.	21	@Kim	other	-40	Unsatisfied
24	10/27/2017 9:20	I'm in Nottingham and the signal keeps dropping in and out and no Wi-Fi I had rehearsals today and couldn't get into building!!		@Jaq	Internet	-33	
25	10/27/2017 11:01	Hi Jaq, we were aware of an issue but it should be fixed, please restart your phone for it to connect back on the network.	24	@TheCompan yName		0	
26	10/27/2017 11:35	Okay It worked.	24	@Jaq		5	Satisfied

Fig. 4. Example of the created social log

The test data set is then used to estimate the accuracy of the generated prediction model. We have tested the model across several classification algorithms, including: Random Forest [37], Naïve Bayes[38], Naïve Bayes Multinomial [39], Sequential Minimal optimization (SMO) [36]. Our evaluation results showed that Sequential Minimal optimization (SMO) has the best results as will be demonstrated next in Sect. 5.1.

---

[4] "Other" means that the person is tweeting by un-meaningful words or not a related tweet indicating a potential problem.

Then, *getting the tweet's thread, detecting customer satisfaction/dissatisfaction*, and *getting the delta time* for each thread steps are done on our dataset (original data set of 8,719 tweet) to create the *Social Log* as illustrated in Sect. 4. A snapshot of the social log is depicted in Fig. 4.

By creating this social log from the raw tweets, the telecommunication company in our running scenario can gain profound insights and can take informed decisions. Aggregated statistical analysis could also be performed to get insights for example of the total number of satisfied/dis-satisfied customer and if they are in line with their strategic plans, and what recovery/correction actions they can take to alleviate any deficiencies.

5.1 Results and Discussion

This section demonstrates the results of the machine learning algorithms we used for testing the classification models we have deployed in our proposed framework. However, before showing the results we have to clarify the meaning of some key terms [40]:

- True Positives (TP): These are the correctly predicted positive values, which means that the value of actual class is positive, and the value of predicted class is positive.
- True Negatives (TN): These are the correctly predicted negative values, which means that the values of both the actual predicted classes are negative.
- False Positives (FP): These cases represent the situation when the actual class is negative and predicted class is positive.
- False Negatives (FN): These cases represent the situation when actual class is positive but the predicted class is negative.
- Accuracy: Accuracy is the most intuitive performance measure and it is simply a ratio of correctly predicted observation to the total observations. One may think that, if we have high accuracy then our model is good enough. Nevertheless, accuracy is a great measure but only when you have symmetric datasets, where values of false positive and false negatives are almost the same. Therefore, you have to look at other parameters to evaluate the performance of your model.

$$Accuracy = TP + TN/TP + FP + FN + TN \qquad (2)$$

- Error rate (Err): The complement of Accuracy is the error rate, which evaluates a classifier by its percentage of incorrect predictions. Accuracy and Err are general measures and can be directly adapted to multiclass classification problems.

$$Err = 1 - Accuracy = FP + FN/TP + FP + FN + TN \qquad (3)$$

- Precision: Precision is the ratio of correctly predicted positive observations to the total predicted positive observations. High precision relates to the low false positive rate.

$$Precision = TP/TP + FP \qquad (4)$$

- Recall (Sensitivity): Recall is the ratio of correctly predicted positive observations to the all observations.

$$Recall = TP/TP + FN \qquad (5)$$

- F1 score: F1 Score is the weighted average of Precision and Recall. Therefore, this score takes both false positives and false negatives into account. Intuitively it is not as easy to understand as accuracy, but F1 is usually more useful than accuracy, especially if you have an uneven class distribution. Accuracy works best if false positives and false negatives have similar cost. If the cost of false positives and false negatives are very different, it is better to look at both Precision and Recall.

$$F1Score = 2 * (Recall * Precision)/(Recall + Precision) \qquad (6)$$

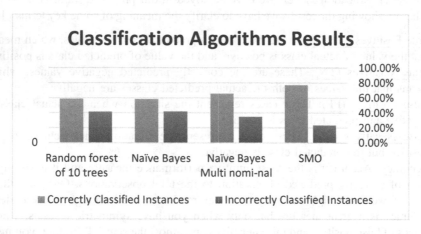

Fig. 5. Classification algorithms results

Figure 5 shows the performance results of the four classification machine-learning algorithms, which are Random Forest, Naïve Bayes, Naïve Bayes Multinomial, SMO (Sequential Minimal Optimization). (The total number of instances here is 5191 tweet) SMO refers to the specific efficient optimization algorithm used inside the SVM Support Vector Machines implementation. From the results, one can see that SMO has the best results. As discussed in [41] SVM are supervised learning classification algorithms which has been extensively used in text classification problems due to the sparse high dimensional nature of the text with few irrelevant features.

Table 1 depicts the detailed accuracy measures by class for SMO classification algorithm, which has the best results in our case study.

Table 1. SMO detailed accuracy by class

TP rate	FP rate	Precision	Recall	F-Measure	ROC area	Classified class
0.2	0.004	0.238	0.2	0.217	0.91	network
0.732	0.067	0.717	0.732	0.725	0.862	customer service
0.526	0.003	0.781	0.526	0.629	0.868	mobile data
0.659	0.018	0.815	0.659	0.728	0.916	financial
0.164	0.002	0.563	0.164	0.254	0.711	sales
0.648	0.033	0.712	0.648	0.679	0.865	phone services
0.69	0.008	0.787	0.69	0.735	0.953	signal
0.918	0.182	0.787	0.918	0.848	0.884	other
0.632	0.011	0.84	0.632	0.721	0.904	internet

6   Conclusions and Future Work

Business processes represent the foundation of all organizations, and as such, organizations are striving for their continuous improvement throughout the complete BP lifecycle. The emergence and the wide adoption of web 2.0 technologies represents a paradigm shift in how individuals collaborate in their businesses and daily lives, enabling organizations to take a customer-centric operating model, and subsequently achieve a competitive advantage. This paradigm shift is known as social business processes or business processes 2.0. This paper contributes with a novel framework that exploits the huge amount of social data (in particular twitter) to enable the identification and resolution of runtime business process disruptions (problems affecting customers' satisfaction in different domains). The main objective of the framework is to inject self-healing capabilities into BPM systems, where the system is autonomously capable of discovering, diagnosing and reacting to disruptions. The paper then proposes a concrete analysis approach by utilizing and integrating text mining techniques (i.e. Twitter LDA), machine learning techniques (i.e. SMO, Naïve Bayes, Naïve Bayes Multinomial, Random Forest classification algorithms using WEKA) to realize the self-diagnosis component.

The proposed analysis approach has been implemented and applied on a real-life case study of a telecommunication company, and our evaluation study revealed that there exists a strong correlation between data analytics of the social side and improving its adherent twin; the business side. The huge amounts of data on the social side can always be utilized to enhance the business side, by removing the cover from many problems and violations that are taking place between users and the organization and can degrade their business if left unnoticed.

Ongoing and future work is going in a number of parallel and complementary directions. This includes:

- Incorporation and integration of other heterogeneous free text social media networks, such as Facebook, to extract more faithful knowledge, for better informed decision-making and better action taking.

- Incorporation of ontologies to capture the semantics of the social and BP worlds, enable their semantic alignment and the integration of heterogeneous social media sources.
- Application of the proposed framework on enterprise social networks [42], which are dedicated private social networks adapted by organizations internally and externally to connect individuals who share similar business interests or activities. Although, we claim that our framework and results are applicable to enterprise social networks, however, this needs a dedicated experimental study for its validation.
- Intensifying the validation and evaluation of the proposed framework by considering other case studies from different industrial sectors, while comparing the domains that are more prone to the adoption of this technology.
- Accommodating with the large volume of today's big data by incorporating a big data platform, such as Hadoop to support the scalability of the proposed framework and its underpinning approaches.

Acknowledgements. We wish to thank Dr. Ahmed Awad, Institute of Computer Science, University of Tartu, Estonia, for providing the essentials of the case study in this paper and for the fruitful discussions and advices.

References

1. Almeida, F.: Web 2. 0 technologies and social networking security fears in enterprises. Int. J. Adv. Comput. Sci. Appl. **3**, 152–156 (2012)
2. Leskovec, J., Adamic, L.A., Huberman, B.A.: The dynamics of viral marketing. ACM Trans. Web. **1**, 39 (2007)
3. Kumar, M., Bala, A.: Analyzing Twitter sentiments through big data. In: 3rd International Conference on Computing for Sustainable Global Development (INDIACom). IEEE (2016)
4. Troussas, C., Virvou, M., Espinosa, K.J., Llaguno, K., Caro, J.: Sentiment analysis of Facebook statuses using Naive Bayes classifier for language learning. In: 2013 Fourth International Information, Intelligence, Systems and Applications (IISA) (2013)
5. Isah, H., Trundle, P., Neagu, D.: Social media analysis for product safety using text mining and sentiment analysis. In: 2014 14th UK Workshop on Computational Intelligence (UKCI). IEEE (2014)
6. Wang, T., Chen, Y.: The power of comments: fostering social interactions in microblog networks. Front. Comput. Sci. **10**, 889–907 (2016)
7. He, W., Zha, S., Li, L.: Social media competitive analysis and text mining: a case study in the pizza industry. Int. J. Inf. Manag. **33**, 464–472 (2013)
8. Moseley, N., Alm, C.O., Rege, M.: Toward inferring the age of Twitter users with their use of nonstandard abbreviations and lexicon. In: IEEE 15th International Conference on Information Reuse and Integration, IRI 2014, pp. 219–226 (2014)
9. Liu, B.: Sentiment Analysis. Cambridge University Press, Cambridge (2015)
10. Shad Manaman, H., Jamali, S., Aleahmad, A.: Online reputation measurement of companies based on user-generated content in online social networks. Comput. Hum. Behav. **54**, 94–100 (2016)

11. Salampasis, M., Paltoglou, G., Giachanou, A.: Using social media for continuous monitoring and mining of consumer behaviour. Int. J. Electron. Bus. **11**, 85 (2013)
12. Younis, E.: Sentiment analysis and text mining for social media microblogs using open source tools: an empirical study. Int. J. Comput. Appl. **112**, 44–48 (2015)
13. Gürsoy, U.T., Bulut, D., Yiğit, C.: Social media mining and sentiment analysis for brand management. Glob. J. Emerg. Trends e-Business, Mark. Consum. Psychol. Online Int. Res. J. **3**, 497–511 (2017)
14. Kontopoulos, E., Berberidis, C., Dergiades, T., Bassiliades, N.: Ontology-based sentiment analysis of twitter posts. Expert Syst. Appl. **40**, 4065–4074 (2013)
15. Ruba, K.V., Venkatesan, D.: Building a custom sentiment analysis tool based on an ontology for Twitter posts. Indian J. Sci. Technol. **8**, 1–5 (2015)
16. Ali, F., Kwak, D., Khan, P., Islam, S.M.R., Kim, K.H., Kwak, K.S.: Fuzzy ontology-based sentiment analysis of transportation and city feature reviews for safe traveling. Transp. Res. Part C Emerg. Technol. **77**, 33–48 (2017)
17. Zehra, S., Wasi, S., Jami, I., Nazir, A., Khan, A., Waheed, N.: Ontology-based sentiment analysis model for recommendation systems. In: 9th International Joint Conference on Knowledge Discovery, Knowledge Engineering and Knowledge Management (KEOD 2017), pp. 155–160 (2017)
18. Ryota, K., Tomoharu, N.: Stock market prediction based on interrelated time series data. In: IEEE Symposium on Computers and Informatics, ISCI 2012, pp. 17–21 (2012)
19. Pagolu, V.S., Challa, K.N.R., Panda, G., Majhi, B.: Sentiment analysis of Twitter data for predicting stock market movements. In: International Conference on Signal Processing, Communication, Power and Embedded System (SCOPES), pp. 1–6 (2016)
20. Pimprikar, R., Ramachandran, S., Senthilkumar, K.: Use of machine learning algorithms and twitter sentiment analysis for stock market prediction. Int. J Pure Appl. Math. **115**, 521–526 (2017)
21. Gaikar, D.D., Marakarkandy, B., Dasgupta, C.: Using Twitter data to predict the performance of Bollywood movies. Ind. Manag. Data Syst. **115**, 1604–1621 (2015)
22. Sumner, C., Byers, A., Boochever, R., Park, G.J.: Predicting dark triad personality traits from Twitter usage and a linguistic analysis of tweets. In: 11th International Conference on Machine Learning and Applications. IEEE (2012)
23. Golbeck, J., Robles, C., Turner, K.: Predicting personality with social media. In: CHI EA 2011 CHI 2011 Extended Abstracts on Human Factors in Computing Systems, pp. 253–262. ACM (2011)
24. Hauder, M.: Bridging the gap between social software and business process management : a research agenda. In: Seventh International Conference on Research Challenges in Information Science (RCIS), pp. 1–6. IEEE (2013)
25. Yunus, M., Moingeon, B., Lehmann-Ortega, L.: Building social business models: lessons from the Grameen experience. Long Range Plan. **43**, 308–325 (2010)
26. Maamar, Z., Sakr, S., Faci, N., Boukhebouze, M., Barnawi, A.: SUPER: social-based business process management framework. In: Toumani, F., et al. (eds.) ICSOC 2014. LNCS, vol. 8954, pp. 413–417. Springer, Cham (2015). https://doi.org/10.1007/978-3-319-22885-3_38
27. Maamar, Z., Burégio, V., Sellami, M.: Collaborative enterprise applications based on business and social artifacts. In: 2015 IEEE 24th International Conference on Enabling Technologies: Infrastructure for Collaborative Enterprises, pp. 150–155 (2015)
28. Brun, Y., et al.: Engineering self-adaptive systems through Feedback loops. In: Cheng, Betty H.C., de Lemos, R., Giese, H., Inverardi, P., Magee, J. (eds.) Software Engineering for Self-Adaptive Systems. LNCS, vol. 5525, pp. 48–70. Springer, Heidelberg (2009). https://doi.org/10.1007/978-3-642-02161-9_3

29. Kephart, J.O., Chess, D.M.: The vision of autonomous computing. IEEE Comput. Soc. **36**, 41–50 (2003). Home Community Technology Leaders
30. IBM: An architectural blueprint for autonomic computing. IBM (2005)
31. Salehie, M., Tahvildari, L.: Self-adaptive software: landscape and research challenges. ACM Trans. Auton. Adapt. Syst. **4**, Article no. 14 (2009)
32. Luckham, D.: The Power of Events: An Introduction to Complex Event Processing in Distributed Enterprise Systems. Addison-Wesley Longman Publishing Co., Inc., Boston (2001)
33. Awad, A., Barnawi, A., Elgammal, A., Elshawi, R., Almalaise, A., Sakr, S.: Runtime detection of business process compliance violations: an approach based on anti patterns. In: Proceedings of the 30th Annual ACM Symposium on Applied Computing, pp. 1203–1210. ACM (2015)
34. Barnawi, A., Awad, A., Elgammal, A., El Shawi, R., Almalaise, A., Sakr, S.: Runtime self-monitoring approach of business process compliance in cloud environments. Clust. Comput. **18**, 1503–1526 (2015)
35. Zou, L., Song, W.W.: LDA-TM: a two-step approach to Twitter topic data clustering. In: IEEE International Conference on Cloud Computing and Big Data Analysis, ICCCBDA 2016, pp. 342–347 (2016)
36. Cortes, C., Vapnik, V.: Support vector networks. Mach. Learn. **20**, 273–297 (1995)
37. Louppe, G.: Understanding random forests from theory to practice (2014)
38. Metzger, A., et al.: Comparing and combining predictive business process monitoring techniques. IEEE Trans. Syst. Man Cybern. Syst. **45**, 276–290 (2015)
39. Xu, S., Li, Y., Wang, Z.: Bayesian multinomial Naïve Bayes classifier to text classification. In: Park, James J.(Jong Hyuk), Chen, S.-C., Raymond Choo, K.-K. (eds.) MUE/FutureTech - 2017. LNEE, vol. 448, pp. 347–352. Springer, Singapore (2017). https://doi.org/10.1007/978-981-10-5041-1_57
40. Costa, E.P., Lorena, A.C., Carvalho, A.C.P.L.F., Freitas, A.A.: A Review of performance evaluation measures for hierarchical classifiers. In: Evaluation Methods for Machine Learning II: Papers from AAAI-2007 Work, pp. 1–6 (2007)
41. Allahyari, M., Trippe, E.D., Gutierrez, J.B.: A brief survey of text mining : classification, clustering and extraction techniques. ArXiv: 1 (2017)
42. Ellison, N.B., Gibbs, J.L., Weber, M.S.: The use of enterprise social network sites for knowledge sharing in distributed organizations. Am. Behav. Sci. **59**, 103–123 (2015)

Semantic IoT Gateway:
Towards Automated Generation
of Privacy-Preserving Smart Contracts
in the Internet of Things

Faiza Loukil[1]([✉]), Chirine Ghedira-Guegan[2], Khouloud Boukadi[3],
and Aïcha Nabila Benharkat[4]

[1] University of Lyon, University Jean Moulin Lyon 3, CNRS, LIRIS, Lyon, France
`faiza.loukil@liris.cnrs.fr`
[2] University of Lyon, University Jean Moulin Lyon 3,
iaelyon School of Management, CNRS, LIRIS, Lyon, France
`chirine.ghedira-guegan@liris.cnrs.fr`
[3] Mir@cl Laboratory, Sfax University, Sfax, Tunisia
`khouloud.boukadi@fsegs.usf.tn`
[4] University of Lyon, INSALyon, CNRS, LIRIS, Lyon, France
`nabila.benharkat@liris.cnrs.fr`

Abstract. The Internet of Things paradigm has brought opportunities to meet several challenges by interconnecting IoT resources, such as sensors, actuators, and gateways on a massive scale. The IoT gateways play an important role in the IoT applications to bridge between sensor networks and the external environment through the Internet. Typically, the IoT gateways collect and send the data collected from sensors and actuators to external platforms where they will be remotely analyzed. However, the users desire a more adapted IoT gateway that can improve the IoT data privacy preservation before sending them to these external platforms. Thus, an IoT gateway that enables a better control over the set of private IoT resources and protects the collected personal data and their privacy is required. For this purpose, we propose a Semantic IoT Gateway that helps implement a dynamic and flexible privacy-preserving solution for the IoT domain. First, it enables to match between the data consumer's terms of service and the data owner's privacy preferences by generating an adapted privacy policy. Second, it converts the privacy policy into a custom smart contract. Finally, it connects a set of private IoT resources to a distributed network using the blockchain technology to host the generated smart contracts. A smart contract is an executable code that runs on top of the blockchain to facilitate, execute and enforce an agreement between untrusted parties without the involvement of a trusted third party. Our proposal, which is highlighted through an example and experimentation on a real-world use-case, has given the expected results.

© Springer Nature Switzerland AG 2018
H. Panetto et al. (Eds.): OTM 2018 Conferences, LNCS 11229, pp. 207–225, 2018.
https://doi.org/10.1007/978-3-030-02610-3_12

1 Introduction

The Internet of Things (IoT) is a novel paradigm, the main strength of which is its high impact on several aspects of everyday's life and behavior of potential users. From the point of view of the private user, the most obvious effects of the IoT introduction will be visible in both working and domestic fields, such as assisted living, e-health, and enhanced learning. Similarly, from the perspective of the business users, the most apparent consequences will be equally visible in fields, such as automation and industrial manufacturing, business/process management, and intelligent transportation of people and goods [3].

Actually, many challenging issues related to the IoT resource characteristics still need to be addressed. In fact, they have a low computation and an energy capacity to protect personal data and the user's privacy. In order to overcome this problem, another IoT resource type is proposed. This is called IoT Gateway, the role of which is to collect and send the collected data from IoT sensors and actuators to external platforms to be remotely analyzed. Moreover, those external platforms gather the IoT data and use them to personalize services, optimize decision-making processes, and predict future trends. However, the IoT data raise security and privacy concerns. In fact, the users have a little or no control over the collected data about themselves [10]. For instance, sharing the collected data by wearable devices with service providers leads to lose the IoT data control and ownership [10]. Moreover, users have no guarantee that the service provider will respect the licensing agreement concerning privacy and security protection [10]. Moreover, the used IoT gateways are generic, with basic settings, and do not preempt the user's requirements especially concerning the privacy issue.

Motivated by the actual basic role and the need to have a more flexible gateway that enables to better control the set of private IoT resources, we propose a Semantic IoT Gateway as a core component of our proposed end-to-end privacy-preserving framework for the IoT data based on the blockchain technology, called PrivBlockchain [9]. The reason behind using the blockchain technology is that the blockchain is an immutable public record of data secured by a network of peer-to-peer participants that hosted smart contracts, which are executable codes that run on top of the blockchain to facilitate, execute and enforce an agreement between untrusted parties without the involvement of a trusted third party. Nevertheless, the Semantic IoT Gateway is intended to convert the data owner's privacy preferences into smart contracts that will be published within the blockchain. Moreover, our Semantic IoT Gateway is based on an IoT privacy ontology, called LIoPY, which is a European legal compliant ontology to preserve privacy for IoT and defined in our previous work [8]. Thanks to LIoPY use, a privacy policy can be inferred according to the data owner's privacy preferences and the data consumer's terms of service. This policy is a set of conditions that the consumer needs to fulfill in order to handle specific shared IoT data. Those conditions are hosted in a smart contract. Thus, the use of a smart contract will prevent any privacy violation attempts by enforcing the data privacy requirements and ensuring that the shared data will be handled as expected in the whole IoT data lifecycle, collection, transmission, storage and processing phases.

This paper is organized as follows. Section 2 deals with the existing researchers who studied how privacy is preserved in the IoT scope. Section 3 presents an overview of the PrivBlockchain framework. Section 4 identifies the framework's core components. Section 5 defines the proposed Semantic IoT Gateway and explains its components and main functionalities. Section 6 validates our solution in a healthcare scenario. Section 7 concludes the paper and presents some future endeavors.

2 Related Work

There are many researchers, who have studied the integration of IoT technology and semantic modeling or blockchain technology for preserving privacy.

Semantic-based privacy preservation solutions are based on ontologies and inference rules for developing smart applications. For instance, Celdran et al. [5] proposed a solution called SeCoMan, in which an ontology is employed to model the description of entities, reason over data to obtain useful knowledge, and define context-aware policies. However, privacy protection is fulfilled in a location-limited level. For their part, Wang et al. [13] proposed an Ontology-based Resource Description Model to describe resources in the IoT environment. They defined a Privacy class that protects the device from illegal access or control. However, ORDM did not offer fine-grained access control to the sensed data.

Blockchain-based privacy preservation solutions are based on the blockchain technology for enabling users to preserve their IoT data privacy while eliminating the need to trust a centralized regulator. For instance, Hashemi et al. [6] proposed a distributed data storage system, which used blockchain to maintain data access control and data storage model. For their part, Zyskind et al. [14] proposed a decentralized personal data management system, which used blockchain to keep track of both data and access transactions. However, the IoT devices have not sufficient resources to store the whole blockchain.

To the best of our knowledge, the combination of semantic modeling and blockchain technology for preserving IoT data privacy has never been explored.

3 PrivBlockchain Overview

Considering the legal rights imposed by the GDPR [11], it is necessary to ensure the privacy requirement compliance to preserve privacy during the whole data lifecycle, covering the collection, transmission, storage and processing phases. In our previous work [9], we have proposed PrivBlockchain, an end-to-end privacy-preserving framework for the IoT data based on the blockchain technology. PrivBlockchain aims at enforcing these privacy requirements and obligations for the IoT environment.

PrivBlockchain is based on the main following principles. **User-driven and transparency:** The user is the master of his own data since he has a full control over the data he shared in the network. **Fairness:** Using the blockchain in our end-to-end privacy-preserving framework improves fairness because nobody could systematically be enforced to lose control over his own data. **Distributed architecture and the lack of a central authority:** Each node in the network directly shares its data with other nodes, without the intervention of any third or trusted entity to manage the whole network. **Fine-granularity:** The use of a smart contract enables the user to implement expressive and granular privacy policies over our framework.

Figure 1 depicts PrivBlockchain, the proposed architecture that includes two types of network: first, the private IoT network, which can be a smart home, smart building, etc. This network includes the IoT resources owned by a data owner, which can be an individual or an organization. The second network is the public IoT network, which represents the external domain of the private IoT network. Moreover, we distinguish three IoT network node types, namely private, public, and storage nodes. Both public and storage IoT nodes belong to the public network. The private node (i.e., Semantic IoT Gateway or private IoT resource) is an IoT node that belongs to both the private and public IoT networks.

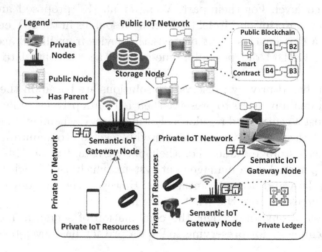

Fig. 1. PrivBlockchain architecture

In the private IoT network, each data owner has one or more high resource devices, known as the "Semantic IoT Gateway", which is responsible for the other owned IoT resources. The communication between the owned IoT resources by the data owner (i.e., the private IoT resources) is stored in a private lockchain called the "private ledger". The communication between the private IoT nodes and the other nodes of the public IoT network is stored in a "public blockchain".

We outline the proposed framework core components in the following section.

4   PrivBlockchain Core Components

This section discusses the main blockchain-based solution components. Indeed, the PrivBlockchain framework consists of nine core components, such as smart contract, transaction, private IoT network, private ledger, Semantic IoT Gateway, local storage, public IoT network, public blockchain, and storage node.

4.1   Smart Contract

Two parties can share a set of conditions by signing a common agreement. This kind of published agreement within the blockchain is known as a smart contract, which contains a code and defines a set of functions. For instance, the smart contract can define the constructor function that enables to create the smart contract itself. The sender of the transaction (i.e., network node) that invokes the constructor function becomes the smart contract owner.

A self-destruct function is another example of the functions that can be defined in a smart contract. Usually, only the smart contract owner can destruct the contract by invoking this function. A smart contract is likely to be a class that contains state variables, functions, function modifiers, events, and structures [4]. Besides, it can even call other smart contracts. We represent the smart contract, which is denoted as SC, as a tuple that has the following form:

$$SC = <states, functions>$$

- **States:** they are variables that hold some data or the owner's Ethereum wallet address (i.e., the address in which the smart contract is deployed). We can distinguish between two state types, namely *constant states*, which can never be changed, and *writable states*, which save states in the blockchain.
- **Functions:** they are pieces of code that can read or modify states. We can distinguish between two function types, namely *read-only functions*, which are marked as constant in the code and do not require *gas* to run and *write functions* that require *gas* because the state transitions must be encoded in a new block of the blockchain.

In order to invoke one smart contract function, a transaction needs to be created.

4.2   Transaction

Communication between IoT resources and network nodes is known as a transaction. In our work, we define a set of transaction types. T_{Add} and T_{Remove} transactions are generated by a Semantic IoT Gateway to add a new private IoT resource or to remove it from the network. $T_{LocalStore}$ transaction is generated by IoT resources to locally store the data. T_{Store} transaction is generated by IoT resources to store data on a Storage Node that can be a cloud storage provider. T_{Access} transaction is generated by an IoT resource, gateway node or IoT network node to access a shared data. $T_{Monitor}$ transaction is generated by an IoT network node to periodically receive near real-time collected data by IoT resources.

Moreover, $T_{GetPermission}$, $T_{GrantPermission}$, and $T_{GetSharedResource}$ transactions are used to ask for permissions to (i) access a specific IoT resource output, (ii) define a set of permissions to a data consumer by a data owner, and (iii) handle the shared data by an allowed data consumer. Each transaction contains a set of parameters, as depicted in Fig. 2, such as the previous transaction identifier to chain transactions, the current transaction identifier, and the transaction type.

Fig. 2. Transaction structure

Lightweight cryptography, such as AES Encryption [7], is used by IoT resources to secure the transactions during the communication. It should be noted that all the transactions between the IoT resources and the Semantic IoT Gateway occur in a private IoT network.

4.3 Private IoT Network

The private IoT network is an area, like a smart home or a smart building, where its owner can control a set of owned IoT resources. Indeed, the private IoT network includes a set of private IoT resources and Semantic IoT Gateway nodes, which are high resource devices that validate communication between the private IoT resources and link these private resources with the public IoT network. In our work, we distinguish between two node types, namely *full nodes*, which process every transaction and store the entire blockchain, and *light nodes*, which only store the relevant information, such as the gateway node and smart contract addresses due to their limited resources. In our private IoT network, the gateway nodes are full nodes while the private IoT resources are light nodes. Each private IoT network maintains a private ledger.

4.4 Private Ledger

A private ledger is a local private blockchain that enables the data owner to control his own IoT resources. This blockchain contains the data owner's private IoT resource communication and has a set of smart contracts that enforce the data owner's privacy preferences on how his IoT resources must behave. Transactions are chained together in a block. Each block in the private ledger contains a block header, which is the hash of the previous block to keep the blockchain immutable. Besides, each block contains a list of transactions (see Fig. 1).

The private ledger is kept and managed by a set of Semantic IoT Gateway nodes.

4.5   Semantic IoT Gateway Node

A Semantic IoT Gateway is a device with high memory and storage capabilities. Each gateway node is responsible for a set of private IoT resources, generates their keys and adds them to the IoT network. In our proposal, IoT resources with low memory and storage capabilities, such as a Beaglebone or an Arduino board, can delegate complicated treatments to the Semantic IoT Gateway. Moreover, it validates the incoming and the outgoing transactions before adding them to the private ledger. On the other hand, the Semantic IoT Gateway is considered as a public node in the public IoT network. In fact, it communicates with both the public and storage nodes, and stores a copy of the public blockchain to benefit from the IoT applications that are offered by the public IoT network nodes. For this purpose, it uses another couple of public and private keys different from the couple used in the private IoT network to reduce the linkability problem.

Furthermore, the Semantic IoT Gateway manages a local storage.

4.6   Local Storage

Local storage is a storing device, which is used to store data locally. It saves the collected data by IoT resources for a long-term storage before sending them to the external storage center, which is the storage node in our case. Each data block is stored using its Data Block ID. In case of a data center failure, the data can be restored from the local storage using the unique data identifiers. The local storage provides an additional capability to the Semantic IoT Gateway to belong as a public node to the public IoT network.

4.7   Public IoT Network

The proposed public IoT network is a peer-to-peer network (P2P) that contains several nodes with different memory and storage capabilities. The public nodes can be a gateway, a storage, or a public node. These nodes require a high memory and storage capabilities to store the public blockchain. Each IoT network node has a unique pair of public (PK) and private (SK) keys. The former, which is known by the other public nodes in the IoT network, is used as a unique node identifier to communicate (send/receive) transactions from the other nodes in the public IoT network while the latter, which is kept secret to the node, is used to sign transactions before sending them. Then, the signature is verified using the node's PK in the transaction. The digital signature, which is the hash of a digital asset (i.e., a transaction), improves transaction sender's authentication (i.e., proves that the transaction sender has the appropriate couple of public key and blockchain address), non-repudiation (i.e., the sender cannot deny having sent a transaction), and integrity (i.e., proves that a transaction is not altered while transmitted). Only valid transactions can be added to the public blockchain and distributed between all the public IoT network nodes. The public IoT network maintains a public blockchain.

4.8   Public Blockchain

The public blockchain can be seen as the history of all the transactions that are sent by the public nodes to access or share IoT data in the public IoT network. In fact, it can ensure auditing functions. Hence, our solution offers a non-repudiation principle compliance, which consists in preventing any public IoT network node from denying actions that are performed by itself. Furthermore, the public blockchain contains smart contracts that enforce the data owner's privacy preferences on how his data must be handled. In fact, the smart contract can be considered as data owner's privacy policy that specifies obligations for handling the shared IoT data. The public blockchain is stored on public and storage nodes.

4.9   Storage Node

The storage node is proposed as a public IoT network node that offers a storing service for both public blockchain and data collected by the IoT resources. For instance, the storage node can be a cloud storage service. Each data owner has the choice whether to use a different storage node for each of his IoT resources or the same storage node to store all of his collected IoT data. It is worth noting that the use of separate storage nodes can reduce privacy hurdles, especially the linkability issue [12]. Thus, separate databases must be created in such a way that common attributes are avoided.

After presenting an overall design architecture of PrivBlockchain, we focus on the core component, which is the Semantic IoT Gateway in the following section.

5   Semantic IoT Gateway

Typically, gateways collect and send the collected data from IoT resources like sensors and actuators to external platforms in order to be remotely analyzed. In order to enable a better control over their private IoT resources, a more flexible gateway is required by the users. For this purpose, we propose a Semantic IoT Gateway that aims at converting the data owner's privacy preferences into smart contracts that will be published within the blockchain to be enforced. As aforementioned, our Semantic IoT Gateway is based on an IoT privacy ontology, called LIoPY and defined in our previous work [8].

The architecture of the Semantic IoT Gateway is shown in Fig. 3. It includes four core components, which are: (i) the Semantic Rule Manager, which aims at matching the data owner's privacy preferences and the data consumer's terms of service in order to generate an adapted privacy policy, (ii) the Smart Contract Factory, which converts the privacy policy into a custom smart contract that will be hosted in the blockchain to enforce the privacy requirements. Moreover, it generates three smart contract types, namely *PrivacyPermissionSetting*, *Ownership*, and *PrivacyPolicy*. The two first smart contracts are published in the private ledger while the third is published in the public blockchain, (iii) the MQTT Client, which enables the Semantic IoT Gateway to subscribe to the data consumer's terms of service and publish the custom smart contract parameters using the MQTT standard, which is a lightweight publish/subscribe messaging protocol, and (iv) the Blockchain Client, which is considered as an access point to the blockchain network to receive a blockchain address and access this latter.

All the Semantic IoT Gateway components interact among them and with the external network in order to preserve the IoT data privacy. We detail below those components, the associated processes/workflows, and an example of a smart contract generation protocol.

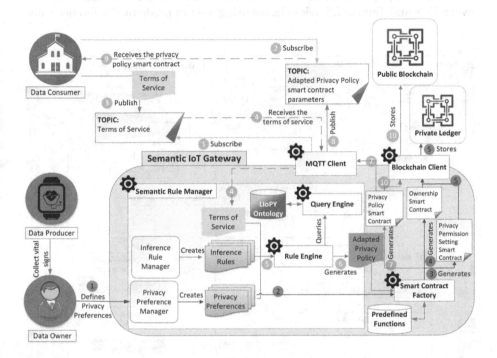

Fig. 3. Architecture of Semantic IoT Gateway

5.1   Semantic Rule Manager: From Privacy Preferences to Privacy Policy

The Semantic Rule Manager provides the Semantic IoT Gateway with semantic capabilities that enable to infer additional knowledge from the defined concepts in the European Legal compliant IoT Privacy-preserving ontologY (LIoPY) [8]. The main purpose of our LIoPY ontology is to enable inferring a privacy policy that aims at protecting privacy during the whole process of collecting, transmitting, storing, and processing the collected data by smart devices.

Figure 4 provides an overview of LIoPY. In order to cover the whole privacy aspects, the LIoPY contains three main modules, namely IoT resource management, IoT description, and IoT resource result sharing management. Each module includes a set of sub-modules. We referred to the data owner's privacy preferences by the Privacy_Rule class and to the data consumer's terms of service by the Terms_of_Service class. Both of these two classes are associated to a set of privacy requirements depicted by the Privacy_Attribute class, which has a set of subclasses, namely Consent, Purpose, Retention, Operation, Condition, and Disclosure. These subclasses specify for what reason, for how long, how, under which conditions the owner's data will be handled, and to whom they can be disclosed. The Semantic Rule Manager is based on our matching algorithm described in [8] to provide the appropriate policy that matches the privacy requirements of both Privacy_Rule and Terms_of_Service classes using a set of predefined inference rules.

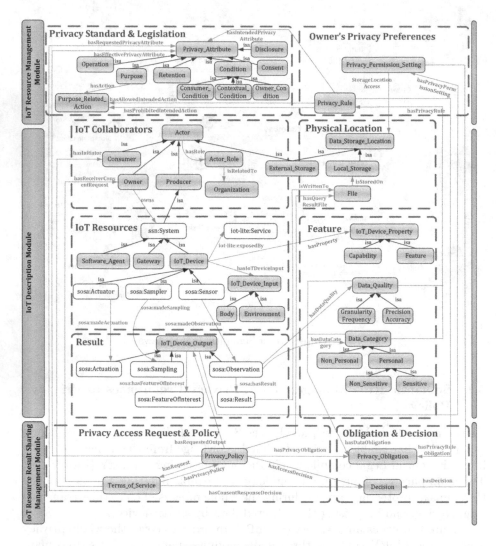

Fig. 4. LIoPY overview

Once a privacy policy is generated by the Semantic Rule Manager, it is converted to a smart contract by the Smart Contract Factory that is detailed below.

5.2   Smart Contract Factory: From Privacy Policy to Smart Contract

As aforementioned, three smart contract types are proposed, namely *PrivacyPermissionSetting*, *Ownership*, and *PrivacyPolicy*. The first and second contracts enforce the data owner's privacy preferences on how his IoT resources must behave according to each data output while the third enforces the data owner's privacy preferences and requirements on how his data must be handled once shared.

PrivacyPermissionSetting Smart Contract. In order to add a new *PrivacyPermissionSetting* smart contract, the Semantic IoT Gateway creates the contract and deploys it in the private ledger. Each IoT resource that knows the smart contract address can use it by invoking its defined functions.

The *PrivacyPermissionSetting* smart contract is designed to store the permission for each IoT resource concerning a specific IoT resource output according to the data owner's privacy preferences. This smart contract defines a set of functions, namely: (i) **LocalStore** function that enables to verify the IoT resource permission to locally store its collected data, (ii) **ExternalStore** function that verifies if the IoT resource has the permission to send the collected data to be stored on an external storage node, (iii) **Read** function that verifies if the IoT resource has the permission to request data from other internal or external IoT resources after verifying the IoT resource permissions, (iv) **Write** function that enables an IoT resource to add and/or modify a requested data collected by other internal or external IoT resources if the IoT resource is permitted, and (v) **Monitor** function that enables to verify the IoT resource permission to receive periodic data from another IoT resource. Furthermore, the *PrivacyPermissionSetting* smart contract includes a self-destruct function. Only the Semantic IoT Gateway can invoke this function to destruct the smart contract in order to revoke the granted privacy permissions for all the IoT resources associated with this contract. It is worth noting that when destructing the smart contract, it will be inoperable, but its history remains in the private ledger.

Ownership Smart Contract. The Semantic IoT Gateway creates an *Ownership* smart contract in order to store its own IoT resource addresses. For each IoT resource, a set of outputs is added. A *PrivacyPermissionSetting* contract is associated with each IoT resource output. Thus, the smart contract address is stored on the *Ownership* smart contract and sent to the appropriate IoT resource according to its data outputs and granted permissions.

The *Ownership* smart contract is designed to enforce data owner's control over his IoT resources and their outputs. It defines a set of functions, namely: (i) **addNewIoTResource** function, which enables to add a new IoT resource by indicating an IoT resource address, an IoT resource output, and the address of the *PrivacyPermissionSetting* smart contract, which is associated with the IoT resource output, (ii) **modifyIoTResource** function, which enables to modify the description of an existing IoT resource except for the set of its outputs, (iii) **removeIoTResource** function, which enables to remove an existing IoT resource, (iv) **addIoTResourceOutput** function, which enables to add a new output to an existing IoT resource by indicating a description output and the associated *PrivacyPermissionSetting* smart contract address, (v) **modifyIoTResourceOutput** function, which enables to modify the description of an existing IoT resource output, and (vi) **removeIoTResourceOutput** function, which enables to remove an existing IoT resource output from an existing IoT resource.

PrivacyPolicy Smart Contract. A *PrivacyPolicy* smart contract is created when a data owner wants to share a new IoT resource output with consumers. A set of subscribers can be added to the allowed consumer's list. This smart contract is designed to enforce the data owner's privacy preferences on how his IoT resource outputs must be handled once shared. The *PrivacyPolicy* smart contract contains many data fields, such as data hash, data path hash (while the data path is sent in a private transaction over HTTPS), creation date, and a set of consumer's addresses. Each consumer is defined with various permissions. The defined **addConsumer** function enables to add a new consumer by indicating its address, and a set of permissions relevant to the privacy requirements, such as the allowed action according to the chosen purpose, operation, retention duration, disclosure limitation, etc. Moreover, a set of conditions can be added to each existing consumer's permission, such as the allowed location consumer's address, time of day for handling the shared data, and the allowed role of the consumer's address. To this end, an **addCondition** function is defined. When the retention duration ends, the consumer's address is automatically removed from the consumer's list by invoking the **removeConsumer** function. Besides, this function is used when the data owner wants to revoke the permissions of a specific consumer. In case of the file content modification, the **updateFile** function needs to be invoked in order to keep consistency between the file hash that is stored on blockchain and the off-blockchain stored file content. Similar to the smart contract owner, a consumer with a *write permission* can invoke this function in order to change the hash of the file content. It should be noted that the use of data hash enables the data integrity.

After presenting the two core components of the Semantic IoT Gateway, we define an example of a smart contract generation in the following subsection.

5.3 *PrivacyPolicy* Smart Contract Generation Protocol

Figure 5 depicts the business process of generating and adding a new *Privacy-Policy* smart contract to the public blockchain. The parameters of such a smart contract are only generated in case of a match between the data consumer's terms of service and the data owner's privacy preferences.

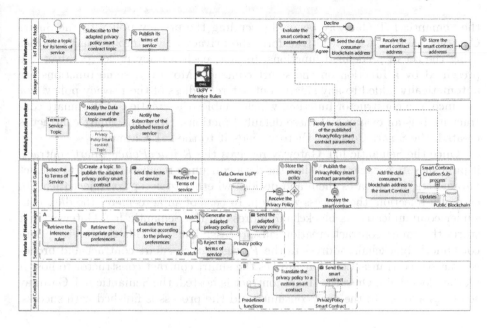

Fig. 5. Adding a new *PrivacyPolicy* smart contract business process notated in BPMN

Each IoT public node (i.e., data consumer in our case) publishes its terms of service by creating the appropriate topic according to the data that will be used. An MQTT broker manages these topic and broadcasts the published messages to the appropriate subscribers. The Semantic IoT Gateway subscribes on the data consumer's terms of service topic and creates a new topic in which it will publish a set of *PrivacyPolicy* smart contract parameters. The Broker notifies the data consumer of the topic creation to subscribe. Once the data consumer publishes new terms of a service, the privacy policy generation sub-process starts (see Fig. 5A). When the Semantic IoT Gateway node receives the terms of service, it communicates with the Semantic Rule Manager, which is responsible for reasoning about the received request and then taking a decision whether to create or not a privacy policy. First, the Semantic Rule Manager retrieves the predefined set of inference rules, which are stored on the storage node and shared between all the involved parties. These inference rules help the Semantic Rule Manager to retrieve the appropriate data owner's privacy preferences from the "Data Owner's LIoPY Instance" base, which is an instance of LIoPY ontology. Then, the Semantic Rule Manager matches the terms of service with the privacy preferences to infer an adapted privacy policy. In case of a match, a privacy policy is created and sent to the Semantic IoT Gateway that will store the privacy policy on the "Data Owner's LIoPY Instance" base and starts the smart contract generation sub-process. Otherwise, the Semantic Rule Manager rejects the terms of service and the process stops. Figure 5B shows the smart contract generation Sub-Process in details. In order to generate a new smart contract, the Semantic IoT Gateway begins by sending the privacy policy to the Smart Contract Factory, which will transform the privacy policy into a smart contract using the set of predefined functions. In fact, each privacy policy parameter is presented by a function on the smart contract. Moreover, some functions are automatically added to any smart contract regardless of the privacy policy. For instance, the constructor function, which enables the creation of the smart contract itself is an example of these default functions. Once the smart contract is created, the Smart Contract Factory sends it to the IoT Gateway.

Once the Semantic IoT Gateway receives the *PrivacyPolicy* smart contract, it publishes its parameters. When the data consumer receives the smart contract parameters, it evaluates them and decides whether to accept or reject them. In case of a rejection, the process will stop. In case of an agreement, the data consumer communicates its blockchain address to the Semantic IoT Gateway that starts the smart contract creation sub-process. Thus, it adds the received data consumer's blockchain address to the *PrivacyPolicy* smart contract and broadcasts a signed transaction that invokes the smart contract constructor to host it on the blockchain. Once the smart contract is hosted, the Semantic IoT Gateway sends its address to the data consumer and the process is finished with success.

6   Prototype and Validation

We implemented our proposed smart contracts using the Solidity language [2] and deployed it to the Ethereum test network. Because our system does not rely on the currency transfer, there is no difference between the real Ethereum network and the Ethereum test network. Ethereum is currently the most common blockchain platform for developing smart contracts [4].

We applied our solution to the following scenario from the healthcare domain in order to validate our solution:

A patient named Alice needs to follow a healthcare protocol, which consists in practicing some sport activities and eating healthy meals. Alice owns a wearable device that collects the user's heart rate. This IoT resource is connected to Alice's Semantic IoT Gateway. Alice regularly goes to a modern gym. During the training, the wearable device collects Alice's vital parameters and sends them to her gateway. The latter receives Alice's sensitive data and decides what information to send to the hospital to be stored on Alice's medical base, which is regularly checked by her doctor. These stored data are analyzed to propose personalized recommendations for patients. Hence, a need for a break or water notifications could be sent to Alice when necessary. Moreover, Alice wants to use a "Healthy Eating" application, which is offered by the gym with the aim of proposing a set of healthy meals according to the needed calories for each specific user.

However, Alice is afraid of losing the control over her data once shared. For this purpose, she uses her Semantic IoT Gateway that checks if the "Healthy Eating" application's terms of service match her privacy preferences. We assume that Alice's gateway and the IoT application are connected to the PrivBlockchain framework and each of them is identified by Ethereum addresses.

The depicted steps by the business process (see Fig. 5) are implemented to validate our proposal. First, the data consumer (i.e., "Healthy Eating" application administrator) creates an MQTT message that includes the requested data, and a set of privacy requirements, such as the use purpose, disclosure, retention duration, the requested operation, etc. The reason behind the use of JSON format for the MQTT message's content is to ensure an easy matching between the data consumer's terms of service and the data owner's privacy preferences, which are defined on RDF format. Listing 1.1 is an example of MQTT message content.

Listing 1.1. MQTT message content notated in JSON's format

```
 1  {
 2          "className": "Terms_of_Service",
 3          "individualName": "eatHealthy_TersmsOfService",
 4          "objectProperties": {
 5                  "hasRequestedPrivacyAttribute": ["Treatment_Purpose",
 6                  "Write", "Retention_180_days", "With_Requester"]
 7          },
 8          "dataProperties": {
 9                  "requested_data_name": "Heartrate"
10          }
11  }
```

When the Semantic IoT Gateway received the "Healthy Eating" application terms of service, it communicated with the Semantic Rule Manager that decided if the received terms of service match Alice's privacy preferences that are already stored on "Alice's LIoPY instance" base. In our case, the Semantic Rule Manager succeeded in inferring an adapted privacy policy that is depicted in Fig. 6.

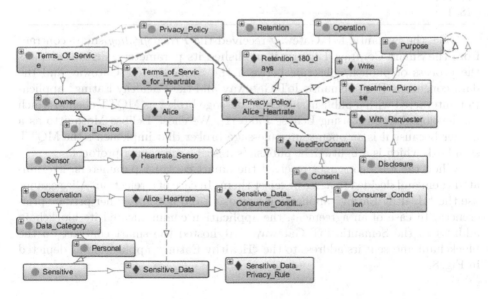

Fig. 6. The generated privacy policy by the Semantic IoT Gateway

Once the privacy policy is created, the Semantic IoT Gateway stored it on "Alice's LIoPY instance" base and sent it to the Smart Contract Factory, which converted the policy into a smart contract using a set of predefined functions. For instance, if in the privacy policy, the permitted operation is 'Write', then the *updateFile* function is added to the custom smart contract. Moreover, the Smart Contract Factory added a modifier, called *AllowedToWrite* that aims at verifying a set of conditions before executing the 'Write' operation. Listing 1.2 shows an example of code that is included in the generated *PrivacyPolicy* smart contract.

Listing 1.2. Example of a *PrivacyPolicy* smart contract including two Solidity predefined functions: *updateFile* function and *AllowedToWrite* modifier

```
 1 contract PrivacyPolicy {
 2 ...
 3 modifier AllowedToWrite(address _account) {
 4         require( owner == _account ||
 5             ( isConsumer[_account] && isAllowedToWrite[_account]
 6             && now < (now + retentionDuration[_account]) ) );
 7         _;
 8     }
 9 function updateFile(string new_data_file_name,
10             string new_data_file_path, bytes new_data_hash) public
11         AllowedToWrite(msg.sender) {
12         require(msg.sender == owner);
13         sharedFile = File(sharedFile.dataFileId, new_data_file_name,
14         new_data_file_path, new_data_hash, sharedFile.createdDate,
15         now, sharedFile.consumers, sharedFile.storageNode);
16     }
17 ...
18 }
```

Once the Semantic IoT Gateway received the *PrivacyPolicy* smart contract from the Smart Contract Factory, it published its parameters. Figure 7 depicts the process of publish/subscribe between the Semantic IoT Gateway and the data consumer. Both Semantic IoT Gateway and the "Healthy Eating" application interacted with a publish/subscribe message broker as MQTT clients, which we developed in Java using Eclipse Paho [1]. We chose Eclipse Mosquitto as a broker because it is an open-source message broker that implements the MQTT standard, which is a lightweight publish/subscribe messaging protocol.

When the data consumer received the smart contract parameters, it evaluated them and decided to agree or reject them. In case of a rejection, Alice cannot use the "Healthy Eating" application because it did not match her privacy preferences. In case of an agreement, the application communicated its blockchain address to the Semantic IoT Gateway that hosted the smart contract on the blockchain and sent its address to the "Healthy Eating" application as depicted in Fig. 8.

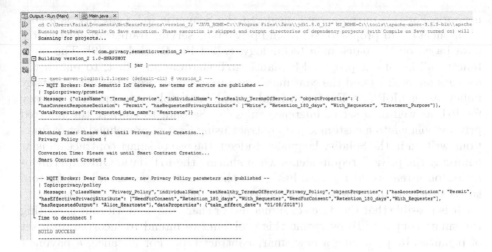

Fig. 7. Prototyping of the Publish/Subscribe process

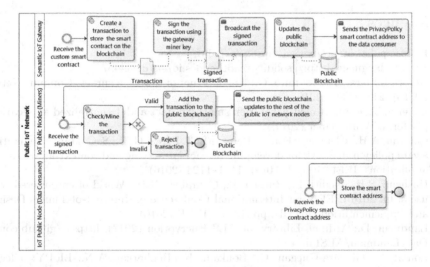

Fig. 8. Storing the *PrivacyPolicy* smart contract on the public blockchain

7 Conclusion

The blockchain technology is a distributed database that records all the transactions that have ever occurred in the network. The main feature of the blockchain is that it allows deploying smart contracts. In fact, a smart contract is an executable code that runs on top of the blockchain to facilitate, execute and enforce an agreement between untrusted parties without the involvement of a trusted third party. Hosting a smart contract in the blockchain can enforce privacy-preserving in the IoT domain. For this purpose, we defined a Semantic IoT Gateway that acts as a bridge between the IoT sensors, actuators and the blockchain network. For this purpose, we considered the Semantic IoT Gateway as a core component of our proposed end-to-end privacy-preserving framework for the IoT data based on the blockchain technology, called PrivBlockchain [9]. The main functionalities of our proposed Semantic IoT Gateway are: first, to match the owner's preferences and the consumer's requirements in order to infer a privacy policy using LIoPY, a European legal compliant ontology to preserve privacy for IoT as well as a set of inference rules [8]. Second, to convert the inferred privacy policy into a custom smart contract using a set of predefined set of functions written in the Solidity language. Indeed, the use of smart contract aims at enforcing the privacy requirements when sharing the IoT data. Our experimentation on a real-world use-case has given the expected results and the custom smart contracts are generated and added to the blockchain with success.

It is possible that the data consumer use or share the data without executing the smart contract. To overcome this problem, we intend to incorporate a set of penalties by proposing a new smart contract type. For instance, a payment would be automatic in case of breaking a contract by sharing data illicitly.

References

1. Eclipse paho. http://www.eclipse.org/paho/
2. Solidity language. http://solidity.readthedocs.io/en/develop/
3. Atzori, L., Iera, A., Morabito, G.: The internet of things: a survey. Comput. Netw. **54**(15), 2787–2805 (2010)
4. Buterin, V., et al.: A next-generation smart contract and decentralized application platform. White paper (2014)
5. Celdrán, A.H., Clemente, F.J.G., Pérez, M.G., Pérez, G.M.: SeCoMan: a semantic-aware policy framework for developing privacy-preserving and context-aware smart applications. IEEE Syst. J. **10**(3), 1111–1124 (2016)
6. Hashemi, S.H., Faghri, F., Rausch, P., Campbell, R.H.: World of empowered IoT users. In: 2016 IEEE First International Conference on Internet-of-Things Design and Implementation (IoTDI), pp. 13–24. IEEE (2016)
7. Landman, D.: Arduino Library for AES Encryption (2017). https://github.com/DavyLandman/AESLib
8. Loukil, F., Ghedira-Guegan, C., Boukadi, K., Benharkat, A.N.: LIoPY: a legal compliant ontology to preserve privacy for the internet of things. In: 2018 IEEE 42nd Annual Computer Software and Applications Conference (COMPSAC), pp. 701–706. IEEE (2018)
9. Loukil, F., Ghedira-Guegan, C., Boukadi, K., Benharkat, A.N.: Towards an end-to-end IoT data privacy-preserving framework using blockchain technology. In: 19th International Conference on Web Information Systems Engineering (WISE) (2018)
10. Maddox, T.: The dark side of wearables: how they're secretly jeopardizing your security and privacy. https://www.techrepublic.com/article/the-dark-side-of-wearables-how-theyre-secretly-jeopardizing-your-security-and-privacy/
11. Regulation, General Data Protection: Regulation (EU) 2016/679 of the European Parliament and of the Council of 27 April 2016 on the protection of natural persons with regard to the processing of personal data and on the free movement of such data, and repealing Directive 95/46. Official Journal of the European Union (OJ), **59**, 1–88 (2016)

12. Spiekermann, S., Cranor, L.F.: Engineering privacy. IEEE Trans. Softw. Eng. **35**(1), 67–82 (2009)
13. Wang, S., Hou, Y., Gao, F., Ma, S.: Ontology-based resource description model for internet of things. In: 2016 International Conference on Cyber-Enabled Distributed Computing and Knowledge Discovery (CyberC), pp. 105–108. IEEE (2016)
14. Zyskind, G., Nathan, O., et al.: Decentralizing privacy: using blockchain to protect personal data. In: 2015 IEEE Security and Privacy Workshops (SPW), pp. 180–184. IEEE (2015)

Crowdsourcing Task Assignment with Online Profile Learning

Silvana Castano$^{(\boxtimes)}$, Alfio Ferrara$^{(\boxtimes)}$, and Stefano Montanelli$^{(\boxtimes)}$

DI, Università degli Studi di Milano, Via Comelico, 39, 20135 Milan, Italy
{silvana.castano,alfio.ferrara,stefano.montanelli}@unimi.it

Abstract. In this paper, we present a reference framework called Argo+ for worker-centric crowdsourcing where task assignment is characterized by feature-based representation of both tasks and workers and learning techniques are exploited to online predict the most appropriate task to execute for a requesting worker. On the task side, features are used to represent requirements expressed in terms of knowledge expertise that are asked to workers for being involved in the task execution. On the worker side, features are used to compose a profile, namely a structured description of the worker capabilities in executing tasks. Experimental results obtained on a real crowdsourcing campaign are discussed by comparing the performance of Argo+ against a baseline with conventional task assignment techniques.

1 Introduction

In the recent years, the crowdsourcing philosophy has gained a lot of attention and many crowdsourcing systems/platforms appeared on the web scene for satisfying the growing need of marketplaces where the offer of requesters providing jobs to execute can meet the work-force provided by the crowd. For requesters, the key point is to get the jobs completed as fast as possible by also considering the tradeoff between the quality of the obtained results and the expenses to sustain [11,14]. For workers, the key point is to find a system where the crowd motivation is triggered and encouraged, as well as the profits are attractive (in terms of experience, advance in human capital, remuneration) [1]. Most of the crowdsourcing platforms (e.g., CrowdFlower - https://www.crowdflower.com/, Amazon Mechanical Turk - https://www.mturk.com/, Crowdcrafting - https://crowdcrafting.org/) are *task-centric*, in that they are designed to support different types of tasks and mechanisms for task result evaluation, mostly characterized by consensus- and/or inference-based techniques [19]. On the opposite, mechanisms for worker evaluation and tools for supervising the worker accuracy/trustworthiness can (strongly) differ from one system to another. Mostly, the worker evaluation is enforced by exploiting the results on the executed tasks: the more a worker provides good/valid results, the more a worker is considered as trustworthy (see for instance [5,27,29,30]). The worker reliability is progressively revealed during task execution and it becomes known at the end of activities.

© Springer Nature Switzerland AG 2018
H. Panetto et al. (Eds.): OTM 2018 Conferences, LNCS 11229, pp. 226–242, 2018.
https://doi.org/10.1007/978-3-030-02610-3_13

However, humans have different knowledge and abilities, thus a crowd worker can be trustworthy on a certain task campaign that is coherent with her/his attitudes, as well as she/he can be inaccurate on another campaign with different topic requirements not compliant with her/his attitudes. In other words, there is the need to switch from a task-centric to a *worker-centric* design paradigm to leverage on the human factors of crowd workers and effectively enforce the crowdsourcing task assignment. A key capability of the worker-centric paradigm is the availability of techniques for learning the worker-profile features during task assignment and execution [2]. We argue that the capability to effectively discover and represent the profile of engaged crowd workers is a strategic asset of future crowdsourcing marketplaces. This way, it becomes possible for a system to predict the worker trustworthiness on considered topics and to selectively choose a qualified and motivated crowd to recruit/involve in a given campaign according to the required knowledge/abilities.

In this paper, we present Argo+, a reference framework for worker-centric crowdsourcing where task assignment is characterized by feature-based representation of both tasks and workers and learning techniques are exploited to online predict the most appropriate worker for a given task. On the task side, features are used to represent requirements expressed in terms of knowledge expertise that are asked to workers for being involved in the task execution. On the worker side, features are used to compose a profile, namely a structured description of the worker capabilities in executing tasks along multiple dimensions (i.e., the features). From the system point of view, the goal of Argo+ is to predict the tasks on which a worker is expected to provide successful results based on the specified task requirements, thus increasing the number of successful results while reducing the number of task executions at the same time. From the worker point of view, the goal of Argo+ is to receive tasks that are compatible with her/his profile, thus leveraging on worker interest, motivation, and finally satisfaction. A further distinguishing aspect of Argo+ is related to the *dynamic* and *adaptive* nature of the worker profiles that are continuously evolving based on the results of executed crowdsourcing activities. Online learning techniques are employed in Argo+ to capture the real worker capabilities by progressively adjusting the features of worker profiles to resemble the features of tasks that have been successfully executed by the worker. In the paper, we focus on discussing an essential implementation of the Argo+ framework characterized by the use of (i) *probabilistic topic modeling* for enforcing task assignment, and (ii) techniques inspired to the *Rocchio relevance feedback* for enforcing worker profile learning. Experimental results are finally provided to analyze the behavior of Argo+ on both a third-party dataset and a real crowdsourcing campaign, by comparing the performance of the proposed framework against a baseline with conventional task routing techniques.

The paper is organized as follows. Section 2 provides the related work. The proposed Argo+ framework is presented in Sect. 3. The Argo+ techniques and the related implementation based on probabilistic topic modeling are illustrated in Sect. 4. Experimental results are discussed in Sect. 5. Concluding remarks are finally provided in Sect. 6.

2 Related Work

The idea to enforce task assignment (a.k.a. *task routing*) by considering specific human factors according to a sort of worker-centric crowdsourcing model is becoming popular in the very-recent crowdsourcing literature [2,16]. In [8], HITs (Human Intelligent Tasks) to execute are assigned to crowd workers by matching keywords extracted from task descriptions and worker preferences extracted from worker profiles on social networks. Limitations still exist due to possible incompatibilities (i.e., use of different keywords) among task features and social-network preferences. A similar approach is presented in [1] where *relevance* and *diversity* measures are introduced to capture the workers that are most appropriate for assignment of a certain task. A *cross-task crowdsourcing* approach (CTC) characterized by a transfer learning method and a bayesian model is proposed in [22] based on the intuition that the history of executed tasks can be exploited to extract knowledge on workers abilities. In CTC, the application target is a large crowdsourcing platform where many tasks are executed by each worker so that learning can be effectively employed. Budget constraints are introduced in [11] to enforce task routing with an incentive-compatible approach. Issues about task routing discussed from the marketplace point-of-view are presented in [14] where a comparative analysis of a set of well-known crowdsourcing platforms is provided. The goal of [14] is to observe the crowdsourcing ecosystem as a whole and to provide insights about possible platform improvements on task design and worker understanding empowerments. An approach based on the specification of Service Level Agreements (SLAs) is proposed in [18] where the task assignment mechanism is dynamically adjusted to fulfill a set of declared requirements. The satisfaction of human skill requirements can be included in the SLA for being exploited by the task scheduler. However, a fixed list of skills is specified in the SLA and it cannot be extended at runtime to catch the real worker attitudes with respect to the executed tasks.

The idea to improve the crowdsourcing performances through learning of worker expertise has been also envisaged. In [10], implicit and explicit learning stages are enforced to capture the predict the degree of worker success in executing specific kinds of tasks. In [24], the worker profile is predicted by observing the results on the executions of *taste* tasks and a *taste-matching* function is proposed to adjust the prevision according to the correct task answer (presuming that such a correct task answer is available). In [9], a warm-up phase is presented for the iCrowd framework to estimate the worker accuracy through the execution of an initial set of tasks with known answer. After warm-up, the tasks assigned to a worker are adaptively chosen according to the estimated worker accuracy based on the quality of provided answers. Other similar approaches are aimed at capturing the worker abilities by relying on cognitive tests in a sort of psycho-analytics approach (e.g., [12]). Sometimes, the cognitive model is based on a predefined taxonomy/ontology of worker skills taken from target crowdsourcing applications or reference skill classifications [13,17]. Learning requires the specification of task types to consider and focused learning tasks to use for training. Online learning techniques have been also proposed for improving the quality of crowdsourcing results. In [20], the problem of crowdsourcing labeling is modeled as a multi-armed bandit (MAB) problem where the goal is to learn the most effective combination of labelers for a given labeling task to execute. However, we stress that this solution is specifically conceived for the labeling problem under majority-voting employment. Other approaches based on (multi-armed) bandits algorithms are presented in [15, 28] where the focus is on optimizing task assignment in relation with predefined budget constraints.

Original Contribution. With respect to the related work discussed above, the Argo+ framework proposed in this paper is characterized by the following original contribution. First, the features used for representing the worker expertise are not predefined in type and number and the degree of worker expertise on a certain feature dynamically changes through learning functions based on the successful rate of executed tasks. Moreover, the specification of an initial worker profile is not mandatory in Argo+, thus allowing the system to start the crowd-sourcing activities without requiring the execution of an extra bulk of profiling tasks/questions devoted to recognize the worker abilities in advance with respect to the participation to real crowdsourcing campaigns. The evolving nature of the Argo+ worker profile captures the worker capabilities that have been concretely employed and shown in task execution, thus allowing to choose the work-force of a task by selecting workers that are really expected to provide a successful contribution/answer.

3 The Argo+ Framework

Consider a *requester*, namely a *campaign manager*, submitting a crowdsourcing campaign C based on a set of tasks $T = \{t_1, \ldots, t_n\}$ for execution on a crowdsourcing system CS by a crowd of workers $W = \{w_1, \ldots, w_k\}$.

A *task* $t \in T$ is defined as $t = \langle id_t, a_t, m_t, d_t, F_t \rangle$, where id_t is the unique *task identifier*, a_t is the *task action*, m_t is the *task modality*, d_t is the *task description*, and F_t is the set of *task-features*. A task action a_t denotes the task target, namely the goal that needs to be satisfied through crowd execution (e.g., picture labeling, movie recognition, sentiment evaluation). A modality m_t represents the kind of worker answer required in task execution. Conventional modalities are (i) *creation* for denoting that the worker is called to generate a free task answer, and (ii) *decision* for denoting that the task answer is chosen by workers within a set of possible alternative options. Further task modalities are possible, such as for example *rating* and *ranking*, and details are provided in [6]. A description d_t represents the task request given to each worker for illustrating what is demanded to her/him in the task execution. For instance, in a task t with action picture labeling and decision modality, the description d_t can be recognize the historical period of the following artwork among the given options. A set of task-features F_t provides a description of task requirements, namely a specification of the capabilities expected from a worker for being involved in the execution of the task t. For each feature $f \in F_t$, a *task-feature weight* $\omega(f)$ is associated to denote the relevance of f within the task-features F_t. A task-feature $f \in F_t$ is a label that denotes an expected worker expertise in a specific domain knowledge. For instance, the task-features of the considered task t can be art history and renaissance to denote required domain knowledge.

A *worker* $w \in W$ is defined as $w = \langle id_w, F_w \rangle$, where id_w is the unique *worker identifier* and F_w is the **worker profile** expressed as a set of worker-features. A worker-feature $f \in F_w$ denotes a worker capability, namely knowledge expertise, and it is associated with a *worker-feature weight* $\omega(f)$ denoting the "degree" of expertise/ability associated with the worker. In Argo+, a worker profile is *adaptive* and *evolving*, meaning that (i) the worker-features F_w of a worker $w \in W$ are dynamically determined based on the executed tasks and (ii) the associated worker-feature weights are progressively adjusted (i.e., learned) based on the quality of executed tasks. For instance, given a set of tasks T with task-features F_t, the more a worker w successfully executes the tasks T, the more the worker-features in F_w and the corresponding worker-feature weights become similar to the task-features of F_t.

The Argo+ framework and the conceptual schema of the underlying worker/task management repository are shown in Fig. 1. The framework is articulated in the following components:

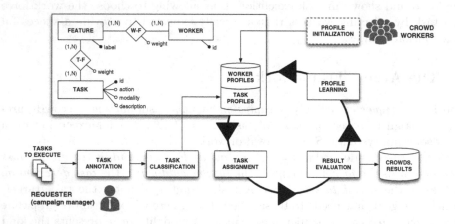

Fig. 1. The Argo+ framework

Task Annotation. This component has the goal to associate a task $t \in T$ with the corresponding set of task-features F_t and related task-feature weights. In a basic scenario, we expect that task annotation is manually performed by the requester, who has the role to choose the set of task-features F_t and to setup the corresponding weights in force of her/his understanding of the crowdsourcing campaign. The requester determines the composition of the set F_t by specifying requirements on the worker domain knowledge. Usually, tasks characterized by a common action, modality, and description are associated with the same set of features F_t. Since a campaign is typically characterized by few different types of task requests, manual task annotation can be considered a viable and effective solution.

Profile Initialization. This component has the goal to associate a worker $w \in W$ with the worker-features F_w and corresponding weights $\omega(f)$. In a crowdsourcing campaign, for each worker, the set of worker-features F_w coincides with the set of task-features F_t associated with the tasks T to execute in the crowdsourcing campaign. A weight $\omega(f)$ is then assigned to each worker-feature $f \in F_w$. Two alternative solutions are envisaged in Argo+ for profile initialization. Similarly to most of the existing crowdsourcing systems, profile initialization can be enforced through an initial *positioning questionnaire* based on ad-hoc questions in which workers self-evaluate their knowledge/abilities before their involvement in task execution (*custom profile initialization*). The questionnaire is usually provided by the campaign manager and it is defined according to the features of tasks T constituting the campaign. In this solution, the result of profile initialization is that task assignment is initially based on worker preferences as determined through the questionnaire answers. As an alternative, the worker profile can be initialized according to a default configuration which is common to any worker, thus avoiding both the preparation and the execution of a positioning questionnaire (*flat profile initialization*). In this solution, each worker profile gradually evolves from the common initialization towards the personal profile as learned from executed tasks. The flat initialization grounds in the idea that the self-evaluation of worker knowledge is subjective and untrustworthy due to possible over-/under-estimation of real worker expertise, thus a flat profile initialization is more effective than a custom one.

Task Classification. This component has the goal to aggregate the tasks T into K classes, so that tasks with similar features F_t are associated with a same class. A number of categorization approaches can be exploited for task aggregation, ranging from distance-based algorithms to probabilistic models based on latent features mined from data [23]. As a general remark, we recommend that the adopted solution for task classification is characterized by *overlap support*, meaning that a task can be associated with more than one class due to different similarities causes with other tasks. This way, it is possible that a task with a plurality of features has multiple associated classes and it can be exploited by workers with different expertise, each one focused on a different class.

Task Assignment. This component has the goal to choose the work-force of each task $t \in T$ and to schedule the related execution. In Argo+, task assignment is expected to follow a *worker-centric* criterion. This means that assignment is triggered by a worker w asking for a task to execute, which is determined "on-the-fly" by the system according to the w profile. Argo+ exploits the results of task classification and it detects the most appropriate task class k for the worker w, then the task t with the highest probability of being associated with k is assigned to w. The most appropriate task class for a worker w is the class k that best fits the w profile, thus maximizing the probability of w to successfully execute a task t in k.

Result Evaluation. For each task $t \in T$, this component has the goal to determine the final task result $\alpha(t)$ on the basis of the answers provided by workers involved in the t execution. Different solutions can be employed by a crowdsourcing system CS for determining $\alpha(t)$. Popular solutions are based on *majority voting* mechanisms where the final task result corresponds to the answer that obtained the majority of preferences by the involved workers [5]. Alternative solutions are characterized by *statistics-based* techniques where the final task result is determined by considering the distribution of collected worker answers through the combination of one or more statistical indicators, like for example arithmetic mean, variance, or deviation [6]. As a result, the final task result $\alpha(t)$ is determined by relying on the adopted solution. Given a task t, a further goal of result evaluation is to assess the quality of each collected worker answer in light of the determined $\alpha(t)$. We say that a worker w provided a *successful contribution* to the task t when the worker answer coincides with (or is equivalent to) $\alpha(t)$ according to the employed techniques for task result evaluation. Otherwise, we say that a worker w provided an *unsuccessful contribution* to the task t. According to this, we define the *worker-task result* $\rho(w, t)$ as follows:

$$\rho(w, t) = \begin{cases} 1 \text{ if } w \text{ provided succ. contrib.} \\ 0 \text{ otherwise} \end{cases}$$

Profile Learning. This component has the goal to update/evolve the worker profiles according to the results of executed tasks. The idea of Argo+ is to progressively learn the knowledge of a worker w by considering the features of tasks and related weights on which w provided successful contributions. For effective profile learning, update/evolution operations on the profile of the worker w should be triggered each time w executes a task t, on the basis of the quality of provided task answer. This way, learning results can affect all the subsequent task assignments for the worker w. However, we need to consider that real crowdsourcing systems are mostly characterized by *group-based task assignment* in which a task t is assigned to a set of independent workers, each one called to autonomously execute t. As a result, the final task result $\alpha(t)$ of a task t can be determined only when a certain number of worker answers are collected. In most situations, this means that, the quality of answer provided by a worker w to a task t cannot be determined immediately after the execution of t by w. A basic solution to this issue is the use of *delayed learning actions*, in which update operations to worker profiles are enforced when the final results of executed tasks become available. A more effective solution is the use of *gang-based learning actions*, in which the quality of worker answers is immediately determined after a task execution according to probability-based predictions determined by observing the behavior of similar workers (that can be considered as a "gang") [7].

4   Techniques for Argo+

In the following, we present a possible implementation of Argo+ based on the use of probabilistic topic modeling for realizing the task classification component. In particular, the proposed solution is characterized by the use of Latent Dirichlet Allocation (LDA) [4] over the task-features, which has the goal to produce a soft task aggregation based on two discrete probability distributions, namely ϕ and θ. ϕ describes the probability distribution of task-features on classes. In particular, ϕ^k denotes the probability of each task-feature f of being associated with the kth class on the K possible classes. θ describes the probability distribution of classes on tasks. In particular, θ_t denotes the probability of the task t of belonging to each class k among the K possible classes. Finally, we denote θ_t^k as the probability of the task t to be associated with the class k. The choice of K, namely the number of classes on which LDA works for task classification, is a configuration parameter and it is discussed in Sect. 5.

The results of task classification has an impact on (i) how to select the tasks for assignment to a requesting worker (task assignment component), (ii) how to update the worker profile according to the task results (profile learning component).

4.1 Assigning Tasks to Workers

Consider a worker w and associated worker profile F_w. When w asks for a task t to execute, the probability distributions (ϕ, θ) created by task classification are exploited. Through ϕ, Argo+ calculates the maximum a posteriori estimation θ_w given the worker features F_w. This is done by using collapsed Gibbs sampling [26] to learn the latent assignment of features to classes given the observed features F_w. In particular, we repeatedly estimate the probability $p(f \mid \phi^k)$ of a feature f to be assigned to a class k and we exploit this to estimate the probability $p(k \mid w)$ of the class k to be the correct assignment for the worker w. This sampling process is repeated until convergence, so that for each class $k \in K$ we finally estimate:

$$\theta_w^k \propto \frac{\sum\limits_{f \in F_w} \omega(f)_k}{\sum\limits_{f \in F_w} \sum\limits_{j \in K} \omega(f)_j}, \tag{1}$$

where $\omega(f)_i$ denotes the weight of features of type f that have been assigned to class i. Then, from the distribution θ_w, we select the class z such that:

$$z = \arg\max_{z \in K} \theta_w^z. \tag{2}$$

Given z as the most relevant task class for the worker w, we assign to w the task t such that:

$$t = \arg\max_{t \in T} \theta_t^z. \tag{3}$$

We stress that a task t is available for assignment until the number of task executions expected by the system is reached, then it is marked as finished and it is excluded from the assignment mechanism.

Example 1. Consider to enforce Argo+ on a system with task classification based on $K = 10$. A worker w asks for a task to execute and the profile F_w is defined by the following features:

$F_w = \langle$ (web search, 0.85), (classification, 0.85), (smartphone, 0.51), (text, 0.34), ... $\rangle$

Starting from F_w, we exploit Eq. 1 in order to classify the worker w with respect to the classes K. The resulting distribution θ_w is:

K	1	2	3	4	5	6	7	8	9	10
θ_w	0.07	0.57	0.02	0.08	0.06	0.02	0.02	0.04	0.02	0.07

From θ_w, we exploit Eq. 2 to select the most relevant class for the worker profile, that is $k = 2$. The top-3 features associated with $k = 2$ in ϕ^2 are: classification, tweets, and web search, which motivates the relevance of the class with respect to the worker profile F_w. Given the class, it is now possible to exploit Eq. 3 in order to select a task t for worker execution. The features F_t of the task selected for assignment to w are:

$$F_w = \langle \text{ (web search, 1.0), (classification, 1.0), (smartphone, 1.0)} \rangle$$

4.2 Learning Worker Profiles

For updating a worker profile, Argo+ relies on learning techniques inspired to the *Rocchio relevance feedback* [21]. When a worker w executes a task t, we associate the worker w with a new set of features $F'_w = F_w \cup F_t$. We denote $\omega(f)^w$ as the weight of the feature f in F_w (possibly being 0 if f was not in F_w) and $\omega(f)^t$ the weight of feature f in F_t (possibly being 0 if f was not in F_t). Then, the new weight $\omega(f)'$ for each feature in F'_w is updated as follows:

$$\omega(f)' = \delta \cdot \omega(f)^w + (1 - \delta) \cdot \theta^z_t \cdot \rho(w, t) \cdot \omega(f)^t, \qquad (4)$$

where δ is a dumping factor in $[0, 1]$ that determines how much of the original weight of the profile features contributes to the new weight, and z is the class chosen for the task assignment. The idea behind profile update is that when a worker profile feature is not included in the task features, its weight is reduced by a factor δ. In the other case, the new profile feature weight $\omega(f)'$ is computed as the weighted sum between the previous profile feature weight $\omega(f)^w$ and the task feature weight $\omega(f)^t$, which contribution is proportional to the relevance θ^z_t of the task t in the class z. The task feature weight $\omega(f)^t$ is forced to be equal to 0 when the worker does not provide a successful contribution on the task (resulting in a reduction of the corresponding profile feature weight).

Example 2. Consider the task assignment of Example 1. The worker w executed t and $\rho(w, t) = 1$. We update F_w by applying Eq. 4. The updated worker profile F'_w is the following (the class-task relevance $\theta^2_t = 0.77$):

$F_w = \langle$ (web search, 0.78), (classification, 0.78), (smartphone, 0.74), (text, 0.03), ... $\rangle$

We note that the three features of t affects the worker profile by changing the relative feature weights. Features web search and classification remain the most relevant, but the weight of smartphone that is a feature of t is increased. On the opposite, the feature text of F_w becomes less relevant in the new worker profile, due to the fact that it is not part of the task feature set F_t. After the profile update, Argo+ will exploit the new worker profile for the subsequent task assignments to w.

5 Evaluation of Learning-Based Task Assignment in Argo+

For evaluation purposes, we rely on the Argo+ implementation presented in Sect. 4 and we consider two different experiments characterized by different task datasets called Amt-dts and Argo-dts, respectively.

Amt-dts contains the results of a third-party crowdsourcing campaign run on Amazon Mechanical Turk and presented in [25]. In Amt-dts, tasks are assigned to workers according to a motivation-based criterion by relying on the intuition that the quality of worker contributions can be improved when tasks are assigned to workers according to their motivation to execute tasks. Through Amt-dts, we consider a third-party dataset with the aim to compare Argo+ against a well-known crowdsourcing system equipped with a task assignment mechanism characterized by totally-different design principles.

Argo-dts, contains the results of a crowdsourcing campaign run through our Argo crowdsourcing system [5]. In Argo-dts, tasks are assigned to workers according to a trustworthiness-based mechanism, in which workers that demonstrate high reliability are rewarded and involved in tasks where commitment is hard to obtain (i.e., complex tasks). Through Argo-dts, the goal is to have a baseline comparison for observing the behavior of Argo+ against a basic version of expertise-based task assignment mechanism.

Our experimental evaluation is organized in two different experiments based on Amt-dts and Argo-dts, respectively. In both the experiments, the task answers contained in the datasets are considered as a baseline for comparison against Argo+ and the goal is to assess whether Argo+ succeeds in improving the task assignments with respect to the considered baseline.

In the following, we first present the experimental setup and associated evaluation metrics, then we discuss the obtained results in the two experiments with Amt-dts and Argo-dts, respectively.

5.1 Experimental Setup and Evaluation Metrics

Both Amt-dts and Argo-dts consist in a set of task answers collected from crowd workers on a given crowdsourcing campaign. Multiple answers are provided for each task by different workers and each worker is involved in the execution of a variable number of tasks.

Given a crowdsourcing execution over a given set of tasks T, we call **stream of answers** $\mathcal{S} = \{a_1, \ldots, a_n\}$ a sequence of worker task answers where $a = \langle w, t, \rho(w, t), r \rangle$ denotes that the worker w executed the task $t \in T$ by providing the worker-task result $\rho(w, t)$. The r value is the *request timestamp* denoting when the worker submitted the task answer and completed the task execution. According to Sect. 3, the worker-task result $\rho(w, t)$ is specified to distinguish the worker answers that represent a successful contribution from the others. In a stream of answers, given a worker w, it is possible to retrace the time-sequence of tasks executed by w which coincides with the task assignments received by w during time.

In both the experiments, the considered dataset, namely Amt-dts or Argo-dts, represents a stream of answers considered as a baseline for comparison against Argo+. Since the set of tasks concretely executed by each worker cannot be changed, our experiments are based on the idea to post-analyze the stream of answers in the baseline and to change the assignment sequence of tasks to workers according to the assignment schedule determined by Argo+. We aim at verifying whether the task assignment of Argo+ succeeds in capturing the worker profile and in improving the time-sequence of executed tasks, so that the tasks successfully executed by a worker w are assigned to w before than other tasks.

Given a stream of answers S, we call $\sigma(r, S)$ the *success rate* of S at the request timestamp r and it is defined as follows:

$$\sigma(r, S) = \frac{1}{r} \sum_{i=1}^{r} \rho(w, t)^i$$

where $\rho(w, t)^i$ is the worker-task result of the ith task execution at the request timestamp r in the stream of answers S. In the experiments, we compare two different streams of answers. One is the baseline stream (i.e., Amt-dts or Argo-dts) and one is the stream of answers of Argo+.

To this end, given two different streams of answers S_1 and S_2, we call *increment value* $I_r(S_1, S_2)$ the ratio between the success rate of S_1 and the success rate of S_2 at the timestamp r. The increment value $I_r(S_1, S_2)$ is defined as follows:

$$I_r(S_1, S_2) = \frac{\sigma(r, S_1)}{\sigma(r, S_2)}$$

Considering a stream of answers S, we call *assignment performance* σ_R^S the comprehensive success rate of S defined as follows:

$$\sigma_R^S = \int_1^R \sigma(r, S) dr$$

where R the overall number of successfully executed tasks in the stream of answers S, namely R is the sum of all the $\rho(w, t) = 1$ in S.

5.2   Results on the Amt-dts Dataset

The Amt-dts dataset contains 707 crowd answers about 22 different kinds of tasks, each one associated with a specific set of tags/keywords taken from a set of 39 thematic tags. The task keywords have been exploited for setting up the task features F_t expected by Argo+. In particular, for each keyword associated with a task t in Amt-dts, a corresponding task feature f is created in Argo+ for the task t with a corresponding weight $\omega(f) = 1$. Moreover, 23 workers are involved in the execution of dataset tasks, each one associated with a static set of featuring keywords. The same set of thematic tags is exploited in Amt-dts for describing both tasks and workers. In the experiment, two different Argo+ configurations are considered. One configuration with a flat worker profile (called Argo+noprofile) where $\omega(f) = 0$ is initially defined for each thematic tag (i.e., worker-feature), and one configuration with a custom worker profile (called Argo+profile) where $\omega(f) = 1$ if the tag/worker-feature is associated with the worker in Amt-dts.

In the experiment, Argo+ has been configured with $K = 10$ classes for task classification and a dumping factor $\delta = 0.3$ for worker profile learning. We note that choosing a high value of K produces classes with few and (usually) precise associated tasks, but the presence of few tasks per class negatively affects the learning mechanism. For instance, consider to choose K so that it corresponds to the number of available tasks. This way, each class is associated with very few tasks (probably just one per class). As a result, once that the task of a class k is assigned to a worker w, learning that w is capable to successfully execute the tasks of k is not useful since the class k is empty for w and subsequent tasks need to be picked up from other classes, thus enforcing a task assignment mechanism that is equivalent to a random choice. On the opposite, choosing a low value of K produces classes with many associated tasks. Given a class k, there are tasks t that are associated with k through a high probability value θ_t^k, but there are also a (usually long) tail of tasks with low probability values. This means that a worker w with a profile fitting the class k is satisfied of the initial tasks taken from k, but subsequent assignments risk to be inappropriate due to the fact that tasks are not so relevant for the class k (i.e., low probability value θ_t^k). In our experiments, the choice of $K = 10$ has been determined experimentally by exploiting *perplexity* which measures the ability of a model to generalize to unseen data [3]. In the experiment, we run the LDA algorithm considering a variable number of K and we calculated the corresponding perplexity value for each execution. Perplexity decreases as K increases and we decided to set K to the minimum value for which perplexity reaches a stable value (i.e., $K = 10$).

The comparison of the baseline Amt-dts ($\mathcal{S}_1$) against Argo+noprofile ($\mathcal{S}_2$) and Argo+profile ($\mathcal{S}_3$) is performed by comparing (i) the success rate $\sigma(r, \mathcal{S}_1)$, $\sigma(r, \mathcal{S}_2)$, and $\sigma(r, \mathcal{S}_3)$ (Fig. 2(a)), and (ii) the increment value $I(\mathcal{S}_2, \mathcal{S}_1)$ and $I(\mathcal{S}_3, \mathcal{S}_1)$ (Fig. 2(b)). We observe that both Argo+noProfile and Argo+Profile succeed in improving the success rate of Amt-dts, since successfully executed tasks are assigned to workers before than others in most cases. For the first 150 requests, the success rate of Argo+noProfile is around 20% better than Amt-dts. It is also interesting to note that at the very beginning of the system execution ($r < 10$) the behavior of Argo+noProfile and Argo+Profile is characterized by an unstable trend. We believe that this behavior is due to the fact that learning has insufficient information for recognizing the appropriate task class for each worker. However, Argo+ quickly learns the worker profile ($r \geq 10$) and this has a positive impact on the assignment of subsequent tasks. The performance of Argo+noProfile becomes similar to Amt-dts after the 300th worker request. This is due to the fact that Argo+ first selects tasks that are highly relevant for the worker profile, but subsequent assignments are about residual tasks of the K classes on which the relevance for the worker profile is weaker.

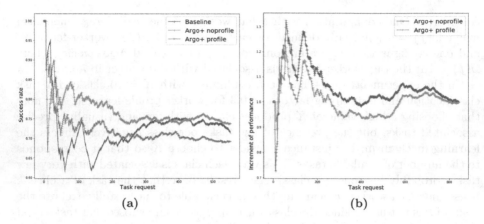

Fig. 2. (a) the success rate on executed tasks, (b) the increment value of Argo+ with respect to Amt-dts

Finally, we compare Amt-dts and Argo+ through the assignment performance measure σ_R and we obtain that $\sigma_R^{\mathcal{S}_1} = 399.59$, $\sigma_R^{\mathcal{S}_2} = 424.66$, and $\sigma_R^{\mathcal{S}_3} = 399.61$. As a result, we observe that the use of a flat initialization of worker profile provides the best performance on the three considered streams of answers (see also Fig. 2(b) on the increment value). This confirms the intuition behind the use of flat profiles which argues that the auto-evaluation of worker knowledge is usually misplaced with respect to the real worker expertise, and thus unreliable and sometimes damaging for the performance of the crowdsourcing system.

5.3   Results on the Argo-dts Dataset

The Argo-dts dataset contains $14,016$ crowd answers about $1,507$ tasks on paintings. The Argo-dts dataset has been collected during a crowdsourcing campaign between November and December 2017 which involved 367 students from the Faculty of Arts and Literature at the University of Milan. The painting tasks are about paintings from 56 different authors spanning from the 13th century to the 20th century. For each task, the worker is asked to examine a painting and to choose the correct author among a set of six possible painters. For painting, both task and worker features are taken from Wikidata[1] and they include the name of the author, the year, and the Wikidata thematic categories available for a painting. An example of task and worker answer is given in Fig. 3. Two different configurations of Argo+ are considered. One called Argo+noprofile is characterized by a flat worker profile where $\omega(f) = 0$ is initially defined for each feature (i.e., worker-feature) taken from Wikidata. One called Argo+profile is characterized by a custom worker profile where $\omega(f) = 1$ for each feature on which the worker has declared an expertise. A self-evaluation questionnaire has been submitted to workers about knowledge of painters and different periods in the art history

---

[1] https://www.wikidata.org.

for collecting the perceived worker expertise before starting the crowdsourcing activities. In the experiment, Argo+ has been configured with a number of classes $K = 30$ for task classification and a dumping factor $\delta = 0.3$ for worker profile learning. The number K has been determined through perplexity by following the same approach discussed in the Amt-dts experiment.

WORKER OPTIONS	TASK FEATURES	EXAMPLE OF WORKER ANSWER
○ Raffaello Sanzio ○ Gustav Klimt ○ Piero della Francesca ○ Francisco Goya ○ Giotto ○ Michelangelo Buonarroti	Raffaello Sanzio 1516 High Renaissance Portrait paintings of cardinals	{ "gold_answer" : "Q5597", "argo_answer" : "Raffaello Sanzio", "worker_answer_id" : "Q5432", "worker_answer" : "Francisco Goya", "task_id" : 1102, "answer_timestamp" : 2017-11-13T14:42:19, "worker_id" : 527, "task_refused" : false }

Fig. 3. Example of painting task in Argo-dts with associated task features and an example of (wrong) worker answer

The comparison of the baseline Argo-dts ($\mathcal{S}_1$) against Argo+noprofile ($\mathcal{S}_2$) and Argo+profile ($\mathcal{S}_3$) is shown for the first 200 tasks requests in Fig. 4(a) and (b), respectively. The Argo-dts experiment confirms that both Argo+noProfile and Argo+Profile succeed in significantly improving the success rate of the baseline. In comparison with the previous Amt-dts, this experiment with Argo-dts show that Argo+noProfile and Argo+Profile follow a very similar trend. This means that, in Argo-dts, the initialization of a custom worker profile does not provide a relevant impact, neither positive or negative, on the success rate of executed tasks. Similarly to the experiment with Amt-dts, it is interesting to note that an unstable trend on the success rate occurs at the beginning of task assignment. In this case, the instability behavior of the two Argo+ streams can be observed within an initial window-frame of around 30–50 tasks. This value is higher than the one observed in Amt-dts and (more or less) coincides with the number K of classes used for task classification (i.e., $K = 30$ in Argo-dts). We argue that this result highly depends on the use of a learning mechanism inspired to the Rocchio relevance feedback, in which the learned worker expertise is immediately exploited for subsequent task assignments. As a result, apart from the perplexity considerations, we argue that the use of a high number of classes for task classification is recommended in large crowdsourcing campaigns. This way, the initial window-frame in which tasks of different classes are assigned to each worker is enlarged and the learning mechanism can discover more worker expertnesses before starting focused assignments based on learned knowledge.

According to the increment values in Fig. 4(b), we note that Argo+Profile provides a slightly better performance than Argo+noProfile. However, the behavior of the two Argo+ configurations is very similar in comparison with the baseline. We argue that this result is due to the training information received by the involved workers before starting the crowdsourcing activities. In particular, workers were recommended to carefully fill out the self-evaluation questionnaire by providing fair degrees about their expertise in Arts and paintings. As a result, we have

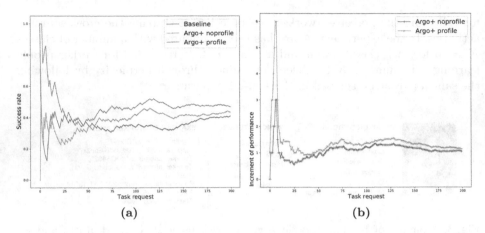

Fig. 4. (a) the success rate on executed tasks, (b) the increment value of Argo+ with respect to Argo-dts

that the use of a trustworthy crowd has a positive impact on the performance of the task assignment mechanism. However, the degree of trustworthiness of a crowd cannot be determined a-priori, before the crowdsourcing activities. Thus, we confirm that the adoption of a flat profile initialization is a more reliable solution that provides the better performance in the general case (see the results of the Amt-dts experiment).

6   Concluding Remarks

In this paper, we presented the Argo+ framework and a related techniques for crowdsourcing task assignment with online learning of worker profiles. An implementation based on techniques inspired to topic modeling and Rocchio relevance feedback is illustrated. Experimental results on a third-party dataset and a real crowdsourcing campaign show promising results by increasing the number of successfully executed tasks in a considered time frame.

Future research activities are devoted to extend the Argo+ framework to consider/support human skills, such as *originality*, *perceptual speed*, and *deductive reasoning*, in addition to knowledge expertise when modeling task features and worker profiles. The possible use of alternative techniques with respect to LDA and Rocchio-inspired learning mechanism will be also considered in future work. In addition, ongoing work are focused on worker profile management over different crowdsourcing campaigns. The idea is that the profile of a worker is saved when a crowdsourcing campaign is terminated, so that it can be eventually exploited in subsequent campaigns. This way, the learned worker experience contributes to positively affect task assignments on startup of crowdsourcing campaigns subsequently joined by a worker. This intuition goes in the direction to consider the crowd as a *permanent* component of a crowdsourcing system where the workforce to involve in a campaign can be dynamically composed according to the expertise degree of each worker on the tasks to execute. As a result, like in a real job-marketplace, it becomes possible that workers can negotiate their participation to crowdsourcing activities based on the contribution they can provide in terms of (profile-certified) expertise.

References

1. Alsayasneh, M., et al.: Personalized and diverse task composition in crowdsourcing. IEEE Trans. Knowl. Data Eng. **30**(1), 128–141 (2018)
2. Amer-Yahia, S., Roy, S.B.: Human factors in crowdsourcing. PVLDB **9**(13), 1615–1618 (2016)
3. Arun, R., Suresh, V., Veni Madhavan, C.E., Narasimha Murthy, M.N.: On finding the natural number of topics with latent Dirichlet allocation: some observations. In: Zaki, M.J., Yu, J.X., Ravindran, B., Pudi, V. (eds.) PAKDD 2010. LNCS (LNAI), vol. 6118, pp. 391–402. Springer, Heidelberg (2010). https://doi.org/10.1007/978-3-642-13657-3_43
4. Blei, D.M., Ng, A.Y., Jordan, M.I.: Latent Dirichlet allocation. J. Mach. Learn. Res. **3**(Jan), 993–1022 (2003)
5. Castano, S., Ferrara, A., Genta, L., Montanelli, S.: Combining crowd consensus and user trustworthiness for managing collective tasks. Futur. Gener. Comput. Syst. **54**, 378–388 (2016)
6. Castano, S., Ferrara, A., Montanelli, S.: A multi-dimensional approach to crowd-consensus modeling and evaluation. In: Johannesson, P., Lee, M.L., Liddle, S.W., Opdahl, A.L., López, Ó.P. (eds.) ER 2015. LNCS, vol. 9381, pp. 424–431. Springer, Cham (2015). https://doi.org/10.1007/978-3-319-25264-3_31
7. Cesa-Bianchi, N., Gentile, C., Zappella, G.: A gang of bandits. In: Proceedings of the 27th Internatioanl Conference on Neural Information Processing Systems, Lake Tahoe, Nevada, USA, pp. 737–745 (2013)
8. Difallah, D.E., Demartini, G., Cudré-Mauroux, P.: Pick-a-crowd: tell me what you like, and I'll tell you what to do. In: Proceedings of the 22nd WWW International Conference, Rio de Janeiro, Brazil (2013)
9. Fan, J., Li, G., Ooi, B.C., Tan, K.l., Feng, J.: iCrowd: an adaptive crowdsourcing framework. In: Proceedings of the ACM SIGMOD International Conference on Management of Data, pp. 1015–1030. ACM, Melbourne (2015)
10. Gadiraju, U., Fetahu, B., Kawase, R.: Training workers for improving performance in crowdsourcing microtasks. In: Conole, G., Klobučar, T., Rensing, C., Konert, J., Lavoué, É. (eds.) EC-TEL 2015. LNCS, vol. 9307, pp. 100–114. Springer, Cham (2015). https://doi.org/10.1007/978-3-319-24258-3_8
11. Goel, G., Nikzad, A., Singla, A.: Allocating tasks to workers with matching constraints: truthful mechanisms for crowdsourcing markets. In: Proceedings of the 23rd WWW International Conference, Seoul, Korea (2014)
12. Goncalves, J., Feldman, M., Hu, S., Kostakos, V., Bernstein, A.: Task routing and assignment in crowdsourcing based on cognitive abilities. In: Proceedings of the 26th WWW International Conference, Perth, Australia (2017)
13. Hassan, U., Curry, E.: A capability requirements approach for predicting worker performance in crowdsourcing. In: Proceedings of the 9th International Conference on Collaborate Computing, Austin, Texas, USA (2013)
14. Jain, A., Sarma, A.D., Parameswaran, A., Widom, J.: Understanding workers, developing effective tasks, and enhancing marketplace dynamics: a study of a large crowdsourcing marketplace. Proc. VLDB Endow. **10**(7), 829–840 (2017)

15. Jain, S., Narayanaswamy, B., Narahari, Y.: A multiarmed bandit incentive mechanism for crowdsourcing demand response in smart grids. In: Proceedings of the 28th AAAI Conference on Artificial Intelligence, Qulébec, Canada, pp. 721–727 (2014)
16. Karger, D.R., Oh, S., Shah, D.: Budget-optimal task allocation for reliable crowdsourcing systems. Oper. Res. **62**(1), 1–24 (2014)
17. Kazai, G., Kamps, J., Milic-Frayling, N.: Worker types and personality traits in crowdsourcing relevance labels. In: Proceedings of the 20th CIKM, Glasgow, Scotland, UK (2011)
18. Khazankin, R., Psaier, H., Schall, D., Dustdar, S.: QoS-based task scheduling in crowdsourcing environments. In: Kappel, G., Maamar, Z., Motahari-Nezhad, H.R. (eds.) ICSOC 2011. LNCS, vol. 7084, pp. 297–311. Springer, Heidelberg (2011). https://doi.org/10.1007/978-3-642-25535-9_20
19. Li, G., Wang, J., Zheng, Y., Franklin, M.J.: Crowdsourced data management: a survey. IEEE Trans. Knowl. Data Eng. **28**(9), 2296–2319 (2016)
20. Liu, Y., Liu, M.: An online learning approach to improving the quality of crowdsourcing. IEEE Trans. Netw. **25**(4), 2166–2179 (2017)
21. Manning, C.D., Raghavan, P., Schütze, H.: Introduction to Information Retrieval, vol. 1. Cambridge University Press, Cambridge (2008)
22. Mo, K., Zhong, E., Yang, Q.: Cross-task crowdsourcing. In: Proceedings of the 19th ACM SIGKDD International Conference, Chicago, Illinois, USA (2013)
23. Müller, E., Günnemann, S., Färber, I., Seidl, T.: Discovering multiple clustering solutions: grouping objects in different views of the data. In: Proceedings of the 28th IEEE ICDE International Conference, Washington, DC, USA, pp. 1207–1210 (2012)
24. Organisciak, P., Teevan, J., Dumais, S.T., Miller, R., Kalai, A.T.: A crowd of your own: crowdsourcing for on-demand personalization. In: Proceedings of the 2nd AAAI HCOMP, Pittsburgh, USA (2014)
25. Pilourdault, J., Amer-Yahia, S., Lee, D., Roy, S.B.: Motivation-aware task assignment in crowdsourcing. In: Proceedings of the 20th EDBT International Conference, Venice, Italy (2017)
26. Porteous, I., Newman, D., Ihler, A., Asuncion, A., Smyth, P., Welling, M.: Fast collapsed gibbs sampling for latent Dirichlet allocation. In: Proceedings of the 14th ACM SIGKDD International Conference, pp. 569–577 (2008)
27. Simpson, E., Roberts, S.: Bayesian methods for intelligent task assignment in crowdsourcing systems. In: Guy, T., Kárný, M., Wolpert, D. (eds.) Decision Making: Uncertainty, Imperfection, Deliberation and Scalability. SCI, vol. 538, pp. 1–32. Springer, Cham (2015). https://doi.org/10.1007/978-3-319-15144-1_1
28. Tran-Thanh, L., Stein, S., Rogers, A., Jennings, N.R.: Efficient crowdsourcing of unknown experts using bounded multi-armed bandits. Artif. Intell. **214**, 89–111 (2014)
29. Tranquillini, S., Daniel, F., Kucherbaev, P., Casati, F.: Modeling, enacting, and integrating custom crowdsourcing processes. ACM Trans. Web **9**(2), 7:1–7:43 (2015)
30. Zheng, Y., Li, G., Li, Y., Shan, C., Cheng, R.: Truth inference in crowdsourcing: is the problem solved? Proc. VLDB Endow. **10**(5), 541–552 (2017)

Distributed Collaborative Filtering for Batch and Stream Processing-Based Recommendations

Kais Zaouali[✉], Mohamed Ramzi Haddad[✉], and Hajer Baazaoui Zghal[✉]

Riadi Laboratory, École Nationale des Sciences de l'Informatique,
University of Manouba, Manouba, Tunisia
zaouali.kais@gmail.com, haddad.medramzi@gmail.com,
hajer.baazaouizghal@riadi.rnu.tn

Abstract. Nowadays, user actions are tracked and recorded by multiple websites and e-commerce platforms, allowing them to better understand their preferences and support them with specific and accurate content suggestions. Researches have proposed several recommendation approaches and addressed several challenges such as data sparsity and cold start. However, the low-scalability problem remains a major challenge when handling large volumes of user actions data. This issue becomes more challenging when it comes to real-time applications. Such constraint requires a new class of low latency recommendation approaches capable of incrementally and continuously update their knowledge and models at scale as soon as data arrives. In this paper, we focus on the user-centered collaborative filtering as one of the most adopted recommendation approaches known for its lack of scalability. We propose two distributed and scalable implementations of collaborative filtering addressing the challenges and the requirements of batch offline and incremental online recommendation scenarios. Several experiments were conducted on a distributed environment using the MovieLens dataset in order to highlight the properties and the advantages of each variant.

Keywords: Distributed recommender systems
Collaborative filtering · Stream data processing
Batch data processing · Incremental learning

1 Introduction

Thanks to recommender systems, users are now assisted by personalized suggestions when searching for goods or content on the internet. Such systems help reduce the information overload problem and enable users to discover suitable choices and make better decisions. These suggestions are inferred by recommendation approaches based on the collected historical data describing users' actions and behaviors. In fact, historical data enable recommender systems and marketers to model users' consumption patterns and therefore better personalize and tailor their communications and offerings to each one based on their

© Springer Nature Switzerland AG 2018
H. Panetto et al. (Eds.): OTM 2018 Conferences, LNCS 11229, pp. 243–260, 2018.
https://doi.org/10.1007/978-3-030-02610-3_14

predicted future interests. Such smart systems are not only beneficial for users but also for service providers since they assist, support and implement their marketing strategy by inferring knowledge from the data they collect.

In the literature, various methods have been used in recommendation systems such as user-centered or item-centered collaborative filtering methods. User-centered collaborative filtering is based on the idea that a user can like items that have already been liked by similar users depending on their action history. The item-centered collaborative filtering is based on the idea that a user can like items similar to items he has previously liked. While collaborative filtering methods only use the users action history, other content-based methods have been used in recommendation systems to exploit the user demographics and the item characteristics. A third method family has been proposed in the literature and it is the hybrid method family that combines the two other method families.

In practice, the use of content-based methods requires user demographics (e.g. age, gender, country, etc.) or item characteristics (e.g. type, price, country of production, etc.) or both together. Since these data are not necessarily available on different recommendation systems, the use of collaborative filtering methods is becoming more and more popular. Thus, in this work we have chosen to focus on collaborative user-centered filtering and the problems that come with its use in current recommendation systems.

The major problems of collaborative filtering are cold-start, data sparsity and low scalability especially for the user-centered variant. Besides, the complexity of collaborative filtering and the fact that it requires processing all the historical data at once in a batch modes makes it resource heavy and leads to long processing times that become excessive with the growth of usage data. This explains the adopted strategy in production environments that resorts to discard all the learned models and reprocess all the available data periodically or when a given volume of usage data is collected. These collaborative filtering constraints makes it inadequate for real-time stream recommendations where suggestions must be provided with low latency in reaction to the latest observed behaviors. In this context, a stream based recommendation approach has the ability to better coordinate and synchronize the service/content provider's marketing and point of sale systems by ensuring they analyze and react to the same up-to-date view of each customer.

On the other hand, companies with common business purposes tend to share their users action history in order to better understand their preferences. Feeding a shared and cooperative recommendation system with user data from different sources further complicates the use of user-centered collaborative filtering techniques since many events (i.e. users actions) may occur from these sources and have to be shared and processed instantly.

In this work, we propose several variants and implementations of the user-centered collaborative filtering in order to make it able to handle the two recommendation scenarios that can face service providers, namely the online scenario for stream based real-time recommendation and the offline one based on a single batch processing of historical data. All the proposed variants' implementations are distributed, resilient, fault tolerant and help reduce the processing time.

The first scenario addresses the continuous online recommendations based on consumption event streams that requires processing new incoming data with low latency in order to rapidly take into account the knowledge it carries. In order to achieve continuous stream recommendation, we resorted to making the collaborative filtering learning algorithm incremental and distributable. Indeed, distributed processing, using stateless procedures, increases the recommender system's horizontal scalability and ability to process more data simply by deploying new replicas of the computational nodes. Moreover, the incremental aspect of the proposed collaborative filtering variant avoids the need to reprocess all the historical data every time a new user action is observed. Therefore, new observations only require updating the part of the model or the subset of knowledge they affect.

The second scenario is related to offline, one time, batch recommendation use cases where large volumes of static data are available and at rest. In this context, we believe that batch recommendation algorithms are more adequate since they allow making several passes over the data which may lead to extract more knowledge when compared with incremental approaches making only one pass. Therefore, we propose a distributed implementation of user-centered collaborative filtering which is more suitable to one time offline learning and recommendation over big datasets.

We should point out that in this work we are careful to respect the privacy and the concerns of those to whom the data relate.

This paper is organized as follows. In Sect. 2, we present an overview of recommender systems and their underlying approaches. In Sect. 3, we present our proposed approach for scalable collaborative filtering with its two variants targeting (1) the online incremental stream processing and (2) the offline batch use cases. In Sect. 4, we evaluate our proposals and discuss the obtained results. Finally, in Sect. 5, we conclude the paper with some possible perspectives.

2   Related Works

Recommender systems are a class of personalization systems whose main objective is to predict users' interests towards the available informational content in the application domain. To achieve this goal, several approaches and methodologies were proposed in the literature [12,18]. Recommender systems based on collaborative filtering techniques use correlations in users' rating patterns in order to predict their interests. Collaborative filtering approaches are mainly classified into two categories: User-Centered Collaborative Filtering (UCCF) and Item-Centered Collaborative Filtering (ICCF).

UCCF is based on the idea that if two users have liked similar items in the past, they will probably have similar preferences and therefore look for the same items in the future.

UCCF first calculates the similarities between each pair of users through based on rating patterns in order to build neighborhood of similar users. From there, the algorithm recommends to the current user the items that are appreciated by his neighborhood [12].

ICCF calculates the similarities between items based on their patterns. Then, items similar to the ones that the current user liked in the past will be recommended to him [15].

The main advantage of collaborative filtering techniques is that they do not require a pinpoint item description. Given that the recommendation process is based on user-item interactions, this technique makes it possible to apply in all application domains in order to recommend complex items without the need to describe or analyze them.

2.1 Scalable Stream Processing of Collaborative Filtering

Highly interactive online services (e.g. VOD websites such as netflix.com, MOD websites such as soundcloud.com) are increasingly popular on the Web. Such services can have a large amount of data characterizing consumers' behaviors since they interact continuously with these services and the items they provide. Since the process of neighborhood construction takes so long, it is then calculated periodically. If the recommendation system begins to analyze users' behaviors at a time t1 and finishes it at t2, the interactions that were observed during this period are not taken into consideration. The larger the amount of data to be analyzed, the greater the time between t1 and t2, and the more the newly collected data are ignored.

On the basis of this observation, several research studies have attempted to provide the necessary solutions to accommodate the continuous data flow problem. Recently, real-time recommendation systems are increasingly attracting researchers' attention [3,7].

Papagelis proposed an approach to solve the scalability problem [14]. His approach is based on incremental updates of the users' similarities, making it possible to calculate the similarities in an online application. Das discussed the problems issued from the large volumes of data and the accelerating content evolution [5]. He then proposed online templates in order to generate recommendations for users on Google News. Koren introduced a new neighborhood model with an improved accuracy [13]. The proposed approach takes into account recent actions. It is then more scalable than the previous method, without compromising recommendations relevance or other desired properties.

Diaz-Aviles and Chen provided users with a real-time thematic recommendation by analyzing social data flows [4,6]. However, the proposed systems have the inconvenience of not being able to support large volumes of data. StreamRec, proposed by Chandramouli, implements a recommendation model based on collaborative filtering in a scalable and incremental way allowing it to handle continuous data flows [2].

Huang was inspired by Papagelis [14] and StreamRec, to build TencentRec [11]. It is a general recommendation system that implements a series of algorithms in order to satisfy the different application requirements. The proposed approach takes into account various types of user action data, including implicit and explicit actions as in StreamRec. Huang has used approximation techniques in this approach. In fact, even though Papagelis' method reduces the necessary computations for the similarities update, these calculations are still very expensive and cannot be done within a real-time delay.

2.2 Distributed Approaches for Batch Collaborative Filtering

Distributed collaborative filtering approaches started by the work of Tveit on UCCF [16]. The author used the Pearson correlation as a similarity measure whose computation was implemented in a peer-to-peer distributed architecture.

The UCCF has been implemented by Han [8,9] using distributed hash tables. Users were distributed over different computation nodes in a way that guarantees that each node handles users having in common at least one item rated similarly. This allowed to build local training sets and to calculate similarities between users belonging to the same computation node and then predict the missing ratings.

Later, more sophisticated approaches were proposed to associate traditional collaborative filtering techniques with other new ideas.

In order to calculate users' similarities, Xie [17] used a schema relying on distributed hash tables (DHT). The number of users considered in the similarity calculation process has been reduced by taking into account only those who have very similar preferences. Furthermore, in order to avoid rating biases, users' ratings have been normalized.

The approach proposed by Castagnos and Boyer in [1] made it possible to take into account the user implicit actions during the training step. Each user is characterized by a profile and an ID. Both allowed users to be categorized using Pearson correlation. In fact, for each user four lists of user ID have been defined: the most similar users, the ones with similarity exceeding the defined similarity threshold, the blacklisted users, and the users who added the active user in their profile.

2.3 Motivation and Objectives

Classic recommendation algorithms are complex and require a heavy processing footprint especially in the neighborhoods construction phase where each pair of items or users is processed in order to measure their similarity degree. Moreover, each time a user's attitude towards an item is observed (e.g. purchase or rating) the user's (resp. item's) neighborhood needs to be updated by recalculating the similarities with all the other users (resp. items).

Offline and batch execution often process large and complex static data and is more concerned with throughput than latency of individual components of the computation. This type of execution addresses the need to predict variables when it comes to making the prediction periodically on data batches.

Real-time recommendation systems are more demanding in terms of latency. They process data streams whose calculation is executed using elementary data or a smallish window of recent data. This calculation is relatively simpler than that of batch data and must be carried out quasi-instantaneously.

To make predictions even simpler, one solution would be to carry out the calculations incrementally in order to reuse the results of the previously carried out calculations and infer the new predictions using fewer computation resources.

3   An Approach for Distributed Collaborative Filtering Supporting Stream and Batch Recommendations

Our proposal includes two implementations of the user-centered collaborative filtering algorithm. First, we detail the approach for processing batches or streams of unary or binary data using incremental calculations. This ensures that the last data collected is taken into account when generating recommendations, which is very well adapted to the real-time context that is found in several applications. Then, we address application domains relying on multi-valued data such as ratings to model users' attitudes by a distributed variant for offline recommendation on big data at rest, which do not plan to handle more user data before the end of the current recommendation process.

3.1   Distributed Incremental Collaborative Filtering for Stream Recommendation (DCFS)

The proposed approach targets use cases where users' behaviors are modeled using unary or binary variables. It is based on a distributed and incremental modeling and prediction of pair of users or items correlation, independently of the other pairs.

A recommendation system based on collaborative filtering has two main steps: users' similarity computation (or the items' one), and ratings' prediction. Since the computations' complexity lies in the similarity step, we have then focused on alleviate these computations in our approach, and for which we have chosen the Jaccard measure. This choice was made by taking into account how to incrementally upgrade the different similarity measures. In fact, the Jaccard measure has the fewest variables to update when it comes to making calculations incrementally, so this is the best measure for our approach.

The Jaccard measure (cf. Eq. 1) is based on calculating two correlations. The first correlation concerns the shared actions (i.e. set of actions made by both users) and the second concerns the union of actions (i.e. set of all actions performed by at least one of the two users). In the case of user-centered collaborative filtering, it can be reduced to merely calculating the number of common items for a pair of users, and the number of items present in their actions history.

$$J(A, B) = \frac{|A \cap B|}{|A \cup B|} \tag{1}$$

The Fig. 1 details the workflow of our proposal which contains two steps. It starts by generating users' correlation tuples in the first step and goes then ahead to calculate similarity degrees in a second step.

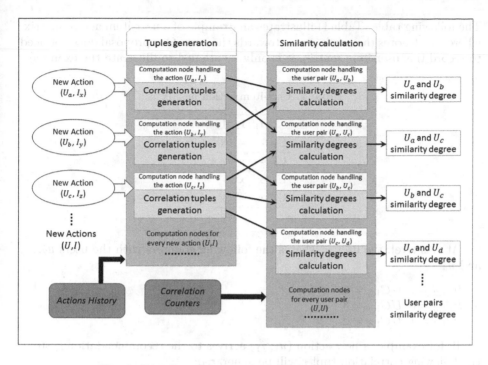

Fig. 1. Distributed incremental collaborative filtering

The calculation of the two correlation counters (intersection counter IC, and union counter UC) for a pair of users can be done in an incremental manner, and independently of the counters of the other pairs. This consists in calculating, for each new action between a user u_x and an item i_i, the correlation between u_x and any other user u_y with respect to i_i: if u_y has a recorded action for i_i, the intersection counter is incremented, otherwise the union counter is incremented.

These increments are carried out by generating correlation tuples in the form (User pair, Intersection counter, Union counter). For example, if the action (u_x, i_i) arrives to the recommendation system, the following correlation tuples will be generated:

- For each user u_y with the same action as u_x towards i_i, we generate the tuple $((u_x, u_y), 1, 0)$ which consists in incrementing the intersection counter.
- For any user u_y that does not have the same action as u_x towards the item i_i, we generate the tuple $((u_x, u_y), 0, 1)$ which consists in incrementing the union counter.

Then, for each user pair, we join their correlation tuples with their intersection and union counters. The correlation tuples of a user pair will allow updating the counters and subsequently infer the new similarity value between the two users:

$$Sim(A, B) = \frac{IC_{A,B}}{UC_{A,B}} \tag{2}$$

The following table (Table 1) illustrates an example of a user-item action matrix, where "x" denotes the user's action towards the item. Our proposal does not need to record this user-item matrix, it is only mentioned to illustrate the example.

Table 1. User-item action matrix

	I_1	I_2	I_3
U_1	-	-	x
U_2	-	x	x
U_3	x	x	-
U_4	x	-	x

At this level, the user u_1 shares the following counters with the users u_2, u_3 and u_4:

- $IC_{1,2} = 1$; $UC_{1,2} = 2$
- $IC_{1,3} = 0$; $UC_{1,3} = 3$
- $IC_{1,4} = 1$; $UC_{1,4} = 2$

If, for example, a new action (u_1, i_1) arrives to the recommendation system, the following correlation tuples will be generated:

- Intersection tuples: Users with the same action as u_1 towards i_1 are u_3 and u_4, therefore we generate the tuples $((u_1, u_3), 1, 0)$ and $((u_1, u_4), 1, 0)$.
- Union tuples: Only the user u_2 does not have the same action as u_1 towards i_1, therefore we generate the tuple $((u_1, u_2), 0, 1)$.

These generated tuples make it possible to update the correlation counters of u_1 with the other users, which will become: $UC_{1,2} = 3$; $IC_{1,3} = 1$; $IC_{1,4} = 2$.

Thus, the similarity degrees between u_1 and other users can be inferred: $Sim_{1,2} = 0,33$; $Sim_{1,3} = 0,33$; $Sim_{1,4} = 1$.

Note that our proposal allows processing static data in a batch mode, with one pass computation, by generating the tuples of correlation for all the actions at the same time. Then, tuples are grouped by user pairs allowing to calculate the sum of the increments' values in order to obtain two final counters for each user, and subsequently deducing the degree of similarity.

In both cases of use, our proposal offers a constant complexity for the similarity calculation after the arrival of a new user action and makes it possible to use the results of the previous calculations in order to perform the future ones.

3.2 Distributed Collaborative Filtering for Batch Recommendation (DCFB)

The incremental variant of our approach used a low complexity similarity measure to better fit the real-time recommendation requirements. Outside the real-time application context, other complex measures can be implemented since they may outperform the Jaccard similarity measure by taking into account all the available data points every time two users need to be compared. Moreover, classic similarity measures makes it possible to handle all types of variables modeling user's actions and attitudes since they aren't restricted to binary or unary variables.

Batch processing allows accessing all the data at hand when a recommendation is needed. This offline processing provides the ability to manage multi-valued static data and calculate the exact degrees of similarity for collaborative filtering when using classic non-incremental measures. In this context, we propose a solution for batch recommendation by incorporating the most used similarity measurements and rating estimators.

The used similarity measures are the Cosine similarity (cf. Eq. 3) and the Pearson correlation (cf. Eq. 4).

$$sim_{x,y} = \frac{\overrightarrow{R_x} \cdot \overrightarrow{R_y}}{||\overrightarrow{R_x}||_2 \times ||\overrightarrow{R_y}||_2} = \frac{\sum_{i \in I} r_{x,i} \times r_{y,i}}{\sqrt{\sum_{i \in I} r_{x,i}^2} \times \sqrt{\sum_{i \in I} r_{y,i}^2}} \tag{3}$$

$$sim_{x,y} = \frac{\sum_{i \in I_x \cap I_y} (r_{x,i} - \bar{r}_x)(r_{y,i} - \bar{r}_y)}{\sqrt{\sum_{i \in I_x \cap I_y} (r_{x,i} - \bar{r}_x)^2} \times \sqrt{\sum_{i \in I_x \cap I_y} (r_{y,i} - \bar{r}_y)^2}} \tag{4}$$

Concerning the rating estimators, the following three estimators were adopted:

- The similarity-weighted average (SW) which predicts the user rating for an item based on the user's neighborhood ratings for this item.

$$r_{x,i} = \frac{\sum_{y \in U} sim_{x,y} \times r_{y,i}}{\sum_{y \in U} |sim_{x,y}|} \tag{5}$$

- The average based on the centered ratings (CR) which is based on biased ratings in order to avoid the differences in user rating behavior.

$$r_{x,i} = \bar{r}_x + \frac{\sum_{y \in U} sim_{x,y} \times (r_{y,i} - \bar{r}_y)}{\sum_{y \in U} |sim_{x,y}|} \tag{6}$$

- The average based on the standard deviation (SD) thereby compensating for users differing in ratings spread as well as mean ratings.

$$r_{x,i} = \bar{r}_x + \sigma_x \frac{\sum_{y \in U} sim_{x,y} \times (r_{y,i} - \bar{r}_y)/\sigma_y}{\sum_{y \in U} |sim_{x,y}|} \tag{7}$$

The general architecture of our proposal is detailed in Fig. 2. The first step allows preparing the necessary statistics for the similarities calculation and ratings predictions. Afterwards, the similarity between each pair of users is calculated according to the chosen similarity measure using the previously calculated statistics. The third step consists on predicting the missing ratings for each user based on his similarity degrees with the rest of the users and using a rating estimator. Finally, after predicting all missing ratings for a given user, a list of the best items is recommended.

In the following, we detail all the steps required to generate recommendations while highlighting their distributed algorithms.

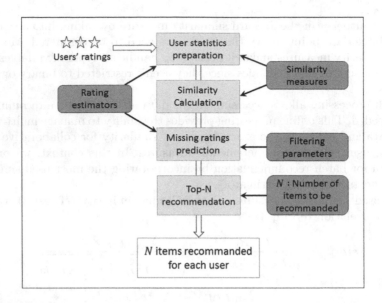

Fig. 2. Distributed collaborative filtering architecture

Users Statistics Preparation: The statistics that need to be computed for each user differ according to the rating estimator and the chosen measure of similarity. They can be the user mean rating $\bar{r}_i$, the Euclidean norm $||\overrightarrow{R_i}||_2$ or the user ratings' standard deviation σ_i. Table 2 summarizes the statistics that need to be calculated according to the used measures of similarity and rating estimators. The calculation is done by grouping ratings per user and then inferring the statistics of each user.

Similarity Calculation: This step focuses on calculating the similarity for each pair of users. First, users' ratings are grouped by items (i.e. ratings concerning the same item are redirected to the same calculation node). Then, for each item, correlation tuples are generated for each pair of users who have rated the item. The correlation tuples are defined as (User pair, T1, T2, T3) tuple.

Table 2. User statistics to be calculated

	Pearson correlation	Cosine similarity
Similarity-weighted average	User mean rating	Euclidean norm
Average based on the centered ratings	User mean rating	User mean rating Euclidean norm
Average based on the standard deviation	User mean rating Standard deviation	User mean rating Euclidean norm Standard deviation

Similarity measures are based on formulas that are of the form of $a/(b \times c)$. The three terms T1 (cf. Eq. 8), T2 (cf. Eq. 9) and T3 (cf. Eq. 10) calculate respectively the elementary values of a, b and c produced from a single item. These terms allow calculating the similarity degree between each pair of users (cf. Eq. 11).

$$T1 = \begin{cases} r_{x,z} \times r_{y,z} & if \; Cosine \\ (r_{x,z} - \bar{r}_x) \times (r_{y,z} - \bar{r}_y) & if \; Pearson \end{cases} \tag{8}$$

$$T2 = \begin{cases} null & if \; Cosine \\ (r_{x,z} - \bar{r}_x)^2 & if \; Pearson \end{cases} \tag{9}$$

$$T3 = \begin{cases} null & if \; Cosine \\ (r_{y,z} - \bar{r}_y)^2 & if \; Pearson \end{cases} \tag{10}$$

$$sim_{x,y} = \begin{cases} \dfrac{\sum T1}{||\vec{R_x}||_2 \times ||\vec{R_y}||_2} & if \; Cosine \\ \dfrac{\sum T1}{\sqrt{\sum T2} \times \sqrt{\sum T3}} & if \; Pearson \end{cases} \tag{11}$$

The following user-item ratings matrix (Table 3) is used to illustrate examples for each component of our proposal for batch data:

Table 3. User-item ratings matrix

	I_x	I_y	I_z	I_w
U_a	$r_{a,x}$	$r_{a,y}$	-	-
U_b	$r_{b,x}$	$r_{b,y}$	$r_{b,z}$	-
U_c	$r_{c,x}$	-	$r_{c,z}$	-
U_d	-	$r_{d,y}$	$r_{d,z}$	$r_{d,w}$

The similarity calculation process for this user-item ratings matrix is described within the Fig. 3.

Missing Ratings Prediction: The third step predicts the ratings that users have not yet assigned. A filtering can be carried out in order to preserve only the most similar users for the active user. This filtering is performed either by pruning the degrees of similarity which are less than a similarity threshold t or by fixing the neighborhood size at n.

The similarity degrees of a same user are then grouped together to generate correlation tuples. For each item that the active user has not rated, and

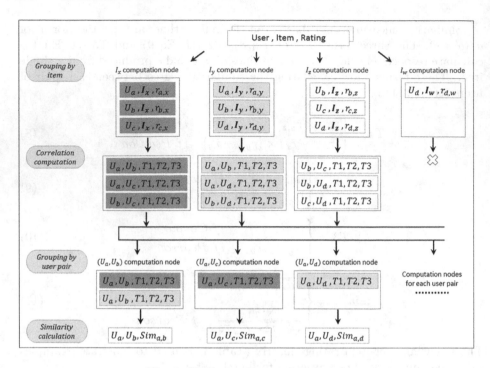

Fig. 3. Similarity degrees calculation

for each similar user who has rated the item, a correlation tuple is generated gathering data about the second user rating and the similarity degree between both users.

These tuples are defined as (Active user, Item not rated, Term T4 (cf. Eq. 12), Term T5 (cf. Eq. 13)) tuple. They allow predicting the missing user ratings through Eq. 14.

The Fig. 4 details the missing ratings prediction process for each user, based on similarity degrees already calculated.

$$T4 = \begin{cases} sim_{x,y} \times r_{y,z} & if \ SW \\ sim_{x,y} \times (r_{y,z} - \bar{r}_y) & if \ CR \\ sim_{x,y} \times \frac{r_{y,z} - \bar{r}_y}{\sigma_y} & if \ SD \end{cases} \tag{12}$$

$$T5 = |sim_{x,y}| \tag{13}$$

$$r_{x,z} = \begin{cases} \frac{\sum T4}{\sum T5} & if \ SW \\ \bar{r}_x + \frac{\sum T4}{\sum T5} & if \ CR \\ \bar{r}_x + (\sigma_x \times \frac{\sum T4}{\sum T5}) & if \ SD \end{cases} \tag{14}$$

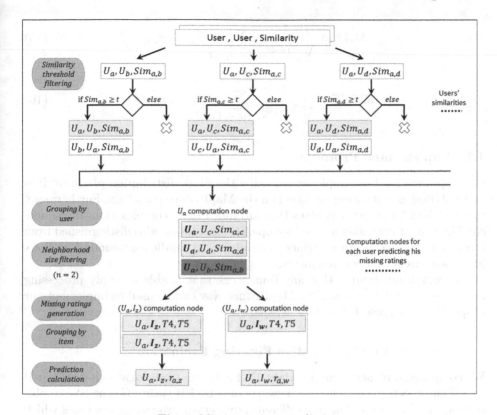

Fig. 4. Missing ratings prediction

4 Experimentation

In this section, we present the experimentation of our approach, detail the targeted application domain, and discuss the obtained results for the different variants that were evaluated.

4.1 Experimentation Setup

In this work, we adopted the MovieLens 1 m dataset [10] since it is the closest to our requirements with regard to the available variables. In fact, this dataset includes ratings assigned by a set of users to the movies they have watched. The dataset used in the following experiments includes 3.900 movies, 6.040 users and 1.000.209 ratings (1 to 5).

In order to evaluate our approaches relevance and performances, we computed the Mean Absolute Error (MAE) and the Root Mean Squared Error (RMSE) as quality measures and the Mean Execution Time (MET) as the evaluation metric of efficiency. The experiments were carried out on a cluster consisting of 16 processing machines, each having one core of 2.4 GHz.

$$MAE = \sqrt{\frac{1}{S_{Test}} \sum_{(u,i) \in S_{Test}} |\hat{r}_{u,i} - r_{u,i}|} \qquad (15)$$

$$RMSE = \sqrt{\frac{1}{S_{Test}} \sum_{(u,i) \in S_{Test}} (\hat{r}_{u,i} - r_{u,i})^2} \qquad (16)$$

4.2 Implemented Prototype

Our approach has been implemented using the Flink distribution platform. It is a distributed execution engine based on the MapReduce paradigm. Furthermore, it allows building execution plans that are much more complex than the simple combination of both map and reduce operations. Flink is also distinguished from other distributed execution engines by its ability to handle continuous data flows in a transparent and controllable way.

It's worth mentioning that any framework that is able to apply processing on a pipeline or data stream in a MapReduce way can be used to implement our proposal (e.g. samza, kafka streams, spark).

4.3 Incremental Collaborative Filtering Evaluation

We compared our incremental variant DCFS to the traditional non-distributed UCCF approach (using Jaccard measure) and DCFB (using Cosine-SW combination). We have experimented different numbers of processing machines which we used to draw the curve shown in Fig. 5. We have to mention that the results in this figure are for the similarity step only. In addition, we have to specify that for the DCFB and the non-distributed Jaccard the similarity of each pair of users was calculated at once (i.e. 18.237.780 similarities were calculated for the 6.040 users), whereas for the DCFS the similarities are calculated progressively as in a real-time recommendation scenario where at the arrival of each new user action the similarities concerned by the new action will be updated (i.e. 6.036.667.952 similarities were calculated or updated for the 6.040 users).

In practice, this means that the results of the DCFS concern the calculations made during the arrival of all the actions of the users, whereas the results of the two other approaches concern a single similarity calculation operation that have been made in order to maintain the accuracy of the similarity values. Then the DCFS calculates 331 times more similarity values compared to the other two approaches. So, to be able to compare these approaches we must divide the DCFS computation times by 331 to obtain the computation times which concern the same number of similarity values calculated by the DFCB or the non-distributed Jaccard.

The mean execution time of our approach is much better than the traditional approach, even when using a single processing machine. Indeed, this is due to the incremental aspect of our approach, which preserves the already performed calculations results and used them to infer the new calculations ones.

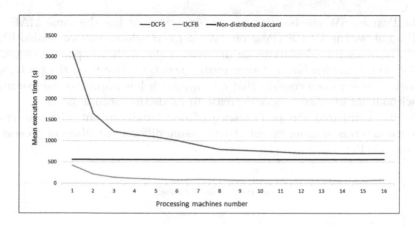

Fig. 5. DCFS mean execution time evaluation

By increasing the number of processing machines, the mean execution time of our approach decreases which validates the motivation behind distributing computations. However, this improvement is not as important as the one coming from the incremental calculations.

4.4   Batch Collaborative Filtering Evaluation

The experiments of the batch variant took into account the six combinations of the similarity measures (Cosine and Pearson) and the rating estimators (SW, CR and SD). These experiments were performed using 16 processing machines and taking into account all users without pruning non-similar users. The Table 4 enumerates MET (in seconds), MAE and RMSE results of these combinations.

The obtained results show that the CR estimator has the poorest prediction quality independently of the similarity measure. The best MAE and RMSE values were obtained respectively by the Cosine-SD and the Pearson-SD combinations. The mean execution time results led to the conclusion that the use of the Pearson correlation measure allows to have better performances than those involving the Cosine similarity. Concerning the rating estimators, SW offers the shortest execution time, followed by SD then CR. Thus, the most efficient combination was the Pearson-SW.

Table 4. DCFB and traditional UCCF evaluation

	P-SW	P-CR	P-SD	C-SW	C-CR	C-SD	UCCF (P-SW)
MAE	0.745	0.811	0.744	0.748	0.787	0.742	0.745
RMSE	0.910	0.998	0.909	0.913	0.971	0.911	0.910
MET	305	314	306	313	318	308	1063

The Pearson-SW combination of our batch variant has the same MAE and RMSE value as the UCCF (Pearson-SW) since it adopts the same algorithm, only distributing its calculations on different nodes with no approximation. In addition, our approach has a shorter mean execution time than the traditional approach. These results confirm that our approach has improved the execution time without information loss or decrease in prediction accuracy.

We also compared the performance of our batch variant with the traditional one in terms of mean execution time using different numbers of processing machines (cf. Fig. 6).

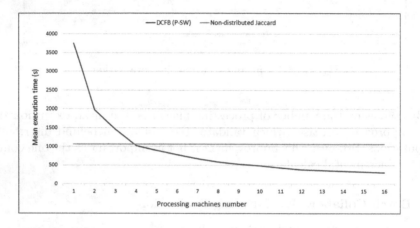

Fig. 6. DCFB mean execution time evaluation

We used in this experiment the Pearson measure and the similarity-weighted average rating estimator. From the obtained results, we can find that starting from the use of 4 processing machines, our approach provides a shorter mean execution time than the traditional approach.

5 Conclusion and Future Works

In this work, we proposed two distributed variants of the collaborative filtering approach for batch and stream processing-based recommendations. Our offline approach resorts to processing distribution in order to manage the large volume of batch data whereas the online one relies on an incremental processing of continuous data streams for online real-time recommendation. The experiments show that the proposed approaches improve the processing time in both scenarios and preserve recommendation quality.

In future works, we will focus on proposing new approximate similarity measures for multi-valued variables which are more adapted to incremental modeling and recommendation. Such measures should improve recommender systems performances without compromising the exactitude of their results or the real-time low latency constraints.

References

1. Castagnos, S., Boyer, A.: Personalized communities in a distributed recommender system. In: Amati, G., Carpineto, C., Romano, G. (eds.) ECIR 2007. LNCS, vol. 4425, pp. 343–355. Springer, Heidelberg (2007). https://doi.org/10.1007/978-3-540-71496-5_32. http://dl.acm.org/citation.cfm?id=1763653.1763695
2. Chandramouli, B., Levandoski, J.J., Eldawy, A., Mokbel, M.F.: StreamRec: a real-time recommender system. In: Proceedings of the 2011 ACM SIGMOD International Conference on Management of Data, SIGMOD 2011, pp. 1243–1246. ACM, New York (2011). http://doi.acm.org/10.1145/1989323.1989465
3. Chang, S., et al.: Streaming recommender systems. In: Proceedings of the 26th International Conference on World Wide Web, WWW 2017, International World Wide Web Conferences Steering Committee, pp. 381–389. Republic and Canton of Geneva, Switzerland (2017). https://doi.org/10.1145/3038912.3052627
4. Chen, C., Yin, H., Yao, J., Cui, B.: TeRec: a temporal recommender system over tweet stream. Proc. VLDB Endow. 6(12), 1254–1257 (2013). https://doi.org/10.14778/2536274.2536289
5. Das, A.S., Datar, M., Garg, A., Rajaram, S.: Google news personalization: scalable online collaborative filtering. In: Proceedings of the 16th International Conference on World Wide Web, WWW 2007, pp. 271–280. ACM, New York (2007). http://doi.acm.org/10.1145/1242572.1242610
6. Diaz-Aviles, E., Drumond, L., Schmidt-Thieme, L., Nejdl, W.: Real-time top-N recommendation in social streams. In: Proceedings of the Sixth ACM Conference on Recommender Systems, RecSys 2012, pp. 59–66. ACM, New York (2012). http://doi.acm.org/10.1145/2365952.2365968
7. Domann, J., Lommatzsch, A.: A highly available real-time news recommender based on apache spark. In: Jones, J.F. (ed.) CLEF 2017. LNCS, vol. 10456, pp. 161–172. Springer, Cham (2017). https://doi.org/10.1007/978-3-319-65813-1_17
8. Han, P., Xie, B., Yang, F., Shen, R.: A scalable P2P recommender system based on distributed collaborative filtering. Expert. Syst. Appl. 27(2), 203–210 (2004). https://doi.org/10.1016/j.eswa.2004.01.003
9. Han, P., Xie, B., Yang, F., Wang, J., Shen, R.: A novel distributed collaborative filtering algorithm and its implementation on P2P overlay network. In: Dai, H., Srikant, R., Zhang, C. (eds.) PAKDD 2004. LNCS (LNAI), vol. 3056, pp. 106–115. Springer, Heidelberg (2004). https://doi.org/10.1007/978-3-540-24775-3_13
10. Harper, F.M., Konstan, J.A.: The movielens datasets: History and context. ACM Trans. Interact. Intell. Syst. 5(4), 19:1–19:19 (2015). http://doi.acm.org/10.1145/2827872
11. Huang, Y., Cui, B., Zhang, W., Jiang, J., Xu, Y.: TencentRec: real-time stream recommendation in practice. In: Proceedings of the 2015 ACM SIGMOD International Conference on Management of Data, SIGMOD 2015, pp. 227–238. ACM, New York (2015). http://doi.acm.org/10.1145/2723372.2742785
12. Kluver, D., Ekstrand, M.D., Konstan, J.A.: Rating-based collaborative filtering: algorithms and evaluation. In: Brusilovsky, P., He, D. (eds.) Social Information Access. LNCS, vol. 10100, pp. 344–390. Springer, Cham (2018). https://doi.org/10.1007/978-3-319-90092-6_10
13. Koren, Y.: Factor in the neighbors: Scalable and accurate collaborative filtering. ACM Trans. Knowl. Discov. Data 4(1), 1:1–1:24 (2010). http://doi.acm.org/10.1145/1644873.1644874

14. Papagelis, M., Rousidis, I., Plexousakis, D., Theoharopoulos, E.: Incremental collaborative filtering for highly-scalable recommendation algorithms. In: Hacid, M.-S., Murray, N.V., Raś, Z.W., Tsumoto, S. (eds.) ISMIS 2005. LNCS (LNAI), vol. 3488, pp. 553–561. Springer, Heidelberg (2005). https://doi.org/10.1007/11425274_57

15. Sarwar, B., Karypis, G., Konstan, J., Riedl, J.: Item-based collaborative filtering recommendation algorithms. In: Proceedings of the 10th International Conference on World Wide Web, WWW 2001, pp. 285–295. ACM, New York (2001). http://doi.acm.org/10.1145/371920.372071

16. Tveit, A.: Peer-to-peer based recommendations for mobile commerce. In: Proceedings of the 1st International Workshop on Mobile Commerce, WMC 2001, pp. 26–29. ACM, New York (2001). http://doi.acm.org/10.1145/381461.381466

17. Xie, B., Han, P., Yang, F., Shen, R.M., Zeng, H.J., Chen, Z.: DCFLA: a distributed collaborative-filtering neighbor-locating algorithm. Inf. Sci. **177**(6), 1349–1363 (2007). https://doi.org/10.1016/j.ins.2006.09.005

18. Zanitti, M., Kosta, S., Sørensen, J.: A user-centric diversity by design recommender system for the movie application domain. In: Companion Proceedings of the The Web Conference 2018, pp. 1381–1389, WWW 2018. International World Wide Web Conferences Steering Committee, Republic and Canton of Geneva, Switzerland (2018). https://doi.org/10.1145/3184558.3191580

Detecting Constraints and Their Relations from Regulatory Documents Using NLP Techniques

Karolin Winter[✉] and Stefanie Rinderle-Ma

Faculty of Computer Science, University of Vienna, Vienna, Austria
{karolin.winter,stefanie.rinderle-ma}@univie.ac.at

Abstract. Extracting constraints and process models from natural language text is an ongoing challenge. While the focus of current research is merely on the extraction itself, this paper presents a three step approach to group constraints as well as to detect and display relations between constraints in order to ease their implementation. For this, the approach uses NLP techniques to extract sentences containing constraints, group them by, e.g., stakeholders or topics, and detect redundant, subsuming, and conflicting pairs of constraints. These relations are displayed using network maps. The approach is prototypically implemented and evaluated based on regulatory documents from the financial sector as well as expert interviews.

Keywords: Compliance · Regulatory documents
Requirements extraction · Text mining · NLP

1 Introduction

Extracting norms and business constraints as well as process models from natural language text is an ongoing challenge since non-compliance to laws or regulatory documents can cost billions of dollars [14]. The growing amount of regulatory documents and the need to constantly update and compare already existing rules is exacerbating the situation.

Several approaches have been presented for supporting this challenge (e.g., [3,5,10,18]), but they either impose restrictions on the input text (e.g., it is assumed that the text does only contain process relevant information) or produce models and rules that are incomplete or contain conflicts [22]. Moreover, the main focus of these approaches is mostly on extracting constraints or mapping them to formal rules. However, users still need to understand the rules and constraints as well as dependencies between them in order to be able to implement them correctly [23]. An identification of redundant, subsumed or conflicting constraints could avoid additional or unnecessary implementation effort as well as implementation errors. Another difficulty is the fact that in large companies not every constraint affects every department or stakeholder, consequently not every person has to always read every part of a document.

© Springer Nature Switzerland AG 2018
H. Panetto et al. (Eds.): OTM 2018 Conferences, LNCS 11229, pp. 261–278, 2018.
https://doi.org/10.1007/978-3-030-02610-3_15

Grouping constraints based on *constraint related subjects*, which are for example topics, stakeholders or departments, as well as displaying relations between constraints should therefore be supported conceptually and by suitable tools.

In [26] we presented a method that is capable of identifying sentences containing constraints based on standard text mining tools as well as grouping of document fragments having similar topics or stakeholders by using term frequencies and k-means clustering. For the presented first case study on ISO security documents it was possible to identify and group fragments dealing with different topics, e.g., measurement and evaluation of ISMS or legal concerns. However, using term frequencies to group documents or sentences, even though this procedure is widely applied in the field of text mining, is rather limited and can lead to vague or incomplete results.

In this paper, we want to overcome this issue by providing means to either integrate additional information such as organizational charts or exploiting the part-of-speech tags of sentences leading to a more purposeful grouping of sentences containing constraints. Moreover, in order to tackle the lack of managing redundancies, subsumptions as well as conflicts between constraints, the method is further extended by an identification and visualization of these relation types.

For this purpose, the following research questions are stated

RQ1. How to group elicited constraints based on constraint related subjects like stakeholders or topics?
RQ2. How to identify relations between constraints?
RQ3. How to display the elicited constraints and the derived relations?

In order to answer these questions, this paper presents a method that makes use of NLP techniques and provides means for integration of additional information whenever it is available. Moreover, a definition of redundancy, subsumption and conflict of constraints with respect to natural language text is stated which can be used to point out potential modelling and implementation errors.

The remainder of the paper is organized as follows. Section 2 describes the method, Sect. 3 the prototypical implementation which is used in Sect. 4 to evaluate the approach on a set of regulatory documents from the financial sector. A short discussion of the approach is presented in Sect. 5, followed by related work in Sect. 6. The paper concludes in Sect. 7 with a summary and outlook of future work.

2   Overall Method

In this section the method is outlined. Since it can be carried out with any NLP framework, details on our prototypical implementation are separately described in the next section. The method (cf. Fig. 1) can be divided into the three typical stages of data mining, **pre-processing**, **processing** ($\mapsto$RQ1 & RQ2) and **post-processing**($\mapsto$RQ3), each of them consisting of several steps. Pre-processing and parts of the processing steps can be viewed as a tool chain since they rely on state-of-the-art NLP techniques and data mining algorithms. The second part of the processing and the post-processing step form the main contribution by providing a definition and application of characteristics of redundant, subsumed, and conflicting constraints with respect to natural language text.

During each stage the elicited sentences containing constraints are maintained as such and not yet mapped to a formal language. The advantage is that non expert users have a better chance of understanding the constraints and their relations. After all relations have been resolved they are visualized using a graph-based representation.

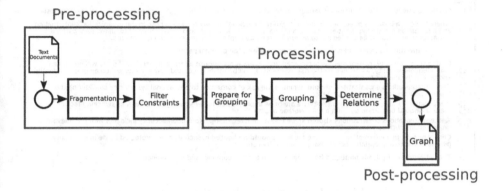

Fig. 1. Overall method

To illustrate the method a running example is provided which is based on parts of two documents from the ISO 27000 security standard family (ISO 27001 and ISO 27011) as depicted in Fig. 2. They were used for a case study in [26]. Sentences 1–9 are taken from ISO 27000 which is an overview document and sentences 10–18 stem from ISO 27011 which outlines security topics for the telecommunications domain.

2.1   Pre-processing

First of all, each document needs to be prepared, i.e., table of contents and references are removed, since these parts do not contain valuable information on constraints. Moreover, since the documents are chunked into sentences these parts of documents would cause errors during the part-of-speech (POS) tagging process. Depending on the documents it might be necessary to remove even more parts that do not contain constraints, e.g., introductions. In addition, it should be checked whether information from tables and pictures is needed and was parsed correctly. A manual inspection of these steps might be required depending on the framework that is used. Another challenge is how to proceed with, e.g., footnotes. If a footnote, contains an explanation or a link to another document, it could be included in the final visualization. Since text passages are often related and depend on each other it might happen that one sentence refers to the subject

ISO 27001

(S1) Managing information security risks requires a suitable risk assessment and risk treatment method which may include an estimation of the costs and benefits, legal requirements, the concerns of stakeholders, and other inputs and variables as appropriate.

(S2) Risk assessments **should** identify, quantify, and prioritize risks against criteria for risk acceptance and objectives relevant to the organization.

(S3) The results **should** guide and determine the appropriate management action and priorities for managing information security risks and for implementing controls selected to protect against these risks.

(S4) Risk assessment **should** include the systematic approach of estimating the magnitude of risks (risk analysis) and the process of comparing the estimated risks against risk criteria to determine the significance of the risks (risk evaluation).

(S5) Risk assessments **should** be performed periodically to address changes in the information security requirements and in the risk situation, e.g. in the assets, threats, vulnerabilities, impacts, the risk evaluation, and when significant changes occur.

(S6) These risk assessments **should** be undertaken in a methodical manner capable of producing comparable and reproducible results.

(S7) The information security risk assessment **should** have a clearly defined scope in order to be effective and **should** include relationships with risk assessments in other areas, if appropriate.

(S8) ISO/IEC 27005 provides information security risk management guidance, including advice on risk assessment, risk treatment, risk acceptance, risk reporting, risk monitoring and risk review.

(S9) Examples of risk assessment methodologies are included as well.

ISO 27011

(S10) Risk assessment **should** be repeated periodically to address any changes that might influence the risk assessment results.

(S11) Where there is a business need for working with external parties that may require access to the organization's information and information processing facilities, or in obtaining or providing a product and service from or to an external party, a risk assessment **should** be carried out to determine security implications and control requirements.

(S12) Controls **should** be agreed and defined in an agreement with the external party.

(S13) If information security management is outsourced, the agreements **should** address how the third party will guarantee that adequate security, as defined by the risk assessment, will be maintained, and how security will be adapted to identify and deal with changes to risks.

(S14) Telecommunications organizations **should** minimize the risk of corruption to operational systems by considering the following guidelines to control changes.

(S15) If applications and operating system software are to be implemented to sensitive systems such as switching facility, the test **should** be carried out with a full coverage of path.

(S16) Telecommunications organizations **should** share information regarding information security incidents with the relevant organizations such as Telecom-ISAC.

(S17) Critical or sensitive information processing facilities **should** be housed in secure areas, protected by defined security perimeters, with appropriate security barriers and entry controls.

(S18) They **should** be physically protected from unauthorized access, damage, and interference.

Fig. 2. Running example–textual input

of a preceding one. In this case determiners or pronouns, e.g., they are used. During the preparation of the documents each of these words must be replaced by the corresponding subject of its preceding sentence. In the running example sentences $S17$ and $S18$ represent such a situation. Here, they in $S18$ must be replaced by information processing facility from $S17$. Another issue is to detect whether multiple subjects are present in one sentence. In this case, the corresponding sentence is split, resulting in two partial sentences. In the running example, $S7$ is not split because the sentence does only contain one subject, i.e., information security risk assessment. The prepared documents form the input, so-called text corpus, for the subsequent steps.

Now, each document in the text corpus is fragmented (chunked) into sentences and POS tagged.[1] Afterwards all sentences containing constraints are filtered out. For this purpose, each sentence is scanned for markers such as shall, should or must. We use markers for deriving constraints, like [10] use markers for detecting BPMN elements. In addition, during the evaluation an expert interview confirmed these assumptions. If a sentence contains at least one of these markers it is tagged as constraint and the following definition can be stated.

Definition 1. *Let* S *be a set of sentences. A constraint is an element* $s \in S$ *such that at least one marker is contained in* s*. The set of all constraints is called* C*.*

---

[1] The POS tags are necessary at a later stage of the method.

In the running example, constraints are the sentences containing words (markers) written in bold font.

Sometimes, constraints are pre- or succeeded by sentences only containing explanatory information but no markers. These sentences are not included in the following steps but can be included in the visualization if necessary. In the running example, these are the sentences containing no word in bold font ($S1$, $S8$, $S9$). During the processing of sentences lemmatized words are used, in order to prevent that, e.g., plural and singular forms of nouns form different groups.[2] The final visualization still contains the original sentences.

2.2 Processing

The processing stage is divided into three steps, the preparation for the grouping, the grouping itself and the determination of relations between pairs of constraints. Consequently, the result of the processing is on the one hand a grouping of constraints and on the other hand the detection of redundant, subsumed and conflicting constraints. Three possibilities for carrying out these steps are included in order to ensure that an analyst can choose the mean that is most suitable for the given collection of documents.

Preparation and Grouping

Term Frequencies: The first option corresponds to unsupervised grouping of sentences using k-means as it was outlined in [26] and should be applied on a large collection of documents. For this, term frequencies need to be determined which can be computed by different measures. If the text corpus contains documents (or in this case sentences) that strongly vary in length, term frequency inverse document frequency (cf. [1]) is recommended resulting in a term-sentence matrix which is used for grouping the sentences.

Besides choosing a suitable term frequency measure another challenge is to determine the appropriate number of groups for k-means. Commonly applied methods for selecting the number of groups are, e.g., elbow or silhouette plots. In order to further improve the approach, we decided to use k-means++ [4]. The result of k-means clustering is a grouping of sentences based on term frequencies.

For the running example the most frequent terms are: `organizations`, `risk(s)`, `assessment`, `telecommunications` and performing k-means++ with $k = 6$ creates the groups schematically displayed in Fig. 3.

The remaining two methods correspond to a supervised grouping of the set of sentences based on a predefined list of terms. So, the labels of groups are given beforehand and each group corresponds to one of the terms that were derived by one of the following techniques.

Structure of Sentences: The second method for grouping the sentences is based on extracting constraint related subjects in an automated way without making use of additional information but by exploiting the structure of sentences and can be applied for small or mid-size document collections. A word is

---

[2] The quality of the outcome of this step relies on the NLP framework that is used.

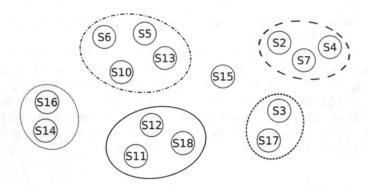

Fig. 3. Running example – grouping by term frequencies

identified as constraint related subject if it is a (compound) subject and followed by a marker, e.g., in sentence $S12$ of the running example the marker is should and the constraint related subject is controls. Based on this, the list of constraint related subjects is created by examining the parse tree of each sentence and searching for the described pattern ((compound) subject + marker). In the running example constraint related subjects are, e.g., information security risk assessment or information processing facility. Since each sentence is processed in its lemmatized form, risk assessment and risk assessments are treated as one grouping subject. Terms like information security risk assessment and risk assessment are not aggregated since these might relate to different things, e.g., risk assessment could be another type of risk assessment than information security risk assessment. For the running example constraint related subjects are risk assessment, information processing facility, telecommunication organization, result, control, agreement, information security risk assessment and application.

Now, each sentence is parsed and checked whether it contains one of the terms from the constraint related subject list. If so, it is shifted to the corresponding group. Figure 4 displays this grouping for the running example.

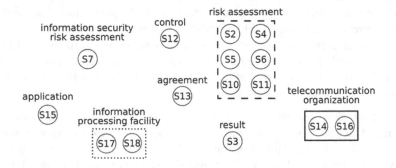

Fig. 4. Running example – grouping by sentence structure

External Information Sources: The third and last processing possibility can be used whenever external information sources, like organizational charts, glossaries, or any other knowledge provided by domain experts that contains information on how to group constraints is available. Based on this information, a list containing possible constraint related subjects is derived. Afterwards, each entry in the list is extended by synonyms. A commonly applied mean to find synonyms is to use a lexical database such as WordNet [17]. Synonyms are relevant in this case because of the diversity of language, e.g., one subject could be represented by several different words. In addition, all subjects should be lemmatized like before. The outcome is again a list containing constraint related subjects. Like in the second method each sentence is shifted into its corresponding group.

For the running example no additional information is available, therefore only the first (word frequencies) and second (sentence structure) method for grouping constraints can be applied. The third method is demonstrated in the evaluation.

Determine Relations

After grouping the set of sentences containing constraints the second part of the processing step is to retrieve dependencies between them in order to detect redundant, subsumed and conflicting constraints. To this end a classification of these types of constraints for natural language text is provided. It is based on [19] which gives a definition of these terms for constraints in a formal language. In order to transfer the characterization to sentences, we first need to define the similarity between pairs of constraint related subjects and tasks. For this Definition 2 is following the one of semantic similarity of text labels in [8]. Constraint related subjects are derived as described before ((compound) subject + marker) while for tasks the techniques of [10] can be applied, e.g., by filtering verbs. In this case we can be more precise since tasks will be preceded by markers.

Definition 2 (Semantic Similarity). *Let C be a set of constraints, $c_1, c_2 \in C$ and let W be the set of all words contained in c_1, c_2. Moreover, let R be the set constraint related subjects of c_1, c_2, $w : R \mapsto P(W)$ be a function that separates an element in R into words. Let $w_1 = w(r_1), w_2 = w(r_2)$ and w_i, w_s be the weights associated with identical words and synonymous words, respectively. The semantic similarity of two constraint related subjects $r_1, r_2 \in R$ is defined as*

$$sem(r_1, r_2) := \frac{2 \cdot w_i \cdot |w_1 \cap w_2| + w_s \cdot (|s(w_1, w_2)| + |s(w_2, w_1)|)}{|w_1| + |w_2|},$$

with $s(w_1, w_2)$ being the set of synonyms of w_1 that appear in w_2.

The semantic similarity of tasks t_1 of c_1, t_2 of c_2 can be defined analogously.

For the running example take $r_1 = $ information security risk assessment and $r_2 = $ information processing facility. It holds $w_1 = $[information, security, risk, assessment] and $w_2 = $[information, processing, facility]. Consequently, with $w_i = 1, w_s = 0.75 : sem(r_1, r_2) = \frac{2 \cdot 1 \cdot 1 + 0.75 \cdot (0+0)}{4+3} \approx 0.286$. So, these constraint related subjects do not have a high similarity.

For defining the targeted relation types of constraints, on the one hand the similarity of sentences and on the other hand a characterization of conflict between sentences is needed. While computing similarity of text is a frequently studied part of natural language processing, to the best of our knowledge, determining conflicting text parts has not been examined very well by now [13, 15]. Mostly, these approaches search for negations or antonyms. Searching for negations might not be that useful when considering constraints since these will not be stated explicitly in a regulatory document. Antonyms in this case correspond to, e.g., constraint related subjects having a low similarity score. Consequently, the following definitions can be stated.

Definition 3 (Constraint Characterization). *Let C be a set of constraints, $c_1, c_2 \in C$ and $sim : C \times C \mapsto I$ be a function that determines the similarity between two constraints with $I \subseteq \mathbb{R}$ an interval. Let $\tau \in I$ be a constant, such that $sim(c_1, c_2) > \tau$. The constraints c_1, c_2 are called*

- *redundant, iff*
  - *they belong to the same group or for constraint related subjects $r_1 \in c_1, r_2 \in c_2$ holds $sem(r_1, r_2) > \eta_1$*
  - *and for two tasks $t_1 \in c_1, t_2 \in c_2$ holds $sem(t_1, t_2) > \eta_2$*
  with $\eta_1, \eta_2 \in I$.

- *subsumed, iff they are redundant and either c_1 or c_2 contains further information related to its task.*
- *conflicting, iff either*
  - *they belong to different groups or for constraint related subjects $r_1 \in c_1, r_2 \in c_2$ holds $sem(r_1, r_2) < \mu_2$*
  - *and for two tasks $t_1 \in c_1, t_2 \in c_2$ holds $sem(t_1, t_2) > \mu_1$*
  with $\mu_1, \mu_2 \in I$ or they are redundant but contain different time spans.

Definition 3 of redundant, subsumed, and conflicting constraint pairs is based on a similarity function *sim* which operates on constraints that are reflected by sentences. The similarity of sentences is computed within and across each group.

For this, all words need to be represented by word vectors. For computing these vectors, several approaches have been proposed during the last years (e.g., [16, 21]). In order to deliver reasonable results mostly large data collections for training the model are needed. Therefore, we suggest to use a pre-trained model or pre-trained word vectors for the language in which the documents are written. The similarity between the word vectors is then computed using a distance measure. For text mining tasks the cosine measure is a common choice [1]. The final outcome is a similarity score which is, in the case of the cosine measure, a value between -1 and 1, with 1 corresponding to absolutely similar, -1 not similar at all, i.e., $I := [-1, 1]$ in this case.

After applying the characteristics set out in Definition 3 the outcome is three lists per method containing redundant, subsumed or conflicting constraints.

For the running example using **term frequencies** results in one pair of subsumed constraints, $(S5, S10)$ with a similarity score of ≈ 0.94031. These are obviously subsumed since the first sentence explains in more detail what needs to be done to address changes. No redundant or conflicting constraints are found which can be easily verified by a manual inspection of the given sentences. Examples of these types are given in the evaluation.

Using the second method, i.e., **sentence structure** delivers one pair of subsumed constraints $(\mathcal{S}5, \mathcal{S}10)$, no redundant and no conflicting constraints.

2.3 Post-processing

To make the derived information available for the user, a suitable visualization is needed. In this approach a graph-based structure (so-called network map, cf. Definition 4) is used but this step can be customized and any other representation could be chosen. In the network map visualization each node corresponds to a sentence and the edges indicate whether sentences are redundant, subsumed, or contradicting. Edges representing redundant and subsumed connections are labeled as r and s while contradicting ones are labeled as c.

Definition 4 (Constraint Network Map). *A network map is a graph $NM = (\mathcal{C}, E)$, with*

- *$\mathcal{C}$ being a set of nodes where each node $c \in \mathcal{C}$ corresponds to one constraint*
- *$E \subseteq \mathcal{C} \times \mathcal{C}$ being the edges.*

Moreover, let $w: E \mapsto RL := \{r, s, c\}$ be a function assigning a label to an edge depending on the corresponding relation between the nodes that span the edge, i.e., redundant (r), subsumed (s), conflicting (c).

Figure 5a displays the network map for the running example based on term frequencies, Fig. 5b the network map for the running example based on the sentence structure. Subsumed constraints are displayed as edges labeled s for subsumed. Note that no redundant or conflicting constraints were found and constraints that are not connected do not have a relation.

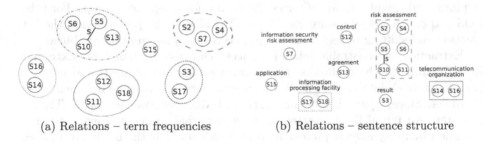

(a) Relations – term frequencies   (b) Relations – sentence structure

Fig. 5. Relations – running example

Another possible post-processing strategy could be to transform each sentence into a formal language in order to construct executable rules or models. Many current approaches are capable of doing this but often conflicts, subsumptions or redundancies are not resolved correctly resulting in incomplete or contradicting rules and models. By first applying the presented approach it might be possible to resolve such clashes at all or at least in a shorter amount of time.

3   Implementation

A prototypical implementation of the method described in Sect. 2 is provided and used for the evaluation in Sect. 4. The prototype is written in Python 3 and integrates the NLP framework Spacy[3], NLTK (cf. [6]) and WordNet[4]. This decision relies on [2], which evaluated several state-of-the-art NLP frameworks.

First of all, the documents are transformed and prepared as described in Sect. 2.1. Due to the variety of document formats it is difficult to provide a generic implementation and further elaborating on this is beyond the scope of the paper.

The first generic step which is carried out using Spacy is the chunking, parsing, POS tagging and lemmatizing of sentences. As recommended, the large model for the English language is used. There are three means on which the grouping can be based on and which are applied after filtering the POS tagged sentences for markers.

The first technique uses clustering based on **term frequencies** in combination with the k-means++ algorithm and can be applied if no additional information is available and the set of documents is large. Since this correlates to parts of the method presented in [26], the corresponding parts of the implementation were migrated from R to Python 3 and integrated into the recent implementation. In the implementation TfIdf as well as k-means++ are taken from the popular scikit-learn library [20].

For the remaining two techniques the grouping relies on a list of constraint related subjects.

The second method exploits the **sentence structure** in order to derive such a list. In particular the annotation attributes of Spacy Tokens are used. Note, that during this process there are some properties that need to be taken care of, e.g., compound terms must be considered as one constraint related subject or plural and singular forms of terms should not form separate groups. For this, POS and dependency tags are considered. To enhance the performance, the list creation and shifting of sentences to their corresponding group is combined.

Extraction of constraint related subjects can also be based on **external information** (in the evaluation an organizational chart is used) and integration of a lexical database for finding synonyms which is in our implementation WordNet. Since, Spacy has no integration of WordNet, we rely on NLTK for this step. An initial list is created from the external source and every term is extended by its synonyms present in WordNet. Now that the list is given, each sentence is processed again and shifted to its corresponding group.

The last step of the processing stage relates to finding relations between pairs of constraints. As outlined in Definition 3, the similarity between sentences needs to be computed. For this task it is either possible to train own word vectors or to use Spacy's similarity function which uses the cosine metric and word vectors that were trained with the word2vec algorithm family [16]. Since

---

[3] https://spacy.io.
[4] https://wordnet.princeton.edu/.

this is a pre-computed model it might happen that not every term has a vector representation, so this needs to be checked and adapted if necessary.

After computing the similarity, all sentences above a certain threshold are further examined whether they have the described characteristics for redundancy, subsumption or conflict stated in Definition 3. Automating the detection and comparison of time spans is the main challenge here. Simple functions like `isdigit()` are not sufficient since digits can also be written out.

In the last step of the method, the results are displayed as network maps. For drawing these graphs the NetworkX (cf. [12]) package is used. Each constraint is integrated and colored based on its group. The edges are drawn whenever a relation between a pair of constraints exists and labeled accordingly. For large documents the result needs to be scalable and it should be possible to display only particular groups.

4   Evaluation

For evaluating the approach a set of documents from the financial sector is used and an expert was consulted in order to estimate the results. The first document is the *BCBS 239*[5] which provides guidelines on risk management of financial institutes. The second document is the *Regulation 2016/867*[6], which specifies guidelines for credit management.

For gaining an overview of the documents and getting to know their structure an expert interview was conducted first. The expert emphasized that constraints always contain markers like `shall`, `should` and `must`. For testing the third processing option (external information sources) an organizational chart of the experts company is integrated.

Before starting the analysis, several questions were stated, e.g., *Did the approach find sentences which do not contain constraints?* or *Were the relations between constraints correctly drawn, i.e., how precise is the approach?*.

The precision can be measured by the ratio between the number of the intersection of relevant sentence pairs and all retrieved sentence pairs divided by the number of retrieved sentence pairs,

$$Precision = \frac{|\{relevant\ pairs\} \cap \{retrieved\ pairs\}|}{|\{retrieved\ pairs\}|}.$$

Relevant in this case means, that a pair is a pair that is indicated by the domain expert to be in the correct group of relations.

Both documents are given in PDF format, and thus first of all transformed into plain text format. Afterwards, the table of contents and references, as well as the introductions are removed and each document is fragmented into sentences and POS tagged. Constraints are filtered out using markers and lemmatized for the processing stage. Each of the three methods is applied on the resulting aggregated set of constraints and the thresholds are set to $\tau = 0.97, \eta_1 = 0.8, \eta_2 = 0.5$.

---

[5] https://www.bis.org/publ/bcbs239.pdf.

[6] https://eur-lex.europa.eu/eli/reg/2016/867/oj.

Term Frequencies: Choosing $k = 15$ results in clusters containing between 6 and 38 constraints.

The number of redundant constraints is 42, among which 10 have a similarity score of 1.0. This is due to the fact that the sentence In the case of natural persons being affiliated with instruments reported to AnaCredit, no record for the natural persons must be reported. appears five times in *Regulation 2016/867* in five different sections. Another example of a redundant pair of constraints is

- If a change takes place, the records must be updated no later than the monthly transmission of credit data for the reporting reference date on which the change came into effect.
- If a change takes place, the records must be updated no later than the monthly transmission of credit data for the reporting reference date on or before which the change came into effect.

with a similarity score of $\approx$0.99897. This pair could also be viewed as subsumed but the difference is so little that the approach detects a redundancy in this case, which is fine according to the consulted domain expert.

The number of subsumed constraint pairs is 10. An example is

- Supervisors should have and use the appropriate tools and re-sources to require effective and timely remedial action by a bank to address deficiencies in its risk data aggregation capabilities and risk reporting practices.
- Supervisors should require effective and timely remedial action by a bank to address deficiencies in its risk data aggregation capabilities and risk reporting practices and internal controls.

with a similarity score of $\approx$0.98628.

In addition, 6 conflicting pairs of constraints are retrieved, e.g.,

- For observed agents that are resident in a reporting Member State, NCBs shall transmit monthly credit data to the ECB by close of business on the 30th working day following the end of the month to which the data relate.
- For observed agents that are foreign branches not resident in a reporting Member State, NCBs shall transmit monthly credit data to the ECB by close of business on the 35th working day following the end of the month to which the data relate.

with a similarity score of $\approx$0.99669. Having a closer look at this pair of constraints and also according to the expert, revealed that this is not a conflicting constraint pair. It rather indicates a decision, i.e., whether an observed agent is resident in a reporting member state or not which must be considered by a user who wants to implement these constraints. This corner case is difficult to detect by automated approaches because the conflict of time intervals is refuted by the opposite subjects indicated by a negation.

Structure of Sentences: The approach delivers 56 constraint related subjects forming also 56 groups which contain between 1 and 22 sentences.

It can be recognized that lemmatization of words did not work out entirely, since, e.g., `bank` and `banks` formed separate groups. In addition, an exceptional case can be seen. `Procedures should be in place to allow for rapid collection and analysis of risk data and timely dissemination of reports to all appropriate recipients. This should be balanced with the need to ensure confidentiality as appropriate.` In this case this refers to the preceding sentence as such.

Redundant pairs of constraints have similar lemmatized constraint related subjects and similar lemmatized tasks. The approach yields 42 of these, e.g.,

- `4.4 The records must be reported no later than the monthly trans- mission of credit data relevant for the reporting reference date on or before which the instrument was registered in AnaCredit.`
- `If a change takes place, the records must be updated no later than the date of the monthly transmission of credit data that is relevant for the reporting reference date on or before which the change came into effect.`

having a similarity score of ≈ 0.97818. The pairs differ slightly from the ones detected by the previous method. The redundant constraint pairs with similarity score 1.0 which were found using term frequencies are not present in this set. The pattern ((compound) subject + word) fails in this case and therefore the sentence is not considered anymore. One possibility to tackle this issue might be to introduce a group "undefined".

Two subsumed pairs of constraints are found, e.g.,

- `Reports should include an appropriate balance between risk data, analysis and interpretation, and qualitative explanations.`
- `Reports should reflect an appropriate balance between detailed data, qualitative discussion, explanation and recommended conclusions.`

having a similarity score of ≈ 0.97119.

The same conflicting pairs of constraints like before are obtained.

External Information Sources: To apply the third method, an organizational chart is used for manually deriving the list of constraint related subjects. It is used for grouping the sentences and consists in this case of 15 terms (before it is extended by synonyms). Such an organizational chart contains a graphical representation of the relation of one official and its department to another within

an organization. Consequently, the grouping is structured among departments.[7]
If a sentence cannot be assigned to one of the terms from the list it is shifted
to a default group. Altogether, 7 groups of size 1 to 131 are received whereupon
the default group is the largest one.

This method delivers 56 redundant constraints, 14 subsumed and the same
6 conflicting constraints as before. The redundant ones are the same when combining the previous two approaches. A difference can be seen regarding the set
of subsumed constraints. In this case four constraints that are not present in the
previous sets are given, e.g.,

- A bank's risk data aggregation capabilities should ensure that
  it is able to produce aggregate risk information on a timely
  basis to meet all risk management reporting requirements.
- Risk management reports should be accurate and precise to
  ensure a bank's board and senior management can rely with
  confidence on the aggregated information to make critical
  decisions about risk.

with a similarity score of ≈ 0.97116.

Finally, the quality of the overall method and of every processing option
needs to be assessed. For the overall method, it can be stated that every sentence that was marked as constraint truly is a constraint, so the approach did
not deliver false positives. What can be taken into account for comparing the
three processing strategies is, e.g., the number of created groups. A large number
of groups enables a differentiated view on the data but can be to fine-granular,
e.g., the second method delivered groups containing only one sentence. On the
other hand, the third method created few groups but these are not very distinc-
tive. Therefore, a good balance between the number of groups and the therein
contained sentences should be targeted. The first method fulfills this criterion
best. For estimating the quality of the derived relations the precision scores for
each method are summed up in Table 1.

Table 1. Precision scores

	Redundancy	Subsumption	Conflict
Term frequencies	69% (77%)	50% (65%)	0%
Sentence structure	54% (60%)	100% (100%)	0%
External information	64% (69%)	50% (61%)	0%

A domain expert checked each detected pair and decided whether it is in the
right category or not. Some sentences were half half, i.e., they can be partly seen
as redundant or subsuming. The score in brackets indicates this by weighting

---

[7] Another possibility is to use a glossary and carry out the grouping based on the
therein contained terms.

these sentences in the computation with 0.5, while the other score counts them as false positives and is therefore a bit lower. The overall outcome is, compared to state-of-the-art approaches fine, when considering that no restrictions were imposed on the input text (for more details cf. [24]). The precision of `conflict` is 0% because of the before mentioned corner case of two conflicting characteristics that cancel out. Moreover, the expert indicated that no conflicts are present in the documents. Conflicts might arise when updated versions of a document are considered, i.e., a new constraint causes a conflict compared to an old one. Evaluating the approach on such a set of documents is planned as future work.

Post-processing
The results for the last method (external information) are schematically visualized in Fig. 6 to demonstrate how a user could benefit from the derived results.

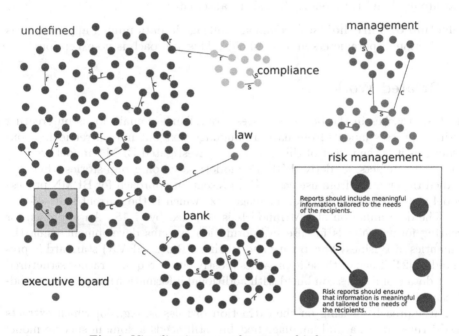

Fig. 6. Visualization – results external information

The grouping of constraints is reflected by the different colors of nodes and their regional proximity. A user can now select the subset of constraints that he wants to have a closer look at. Moreover, he can view the relations and sentences in more detail as indicated by the box in the right lower corner.

5   Discussion and Limitations

Ambiguity of Language: Since natural language can be versatile the completeness of markers is hard to estimate. Also, extracting synonyms can be a challenge since meanings of words differ when the context is changing. Using domain specific ontologies might overcome this issue. During the evaluation we also observed that, e.g., searching for time spans is not straight forward and several iterations of implementations are needed in order to get reasonable results.

Choosing a NLP Framework: There are lots of NLP frameworks available and many of them provide different features. The quality of the results relies on their capabilities to parse information correctly. As it could be demonstrated in the evaluation, lemmatization was not carried out correctly for each case. (Manual) adoptions tailored to the regarded document collection might be necessary.

Integration of External Information: Another task that requires manual inspection is the integration of external information for deriving the list of constraint related subjects. Again, this step relies on the input format and tools that are used and is therefore difficult to automate.

Selection of Thresholds: A common challenge in data mining applications is the selection of parameters and thresholds. This approach is no exception.

6   Related Work

Most approaches in the business process compliance domain focus on creating business process models from natural language text but not on retrieving constraints as it is the target of this paper. [11] investigated BPMN model creation from text artefacts, [3] derived BPMN models based on group stories while [25] studied the creation from use cases. [10] present an approach for BPMN process model generation from natural language text which is the current state-of-the-art. The determination of UML models is targeted by [7,18]. An approach for creating formal models for use in information systems development using the Semantics of Business Vocabulary and Business Rules (SBVR) standard is presented in [24]. Each of these approaches mostly either requires rather structured input data (sometimes combined with additional information) or produces models that lack precision.

An approach focusing on the extraction of rules is, e.g., [5] which extracts SBVR rules from natural language text but still needs a domain specific model is needed. Our approach does not require such information. [9] outline a method for extracting rules from legal documents by using logic-based as well as syntax-based patterns.

Resolving relations between sentences containing constraints is not discussed in any of the mentioned approaches but might help to improve derived business rules and process models.

7   Conclusion and Future Work

In this paper an approach for grouping sentences containing constraints and resolving relations between them was presented. Relations could be resolved based on a characterization of redundancy and conflict. A state-of-the-art NLP framework as well as common data mining algorithms were used for implementing the method. The evaluation was carried out on a set of documents from the financial sector and the results were assessed by a domain expert.

The most crucial target of future work is to evaluate to what extent our method can resolve the lack of precision generated by state-of-the-art approaches for process rule and model extraction from natural language text. Besides that, we plan to further extend the evaluation in order to improve the implementation by covering more exceptional cases. Another interesting point is to consider sets of documents that consist of updated versions of one document and to retrieve examples of constraint pairs that changed during the versions. This might help to manage and update business rules accordingly. Creating an interactive visualization that integrates the original documents is also envisaged.

Acknowledgment. This work has been funded by the Vienna Science and Technology Fund (WWTF) through project ICT15-072.

References

1. Aggarwal, C.C., Zhai, C.: Mining Text Data. Springer, New York (2012). https://doi.org/10.1007/978-1-4614-3223-4
2. Al Omran, F.N.A., Treude, C.: Choosing an NLP library for analyzing software documentation: a systematic literature review and a series of experiments. In: Proceedings of the 14th International Conference on Mining Software Repositories, pp. 187–197. IEEE Press (2017)
3. de AR Goncalves, J.C., Santoro, F.M., Baiao, F.A.: Business process mining from group stories. In: 13th International Conference on Computer Supported Cooperative Work in Design, CSCWD 2009, pp. 161–166. IEEE (2009)
4. Arthur, D., Vassilvitskii, S.: k-means++: the advantages of careful seeding. In: Proceedings of the Eighteenth Annual ACM-SIAM Symposium on Discrete Algorithms, pp. 1027–1035. Society for Industrial and Applied Mathematics (2007)
5. Bajwa, I.S., Lee, M.G., Bordbar, B.: SBVR business rules generation from natural language specification. In: AAAI Spring Symposium: AI for Business Agility, pp. 2–8 (2011)
6. Bird, S., Klein, E., Loper, E.: Natural Language Processing with Python: Analyzing Text with the Natural Language Toolkit. O'Reilly Media, Inc., Sebastopol (2009)
7. Deeptimahanti, D.K., Babar, M.A.: An automated tool for generating UML models from natural language requirements. In: Proceedings of the 2009 IEEE/ACM International Conference on Automated Software Engineering, pp. 680–682. IEEE Computer Society (2009)
8. Dijkman, R., Dumas, M., Van Dongen, B., Käärik, R., Mendling, J.: Similarity of business process models: metrics and evaluation. Inf. Syst. **36**(2), 498–516 (2011)
9. Dragoni, M., Villata, S., Rizzi, W., Governatori, G.: Combining NLP approaches for rule extraction from legal documents. In: 1st Workshop on MIning and REasoning with Legal texts (MIREL 2016) (2016)
10. Friedrich, F., Mendling, J., Puhlmann, F.: Process model generation from natural language text. In: Mouratidis, H., Rolland, C. (eds.) CAiSE 2011. LNCS, vol. 6741, pp. 482–496. Springer, Heidelberg (2011). https://doi.org/10.1007/978-3-642-21640-4_36

11. Ghose, A., Koliadis, G., Chueng, A.: Process discovery from model and text artefacts. In: 2007 IEEE Congress on Services, pp. 167–174. IEEE (2007)

12. Hagberg, A., Swart, P., Schult, D.: Exploring network structure, dynamics, and function using networkx. Technical report, Los Alamos National Laboratory (LANL), Los Alamos, NM (United States) (2008)

13. Harabagiu, S., Hickl, A., Lacatusu, F.: Negation, contrast and contradiction in text processing. In: AAAI, vol. 6, pp. 755–762 (2006)

14. Hashmi, M., Governatori, G., Lam, H.P., Wynn, M.T.: Are we done with business process compliance: state of the art and challenges ahead. Knowl. Inf. Syst. **57**, 1–55 (2018)

15. de Marneffe, M.C., Rafferty, A.R., Manning, C.D.: Identifying conflicting information in texts. In: Handbook of Natural Language Processing and Machine Translation: DARPA Global Autonomous Language Exploitation (2011)

16. Mikolov, T., Yih, W.T., Zweig, G.: Linguistic regularities in continuous space word representations. In: Proceedings of the 2013 Conference of the North American Chapter of the Association for Computational Linguistics: Human Language Technologies, pp. 746–751 (2013)

17. Miller, G.A., Beckwith, R., Fellbaum, C., Gross, D., Miller, K.J.: Introduction to wordnet: an on-line lexical database. Int. J. Lexicogr. **3**(4), 235–244 (1990)

18. More, P., Phalnikar, R.: Generating UML diagrams from natural language specifications. Int. J. Appl. Inf. Syst. **1**(8), 19–23 (2012)

19. Nguyen, T.A., Perkins, W.A., Laffey, T.J., Pecora, D.: Checking an expert systems knowledge base for consistency and completeness. In: IJCAI, vol. 85, pp. 375–378 (1985)

20. Pedregosa, F., et al.: Scikit-learn: machine learning in python. J. Mach. Learn. Res. **12**(Oct), 2825–2830 (2011)

21. Pennington, J., Socher, R., Manning, C.: Glove: global vectors for word representation. In: Proceedings of the 2014 Conference on Empirical Methods in Natural Language Processing (EMNLP), pp. 1532–1543 (2014)

22. Riefer, M., Ternis, S.F., Thaler, T.: Mining process models from natural language text: a state-of-the-art analysis. In: Multikonferenz Wirtschaftsinformatik (MKWI 2016), pp. 9–11, March 2016

23. Rinderle-Ma, S., Ma, Z., Madlmayr, B.: Using content analysis for privacy requirement extraction and policy formalization. In: Enterprise Modelling and Information Systems Architectures, pp. 93–107 (2015)

24. Selway, M., Grossmann, G., Mayer, W., Stumptner, M.: Formalising natural language specifications using a cognitive linguistic/configuration based approach. Inf. Syst. **54**, 191–208 (2015)

25. Sinha, A., Paradkar, A.: Use cases to process specifications in business process modeling notation. In: 2010 IEEE International Conference on Web Services (ICWS), pp. 473–480. IEEE (2010)

26. Winter, K., Rinderle-Ma, S., Grossmann, W., Feinerer, I., Ma, Z.: Characterizing regulatory documents and guidelines based on text mining. In: Panetto, H. (ed.) OTM 2017. LNCS, vol. 10573, pp. 3–20. Springer, Cham (2017). https://doi.org/10.1007/978-3-319-69462-7_1

Empirical Analysis of Sentence Templates and Ambiguity Issues for Business Process Descriptions

Thanner Soares Silva[1]($\boxtimes$)(iD), Lucinéia Heloisa Thom[1](iD), Aline Weber[1](iD),
José Palazzo Moreira de Oliveira[1](iD), and Marcelo Fantinato[2](iD)

[1] Department of Informatics, Federal University of Rio Grande do Sul, UFRGS,
Porto Alegre, Brazil
{thanner.silva,lucineia,aline.weber,palazzo}@inf.ufrgs.br
[2] School of Arts, Sciences and Humanities, University of São Paulo,
São Paulo, Brazil
m.fantinato@usp.br

Abstract. Business process management has become an increasingly present activity in organizations. In this context, approaches that assist in the identification and documentation of business processes are presented as relevant efforts to make organizations more competitive. To achieve these goals, business process descriptions are considered as a useful artifact in both identifying business processes and complementing business process documentation. However, approaches that automatically generate business process descriptions do not explain how the sentence templates that compose the text were selected. This selection influences the quality of the text, as it may produce ambiguous or non-recurring sentences, which could make it difficult to understand the process. In this work, we present an empirical analysis of 64 business process descriptions in order to find recurrent sentence templates and filter them for ambiguity issues. The analysis made it possible to find 101 sentence templates divided into 29 categories. In addition, 13 of the sentence templates were considered to have ambiguity issues based on the adopted criteria. These findings may support other approaches in generating process descriptions more suitable for process analysts and domain experts.

Keywords: Business Process Model and Notation
Natural language processing · Sentence template · Ambiguity

1 Introduction

In order to stay competitive, organizations need to document and manage their business processes. Approaches related to process descriptions and Business Process Model and Notation (BPMN) can help them to achieve these goals. The combination of distinct approaches can improve understanding of the process as

© Springer Nature Switzerland AG 2018
H. Panetto et al. (Eds.): OTM 2018 Conferences, LNCS 11229, pp. 279–297, 2018.
https://doi.org/10.1007/978-3-030-02610-3_16

they provide different perspectives on it [21,23]. Several proposals have been presented with the goal of contributing to the relationship between texts and process models in different scenarios, such as: automatic generation of process descriptions from process models [17,18,20], automatic generation of process models from process descriptions [8,14,15,25], process mining from natural language text [5,6,10,19], identification of business process elements in natural language texts [12], integration between texts and process models [16] and verification of conformity between texts and process models [1,3,24].

A number of the proposed approaches related to this context use sentence templates to generate or transform process descriptions. However, the corresponding related works do not explain how the sentence templates that compose the process description were selected and this information is important, as it directly interferes with the quality of the text. Sentence templates not carefully selected may produce sentences with ambiguity problems that may not be understood by the stakeholders of the business processes, such as process analysts and domain experts.

Furthermore, this work aims to help in the development of an approach that generates process-oriented texts from natural language texts. In the context of this approach, a process-oriented text is defined as a text that is both structured and capable of maintaining the maximum information related to the business process. In addition, it is expected to verify that the business process described in the text conforms to the BPMN specification. The approach consists of five stages: input data, text reading, BPMN verification, text writing and text output. Firstly, a natural language text is given as input to the input data (i.e., input data stage). Then the approach reads the natural language text and produces an intermediate structure (i.e., text reader stage). Then, the intermediate structure is used to verify the process described by the text (i.e., BPMN verification stage) and to generate the text writer (i.e., text writer stage). Finally, the verification and the structured text are combined for the generation of the process-oriented text (i.e., output text stage). Figure 1 shows the approach with its respective stages. This paper seeks to contribute with the sentence templates repository presented in stage "4. Text writer".

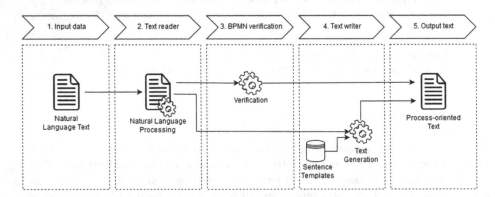

Fig. 1. Process-oriented text generation.

In this context, this work proposes an analysis to identify sentence templates that are common in processes descriptions and that causes less ambiguity problems, so the findings can be useful in writing new process descriptions. For this, an empirical analysis of 64 texts describing business processes was performed. These texts were taken from the book Fundamentals of Business Process Management [9] and Friedrich [13]. Moreover, an investigation of ambiguity issues in process descriptions was carried out based on the literature and on the sentence templates identified. This paper presents the results of these analyses.

A total of 101 sentence templates was found, divided into 29 categories based on three criteria, namely: source, target and relationship. Six types of ambiguities were identified and, when compared with the found sentence templates, enabled us to define 13 templates related to ambiguity issues.

The remainder of this paper is organized as follows: Sect. 2 presents BPMN as background of the presented research. Section 3 presents the method used to identify and classify sentence templates as well as the identified ambiguity issues. Section 4 reports the results analysis. Section 5 contemplates the applications of this work, as well as discusses the results obtained. Section 6 presents the related work. Finally, Sect. 7 presents the final conclusions.

2   Background

The BPMN is a standard for process modeling maintained by the Object Management Group (OMG) [22]. BPMN includes five elements categories: flow objects (activities, events and gateways), data (data objects, data inputs, data outputs and data stores), connecting objects (sequence flows, message flows, associations and data associations), swimlanes (pools and lanes) and artifacts (groups and text annotations). In this work, the focus is on flow objects, swimlanes and connecting objects because they are used for the identification and classification of sentence templates.

Regarding flow objects, activities can be defined as a task that a company performs in a process. An activity can be atomic or non-atomic and is represented as rounded boxes. Events are represented as circles and indicate where a particular process starts (start event) or ends (end event). Moreover, there are events that can occur between a start event and an end event (intermediate event) which can affect the flow of the process but cannot start it or end it. Finally, gateways are represented as a diamond shape and are responsible for controlling divergence (split) and convergence (join) of sequence flows in a process. There are six different types of gateways which differ in both the logic that they execute and the representation placed within the gateway diamond. Among them, it can be highlighted: exclusive gateway (XOR, represented with or without a "X" marker), where the decision making leads to the execution of exactly one path; parallel gateway (AND, represented with a "+" marker), where all possible paths must be executed; and inclusive gateway (OR, represented with a "O" marker), where decision making leads to the execution of at least one path.

In relation to swimlane, a pool represents a participant in a business process. A pool is graphically represented as a container that partitions a process from other participants. If a pool does not contain a process, it is considered as a black box. Lane, on the other hand, are the partitions used to organize and categorize activities within a pool. Lanes are often used for representing internal roles (e.g., Manager, Associate), systems (e.g., an enterprise application), or an internal department (e.g., shipping, finance).

Regarding connecting objects, sequence flows are used to show the order of flow objects in a process. A sequence flow is represented as a solid single line with a solid arrowhead. A message flow represents the flow of messages between two different participants and is represented as a dashed single line with an open circle line start and an open arrowhead line end. In addition, an association is used to link information and artifacts with flow objects. Data associations, on the other hand, are used to relate data objects and activities. Both association and data association are represented as a dotted single line.

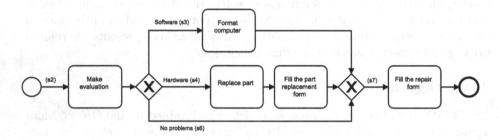

Fig. 2. Example of BPMN model: computer repair.

Figure 2 presents an example of a BPMN process model composed by one start event, five activities, one exclusive decision gateway (XOR-split), one exclusive merge gateway (XOR-join) and one end event. After the process starts, an activity is executed (called "Make evaluation"). Then, there is a decision making in which only one of three possible paths can be followed. After one path is followed, the process returns to the main path, another activity is performed and the process ends. A possible description of the process shown in Fig. 2 can be seen in Fig. 3. The relationships between the text and the model are evidenced through s_x, where x refers to the sentence number in the text.

3   Sentence Templates and Ambiguity Issues

This section presents the method used to identify the sentence templates.

The process descriptions used in the analysis came from two different sources. Only process descriptions in English were considered, as templates are very sensitive to the language. Firstly, 47 process descriptions present in Friedrich [13] were identified. From this first source, 17 process descriptions were disregarded for the following reasons: they were translated from another language through machine translation services (14 texts), were duplicated (2 texts) or had a description

---

(s_1) The repair department of the company X performs repairs of computers and printers. (s_2) Once a computer with problems arrives, the technician must perform an evaluation. (s_3) In case it is a software problem, the technician must format the computer. (s_4) For the case that it is a hardware problem, the technician must replace the part and fill out the part replacement form. (s_5) This form must contain the part identification code and the technician's signature. (s_6) On the other hand, if no problem is found, no modification should be made to the computer. (s_7) The process finishes after the technician completes the repair form.

Fig. 3. Example of process description: computer repair.

format based on enumeration (1 text). Secondly, 34 process descriptions from the book Fundamentals of Business Process Management [9] were identified. The final set of 64 process descriptions, as well as their respective sources and types are presented in Table 1.

Table 1. Data sources.

Source	Amount	Type
HU Berlin	4	Academic
TU Berlin	2	Academic
QUT	8	Academic
TU Eindhoven	1	Academic
Vendor Tutorials	4	Industry
inubit AG	3	Industry
BPM Practitioners	1	Industry
BPMN Prac. Handbook	3	Textbook
BPMN M&R Guide	4	Textbook
Fundamentals of BPM	34	Textbook
Total	64	–

The following subsections present the procedures followed to: (Sect. 3.1) prepare the sentences, (Sect. 3.2) identify and classify the sentence templates and (Sect. 3.3) address the ambiguity issues.

3.1   Preparation of Sentences

In this first stage, the sentences are prepared for the identification and classification of sentence templates. For this, the sentences of a process description are modified to become more generic.

A business process description may contain snippets of text that are directly related to the process context. As an example, the sentences "The manager must sell the product" and "The salesman must sell the product" are identical, except by who carries out the activity of selling the product. This difference can hinder the identification and classification of sentence templates, since these sentences can be considered as different sentence templates. In this sense, a term capable of representing both "manager" and "salesman" could be used in order to make these two sentences equal and, consequently, to define both as the same sentence template. Thereby, the process descriptions were previously analyzed and four different placeholders were created with the goal to replace in the sentences the snippets related to the context by more generic information. The created placeholders are: *role, condition, number* and *object*.

The placeholder *role* is associated with the role responsible for performing a particular action. In relation to the business process model, a role could be considered as a participant. As an example, in the sentence "The process finishes after the technician completes the repair form.", once the technician is the one performing the action, the word "technician" can be replaced by the placeholder *role*. Therefore, the sentence after the modification would be written as "The process finishes after the *role* completes the repair form".

The placeholder *condition* aims to define some condition that needs to be satisfied for a given flow to occur. Normally, the condition appears in a business process model as a label that tracks the output sequence flow of an exclusive or inclusive gateway. For instance, in the sentence "In case it is a software problem, the technician must format the computer." it is possible to observe that to be done the activity of formatting the computer must exist before the condition "it is a software problem". Therefore, this condition will be replaced in the text by the placeholder *condition*. Moreover, the technician can also be replaced in this sentence by the placeholder *role*.

The placeholder *number* is used to represents a certain amount of process elements or paths in a process model. As an example, in the sentence "After all five activities are completed, the process ends.", the amount "five" can be replaced by the placeholder *number*.

Finally, the placeholder *object* can represent the business object to which the sentence refers. For instance, in the sentence "The car can be sold by the manager or the seller", the business object "car" can be replaced by the placeholder *object*. In addition, the placeholders *role 1* and *role 2* could be created to represent the manager and seller respectively.

After the preparation stage, the modified sentences containing placeholders will be used for the identification and classification of sentence templates.

3.2   Identification and Classification of Sentence Templates

In the context of this work, a sentence template was considered as each pattern present in a sentence that is able to describe one or more process elements. These process elements appear in the template as placeholders to be replaced. For the scope of this paper, a reduced set of elements is taken into account to find sentence templates, being: activity (A_c), AND-split (G_{+s}), AND-join (G_{+j}), XOR-split (G_{Xs}), XOR-join (G_{Xj}), OR-split (G_{Os}), OR-join (G_{Oj}), start event (E_s), intermediate event (E_i), end event (E_e). In addition, "empty" is used to define paths without elements (e.g., Fig. 2, s_6).

In order to identify a sentence template, it is necessary to identify beforehand the process elements in the text. Although there are works that contribute to the automated identification of process elements in texts, to the best of our knowledge, there is no approach capable of extracting the process elements in a textual description with complete precision [10,12,14]. In addition, automated identification approaches can draw incorrect conclusions about a process by making assumptions about texts that allow for multiple interpretations [2]. Therefore, an automated analysis of sentence templates could be compromised by the selected approach of extracting process elements, so the identification of the sentence templates was carried out manually.

The identification and classification of sentence templates were carried out in parallel. In the context of this paper, each sentence template is considered as composed by the following elements:

- **Target:** the set of process elements described by the sentence template. A target must appear in the sentence, even if implicitly (i.e., without a placeholder to fill with the process element).
- **Relationship:** how the process elements in the sentence are associated to each other: none (R_N), composed by 0 or 1 process element; sequential (R_S), one element follows the other; exclusive (R_X); inclusive (R_O); and parallel (R_+).
- **Source:** the process element that occurs immediately before the analyzed sentence. As in the BPMN specification [22], the source can be understood as the element prior to the currently described element connected by a sequence flow. A source may or may not be evidenced in the sentence.

As an example, for the process description presented in Fig. 3 (corresponding to the BPMN model depicted in Fig. 2), five sentence templates were identified, two of which has target with the sequential relationship (s_4, s_7) and three has target with the none relationship (s_2, s_3, s_6). In the sentence s_2, it is possible to define the sentence template "Once E_s, the *role* must A_c", where E_s is the source evidenced in the sentence template, A_c is a placeholder for an activity described in the target and *role* refers to some participant in the process that performs the activity A_c. The sentences s_3, s_4 and s_6 have as source a XOR-split gateway not evidenced in the sentence template. In addition, the sentence s_7 has as its source a XOR-join gateway and as target an activity (A_c) and an end event (represented implicitly by "The process finishes after"). Not all sentences in a text are necessarily mapped to a sentence template, since process descriptions can be composed by other information, such as statements that contextualize the process (s_1) and statements that detail an activity or business rule (s_5).

In order to identify the sentence templates, each process description was inserted into a spreadsheet, as illustrated in Table 2. In the spreadsheet, each line represents one sentence and the columns represent the following attributes:

sentence, sentence template ID (i.e., the ID of the sentence template that can be a number or "none") and sentence template.

Table 2. Example of identification of sentence templates.

Process description: computer repair		
Sentence	Sentence template ID	Sentence template
The repair department of the company X performs repairs of computers and printers	None	
Once a computer with problems arrives, the technician must perform an evaluation	3	Once E_s, the *role* must A_c
In case it is a software problem, the technician must format the computer	10	In case *condition*, the *role* must A_c
For the case that it is a hardware problem, the technician must replace the part and fill out the part replacement form	25	For the case *condition*, the *role* must A_c and A_c
This form must contain the part identification code and the technician's signature.	None	
On the other hand, if no problem is found, no modification should be made to the computer	1	On the other hand, if *condition*, *empty*
The process finishes after the technician completes the repair form	8	The process finishes after the *role* A_c

After all the sentence templates were identified, they were grouped into categories based on source, target and relationship. As a result, each category is composed by sentence templates that can be replaced in a process description and represent the same information. As an example, the sentence s_3 is defined as a sequential relationship between a XOR-split (source) and an activity (target). This sentence can be rewritten by another sentence template that have the same properties, therefore the same category, such as: "Once *condition*, the *role* needs to A_c".

The analysis of sentence templates was done in two different manners, namely atomic level analysis and group level analysis. At the atomic level analysis, it is considered that if two sentences have the same text, but represent different process elements in the source, they are defined as two distinct sentence templates. For example, sentences "When a computer with problems arrives, the technician must perform an evaluation." and "When performing a repair, the technician must perform an evaluation." are defined as different sentence templates because they have different process elements in the source, being respectively: "When E_s, A_c." and "When A_c, A_c.". On the other hand, at the grouped level it is considered that different possible process elements can be translated as the same sentence template. In this case, the two sentence templates described above can be viewed as a single sentence template (i.e., "When $(A_c$ or $E_s)$, A_c."), capable to have as source either an activity or a start event.

To facilitate the categorization of sentence templates, a notation was created based on the previously defined criteria. $St_i = R_s(source, target)$ can be interpreted as: there is a sentence template St_i that starts from a $source$, can describe a $target$ and is associated through a sequential relationship R_s. In the case of atomic level analysis, a $source$ is a process element. On the other hand, in the case of group level analysis a $source$ is a set of possible process elements (e.g., $A_c|G_{Xs}|G_{+s}$). A target can be described as $target = R_x(component_1, ..., component_n)$, i.e., a target is a set of components that relate to each other through a relationship R_x. Finally, a component can be a process element, $empty$, or another target. Thus, two different sentence templates belong to the same category if they share the same notation, which means to start from the same $source$ and reach the same $target$.

3.3   Ambiguity in Sentence Templates

After the identification and classification of the sentence templates, they were analyzed in relation to ambiguity issues. A sentence template was considered ambiguous when it allows multiple interpretations of the process. To identify common ambiguity issues in process descriptions, two approaches were carried out: analysis of the literature and analysis of the sentence templates.

As for the analysis of the literature, works related to ambiguity in process descriptions were investigated. Although some works related to this subject were found, only a few of them [1–4] presented cases of ambiguity. This analysis of the literature made it possible to find eight ambiguity problems that were categorized into five different types of ambiguity.

In terms of the analysis of the sentence templates, two independent tasks were conducted. In the first part, an analysis of each sentence template was carried out individually in order to identify ambiguity issues. In the second part, an analysis was carried out involving the combination of sentence templates. For the latter case, sentence templates that share the same description, but do not have the same classification (i.e., source, target or relationship) were considered candidates for ambiguity issues.

Table 3 presents the six different types of ambiguities that were identified in this work, with their respective identifiers $(Ambi_{ID})$, descriptions, examples and source.

4   Results and Analysis

After analyzing the process descriptions in an atomic level, it was possible to obtain a set of 101 sentence templates for 29 categories. Of these, 13 sentence templates were classified as having one of the six ambiguity issues. Tables 4,

Table 3. Identified ambiguity issues.

$Ambi_{ID}$	Description	Example	Source
$Ambi_1$	The term "and" can have different meanings, such as: sequence, dependence, parallelism, contrast	*The employee must update the document **and** prepare the product for shipping*	- Akbar *et al.* [4] - van der Aa *et al.* [2,3]
$Ambi_2$	The terms "or" and "sometimes" may raise doubts whether it includes or is mutually exclusive to the different alternatives	*(1) The document is accepted **or** denied. (2) The bicycle can be mounted **or** painted*	- Akbar *et al.* [4] - this paper
$Ambi_3$	The term "latter" usually does not make clear to what previous activities it refers	*In parallel to the **latter** steps...*	- van der Aa *et al.* [2,3]
$Ambi_4$	The terms "meanwhile", "concurrently", "meantime", "in the meantime" and "at the same time" make it difficult to specify which sets of activities they refer to	***In the meantime**, the sales department must prepare the receipt*	- van der Aa *et al.* [3] - this paper
$Ambi_5$	Repetitions usually not clear what activities should be performed again	*The previous steps must be **repeated***	- van der Aa *et al.* [1–3]
$Ambi_6$	The term "while" can mean simultaneity or concession	***While** it is true that the company has the money, they can't build the houses*	- this paper

5 and 6 show the sentence templates for each one of the 29 categories (C_{ID}), with their respective category notation. Each sentence template has a specific identifier presented in the "St_{ID}" column. In addition, the number of times each sentence template appeared in the process descriptions analyzed is presented in the "N" column. Moreover, the ambiguity issues identified for each sentence template, when identified, is presented in the "$Ambi_{ID}$" column, based on the elements in Table 3.

Of the identified sentence templates, the most recurrent is "If *condition*, A_c." (St_{71}), from category C_{18}, which appeared 81 times. It is possible to observe that this sentence template is fairly recurrent in process descriptions because the two sentence templates that appear the most after this first (i.e., St_1 and St_{36}) were identified only 15 times. Moreover, the category that presented the largest diversity of sentence templates is C_1, with 19 distinct sentence templates, followed by C_{18} (with 12) and C_3 (with 10).

In terms of ambiguity, the type that appeared most in the sentence templates is related to the term "and" ($Ambi_1$), having occurred five times. Moreover, $Ambi_2$ appeared three times, followed by both $Ambi_4$ and $Ambi_6$ (2 times), and $Ambi_5$ (1 time). In addition, in the identified sentence templates no case was found related to the ambiguity $Ambi_3$.

Table 4. Atomic sentence templates by category – 1.

C_{ID}	Notation	St_{ID}	Sentence template	N	$Ambi_{ID}$
C_1	$R_S(A_c, R_N(A_c))$	St_1	Once A_c, A_c	15	
		St_2	Then A_c	13	
		St_3	When A_c, A_c	11	
		St_4	After A_c, A_c	10	
		St_5	Next A_c	5	
		St_6	Afterwards, A_c	4	
		St_7	As soon as A_c, A_c	3	
		St_8	Subsequently A_c	3	
		St_9	The *role* then A_c	3	
		St_{10}	Upon A_c	3	
		St_{11}	After that A_c	2	
		St_{12}	Likewise A_c	2	
		St_{13}	A_c, after which A_c	1	
		St_{14}	Immediately after A_c, A_c	1	
		St_{15}	In addition to A_c, A_c	1	
		St_{16}	In the following A_c	1	
		St_{17}	Moreover, A_c	1	
		St_{18}	Thereafter A_c	1	
		St_{19}	Therefore A_c	1	
C_2	$R_S(A_c, R_N(E_e))$	St_{20}	After A_c, the process ends	1	
		St_{21}	After all *number* activities are completed the process ends	1	
		St_{22}	For *role* the process ends then	1	
		St_{23}	The process ends here	1	
		St_{24}	The process finishes only once A_c	1	
C_3	$R_S(A_c, R_N(G_{Xs}))$	St_{25}	After A_c, the *object* may lead to *number* possible outcomes: *condition* or *condition*	2	$Ambi_2$
		St_{26}	After A_c, it is checked whether *condition*	1	
		St_{27}	After A_c, the *role* can either *condition*, *condition* or *condition*	1	
		St_{28}	After A_c, the *role* investigates whether *condition* or *condition*	1	
		St_{29}	After A_c, the *role* may either *condition* or *condition*	1	
		St_{30}	One of the *number* alternative process paths may be taken	1	
		St_{31}	The *role* can either *condition* or *condition*	1	
		St_{32}	The *role* can then *condition* or *condition*	1	$Ambi_2$
		St_{33}	This procedure is repeated for each *condition*	1	$Ambi_5$
		St_{34}	When A_c, it is first checked whether *condition*	1	

Table 5. Atomic sentence templates by category – 2.

C_{ID}	Notation	St_{ID}	Sentence template	N	$Ambi_{ID}$
C_4	$R_S(A_c, R_N(G_{Xj}))$	St_{35}	Then the process continues normally	1	
C_5	$R_S(A_c, R_S(A_c, A_c))$	St_{36}	A_c and A_c	15	$Ambi_1$
		St_{37}	Also A_c and A_c	1	$Ambi_1$
C_6	$R_S(A_c, R_S(A_c, A_c, A_c))$	St_{38}	A_c and A_c and A_c	1	$Ambi_1$
		St_{39}	First A_c, then A_c, and finally A_c	1	
C_7	$R_S(A_c, R_S(A_c, E_e))$	St_{40}	A_c, which ends the process	2	
		St_{41}	Finally, A_c	2	
		St_{42}	The process completes with A_c	2	
		St_{43}	A_c, then the process ends	1	
		St_{44}	After A_c, this process path ends	1	
		St_{45}	Afterwards, A_c and finishes the process instance	1	
		St_{46}	The process finishes when A_c	1	
C_8	$R_S(A_c, R_S(A_c, G_{Xs}))$	St_{47}	The *role* A_c and may decide to either A_c or A_c	1	$Ambi_1$
C_9	$R_S(A_c, R_+(A_c, A_c))$	St_{48}	While A_c, A_c	2	$Ambi_6$
		St_{49}	Next, A_c while A_c	1	$Ambi_6$
		St_{50}	Once A_c, A_c and meantime A_c	1	
C_{10}	$R_S(A_c, R_X(A_c, A_c))$	St_{51}	After A_c, the *role* either A_c or A_c	1	
		St_{52}	When A_c, the *role* may either A_c or A_c	1	
C_{11}	$R_S(A_c, R_X(A_c, A_c, A_c))$	St_{53}	Sometimes A_c, sometimes A_c and sometimes A_c	1	$Ambi_2$
C_{12}	$R_S(A_c, R_O(A_c, A_c))$	St_{54}	*object* may be A_c from either *role 1* or *role 2* or from both	1	
		St_{55}	The *role* may either A_c or also A_c	1	
C_{13}	$R_S(E_i, R_N(E_e))$	St_{56}	After *role* E_i, the process flow ends	1	
C_{14}	$R_S(E_s, R_N(A_c))$	St_{57}	The process starts with A_c	7	
		St_{58}	The process starts when A_c	6	
		St_{59}	First, A_c	3	
		St_{60}	The process starts by A_c	2	
		St_{61}	When E_s, A_c	2	
		St_{62}	Whenever E_s, A_c	2	
		St_{63}	After the process starts, A_c	1	
		St_{64}	The process is triggered by A_c	1	
		St_{65}	The process starts once A_c	1	
C_{15}	$R_S(E_s, R_S(A_c, A_c, A_c))$	St_{66}	A_c and A_c and A_c	1	$Ambi_1$
C_{16}	$R_S(G_{+s}, R_N(A_c))$	St_{67}	In the meantime, A_c	2	$Ambi_4$
		St_{68}	At the same time, A_c	1	$Ambi_4$
C_{17}	$R_S(G_{+j}, R_N(A_c))$	St_{69}	Afterwards, A_c	2	
		St_{70}	After each of these activities, A_c	1	

Table 6. Atomic sentence templates by category − 3.

C_{ID}	Notation	St_{ID}	Sentence template	N	$Ambi_{ID}$
C_{18}	$R_S(G_{Xs}, R_N(A_c))$	St_{71}	If *condition*, A_c	81	
		St_{72}	Otherwise, A_c	10	
		St_{73}	In this case A_c	7	
		St_{74}	In case *condition*, A_c	3	
		St_{75}	For the case *condition*, A_c	2	
		St_{76}	*condition*, in which case A_c	1	
		St_{77}	*condition*, otherwise A_c	1	
		St_{78}	However, if *condition*, A_c	1	
		St_{79}	In that case, A_c	1	
		St_{80}	In the latter case, A_c	1	
		St_{81}	On the other hand, if *condition* A_c	1	
		St_{82}	Sometimes *condition*, then A_c	1	
C_{19}	$R_S(G_{Xs}, R_N(E_e))$	St_{83}	If *condition* the process will end	1	
		St_{84}	In the former case, the process instance is finished	1	
C_{20}	$R_S(G_{Xs}, R_N(G_{+s}))$	St_{85}	This action consists of *number* activities, which are executed in an arbitrary order	1	
C_{21}	$R_S(G_{Xs}, R_N(G_{Xs}))$	St_{86}	If *condition*, this results in either *condition* or *condition*	1	
C_{22}	$R_S(G_{Xs}, R_S(A_c, A_c, A_c))$	St_{87}	If *condition*, *role* may need to first A_c, then A_c and finally A_c	1	
C_{23}	$R_S(G_{Xs}, R_+(A_c, A_c))$	St_{88}	Once *condition*, A_c and meantime A_c	3	
C_{24}	$R_S(G_{Xs}, R_X(A_c, A_c))$	St_{89}	If *condition*, A_c, otherwise A_c	4	
		St_{90}	In case *condition*, A_c, otherwise A_c	1	
		St_{91}	*role* either A_c or A_c	1	
C_{25}	$R_S(G_{Xs}, R_X(A_c, E_e))$	St_{92}	If *condition* A_c, otherwise the process is finished	1	
C_{26}	$R_S(G_{Xs}, R_X(A_c, empty))$	St_{93}	If *condition*, A_c, except if *condition*	1	
		St_{94}	In case *condition*, A_c otherwise the process continues	1	
C_{27}	$R_S(G_{Xj}, R_N(A_c))$	St_{95}	In any case, A_c	4	
		St_{96}	In either/any case, A_c	1	
		St_{97}	Once one of these *number* activities is performed, A_c	1	
		St_{98}	The process then continues with A_c	1	
C_{28}	$R_S(G_{Xj}, R_S(A_c, E_e))$	St_{99}	Afterwards, A_c and the process completes	1	
		St_{100}	Finally, A_c	1	
C_{29}	$R_S(G_{Xj}, R_+(A_c, A_c))$	St_{101}	Then, two current activities are triggered, A_c and A_c	1	

Furthermore, it is possible to notice that not all relationships between process elements are explored in Tables 4, 5 and 6. This occurs because some relationships that occur in the model are not explicitly transformed into sentences. Also, there are some relationships that appear less frequently than others in the texts considered.

In the group level analysis, the atomic level sentence templates that share the same target but presents different process elements as sources were grouped. From the data collected in the atomic level analysis, it was possible to identify 8 sentence templates that were transformed into four grouped sentence templates. Table 7 presents the grouped sentence templates. In this table are presented the new notations able to represent the grouped sentence templates, as well as the new sentence templates. In addition, the identifier of the atomic sentence templates used in each grouped sentence template are presented in the "St_{ID}" column. Finally, as in atomic level analysis, the number of times each grouped sentence template appeared in the process descriptions analyzed is presented in the "N" column.

Table 7. Grouped sentence templates.

Notation	Sentence template	St_{ID}	N
$R_s(A_c\|E_s, Rn(A_c))$	When $A_c\|E_s$, A_c	St_3, St_{61}	13
$R_s(A_c\|G_{+j}, Rn(A_c))$	Afterwards, A_c	St_6, St_{69}	6
$R_s(A_c\|G_{Xj}, R_s(A_c, E_e))$	Finally, A_c	St_{41}, St_{100}	3
$R_s(A_c\|E_s, R_s(A_c, A_c, A_c))$	A_c and A_c and A_c	St_{38}, St_{66}	2

5   Discussion

The identified sentence templates can help approaches for identification of process elements in natural language texts and for automated creation of business process descriptions. For identification of process elements in natural language texts, the approaches can use the identified sentence templates as patterns to be sought in texts. In this case, sentence templates could be searched in the sentences of a process description. By finding a sentence corresponding to a sentence template, the process elements present in the sentence could be identified.

For automated creation of process descriptions, the approaches can choose to use the most recurring sentence templates, or take advantage of the variety of sentence templates in each category to make the text more diversified. As a demonstration, the text described in Fig. 3 could with the use of the identified sentence templates be rewritten as presented by Fig. 4. In this new process description, the sentence s_1 remained the same because it only presents context information. In addition, the sentence s_2 (belonging in the category C_{14}) was modified by the sentence template "The process starts with A_c" (St_{57}) because it is recurrent in the process descriptions analyzed. Moreover, the sentences s_3

and s_6 (both belonging to category C_{18}) were rewritten using the sentence template "If *condition*, A_c" (St_{71}) because this sentence template appears recurrent in describing activities starting from an XOR-split. The sentence s_5 also has not been modified because this sentence only details an activity. Furthermore, the sentence s_6 has an ambiguity problem related to "and" ($Ambi_1$). Although the sentence templates collected do not present any candidate capable of removing the problem, the results help indicate that there is a problem in the process description. Among possible solutions, the sentence could separate into two new sentences or seek for a different sentence capable of avoiding the problem of ambiguity, such as: "If it is a hardware problem, the technician must replace the part and then fill out the part replacement form.". Finally, the sentence s_7 (belonging to category C_7) was modified by the sentence template "Finally, A_c" (St_{41}) because it appears to be most recurrent in the identified sentence templates.

(s_1) The repair department of the company X performs repairs of computers and printers. (s_2) The process starts with the technician performing an evaluation. (s_3) If it is a software problem, the technician must format the computer. (s_4) If it is a hardware problem, the technician must replace the part and then fill out the part replacement form. (s_5) This form must contain the part identification code and the technician's signature. (s_6) If no problem is found, no modification should be made to the computer. (s_7) Finally, the technician must complete the repair form.

Fig. 4. Example of rewritten process description: computer repair.

Regarding categorization of sentence templates, although the classification based on source, target and relationship helps in the task of grouping sentence templates that have the same characteristics, in some cases an analysis of the context can help to select a sentence template that more fits the text and, consequently, to produce a text more suitable for process analysts and domain experts. For instance, the sentence templates "If condition, A_c" (St_{71}) and "Otherwise, A_c" (St_{72}) are related in the same category (C_{18}) and are both able to represents an activity being performed after an XOR-split. However, the second sentence template could produce a disconnected text by referring to the first possible path of a decision making in a process description.

6 Related Work

The work presented here relates to two different streams of research: generation of process descriptions and ambiguities present in process descriptions.

For the generation of process descriptions context, Leopold *et al.* [18] proposed a technique to generate natural language texts from business process models. The authors used sentence templates to transform business process models into sentences that compose the text. Furthermore, Aysolmaz *et al.* [7] defined a semi-automated approach to generate natural language requirements documents based on business process models. The authors adopted a template filling technique, in which sentence templates are defined containing gaps that must be filled with information from a requirements model. In addition, Caporale [8] suggested a method that allows generating process models from process descriptions. To achieve this, the author proposed that the process descriptions should be specified with a controlled natural language, based on sentence templates, in order to facilitate the extraction of information necessary to generate the models. Moreover, Ghose *et al.* [15] proposed a framework and prototype tool that can query information resources (e.g. corporate documentation, web-content, code) for construct models to be incrementally adjusted to correctness by an analyst. One of the techniques used by authors to extract information from text is based on template extraction. In this technique, the authors created templates from textual structures that are commonly used in describing processes and used these templates to extract knowledge from text documents.

Regarding the ambiguity present in process descriptions, Ferrari *et al.* [11] conducted a literature review and a set of interviews with different public institutions aiming at improving the process descriptions to be used in public administrations. The authors concluded that ambiguity is one of the macro-areas of research in which computer scientists can contribute towards more quality in business process descriptions. Moreover, van der Aa *et al.* [1] presented an approach to automatically detect inconsistencies between process model and the corresponding textual description. The authors identified that a technique to detect inconsistencies must deal with ambiguity issues present in natural language. In another work, van der Aa *et al.* [2] proposed to deal with ambiguity in textual process descriptions introducing the behavioral space concept. The behavioral space captures all possible behavioral interpretations of a textual process description. Furthermore, van der Aa *et al.* [3] presented an approach to verify the compliance between a process and a process description, considering the ambiguity present in texts. To handle the ambiguity issues, they used the concept of behavioral space previously proposed in [2].

As can be observed, the first stream of research makes use of sentence templates for generation of business process descriptions from process models and extraction of text information for the generation of process models. Moreover, the second stream discusses the problem of ambiguity in process descriptions and presents some cases of ambiguity. This work is distinguished by the fact that it presents an empirical analysis in business processes descriptions, in order to find recurrent sentence templates and to highlight ambiguity issues in these sentence templates.

7   Conclusion

In this work, an empirical analysis of business process descriptions was carried out in order to discover the most recurrent sentences used to describe BPMN process models. In addition, an analysis was performed in order to find sentences with ambiguous meaning. The analysis consisted of three different steps. First, a set of 64 process descriptions was selected and their sentences were prepared for the identification of sentence templates. Then, an identification and a classification of sentence templates was performed in the prepared sentences. Finally, the sentence templates were marked as having or not ambiguity issues. Among all, 101 sentence templates were found and they were classified into 29 different categories. Of these, 13 sentence templates were considered as having ambiguity issues.

This work aims to contribute to the description of business processes in a way that is closer to a pattern and with less ambiguity issues. It can be useful for creating process descriptions more suitable for process analysts and domain experts. In addition, this analysis can be used by tools that automate the creation of process descriptions. As future work, it is intended to increase this analysis for other elements in BPMN, expand the sample of textual descriptions and construct a technique that uses these sentence templates to produce process descriptions.

Acknowledgments. We are grateful to the national research funding agencies CNPq (National Council for Scientific and Technological Development) and CAPES (CAPES Foundation) for financial support. We also acknowledge the support we have been receiving from the Graduate Program in Computer Science as well as the Institute of Informatics, UFRGS.

References

1. van der Aa, H., Leopold, H., Reijers, H.A.: Detecting inconsistencies between process models and textual descriptions. In: Motahari-Nezhad, H.R., Recker, J., Weidlich, M. (eds.) BPM 2015. LNCS, vol. 9253, pp. 90–105. Springer, Cham (2015). https://doi.org/10.1007/978-3-319-23063-4_6
2. van der Aa, H., Leopold, H., Reijers, H.A.: Dealing with behavioral ambiguity in textual process descriptions. In: La Rosa, M., Loos, P., Pastor, O. (eds.) BPM 2016. LNCS, vol. 9850, pp. 271–288. Springer, Cham (2016). https://doi.org/10.1007/978-3-319-45348-4_16
3. van der Aa, H., Leopold, H., Reijers, H.A.: Checking process compliance against natural language specifications using behavioral spaces. Inf. Syst. (2018)
4. Akbar, S., Bajwa, I.S., Malik, S.: Scope resolution of logical connectives in NL constraints. In: Eighth International Conference on Digital Information Management, pp. 217–222. IEEE (2013)
5. de AR Goncalves, J.C., Santoro, F.M., Baiao, F.A.: Business process mining from group stories. In: 13th International Conference on Computer Supported Cooperative Work in Design, CSCWD 2009, pp. 161–166. IEEE (2009)
6. de AR Gonçalves, J.C., Santoro, F.M., Baião, F.A.: Let me tell you a story-on how to build process models. J. UCS **17**(2), 276–295 (2011)
7. Aysolmaz, B., Leopold, H., Reijers, H.A., Demirörs, O.: A semi-automated approach for generating natural language requirements documents based on business process models. Inf. Softw. Technol. **93**, 14–29 (2018)

8. Caporale, T.: A tool for natural language oriented business process modeling. In: ZEUS, pp. 49–52 (2016)
9. Dumas, M., La Rosa, M., Mendling, J., Reijers, H.A., et al.: Fundamentals of Business Process Management, vol. 1. Springer, Heidelberg (2013). https://doi. org/10.1007/978-3-642-33143-5
10. Epure, E.V., Martín-Rodilla, P., Hug, C., Deneckère, R., Salinesi, C.: Automatic process model discovery from textual methodologies. In: 2015 IEEE 9th International Conference on Research Challenges in Information Science (RCIS), pp. 19–30. IEEE (2015)
11. Ferrari, A., Witschel, H.F., Spagnolo, G.O., Gnesi, S.: Improving the quality of business process descriptions of public administrations: resources and research challenges. Bus. Process Manag. J. **24**, 49–66 (2017)
12. Ferreira, R.C.B., Thom, L.H., de Oliveira, J.P.M., Avila, D.T., dos Santos, R.I., Fantinato, M.: Assisting process modeling by identifying business process elements in natural language texts. In: de Cesare, S., Frank, U. (eds.) ER 2017. LNCS, vol. 10651, pp. 154–163. Springer, Cham (2017). https://doi.org/10.1007/978-3-319-70625-2_15
13. Friedrich, F.: Automated generation of business process models from natural language input. M.Sc., School of Business and Economics. Humboldt-Universität zu Berli (2010)
14. Friedrich, F., Mendling, J., Puhlmann, F.: Process model generation from natural language text. In: Mouratidis, H., Rolland, C. (eds.) CAiSE 2011. LNCS, vol. 6741, pp. 482–496. Springer, Heidelberg (2011). https://doi.org/10.1007/978-3-642-21640-4_36
15. Ghose, A., Koliadis, G., Chueng, A.: Process discovery from model and text artefacts. In: 2007 IEEE Congress on Services, pp. 167–174. IEEE (2007)
16. Leopold, H., van der Aa, H., Pittke, F., Raffel, M., Mendling, J., Reijers, H.A.: Integrating textual and model-based process descriptions for comprehensive process search. In: Schmidt, R., Guédria, W., Bider, I., Guerreiro, S. (eds.) BPMDS/EMMSAD -2016. LNBIP, vol. 248, pp. 51–65. Springer, Cham (2016). https://doi.org/10.1007/978-3-319-39429-9_4
17. Leopold, H., Mendling, J., Polyvyanyy, A.: Generating natural language texts from business process models. In: Ralyté, J., Franch, X., Brinkkemper, S., Wrycza, S. (eds.) CAiSE 2012. LNCS, vol. 7328, pp. 64–79. Springer, Heidelberg (2012). https://doi.org/10.1007/978-3-642-31095-9_5
18. Leopold, H., Mendling, J., Polyvyanyy, A.: Supporting process model validation through natural language generation. IEEE Trans. Softw. Eng. **40**(8), 818–840 (2014)
19. Li, J., Wang, H.J., Zhang, Z., Zhao, J.L.: A policy-based process mining framework: mining business policy texts for discovering process models. Inf. Syst. E-Bus. Manag. **8**(2), 169–188 (2010)
20. Meitz, M., Leopold, H., Mendling, J.: An approach to support process model validation based on text generation. In: EMISA Forum, vol. 33, pp. 7–20 (2013)
21. Nawrocki, J.R., Nedza, T., Ochodek, M., Olek, L.: Describing business processes with use cases. In: BIS, pp. 13–27 (2006)
22. OMG: Business process modeling notation (BPMN) version 2.0.2 (2013). https:// www.omg.org/spec/BPMN/
23. Ottensooser, A., Fekete, A., Reijers, H.A., Mendling, J., Menictas, C.: Making sense of business process descriptions: an experimental comparison of graphical and textual notations. J. Syst. Softw. **85**(3), 596–606 (2012)

24. Sànchez-Ferreres, J., Carmona, J., Padró, L.: Aligning textual and graphical descriptions of processes through ILP techniques. In: Dubois, E., Pohl, K. (eds.) CAiSE 2017. LNCS, vol. 10253, pp. 413–427. Springer, Cham (2017). https://doi.org/10.1007/978-3-319-59536-8_26
25. Sinha, A., Paradkar, A.: Use cases to process specifications in business process modeling notation. In: 2010 IEEE International Conference on Web Services (ICWS), pp. 473–480. IEEE (2010)

Evolution of Instance-Spanning Constraints in Process Aware Information Systems

Conrad Indiono[1(✉)], Walid Fdhila[2], and Stefanie Rinderle-Ma[1]

[1] Faculty of Computer Science, University of Vienna, Vienna, Austria
{conrad.indiono,stefanie.rinderle-ma}@univie.ac.at
[2] SBA Research, Vienna, Austria
walid.fdhila@univie.ac.at

Abstract. Business process compliance has been widely addressed resulting in works ranging from proposing compliance patterns to checking and monitoring techniques. However, little attention has been paid to a specific type of constraints known as instance spanning constraints (ISC). Whereas traditional compliance rules define constraints on process models, which are checked separately for each instance, ISC impose constraints that span multiple instances. This paper focuses on ISC evolution and its impact on process compliance. In particular, ISC change operations, as well as change strategies are defined, and the impact on both the ISC monitoring engine and the process instances during run time are analyzed. The concepts are prototypically implemented.

1 Introduction

Constraints imposed on business processes evolve constantly, for example, when new constraints are added or existing constraints are updated. Whereas business process changes have been investigated in detail (e.g., [17,18]), constraint changes – also in interplay with process changes – have lacked attention so far. Some approaches address the impact of business process change on constraint checking [12], but concepts for the evolution of the constraints and the impacts on other constraints and business processes are missing. This holds particularly true for so called instance-spanning constraints (ISC), i.e., constraints that span multiple instances of one or several business processes [4]. In this work we revisit change strategies as formulated for business process evolution, i.e., versioning, migration, and clean state [1] for ISC changes in an abstract manner. Additionally, a concrete versioning approach is discussed and we investigate the related change impacts culminating in the following research questions:

RQ1. How can ISC changes be handled ($\mapsto$ change strategies)?
RQ2. How to visualize and measure the impact of ISC changes?
RQ3. How to realize versioning in the context of ISC change?

© Springer Nature Switzerland AG 2018
H. Panetto et al. (Eds.): OTM 2018 Conferences, LNCS 11229, pp. 298–317, 2018.
https://doi.org/10.1007/978-3-030-02610-3_17

Answering the questions is challenging due to the complexity of the ISC: ISC consist of structural patterns referring to the underlying processes and conditions concerning data, time, and resource aspects [14]. Whereas these parts can be also found for intra-instance constraints, i.e., independent constraints that can be verified for each process instance in a separated way, ISC can also contain trigger and action parts. These parts define the "active" components of an ISC such as putting the instance execution to a suspend state for synchronization [4,13]. Moreover, ISC might "share" data and resources. Hence, changing one ISC might affect other ISC even over different versions. Finally, ISC change and impact have to be considered during both, design and runtime.

This work tackles RQ1–RQ3 as follows: At first, ISC change operations are defined ($\mapsto$ RQ1) in accordance with business process change patterns [17] and constraint changes proposed in literature [10], i.e., ISC change operations for adding, deleting, and updating ISC are proposed. The complexity of ISC adds several elements to defining change operations as each of the parts concerning structure, data, time, resources, and trigger/actions might be adapted. Established process change strategies such as versioninig, migration, and clean state are transferred to ISC change ($\mapsto$ RQ1). This is also connected with a motivation on which ISC formalism and inference techniques can be selected, in this work Event Calculus and RETE. The impact of ISC changes on the ISC themselves as well as on associated process instances is systematically studied at an abstract level and implemented using the RETE matching algorithm ($\mapsto$ RQ2). As the field of ISC evolution is entirely new, a first ISC versioning algorithm is proposed and a prototype proof of concept is presented ($\mapsto$ RQ3).

2   Preliminaries

While process instances reflect the actual execution of a business process, ISC are means to define and enforce restrictions over multiple process instances. As an illustrative example, we use the process scenario of Fig. 1 [8]. It depicts an integrated energy management solution to deliver end-to-end advanced metering. Assume that the provider aims at ensuring that 99% of all readouts (of different instances) are performed within 6 hours and the aggregate read out value does not exceed x. This constraint is considered as an ISC since it imposes restrictions over multiple instances. The ISC meta model follows the IUPC structure [14], which is a tuple ISC(context$^+$, connection$^+$, condition*, behavior$^+$) where:

- a context refers to the process/process instances subjected by an ISC.
- a connection defines the events required to check an ISC (e.g., activity started or time trigger event).
- a condition is a constraint on either resources, time or process/rule data.
- a behavior is an action triggered when all conditions are met (e.g., suspend a task, resume a task).
- $^+$ denotes an at-least-one, while * denotes an optional quantifier.

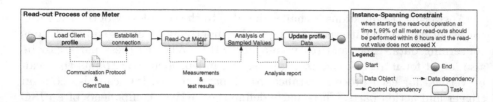

Fig. 1. Process example from the energy domain, taken from [8]

Example 1 (ISC). *The ISC described in Fig. 1 can be defined as follows:*
ISC(Context(*readout_process*), **Connection**(*readout_meter, global_readout_start*), **Condition**(*total_readout > x*), **Condition**(*at global_readout_start.t +* 6 less than 99% of total meters are finished), **Behavior** (*send alert*)). *The event* (*id, global_readout_start, t*) *is a time trigger to launch all read-out instances at the same time t.*

The ISC representation employed in this paper is simplified and needs to be converted into a rule engine language in order to be executed and monitored [8] by an ISC monitor. ISC formalization using, e.g., Event Calculus (EC) is proposed in [4,8], which we implement on top of a Rete rule engine acting as the ISC monitor. EC is a logic programming approach to model time and change. It uses first order predicate logic (FOL) as the basis and introduces fluents and domain-independent predicates to assert fluents for the ability to model time-varying state. Section 4.2 gives an overview of ISC implementation based on EC.

Definition 1 (Event). *An event represents the occurrence of an action (e.g., execution of a process task) at a given time and is defined as a tuple (event_id, type, timestamp, payload*) where type is the event name, timestamp is the time of its occurrence and payload holds the event related data.*

Example 2 (Event). (*id, readout_meter, time, meter, readout_value*) *is an event of type readout_meter that provides the value of the readout of a given meter.*

3 Atomic ISC Change Operations

ISC changes can range from deleting an ISC attribute (e.g., condition or connection) to adding or updating new or existing attributes respectively. Similar to process change operations [17], three main change operation groups can be identified, i.e., delete, add, and update. Each of these groups include various change operations with different impacts on both the process instances and the ISC monitor, cf. Definition 2.

DELETE	DELETE - BEFORE	DELETE - AFTER
CONTEXT	For both **gas** and **electricity**, 99% of all meters should be readout within 6 hours.	For **electricity**, 99% of all meters should be readout within 6 hours.
CONNECTION	At **12:00** and **14:00**, the average readout of all meters should have a value less than X	At **12:00**, the average readout of all meters should have a value less than X
CONDITION	When starting the read-out at 00:00, 99% of all meters should be read out within 6 hours and **readout value should not exceed X.**	When starting the read-out of 00:00 values, 99% of all meters should be read out within 6 hours.
BEHAVIOR	For 100 (simultaneous) ad hoc readouts , if 10 meter checks exceed 6 hours then send an alert and **stop the readouts.**	For 100 (simultaneous) ad hoc readouts , if 10 meter checks exceed 6 hours then send an alert.
ISC	When starting the read-out of 00:00 values 99% of all meters should be read out within 6 hours.	-
ADD	ADD - AFTER	ADD - BEFORE

Events: **e1** (id, Readout_meter ,time, meter, value) - **e2** (id, GlobalReadoutStart ,time)

Fig. 2. Change operations: delete and add examples

Definition 2 (Change Operations)

$Change\_operation :: = Operation\_type(ISC, Component)|Delete(ISC)$

$\qquad |Add(ISC, Context^+, Connection^+, Condition^*, Behavior^+)$

$Operation\_type :: = Delete|Add|Update$

$\qquad Component :: = Context|Connection|Condition|Behavior$

While the first change operation acts on a specific attribute of an existing ISC using one of the three change types (i.e., Delete, Add and Update), the second change operation deletes an entire ISC and, the third adds a new ISC. Based on Fig. 2 we selectively illustrate some *delete* change operations, while omitting the description for *add* operations due to space constraints. Examples for *add* operations are illustrated in the same Fig. 2 by reading from right to left. Note that the action part is considered as "send alert" by default.

Delete Context. A context represents the process model to which an ISC refers. Monitoring an ISC requires execution events of the corresponding process. Multiple contexts might be defined within the same ISC. Deleting a context implies that all related events are no longer required for its monitoring, and automatically removes all corresponding linkages, i.e., (context, connection). In Fig. 2 (CONTEXT), the ISC has changed from considering both processes of electricity and gas meter readouts respectively, to only electricity.

Delete Connection. A connection refers to the events necessary for checking an ISC. Within a single ISC, multiple connections can be specified, which might refer to different contexts, i.e., events from different process executions. Therefore, deleting a connection reduces the number of event types to be checked by an ISC. Note that deleting all connections of a same context implies the deletion of the latter. In Fig. 2 (CONNECTION), it is checked at both times 12:00 and 14:00 whether the aggregate readout value of all meters is less than the threshold x. Deleting trigger time 12:00 means that the monitor will still receive the readout events, but will check the threshold condition at 14:00 only.

Delete Condition. An upcoming event or set of events, in order to be considered for an ISC, needs to match its conditions. Conditions are defined as constraints on the data associated with the events; e.g., process data, resources, time. Multiple conditions might be combined for the same ISC, and therefore, deleting a condition releases a restriction on the events used for the ISC. In Fig. 2 (CONDITION), the condition on the threshold value was removed, which means that the latter will not be checked when the change becomes effective.

Delete Behavior. As aforementioned, a behavior is an action that is executed if an ISC has fired; e.g., a stop or wait task. Deleting a behavior reduces the number of actions to be enacted when all conditions evaluate to true. In Fig. 2 (BEHAVIOR), the action stop is removed. Consequently, only the action send alert is executed when the ISC fires. This might have impacts on the process instances that are stopped before the change. Compensation actions might be required in order to continue those instances.

4   ISC Evolution Approach

This section introduces the general approach for performing ISC Evolution, starting with an overview of the concepts applied.

4.1   General ISC Evolution Strategies

Inspired by the strategies for process evolution in [1], we introduce three general change strategy approaches for managing ISC evolution. These are versioning, migration, and clean state. Figure 3 illustrates the differences between these strategies by comparing how the same change sequence is handled in each case. The illustrative case shows how an initial set of ISC, i.e., $\{ISC1_0.ISC2_0\}$ are changed over time, abstracting from the concrete atomic change operation applied. At time t $ISC1_0$ is changed into $ISC1_1$. At time $t+1$ there are three changes: (1) the previously changed $ISC1_1$ transformed to $ISC1_2$, (2) the change of $ISC2_0$ to $ISC2_1$, and (3: only in the versioning case) the original $ISC1_0$ change to $ISC1_3$. Critical to the evolution of ISC is state management, where state refers to any part of an *ISC(context, connection, condition, behaviour)* being shared among different ISC. For example, state $S1$ is the state shared between two ISC: $ISC1_0$ and $ISC2_0$, which in the running example (Fig. 1) might represent the aggregate read-out value from all smart meters, e.g., parts of the condition attribute of both ISC. State management is performed differently among the three strategies.

Migration. In the migration case, a domain-specific abstract transformation function f is applied to $S1$ at time t. A concrete function f needs to be determined on a case by case basis for each change scenario. Furthermore, in this case f is bound by the change of $ISC1_0$ to ISC_1, as well as by the actions that $ISC1_0$ has already performed. In the latter case, compensation actions inverse to the original ones may need to be executed. An example could be in a medical scenario where 10 patients are pre-approved for a novel operation, where new test results suggest the requirement to reevaluate these patients. The change

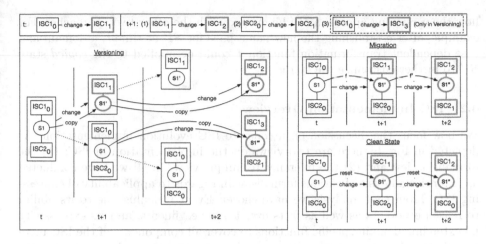

Fig. 3. General ISC evolution strategies

from $ISC1_0$ to $ISC1_1$ represents the change in verification of the approval process. The concrete function f in this case reduces the pre-approved 10 patients using the now correct approval logic, e.g., 6 pre-approved patients. Similarly, at time $t + 1$ f' is applied for migrating state $S1'$ to $S1''$, which is bound by both changes $ISC1_1$ to $ISC1_2$ and $ISC2_0$ to $ISC2_1$. In that way, the migration transformation function f and f' are domain and case dependent.

Clean State. The clean state strategy can be seen as an instance of the migration strategy where the function f is fixed as the function *reset*. This state management function, as the name implies, resets all the information within a state to their default values, which depend on their data types. In the running example (Fig. 1), this could be the resetting of the aggregate read-out values to the default value: 0. This *reset* function does not depend on the changes being performed, but only on the data types being modified within the states. This makes the *reset* function domain and case independent.

Versioning. The versioning strategy also has a domain and case independent state transformation function: *copy*. Also unique to the versioning strategy is the concept of namespacing ISC and associated states, to allow the differing ISC versions to coexist. At time t, the versioning strategy creates a new namespace for each changed ISC to occupy, together with any associated state. For example, at time t, the change of $ISC1_0$ to $ISC1_1$ leads to the creation of a new namespace for $ISC1_1$ and its associated copied state $S1'$. Notice that the original namespace spanning $ISC1_0$, $ISC2_0$, and $S1$ are still maintained, representing the previous version of $ISC1$. At t+1: while case (1) leads to the usual namespacing for the new $ISC1_2$, cases (2) and (3) lead to the creation of a common namespace due to the common state $S1'''$. Namespacing is directly tied to the state being copied, establishing a separate context for the newly changed ISC, allowing them to be processed independently from coexisting ISC versions. It can be imagined how the versioning approach can be merged with the other two approaches. For example, after establishing the new namespace for $S1'$ at time t, a domain and case dependent transformation function f could be applied to the *copied* state $S1'$ to further customize the state.

4.2   ISC Implementation Overview

We now focus the discussion on a concrete ISC evolution implementation. As depicted in Fig. 4, there are two views on the implementation of ISC: (1) the formalism view and (2) the inference technique view. In [4] we have conducted an extensive analysis of various formalisms in regards to applicability of expressing ISC. From that analysis we have chosen Event Calculus due to its ability to reason on events as well as facts over time, i.e., fluents, and its expressivity by extending domain-specific functions to cover all components of the ISC meta model. For the purpose of runtime checking using EC, we have adapted the RETE algorithm [8] as the inference technique, specifically due to its forward chaining, reactive nature of dealing with events. Additionally, the visualization of the RETE network helps with understanding the impact of the atomic ISC change operations. Alternatives to RETE are available, e.g., in the form of Mobucon EC [15], which uses an embedded prolog inference technique combined with cached fluents for improved performance. In this case, EC is compiled to embedded prolog as the target language. Alternatively, CEP implemented via Drools could be employed. While rule management and offline versioning is supported through a plugin (Guvnor), runtime versioning is not available due to lack of state management. In all cases, the general algorithms proposed in this paper need to be specialized for the chosen inference technique to support runtime ISC evolution.

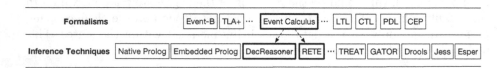

Fig. 4. Formalisms and inference techniques

Internally, a RETE rule engine represents an ISC in a specific structure (e.g., Rete graph), which reflects all its attributes (e.g., connections, conditions, behavior). While a process instance represents the actual execution of a process model (within the process engine), an ISC instance represents the actual execution of an ISC structure (within the rule engine). This implies that the conditions and fact evaluations (e.g., aggregate readout value) of an ISC define its state at runtime, i.e., ISC instance. Each event occurrence might change the state of an ISC instance and consequently the evaluation of a condition or a variable over time. An ISC instance terminates when all its actions are enacted. Changing an ISC not only impacts its structure (i.e., static impacts), but also the related process instances as well as the ISC instance itself (i.e., dynamic impacts).

Additionally, ISC can share state (i.e., conditions, connections, or contexts). This means that deleting a connection or condition from an ISC does not necessarily result in its removal from the ISC monitor as it can still be used by another ISC. Similarly, adding a new condition to an ISC does not necessarily result in its implementation in Rete as it might have been already implemented by another ISC that shares the same condition. Deploying a new ISC does not always result in a completely new structure (e.g., Rete graph) inside the monitor. The monitor will only add those parts of the structure that did not exist before.

Another challenging problem is that multiple ISC are running simultaneously, and consequently changing an ISC might cause conflicts with other ones. Therefore, it is necessary to ensure that no conflicts among ISC are generated as a result of a change. Furthermore, ISC generally refer to events related to task executions or process context data at run-time. As such, it is primordial to check whether a change affects the ISC compliability (cf. [3]) with the corresponding process models (e.g., referring to events that are not produced by the process execution). As discussed in a previous work [4], ISC can be checked at both design and runtime. While the former focuses on verifying ISC compliability with process models as well as detecting conflicts between multiple ISC, runtime checking aims at identifying ISC violations by monitoring execution events. A priori checking of ISC using Event Calculus (EC) requires both process models and ISC to be transformed into EC and then fed into an EC solver (e.g., DecReasoner[1]) to detect either conflicting constraints or incorrect specification. As mentioned in [4], design time checking is not always decidable due to loops or quantification over infinite sets (e.g., arbitrary data objects). In this work, we assume that all ISC as well as change operations are specified correctly, which means that the application of a change operation does not result in inconsistencies within the ISC structure, e.g., adding a condition that refers to connections not specified in the changed ISC. Analyzing the correctness of ISC is not the focus of this paper. For more details on how ISC are modeled and checked with EC, the reader may refer to [4].

4.3 Rete-Based ISC Monitoring

Having established EC as the chosen formalism and RETE as the inference technique for executing EC, we now introduce the fundamentals of ISC monitoring based on the RETE algorithm [8] on which we specialize and implement the ISC evolution strategies introduced in Sect. 4.1. In this context, the paper shows only the fundamentals of the RETE algorithm for discussing ISC evolution. To see how EC can be mapped to RETE for complete runtime execution, please refer to [8]. The ISC Monitor listens to a stream of events sent by the process engine and performs actions defined by the ISC, e.g., suspend/continue activities. This process follows the knowledge-based world view, where each event structure is

---

[1] http://decreasoner.sourceforge.net.

deconstructed as individual facts and fed to a knowledge base. From there, the inference engine applies the Rete pattern matching algorithm [6] to derive the updated states based on the newly submitted facts. We use a variation of the Rete algorithm as outlined in [8] to improve the matching performance for ISC. An ISC is *fired* when all of its conditions match some subset of facts from the knowledge base and the associated behaviour is to be executed. This behaviour can consist of one or many actions and may affect the process engine (e.g., when suspending a running process instance). Figure 5 shows a sample Rete network representing parts of the ISC (cf. Fig. 1) for verifying that (1) all meters are read out within six hours from the read out event starting at 00:00 and (2) the aggregated value of these read out values does not exceed a certain value. A Rete network consists of three main components: knowledge base, alpha network and beta network.

Knowledge Base. The knowledge base collects the set of facts previously submitted in the form of events sent by a governing process engine. These facts, also called *working memory elements* (WME) are defined as follows.

Definition 3 (Fact). *A fact is an element that is decomposed from an event structure (cf. Definition 2): (id, attribute, value), where value represents the value of attribute and (id, attribute) is a composite key.*

An event occurrence might change the valuation of one or multiple facts. For example, an incoming event holding a single *readout_value* as the payload of a single meter, might affect the fact representing the aggregated value: *accumulated_values*. We define the id part to be a tuple of (event$_{id}$, instance$_{id}$). Whereas the event$_{id}$ is a globally unique identifier for the specified event (e.g., completion of activity "Readout Value"), the instance$_{id}$ is a globally unique identifier referring to a process instance.

Alpha Network. The alpha network is a projection network to match facts from the knowledge base and store them inside *alpha nodes*. A pattern is associated to each alpha node where the triple structure for facts is reused for matching: (*id const* | *?id, attribute const* | *?attribute, value const* | *?value*). Each part of the triple structure can be either a constant value or a variable. The latter is marked with a prepended *?* to its name. For example, the triple (*?id*, type, "read-out meter end") is a pattern that matches all facts having *any* value inside the id part, the exact value *type* in the *attribute* part and the exact value "read-out meter end" inside the *value* part. Thus a fact ((event$_1$, instance$_1$), type, "read-out meter end") represents the event which has been emitted by the process instance id = 1 and event id = 1 where the activity "read-out meter" has completed. In Fig. 5 this fact would be stored as *f1* in the alpha node *A1*.

Beta Network. The beta network is responsible for joining together facts that match certain conditions. This is accomplished by *join nodes*, each one connecting a single alpha node with its parent beta node. Attached to join nodes can be an arbitrary number of *join tests*, matching facts stored in alpha nodes with tokens stored in beta nodes. The default join test behaviour is to check

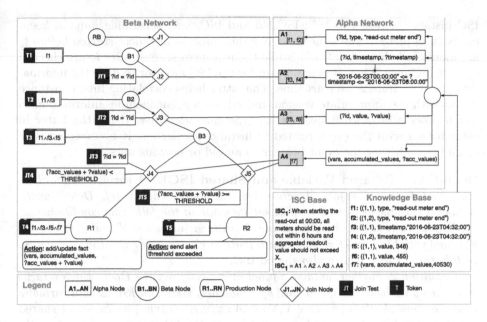

Fig. 5. ISC monitor: inside view of the rete network (adapted from [8])

whether the *id* part of the facts contained in the beta and alpha nodes are equal, i.e., $(\text{event}_{id}^{alpha}, \text{instance}_{id}^{alpha}) = (\text{event}_{id}^{beta}, \text{instance}_{id}^{beta})$. This enables joining together the string of facts originating from the same instance as well as event, which is simplified as $?id = ?id$ in Fig. 5. Arbitrary complex join tests at this level can be employed to relate two facts in various ways. For example, time-based comparison operators would be employed here. In the case of a successful join test, tokens referencing the matching WMEs are stored to all children beta nodes, allowing subsequent join tests to be performed. Following this series of successful join tests until the end will lead to the firing of *production nodes* which contain the sequence of actions that need to be executed.

4.4 Change Strategy: Versioning

We now discuss the specialization of the ISC evolution strategies using the RETE-based ISC monitor. In Sect. 4.1 we have discussed the differences between the three ISC evolution strategies, as well as the necessary operations required to conduct state management. In this section we will specialize the versioning change strategy and propose a concrete versioning algorithm, as well as define proper state management for RETE-based ISC. First we discuss the role of the time of change t_c. Even after employing the necessary versioning techniques (namespacing and state copy), events need to be relayed to the correct version of the ISC at runtime. For this purpose, the time of change t_c becomes critical and necessitates the concept of a **router**, that routes events to the correct

ISC instance. Given two ISC: ISC_old and ISC_new, versioning aims at keeping both in place at the same time. Notably, process instances started before t_c are monitored through ISC_old, while the instances started after t_c are checked against ISC_new. Thus, it is important for the ISC monitor to have the information about each instance start time. The latter helps correlating future instance events with the appropriate version. Indeed, each event includes information to which process instance it belongs, i.e., instance identifier. Using the latter in combination with the event related to instance start time, it becomes possible to find out whether the event belongs to an old or new instance.

Definition 4 (Shared Variable and Shared ISC). *A variable is an alpha node that follows the fact structure (id, attribute, value) (cf. Definition 3). A shared variable is a variable that satisfies one of the following conditions:*

Condition 1: Given two ISC Δ and Δ' with variables $v \in V$ and $v' \in V'$, v is a shared variable iff $v.id = v'.id \wedge v.attribute = v'.attribute \wedge v.value = v'.value$. In this case Δ and Δ' are also shared ISC.

Condition 2: Further, assume a function $\pi : (var, act) \mapsto Bool$, which returns true when a variable var is modified by action act. Given an ISC Δ with variables V and actions $\mathcal{A}$, then $\{v \mid \forall v \in V, \forall \alpha \in \mathcal{A} : \pi(v, \alpha) = true\}$ is the set of shared variables for ISC Δ due to being modified in the behaviour component of the ISC.

Isolation of Versioned ISC. *Shared variables* become a source of complexity when versioning is considered. Imagine an ISC (ISC_old) from the energy domain (cf. Fig. 1), implementing the readout example, which alerts whenever a certain threshold is exceeded. Furthermore, consider a new version of the ISC (ISC_new), which changes the threshold value to be higher before triggering an alert. The usage of the same *shared variable* causes both versions to be evaluated, and in consequence, both ISC_old and ISC_new could be triggered, which is not the intended behaviour. Since we want all process instances that have started after the time of change ($>t_c$) to fall under ISC_new, we can introduce isolation of *shared variables* to achieve the intended behaviour. This isolation can be conducted by creating a copy of the original *shared variable* and storing it under a unique *namespace* intended for ISC_new. *Namespacing* can be realized in Rete by defining the *shared variables* to have a unique prefix in the id part of the fact triple (cf. Definition 3). Furthermore, the production node needs to know which namespace of the *shared variable* it is supposed to access (i.e. when updating the aggregated read out value after each successful readout event). Namespacing thus affects both the alpha nodes (representing the *shared variables*) as well as the production nodes.

Definition 5 (Namespacing of Shared Variables). *Given a shared variable v, $\eta(v)$ returns a new variable v', such that $\neg \exists v'' \in V : v''.id = v'.id$ and thus does not satisfy Condition 1 of Definition 4. Furthermore, any action that modifies a shared variable, $\alpha \in \mathcal{A} : \pi(v, \alpha) = true$, needs to be namespaced as well $\eta(\alpha, v') \mapsto \alpha'$, such that $\alpha.id \neq \alpha'.id \wedge v'.id = \alpha'.id$.*

Now that all *shared variables* are namespaced uniquely for ISC_new, fact evaluation and ISC triggering happen in isolation from ISC_old. What happens if there is a *shared ISC* (cf. Definition 4), ISC_shared, connected to the same *shared variable*, having unrelated behaviour to both ISC_old and ISC_new? At the current state, due to ISC_new being isolated using its unique *namespace*, anytime ISC_old evaluates the fact related to the *shared variable*, the *shared ISC* will also be affected. ISC_shared needs to independently evaluate the *shared variable* from both ISC_old and ISC_new. Therefore, both ISC_old and ISC_new need to be *namespaced*, leading to three distinct namespaces for the same initial *shared variable*. Creating isolated namespaces for ISC_old and ISC_new raises the question of how the *shared variables* should be initialized. Two options are available: 1) copying the original value of the *shared variable* (default) or 2) applying a custom function f to re-initialize the *shared variable*, which could be just resetting it to the default value, depending on its data type. For example, in the case where the *shared variable* is an integer variable, re-initializing could be set to the default value 0. It is part of the versioning specification to deal with the *initialization* of namespaced *shared variables*.

Definition 6 (Initializing Shared Variable). *Given the set of options ops =* $\{copy, reinit\}$, *a* shared variable *s, a namespaced* shared variable *s' (where* $\eta(s) = s'$), *and a custom transformation function that performs a domain-specific action to re-initialize s:* f, *the initialization of the* shared variable *s' is defined as*

$$init(s, s', op, f) = \begin{cases} s.value = s'.value, & \textit{if } op = copy \\ s.value = f(s.value), & \textit{otherwise} \end{cases}$$

Preserving Previously Evaluated Facts. Namespacing an ISC Δ transforms it into a new ISC Δ' to be added to the Rete graph. Since only one of those ISC two can exist at any one time, the original ISC Δ needs to be removed, while at the same time trying to avoid loss of previously evaluated facts (i.e. tokens inside the beta nodes). Simply removing the original ISC Δ causes all related nodes within the beta network to be removed, including the tokens which represent the previously evaluated facts. A safe delete of the original ISC Δ needs to be performed which ensures that such tokens shared by both Δ and Δ' are not removed when deleting Δ.

Definition 7 (Safe Deletion of an ISC). *Given an ISC Δ and a shared ISC Δ^s, a* $safe\_delete(\Delta, \Delta^s)$ *operation ensures that the structurally shared beta and join nodes, which hold the previously evaluated facts (i.e. tokens) remain intact when deleting Δ. A safe delete can be realized by recursively deleting from the bottom production node (representing the ISC Δ), up to the root beta node, while only deleting those nodes which do not reference Δ^s.*

ISC Versioning Algorithm. We are now able to define an ISC Versioning Algorithm (cf. Algorithm 1) that utilizes the Rete graph structure for unbounded

---

Algorithm 1. Unbounded Versioning Algorithm for Rete-based ISCs

Input: $\{\Delta_1, ..., \Delta_n\}$, Δ', t_c, op_{old}, op_{new}, f_{old}, f_{new}

1 **Begin**
2   // (1) adjust the last ISC for proper routing based on t_c, if necessary
3   **if** $length(\{\Delta_1, ..., \Delta_n\}) = 0$ **then**
4 | // no previous ISC to adapt, Δ' is submitted as is
5 | $rete\_add(\Delta')$; **return**
6   **else if** $length(\{\Delta_1, ..., \Delta_n\}) = 1$ **then**
7 | // This is the first time the ISC is versioned, add router condition
8 | $\Delta_{last} = last(\{\Delta_1, ..., \Delta_n\})$
9 | $\Delta'_{last} = \Delta_{last}.alpha\_nodes \cup \{?id, instance\_start\_time, ?ist, \{?ist < t_c\}\}$
10   **else if** $length(\{\Delta_1, ..., \Delta_n\}) > 1$ **then**
11 | // Router condition exists, add proper join test to this condition
12 | $\Delta_{last} = last(\{\Delta_1, ..., \Delta_n\})$
13 | $\Delta'_{last} = \Delta_{last}.router.join\_tests \cup \{?ist < t_c\}$
14   // (2) adapt the new ISC for proper routing
15    $\Delta' = \Delta'.alpha\_nodes \cup \{?id, instance\_start\_time, ?ist, \{?ist >= t_c\}\}$
16   // (3) Namespacing and Initialization of shared variables (cf. Def. 5 and Def. 6)
17    $\Delta'_{last}.S = \{init(s, \eta(s), op_{old}, f_{old}) \mid \forall s \in \Delta_{last}.S, s' \in \Delta'.S : s' = s\}$
18    $\Delta'_{last}.A = \{\eta(\alpha, s) \mid \forall \alpha \in \Delta'_{last}.A, s \in \Delta'_{last}.S : \pi(s, \alpha) = true\}$
19    $\Delta'.S = \{init(s, \eta(s), op_{new}, f_{new}) \mid \forall s \in \Delta_{last}.S, s' \in \Delta'.S : s' = s\}$
20    $\Delta'.A = \{\eta(\alpha, s) \mid \forall \alpha \in \Delta'.A, s \in \Delta'.S : \pi(s, \alpha) = true\}$
21   // (4) ISC Fact evaluation preserving change from Δ_{last} to Δ'_{last}(cf.Def.7)
22    $rete\_add(\Delta'_{last})$; $safe\_delete(\Delta_{last}, \Delta'_{last})$; $rete\_add(\Delta')$

---

versioning of an ISC. The algorithm deals with the three previously discussed aspects: (1) utilizing the *router* concept to detect which version of the ISC should handle the incoming event, (2) *Namespacing* (cf. Definition 5) and the subsequent *Initialization* (cf. Definition 6) of *shared variables* to isolate the different versions of an ISC, and (3) the preservation of previously evaluated facts (i.e. tokens) using *safe delete* (cf. Definition 7). As input, the algorithm requires the list of previous versions of the ISC $\{\Delta_1, ..., \Delta_n\}$, the modified ISC itself Δ', where $\gamma(\Delta_n) \mapsto \Delta'$ is assumed, the time of change t_c and the type of operation for initializing the shared variables op_{old} and op_{new}. Whereas op_{old} is the operation to be performed when initializing the *shared variables* in Δ_n, op_{new} is responsible for initializing Δ'. Both are assumed to be one of $\{copy, reinit\}$ (cf. Definition 6).

For illustrating the algorithm, we take the original ISC defined in Fig. 5, which does not include the 99% of all smart meters condition, and apply it using the *add condition* operation combined with the *versioning* change strategy. This change is visualized in Fig. 6. The color of the nodes identifies the associated ISC: white being the common path for shared attributes (i.e., same context, connection and condition). Orange nodes and edges represent the old ISC, whereas the blue elements representing the new ISC.

Lines 1–16 of Algorithm 1 deal with the *router* aspect (1). Here three subcases are handled. The simplest case is where Δ' has never been versioned before, signaled by an empty $\{\Delta_1, ..., \Delta_n\}$. In this case neither routing, nor isolation

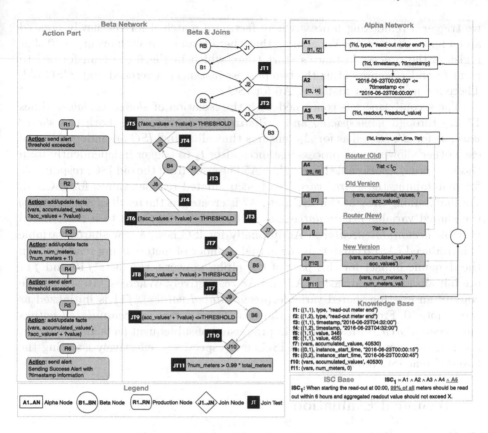

Fig. 6. Change impact evaluation: adding condition + versioning (Color figure online)

of shared variables, nor the preservation of previous facts are necessary. In the case where Δ' represents the first versioning event of Δ_n, the router needs to be setup for the first time. This is accomplished by adding the pattern *(?id, instance_start_time, ?ist)* to the alpha nodes and linking it to both Δ_n and Δ'. As previously mentioned, the process instance's *starting time* is a unique discriminator to identify which ISC version is responsible for handling an event. Here Δ_n is picked for dealing with cases where the process instance's *starting time* is smaller than t_c, otherwise Δ' is responsible for handling the event. The appropriate *join tests* are added to the routing pattern (e.g. $\{?ist < t_c\}$ for Δ_n, or $\{?ist >= t_c\}$ for Δ'). In the case where the routing pattern already exists, which happens when the versioning algorithm has been applied once, then the same routing pattern can be reused in the alpha nodes. Only the new *join tests* are added in that case. Applying the algorithm recursively in this fashion allows unbounded versioning of the same ISC. In the illustration, the routing behaviour is reflected in the creation of the alpha nodes A4, representing old process instances whose instance start time is $< t_c$, and correspondingly A6, representing new process instances started after t_c. On the dynamic impact level, we trigger a reindexing process where facts are matched for the newly created alpha nodes A4 and A6. We assume that there is an event with $event_{id} = 0$ that registers the process instance's start time, which in Fig. 6 are transformed to facts $f8$ and $f9$ representing the two process instances goverened under ISC_old. There are no facts yet matching A6 for ISC_new.

The second aspect is concerned with the isolation of *shared variables* (lines 17–22) through *namespacing* and *initialization*. *Namespacing* both the *shared variable* for Δ', as well as for Δ_n, ensures that all *shared ISC* are independent of each other allowing new process instances after t_c to trigger independently from those before t_c. Noticeable here is that A5 is reused from the old ISC, responsible for maintaining the *shared variable* of accumulated readout values for ISC_old. For maintaining the state for ISC_new, A7 is created as the result of *namespacing* the shared variable *accumulated_values* by *copying* its value to a new shared variable *accumulated_values'*. Additionally for ISC_new, a new shared variable is introduced (A8) to maintain the actual number of meters being read out. On the dynamic impact for this part of the alpha network, the facts $f10$ and $f11$ are initialized and matched to A7 and A8 respectively. Whereas $f10$ is a simple copy of an existing shared variable (*accumulated_values*), $f11$ is initialized as the counter 0. From this point on the two *shared variables accumulated_values* and *accumulated_values'* diverge in processing of subsequent facts and represent the ISC instances of ISC_old and ISC_new respectively, effectively isolating the two different ISC versions.

5   Technical Evaluation

We implemented the concepts introduced in this paper as a prototypical proof-of-concept, extending the ISC Monitor [8][2]. For conducting this technical evaluation we followed the methodology outlined in Fig. 7 in order to tackle research questions RQ1 to RQ3. Concretely, we (1) analyse the correctness of the ISC versioning algorithm, (2) observe the effectiveness of *safe_delete* for avoiding costly reindexing of facts during ISC versioning and (3) highlight the change impact on the Rete graph aggregated by change operation type.

Collected Dataset. In [4] and [19] we collected ISC examples from five different domains (i.e., health care, security, transport/logistics, manufacturing and energy). Some parts of these example scenarios have been simulated on the Cloud Process Execution Engine (CPEE) [20] and subsequently logged as event logs in the Extensible Event Stream (XES) format. Through this collection phase we can already map the common ISC types related to the domain. Consequently, the most likely change operation for the ISC operating under certain domains can be classified. For example, 53% of the collected ISC examples are classified as single-context, meaning that for those cases an *add context* operation is appropriate.

---

[2] The full source code, as well as supplementary material can be found under http://gruppe.wst.univie.ac.at/projects/crisp/index.php?t=iscevolution.

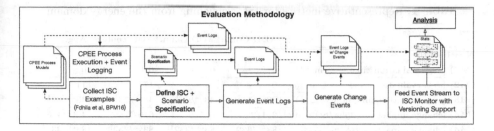

Fig. 7. Evaluation methodology

Scenario Specification. In addition to the modeled examples, we aim to extend the number of event logs to test multitude of realistic scenarios in various domains. For this purpose, we have implemented a *scenario generator* that takes as input a *scenario specification*. The latter is a formal specification of the scenario, which is used to generate event logs. The specification allows the realistic definition of structural control flow as well as data elements. In addition, care has been taken to generate realistic timestamps which follow the known probabilistic distribution from collected scenarios. In this fashion we have specified in total 10 scenarios, each varying in the number of process instances being generated [10, 100, 1000, 10K]. Along with the *scenario specification* we associate the corresponding ISC that governs the monitoring.

Generate Change Events. In order to study the impacts of ISC changes, as well as verify the change process, we embed change events into the event logs. We perform this embedding step randomly over the event logs, following the probability distribution of change operations that could occur depending on the scenario domain (e.g. a change on condition referring to *resource* occurs only very rarely in scenarios in the energy domain. Most ISC in that domain refer to *execution data*). The change events are not completely random, as to avoid invalid change operations. We thus specify the valid base of change operations per scenario, and limit the randomization based on these operations. To ensure an equal distribution of ISC_old and ISC_new events, we automatically pick t_c for each scenario based on the generated event stream, such that the difference in number of events between the ISC versions remains within a max. deviation of 5%.

Feed Event Streams to the ISC Monitoring Engine. Finally, the generated event streams for each domain are fed into the Rete-based ISC Monitoring Engine, which implements the proposed ISC versioning algorithm. Whenever a change event is met, the versioning algorithm is applied. Any other event is fed normally for processing. Key metrics (cf. Table 1) are determined per domain allowing us to analyse various aspects of the algorithm. Additionally, we also track (**M1**) the actual number of ISC activations for each version to ensure correctness of the ISC versioning algorithm.

Table 1. Representative metrics for ISC versioning from the energy domain

	ADD Cond.	ADD Conn.	ADD context	Avg.
$N = 1000$				
(**M2**) [avg. evaluation time $< t_c$ in ms]	0	0	0	0
(**M2**) [avg. evaluation time $>= t_c$ in ms]	2	14	17	11
(**M3**) [avg. evolution time in ms ($\triangle$/†)]	$39^{\triangle}/71^{\dagger}$	$43^{\triangle}/64^{\dagger}$	$48^{\triangle}/73^{\dagger}$	$43^{\triangle}/69^{\dagger}$
(**M4**) [change impact on alpha/beta/join]	2/1/2	5/5/6	6/6/7	4.33/4/5
(**M4**) [change impact on tokens ($\triangle$/†)]	$83^{\triangle}/332^{\dagger}$	$83^{\triangle}/332^{\dagger}$	$83^{\triangle}/332^{\dagger}$	$83^{\triangle}/332^{\dagger}$
	DEL Cond.	DEL Conn.	DEL context	Avg.
$N = 1000$				
(**M2**) [avg. evaluation time $< t_c$ in ms]	1	1	1	1
(**M2**) [avg. evaluation time $>= t_c$ in ms]	27	10	9	15.33
(**M3**) [avg. evolution time in ms ($\triangle$/†)]	$48^{\triangle}/71^{\dagger}$	$41^{\triangle}/64^{\dagger}$	$42^{\triangle}/65^{\dagger}$	$43^{\triangle}/66^{\dagger}$
(**M4**) [change impact on alpha/beta/join]	6/5/6	4/1/2	5/2/3	5/2.66/3.66
(**M4**) [change impact on tokens ($\triangle$/†)]	$83^{\triangle}/175^{\dagger}$	$83^{\triangle}/332^{\dagger}$	$83^{\triangle}/332^{\dagger}$	$83^{\triangle}/279.66^{\dagger}$

$\triangle$ = with *safe_delete*, † = without *safe_delete*

Analysis: ISC Versioning Algorithm Correctness. The first question to tackle is the correctness of the ISC versioning algorithm. Towards that goal we split up three different ISC sets to the same event stream for a given t_c. The first set (**S1**) consists solely of the original ISC (ISC_old), with the additional pattern that it only handles process instances with *starting time* $< t_c$. Similarly, the second set (**S2**) consists of ISC_new handling instances with *starting time* $>= t_c$. The third set (**S3**) utilizes the ISC versioning algorithm to maintain both ISC_old and ISC_new. **M1** confirms that the number of ISC activations in **S1** equals the ISC_old activations of **S3**, and the same for **S2** for ISC_new in **S3**. This confirms that the *router* logic, *namespacing* and *initialization* logic of *shared variables* work correctly as intendend.

Analysis: Change Impact by Change Operation Type. Table 1 shows the collected metrics **M2-M4** for scenarios within the energy domain. Generally we can observe that the lowest cost (in terms of evaluation time, evolution time and number of nodes affected) is by performing an *ADD Condition* operation, which only affects a single alpha node (twice due to *namespacing*). The next level in complexity is the operation *ADD Connection*, which can be explained due to the necessity of adding facts related to another activity in the same event stream. Finally, *ADD Context* can be interpreted as an additional alpha node, representing the new process model, being added on top of a *ADD Connection*. The effectiveness of utilizing *safe_delete* in terms of evolution time and reduced fact reindexing can be seen through metrics **M3+M4**. **M3** shows that the average evolution time *with safe_delete* for both *add* and *delete* is nearly 37% faster compared to the variant *without safe_delete*, i.e., 43ms vs 69ms. Similary, only 25% of the tokens need to be processed when using *safe_delete*, compared to without, i.e., 83 vs 332 (**M4**).

6 Related Work

Change and evolution in PAIS have been research topics for many years, focusing on the definition, soundness, and realization of process schema and instance changes [17]. The impact of process schema and instance changes on compliance constraints has been addressed in [12] by reducing the effort of compliance verification to those compliance constraints that are affected by the changes. Different change scenarios have been considered ranging from ad hoc instance changes to process evolution with concurrent instance changes. Compliance of constraints after ad hoc instance changes has been also subject to the work presented in [11]. Here three states can be distinguished after an instance change, i.e., valid, partially valid, and invalid. Changes of compliance constraints for central processes have been addressed in [10]. The work describes a unified compliance management framework that also considers changes of constraints (adding, changing, and deleting). Some approaches address change and compliance in distributed processes. In [9], algorithms are proposed in order to detect the effects of changing private processes on global compliance rules. [5,9] follows up by stating challenges for the evolution of collaborative processes and their compliance rules. In a similar direction, [2] aims at finding alignments between the monitoring of business networks and compliance at the presence of change at both levels. If a change occurs, a set of actions is determined in order to maintain the ability to monitor the business network. Despite the interest in studying the impacts of process and compliance change, evolving ISC and their impacts on processes and vice versa have not been considered yet. In general, supporting ISC in PAIS is an emerging topic where some approaches have dealt with ISC selectively in the area of, for example, batching [16] or security [21]. In [4], the relevance and modeling of ISC has been addressed in a general way. [8] has proposed techniques for checking and monitoring ISC. A visual notation for ISC has been introduced in [7]. However, none of these approaches has dealt with evolving ISC and processes.

7 Conclusion

This paper addresses ISC evolution in process aware information systems by first revisiting general ISC evolution change strategies: *versioning, migration* and *clean state*, as well as identifying atomic change operations (R1). State management is identified as a critical component when employing these high-level change strategies, which are handled differently depending on the strategy. One strategy, *versioning*, has been concretely defined and implemented on the basis of EC as the ISC formalism and Rete as the inference technique. The prototype is implemented and evaluated (RQ2+RQ3), showing the correctness, efficiency considerations, and the change impact of the proposed ISC versioning algorithm. Future work will address the effects of process changes on ISC.

Acknowledgment. This work has been funded by the Vienna Science and Technology Fund (WWTF) through project ICT15-072.

References

1. Casati, F., Ceri, S., Pernici, B., Pozzi, G.: Workflow evolution. DKE **24**(3), 211–238 (1998)
2. Comuzzi, M.: Aligning monitoring and compliance requirements in evolving business networks. OTM 2014. LNCS, vol. 8841, pp. 166–183. Springer, Heidelberg (2014). https://doi.org/10.1007/978-3-662-45563-0_10
3. Fdhila, W., Knuplesch, D., Rinderle-Ma, S., Reichert, M.: Change and compliance in collaborative processes. In: Services Computing (2015)
4. Fdhila, W., Gall, M., Rinderle-Ma, S., Mangler, J., Indiono, C.: Classification and formalization of instance-spanning constraints in process-driven applications. In: La Rosa, M., Loos, P., Pastor, O. (eds.) BPM 2016. LNCS, vol. 9850, pp. 348–364. Springer, Cham (2016). https://doi.org/10.1007/978-3-319-45348-4_20
5. Fdhila, W., Rinderle-Ma, S., Knuplesch, D., Reichert, M.: Change and compliance in collaborative processes. In: Services Computing, pp. 162–169 (2015)
6. Forgy, C.: Rete: a fast algorithm for the many patterns/many objects match problem. Artif. Intell. **19**(1), 17–37 (1982)
7. Gall, M., Rinderle-Ma, S.: Visual modeling of instance-spanning constraints in process-aware information systems. In: Dubois, E., Pohl, K. (eds.) CAiSE 2017. LNCS, vol. 10253, pp. 597–611. Springer, Cham (2017). https://doi.org/10.1007/978-3-319-59536-8_37
8. Indiono, C., Mangler, J., Fdhila, W., Rinderle-Ma, S.: Rule-based runtime monitoring of instance-spanning constraints in process-aware information systems. In: Debruyne, C. (ed.) OTM 2016. LNCS, vol. 10033, pp. 381–399. Springer, Cham (2016). https://doi.org/10.1007/978-3-319-48472-3_22
9. Knuplesch, D., Fdhila, W., Reichert, M., Rinderle-Ma, S.: Detecting the effects of changes on the compliance of cross-organizational business processes. In: Johannesson, P., Lee, M.L., Liddle, S.W., Opdahl, A.L., López, Ó.P. (eds.) ER 2015. LNCS, vol. 9381, pp. 94–107. Springer, Cham (2015). https://doi.org/10.1007/978-3-319-25264-3_7
10. Koetter, F., Kochanowski, M., Renner, T., Fehling, C., Leymann, F.: Unifying compliance management in adaptive environments through variability descriptors. In: Service-Oriented Computing and Applications, pp. 214–219 (2013)
11. Kumar, A., Yao, W., Chu, C.: Flexible process compliance with semantic constraints using mixed-integer programming. INFORMS J. Comput. **25**(3), 543–559 (2013)
12. Ly, L.T., Rinderle, S., Dadam, P.: Integration and verification of semantic constraints in adaptive process management systems. DKE **64**(1), 3–23 (2008)
13. Mangler, J., Rinderle-Ma, S.: Rule-based synchronization of process activities. In: Commerce and Enterprise Computing, pp. 121–128 (2011)
14. Mangler, J., Rinderle-Ma, S.: IUPC: identification and unification of process constraints. CoRR abs/1104.3609 (2011). http://arxiv.org/abs/1104.3609
15. Montali, M., Maggi, F., Chesani, F., Mello, P., van der Aalst, W.: Monitoring business constraints with the event calculus. ACM TIST **5**(1), 17 (2013)
16. Pufahl, L., Herzberg, N., Meyer, A., Weske, M.: Flexible batch configuration in business processes based on events. In: Franch, X., Ghose, A.K., Lewis, G.A., Bhiri, S. (eds.) ICSOC 2014. LNCS, vol. 8831, pp. 63–78. Springer, Heidelberg (2014). https://doi.org/10.1007/978-3-662-45391-9_5
17. Reichert, M., Weber, B.: Enabling Flexibility in Process-Aware Information Systems - Challenges, Methods, Technologies. Springer, Heidelberg (2012). https://doi.org/10.1007/978-3-642-30409-5

18. Rinderle, S., Reichert, M., Dadam, P.: Correctness criteria for dynamic changes in workflow systems - a survey. Data Knowl. Eng. **50**(1), 9–34 (2004)
19. Rinderle-Ma, S., Gall, M., Fdhila, W., Mangler, J., Indiono, C.: Collecting examples for instance-spanning constraints. Technical report, arXiv:1603.01523 arXiv (2016)
20. Stuermer, G., Mangler, J., Schikuta, E.: Building a modular service oriented workflow engine. In: 2009 IEEE International Conference on SOCA, pp. 1–4 (2009)
21. Warner, J., Atluri, V.: Inter-instance authorization constraints for secure workflow management. In: SACMAT, pp. 190–199 (2006)

Process Histories - Detecting and Representing Concept Drifts Based on Event Streams

Florian Stertz[✉] and Stefanie Rinderle-Ma

Faculty of Computer Science, University of Vienna, Vienna, Austria
{florian.stertz,stefanie.rinderle-ma}@univie.ac.at

Abstract. Business processes have to constantly adapt in order to react to changes induced by, e.g., new regulations or customer needs resulting in so called concept drifts. By now techniques to detect concept drifts are applied on process execution logs ex post, i.e., after the process is finished. However, detecting concept drifts during run-time bears many benefits such as instant reaction to the concept drift. Introducing *process histories* as a novel way to detect and represent incremental, sudden, recurring, and gradual concept drifts through mining the evolution of a process model based on an event stream will face this challenge. Therefore, a formal definition of process histories is given, the concept of process histories is prototypically implemented and compared with existing approaches based on a synthetic event log.

Keywords: Process mining · Event streams · Runtime
Concept drift · Process histories

1 Introduction

Business processes have to constantly adapt in order to react to changes [15] induced by, for example, new regulations or customer needs resulting in so called concept drifts [5]. A recent example for new regulations possibly forcing business process changes is the General Data Protection Regulation as non-compliance with these regulations can cause fines up to €20 million [8]. Process changes are explicitly defined and stored in so called change logs [16] and hence are known to the company. Contrary, concept drifts are happening as the process evolves and hence are to be detected from process execution logs, i.e., logs that store events of executing process instances such as starting or completing process tasks. By now techniques to detect concept drifts in business processes work on process execution logs and are hence applied ex post, i.e., after the process is finished. However, detecting concept drifts during run-time bears many benefits such as being able to instantly react to the concept drift.

Run-time detection of concepts drifts works on event streams rather than on process execution logs. As event streams are infinite, online concept drift

© Springer Nature Switzerland AG 2018
H. Panetto et al. (Eds.): OTM 2018 Conferences, LNCS 11229, pp. 318–335, 2018.
https://doi.org/10.1007/978-3-030-02610-3_18

detection faces the following challenges: start and end event of the stream are unknown; it is not known how many events belong to a trace; it is not known which future events will occur and when they will occur. Moreover, different kinds of concept drift are to be distinguished, i.e., incremental, sudden, recurring, and gradual drifts [5]. So far, the focus has been put on incremental and sudden drifts only, however, detecting recurring and gradual drifts can be important for many application domains as well. These challenges will be tackled along the following research questions:

RQ1 How to detect and reflect process model evolution based on event streams?
RQ2 How to detect incremental, sudden, recurring, and gradual concept drifts based on event streams?

For addressing RQ1, *process histories* are introduced. A process history reflects viable models that are discovered for a process based on an event stream. Process histories provide a novel way to detect and represent concept drifts through mining the evolution of a process model based on an event stream. The challenging question is when a new model is created, i.e., which event or sequence of incoming events triggers the creation of a new model in the history. We present two new algorithms. The first algorithm creates the process history and discovers new viable models. The detection of a viable model, is based on conformance [17] and the "age" of the event information using the sliding window approach, i.e., older process instances do have no impact on the current business process logic. The second algorithm determines concept drifts based on the synthesised process histories (RQ2) and enables the detection of incremental, sudden, recurring, and gradual drifts. The evaluation comprises a prototypical implementation as well as a comparison with existing approaches on detecting concept drifts based on synthetic and real-life logs. In summary, this work provides means to detect incremental, sudden, recurring, and gradual concept drifts based on event streams and the concept of process histories during run-time.

The paper is structured as follows: In Sect. 2, the required definitions and techniques for this work are described. Section 3 features the main contribution of this work, followed by an evaluation, Sect. 4, based on a synthetic log created using the process models of [4]. The related work is presented in Sect. 5 and an outlook and summary is provided in the last section, Sect. 6.

2   Fundamentals

This section introduces fundamentals on business processes and process mining, defines process histories and discusses event streams in comparison with process execution logs.

Process Execution Log. Every time a business process is executed, the information on the execution is usually recorded using the XES[1] [1] format. In short,

---

[1] Extensible Event Stream.

a business process corresponds to a **log** node in a XES file. One log can have zero to many **trace** elements. A trace corresponds to an executed process instance of the business process and it contains information about the process instance, like an id, Runtime and has zero to many **event** elements. An event symbolises an executed activity in the process instance and contains information related to this event and the id of the log as well. In a stream based architecture, the last part is important to relate specific events to their corresponding log.

These log files, allow for the three main types of process mining [18].

- *Process Model Discovery.* This technique is used for finding a fitting process model for the log file.
- *Process Conformance Checking.* This technique takes an event log and an already discovered process model, and checks if the log fits the process model.
- *Process Enhancement.* This technique allows to change and improve the already discovered process model with a new log file.

Process History. A process history contains every process model for one business process and is defined as follows:

Definition 1 (Process History). *Let P be a business process. A process history H_P is a list of viable process models $M_n, n \in \mathbb{N}$ that have been discovered for P with M_n being the current model for P, formally:*

$$H_P := < M_0, M_1, ..., M_{n-1}, M_n, .. > \tag{1}$$

For synthesising a process history, process model discovery and process conformance checking techniques are applied in Algorithm 1 in Sect. 3 to find, check and adapt the current process model.

Since log files are static and created after the execution of process instances, process mining approaches are often applied offline, i.e., ex-post. Our contribution aims at synthesising a process history for a business process online, i.e., at run-time. To achieve this goal, an event stream, instead of log files is used [23].

Event Stream. An event stream represents a continuous flow of events produced by process instances of a business process. To help identify which events belong to which trace or event log, a unique identifier is embedded in the event itself. There are at least two main differences between a log file and an event stream. First, a log file is finite meaning that the information on the number of events per trace and which events appear is available. Second, a log file is also complete, so the specific end and start events of a business process are known. In an event stream it is not guaranteed, that there are no more events for a trace coming in. It is unlikely as well, that we listen to the stream from the time when the first event of the first process instance has occurred, until the last event of the last process instance, because of main memory issues and of course usability. It should be possible to start listening to an event stream at any time. For the implementation of Algorithm 1, special data structures are required, to discover a process model.

Data Structures: A process history covers all viable process models that have been executed for a specific process. Any time an event is sent to the process execution engine, it will be processed using the three types of process mining. For process model discovery, there is a plethora of algorithms available to mine process models, for example, the α-miner [19] or the inductive miner [10].

For process model discovery, we are using an adapted version of the stream-based abstract representation (S-BAR) [23]. S-BAR introduces an abstract representation of the directly follows set of events. This set of events consists of every observed pair of subsequently executed events. This is achieved by creating two maps. In this case, a map relates to the well-known data structure of a *hash table* [7], consisting of keys and their corresponding values.

The first map, **trace_map**, is built using the trace id of a trace as key. The corresponding value to a trace id, is the whole trace. In an event stream, one event at a time is processed. After processing the event, is put into the trace_map. To cope with memory issues and to help determine active traces, the point in time when the first and currently last event of a trace is being processed is also stored. The second map, **directly_follows_map** represents the directly follow relations of all events w.r.t. the trace_map. As key, the preceding event is being stored, with the following event as the corresponding value.

The usage of these maps is explained in detail in Sect. 3. Figure 1, shows the trace_map and directy_follows_map for the traces [A, B, C] and [A, C, B].

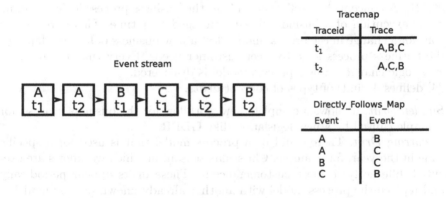

Fig. 1. Event stream containing two traces with different order of events.

For using the inductive miner, an assumption of specific start and end events is required. Since we look at an event stream, it cannot be guaranteed to identify the correct end or start events. E.g., a set of collected start events of each trace can be taken as input. The example in Fig. 1, shows two traces that both have the same starting event "A", so this event would be the only start event for the inductive miner. For the end events, we can only consider the last known events of already known traces. The traces in Fig. 1 provide two possible end events. The first trace has as an end event "C", while the second trace has as an end event "B". This results in a set containing two end events marked for the inductive miner, namely "B" and "C".

Conformance Checking: Every time a new event is processed, we check if this newly extended trace is fitting an already mined process model using conformance checking [17]. Conformance checking replays a trace on a given process model and tries to align it as good as possible. The fitness of a trace in its most basic way, is calculated using costs for inserting events into the model if there are too many events in the log and inserting events in the log, if there are not enough events in the log to fit the model.

Sliding Window: The two maps, trace_map and directly_follows_map, contain the necessary information about every processed event. Since business processes change, already finished or older traces may be part of a preceding version of the business process. Another important factor is, that not every trace can be saved in the main memory, because of capacity issues. To resolve that, the sliding window approach is used. The sliding window only stores k data entries, here keys in the trace_map. If there are already k entries stored, the oldest one is removed before storing the new entry. If the trace_map would be exceeding k, the key value pair with the oldest currently known end event is removed. This method ensures, that only currently active and newer traces are taken into account while discovering a new process model and checking its fitness.

Concept Drift: The last important concept for process histories are concept drifts [21]. A concept drift reflects a shift in the business process logic, meaning that the execution of a business process changed over time. There are several reasons for a change in the process model, like a new business policy or adaptions in the business process logic to meet customer needs. Every time the business process logic changes, a new process model is discovered.

[4] defines 4 kinds of types of concept drifts.

- *Sudden Drift.* It shows a complete new workflow for the business process, for example caused by a new legislation, like GDPR.
- *Recurring Drift.* There could be a process model that is used for a specific time in the year, for example, Christmas season, in which workflows are executed differently to meet customer needs. These drifts appear periodically and replace the process model with another already known process model.
- *Incremental Drift.* Describes small changes, that are natural in the evolution of a business process. Especially in the beginning, the process history will be often extended, because a new process model is discovered after each new event in the event stream. This results in many sub process models.
- *Gradual Drift.* The process got changed and all process instances since the change point have a different process model. M_n and M_{n-1} coexist, as long as already started process instances of M_{n-1} are still running.

It is to be noted, that recurring drifts and incremental drifts, can also be gradual drifts, since process instances of the PM_{n-1} could still be executed.

A process history enables the detection of each of these drifts and is defined in the next section.

3   Contribution

This section describes the contribution of this paper, synthesising process histories and detecting concept drifts.

3.1   Synthesising a Process History

For tackling RQ1, a process history, H_P, is synthesised. The process history contains a list of already known process models, M_i, for a process P, where M_0 is the first known process model for P and M_n is the last known and currently used model for P. With the list of process models, all historical changes of P's logic are represented in H_P, and show the evolution of P.

These models are discovered using an event stream.

The developed Algorithm 1 synthesises a process history and is described in the remainder of this subsection. As input an event stream, ES, a window limit k and the thresholds ϕ and σ are required. The thresholds are described in detail in the following paragraphs.

At the beginning the process history, H_P, is an empty list and does not contain any process models. The trace_map, explained in Sect. 2, contains no items in the beginning. The directly_follows_map is created after an unfit trace is detected.

For usage of the sliding windows approach, the window size k must be defined. Only k items are possible in the trace_map. Every time a new event is processed, it is checked, if its trace id is already existing in the trace_map. If it does not exist and the size of the map is smaller than k, the trace id is used as key and as a value, the event is used as the starting event of the corresponding trace. If the map has already k items, the oldest trace is removed from the trace_map. If the trace id is found in the trace_map, this event will be appended to the trace.

Afterwards, if there is already at least one model in the process history, the fitness of the active trace is checked. For this purpose, we use common conformance checking techniques. Conformance checking returns the fitness value for a trace for a process model by replaying and aligning the trace to the model [2]. For the alignment costs of the trace, two different costs are calculated. The costs for a move in a log, describe if an event is found in the log but not in the model at this position. On the other hand, costs for a move in a model, describe if an event is found in the model but not in the log. For our purposes, only moves in a log are considered, because in an online environment, it is not known, if a process instance reached its end event yet, which means, that the trace can still fit the model. The fitness value of a trace for a model ranges between 0, does not match at all, and 1, matches the model perfectly.

The model in Fig. 2 for example, is our last known model in the process history. The two traces that would match completely would be [A, B, C, D] and [A, C, B, D]. Since we only take the moves in the log into account, the two traces [A, B, C] and [A, C, B] receive a perfect score, and it is assumed, that those process instances are still being executed and the end event "D" has not been processed at the moment.

Input: Event Stream ES (a series of events)

        k (Limit for number of trace_map items)

        σ (Threshold for the fitness of a trace for a model, [0,1])

        ϕ (Threshold for distinction of a new viable model [0,1])

Result: **Process History** H_P (contains all viable process models in

        chronological order.)

$H_P = [\,]$

$M\_duration = 0$

trace_map<trace_id,trace> = 0

for e *in ES* **do**

    if *trace_map contains_key e.trace_id* **then**

       | trace_map['e.trace_id'].append(e)

    else

       if *trace_map.size $\geq$ k* **then**

         | trace_map.delete_oldest

       trace_map.insert(e.trace_id,e)

    if *H.size $\neq$ 0 and conformance_checking(traces[e.trace_id],H_P.last) < σ*

    then

       directly_follows_map<event,event> = 0

       for t *in trace_map.values* **do**

         if *conformance_checking(t,H_P.last) < σ* **then**

           for i *in t.size* **do**

             if *i != 0* **then**

               | directly_follows_map.insert(t[i-1],t[i])

       Model = inductive_miner(directly_follows_map)

       fitting_traces_counter = 0

       durations = [\,]

       for t *in trace_map.values* **do**

         if *conformance_checking(t,Model)$\geq$ σ* **then**

           fitting_traces_counter+= 1

           if *t.end_event in Mode.end_events* **then**

             | durations.append(t.end_event.time-t.start_event.time)

       Score$_{Model,trace\_map.values}$ s= fitting_traces_new / trace_map.values.size

       if *s $\geq$ ϕ* **then**

         H_P.append(Model)

         $M\_duration$ = durations.average + durations.std_deviation

         unfinished_traces =

         trace_map.get_unfinished($H_P[H_P$.size-1],trace_map, $M\_duration$)

         detect_concept_drift(trace_map.values,unfinished_traces,H_P,ϕ,δ)

    if *$|H_P| = 0$* **then**

       | H_P.append(inductive_miner(e))

The last trace [A, D, B], received a lower fitness score, based only on moves in the log. The second event "D" is not expected this early in the process model and cannot be aligned in a perfect way, so it is moved in the log.

To define if a trace fits the model, a threshold, σ is introduced, ranging from 0 to 1. While 0, would result in any trace fitting any model, 1 would only consider perfectly matching traces as fitting. For the purpose of detecting viable process

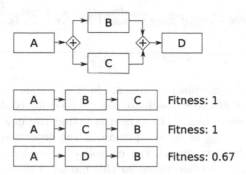

Fig. 2. A process model with one parallel gateway and 3 related traces. While the Move-Log fitness is perfect for the first two traces, the last trace contains an additional event and receives a lower score

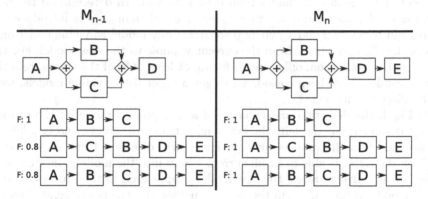

Fig. 3. Model M_{n-1} is only fitting one the first trace perfectly. Model M_n is fitting all traces. M_n is now the new model.

models, a high threshold like 1 is suggested. This guarantees to only consider perfectly matching traces for the distinction.

If the trace of the currently processed event, does not fit the last known process model of the process history, a new model is mined, using the inductive miner. As input for the inductive miner, the abstract representation of the directly follows relation is sufficient. Only unfitting traces in the current window are used for discovering the new process model. The inductive miner always produces sound workflow nets and suffers from less instabilities like the α-miner, which does not detect short loops for example.

To distinguish between viable new process models and anomalous process instances, a score for a process model for a set of traces is defined as:

Definition 2 (Model score). *Let T be a given set of traces, M a process model, and $\phi \in (0,1]$ be a threshold. Moreover let $A \subseteq T$ be the set of all traces having a fitness score greater or equal than ϕ and let $\chi_A(t)$ be the indicator function, returning 1 if $t \in T$ is in A, 0 otherwise.*

Then $\mathcal{S}_{M,T}$, the model score of process model M w.r.t. T, is defined as

$$\mathcal{S}_{M,T} = \frac{\sum_{t \in T} \chi_A(t)}{|T|}.$$

In the algorithm, for calculating the score for the new model using the current trace_map, the values of the whole map, are checked for conformance with the newly discovered process model. If a trace is fitting the new model, a counter is increased by 1, starting at 0. In addition, if the trace's end event is one of the end events of the new model, the complete execution time is calculated and stored in a list of execution times for this model. The variable $M\_duration$ describes the average execution time of M plus the standard deviation. The number of fitting traces is then divided by the number of all possible traces from the trace_map, which results in the score for the new model and the trace_map. The score for this model, ranges from 0 to 1 as well. To determine if the new model is viable, the score must be greater or equal than ϕ. ϕ is introduced as a threshold between 0 and 1, where 0 considers any model as viable and 1 only models that fit every trace from the current window to the new model. For the history in the evaluation, only models fitting at least 90% of the traces from the current window have been considered to get a strict list of viable models, with results discussed in Sect. 4.

In Fig. 3, the detection and creation of a new process model in the process history is shown. On the left, the two longer traces do not fit the last known process model in the process history. The newly discovered model, visible in Fig. 3 on the right, is able to fit all current traces into the model. Since the new model fits all current traces, the new model is appended to the process history. The old model is now M_{n-1} in the process history and the newly created model is now the last and current model in our process history, M_n.

Every time a new model is discovered and appended to H_P, a concept drift is detected. To determine the type of the concept drift, unfinished traces for M_{n-1} from the trace_map need to be collected. A trace is likely to be unfinished if its end event is not part of the end events of M_{n-1} and its current execution time is lower than the execution time stored in $M\_duration$. If its execution time is larger, the process instance is likely to be cancelled.

3.2   Concept Drift Distinction

Algorithm 1 synthesises a process history for a specific process. Every time a new process model is appended to the process history, a concept drift is detected. The 4 types of concept drifts, in relation to a process history, can be defined formally as follows:

Definition 3 (Concept Drift Types). *Let T be a given set of traces and U be a given set of unfinished traces. Moreover let H be a process history for a process P and $\delta \in [0,1]$, $\epsilon \in [0,1]$ be thresholds and the function fitness, defined for one trace and a model, ranging from 0 to 1. The following drift types are defined as follows:*

- *Incremental Drift if* $|H| \geq 2 \wedge \exists (t \in T, fitness(t, M_{n-1}) \geq \delta \wedge fitness$
  $(t, M_n) \geq \delta)$
- *Recurring Drift if* $|H| \geq 3 \wedge \neg IncrementalDrift \wedge \exists m \in \mathbb{N}, 2 \leq m \leq n,$
  $|S_{M_n,T} - S_{M_{n-m},T}| \leq \epsilon$
- *Gradual Drift if* $U \neq \varnothing$
- *Sudden Drift if* $\neg IncrementalDrift \wedge \neg RecurringDrift \wedge \neg GradualDrift$

As a fitness function, this work is using again conformance checking with only considering moves in the log [2].

It is to be noted, that an incremental drift and a recurring drift can be a gradual drift as well. This approach allows to detect concept drifts and identify the type of the concept drift with the use of Algorithm 2 and tackle RQ2.

As input parameters a list of traces T, the traces from the trace_map, a list of unfinished traces U for M_{n-1}, collected by Algorithm 1, for detecting gradual drifts, a process history H_P, δ for determining fit traces and ϵ are required. ϵ describes the maximum error that is allowed between two model scores to be equally viable for T and ranges from 0 to 1, where 0 only determines equal scores to be similar viable and 1 determines any scores to be similar viable.

If there are less than two process models in the process history, it can be concluded that there is no concept drift, since a drift appears when the business process logic changes and a new model is discovered.

For every process model of H the model score is calculated using traces from T, like described in Algorithm 1. The variable `Incremental` is calculated during the calculation of the scores to save execution time. If "Incremental" equals 1 an incremental drift is detected, otherwise not.

If there are traces out of T that fit the preceding model and the current model, an incremental drift is detected. As as long as U is not empty, the incremental drift is a gradual drift as well. Otherwise it is a sudden incremental drift.

For recurring drifts, the score of any model from M_{n-2} to M_0 is calculated. If there is at least one model M_m, where the difference between $S_{M_n,T}$ and $S_{M_m,T}$ is less or equal ϵ, a recurring drift is detected. Then it is again distinguished between a gradual recurring drift and a sudden recurring drift, using the same approach as before.

If it is not a recurring drift or an incremental drift, it number of elements in U is checked. If there is at least one trace, a gradual drift is detected. Otherwise it is not a gradual drift and a sudden drift is detected, since it is already concluded that it is not an incremental or recurring drift either.

The return value is a vector with 4 items corresponding to `Incremental Drift`, `Recurring Drift`, `Gradual Drift` and `Sudden Drift`. E.g., a gradual recurring drift return [0, 1, 1, 0], while a sudden drift returns [0, 0, 0, 1].

In Fig. 4, a complete process history is shown with $\epsilon = 0.05$ and $\delta = 1$. The first concept drift from M_0 to M_1 is detected with T containing t_1 [A, B, C, D, E] and t_2 [A, C, B, D, E]. An incremental drift can be detected between M_0 and M_1, since there is only a new event, E, added to the end of the process. The same traces that fit M_0, fit M_1 as well.

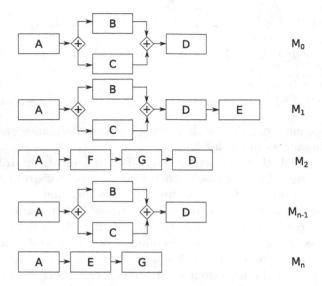

Fig. 4. Complete process history of a single business process

Let $t_{2,3}$ [A, B, C, D], $t_{4,5}$ [A, C, B, D] and $t_{6,...,9}$ [A, F, G, D] be new traces in the event stream. With a small window size k, e.g., 5, M_2 is mined. Only $t_{6,...,9}$ are considered for the model, since they are not fitting M_1. The difference between $S_{M_2,t_{5-9}}$ and $S_{M_1,t_{5-9}}$ or $S_{M_0,t_{5,...,9}}$ is greater than ϵ, so it is not a recurring drift. There are no traces fitting M_2 and M_1 as well, so an incremental drift is not possible. Since t_5 is likely to be not finished for M_1, a gradual drift is detected.

Assume the next traces in the event stream are t_{10-12} [A, B, C, D] and t_{13} [A, C, B, D]. This results in M_{n-1}. The difference between $S_{M_0,t_{9,...,13}}$ and $S_{M_{n-1},t_{9,...,13}}$ is 0. A recurring drift is detected with the recurrence of M_0. Since t_9 s already finished, it is not a gradual drift as well.

Let the next traces be, $t_{14,...,17}$ [A, E, G], which result in M_n. There is no equally similar score to $S_{M_n,t_{13,...,17}}$ and there are no unfinished traces.

In the next section, the two algorithms are evaluated on a synthesised log, following the insurance example used in [4].

4   Evaluation

This section describes a tested example of process histories. Process Execution log files have been synthesised, transformed into an event stream and the process history discovered. The business process depicts an insurance process, first described in [4]. Since not all concept drifts have been integrated in the source, additional concept drifts have been added. With this process history, the type of concept drifts has been detected. The first part of this section covers the implementation of the algorithms and the framework for the evaluation. The second part shows the execution and results. The implementation is available at http://gruppe.wst.univie.ac.at/projects/ProcessHistory.

Input: Traces traces (list of traces), Traces u (list of unfinished traces), H
(Process History),

   δ (Threshold for fitting models $\in$ [0,1]

   ϵ (maximum error between similar process models)

**Result: type_vector[0,0,0,0] (Positions represent Drifts
   [Inc,Rec,Grad,Sudden], 1 represents this type of drift
   occurred.)**

if *H.size* <= *1* **then**
| return "Error: No drift"

Scores = [ ]

Incremental = 0

for *M in H* **do**
| model_score = 0
| **for** *t in traces* **do**
| | **if** *conformance_checking(t,PM)*>= δ **then**
| | | model_score += 1
| | | **if** *M == M_{n-1} and conformance_checking(t,M_n)*>= δ **then**
| | | | Incremental = 1
| Scores.append(model_score/traces.size)

Scores = Scores.reverse // Reverse order so Scores[0] == M_n

if *(Incremental == 1)* **then**
| **if** *u.size $\neq$ 0* **then**
| | return [1,0,1,0] //Incremental Gradual Drift
| **else**
| | return [1,0,0,0] // Incremental Drift

for *i in Scores.size* **do**
| //Start with 0 **if** *i $\leq$ 1* **then**
| | next
| **if** *($|Scores[0]-Score| \leq \epsilon$)* **then**
| | **if** *u.size $\neq$ 0* **then**
| | | return [0,1,1,0] //Recurring Gradual Drift
| | **else**
| | | return [0,1,0,0] // Recurring Drift

if *u.size $\neq$ 0* **then**
| return [0,0,1,0] // Gradual Drift

return [0,0,0,1] // Sudden Drift

4.1   Implementation

A tool to synthesise process execution log files, has been implemented in Ruby
[12]. A web service has been implemented as well, to transform static log files
into an event stream. The generated log files have time stamps in every event
as information. The web service extracts the list of events from the log files
and orders them, based on time stamps. This results in a chronological correct
event stream. Every event is then sent to the main web service. This web service,
written in Ruby as well, processes each event and runs both algorithms described
in Sect. 3. The process history is constantly adapted and provided through a
REST interface. For the mining algorithm we are using the inductive miner,

implemented in ProM called from extension RapidProm [3] of Rapidminer. The output is then retrieved using the REST interface.

4.2   Evaluation

For the evaluation, we synthesised process execution log files, based on the process models used in [5]. Small modifications have been applied, because only one path of some decisions showed concept drifts. For every process model 100 process instances were created. Since not all types of drifts are detectable in these models, we added new process instances to find every type of concept drift. For the creation of the process history, k was set to 50, δ to 1 and σ to 0.9.

For the distinction of a concept drift, ϵ was set to 0.05 and σ to 1. As is often the case in data mining approaches, the values for the parameters are highly dependent on the domain of the business processes. The parameters in this evaluation are fairly strict and only consider perfect fitting models as viable if at least 90 % of the current traces are fitting the new model.

The first 100 process instances consisted of "Register", "Decide High/Low", , "High Insurance Check", "High Med.History Check", "Contact Hospital", "Prepare Notification", "By Phone", "By Email", "By Post" and "Notification Sent". The order of "High Insurance Check", "High Med. History Check" and the order and existence of "By Phone", "By Email", "By Post" have been randomised, so that the inductive miner is able to detect the parallel paths and decisions. The first models produced can vary a lot, depending on the order of events in the event stream. Figure 5 shows the first discovered viable process models in the process history. M_0 consists of only one event. During the first 100 instances, the process model evolves and, depending on the order of the execution of the process instances, the first part of the first parallel gateway can be seen in M_4. Algorithm 2 detects for the first process models in the history only incremental drifts, as expected. This can be reasoned because, every time a new event is found at the end of a trace or a new parallel order instead of sequence is mined, all other traces from the previous model are fitting the new model, e.g., the trace ["Register", "Decide High/Low", "High Insurance Check", "Contact Hospital"] and the trace ["Register", "Decide High/Low", "Contact Hospital", "High Insurance Check"] are both fitting M_5.

After each possible combination is executed, M_{n-4} (Fig. 7), is discovered. To create a gradual drift, the next 50 instances are fitting M_{n-4}, but did not finish before the next 100 instances started in the stream, containing small adaptations. Instead of a parallel gateway for the medical checks, cheaper checks, like "High Insurance Check", are done at the beginning. If this check fails, the other checks are automatically skipped. Unfit traces now contain only a subset of the 3 events. After 45 instances, σ and the score of the new process model M_{n-3} for the traces of the trace_map are equal, so the model is appended to the process history. Since there are 5 instances for M_{n-4} not finished as well, we detected a gradual drift. An incremental drift has been detected as well, since some traces fit M_{n-4} and M_{n-3} perfectly.

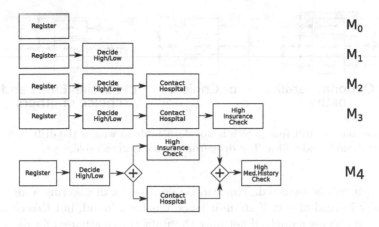

Fig. 5. Process Models containing concept drifts.

All the concept drifts from [4] cannot be detected with our approach. As can be seen in Fig. 6, the first process model includes the events "By Phone", "By Email", "By Post" in optional parallel paths. The first concept drift described in [4] changes the parallel gateway to a decision, where only one event is chosen (Fig. 6(b)). Since all paths are optional anyway, all traces fit, even if only one event is present. To negate this, a periodical model could be mined, using all traces in the trace_map to detect a stricter model fitting all traces. The other model containing again a subset of choices already possible in the parallel optional model, suffers from the same problem. The concept drift from Fig. 6 (b) to (c) could be detected, but only if b is discovered. The drift from (a) to (c) cannot be detected.

Another 100 instances have been put into the event stream, representing a new legislation. The split of high and low insurance claims has been removed, i.e., every claim is treated the same way. The notifications are only allowed to be sent per post as well. After 45 instances, the model M_{n-2} is discovered. This model varies vastly from M_{n-3}, since the score from M_{n-2} is 0.9 and the score of M_{n-3} is 0.1. No traces from M_{n-3} match M_{n-2}. Also M_{n-2} does not conform to any other known model in the process history. A sudden drift is detected.

In the next 100 instances, a new event at the end has been discovered. "Receive delivery confirmation" is appended to the end of the new process instances. Again after 45 instances, M_{n-1} is discovered. Since the 5 oldest traces still fit M_{n-2} and M_{n-1} an incremental drift is detected.

Afterwards the first 100 instances have been put into the stream again with modified time stamps. As expected, after 45 instances, M_n is discovered, which is identical to M_{n-4} and both have the same score of 0.9. A recurring drift has been detected.

All four types of concept drifts can be detected. A problem occurs, if the process model after the concept drift is just a stricter model. This means if new traces fit the current model perfectly, no new model will be discovered and no

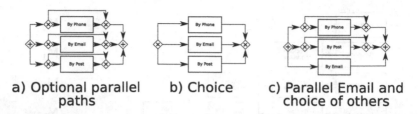

a) Optional parallel paths   b) Choice   c) Parallel Email and choice of others

Fig. 6. The concept drift from a to b is not detectable as well as the drift from a to c, since traces from b and c fit a. The drift from b to c is detectable.

concept drift will be detected. This can be negated by discovering a new model periodically instead of only if an unfit trace has been found, but this could lead to big mixed process models, if not only the unfit traces are used for discovering a new model. E.g., the sudden drift in Fig. 7 from M_{n-3} to M_{n-2}, could also be interpreted with a decision after the "Register" event, which leads to the path from M_{n-3} or the path from M_{n-2}.

Section 5 covers the related work in this field.

5   Related Work

A plethora of algorithms use XES files for discovering process model. The most prominent mining algorithm is the α-miner [13]. This mining technique transforms a directly follow abstraction [23] into a Petri Net [14]. These XES files do not change while they are used for process mining.

Online process mining: An online setting using abstract methods is described in [23] working with directly follows relation, [6] using the heuristics net miner [20] or [11] detecting concept drifts in ltl declared models. The requirements for an online setting, are (a) finite memory. Process execution logs tend to get larger and larger. The size of these files gets so big, that mining the entire XES file at once is not possible, because there is not enough main memory available. Since the files get larger, there is also more data to process. For an online setting, the calculations need to be finished at run-time (b), therefore there are run-time constraints. To cope with these requirements in an offline setting, many process mining approaches, create an abstract representation of an event log to retrieve a process model. The S-BAR approach complies with the following principles. It reuses existing approaches for finding the process model. For the usage of existing techniques, an abstract representation is built.

Concept Drift in Process Mining: A sudden shift, induced by a new legislation for example, can cause a change in a business process. This is followed by a change in the process execution and the process execution logs. The already mined process model might be not suitable any more for conformance checking, a concept drift [21]. In [4], an approach to find concept drifts in process execution logs is discussed. Features to interpret the relationships of events are introduced.

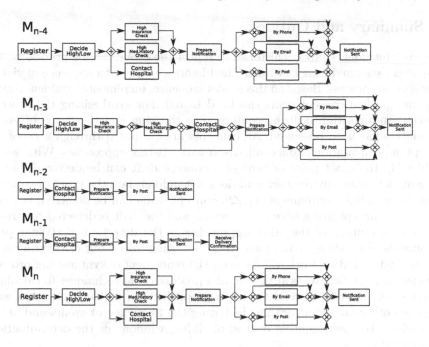

Fig. 7. Process Models containing concept drifts.

- **Relation Type Count.** Defines a vector for every event, containing the number of events that always, sometimes and never follow a specific event.
- **Relation Entropy.** The average rate at which a specific relation is being created.
- **Window Count.** The count is defined for a specific relation, like the follows relation, on a give length.
- **J-measure.** Originally proposed by Smyth and Goodman [9], to calculate the goodness of a rule, like b follows a. This is done with cross-entropy of two events and a specific windows size.

The first two features are calculated using the whole log, while the other two features are calculated on each trace. Concept drifts, are detected by splitting the log in smaller sub-logs and finding the point of change through statistical tests, like the **Kolmogorov-Smirnov test** and the **Mann-Whitney U** test.

The related work, offers ways to retrieve process models from process execution logs and detect concept drifts in an offline environment. While it is possible to detect the exact point in time when the drift is happening, there is no differentiating of types of concept drifts. Also since it is done offline, the results are ex-post. The online mining approaches use a similar strategy to discover process models at run-time. Concept drifts can be detected, but are not differentiated and not all types of concept drifts can be detected.

6   Summary and Outlook

This work introduces process histories to reflect the evolution of a process based on an event stream during run-time. The histories consist of a sequence of viable models of this process. Based on this model sequence, incremental, sudden, recurring, and gradual concept drifts can be detected. For synthesizing the process histories, an algorithm utilizing conformance checking and the "age" of event information has been presented. All concepts are evaluated through a proof-of-concept implementation and a comparison with existing approaches. With static log files [4], the exact point of time of a concept drift can be detected, but is not using an online environment and does not differentiate the types of concept drifts. In an online environment [11,22], concept drifts can be detected, but not all types of concept drifts have been covered and the drift is detected relatively late. The advantage of the other approaches is the detection of stricter process models, since they are not focused on detecting drifts, but discovering new process models. Future work will focus on the refinement of synthesizing process histories, i.e., parallel events at the end of a process, other techniques to calculate the fitness of a specific trace, detecting concept drifts in a stricter model as well as testing other mining algorithms including the frequency of events and other approaches like lossy counting instead of sliding window for the determination of impactful traces.

Acknowledgment. This work has been funded by the Vienna Science and Technology Fund (WWTF) through project ICT15-072.

References

1. IEEE standard for extensible event stream (XES) for achieving interoperability in event logs and event streams. IEEE Std 1849–2016, pp. 1–50 (Nov 2016)
2. Van der Aalst, W., Adriansyah, A., van Dongen, B.: Replaying history on process models for conformance checking and performance analysis. Wiley Interdiscip. Rev.: Data Min. Knowl. Discov. **2**(2), 182–192 (2012)
3. van der Aalst, W.M., Bolt, A., van Zelst, S.J.: RapidProM: mine your processes and not just your data. arXiv preprint arXiv:1703.03740 (2017)
4. Bose, R.P.J.C., van der Aalst, W.M.P., Žliobaitė, I., Pechenizkiy, M.: Handling concept drift in process mining. In: Mouratidis, H., Rolland, C. (eds.) CAiSE 2011. LNCS, vol. 6741, pp. 391–405. Springer, Heidelberg (2011). https://doi.org/10.1007/978-3-642-21640-4_30
5. Bose, R.J.C., Van Der Aalst, W.M., Zliobaite, I., Pechenizkiy, M.: Dealing with concept drifts in process mining. IEEE Trans. Neural Netw. Learn. Syst. **25**(1), 154–171 (2014)
6. Burattin, A., Sperduti, A., van der Aalst, W.M.: Heuristics miners for streaming event data. arXiv preprint arXiv:1212.6383 (2012)
7. Cormen, T.H., Leiserson, C.E., Rivest, R.L., Stein, C.: Introduction to Algorithms. MIT Press (2009)
8. Drolet, M.: How much will non-compliance with GDPR cost you? CSO, October 2017. https://www.csoonline.com/article/3234685/data-protection/how-much-will-non-compliance-with-gdpr-cost-you.html

9. Goodman, R.M., Smyth, P.: Rule induction using information theory. G. Piatetsky (1991)
10. Leemans, S.J.J., Fahland, D., van der Aalst, W.M.P.: Discovering block-structured process models from event logs - a constructive approach. In: Colom, J.-M., Desel, J. (eds.) PETRI NETS 2013. LNCS, vol. 7927, pp. 311–329. Springer, Heidelberg (2013). https://doi.org/10.1007/978-3-642-38697-8_17
11. Maggi, F.M., Burattin, A., Cimitile, M., Sperduti, A.: Online process discovery to detect concept drifts in LTL-based declarative process models. In: Meersman, R., et al. (eds.) OTM 2013. LNCS, vol. 8185, pp. 94–111. Springer, Heidelberg (2013). https://doi.org/10.1007/978-3-642-41030-7_7
12. Matsumoto, Y., Ishituka, K.: Ruby programming language (2002)
13. Alves de Medeiros, A., Van Dongen, B., Van Der Aalst, W., Weijters, A.: Process mining: extending the alpha-algorithm to mine short loops. Technical report, BETA Working Paper Series (2004)
14. Peterson, J.L.: Petri net theory and the modeling of systems (1981)
15. Reichert, M., Weber, B.: Enabling Flexibility in Process-Aware Information Systems: Challenges, Methods, Technologies. Springer, Heidelberg (2012). https://doi.org/10.1007/978-3-642-30409-5
16. Rinderle, S., Reichert, M., Jurisch, M., Kreher, U.: On representing, purging, and utilizing change logs in process management systems. In: Dustdar, S., Fiadeiro, J.L., Sheth, A.P. (eds.) BPM 2006. LNCS, vol. 4102, pp. 241–256. Springer, Heidelberg (2006). https://doi.org/10.1007/11841760_17
17. Rozinat, A., Van der Aalst, W.M.: Conformance checking of processes based on monitoring real behavior. Inf. Syst. 33(1), 64–95 (2008)
18. van der Aalst, W., et al.: Process mining manifesto. In: Daniel, F., Barkaoui, K., Dustdar, S. (eds.) BPM 2011. LNBIP, vol. 99, pp. 169–194. Springer, Heidelberg (2012). https://doi.org/10.1007/978-3-642-28108-2_19
19. Van Der Aalst, W., Van Hee, K.M., van Hee, K.: Workflow Management: Models, Methods, and Systems. MIT Press (2004)
20. Weijters, A., van Der Aalst, W.M., De Medeiros, A.A.: Process mining with the heuristics miner-algorithm. Technische Universiteit Eindhoven, Technical report WP 166, pp. 1–34 (2006)
21. Widmer, G., Kubat, M.: Learning in the presence of concept drift and hidden contexts. Mach. Learn. 23(1), 69–101 (1996)
22. van Zelst, S.J., Bolt, A., Hassani, M., van Dongen, B.F., van der Aalst, W.M.: Online conformance checking: relating event streams to process models using prefix-alignments. Int. J. Data Sci. Anal. 1–16 (2017)
23. van Zelst, S.J., van Dongen, B.F., van der Aalst, W.M.: Event stream-based process discovery using abstract representations. Knowl. Inf. Syst. 54(2), 407–435 (2018)

Lifecycle-Based Process Performance Analysis

Bart F. A. Hompes[1,2(✉)] and Wil M. P. van der Aalst[1,3]

[1] Department of Mathematics and Computer Science,
Eindhoven University of Technology, Eindhoven, The Netherlands
[2] Philips Research, Eindhoven, The Netherlands
`b.f.a.hompes@tue.nl`
[3] Lehrstuhl für Informatik 9/Process and Data Science,
RWTH Aachen University, Aachen, Germany
`wvdaalst@pads.rwth-aachen.de`

Abstract. Many business processes are supported by information systems that record their execution. Process mining techniques extract knowledge and insights from such process execution data typically stored in event logs or streams. Most process mining techniques focus on process discovery (the automated extraction of process models) and conformance checking (aligning observed and modeled behavior). Existing process performance analysis techniques typically rely on ad-hoc definitions of performance. This paper introduces a novel comprehensive approach to process performance analysis from event data. Our generic technique centers around business artifacts, key conceptual entities that behave according to state-based transactional lifecycle models. We present a formalization of these concepts as well as a structural approach to calculate and monitor process performance from event data. The approach has been implemented in the open source process mining tool ProM and its applicability has been evaluated using public real-life event data.

Keywords: Process mining · Performance analysis
Business artifacts · Transactional lifecycle models

1 Introduction

Business processes generate most of the cost of any business. Improving efficiency and efficacy in any organization therefore generally necessitates improving its processes [11]. Nowadays, processes are typically supported by *Process-Aware Information Systems (PAIS)* such as ERP, CRM, and BPM systems, that record data about the execution of the process. Process mining is a discipline that deals with extracting knowledge and non-trivial insights from such event data. Existing process mining techniques have focused on three main areas: process discovery (the automated extraction of process models from event data), conformance checking (aligning observed with modeled behavior), and process enhancement (the extension of existing a priori process models with recorded information).

© Springer Nature Switzerland AG 2018
H. Panetto et al. (Eds.): OTM 2018 Conferences, LNCS 11229, pp. 336–353, 2018.
https://doi.org/10.1007/978-3-030-02610-3_19

Performance analysis focuses on the time perspective, and typically relates to frequencies of occurrence and differences in time between some special events. From most event logs, we can obtain information about process performance characteristics such as waiting times, throughput times, and utilization rates. Performance characteristics such as these are often used in *Key Performance Indicators (KPIs)* and *Service Levels*. However, relatively little research has gone into the structural analysis of process performance from event data. Most performance-oriented techniques rely on ad-hoc definitions of performance [11], and are usually limited to cases (i.e. process instances) and control-flow activities of a process. Often, however, organizations are also interested in the performance of other process entities such as resources, work-items, and other artifacts. This simultaneous analysis from multiple perspectives is usually not supported.

In this paper, we introduce a novel structural approach to analyzing business process performance. We follow the notion of business artifacts, which have been defined as key conceptual entities that are central to the operation of part of a business and who's lifecycles define the overall business process [3,18]. Given process execution data in terms of an event log or stream, our approach measures performance by correlating events to state transitions in business artifact lifecycle models represented by finite state machines (FSMs). Performance metrics are expressed in terms of generic measures defined on these lifecycle models, such as the time spent in a state or the number of times a transition was executed. Events may signal transitions in the lifecycles of multiple artifact instances, and multiple lifecycle models can be identified for the different artifact types. As such, we allow for the simultaneous analysis of multiple artifacts on different levels. An example is shown in Fig. 1. Tool support is provided as well.

The remainder of this paper is structured as follows. Section 2 introduces preliminary definitions. Section 3 shows how events need to be correlated to business artifacts. Section 4 formalizes lifecycle models and explains how events are linked to lifecycle transitions. Section 5 defines performance measures and metrics. The implementation of the approach in the open source process mining tool ProM is discussed in Sect. 6. We evaluate the approach on a publicly available real-life dataset in Sect. 7. Related work is discussed in Sect. 8. Section 9 summarizes the paper and concludes with views on future work.

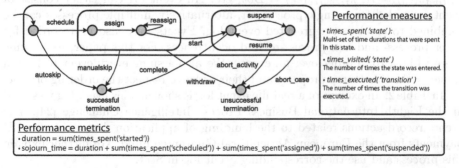

Fig. 1. Example of the artifact-centric performance analysis approach. Basic performance measures are used to form more complex process metrics.

2   Preliminaries

Event logs serve as input for any process mining technique. These techniques typically reason about a number of *business artifacts*. Typically, the execution of an *activity* for a *case* is recorded by one or more *events* that are associated with the lifecycle state *transitions* for the activity *instance*. As we will show later, different lifecycle models can be used for different artifacts. For each event, the *resource* that executed the *activity* is usually recorded as well.

Definition 1 (Universes). $\mathcal{T}$ *is the universe of time stamps.* $\mathcal{B}$ *is the universe of business artifact instances.* $\mathcal{E}$ *is the event universe, i.e. the set of all possible event identifiers. We assume $\mathcal{E}$ to be a totally ordered set.* $\mathcal{P}$ *is the universe of properties.* $\mathcal{V}$ *is the universe of property values.*

Definition 2 (Intervals). $\mathcal{I} = \{(t_1, t_2) \in \mathcal{T} \times \mathcal{T} \mid t_1 \leq t_2\}$ *is the universe of intervals. The function $\delta \in \mathcal{I} \to \mathbb{R}_0^+$ with $\delta(t_1, t_2) = t_2 - t_1$ for $(t_1, t_2) \in \mathcal{I}$ defines the duration of an interval[1].*

Definition 3 (Events, Event properties). *Function $\pi \in \mathcal{P} \to (\mathcal{E} \nrightarrow \mathcal{V})$ is the event property function π. For any property $p \in \mathcal{P}$, $\pi(p)$ (denoted π_p) is a partial function mapping events onto values. If $\pi_p(e) = v$, then event $e \in \mathcal{E}$ has a property $p \in \mathcal{P}$ and the value of this property is $v \in \mathcal{V}$. If $e \notin dom(\pi_p)$, then event e does not have property p and we write $\pi_p(e) = \perp$. We assume all events to have a value for the property 'timestamp', i.e. $dom(\pi_{timestamp}) = \mathcal{E}$. Furthermore, $\forall e_1, e_2 \in \mathcal{E} : e_1 < e_2 \implies \pi_{timestamp}(e_1) \leq \pi_{timestamp}(e_2)$, and $e_1 = e_2 \implies \pi_{timestamp}(e_1) = \pi_{timestamp}(e_2)$, i.e. time stamps respect the total order on events.*

Definition 4 (Event logs). $\mathcal{L}$ *is the universe of event logs. An event log $L = (E, P) \in \mathcal{L}$ defines a set of events $E \in \mathcal{E}$ and a set of event properties $P \in \mathcal{P}$.*

Note that $e \in \mathcal{E}$ is only a unique identifier and function π is needed to attach meaning to e. The set E contains individual events, and P is the set of properties that events in the event log may or may not have. An example event log L_1 is shown in Table 1. $P = \{id, timestamp, case, age, type, activity, activity\ instance, transition, resource\}$ corresponds to the properties recorded in L_1 and $\pi_{timestamp}(e_6) = $ 2018-01-06 10:43, $\pi_{case}(e_6) = c_2$, $\pi_{age}(e_6) = 27$, $\pi_{activity}(e_6) = A$, $\pi_{resource}(e_6) = Bob$, etc. An event log can either consist of events recorded for a single process, or, alternatively, multiple processes can be combined to create an aggregated event log. Events may also be recorded for other process- and non process-related occurrences. For this paper, we assume there to be a total order over all events. We also assume every event to have a value for the timestamp property and their order to respect the ordering of time.

In Table 2, an excerpt of a real-life event log is shown. This event log was used in the Eighth International Business Process Intelligence Challenge [24]. The events record actions related to the handling of application documents for EU payments from the European Agricultural Guarantee Fund. We further explain this process and use the corresponding event log in Sect. 7.

---

[1] $\mathbb{R}_0^+ = \{r \in \mathbb{R} \cup \{0\} \mid r \geq 0\}$.

Table 1. Example event log L_1. Events are characterized by multiple properties.

Event ID	Event properties							
	Timestamp	Case	Age	Type	Activity	Activity instance	Transition	Resource
e_1	2018-01-04 08:00	c_1	33	Gold	A	a_1	start	John
e_2	2018-01-04 09:15	c_1	33	Gold	A	a_1	complete	John
e_3	2018-01-04 10:12	c_1	33	Gold	B	a_2	complete	Bob
e_4	2018-01-04 14:00	c_1	33	Gold	C	a_3	start	Sue
e_5	2018-01-04 14:05	c_1	33	Gold	C	a_3	complete	Sue
e_6	2018-01-06 10:43	c_2	27	Bronze	A	a_4	start	Bob
e_7	2018-01-06 11:00	c_2	27	Bronze	A	a_4	complete	Bob
e_8	2018-01-07 09:33	c_2	28	Bronze	B	a_5	complete	$\perp$
e_9	2018-01-07 09:35	c_2	28	Bronze	C	a_6	start	Sue
e_{10}	2018-01-07 09:37	c_2	28	Bronze	C	a_6	complete	Sue
e_{11}	2018-01-09 09:27	c_3	18	Gold	A	a_7	start	John
e_{12}	2018-01-09 10:40	c_3	18	Gold	A	$\perp$	complete	John
e_{13}	2018-01-09 15:03	c_3	18	Gold	B	a_8	complete	Bob

Table 2. Excerpt from a real-life event log containing multiple artifacts [24].

Event ID	Event properties				
	Timestamp	Application	DocType	DocID	Activity
...	...	...	...	...	...
6062	2015-07-20 13:07:28	8b99873a	Parcel document	5160	Finish editing
7139	2015-07-24 21:10:08	8b99873a	Ref. alignment	7142	Initialize
7131	2015-07-24 21:10:08	8b99873a	Ref. alignment	7142	Performed
7192	2015-09-15 19:31:11	8b99873a	Dep. contr. parc.	7195	Performed
3398	2015-10-26 13:27:10	8b99873a	Ent. application	3397	Initialize
5785	2015-10-31 05:10:31	8b99873a	Ent. application	3397	Begin editing
5307	2015-10-31 05:10:35	8b99873a	Ent. application	3397	Calculate
2108	2015-11-05 11:01:26	8b99873a	Ent. application	3397	Save
1924	2015-11-05 11:01:37	8b99873a	Ent. application	3397	Calculate
1863	2015-11-05 11:01:39	8b99873a	Ent. application	3397	Finish editing
...	...	...	...	...	...

3 Correlating Events

Most existing process mining techniques assume events in an event log to be grouped per case. This seemingly simple requirement can be quite challenging as it requires *event correlation*, i.e. events need to be related to each other. Event correlation is one of the primary challenges when extracting event logs from information systems [22]. Few approaches have been proposed to deal with this issue, and most process mining techniques assume event logs to contain the necessary correlation information. To illustrate the issue, consider the following.

Events are often referred to by the name of the activity that generated them. As such, many events may refer to the same activity name. They may refer to the same activity instance (e.g. lifecycle transitions) or to different activity instances (e.g. in a loop). Consider the scenario in which the same activity is started twice, i.e. two activity instances are running in parallel. Given the footprint of two "starts" followed by two "completes", there are two possible scenarios as illustrated by Fig. 2. In the first scenario, the time between related events is similar for both activity instances, whereas in the second scenario, this is not the case. It is clearly important to be able to correlate events, as timing information recorded by them is used to measure performance. Different strategies can be used to correlate events to activity instances a posteriori. For example, heuristics can be used. In Fig. 2 one could assume a first-in-first-out order and pick the first scenario. Moreover, one could introduce domain-specific rules such as timeouts to correlate events to activity instances [22]. Additional information from process models such as whether or not there are loops or parallelism can be used as well.

In [10], the authors show how events can be related to cases (i.e. process instances) without a priori information. Relating events to cases is described as the "primary correlation problem", and relating events within the same case as the "secondary correlation problem". However, events may relate to multiple business artifacts. As such, events can be correlated to each other on different perspectives at the same time. For example, events can be grouped per activity, by the resource that executed it, per day of the week, or by all of the above. Given more complex artifact lifecycle models with additional states and transitions, many ordering scenarios are possible and correlating events a posteriori becomes even more challenging. A complete overview of event correlation techniques is outside the scope of this paper. Consequently, for the remainder, we will assume event correlation information to be available in the event log.

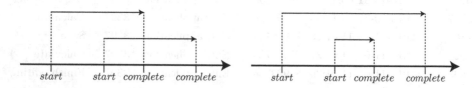

Fig. 2. Two scenarios involving two activity instances leaving the same footprint in the event log. Correlating events is paramount for accurate performance analysis.

4   Business Artifact Lifecycle Models

Traditional process mining techniques are based on activity flows for cases (process instances). Additionally, the data aspect is often considered only as an afterthought. In contrast, an emerging family of approaches uses (business) artifacts that combine both data aspects and process aspects into a holistic unit.

Artifacts have been defined as key conceptual entities that are central to the operation of part of a business and who's lifecycles define the overall business process [3, 18]. Several different artifact lifecycle models have been used in literature, tools, and PAIS. For example, the recently accepted IEEE standard for eXtensible Event Streams (XES) [1] contains two distinct lifecycle models for activities by default. Generally speaking however, only activity lifecycle models are considered and the lifecycles of cases are expressed using process models.

In this paper, we consider the general case in which business artifacts behave according to state-based transactional lifecycle models, i.e. the behavior of activities, cases, resources, and the process as a whole can be described using lifecycle models. Note that this list of artifacts is not exhaustive. It may be extended to include any concept that is relevant for the process under investigation. A typical example artifact-centric process is a build-to-order process where several customer orders are collected, and based on the order, materials need to be ordered from suppliers. Customer orders can relate to one or more supplier orders which contain goods related to multiple customer orders [19]. Other domains where business artifacts play an important role are healthcare [25], order-to-cash [16], and logistics and transportation. Furthermore, business artifacts may relate to multiple lifecycle models, to enable analysis on multiple levels of abstraction.

4.1 Formalizing Lifecycle Models

A key concept in any lifecycle model is that state transitions can occur when time evolves, and that every instance of the model can only be in a single state at any given point in time. As such, we formalize lifecycle models as deterministic finite state machines in which there is exactly one initial state.

Besides manual definition, lifecycle models may be automatically discovered from event data. Several transition system discovery techniques have been proposed [16, 19, 23, 25]. Such discovered models can be easily converted to lifecycle models and used in the analysis presented in this paper. For example, in case more than one natural initial state is present, an artificial initial transition can be added that has an outgoing transition to each of the original initial states.

Definition 5 (Lifecycle models). S *is the universe of states,* A *is the universe of actions, and* M *is the universe of lifecycle models. A lifecycle model* $M = (S, A, T) \in M$ *is a deterministic finite state machine, where* $S \subseteq S$ *is the finite set of states,* $A \subseteq A$ *is the finite set of actions,* $T \subseteq S \times A \times S$ *is the finite set of transitions,* $S^{initial} \in S$ *is the initial state, and* $S^{final} \subseteq S$ *is the set of final states. A transition* $(s, a, p) \in T$ *is also denoted as* $s \xrightarrow{a} p$. *Furthermore,* $\forall t_1 = s \xrightarrow{a} p, t_2 = s' \xrightarrow{a'} p' \in T : s = s' \land a = a' \implies t_1 = t_2$ *(determinism).*

The lifecycle model shown in Fig. 1 models an activity lifecycle, and can be modeled as a finite state machine $M_1 = (S_1, A_1, T_1)$ where $S_1 = \{initial, scheduled, assigned, started, suspended, completed\ successfully, completed\ unsuccessfully\}$, $S_1^{initial} = \{initial\}$, $S_1^{final} = \{completed\ successfully, completed\ unsuccessfully\}$,

$A_1 = \{schedule, assign, reassign, start, suspend, resume, autoskip, manualskip,$
$\ldots\}$, and $T_1 = \{initial \xrightarrow{schedule} scheduled, scheduled \xrightarrow{assign} assigned, assigned$
$\xrightarrow{reassign} assigned, assigned \xrightarrow{start} started, started \xrightarrow{suspend} suspended, suspended$
$\xrightarrow{resume} started, initial \xrightarrow{autoskip} completed\ successfully, scheduled \xrightarrow{manualskip}$
$completed\ successfully, \ldots\}$. Note that here, transitions are drawn between groups
of states in order to improve readability of the model. A transition from resp. to a
group of states represents that such a transition is present from resp. to any state
in the group. Groups can be considered to represent higher-level states.

4.2   Linking Events to Lifecycle Transitions

Events are often referred to by the name of the activity that generated them.
However, as mentioned in Sect. 3, this is technically incorrect as there may be
many events referring to the same activity name. Additionally, events might
be recorded that do not refer to any activity in the process, but do refer to
other artifacts. As such, we use event properties to correlate events to lifecycle
transitions of arbitrarily many business artifacts. The general idea is that events
may relate to zero or more lifecycle moves. A lifecycle move records the execution
of an action in a lifecycle model for a single instance of an artifact. Formally,
lifecycle moves can be defined as follows.

Definition 6 (Lifecycle moves). $\mathcal{N}$ *is the universe of lifecycle moves. A life-*
cycle move $n = (M, i, a) \in \mathcal{N}$ *is a tuple where* $M = (S, A, T) \in \mathcal{M}$ *refers to a*
lifecycle model, $i \in \mathcal{B}$ *to a business artifact instance, and* $a \in A$ *to an action*
in M. *We also define a corresponding property* $lifecycle\_moves \in \mathcal{P}$ *such that*
$\pi(lifecycle\_moves) \in \mathcal{E} \to \mathbb{P}(\mathcal{N})^2$.

Note that there can be multiple transitions in a model that refer to the same
action. Logically, each event can only relate to a single action per model, per
artifact instance. In Table 3, event log L_1 of Table 1 is restructured to show the
lifecycle moves of its events. For this example, two (traditional) business artifacts
are considered: *case* and *activity*. Both artifacts follow the same simple lifecycle
model $M_2 = (S_2, A_2, T_2)$ where $S_2 = \{initial, started, completed\}$, $S_2^{initial} =$
$initial$, $S_2^{final} = \{completed\}$, $A_2 = \{start, complete\}$, and $T_2 = \{initial \xrightarrow{start}$
$started, started \xrightarrow{complete} completed\}$. The first event recorded for each case
signals the *start* action, and events signaling the *complete* action of activity C
are considered to also signal the *complete* action of the related case. For example,
$\pi_{lifecycle\_moves}(e_1) = \{(M_2, a_1, start), (M_2, c_1, start)\}$. In this example, relating
the lifecycle moves in L_2 to the transitions in M_2 is trivial, as each action is only
referred to by in a single transition.

---

[2] $\mathbb{P}(X)$ denotes the powerset of a set X, i.e. $Y \in \mathbb{P}(X) \iff Y \subseteq X$.

Table 3. Event log L_2. Events are correlated to zero or more lifecycle moves.

Event ID	Event properties					
	Timestamp	Activity	Artifact	Lifecycle model	Artifact instance	Action
e_1	2018-01-04 08:00	A	activity	M_2	a_1	start
			case	M_2	c_1	start
e_2	2018-01-04 09:15	A	activity	M_2	a_1	complete
e_3	2018-01-04 10:12	B	activity	M_2	a_2	complete
e_4	2018-01-04 14:00	C	activity	M_2	a_3	start
e_5	2018-01-04 14:05	C	activity	M_2	a_3	complete
			case	M_2	c_1	complete
e_6	2018-01-06 10:43	A	activity	M_2	a_4	start
			case	M_2	c_2	start
e_7	2018-01-06 11:00	A	activity	M_2	a_4	complete
e_8	2018-01-07 09:33	B	activity	M_2	a_5	complete
e_9	2018-01-07 09:35	C	activity	M_2	a_6	start
e_{10}	2018-01-07 09:37	C	activity	M_2	a_6	complete
			case	M_2	c_2	complete
e_{11}	2018-01-09 09:27	A	activity	M_2	a_7	start
			case	M_2	c_3	start
e_{12}	2018-01-09 10:40	A	activity	M_2	$\perp$	complete
e_{13}	2018-01-09 15:03	B	activity	M_2	a_8	complete

4.3 Data Quality and Conformance

Above, we have shown how events can be related to actions in lifecycle models. This generally works well when the recorded event data is of good quality. However, in real-life settings, data quality is imperfect and often contains noise. For example, it is possible that not every lifecycle transition for every lifecycle model for every artifact instance was recorded, or that data is missing or otherwise corrupted. Case in point, e_{12} in L_2 is missing the *artifact instance* value of its one associated lifecycle move. Additionally, recorded event data only stores which lifecycle action occurred, rather than the exact lifecycle transition, and there might be multiple transitions related to the same action. As such, missing event properties or incorrect event ordering could lead to problems when relating events to lifecycle transitions a posteriori. Specific conformance checking techniques need to be developed to deal with non-conformance of event data related to business artifact lifecycle models. This is part of the larger event correlation problem discussed in Sect. 3. Handling non-conformance and other data quality issues is outside the scope of this paper, and for the remainder, we assume all necessary lifecycle information to be available in the event data. Nevertheless, it should be relatively straight-forward to combine the ideas in this paper with existing approaches for dealing with noise and non-conformance.

5   Analyzing Process Performance

Most process performance analysis techniques use ad-hoc definitions of performance, and, to the best of our knowledge, no structural approach for the analysis of business process performance currently exists for processes involving multiple intertwined business artifacts. However, as performance is usually specified in terms of events generated by business artifacts, in our approach, we utilize the artifact lifecycle models defined earlier as the basis for performance analysis. For each lifecycle model related to every instance of an artifact, basic performance measures are calculated. These measures are the same for any lifecycle model, and independent of the model's structure. More complex, model-specific performance metrics can then be defined in terms of these basic measures.

5.1   Performance Measures

Events in an event log are per definition atomic (i.e. they do not have a duration). As described in Subsect. 4.2, events are related to transitions in the lifecycles of any number of business artifacts. As a result, we can generate performance measurements by "replaying" the event log on the lifecycle models.

Events signal the occurrence of lifecycle transitions at certain time stamps and consequently indicate when the source state is left and the target state is entered. State visits are intervals that indicate a single visit to the state, i.e. the times at which a state was entered and exited. As states can be visited multiple times, a multi-set of state visits is kept for each state. For each transition, we keep track of a multi-set of transition executions (timestamps). In order to reason over these basic performance measures, we formally define them as follows.

Definition 7 (Transition executions). *Let $M = (S, A, T) \in \mathcal{M}$ be a lifecycle model, let $L \in \mathcal{L}$ be an event log, and let $i \in \mathcal{B}$ be an artifact instance. The function $\chi_{M,L,i} \in T \to \mathbb{B}(\mathcal{T})$ maps transitions to multi-sets of time stamps*[3].

Definition 8 (State visits). *Let $M = (S, A, T) \in \mathcal{M}$ be a lifecycle model, let $L \in \mathcal{L}$ be an event log, and let $i \in \mathcal{B}$ be a business artifact instance. The function $\gamma_{M,L,i} \in S \to \mathbb{B}(\mathcal{I})$ maps states to multi-sets of intervals named state visits. The function $\rho_{M,L,i} \in S \to \mathbb{B}(\mathbb{R}_0^+)$ maps states to multi-sets of state visit durations where $\rho_{M,L,i}(S) = [\delta(sv) \mid sv \in \gamma_{M,L,i}(S)]$.*

Note that the functions χ, γ, and ρ are defined only in the context of a lifecycle model, an event log, and a business artifact instance.

Using M_2 and L_2, we obtain the following performance measures:

- $\chi_{M_2,L_2,c_1}(initial \xrightarrow{start} started) = [\text{2018-01-04 08:00}]$
- $\chi_{M_2,L_2,c_1}(started \xrightarrow{complete} completed) = [\text{2018-01-04 14:05}]$
- $\gamma_{M_2,L_2,c_1}(started) = [(\text{2018-01-04 08:00, 2018-01-04 14:05})]$
- $\rho_{M_2,L_2,c_1}(started) = [\{365 \min\}]$

---

[3] $\mathbb{B}(X)$ denotes the set of multi-sets (bags) over a set X.

- $\chi_{M_2,L_2,a_3}(initial \xrightarrow{start} started) = [2018\text{-}01\text{-}04\ 14:00]$
- $\chi_{M_2,L_2,a_3}(started \xrightarrow{complete} completed) = [2018\text{-}01\text{-}04\ 14:05]$
- $\gamma_{M_2,L_2,a_3}(started) = [(2018\text{-}01\text{-}04\ 14:00, 2018\text{-}01\text{-}04\ 14:05)]$
- $\rho_{M_2,L_2,a_3}(started) = [\{5\ \text{min}\}]$

From the transition executions and state visits, we can easily compute aggregate statistics such as the number of times a transition t was executed ($|\chi_{M,L,i}(t)|$), the total time spent in a state s ($sum(\rho_{M,L,i}(s))$), or the number of times state s was visited ($|\gamma_{M,L,i}(s)|$). The *times_executed*, *times_spent*, and *times_visited* functions in Fig. 1 refer to the performance measures defined above.

5.2 Performance Metrics

From the basic performance measures, we can derive more complex performance metrics. Performance metrics are therefore represented as expressions in terms of performance measures. As the definition of specific metrics depends on the artifacts in the domain under investigation, and the lifecycle models they correspond to, we define the signature of performance metric functions. In a process performance analysis project, investigators can then define domain-specific performance metrics as functions of this signature.

Definition 9 (Performance metrics). $\mathcal{C} = \mathcal{M} \nrightarrow (\mathcal{L} \times \mathcal{B} \to \mathbb{R}_0^+)$ *is the universe of performance metrics. For any performance metric $c \in \mathcal{C}$, and any lifecycle model $M \in \mathcal{M}$, $c(M)$ (denoted c_M) maps an event log and business artifact instance to a real number. If $c_M(L, i) = r$, then performance metric $c \in \mathcal{C}$ is defined for lifecycle model $M \in \mathcal{M}$ and returns value $r \in \mathbb{R}_0^+$ for event log $L \in \mathcal{L}$ and business artifact instance $i \in \mathcal{B}$. If $M \notin dom(c)$, then performance metric c is undefined for lifecycle model M.*

Formally, every performance metric is only defined in the context of a single lifecycle model. This is necessary, as metrics with the same name can be defined differently for different lifecycle models. For example, for M_1 in Fig. 1, there are two performance metrics defined: *duration*, and *sojourn time*. The definition for *sojourn time* uses the definition for *duration*, and both depend solely on the total time spent in a selection of states. Using the formalizations given above, we can define *duration* as follows: $duration_{M_1}(L, i) = sum(\rho_{M_1,L,i}(started))$.

5.3 Aggregating Performance

So far, the performance measures and metrics we have defined are calculated for individual artifact instances in the context of a lifecycle model and an event log. Note that events that signal the lifecycle transitions are correlated with individual artifact instances by lifecycle moves. Often however, analysis on the artifact instance level is too fine-grained. We are generally not interested in the duration of a single instance of an activity, case, order, patient visit, etc. In many situations, aggregate statistics such as minimum, maximum, and average values provide more insight. To this end, we provide a means to aggregate performance characteristics over multiple artifact instances using properties.

Definition 10 (Business artifact properties). *Function $\lambda \in \mathcal{P} \to (\mathcal{B} \nrightarrow \mathcal{V})$ is the business artifact property function λ. For any property $p \in \mathcal{P}$, $\lambda(p)$ (denoted λ_p) is a partial function mapping business artifact instances onto values. If $\lambda_p(i) = v$, then artifact instance $i \in \mathcal{B}$ has a property $p \in \mathcal{P}$ and the value of this property is $v \in \mathcal{V}$. If $i \notin dom(\lambda_p)$, then artifact instance i does not have property p and we write $\lambda_p(e) = \perp$.*

As defined in Definition 4, events may have many property values. The same applies to business artifacts (i.e. correlated events). The value of a property for an artifact instance might be derived from the events related to the instance. Alternatively, more advanced techniques may be used to assign property values to artifact instances.

Definition 11 (Projection function). *Function $\lceil \in \mathcal{P} \times \mathcal{V} \to (\mathbb{P}(\mathcal{B}) \to \mathbb{P}(\mathcal{B}))$ is the business artifact projection function $\lceil$. For any property $p \in \mathcal{P}$, value $v \in \mathcal{V}$, and any set of artifact instances $B \in \mathcal{B}$, $\lceil (p, v)$ (denoted $\lceil_{p,v}$) is a function mapping a set of artifact instances onto another set of artifact instances such that $\lceil_{p,v} (B) = \{i \in B \mid \lambda_p(i) = v\}$.*

Given the projection function $\lceil$, we can select all artifact instances that share a value for a property. $B = \left\{ i \mid (M, i, a) \in \bigcup \{\pi_{lifecycle\_moves}(e) \mid e \in E_2\} \right\}$ is the set of all artifact instances recorded in $L_2 = (E_2, P_2)$. Consider the event property $activity \in P_2$ to be copied to the related artifact instances. The set of artifact instances with value A for property $activity$ is $\lceil_{activity,A} (B) = \{a_1, a_4, a_7\}$. Additionally, multiple projections can be applied sequentially.

Definition 12 (Aggregate performance metrics). $\mathcal{G} = \mathcal{M} \nrightarrow (\mathcal{L} \to \mathbb{R}_0^+)$ *is the universe of aggregate performance metrics. For any aggregate performance metric $g \in \mathcal{G}$, and any lifecycle model $M \in \mathcal{M}$, $g(M)$ (denoted g_M) maps an event log to a real number. If $g_M(L) = r$, then aggregate performance metric $g \in \mathcal{G}$ is defined for lifecycle model $M \in \mathcal{M}$ and returns value $r \in \mathbb{R}_0^+$ for event log $L \in \mathcal{L}$. If $M \notin dom(g)$, then aggregate performance metric g is undefined for lifecycle model M.*

Given a set of artifact instances it is easy to obtain aggregate performance metrics. Applying a metric on each individual instance leads to a multi-set of values, over which we can compute aggregate statistics. For example, take performance metric $duration_{M_2}(L, i) = sum(\rho_{M_2,L,i}(started))$, and event log L_2. Note that $duration(M_2) \neq duration(M_1)$, i.e. the performance metric $duration$ is defined differently for the two different lifecycle models. In order to aggregate the durations, we can define the aggregate performance metric $avgDuration \in G$ as follows. $avgDurtation_{M_2}(L_2) = avg([duration_{M_2}(L_2, i) \mid i \in B])$. This aggregated metric measures the average duration of all business artifact instances in event log L_2 that follow M_2. If we are interested in only a selection of artifact instances, we can use the projection function. For example $avgDurationA_{M_2}(L_2) = avg([duration_{M_2}(L_2, i) \mid i \in \lceil_{activity,A} (B)])$ measures the average duration of all artifact instances related to activity A.

6 Implementation

The IEEE XES standard defines an XML-based grammar that enables a unified and extensible methodology for capturing and interchanging event data [1]. By default, XES defines a set of extensions, which describe special properties on the event, case, or log level. One such extension provided by default is the *lifecycle extension*, which specifies for events the lifecycle transition they represent in a transactional model of their generating activity. This model can be arbitrary, and two standard transactional models for activities are included. However, the standard XES lifecycle extension is limited: only activity artifacts are considered, only one lifecycle model can be specified for the entire event log, and only a single lifecycle move can be specified per event. We developed a new XES extension that solves these problems and matches the definitions provided in this paper (defined in Sect. 4). It allows events to relate to transitions in multiple lifecycle models of multiple business artifacts used concurrently. The extension is currently under review for approval by the XES working group.

The process performance analysis approach presented in this paper has been implemented in ProM, which is an open source process mining tool that supports a wide variety of process mining techniques in the form of plug-ins[4]. We provide a graphical editor that can be used to create and modify lifecycle models (as defined in Subsect. 4.1. Starting from an existing (automatically discovered) transition system is possible as well. An expression editor is provided that allows for the specification of performance metrics (as defined in Subsect. 5.2). Furthermore, lifecycle models can be imported from and exported to an XML-based format. The tool provides functionality to link events to lifecycle moves. It can be configured to use existing event property values for the lifecycle model, artifact instance, and action properties in lifecycle moves (as defined in Subsect. 4.2). Alternatively, the standard XES lifecycle extension properties can be converted to the artifact-centric format. Finally, the tool measures process performance using event data based on artifact lifecycle models. Next to an event log or stream, it expects as input a set of lifecycle models that include definitions for performance metrics. As optional parameters, a selection of properties can be given that will be used to aggregate the performance information upon (as defined in Subsect. 5.3). For each event, the associated lifecycle moves are evaluated and their effect on the lifecycle models are calculated, resulting in a collection of performance measure values (as defined in Subsect. 5.1). The performance metrics are evaluated using the computed measures and shown to the user in an interactive interface. Figure 3 contains a screenshot of this interface, showing general information and the results of a performance analysis using a synthetic event log concerning a mobile phone repair process. Here, a single artifact *activity* was considered that follows a lifecycle model M_2 (as defined earlier). Two aggregations are selected, such that the aggregated performance metric shows aggregate statistics for the duration of the *test repair* activity instances where the defect was fixed.

---

[4] See http://promtools.org and the *LifecyclePerformance* package for more details.

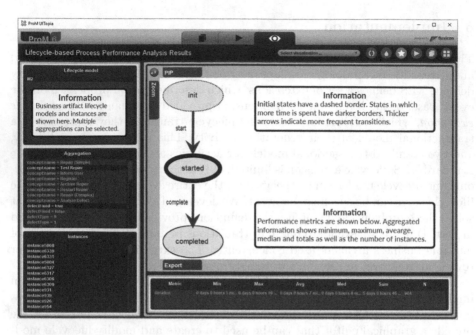

Fig. 3. Screenshot of the ProM plug-in interface showing the results of the lifecycle-based process performance analysis. Multiple aggregations can be selected per lifecycle model, and individual artifact instances can be included or excluded.

7   Evaluation

In this section we evaluate the approach by applying it on real-life public event data used in the Eighth International Business Process Intelligence Challenge (BPIC) [24]. The events record actions related to the handling of applications sent to the European Agricultural Guarantee Fund. The process operates in terms of documents, where each document has a state that allows for certain actions. For each application document several other documents are created[5]. Consequently, the application documents can be considered business artifacts of this process.

The dataset contains nine individual event logs for the different document types involved in the process. It also includes a combined event log, which in total, contains $2,514,266$ events for $43,809$ applications and data has been recorded for a period of three years. The applications contain between 24 and $2,973$ events, with an average of 57. In the event log, several data attributes are recorded, such as the department ID, the resource involved, the year of application, the number and total area (in m^2) of parcels, the type of payment, any penalties, risk factors, etc. An excerpt of the combined event log is shown in Table 2.

---

[5] For more information on the process and the data, see http://www.win.tue.nl/bpi/.

Documents are typically created by an *initialize* activity, after which they can be edited, recorded by the *begin editing* and *finish editing* activities. While editing, amongst other events, calculations can be made and the documents can be saved. For our analysis, we are interested in the time spent editing a document. As such, we created the lifecycle model M_{doc} and a performance metric *et*, defined as the sum of the durations of the state visits to the state *editing*. More formally, $et_{M_{doc}} \in C$, where $et_{M_{doc}}(L, i) = sum(\rho_{L,i}(editing))$ for any event log $L \in \mathcal{L}$ and artifact instance $i \in \mathcal{B}$. Using the recorded event properties, we add aggregations. Additionally, we add aggregated performance metrics to show the minimum, maximum, average, median, and sum of *et* of the selected document instances as well as how many documents are in the selection.

Figure 4 shows the results of applying our approach to the event log. The left of the screen shows the lifecycle models corresponding to the nine different document types as well as the aggregation options and individual documents (artifact instances). The right of the screen shows the annotated lifecycle model of the selected artifacts. More time is spent in the *created* state compared to the *editing* state, and documents are edited multiple times. By selecting different combinations of documents and aggregations, we can answer a multitude of performance questions, such as those posed in the BPIC. We highlight some interesting insights for the selected document type *Geo parcel document*. In 2016, Geo parcel documents handled by department 6b spent on average 33 days and 10 h being edited, while in 2017 the average dropped to 23 days and 10 h. Other departments also spent less time editing Geo parcel documents in 2017 compared with 2016, as can be seen in Table 4. Furthermore, the 29,030 Geo parcel documents labeled successful spent on average 25 days editing, compared to 46 days for the 29 documents that were not successful. Naturally, more complex lifecycle models and performance metrics can be defined to analyze the process in greater detail. Using our approach, we can quickly gain interesting insights into complex performance metrics without the need for programming.

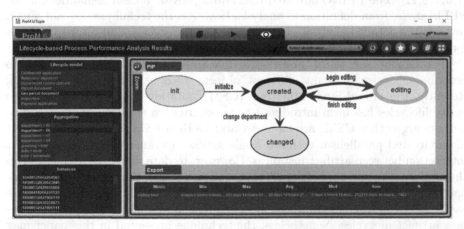

Fig. 4. Screenshot of the ProM interface showing the results of the lifecycle-based process performance analysis. On average, in 2016 and 2017, the 7463 *Geo parcel documents* handled by department 6b spent 28 days and 10 h being edited.

Table 4. Average time spent editing Geo parcel documents for different departments and years. All departments spent less time in 2017 compared to 2016.

Department	2016	2017
4e	25 days 13 h 26 min	17 days 14 h 25 min
6b	33 days 10 h 5 min	23 days 10 h 47 min
e7	33 days 10 h 5 min	25 days 13 h 26 min
d3	33 days 10 h 5 min	25 days 13 h 26 min

8 Related Work

As discussed, in general, little research has gone into the structural specification and analysis of process performance and most performance-oriented techniques use ad-hoc definitions of performance measures and metrics [11]. In [20], the authors define a specification language for performance indicators and their relations based on predicate logic. Furthermore, the technique allows for an automatic verification of the requirements of performance indicators. Although formal specification is used, the semantics of the language are not described and it is unclear how performance can be analyzed using event data. In [4,6], del-Río-Ortega et al. define a metamodel for Process Performance Indicators (PPIs) called PPINOT which allows for the definition of PPIs, and provides an automatic mapping from the metamodel to Description Logic (DL) such that DL reasoners can be used for design-time analysis. A key benefit of the PPINOT approach is that individual PPIs can be traced back to their related business process elements and that a scope can be defined that acts as a filter for the relevant element instances, much like the projection function for artifact instances in this paper. The authors further extend their work along several directions [4,5,7,8,21]. The PPINOT-related techniques provide a clear semantics for the PPI lifecycle, from definition to analysis. However, the techniques are not aimed towards the detailed analysis of business artifact lifecycles. Our approach therefore can be seen as complementary to the PPINOT-based techniques.

Throughout the years, finite state machines (FSMs) have been used to describe lifecycles of many different objects such as business artifacts [9]. Recently, the Guard-Stage-Milestone (GSM) notation for specifying business entity lifecycles has been introduced as an alternative notation in [14,15]. The authors argue that GSMs are more declarative than FSM variants and support hierarchy and parallelism within a single artifact instance. GSMs also support interaction between artifact instances. However, to date, little research exists on the applicability and analysis of GSM models in general. Conversely, ample work has been done on the analysis of transition systems. As such, in this work, we use FSMs (a subclass of transition systems) to reason about and represent business artifact lifecycles. Nonetheless, the technique presented in this paper may be extended to support GSM models and the specific advantages they bring in future work.

Recently, techniques have been proposed that automate the discovery of business artifact lifecycle models from event data. For example, [19], discovers lifecycle models by decomposing the discovery problem such that existing process discovery techniques can be used. Then, the resulting process models are converted to GSM notation. This technique is limited to the discovery of individual models, and no interaction between artifacts can be discovered. In [16] a technique is proposed that can discover interacting artifacts from ERP systems. The Composite State Machines defined in [25] combine the states of FSMs of different process perspectives and can automatically discover such lifecycle models from event data. Automatic lifecycle model discovery can be easily paired with our technique to analyze process performance using more complex lifecycle models.

9 Conclusions and Future Work

This paper introduced a novel generic approach for the structural analysis of business process performance. It uses finite state machines (FSMs) that represent the lifecycles of key conceptual process entities known as business artifacts. Using event data we compute basic performance measures defined on the states and transitions of the artifact lifecycle models. These measures serve as input to more complex, user-defined performance metrics such as durations, waiting times, and utilization rates. By using business artifacts, the approach allows for the definition and analysis of the performance of traditional concepts such as cases, activities, and resources, as well as that of domain-specific artifacts potentially relating to multiple process instances. Lifecycle models automatically discovered from event data may be used in lieu of their manually modeled counterparts in order to calculate performance measures and metrics for more complex business artifacts. Aggregate performance information for selections of artifact instances can easily be computed. The presented approach was implemented in the open source process mining tool ProM and evaluated using real-life public event data, demonstrating its capacity to define, compute, and analyze process performance characteristics for any business artifact and lifecycle model using event data. Using our approach and tool, complex performance analysis questions can quickly be answered without the need for implementation. Furthermore, this approach enables the application of our previous work on the automated analysis of contextual aspects [12] and causal factors [13] to performance characteristics defined for arbitrary business artifacts and domains.

In Sect. 3, we showed that event correlation (relating events to the correct lifecycle transitions) and data quality issues resulting in non-conformance are two interesting open challenges. Future work should investigate how event correlation can be performed without a priori information, and how non-conformance can consequently be dealt with adequately. A promising direction for the latter challenge are so-called alignments between event data and lifecycle models, which have recently gained interest in process conformance and compliance checking research [2, 17]. Using these techniques, an alignment can be found between lifecycle moves recorded in the event data and transitions in lifecycle models.

References

1. IEEE Standard for extensible event stream (XES) for achieving interoperability in event logs and event streams. IEEE Std 1849–2016, pp. 1–50, November 2016
2. Carmona, J., van Dongen, B., Solti, A., Weidlich, M.: Conformance Checking - Relating Processes and Models. Springer, Heidelberg (2018). https://doi.org/10.1007/978-3-319-99414-7
3. Cohn, D., Hull, R.: Business artifacts: a data-centric approach to modeling business operations and processes. IEEE Data Eng. Bull. **32**(3), 3–9 (2009)
4. del-Río-Ortega, A., Cabanillas, C., Resinas, M., Ruiz-Cortés, A.: PPINOT tool suite: a performance management solution for process-oriented organisations. In: Basu, S., Pautasso, C., Zhang, L., Fu, X. (eds.) ICSOC 2013. LNCS, vol. 8274, pp. 675–678. Springer, Heidelberg (2013). https://doi.org/10.1007/978-3-642-45005-1_58
5. del-Río-Ortega, A., Resinas, M., Cabanillas, C., Ruiz Cortés, A.: Defining and analysing resource-aware process performance indicators. In: Proceedings of the CAiSE 2013 Forum at the 25th International Conference on Advanced Information Systems Engineering (CAiSE), Valencia, Spain, 20th June 2013, pp. 57–64 (2013)
6. del-Río-Ortega, A., Resinas, M., Cabanillas, C., Ruiz Cortés, A.: On the definition and design-time analysis of process performance indicators. Inf. Syst. **38**(4), 470–490 (2013)
7. del-Río-Ortega, A., Resinas, M., Durán, A., Bernárdez, B., Ruiz-Cortés, A., Toro, M.: Visual PPINOT: a graphical notation for process performance indicators. Bus. Inf. Syst. Eng. 1–25 (2017). https://doi.org/10.1007/s12599-017-0483-3
8. del-Río-Ortega, A., Resinas, M., Durán, A., Ruiz Cortés, A.: Using templates and linguistic patterns to define process performance indicators. Enterp. IS **10**(2), 159–192 (2016)
9. Ebert, J., Engels, G.: Specialization of object life cycle definitions. Technical report (1997)
10. Ferreira, D.R., Gillblad, D.: Discovering process models from unlabelled event logs. In: Dayal, U., Eder, J., Koehler, J., Reijers, H.A. (eds.) BPM 2009. LNCS, vol. 5701, pp. 143–158. Springer, Heidelberg (2009). https://doi.org/10.1007/978-3-642-03848-8_11
11. González, L., Rubio, F., González, F., Velthuis, M.: Measurement in business processes: a systematic review. Bus. Process Manag. J. **16**(1), 114–134 (2010)
12. Hompes, B., Buijs, J., van der Aalst, W.: A generic framework for context-aware process performance analysis. In: On the Move to Meaningful Internet Systems: OTM 2016 Conferences - Confederated International Conferences: CoopIS, C&TC, and ODBASE 2016, Rhodes, Greece, 24–28 October 2016, Proceedings,pp. 300–317 (2016)
13. Hompes, B., Maaradji, A., La Rosa, M., Dumas, M., Buijs, J., van der Aalst, W.: Discovering causal factors explaining business process performance variation. In: Advanced Information Systems Engineering - 29th International Conference, CAiSE 2017, Essen, Germany, 12–16 June 2017, Proceedings, p. 177–192 (2017)
14. Hull, R., et al.: Business artifacts with guard-stage-milestone lifecycles: managing artifact interactions with conditions and events. In: Proceedings of the Fifth ACM International Conference on Distributed Event-Based Systems, DEBS 2011, New York, NY, USA, 11–15 July 2011, pp. 51–62 (2011)
15. Hull, R., et al.: Introducing the guard-stage-milestone approach for specifying business entity lifecycles. In: Bravetti, M., Bultan, T. (eds.) WS-FM 2010. LNCS, vol. 6551, pp. 1–24. Springer, Heidelberg (2011). https://doi.org/10.1007/978-3-642-19589-1_1

16. Lu, X., Nagelkerke, M., van de Wiel, D., Fahland, D.: Discovering interacting artifacts from ERP systems. IEEE Trans. Serv. Comput. **8**(6), 861–873 (2015)
17. Munoz-Gama, J.: Conformance Checking and Diagnosis in Process Mining: Comparing Observed and Modeled Processes. Lecture Notes in Business Information Processing, vol. 270. Springer, Heidelberg (2016). https://doi.org/10.1007/978-3-319-49451-7
18. Nigam, A., Caswell, N.: Business artifacts: an approach to operational specification. IBM Syst. J. **42**(3), 428–445 (2003)
19. Popova, V., Fahland, D., Dumas, M.: Artifact lifecycle discovery. Int. J. Coop. Inf. Syst. **24**(1), 1550001 (2015)
20. Popova, V., Treur, J.: A specification language for organisational performance indicators. Appl. Intell. **27**(3), 291–301 (2007)
21. van der Aa, H., Leopold, H., del-Río-Ortega, A., Resinas, M., Reijers, H.: Transforming unstructured natural language descriptions into measurable process performance indicators using hidden Markov models. Inf. Syst. **71**, 27–39 (2017)
22. van der Aalst, W.: Process Mining: Data Science in Action. Springer, Heidelberg (2016). https://doi.org/10.1007/978-3-662-49851-4
23. van der Aalst, W., Rubin, V., Verbeek, H., van Dongen, B., Kindler, E., Günther, C.: Process mining: a two-step approach to balance between underfitting and overfitting. Softw. Syst. Model. **9**(1), 87–111 (2010)
24. van Dongen, B., Borchert, F.: BPI Challenge 2018. Eindhoven University of Technology. Dataset (2018). https://doi.org/10.4121/uuid:3301445f-95e8-4ff0-98a4-901f1f204972
25. van Eck, M.L., Sidorova, N., van der Aalst, W.M.P.: Discovering and exploring state-based models for multi-perspective processes. In: La Rosa, M., Loos, P., Pastor, O. (eds.) BPM 2016. LNCS, vol. 9850, pp. 142–157. Springer, Cham (2016). https://doi.org/10.1007/978-3-319-45348-4_9

A Relevance-Based Data Exploration Approach to Assist Operators in Anomaly Detection

Ada Bagozi, Devis Bianchini$^{(\boxtimes)}$, Valeria De Antonellis, and Alessandro Marini

Department of Information Engineering, University of Brescia,
Via Branze, 38, 25123 Brescia, Italy
devis.bianchini@unibs.it

Abstract. Data is emerging as a new industrial asset in the factory of the future, to implement advanced functions like state detection, health assessment, as well as manufacturing servitization. In this paper, we foster Industry 4.0 data exploration by relying on a relevance evaluation approach that is: (i) flexible, to detect relevant data according to different analysis requirements; (ii) context-aware, since relevant data is discovered also considering specific working conditions of the monitored machines; (iii) operator-centered, thus enabling operators to visualise unexpected working states without being overwhelmed by the huge volume and velocity of collected data. We demonstrate the feasibility of our approach with the implementation of an anomaly detection service in the Smart Factory, where the attention of operators is focused on relevant data corresponding to unusual working conditions, and data of interest is properly visualised on operator's cockpit according to adaptive sampling techniques based on the relevance of collected data.

Keywords: Data exploration · Data relevance · Data summarisation
Clustering · Big data · Anomaly detection · Industry 4.0

1 Introduction

Big data management is an ever-growing research topic given the emerging data-intensive applications of the Smart Factory. In order to improve operation process performance, monitoring, control and health assessment [11], big data streams, generated by embedded systems (RFID technology, sensors, mobile and wearable devices) are collected and processed in the cyber space (edge and cloud computing). In this context, human operators still play a crucial role to recognise critical situations that have not been encountered before, based on their long-term experience, but they must be supported in the identification of relevant data without being overwhelmed by the huge amount of information. In the so-called "Human in the Loop Cyber Physical Systems (CPS)", human actions and machine actuations go hand-by-hand and can often complement each other [14].

© Springer Nature Switzerland AG 2018
H. Panetto et al. (Eds.): OTM 2018 Conferences, LNCS 11229, pp. 354–371, 2018.
https://doi.org/10.1007/978-3-030-02610-3_20

As an example of CPS, let's consider a multi-spindle machine, designed to perform flexible manufacturing tasks. The machine is equipped with multiple spindles (e.g., from three to five), that work independently each other on the raw material. Spindles use different tools (that are selected according to the instructions specified in the part program executed by the numerical control of the machine) in distinct steps of the manufacturing process. Spindle precision, working performance, as well as minimisation of tool breaks and machine downtimes are critical factors in these kinds of systems. Therefore, monitoring activities might be very complex, checking several kinds of events in multiple conditions in order to identify anomalies. Anomalies can be discovered when incoming data goes beyond or below an expected range or with the occurrence of unexpected data patterns [13]. Traditional anomaly detection solutions (e.g., [8,9]) apply machine learning techniques to train proper models using historical data and use them to predict the future behaviour of monitored systems. The occurrence of unknown working states, never used before to train machine learning models, can be recognised and managed by operators according to their expertise. To this aim, operators must be supported in the effective exploration of data streams. For example, in the multi-spindle machine the 'spindle rolling friction torque increase' and the 'tool wear' should be promptly detected and avoided. The former one may happen for lack of lubrication or other mechanical wears like bearings damage. The latter one may lead to long downtimes as well and is managed through tool usage optimisation in order to balance the trade-off between the tools wear and the risk of tool breaking during manufacturing. Several working conditions must be considered, with a high likelihood of finding behaviours never met before. On the other hand, the increasing importance of human-machine interactions [7] calls for new models and techniques to organise collected data according to different exploration perspectives and to attract the attention of operators on relevant data only.

In this paper, we propose a novel approach where multi-dimensional data modelling, data summarisation and relevance evaluation techniques are proposed to implement big data exploration and anomaly detection based on data streams. In particular: (i) collected data are organised according to different dimensions, in order to meet distinct system monitoring requirements; (ii) a clustering algorithm for big data streams is applied to provide a comprehensive view over collected data and to enable data exploration using a reduced amount of information; (iii) data relevance techniques focus the attention of operators on relevant data only, thus increasing the effectiveness and efficiency of the data exploration process. The proposed model and techniques have been tested in the Smart Factory context for anomaly detection. Nevertheless, they have to be intended as a general approach for Big Data exploration. Proposed model and techniques aim at preparing data to address "Human in the Loop" issues.

The proposed approach relies on the IDEAaS (Interactive Data Exploration As-a-Service) framework [3]. Specifically, we extend the research presented in [3] with the following novel contributions:

(i) we introduce a mechanism to adapt data monitoring (e.g., for anomaly detection) based on the relevance evaluation;
(ii) we address relevance-driven adaptive sampling for visualisation purposes on the operator's cockpit;
(iii) we expand the experimental results, performing additional experiments to test effectiveness and response times of data relevance evaluation.

The paper is organised as follows: in Sect. 2 we introduce the research challenges; in Sect. 3 we provide an overview of the relevance-based data exploration approach and of the IDEAaS framework; Sect. 4 contains the description of the multi-dimensional model on which the approach relies; in Sects. 5 and 6 relevance-based techniques and adaptive data visualisation are described; Sect. 7 presents experimental evaluation; in Sect. 8 related work are discussed; finally, Sect. 9 closes the paper.

2   Research Challenges

To support big data exploration in dynamic contexts of interconnected systems, such as the considered application scenario, several research challenges raise and must be addressed.

Flexibility. Exploration depends on different analysis requirements. For example, in the considered application scenario the 'spindle rolling friction torque increase' and 'tool wear' events must be monitored to manage maintenance activities and purchase of new tools. Since many unknown situations may occur, due to the complexity of monitored system, analysts and operators must be supported in the identification of possible invisible problems [12]. Multi-dimensional data modelling represents a powerful mean to enable organisation of data according to different perspectives, in turn related to distinct observed problems and requirements. Data modelling according to "facets" or "dimensions", either flat or hierarchically organised, has been recognised as a factor for easing data exploration, since it offers the opportunity of performing flexible aggregations of data [18]. Moreover, a definition of *relevance* is required to attract the operator's attention on relevant data only, corresponding to an unexpected status. Also the concept of *unexpected status* must be defined as well.

Context-Awareness. The detection of relevant data may also depend on the specific working conditions of the observed system. For example, the machine performance may change with respect to the specific part program that is being executed. In different conditions, the range of tolerance for a given measure may be different. Relevance evaluation algorithms and visualisation tools must reflect this difference.

Operator-Centered Visualisation. Operators must be able to visualise unexpected working states and relevant data without being overwhelmed by the huge volume and velocity of collected data. The ability of providing a compact view over data is strongly required. Data summarisation and sampling techniques are recommended, where data is processed and observed in an aggregated way, instead of monitoring each single record [1].

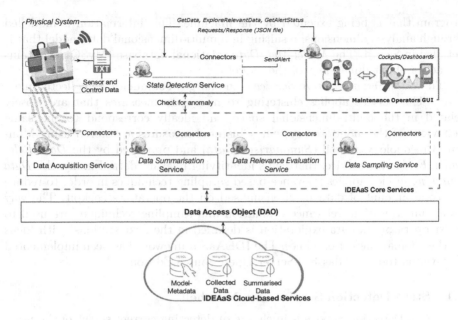

Fig. 1. The IDEAaS framework architecture.

3   Approach Overview

Figure 1 presents the IDEAaS framework modular architecture. The framework is implemented according to a service-oriented architecture, where *Core Services* implement data acquisition, data summarisation, sampling and relevance evaluation, and extensible services, built upon core ones, implement data-intensive functionalities for different application domains, such as the Industry 4.0 one. Among these data-intensive functionalities, in this paper we describe the *State Detection Service*.

As shown in the figure, data coming from the physical system, collected through sensors and IoT technologies, is sent to the *Data Acquisition Service* to be stored in the cyber space. Data is collected according to a set of features. Examples of features for the considered multi-spindle machines are spindle velocity (nm/min), the absorbed electric current (Amp) on X, Y and Z axes, the spindle rotation speed (rpm) and the percentage of absorbed power (%). We refer to measures as the values collected for each feature, associated with a given timestamp. Let's denote with $F = \{F_1, F_2 \ldots F_n\}$ the overall set of features. We formally define a measure for the feature F_i as a scalar value $X_i(t)$, expressed in terms of the unit of measure u_{F_i}, taken at the timestamp t. The Data Acquisition Service operates in order to minimise time spent for data acquisition. Specifically, measures are first saved as JSON documents within a NoSQL database (*Collected Data*), using MongoDB technology. Measures are associated with other information about the physical system and the working conditions in which measures have been collected, for example, the tool used for manufacturing or the part program that is being executed by the machine. This information is modelled through analysis dimensions, resulting in a multi-dimensional data model that is detailed in Sect. 4. The *Model-MetaData* relational database (MySQL) contains metadata about dimensions of the model.

The *Data Summarisation Service* is in charge of summarising collected measures. This service applies clustering to aggregate measures that are closely related in the multi-dimensional space and ideally correspond to the same behaviour of the monitored system. Clustered measures are stored using MongoDB technology as well (*Summarised Data*) and processed by the *Data Relevance Evaluation* service, that helps identifying relevant data. Finally, the *Data Sampling* service applies relevance-based sampling techniques in order to reduce the total amount of data to be visualised on the operator's cockpit. The way data summarisation, relevance evaluation and sampling techniques are used to assist operators in data exploration is detailed in the next sections, with focus on the anomaly detection issues. The IDEAaS framework has been implemented in Java, on top of a Glassfish Server Open Source Edition 4.

3.1   State Detection Service in a Nutshell

The State Detection Service is in charge of detecting current status of the monitored system and managing the interaction with visualisation tools, such as cockpits and dashboards, on which operators can explore data.

We consider four different values for the *status* of the monitored system, (a) ok, when the system works normally; (b) changed, when the system behaviour changed with respect to the normal one, but no anomalies have been detected yet; (c) warning, when the system works in anomalous conditions that may lead to breakdown or damage; (d) error, when the system works in unacceptable conditions or does not operate. The changed and warning status are used to perform an early detection of a potential deviation towards an error status. The warning or error status occurs when one or more features exceed a given bound. Besides defining *features* bounds, we introduced the notion of *contextual bounds*. A contextual bound represents the limit of a feature within specific conditions (e.g., determined by the tool used and/or the part program that is being executed) in which the feature is measured. The rationale is that, in specific conditions, a feature should assume values within a specific range, that might be different from the overall physical limits for the same feature disregarding the working conditions. If the measure overtakes warning bounds, but not the error ones, then the feature status is warning, otherwise the feature is in the error status. Features (contextual) bounds are fixed by domain experts, for instance through to the FMEA/FMECA analysis. The operators can monitor state changes in order to revise features and contextual bounds for specific working conditions.

The *State Detection Service* includes data relevance evaluation techniques to attract the operator's attention on every state change. In fact, the State Detection Service provides the following methods, as remarked in Fig. 1:

- **SendAlert** sends asynchronous notifications about detected changes of the working status in the monitored system, based on Summarised Data; to this

aim, this method relies on the Data Relevance Evaluation Service and adapts the anomaly detection frequency according to the data relevance, as detailed in Sect. 5.2;
- `GetAlertStatus` sends a summary report on the current status of the monitored system; this service is required to synchronise visualisation tools to the current status of the physical system, when external cockpits and dashboards get connected with the State Detection Service.

Data visualisation must take into account the high volume of information to be visualised and facilitate the interaction of operators with the Graphical User Interface (GUI) of the visualisation tool. To this purpose, the following additional methods are exposed by the State Detection Service:

- `ExploreRelevantData` sends relevant data, by relying on the Data Relevance Evaluation service; data is transferred as clusters of aggregated measures (as shown in Sect. 4) and visualised according to the multi-dimensional model described in the next section; this method has been designed to support operators to focus on relevant data only, without specifying any data search and filtering criteria, since operators do not have any a-priori knowledge about which data can be considered as relevant;
- `GetData` sends data within a given time interval and/or for specific search and filtering criteria expressed on dimensions of the multi-dimensional model; this functionality can be used, for example, once relevant summarised data has been identified; since sent data may reach a massive size, sampling techniques are applied; hence, sampling takes into account the relevance of data that is being transmitted, by adapting the sampling ratio to the data relevance, as described in Sect. 6.

4   Clustering Based Multi-dimensional Model

In the multi-dimensional model used within the IDEAaS framework, *measures* are organised through the feature spaces and the domain-specific dimensions.

A feature space conceptually represents a set of related features, that are jointly measured to observe a physical phenomenon. In the example domain, the set composed of spindle power absorption and rpm features is a feature space used to monitor spindle rolling friction torque increase. In fact, spindle rolling friction torque increase may be identified when the rpm value decreases and, at the same time, the power absorption increases. Therefore, these two features must be monitored jointly. Given a feature space $FS_j = \{F_1, \ldots F_h\}$, we denote with $\mathbf{X}_j(t)$ a record of measures $\langle X_1(t), \ldots X_h(t) \rangle$ for the features in FS_j, synchronised with respect to the timestamp t.

Domain-specific dimensions organise records according to different "facets", such as the observed machine, the tool used during manufacturing, the part program that is being executed by the numerical control of the monitored system. Domain-specific dimensions can be organised in hierarchies: tools can be aggregated into tool types, while monitored physical components (e.g., spindles) can be aggregated into the machines they belong to, in turn organised into plants and enterprises. Therefore, a record $\mathbf{X}_j(t)$ is always associated with: (i) the timestamp at which measures in the record have been collected; (ii) the monitored feature space FS_j; (iii) the values of domain-specific dimensions. Once the feature space and domain-specific dimensions have been fixed, the stream of records over time can be used to monitor the evolution of the feature space for the considered dimensions.

Data summarisation is used here to provide an overall view over a set of records using a reduced amount of information and allows to depict the behaviour of the system better than single records, that might be affected by noise and false outliers. In our approach, data summarisation is based on clustering-based techniques. The application of the clustering algorithm to the stream of records incrementally produces a set of *syntheses* $S = \{s_1, s_2, \ldots, s_n\}$, providing a lossless representation of records.

A synthesis conceptually represents a working behaviour of the monitored system, corresponding to a set of records, with close values for each feature. Please refer to [4] for more details about the incremental clustering algorithm. Formally, we define a synthesis of records as:

$$s_i = \langle id_i, N_i, \mathbf{LS}_i, SS_i, \mathbf{X}_i^0, R_i \rangle \tag{1}$$

where: (i) id_i is the unique identifier of s_i; (ii) N_i is the number of records included into the synthesis; (iii) $\mathbf{LS}_i$ is a vector representing the linear sum of measures in s_i; (iv) SS_i is the quadratic sum of points in s_i for each feature; (v) $\mathbf{X}_i^0$ represents the centroid of the synthesis in the feature space; (vi) R_i is the radius of the synthesis.

The clustering algorithm at a given time t produces a set of syntheses $S(t)$ starting from records collected from timestamp $t - \Delta t$ to timestamp t and built on top of the previous set of syntheses $S(t - \Delta t)$ for a given feature space FS_j and domain-specific dimensions. Therefore, we formally define the multi-dimensional model as a set $\mathcal{V}$ of nodes within an hypercube structure, where time, feature spaces and domain-specific dimensions represent hypercube axes and each node $v \in \mathcal{V}$ is described as

$$v = \langle S(t), FS_j, d_1, d_2, \ldots d_p \rangle \tag{2}$$

where $S(t)$ is the set of syntheses at time t, for the feature space FS_j and the values $d_1, d_2, \ldots d_p$ of domain-specific dimensions $\mathcal{D}_1, \ldots \mathcal{D}_p$.

For example, an arbitrary node $v_A = \langle S(t_1), FS_1, m_1, c_2, u_2, pp_a \rangle$, represents the set of syntheses obtained by summarising records collected from time $t_1 - \Delta t$ to t_1 for machine m_1 (spindle c_2), while using tool u_2 and executing part program pp_a, considering features in the feature space FS_1. Data exploration is performed over dimensions and is guided by data relevance evaluation techniques as described in the following.

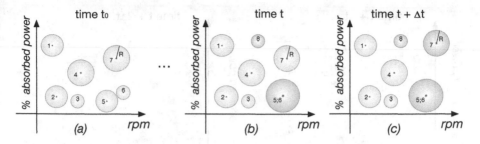

Fig. 2. Evolution of summarised data (syntheses) over time. Feature space and domain-specific dimensions are fixed and not shown here.

5   Relevance-Based Data Exploration

We define data relevance as the *distance* of the physical system behaviour from an *expected status*. This status corresponds to the normal working conditions of the system and is represented by the set of syntheses $\hat{S}(t_0)$. $\hat{S}(t_0)$ can be tagged by the domain experts while observing the monitored system when operates normally. Data relevance at time t is based on the computation of *distance* between the set of syntheses $S(t) = \{s_1, s_2, \ldots, s_n\}$ and $\hat{S}(t_0) = \{\hat{s}_1, \hat{s}_2, \ldots, \hat{s}_m\}$, where n and m represent the number of syntheses in $S(t)$ and $\hat{S}(t_0)$, respectively, and n and m do not necessarily coincide. We denoted this distance with $\Delta(S(t), \hat{S}(t_0))$, computed as:

$$\Delta(\hat{S}(t_0), S(t)) = \frac{\sum_{\hat{s}_i \in \hat{S}(t_0)} d(\hat{s}_i, S(t)) + \sum_{s_j \in S} d(\hat{S}(t_0), s_j)}{m + n} \qquad (3)$$

where $d(\hat{s}_i, S(t)) = min_{j=1,\ldots n} d_s(\hat{s}_i, s_j)$ is the minimum distance between $\hat{s}_i \in \hat{S}(t_0)$ and a synthesis in $S(t)$. Similarly, $d(\hat{S}(t_0), s_j) = min_{i=1,\ldots m} d_s(\hat{s}_i, s_j)$. To compute the distance between two syntheses $d_s(\hat{s}_i, s_j)$, we combined different factors: (i) the euclidean distance between syntheses centroids $d_{X_0}(\hat{s}_i, s_j)$, to verify if s_j moved with respect to $\hat{s}_i$ and (ii) the difference between syntheses radii $d_R(\hat{s}_i, s_j)$, to verify if there has been an expansion or a contraction of synthesis s_j with respect to $\hat{s}_i$. Formally:

$$d_s(\hat{s}_i, s_j) = \alpha d_{X_0}(\hat{s}_i, s_j) + \beta d_R(\hat{s}_i, s_j) \qquad (4)$$

where α, $\beta \in [0, 1]$ are weights such that $\alpha + \beta = 1$, used to balance the impact of terms in Eq. (4). Weights α and β can be set by operators according to their domain knowledge. For preliminary experiments we equally weighted the two terms of Eq. (4), that is, $\alpha = \beta = \frac{1}{2}$. Future efforts will be devoted to automatically identify the best values to set-up α and β.

Roughly speaking, the relevance techniques allow to identify what are the syntheses that changed over time (namely, appeared, have been merged or removed) for a specific feature space and given values of domain-specific dimensions. Let's denote with $\overline{S}(t) = \{\overline{s}_1, \overline{s}_2, \ldots, \overline{s}_k\}$ such syntheses at time t, where

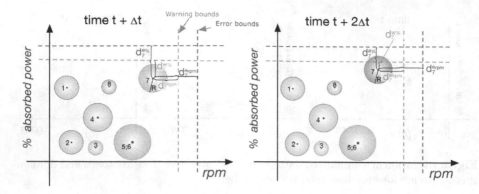

Fig. 3. Anomaly detection through data exploration based on relevance evaluation: data relevance techniques detect changes in syntheses set due to spindle rolling friction torque increase, that may be identified when the rpm value decreases and, at the same time, the power absorption increases.

$k \leq n$ and n is the number of syntheses $\in S(t)$. These syntheses are considered as relevant and will be proposed to the operators to start the exploration. For example, let's consider Fig. 2. Figure 2(a) corresponds to the normal working conditions, as labelled by domain experts according to their expertise, therefore $\hat{S}(t_0) = \{\hat{s}_1, \hat{s}_2, \ldots, \hat{s}_7\}$. At time t, shown in Fig. 2(b), a new synthesis 8 is identified while syntheses 5 and 6 have been merged, that is, $S(t) = \{s_1, s_2, \ldots, s_{[5,6]}, s_7, s_8\}$ and $\overline{S}(t) = \{\overline{s}_{[5,6]}, \overline{s}_8\}$. Finally, in Fig. 2(c) the synthesis 7 moved and $\overline{S}(t + \Delta t) = \{\overline{s}_{[5,6]}, \overline{s}_7, \overline{s}_8\}$.

5.1   Relevance-Based Data Exploration for Anomaly Detection

For anomaly detection purposes, for each synthesis $\overline{s}_c \in \overline{S}(t)$, the distance of synthesis centroid from the warning and error bounds is computed. In the following, we will consider features bounds, but the same considerations hold for the contextual ones. We denote with $\mathbf{d}_c^w$ the record of distances between the centroid of the synthesis $\overline{s}_c$ and the warning bounds and with $\mathbf{d}_c^e$ the record of distances between the centroid of $\overline{s}_c$ and the error bounds. The State Detection Service uses $\mathbf{d}_c^w$ and $\mathbf{d}_c^e$ to perform anomaly detection, by distinguishing among ok, **warning** and **error** status. Both $\mathbf{d}_c^w$ and $\mathbf{d}_c^e$ are records having as components the distance for each feature. For example, $d_7^{e\%}$ represents the distance of the centroid of the synthesis 7 from the error bound of the percentage of absorbed power (see Fig. 3). Each relevant synthesis in $\overline{s}_c$ is described as:

$$\overline{s}_c = \langle id_c, N_c, \mathbf{LS}_c, SS_c, \mathbf{X}_c^0, R_c, \mathbf{d}_c^w, \mathbf{d}_c^e \rangle \tag{5}$$

Every Δt seconds, when the syntheses set $S(t)$ is updated, data is analysed to check for anomalies.

For example, in Fig. 3 synthesis 7 moved over time getting closer to the boundaries. Note that distance also helps to detect *potential* state changes. In fact, at time $t + \Delta t$ synthesis 7 still remains inside the wealth zone (ok status), but its movement is detected through relevance-based techniques. Therefore, synthesis 7 is recognised as relevant and monitored to promptly detect potential warning or error status occurrences. After Δt seconds, synthesis 7 moved again and crosses the warning bound of the percentage of absorbed power feature, causing a warning alert. The warning status is assigned to the feature and is propagated to the feature space and over the hierarchy of monitored system according to the following rules: (i) the status of a feature space corresponds to the worst one among its features; (ii) similarly, the status of a physical component (e.g., the spindle) corresponds to the worst one among monitored feature spaces on that component and the status of composite systems (e.g., the multi-spindle machine) corresponds to the worst one among its components. Figure 3 also shows that it is possible to identify the feature with respect to the warning or error bound that has been exceeded (e.g., among rpm and percentage of absorbed power). When a synthesis moved closer to bounds, the IDEAaS framework reacts by reducing the interval time Δt to check data for anomalies as described in the following.

5.2   Adaptive Relevance Evaluation

The State Detection Service checks the system status by relying on Data Relevance Evaluation Service and after the application of the Data Summarisation Service. If the relevance evaluation detects changes in data compared to the expected working behaviour, the State Detection Service identifies the new status of the system. If a warning or error status is detected, the State Detection Service notifies an alert message to the cockpit with the new status, using the SendAlert method. This check is performed every Δt seconds.

Therefore, setup of Δt parameter influences the performances of the system. Small Δt values increase the promptness in identifying relevant syntheses, in order to attract the attention of the operators on them. On the other hand, response times of data acquisition and clustering may not be able to face small Δt values (see experimental evaluation in Sect. 7). The rationale behind our approach is to change Δt as syntheses get closer to warning and error bounds, since they correspond to potentially critical situations that must be monitored at finer granularity.

To this aim, Δt value is changed according to the distance of relevant synthesis $\bar{s}_c \in \overline{S}(t)$ that is closer to warning and error bounds. We denote with $d_c^{w\_min}$ (resp., $d_c^{e\_min}$) the component of $\mathbf{d}_c^w$ (resp., $\mathbf{d}_c^e$) that presents the minimum distance from the warning bounds (resp., the error bounds). The interval time Δt is updated as follows:

- if $\frac{d_c^{w\_min}}{R} > 1$, the feature status is set to ok (see for example synthesis 7 in Fig. 3 at time $t + \Delta t$), Δt is set to a default value defined by the domain expert according to his/her knowledge about the monitored system;
- if $\frac{d_c^{w\_min}}{R} <= 1$ and $\frac{d_c^{e\_min}}{R} > 1$ the synthesis centroid is between warning bounds and error bounds (see for example synthesis 7 in Fig. 3 at time $t+2\Delta t$),

the feature status is set to `warning`, Δt is reduced as $\Delta t = \Delta t(\frac{d_c^{e\_min}}{R} - 1)$ until $\Delta t = $ minimum value supported by the framework (see experimental evaluation in Sect. 7);

- if $\frac{d_c^{e\_min}}{R} <= 1$ the synthesis centroid is beyond error bounds, the feature status is set to `error`, Δt is set to the minimum supported value (that is, checks are made as more frequently as possible).

6 Adaptive Sampling for Data Visualisation

An effective visualisation of an unexpected working status and related data on operator's cockpit must consider the impact of data volume and velocity, to avoid operators be overwhelmed by the huge amount of data. To this purpose, data sampling techniques are usually applied, where sampling is performed taking into account the size and capacity of the cockpit interface, independently of the specific conditions which visualised data refers to. In our approach, clustering and relevance evaluation techniques are used to implement adaptive sampling for data visualisation. To this purpose, `ExploreRelevantData` and `GetData` methods of the State Detection Service have been implemented.

Request for Relevant Data. When the operator at time t requests for relevant data, the method `ExploreRelevantData` is invoked. This method relies on relevance evaluation techniques to recognise the most recent relevant syntheses set $\overline{S}(t_i)$, processed at time t_i ($t_i <= t$). Each synthesis $\overline{s}_c \in \overline{S}(t_i)$ is marked with the corresponding status and with additional information about whether the synthesis moved, changed (expansion or contraction) or has been removed. All syntheses in $\overline{S}(t_i)$ recognised as anomalous are visualised as shown in Fig. 3.

Exploration of Relevant Syntheses. Once relevant syntheses have been identified, the operator may request to explore in detail records that have been clustered within relevant syntheses. These records are returned by invoking the `GetData` method. Records may correspond to a time-window h, and for specific values of analysis dimensions, the amount of extracted data may be really large and difficult to visualise. In order to enable data visualisation, a classical adaptive sampling technique has been designed. Nevertheless, in our approach sampling frequency varies according to data relevance evaluation. Considering max_n as the maximum number of data supported by the visualisation tool and n as the number of data extracted from the database, when $n >> max_n$ a sampling technique is applied selecting only max_n data among the n data ready for visualisation. Sampling rate is adaptively modified by a factor that depends on the detected status (`warning` or `error`) within the time-window. When data is not recognised as critical, the sampling rate is set to the minimum value. In the case all data in the interval is not relevant, or is equally relevant, the sampling frequency is set to $\frac{max_n}{t-h}$. This strategy facilitates the cooperation between operators who acts remotely on powerful visualisation interfaces and on-site operators, who may need data visualisation on less powerful HMI embedded in or close to the monitored machine, by setting different values of max_n.

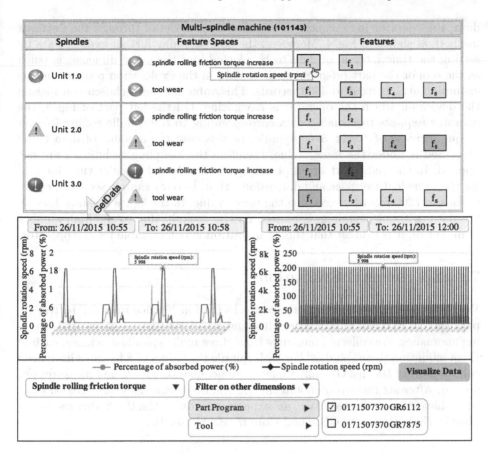

Fig. 4. Visualisation of relevant data on operator's cockpit in the anomaly detection application scenario (`GetData` method).

Figure 4 shows an implementation of remote visualisation cockpit. The cockpit guides data exploration through analysis dimensions in the considered domain, therefore it first considers the monitored system, along with the relevant feature spaces. Figure 4 shows an overview of the data of the multi-spindle machine with ID 101143 and its status. In the overview, the operator can visualise the status of the three spindles of multi-spindle machine, denoted with "Unit 1.0", "Unit 2.0" and "Unit 3.0". Indeed spindle "Unit 1.0" is working correctly with respect to all the observed feature spaces, while spindle "Unit 2.0" is in warning status. In particular, syntheses calculated for features "f4" and "f5" are detected as relevant and associated to the warning status. Therefore, the warning status is propagated to the "tool wear" feature space as well. Finally, spindle "Unit 3.0" is in error status. In fact, even if the "tool wear" warning status has been detected, a more critical status is identified for feature space "spindle rolling friction torque increase". Starting from relevant data, the operator may request to visualise data in detail through the `GetData` method, as shown in Fig. 4. Moreover, the operator may further explore data by setting the time interval of data to be plotted and the other dimensions (such as the tool or the part program) to filter data in the exploration process. In this example max_n is fixed to 3600 records. This value has been chosen considering the device on which the operator is navigating. On the left part of Fig. 4, the operator requests to visualise data corresponding to the spindle rolling friction torque increase of "`Unit 3.0`" spindle. In this case the amount of data to be visualised is under the max_n value, therefore the sampling techniques are not applied. In the right part of Fig. 4 the operator selected a wider time interval for the same feature space and dimensions, that, in our example scenario, corresponds to 7200 records, exceeding the max_n value. In the figure is shown how all the data, without sampling, is plotted on the cockpit: due to the high number of measures, it is evident that this visualisation is not valuable for the operator.

7   Experimental Evaluation

We performed experiments on the State Detection Service in order to test its performance in terms of processing time and its effectiveness in promptly detecting anomalies. We collected measures from three multi-spindle machines, each of them mounting three spindles. For each spindle the values of 8 features have been collected every 500 ms. Globally we faced an acquisition rate of 144 measures per second. After six months of monitoring on the three machines 630,720,000 measures have been collected. We run experiments on a MacBook Pro mounting MacOS High Sierra, 2.8 GHz Intel Core i7, RAM 16 GB.

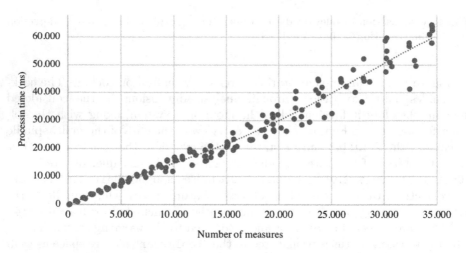

Fig. 5. Response times of the State Detection Service with respect to the number of processed measures.

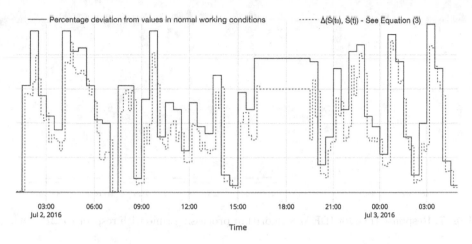

Fig. 6. Correlation between the value of the percentage deviation from the values of rpm and absorbed power features in normal working conditions (black line) and the value of $\Delta(\hat{S}(t_0), S(t))$ computed according to Eq. (3) (dashed red line) (Color figure online).

Figure 5 plots response times with respect to the number of analysed measures. As evident in the figure, response times proportionally (but not exponentially) increase with the number of processed measures. As shown in Fig. 5 our State Detection Service can process 35000 measures in 60 s on average, corresponding to $\sim$583 measures per second. Therefore, our State Detection Service can successfully cope with the acquisition rate.

To test effectiveness of the service to detect anomalies, we artificially introduced a percentage of values for rpm and absorbed power features with respect to their value in normal working conditions. Further evaluation in an actual production environment with real faults is being performed. Figure 6 shows how our relevance evaluation techniques promptly react to the introduced variations. For this experiment, we set the weights $\alpha = \beta = \frac{1}{2}$ in Eq. (4).

In order to quantify the correlation between the two curves in Fig. 6, we used the Pearson Correlation Coefficient (PCC) $\in [-1, +1]$. In the experiment, the value of PCC is higher than 0.85, that represents a strong correlation.

Figure 7 shows the average time required by the IDEAaS techniques to process a single record for different Δt values. In figure is shown how lower Δt values require more time to process data. In fact, every time clustering is applied, some initializations have to be performed (e.g., opening/closing connection to database, access to the set of syntheses previously computed). Therefore, lower Δt values lead to more frequent initializations. On the other hand, higher Δt values decrease the promptness in identifying anomalous situations, as shown in Fig. 6.

As a final remark, for what concerns the efficacy of the cockpit to support domain experts during data exploration, sampling techniques offer doubtless

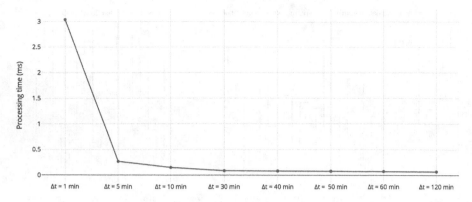

Fig. 7. Response time for IDEAaS algorithm processing time with respect to Δt value.

advantages to ease exploration of data through the proposed implementation of visualisation cockpit. It is straightforward that visualising all the data, without adaptive sampling techniques, is not valuable for the operators and will prevent them to easy inspect and identify incoming anomalies.

8   Related Work

The IDEAaS approach we described in this paper can be classified among approaches that have been proposed to address anomaly detection in presence of big data streams (please refer to [16] for a comprehensive survey). These approaches differ from those based on static data, since all the observations are not available at once and measures are collected and processed incrementally. Moreover, the IDEAaS framework also differs from solutions for anomaly detection in presence of evolving graphs [10,15], that are characterized by causal/non-causal relationships between measurements.

Among the approaches for anomaly detection on evolving data, the authors in [16] focused on unsupervised proposals, since supervised and semi-supervised scenarios are rare to happen in real-world applications, due to the lack of label information regarding the anomalies that could be detected in collected observations. Unsupervised approaches can be in turn classified into statistical-based, nearest neighbors-based and clustering-based. Statistical-based approaches usually require a priori knowledge about the underlying distribution of the measures, that is almost always unavailable when data is collected incrementally. In [8] an approach based on in-memory big data processing is described. A preparation phase is used to generate a model for the "usual state" of the system, by applying machine learning (pre-training) on stored data. An operation phase compares real-time incoming data with the "usual state" to identify anomalies. Similarly, in [9] machine learning is used to train data collected during regular execution of the manufacturing process in order to learn a probabilistic "normal model".

Authors in [2] applies Hierarchical Temporal Memory (HTM) to anomaly detection, by performing two post-processing steps over the output of HTM system: (i) computing the prediction error; (ii) computing the anomaly likelihood.

Nearest neighbors-based approaches rely on the assumption that a measure can be considered as an anomaly if its distance from a significant portion of other measures is greater than a given threshold [5, 19]. In clustering-based approaches, anomalies are discovered either: (a) since they are assumed to fall into clusters with small number of data points or low density; (b) based on their distance from nearest clusters centroids. The approach in [17] operates in two steps: (i) learning of the normal behaviour of the system (based on past data), using a clustering technique (K-means algorithm); (ii) detecting at real-time an anomalous behaviour when new data does not belong to previously detected clusters. The approach in [6] builds a cluster model using Gaussian clustering, that is updated as incoming data arrives. Clustering is performed over a time window. As a new data arrives, the algorithm tries to assign it to an existing cluster. If this is not possible, the evaluation on new data is suspended. When the time window expires, a batch clustering algorithm (e.g. DBScan) is performed, in order to check if suspended data is an anomaly or can be recognized as a new cluster.

Although our approach is cluster-based, it is focused on the evolution of summarised data over time in order to detect anomalies. Indeed, we rely on summarisation techniques as a basis on which to apply relevance evaluation. Moreover, exploration is performed over the multi-dimensional model. This distinguishes the IDEAaS framework from the approaches described in [16] and from traditional Complex Event Processing (CEP) approaches, that are mainly based on pre-defined queries and event detection rules.

9   Conclusions and Future Work

In this paper, we proposed a general-purpose framework that relies on relevant-based data exploration to support domain experts in the inspection and identification of critical situations, out from the large amount of available measure taken from a monitored system. In particular, the framework relies on the combined use of different techniques: (i) an incremental clustering algorithm, to provide summarised representation of collected data; (ii) data relevance evaluation techniques, to attract the experts' attention on relevant data only; (iii) a multi-dimensional organisation of summarised data and adaptive sampling, to enable effective visualisation of data for operators.

The proposed framework has been tested in the Smart Factory context for anomaly detection. Nevertheless, it must be intended as a general approach for Big Data exploration. In fact, the framework can be generalised by defining the dimensions of the multi-dimensional model for different case studies and domains. Summarisation and data relevance evaluation techniques are designed to be applied in any domain that is based on numeric measures collected from a monitored system.

Although preliminary experiments are promising, future development will be focused on further improving the approach using technologies for streaming and parallel processing, such as Spark/Storm. Moreover, the State Detection Service, on which we focused to test the relevance-based data exploration, will be enhanced by introducing pattern recognition techniques to learn from the syntheses evolution. Further usability studies are being performed on the operator's cockpit. This would in principle enable the implementation of health assessment strategies, on top of the ecosystem of services and techniques described in this paper.

References

1. Agrawal, R., Kadadi, A., Dai, X., Andres, F.: Challenges and opportunities with big data visualization. In: Proceedings of the 7th International Conference on Management of Computational and Collective intElligence in Digital EcoSystems (MEDES), pp. 169–173 (2015)
2. Ahmad, S., Lavin, A., Purdy, S., Agha, Z.: Unsupervised real-time anomaly detection for streaming data. Neurocomputing **262**, 134–147 (2017)
3. Bagozi, A., Bianchini, D., De Antonellis, V., Marini, A., Ragazzi, D.: Big data summarisation and relevance evaluation for anomaly detection in cyber physical systems. In: Panetto, H. (ed.) OTM 2017. OTM 2017 Conferences, vol. 10573, pp. 429–447. Springer, Cham (2017). https://doi.org/10.1007/978-3-319-69462-7_28
4. Bagozi, A., Bianchini, D., De Antonellis, V., Marini, A., Ragazzi, D.: Summarisation and relevance evaluation techniques for big data exploration: the smart factory case study. In: Dubois, E., Pohl, K. (eds.) CAiSE 2017. LNCS, vol. 10253, pp. 264–279. Springer, Cham (2017). https://doi.org/10.1007/978-3-319-59536-8_17
5. Cai, L., Thornhill, N.F., Kuenzel, S., Pal, B.C.: Real-time detection of power system disturbances based on k -nearest neighbor analysis. IEEE Access **5**, 5631–5639 (2017)
6. Chenaghlou, M., Moshtaghi, M., Leckie, C., Salehi, M.: Online clustering for evolving data streams with online anomaly detection. In: Phung, D., Tseng, V.S., Webb, G.I., Ho, B., Ganji, M., Rashidi, L. (eds.) PAKDD 2018. LNCS (LNAI), vol. 10938, pp. 508–521. Springer, Cham (2018). https://doi.org/10.1007/978-3-319-93037-4_40
7. Gorecky, D., Schmitt, M., Loskyll, M., Zuhlke, D.: Human-machine interaction in the industry 4.0 era. In: IEEE International Conference on Industrial Informatics (INDIN), pp. 289–294 (2014)
8. Hanamori, T., Nishimura, T.: Real-time monitoring solution to detect symptoms of system anomalies. FUJITSU Sci. Tech. J. **52**, 23–27 (2016)
9. Huber, M., Voigt, M., Ngomo, A.C.N.: Big data architecture for the semantic analysis of complex events in manufacturing, pp. 353–360 (2016)
10. Koutra, D., Shah, N., Vogelstein, J.T., Gallagher, B., Faloutsos, C.: DELTACON: principled massive-graph similarity function with attribution. ACM Trans. Knowl. Discov. Data **10**(3), 28:1–28:43 (2016)
11. Lee, J., Ardakani, H., Yang, S., Bagheri, B.: Industrial big data analytics and cyber-physical systems for future maintenance and service innovation. In: Proceedings of Conference on Intelligent Computation in Manufacturing Engineering (CIRP), vol. 38, pp. 3–7 (2015)

12. Lee, J., Lapira, E., Bagheri, B., Kao, H.: Recent advances and trends in predictive manufacturing systems in big data environment. Manuf. Lett. **1**(1), 38–41 (2013)
13. Lopez, F., et al.: Categorization of anomalies in smart manufacturing systems to support the selection of detection mechanisms. IEEE Robot. Autom. Lett. **2**(4), 1885–1892 (2017)
14. Nunes, D., Silva, J.S., Boavida, F.: A Practical Introduction to Human-in-the-Loop Cyber-Physical Systems. Wiley IEEE Press, Hoboken (2018)
15. Rashidi, L., et al.: Node re-ordering as a means of anomaly detection in time-evolving graphs. In: Frasconi, P., Landwehr, N., Manco, G., Vreeken, J. (eds.) ECML PKDD 2016. LNCS (LNAI), vol. 9852, pp. 162–178. Springer, Cham (2016). https://doi.org/10.1007/978-3-319-46227-1_11
16. Salehi, M., Rashidi, L.: A survey on anomaly detection in evolving data: [with application to forest fire risk prediction]. SIGKDD Explor. Newsl. **20**(1), 13–23 (2018)
17. Stojanovic, L., Dinic, M., Stojanovic, N., Stojadinovic, A.: Big-data-driven anomaly detection in industry (4.0): an approach and a case study. In: 2016 IEEE International Conference on Big Data (Big Data), pp. 1647–1652 (2016)
18. Wongsuphasawat, K., Moritz, D., Anand, A., Mackinlay, J., Howe, B., Heer, J.: Voyager: exploratory analysis via faceted browsing of visualization recommendations. IEEE Trans. Vis. Comput. Graph. **22**(1), 649–658 (2016)
19. Zhang, L., Lin, J., Karim, R.: Adaptive kernel density-based anomaly detection for nonlinear systems. Knowl.-Based Syst. **139**, 50–63 (2018)

Exploiting Smart City Ontology and Citizens' Profiles for Urban Data Exploration

Devis Bianchini[✉], Valeria De Antonellis, Massimiliano Garda,
and Michele Melchiori

Department of Information Engineering, University of Brescia,
Via Branze 38, 25123 Brescia, Italy
{devis.bianchini,valeria.deantonellis,
m.garda001,michele.melchiori}@unibs.it

Abstract. Smart Cities are complex systems, collecting together huge amounts of heterogeneous data mainly concerning energy consumption, garbage collection, level of pollution, citizens' safety and security. In the recent years, several approaches have been defined to enable Public Administration (PA), utility and energy providers, as well as citizens, to share and use information in order to take decisions about their daily life in Smart Cities. Research challenges concern the study of advanced techniques and tools to enable effective urban data exploration. In this paper, we describe a framework that combines ontology-based techniques and citizens' profiles in order to enable personalised exploration of urban data. Ontologies may provide a powerful tool for semantics-enabled exploration of data, by exploiting the knowledge structure in terms of concepts organised through hierarchies and semantic relationships. Smart City indicators are used to aggregate data that can have different relevance for target users, the activities they are performing and their role (e.g., PA, utility and energy providers, citizens) within the Smart City. Ontologies combined with users' profiles enable effective and personalised recommendation and exploration of urban data.

Keywords: Urban data exploration · Smart city indicators
Ontologies · Smart cities

1 Introduction

In a Smart City, investments in human and social capital and modern ICT infrastructure fuel sustainable economic growth and a high quality of life, with a wise management of natural resources, through citizens' participation [7]. In this context, techniques and tools for sharing and exploring urban data about energy consumption, garbage collection, level of pollution, citizens' safety and security are gaining momentum to enable the citizens to take decisions about their daily

© Springer Nature Switzerland AG 2018
H. Panetto et al. (Eds.): OTM 2018 Conferences, LNCS 11229, pp. 372–389, 2018.
https://doi.org/10.1007/978-3-030-02610-3_21

life, depending on different roles they play in the Smart City. Public Administration (PA), utility and energy providers may need to explore urban data at city level, in order to take actions for improving citizens' life. Building managers can use urban data concerning the administered buildings to take decisions for their daily activities (e.g., they can inspect information on energy consumption in order to implement energy saving actions). Citizens need new tools to explore open urban data about pollution, garbage collection, safety and security to take autonomously decisions on activities they may perform or to compare their own energy consumption against average values at building, district and city level to take virtuous behaviours. Let's consider two data exploration scenarios focused on different types of users. John is the manager of several buildings located in different districts of the Smart City. John monitors electrical consumptions of the buildings, in order to implement energy saving policies (e.g., introduction of LED lamps in common spaces or planning renovation work to increase the energy efficiency class). Alice is a citizen who is enthusiastic about bicycle and wants to inspect whether environmental status (in terms of pollution and air quality) is suitable for practising outdoor leisure activities. Challenging issues are related to the capability of John and Alice to fruitfully exploit available information.

In this context, we propose an ontology-based framework for enabling personalised exploration of urban data. In the framework, different categories of end users can be defined and their profiles are taken into account to suggest fruitful exploration, proactively supporting users for making decisions according to their interests. Through the framework, users with different roles may share, access and possibly update urban data coming from different types of data sources, guided by the knowledge structure provided by the ontology and according to their own interests. The framework supports the exploration of indicators that aggregate urban data from various data sources about energy consumption, garbage collection, level of pollution, citizens' safety and security. Actually, indicators provide a comprehensive view over underlying data according to several perspectives without being overwhelmed by the data volume [13]. Indicators are computed at the levels of building, district or city, aggregating data that can have different relevance with respect to the various target users, also considering the activities they perform and their role in the Smart City. Our framework relies on both a Smart City Ontology (SCO), which provides a powerful tool for semantics-enabled exploration of urban data, and on information associated with users' profiles. On the one hand, the SCO is used to properly represent indicators in terms of concepts, hierarchies and semantic relationships. The resulting knowledge organisation can be used for the selection and exploration of the most suitable indicators for a specific user and request. Moreover, inheritance in the ontology hierarchies allows exploration at different granularity levels. On the other hand, users' profiles are exploited to refine the list of suggested indicators. With reference to the examples given above, the framework may allow John to monitor electrical consumption of administered buildings, by exploiting the indicators hierarchy in the ontology to distinguish electrical consumption according to different perspectives (e.g., consumption in common spaces, consumption of elevators), and to compare average values of consumption with other buildings at district or city level. Furthermore, Alice may use the framework to make decisions about her activities, by observing specific indicators (e.g., to avoid sport activities when pollution levels overtake tolerance thresholds).

Novel contribution of the framework relies on the two-fold nature of the approach: a recommendation step, which selects candidate indicators, is followed by and interleaved with an interactive exploration step, which permits to refine the set of indicators. These steps are based on the Smart City Ontology and users' profiles in a combined way. Firstly, indicators are selected by filtering available indicators, exploiting the information in the request issued by the user, as well as his/her profile. Once a set of candidate indicators has been identified, semantics-enabled data exploration is enabled, where concepts hierarchies and semantic relationships in the ontology are used to further refine and rank the indicators of interest. A preliminary validation of the framework has been performed in the context of the Brescia Smart Living (BSL) Italian project[1], which promotes a holistic view of the city where different types of data must be collected and properly explored to provide new services to several city stakeholders and, in particular, educational indications to citizens in order to promote virtuous behaviours.

The paper extends the research presented in [10] with the following innovative contributions: (i) a formalisation of the candidate indicators selection algorithm; (ii) a ranking function, applied to the set of candidate indicators, and (iii) a refinement of the framework architecture, providing additional details about the services and their constituent modules, within each layer.

This paper is organized as follows: in Sect. 2 we compare our approach against related work in literature; Sect. 3 provides an approach overview and describes the Smart City Ontology on which the approach relies; Sect. 4 describes the candidate indicators recommendation step, while in Sect. 5 we introduce the urban data exploration; Sect. 6 discusses the framework implementation; in Sect. 7 we present preliminary experiments; finally, Sect. 8 closes the paper.

2   Related Work

Our approach focuses on an ontology-based data exploration perspective for urban data, properly aggregated in the form of Smart City indicators, considering users' preferences and novel search interests and exploiting additional knowledge provided by the SCO. According to this perspective, our approach differs from *Ontology-Based Data Access* (OBDA), coping with the integration of heterogeneous data sources inside the Smart City [1,6,9,17], and from general purpose recommender systems as described in recent surveys (e.g., [11]), where the explorative viewpoint is not explicitly addressed. Furthermore, compared to approaches focused on Ontology-Based Data Warehouses (OBDW), which store analytical data, indicators, requirements and their semantics [14,15,23],

---

[1] http://www.bresciasmartliving.eu/.

our data exploration framework exploits indicators hierarchy and considers users' profiles to enrich the exploration experience, also considering the influence that the knowledge about indicators might have on users' activities. For what concerns the use of Semantic Web technologies in existing Smart City projects, ontologies have been used for *energy management*, where diagnostic models are defined to discover energy losses [25], or to perform optimisations for cost saving [2,5,12,21]; for *facility discovery*, to search for city facilities and services [4,12]; for *events monitoring and management* [3,19]. Existing approaches proposed recommendation of specific kinds of urban data, e.g., advising environmental recommendations to improve the quality of life of people [20], ontology-based exploration of crime data, that relies on Association Rule Mining [8], ontology-based visualisation of data or mobility [24]. The approach presented in [16] focuses on the ontology development phase. It defines a set of high level concepts, mapped to the ones of ontologies underneath twenty Smart City applications. In [18] an ontology, which models a Smart City as a composition of information objects, agents and measures, is proposed. In [23] a semantic characterisation of Smart City indicators is provided.

Differently from the aforementioned solutions, our proposal uses the concept of indicator to provide a holistic view over the entire Smart City domain, including a wider heterogeneous spectrum of urban data, such as data on energy consumption, environmental conditions, safety and security. Moreover, we foster personalised exploration of data for different categories of users (e.g., citizens, building managers, PA).

3   Approach Overview

The proposed approach is based on both ontology-based descriptions of Smart City indicators and users' profiles, as detailed in the following. According to the intrinsic modular nature of ontologies, the Smart City Ontology may be reused within other Smart City projects, after being properly extended to include the spectrum of concepts and relationships of the considered context. We do not discuss how to compute indicators from heterogeneous data sources, as this is out of the scope of the paper.

3.1   The Smart City Ontology

Figure 1 reports the main concepts and relationships of the Smart City Ontology[2] (SCO). The SCO provides a formal representation of Smart City indicators, with reference to the kinds of activities and users' categories for which indicators can provide relevant information. In particular, the SCO contains the definitions of the following main concepts and mutual relationships between them.

Indicators. Indicators represent an aggregation of urban data of interest for the citizens of the Smart City. Through the SCO, indicators are specified as

---

[2] The TBox of the ontology can be found at https://tinyurl.com/sco-onto (a free Web Protégé account is required).

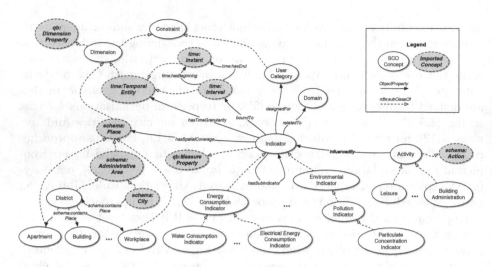

Fig. 1. A portion of the Smart City Ontology, containing main concepts and semantic relationships.

individuals of the `Indicator` concept or one of its sub-concepts in the indicators hierarchy. For instance, `ElectricalEnergyConsumptionIndicator` is defined as a sub-concept of `EnergyConsumptionIndicator` in Fig. 1. We denote with $\mathcal{I}$ the overall set of individuals of available indicators. In the SCO, an indicator is `relatedTo` a domain and is further specified through a set of `Constraint` individuals in the ontology.

Domains. A domain represents a concept used to limit the scope of an indicator, among the types of urban data that can be explored (e.g., environment, safety, energy, mobility). Therefore, an indicator $i \in \mathcal{I}$ is associated with a set of individuals of the `Domain` concept D_i (`relatedTo` relationship).

Constraints. In the SCO a constraint $c_i \in C_i$ can be either an user's category (e.g., citizen, building manager) or a dimension (i.e., time or space). On the one hand, an indicator is designed for specific users' categories (`designedFor` relationship). For instance, the indicator on electrical consumption of buildings is designed to be browsed and explored by the buildings manager. The user's category will enable personalised indicators filtering. On the other hand, an indicator can be `boundTo` a time interval (e.g., values of electrical consumption are available for the year 2017), may have a time granularity (`hasTimeGranularity` relationship) and may be defined at the city, street, district or more specific levels, such as buildings, workplaces and private apartments (`hasSpatialCoverage` relationship).

Activities. This concept is used to represent users' activities (e.g., leisure, building administration) that can be influenced by the knowledge provided by accessing an indicator. For instance, the `EnergyConsumptionIndicator` may provide useful insights for a building manager for implementing energy saving activities in the administered buildings. Similarly, pollution indicators may prevent citizens from practising outside sport activities. To this aim, the `Activity` concept is connected in the SCO to the `Indicator` concept through the `influencedBy` property.

Other SCO Concepts. In defining the SCO, pivotal concepts from available foundation ontologies are exploited to: (i) represent geospatial concepts of the city (e.g., district, street) and users' activities (Schema.org[3] ontology), (ii) define temporal entities, used as analysis dimensions (Time[4] ontology) and (iii) characterise indicators as analytical data entities (Data Cube[5] ontology).

Formally, we can summarise an indicator i as a tuple $\langle ID_i, T_i, D_i, C_i \rangle (\forall i = 1, \ldots, N_{\mathcal{I}})$, where ID_i is a unique identifier (i.e., an URI), T_i is the indicator type (e.g., `ElectricalEnergyConsumptionIndicator`), D_i is the set of domains individuals, C_i is the set of constraints, and where $N_{\mathcal{I}}$ is the number of indicators in $\mathcal{I}$.

3.2   Users' Profiles

Users are described according to a category (e.g., citizen, building manager), activities of interest, defined as individuals of the concept `Activity` or some of its sub-concepts, the types of indicators explored by the user through the interactions with the framework, defined through the concept `Indicator` or its sub-concepts. Different users may have access to different indicators: a citizen can select indicators concerning his/her apartment only, building managers can select indicators on their administered buildings only, energy managers can select indicators that only concern the workplaces they are responsible for, etc.

To this aim, after selecting the activities of interests (if any) during the registration to the framework, citizens, building managers and other categories of users have also to specify the places (e.g., an apartment, a building) they act in. Only urban data that has been aggregated within indicators associated with that places in the SCO (`hasSpatialCoverage` relationship) will be displayed. This has a two-fold advantage: (i) it enables data privacy preservation, for instance preventing building managers to visualise data on buildings they do not administer; (ii) it will be used to personalise indicators selection and data exploration, as explained in the next sections.

Formally, the profile $p(u)$ of a user $u \in \mathcal{U}$ can be summarised as $p(u) = \langle ID_u, cat_u, \mathcal{I}_u, A_u, P_u \rangle$, where ID_u is the identifier associated with the user's account, cat_u is the user's category, $\mathcal{I}_u$ is the set of individuals representing indicators that have been selected and explored by u in previous interactions with the framework, A_u is the set of activities of interest for the user, P_u is the list of individuals of concepts representing specific places where the user takes actions, namely individuals of `Building` for building managers, of `Apartment` for citizens, of `Workplace` for energy managers, etc. A registration wizard, starting

---

[3]   http://schema.org/.
[4]   http://www.w3.org/2006/time.
[5]   http://purl.org/linked-data/cube.

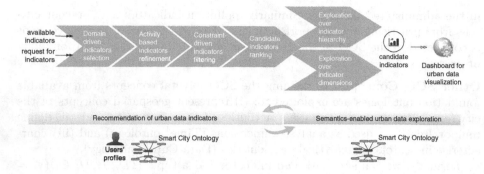

Fig. 2. The steps of semantics-enabled personalised data exploration.

from the user's category, prompts to the user proper masks to insert the instances defining the set P_u and associate them with districts. Such instances will be also inserted in the SCO to enable semantics-enabled urban data exploration, while only districts and upper level places (e.g., the city districts) are inserted in the ontology by domain experts. This reduces the complexity of ontology population and maintenance for platform administrators.

3.3   Urban Data Exploration Steps

The semantics-enabled approach proposed here for urban data exploration is articulated over two main steps, as shown in Fig. 2: (i) the overall set of available indicators is properly pruned, by taking into account both explicit requirements of the user as expressed in a search request and the user's profile, and then ranked (*recommendation of urban data indicators*); (ii) the list of selected indicators is presented to the user as starting point to enable a semantics-enabled personalised exploration of urban data (*semantics-enabled urban data exploration*).

Urban data exploration starts from a request formulated by the user u, denoted with $r(u)$, that contains the domains and indicators of interest, specified as individuals of `Domain` and `Indicator` concepts or sub-concepts of the SCO, respectively. Nevertheless, in order to provide support to the request formulation without demanding a detailed knowledge of ontology concepts and individuals, the framework allows the user to specify a set of keywords $K_r = \{k_{r1}, k_{r2}, \ldots, k_{rn}\}$. The set K_r is processed according to techniques aimed to match the keywords with ontology terms [22]. The adopted disambiguation procedure relies on WordNet[6] to retrieve synonyms, hypernyms and hyponyms of the keywords, and identifies a mapping between the input list of keywords and ontology individuals using probabilistic techniques. Following this approach, requests are processed in a more flexible way, to deal with the different levels of expertise (i.e., knowledge of the terminology and lexicon) users have. Beyond the domains and/or indicators explicitly indicated by the user in K_r, the user's

---

[6] https://wordnet.princeton.edu/.

profile is exploited to take into account other elements for indicators selection, namely the user's category cat_u, his/her activities of interest $\mathcal{A}_u$ and the set $\mathcal{I}_u$ of indicators explored in the past by the user.

Formally, we represent the request submitted by user u as follows: $r(u) = \langle D_r, \mathcal{I}_r, p(u) \rangle$, where D_r is the set of desired domains, $\mathcal{I}_r$ is the set of indicators of interest, $p(u)$ is the user's profile. D_r and $\mathcal{I}_r$ are the output of the WordNet-based disambiguation procedure.

Recommendation of urban data indicators is further organised in four sub-steps, namely:

- *domain-driven indicators selection*, in which a preliminary pruning of candidate indicators is performed based on the domains and indicators of interest specified in the request $r(u)$;
- *activity-based indicators refinement*, where the set of candidate indicators is further enriched considering the activities in the user's profile and the influencedBy ontological relationship;
- *constraint-driven indicators filtering*, where user's category and spatial constraints are exploited to further filter candidate indicators;
- *candidate indicators ranking*, performed by considering both the user's request and the history of indicators in the user's profile.

Urban data exploration, starting from the output of the recommendation of urban data indicators, is supported through the indicators hierarchy and semantic relationships of the SCO, enabling the users to browse the set of available indicators starting from the recommended ones, as explained in Sect. 5. Exploration actions may require to revise candidate indicators recommendation.

Finally, the user can visualise the actual values of one or more indicators in a numerical or graphical way by means of a suitable web-based dashboard. Even if we have implemented the web-based dashboard in the context of the BSL project, in the following the focus will be on the recommendation and exploration steps.

4   Recommendation of Urban Data Indicators

Candidate Indicators Selection. The selection process of candidate indicators is described by Algorithm 1. It takes as input the set $\mathcal{I}$ of all available indicators, the request $r(u)$ and the SCO. The output of the selection process is a set of candidate indicators, namely $\mathcal{I}_{cand}$, containing indicators that are compliant with the request.

Starting from the set $\mathcal{I}_r$ of indicators of interest as specified in the request $r(u)$ (see line 1), the selection process performs the domain-driven indicators selection (lines 2–6), the activity-based indicators refinement (lines 7–10), the constraint-driven indicators filtering (lines 11–18).

In the domain-driven indicators selection, the set D_i of individuals of the Domain concept, associated with each indicator $i \in \mathcal{I}$, is retrieved by considering

---

Algorithm 1. Candidate indicators selection

Input : Request $r(u) = \langle D_r, \mathcal{I}_r, p(u) \rangle$, set $\mathcal{I}$ of available indicators, the Smart
        City Ontology SCO
Output: The set of candidate indicators $\mathcal{I}_{cand} \subseteq \mathcal{I}$

1   $\mathcal{I}_{cand} \leftarrow \mathcal{I}_r$;
2   **if** $(D_r \neq \emptyset)$ **then**
3   **foreach** *indicator* $i \in \mathcal{I}$ **do**
4          $D_i \leftarrow \{domain \mid \texttt{relatedTo}(i, domain)\}$;
5   **if** $(D_i \cap D_r \neq \emptyset)$ **then**
6             $\mathcal{I}_{cand} \leftarrow \mathcal{I}_{cand} \cup \{i\}$;

7   **if** $(A_u \neq \emptyset)$ **then**
8   **foreach** *activity* $a \in A_u$ **do**
9          $\mathcal{I}_a \leftarrow \{indicator \mid \texttt{influencedBy}(a, indicator)\}$;
10        $\mathcal{I}_{cand} \leftarrow \mathcal{I}_{cand} \cup \mathcal{I}_a$;

11   $C \leftarrow P_u \cup cat_u$, both extracted from $p(u)$;
12   **if** $(C \neq \emptyset)$ **then**
13   **foreach** $(c \in C)$ **do**
14   **switch** c **do**
15   **case** $\texttt{UserCategory}(c)$
16                $\mathcal{I}_{cand} \leftarrow \mathcal{I}_{cand} \setminus \{i \mid i \in \mathcal{I}_{cand} \wedge \neg \texttt{designedFor}(i, c)\}$;
17   **case** $\texttt{schema:Place}(c)$
18                $\mathcal{I}_{cand} \leftarrow \mathcal{I}_{cand} \setminus \{i \mid i \in \mathcal{I}_{cand} \wedge \neg \texttt{hasSpatialCoverage}(i, c)\}$;

19   **if** $(\mathcal{I}_{cand} == \emptyset)$ **then**
20      $\mathcal{I}_{cand} \leftarrow \mathcal{I}_u$;

21   **return** $\mathcal{I}_{cand}$;

---

the `relatedTo` relationship in the SCO. If there is an overlapping between the sets D_i and D_r, then the indicator i is added to the set of candidates $\mathcal{I}_{cand}$.

In the activity-based indicators refinement, for each activity $a \in \mathcal{A}_u$ extracted from the user's profile, the `influencedBy` relationship in the SCO is used to retrieve additional candidate indicators.

During the constraint-driven indicators filtering, each indicator $i \in \mathcal{I}_{cand}$ is analysed to filter out candidate indicators that are not compliant with the user's category cat_u and are not available for the places P_u where the user takes actions, which are both extracted from the user's profile as a set of constraints C (line 11). Considering the `designedFor` relationship in the SCO to get the individuals of concept `UserCategory` that are semantically related to i, cat_u is used to filter out non relevant indicators from $\mathcal{I}_{cand}$ (lines 15–16). Similarly, indicators that are not available for the places P_u are discarded considering the `hasSpatialCoverage` relationship (lines 17–18). If the set of candidate indicators is empty, then it is populated with indicators included in $p(u)$ (lines 19–20). The rationale behind this choice is that the set $\mathcal{I}_u$ traces past exploration history of the user providing known, albeit not novel, candidate indicators.

Candidate Indicators Ranking. Once the set $\mathcal{I}_{cand}$ has been identified, candidate indicators are properly ranked by combining two different criteria: (a) the similarity with the user's request $r(u)$ (denoted with Sim_{req}) and (b) the similarity with indicators selected by the user in past exploration activities (denoted with Sim_{past}). The rationale is that the latter criterion is used to maintain a certain compliance with usual interests of the user, as represented through the set $\mathcal{I}_u$ of indicators in his/her profile. On the other hand, only relying on past choices of the user would penalise user's new interests. This can be viewed as a variation of the cold start problem [11] that affects recommendation systems and is balanced by the first criterion.

Specifically, for what concerns (a), the similarity measure is calculated by estimating the overlap between the set D_r of domains, included in $r(u)$, and the domain(s) associated with each candidate indicator $i_{cand} \in \mathcal{I}_{cand}$. The second kind of similarity, that is based on the proximity of each i_{cand} with respect to indicators in $\mathcal{I}_u$, is computed by measuring the similarity between places and domains that are shared by the computed indicators.

The ranking function $\rho : \mathcal{I}_{cand} \longmapsto [0, 1]$ is computed as follows:

$$\rho(i_{cand}) = \alpha \cdot Sim_{req}(D_r, D_{i_{cand}}) + \beta \cdot Sim_{past}(\mathcal{I}_u, i_{cand}) \tag{1}$$

where $\alpha + \beta = 1$ (in our preliminary experiments we set $\alpha = \beta = \frac{1}{2}$), and $D_{i_{cand}}$ is the set of domains associated with $i_{cand} \in \mathcal{I}_{cand}$.

The term Sim_{req} is computed by applying the Dice coefficient, which estimates the overlapping between two sets, as follows:

$$Sim_{req}(D_r, D_{i_{cand}}) = 2 \cdot \frac{|D_r \cap D_{i_{cand}}|}{|D_r| + |D_{i_{cand}}|} \tag{2}$$

Regarding Sim_{past}, this similarity is evaluated as:

$$Sim_{past}(\mathcal{I}_u, i_{cand}) = MAX_{i_u \in \mathcal{I}_u} Sim(i_u, i_{cand}) \tag{3}$$

where $Sim()$ combines the similarity coefficients, equally weighted, between the two indicators according to their domains and their places:

$$Sim(i_u, i_{cand}) = \frac{|P_{i_u} \cap P_{i_{cand}}|}{|P_{i_u}| + |P_{i_{cand}}|} + \frac{|D_{i_u} \cap D_{i_{cand}}|}{|D_{i_u}| + |D_{i_{cand}}|} \tag{4}$$

where P_{i_u} (resp., D_{i_u}) is the set of places (resp., domains) associated with the indicator i_u and $P_{i_{cand}}$ (resp., $D_{i_{cand}}$) is the set of places (resp., domains) associated with the indicator i_{cand}. Please note that all these similarity coefficients are in the range $[0, 1]$.

5   Semantics-Enabled Personalised Urban Data Exploration

The SCO is used to support both the candidate indicators selection and ranking, as discussed above, and the exploration of indicators. The latter one is meant as an iterative process guided by the SCO organisation of concepts and relationships, and including exploration of both semantic description of indicators and of actual values accessible through the web-based dashboard.

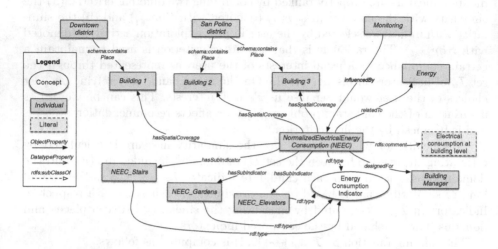

Fig. 3. Example of candidate indicator and related properties.

Starting from candidate indicators returned through the selection step, the users of the platform can further explore other indicators being guided by the semantic relationships in the SCO. Exploration can be performed according to different perspectives, given the knowledge structure in the ontology: (a) exploration over the indicators hierarchy; (b) personalised exploration over the indicators dimensions. Let us explain how this can be done in our framework by a simple demonstration scenario.

Let's consider the user John in the motivating example, who is the manager of three buildings (namely `Building 1`, `Building 2` and `Building 3`) located in two districts of the city. In particular, `Building 1` is located in the city downtown, while `Building 2` and `Building 3` are located in the modern district of San Polino. Since John is interested in monitoring buildings, during the registration to the platform he specifies the activity `Monitoring` in his profile. Moreover, he specifies what are the administered buildings and associates them with the districts they are located in, as part of his profile. Buildings are also inserted into the SCO and linked to the districts by means of the `schema:containsPlace` relationship. In order to have an insight on the status of the buildings he administers, for instance to evaluate whether replacing standard lamps with less energy-demanding LED ones, John logs in to the platform and asks for consumption indicators, specifying the keywords $K_r = \{$`energy, consumption`$\}$. The platform processes the request as explained in the previous section and returns, among the others, the indicator `NormalizedElectricalEnergyConsumption` (`NEEC`), which reports electrical consumption normalised with the number of apartments in the building. The indicator is selected because it is both compatible with the keywords given in the request and associated with the activity `Monitoring` in the ontology. Semantic description of the indicator `NEEC` is shown in Fig. 3.

If John decides to explore the indicators hierarchy, he may select the `NEEC` indicator and the framework suggests him more specific indicators `NEEC_Stairs`, `NEEC_Elevators` and `NEEC_Gardens`, which are related to the `NEEC` indicator through the `hasSubIndicator` relationship in the ontology. Since John's focus is on evaluating the electrical consumption of the lighting plants of stairs, he selects `NEEC_Stairs`.

Personalised exploration over the indicators dimensions exploits the semantic relationship `schema:containsPlace` that relates each others individuals of `schema:Place` concept or its sub-concepts. In particular, knowledge on the spatial coverage of indicators is obtained through the `hasSpatialCoverage` relationship. Starting from indicators previously selected for the John's building, either `NEEC` or one of its sub-indicators, the containment relationship that relates John's buildings with districts is exploited. Therefore, John could choose to visualise the average consumption provided by the indicators for the buildings of the districts, in order to compare his buildings against other ones having similar characteristics or using different lighting solutions. Similarly, indicators for John's buildings could also be suggested over several years (`boundTo` relationship) or over different time granularities (e.g., years, months, days), according to the `hasTimeGranularity` relationship. Comparison between indicators may stimulate John to consider the replacement of energy consuming light bulbs with modern LED lamps in shared spaces, after analysing the affordability of the expenses, with respect to the ones sustained by other similar buildings.

6   Three-Layered Framework Architecture

Figure 4 shows an overview of the semantics-enabled data exploration framework architecture. The framework is developed with web-based technologies and is organised over multiple layers. It has been preliminarily applied in the context of the Brescia Smart Living (BSL) Italian project. Data on field, collected from domain-specific platforms through IoT technologies, as well as data from sources external to the BSL project (weather data, pollution data, etc.) are loaded into the BSL platform database (*BSL Platform Layer*). Data is transferred on the BSL platform using RESTful services, SOAP-based services and MQTT Agents. The *Semantic Layer* enables personalised urban data exploration, as described in Sects. 4 and 5. The *User Access Layer* includes a web-based Smart City Dashboard to be used by citizens, PA and other users to explore urban data.

Using the web browser, users can register themselves and update their profile. Within each layer of the platform, proper Web Services, implemented in Java and deployed under the Apache TomEE[7] application server, elaborate the incoming requests from users to retrieve the set of recommended indicators. The Smart

---

[7] http://tomee.apache.org/.

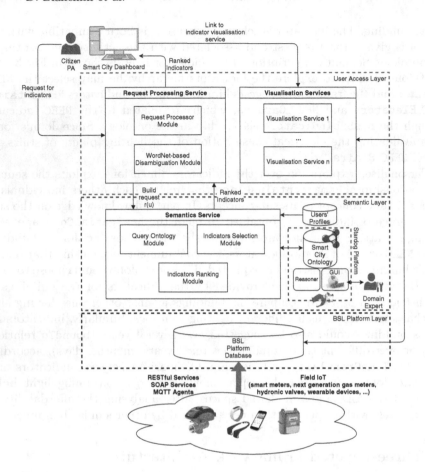

Fig. 4. Web-based architecture of the semantics-enabled data exploration framework.

City Ontology is deployed in OWL using Stardog[8], a NoSQL graph database based on W3C Semantic Web standards. The Stardog Platform supports domain experts in order to maintain the ontology (concepts, relations and individuals, including the insertion of new indicators individuals), interacting with the web-based administration console provided by the platform.

In this section, we introduce the main Web Services, located within the layers of the framework, invoked to process users' requests for indicators recommendation and exploration. Here, we present only the services and their composing modules at a high level, without lingering much on technical details, showing the interaction flow triggered by the user (either a citizen or PA) when issuing a request to the framework. The sequence diagram of Fig. 5 illustrates the interactions (i.e., invocation of methods and exchanged parameters) between the main

---

[8] https://www.stardog.com/.

modules of the framework in order to perform the indicators recommendation step. These modules are implemented as services described in the following.

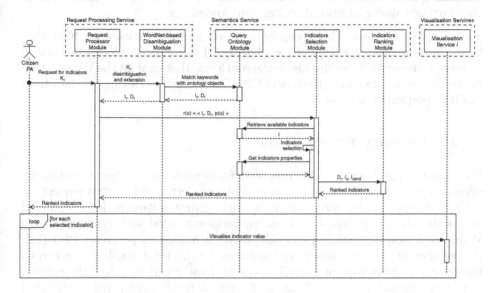

Fig. 5. Interactions between the main modules of the architecture.

Request Processing Service. This service is in charge of processing the incoming users' requests, starting from the set of keywords K_r and the user's profile $p(u)$. It is composed of the following modules: (i) the *Request Processor Module*, that routes the requests to the other services of the platform, handles partial results and returns the set of recommended indicators to the user; (ii) the *WordNet-based Disambiguation Module*, that looks up WordNet database to accomplish the semantic enrichment and disambiguation of the keywords contained in the query.

Semantics Service. The modules embedded in this service operate on the Smart City Ontology, to retrieve and select the candidate indicators, compliant with the user's request. Specifically: (i) the *Query Ontology Module* queries the ontology through the API provided by the Stardog Platform; (ii) the *Indicators Selection Module*, performs the selection of candidate indicators; (iii) the *Indicators Ranking Module* implements the techniques exposed in Sect. 4 to rank the set of candidate indicators. The Query Ontology Module, embedded into the Semantics Service, is in charge of issuing SPARQL queries over the SCO; specifically, it handles two different types of queries: (i) *selection* queries and (ii) *boolean* queries. The former ones (also known as *SELECT* queries) are used to retrieve individuals from the ontology (e.g., to collect the set of indicators apt to citizens) whereas the latter ones (also known as *ASK* queries) are useful to query the ontology to get true/false answers (e.g., to check whether the NEEC indicator is available for citizens). SPARQL queries can benefit from the underlying reasoning engine, provided by the Stardog Platform, to enrich their results when computing hierarchies (e.g., to build the lineage of an indicator, up to its topmost ancestor) or to infer additional background knowledge.

Visualisation Services. This group of services comprises legacy modules that lie in the visualisation engine of the platform. Each indicator has a link which triggers a proper Web Service inside the platform, that enables the visualisation of the indicator values onto the Smart City Dashboard, retrieving the data from the BSL platform database.

7   Preliminary Experiments

Preliminary experiments on the proposed framework aim at demonstrating its effectiveness in supporting candidate indicators selection for a given request r.

To this aim, we compared our framework against a baseline keyword-based approach, where keywords have been properly expanded with synonyms using WordNet lexical system. In particular, given a request r, precision $P(r)$ (i.e., the number of relevant indicators compared to the total number of returned indicators, in the context of a search operation) and recall $R(r)$ (i.e., the number of relevant indicators returned among the search results, compared to the total number of relevant indicators) are measured for quantifying the effectiveness of the candidate indicators selection. To measure precision and recall, we used a SCO composed of 57 concepts (and, among them, 30 indicators), 104 individuals, 207 object and datatype properties. Table 1(a) reports average precision and recall values of our approach compared to a keyword-based search. Precision and recall values have been computed on two kinds of requests: (a) requests where the user specified a set of keywords to identify desired domains and indicators, and the user's profile does not contain any activity or preferential indicator (r_A); (b) requests where the user presents a richer profile (containing category, activities and preferential indicators), but specifies a few keywords in the keyword set K_r, that only correspond to individuals of the Domain concept (r_B). Five requests for each type have been issued and average values have been computed.

The second kind of request is used to demonstrate how relationships within the SCO are effective in improving precision and recall for indicators selection. In fact, with respect to the keyword-based approach, our framework enables a better precision by refining the set of candidate indicators based on the user's category, the specified domain(s) and other ontological relationships. On the other hand, recall is increased by exploiting the relationships between other elements of the user's profile (i.e., activities and preferential indicators) and the available indicators in the ontology, thus including among search results candidate indicators that are not described with the keywords specified in the user's request or with keyword synonyms as extracted from WordNet. Since both the compared approaches use WordNet to perform keywords disambiguation and the same keywords across the approaches have been used during tests, difference

Table 1. (a) Average precision and recall values obtained for the preliminary evaluation; (b) average execution time (in msec) for indicators recommendation.

	Keyword-based search	Ontology-based search
$P(r_A)$	0.49	0.99
$R(r_A)$	0.97	0.98
$P(r_B)$	0.33	0.94
$R(r_B)$	0.27	0.93

(a)

Request type	Indicators selection	Keywords disambiguation	Ranking	Total
r_A	1454	1105	805	3361
r_B	269	1056	880	2205

(b)

in average precision and recall is due to the knowledge structure provided by the ontology.

The formulation of the request as a set of keywords, instead of asking the user to specify required properties and constraints, enables more flexibility, since it does not demand for a detailed knowledge of the ontology, its concepts and relationships. Furthermore, the processing time required to expand keyword sets with the use of WordNet is affordable and acceptable for the considered exploration scenarios. Table 1(b) contains average execution time (in msec) for indicators recommendation (including time spent for the WordNet-based disambiguation of keywords and for ranking) in the preliminary evaluation. Tests have been performed on a Windows-based machine equipped with an Intel i7 2.00 GHz CPU, 8 GB RAM, SSD storage.

Usability tests are being performed to check the capability of the framework in facilitating user's access to urban data through the suggestion of candidate indicators. To perform usability tests, we considered a population of users using metrics such as the number of exploration steps needed to obtain desired data, number of fails, number of successful explorations. Usability experiments are being carried on within the Brescia Smart Living project until September 2018. Currently, the framework is being tested, with satisfaction, by a sample of users in two districts, a modern one (San Polino), where new generation smart meters have been installed, and a district in city downtown, more densely populated and presenting older buildings. The framework will be also used by other partners involved in the project as representatives of PA (in particular, the Municipality of Brescia, Italy), utility and energy providers.

8   Conclusions and Future Works

In this paper, we described a framework for enabling personalised exploration of urban data. The framework relies on the knowledge structure provided by a

Smart City Ontology (SCO) and the information contained within the citizens' profiles. The SCO is used to properly represent indicators in terms of concepts, hierarchies and semantic relationships, thus they can be used to facilitate exploration by exploiting the knowledge structure. Moreover, the concept hierarchies in the ontology allow exploration at different granularity levels. Candidate indicators are recommended if compliant with user's profile, ensuring a personalised selection and exploration over the set of available Smart City indicators. The semantics-enabled personalised urban data exploration is articulated over two steps: (i) indicators are properly recommended taking into account both explicit requirements of the user as expressed in a search request and the user's profile; (ii) recommended indicators are exploited as starting point to set up interactive exploration of urban data. Future efforts will be devoted to perform further experiments (e.g., comparing [22] with other matching techniques) and to extend the set of semantic relationships in the SCO as follows: (a) further relationships between indicators will be identified (e.g., to assert that two or more environmental indicators must be jointly monitored due to their harmful impact on the ecosystem); (b) strategies to promote the users' virtuous behaviours will be studied and implemented on top of the relationships, providing advices for healthy activities that should be practised by users. This will be accomplished by collecting and formalising additional knowledge about users' lifestyle, and then enriching the SCO with specific background semantics. Finally, support to the insertion of indicators individuals, that is currently performed by domain experts given their skill in the application context, will be further developed as well.

References

1. The GrowSmarter project. http://www.grow-smarter.eu/home/
2. OPTIMising the energy USe in cities with smart decision support systems. http://optimus-smartcity.eu/
3. The Res Novae project. http://resnovae-unical.eu
4. The San Francisco Park project. http://sfpark.org
5. The BESOS project: Building Energy decision Support systems fOr Smart cities. http://besos-project.eu
6. The ROMA project: Resilience enhancement Of a Metropolitan Area. http://www.progetto-roma.org
7. Anttiroiko, A.V.: City-as-a-platform: the rise of participatory innovation platforms in finnish cities. Sustainability 8(9), 922 (2016)
8. Balasubramani, B.S., Shivaprabhu, V.R., Krishnamurthy, S., Cruz, I.F., Malik, T.: Ontology-based urban data exploration. In: Proceedings of the 2nd ACM SIGSPATIAL Workshop on Smart Cities and Urban Analytics (UrbanGIS), pp. 10:1–10:8 (2016)
9. Bellini, P., Benigni, M., Billero, R., Nesi, P., Rauch, N.: Km4City ontology building vs data harvesting and cleaning for smart-city services. J. Vis. Lang. Comput. 25(6), 827–839 (2014)
10. Bianchini, D., De Antonellis, V., Garda, M., Melchiori, M.: Semantics-enabled personalised urban data exploration. In: Proceedings of 19th International Conference on Web Information System Engineering (WISE) (2018). Accepted for publication

11. Bobadilla, J., Ortega, F., Hernando, A., Gutirrez, A.: Recommender systems survey. Knowl.-Based Syst. **46**, 109–132 (2013)
12. Brizzi, P., Bonino, D., Musetti, A., Krylovskiy, A., Patti, E., Axling, M.: Towards an ontology driven approach for systems interoperability and energy management in the smart city. In: International Conference on Computer and Energy Science (SpliTech), pp. 1–7 (2016)
13. Chauhan, S., Agarwal, N., Kar, A.: Addressing big data challenges in smart cities: a systematic literature review. Info **18**(4), 73–90 (2016)
14. Fox, M.S.: PolisGnosis project: representing and analysing city indicators. In: Enterprise Integration Laboratory, University of Toronto Working paper (2015)
15. ISO: Sustainable development of communities - Indicators for city services and quality of life. Standard, International Organization for Standardization (2014)
16. Komninos, N., Bratsas, C., Kakderi, C., Tsarchopoulos, P.: Smart city ontologies: improving the effectiveness of smart city applications. J. Smart Cities **1**(1), 1–16 (2015)
17. Lopez, V., Stephenson, M., Kotoulas, S., Tommasi, P.: Data access linking and integration with DALI: building a safety net for an ocean of city data. In: Arenas, M., et al. (eds.) ISWC 2015. LNCS, vol. 9367, pp. 186–202. Springer, Cham (2015). https://doi.org/10.1007/978-3-319-25010-6_11
18. Psyllidis, A.: Ontology-based data integration from heterogeneous urban systems: a knowledge representation framework for smart cities. In: Proceedings of the 14th International Conference on Computers in Urban Planning and Urban Management (2015)
19. Rani, M., Alekh, S., Bhardwaj, A., Gupta, A., Vyas, O.P.: Ontology-based classification and analysis of non-emergency smart-city events. In: 2016 International Conference on Computational Techniques in Information and Communication Technologies (ICCTICT), pp. 509–514 (2016)
20. Riga, M., Kontopoulos, E., Karatzas, K., Vrochidis, S., Kompatsiaris, I.: An ontology-based decision support framework for personalized quality of life recommendations. In: Dargam, F., Delias, P., Linden, I., Mareschal, B. (eds.) ICDSST 2018. LNBIP, vol. 313, pp. 38–51. Springer, Cham (2018). https://doi.org/10.1007/978-3-319-90315-6_4
21. Rossello-Busquet, A., Brewka, L.J., Soler, J., Dittmann, L.: OWL ontologies and SWRL rules applied to energy management. In: Proceedings of the 2011 UkSim 13th International Conference on Computer Modelling and Simulation, pp. 446–450 (2011)
22. Royo, J.A., Mena, E., Bernad, J., Illarramendi, A.: Searching the web: from keywords to semantic queries. In: Third International Conference on Information Technology and Applications (ICITA), pp. 244–249 (2005)
23. Santos, H., Dantas, V., Furtado, V., Pinheiro, P., McGuinness, D.L.: From data to city indicators: a knowledge graph for supporting automatic generation of dashboards. In: Blomqvist, E., Maynard, D., Gangemi, A., Hoekstra, R., Hitzler, P., Hartig, O. (eds.) ESWC 2017. LNCS, vol. 10250, pp. 94–108. Springer, Cham (2017). https://doi.org/10.1007/978-3-319-58451-5_7
24. Sobral, T., Galvão, T., Borges, J.: Semantic integration of urban mobility data for supporting visualization. Transp. Res. Procedia **24**, 180–188 (2017)
25. Tomašević, N.M., Batić, M.C., Blanes, L.M., Keane, M.M., Vraneš, S.: Ontology-based facility data model for energy management. Adv. Eng. Inform. **29**(4), 971–984 (2015)

Design and Performance Analysis of Load Balancing Strategies for Cloud-Based Business Process Management Systems

Michael Adams[1]([✉]), Chun Ouyang[1], Arthur H. M. ter Hofstede[1], and Yang Yu[2]

[1] Queensland University of Technology, Brisbane, Australia
{mj.adams,c.ouyang,a.terhofstede}@qut.edu.au
[2] Sun Yat-sen University, Guangzhou, China
yuy@mail.sysu.edu.cn

Abstract. Business Process Management Systems (BPMS) provide automated support for the execution of business processes in modern organisations. With the advent of cloud computing, the deployment of BPMS is shifting from traditional on-premise models to the Software-as-a-Service (SaaS) paradigm with the aim of delivering *Business Process Automation as a Service* on the cloud. To cope with the impact of numerous simultaneous requests from multiple tenants, a typical SaaS approach will launch multiple instances of its core applications and distribute workload to these application instances via *load balancing* strategies that operate under the assumption that tenant requests are stateless. However, since business process executions are *stateful* and often long-running, strategies that assume statelessness are inadequate for ensuring a uniform distribution of system load. In this paper, we propose several new load balancing strategies that support the deployment of BPMS in the cloud by taking into account (a) the workload imposed by the execution of stateful process instances from multiple tenants and (b) the capacity and availability of BPMS workflow engines at runtime. We have developed a prototypical implementation built upon an open-source BPMS and used it to evaluate the performance of the proposed load balancing strategies within the context of diverse load scenarios with models of varying complexity.

Keywords: Business Process Management System
Software-as-a-Service · Load balancing · Workflow engine · Scalability

1 Introduction

As a leading exemplar of process-aware information systems, Business Process Management Systems (BPMS) are dedicated to providing automated support for the execution of business processes in modern organisations. In recent times, the advantages offered by cloud computing have triggered an increased demand for the deployment of BPMSs to shift from traditional on-premise models to the Software-as-a-Service (SaaS) paradigm.

© Springer Nature Switzerland AG 2018
H. Panetto et al. (Eds.): OTM 2018 Conferences, LNCS 11229, pp. 390–406, 2018.
https://doi.org/10.1007/978-3-030-02610-3_22

Traditionally, BPMS are installed on-site to serve a single organisation. In a multi-tenant cloud environment, scaling up a discrete BPMS to serve simultaneous demands from multiple organisations is challenging given the limited capacity of a single deployment. Thus it has become necessary to reshape the architectures of BPMS so that better support can be offered to clients seeking the benefits offered by cloud-based implementations. A review of the existing efforts towards a generic architecture for a scalable BPMS in the cloud reveals that this aim has not yet been achieved.

In cloud computing generally, to deal with the increasing load from multiple clients, a SaaS approach often deploys multiple instances of its core applications and distributes workload (of tenant requests) to these application instances via a load balancing strategy [4]. This approach assumes that the tenant requests are stateless, but a business process execution is *stateful* and often long-running, and so existing load balancing strategies designed for handling stateless requests are not suitable for handling the execution of process instances. Rather, load balancing strategies need to take into account the work currently being performed statefully on each application instance.

In this paper, we propose a design for a multi-tiered approach to developing load balancing strategies that can better support the stateful requests and responses associated with the execution of process instances for a multi-tenant cloud environment, with a focus on supporting scalability. We define a dual-faceted metric to measure current process engine load, the first component based on the complexity of a process definition *from the perspective of the computational execution engine*, and the second on measures taken in real time from the operational environment. To validate our approach, a prototype has been developed that deploys multiple instances of a traditional open-source BPMS in a cloud environment, using a stateful load balancing middleware component that implements a number of load balancing strategies at various levels of complexity and suitability. A realistic process load is generated via a simulation tool to demonstrate and validate the applicability of our approach, and demonstrates improved capabilities for supporting large volumes of work in a multi-tenanted cloud environment.

The rest of the paper is organised as follows. Section 2 provides the background and sets out the research problem to address. Section 3 proposes the design of a suite of load balancing strategies. Section 4 discusses a prototypical implementation of a scalable BPMS and the load balancing component, and presents a validation of the approach. Section 5 reviews related work. Finally, Sect. 6 concludes the paper and provides an outline of future work.

2   Background: A Cloud-Based BPMS Architecture

Business Process Management Systems (BPMS) are systems that aim to support business process execution by coordinating the right work to the right person (or application) at the right time. A generic architecture of *traditional* BPMS, as depicted in Fig. 1(a), follows the *workflow reference model* [10] proposed by the

Workflow Management Coalition. The core component is the *workflow engine*, which is responsible for creating, running, and completing execution instances of business processes. A process instance is also called a *case*. Each case consists of a sequence of *work items* and these are managed by the *worklist handler* through which the end users (e.g. staff in an organisation) interact with their own list of work items, e.g. to check out a work item to start the work, or to check in a work item when the work is completed. Other operational matters in a BPMS such as resource administration (e.g. managing the access control of individual end users) and process monitoring (e.g. tracking the progress of each running case) are taken care of by the *administration and monitoring tools*.

In addition, there are also two important data repositories. One is the *process model repository*. Before a process can be executed in the workflow engine, it needs to be defined in the form of a process model using a proper modelling language. A process modelling tool is used to create process models. It is also part of a BPMS but is not further considered in this paper, since we are interested in business process execution (rather than process modelling). Hence, we assume that a process model has always already been deployed to the workflow engine from the process model depository. Next, the workflow engine records the (step-by-step) execution of a process and exports the relevant data in the form of *execution logs*. Such data are valuable for process analysis and monitoring.

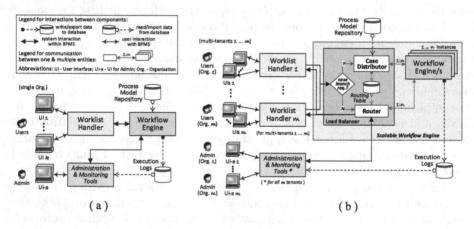

Fig. 1. BPMS architectures: (a) a traditional BPMS deployed on-premise for a single organisation *vs.* (b) a cloud-based BPMS for multiple tenant organisations

However, a traditional BPMS is usually deployed on-premise to serve a single organisation, and as such it is not designed to address simultaneous demands from multiple organisations. When deploying a BPMS in a cloud environment to provide a process automation service to multiple clients, it is necessary to redesign its underlying architecture so that the resulting system is capable of scaling up to cope with large volumes of work from these organisations. Figure 1(b) depicts a generic architecture for cloud-based BPMS. It is a redesign of the traditional BPMS architecture shown in Fig. 1(a) following the principle of a well-established SaaS maturity model [4]. A detailed proposal of the design of this architecture is presented in our previous work [14].

As Fig. 1(b) depicts, the key component to handle simultaneous requests from multiple organisations is the so-called *scalable workflow engine*, which consists of *multiple instances* of workflow engines coordinated via a *load balancer*. The load balancer further comprises a *case distributor*, a *router* and a *routing table*. The case distributor is responsible for handling a request for launching a case. It allocates the case to one of the workflow engines based on a *load balancing strategy* (e.g. to identify the least occupied engine) (see Sect. 3.4). Once a case is distributed to a workflow engine, the case will be executed in a *stateful* manner, meaning that all the work items belonging to the case will be handled by the *same* workflow engine until the case is completed. After a case is launched, the router takes over the responsibility of communication to direct work items between worklist handlers and workflow engines. The routing table records the necessary information in coordinating each work item between the right workflow engine and the right worklist handler.

Figure 2 shows a UML sequence diagram capturing the sequence of interactions between a worklist handler, a case distributor and the workflow engines for launching a case. Upon receiving a request to launch a case, the case distributor will (1) ask each workflow engine for its current busyness status in a certain time period, (2) invoke a load balancing algorithm to decide on an appropriate engine to allocate the case, (3) communicate with that engine to launch the case, and (4) record in the routing table the information necessary for correctly routing the work items belonging to the case.

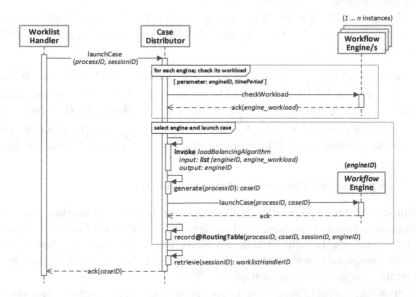

Fig. 2. A UML sequence diagram capturing the interactions of launching a case

3   Design of Load Balancing Strategies

This section focusses on the design of several load balancing strategies that may be applied to handle stateful management of process instances and to achieve an even distribution of the load across an array of process engines in a cloud-based BPMS. It covers a definition of the load balancing problem to address within the scope of this study, an initial proposal for measuring load complexity and resource occupation, and a specification of different levels of capabilities to be achieved for a complete load balancing strategy.

3.1   Problem Definition

A key objective of our load balancing strategies is to ensure a uniform distribution of system load among multiple instances of workflow engines deployed in a cloud-based BPMS. We assume that the number of workflow engines in the system is fixed and known *a priori*, and also that all these engines are of similar (computational) efficiency and capacity. Below, we define the problem of load balancing among multiple workflow engines.

Definition 1 (Balanced workload distribution among engines). *For a cloud-based BPMS, $\mathcal{G}$ denotes a set of workflow engines deployed in the system, $\mathcal{C}$ a set of cases executed in the system, $\mathcal{T}$ a set of timestamps, and $\mathcal{TD}$ a set of time durations. Let $\mathcal{W} \subseteq \mathbb{R}^+ \cup \{0\}$ be a set of values representing the amount of workload.*

- *$\forall g \in \mathcal{G}$, $\omega_g : \mathcal{T} \to \mathcal{W}$ specifies the amount of workload carried by a given workflow engine g at any point of time.*
- *$\gamma : \mathcal{C} \to \mathcal{G}$ specifies the workflow engine to which a case is allocated, i.e. each case $c \in \mathcal{C}$ is allocated to only one workflow engine to support the stateful execution of the case.*

Given a point of time $t \in \mathcal{T}$ and a time duration $d \in \mathcal{TD}$, $[t, t + d]$ defines a time interval, and $\int_t^{t+d} \omega_g(t)$ represents the accumulated amount of workload of a workflow engine g during the time interval $[t, t+d]$. Hence, let $\mathcal{G} = \{g_1, ..., g_n\}$, given any time period $[t, t + d]$, $\int_t^{t+d} \omega_{g_1}(t) \approx ... \approx \int_t^{t+d} \omega_{g_n}(t)$ demonstrates a balanced distribution of work load among the group of engines (specified by $\mathcal{G}$).

3.2   Complexity of Process Instances Measure

One important fact to consider is that process instance definitions may have varying computational complexity, thus imposing workloads of varying capacities on process engines. In light of this, we introduce a new case attribute, namely the *Case Complexity Indicator* (*CCI*), as a quantifiable measure to capture the complexity of a process specification.

Interestingly, while there have been many complexity metrics proposed for process models (see [16] for a survey), every one of them is aimed at measuring complexity from the perspective of a human reader, i.e. as a tool for communication and/or manual analysis. There have been no studies or definitions of complexity metrics of process model instances at runtime from the perspective of the execution engine, to our knowledge. Nevertheless, for this work we have made an initial attempt to consider what may be useful ways to measure how complex a process engine may find a particular case during its execution, taking into account each of the three main process perspectives: control-flow, data and resourcing.

As a starting point, we use the *Extended Cardoso Metric (ECaM)* [11] to inform the value of the CCI. ECaM can be used to measure the structural complexity of a process model and it is built on the classic Cardoso metric [3] which is applicable to all processes (or process languages) that support XOR-, OR-, and AND-splits. ECaM generalises and improves the classic Cardoso metrics so that they can be applied to all Petri net aligned languages.

However, the ECaM on its own is an insufficient measure for execution complexity. We added consideration for the number of tasks defined within a model, since that frequency has a direct effect on the frequent calculation of process state space during execution. These two measures are a first attempt to define control-flow complexity at runtime.

For the data perspective, we consider the number of user-defined data types, and the number of data assignments to input and output variables of tasks, to be indicators of runtime complexity. These attributes are considered to be useful indicators of data load, because each value assignment invokes a transposition that often requires a (re)construction of an explicit data structure to store instance values. Finally, for the resource perspective, the number of roles assigned to tasks is considered, since role groups have to be unpacked at runtime to the contained set of its individual participants, so that work item allocation to individuals can be achieved.

Using these five input measures, we define a runtime complexity metric below.

Definition 2 (Runtime complexity metric). *A runtime complexity metric RCM of a process model is a tuple* $(C, T, U, V, R, W_C, W_T, W_U, W_V, W_R)$ *such that:*

- *C is the extended Cardoso metric*
- *T is the number of tasks*
- *U is the number of user-defined types*
- *V is the number of input and output variables*
- *R is the number of resource roles assigned to tasks*
- *W_C is a weighting factor for the extended Cardoso metric*
- *W_T is a weighting factor for the number of tasks metric*
- *W_U is a weighting factor for the number of user-defined data types metric*
- *W_V is a weighting factor for the number of input and output variables metric*
- *W_R is a weighting factor for the number of roles metric*

A weighting applied to each measure provides the capacity for an overall, balanced complexity metric calculation that can be effectively used to contribute to the measure of load placed on a process engine by a process instance definition at runtime.

3.3   Engine Busyness Measure

The second component of our measure of engine load is derived for four key runtime data points related to the operational environment. We define this component below.

Definition 3 (Operational busyness). *A process engine's operational busyness measure is a tuple* $(R, P, I, T, W_R, W_P, W_I, W_T, L_R, L_P, L_I, L_T)$ *such that:*

- *R the number of requests processed per second*
- *P the average processing time per processed request (in milliseconds)*
- *I the number of work item starts and completions per second*
- *T the number of worker threads currently executing in the engine's container*
- *W_R is a weighting factor for the number of requests processed per second metric*
- *W_P is a weighting factor for the average processing time per processed request metric*
- *W_I is a weighting factor for the number of work item starts and completions per second metric*
- *W_T is a weighting factor for the number of worker threads metric*
- *L_R is an upper limit on the number of requests processed per second, representing a maximum comfortable load for an engine*
- *L_P is an upper limit on the average processing time per processed request, representing a maximum comfortable load for an engine*
- *L_I is an upper limit on the number of work item starts and completions per second, representing a maximum comfortable load for an engine*
- *L_T is an upper limit on the number of worker threads currently executing, representing a maximum comfortable load for an engine*

The upper limit attributes should be configurable, to account for different hardware infrastructures, while the weightings can be set for each so that an accurate factor for any platform can be established. The final busyness factor calculation (B) for an engine is given as:

$$B = (((((R/L_R) \cdot W_R) + ((P/L_P) \cdot W_P) + ((I/L_I) \cdot W_I) + ((T/L_T) \cdot W_T))/4) \cdot 100$$

A final metric for the execution load a process engine is currently experiencing can be defined as the total of all process complexity metrics for currently executing cases combined with the total operational busyness factor for the engine.

3.4   Capability Requirements Analysis

We define a set of specific requirements for a load balancer so that it can reach different levels of capability in distributing the load of process instances to the appropriate process execution engine at run-time. The even distribution of load across all active engines in an array can be supported through the dynamic calculation of the *busyness* of each engine using different strategies, leading to the following four capability levels.

- *Level 0:* Applying a general work scheduling or resource allocation mechanism (e.g. random choice) *without* considering how busy each engine is. This is essentially equivalent to a default stateless load distribution.
- *Level 1:* Considering each engine's busyness *at the exact instant of time* when a new case is required to be distributed to an engine for execution.
- *Level 2:* Considering each engine's busyness *within the time period* of a sliding window looking *backward* (i.e. a number of time units in the past) from the time of case distribution.
- *Level 3:* Considering each engine's busyness *within the time period* of a sliding window looking *forward* (i.e. a number of time units in the future) from the time of case distribution.

4   Performance Analysis and Validation

4.1   Proof of Concept

A prototype that implements the conceptual design shown in Fig. 1(b) has been realised in the YAWL environment [9]. YAWL was selected as the implementation platform because it is open-source, stable, and offers a service-oriented architecture, allowing the new load balancing component to be implemented independent to the existing components. Importantly, absolutely no changes were required to be made to the YAWL environment itself in enabling support for an array of YAWL engines in a cloud environment.

The load balancer component manages a set of available engines. It is situated as a middleware layer that captures all interface calls from external worklist endpoints and redirects them to the appropriate process engine for processing. This redirection is achieved by a trivial configuration change in each worklist so that rather than having a single engine as its remote endpoint, that endpoint instead refers to the load balancing component. The prototype component currently supports all four load distribution capability levels (0–3) described in the previous section. The component is extensively configurable, and changes to configuration values are applied in real time.

Each time the load balancing component receives a request to launch a new process instance, it will poll each engine for its current busyness factor, in terms of the currently configured capability level, and then pass the request to the least busy engine, except when configured to distribution capability level 0 when a random choice is made. For all other (non-launch case) requests, the routing table is queried and the request directed to the engine handling the relevant process instance.

4.2   Implementation

Whenever the load balancing component receives a request to launch a new case, it uses a selection algorithm to select the *idlest* (least busy) engine, based on the currently configured capability level. The method for engine selection on case launch is shown in Algorithm 1. The procedure begins with the current capability level (*mode*) and the set of active engines (*eSet*) (line 1). A reference to the current least busy engine is stored in the *idlest* attribute, and the lowest busyness value is stored in the *lowb* attribute (primed on line 6). Each engine's busyness factor is derived, depending on the current capability level (line 8). If it is less than the current value of *lowb* attribute, then that value is stored in *lowb* (line 10) and a reference to that engine is stored in the *idlest* attribute (line 11). Once all engines have been processed, the idlest engine reference is returned.

Algorithm 1. Selection of Idlest Engine

```
 1: procedure IDLESTENGINE(mode, eSet)
 2:     if mode.random then                          ▷ Baseline random distribution
 3:         return RANDOMITEM(eSet)
 4:     end if
 5:     idlest ← nil
 6:     lowb ← max                                   ▷ Prime lowest load attribute
 7:     for all e in eSet do                         ▷ For each engine
 8:         b ←GETBUSYNESS(mode, e)                  ▷ Get engine's load
 9:         if b < lowb then
10:             b ← lowb
11:             idlest ← e                           ▷ This e is current idlest
12:         end if
13:     end for
14:     return idlest
15: end procedure
```

A different method is used to find the idlest engine at each capability level on a case launch request, as detailed below.

Capability Level 0. At level 0, a random number, limited by the number of available engines, is generated in the programming environment and then used to select an engine from the set.

Capability Level 1. At level 1, each engine is immediately polled for a busyness factor, comprising the various measures described in Sect. 3.3 that exist for the engine at that instant, the total complexity metric of all the cases currently running on the engine, and the configured weightings for each.

Capability Level 2. When the load balancer is set to distribution capability level 2, a backward sliding window to smooth factor readings is applied. Level 2 uses an exponentially weighted moving average to calculate engine busyness, which allows for a configurable weighting factor to be applied in an exponentially decreasing manner to older readings.

It is calculated recursively:

$$S_t = \begin{cases} Y_1, & t = 1 \\ \alpha \cdot Y_t + (1 - \alpha) \cdot S_{t-1}, & t > 1 \end{cases}$$

where:

- α is the weighting factor (a.k.a. the forget factor), between 0 and 1, representing the degree of decrease for older readings, which diminish more quickly as values of alpha approach 1.
- Y_t is the engine busyness reading at time instance t.
- S_t is the weighted moving average at time instance t.

Capability Level 3. Level 3 dynamically invokes one of three algorithms, depending on the current busyness values stored as a time series, which attempt to predict near-term future values of the series. The three algorithms used are: single variable regression, single variable polynomial regression, and multiple variable linear regression[1]. Each algorithm seeks to extrapolate from the previous set of busyness measures for each engine a trend line or curve that best fits the data, then uses that trend to forecast upcoming busyness values. If a threshold of time series data values has not yet been met, this strategy reverts to capability level 2.

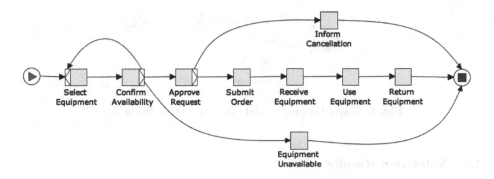

Fig. 3. *Rental Equipment* model used for testing

---

[1] The java-based OpenForecast libraries are used to dynamically select the best-fit algorithm and to perform the calculations (www.stevengould.org/software/openforecast/).

4.3   Validation Environment

We installed the load balancing component and four YAWL engines on the QUT High Performance Computing platform, using a discrete virtual machine for each engine, and another for the load balancing component. Each VM ran on a single core of an Intel Xeon CPU E5-2680 @ 2.70 GHz, and 4 Gb of RAM was available to each.

A simulation tool was used to launch a new instance of a given process specification every 3 s for a total of 200 instances. The tool supports the entire execution of each case by processing work items with resource 'robots', i.e. automated agents that mimic the role of human resources. These agents were configured to simulate processing of each work item for a randomised period within upper limit of between 100 and 3000 ms.

Two process specifications were used for testing. The first is a mostly linear model of an equipment rental process, with a process loop included between the *Select Equipment* and *Confirm Availability* tasks (Fig. 3) to add some variation between cases. An XQuery function was applied to each XOR-split predicate to simulate actual behaviour, with the loop actuating on approximately 40% of *Confirm Availability* task completions, and each cancellation task occurring in approximately 10% of cases.

The second test specification was based on the example found in [1], and models a travel booking process (Fig. 4). Again, an XQuery function was applied to the OR-split predicates following the *register itinerary* task so that for each instance of the specification, one, two or all of the subsequent tasks were executed.

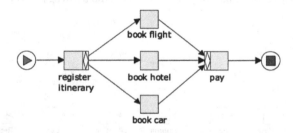

Fig. 4. Travel booking model used for testing (from [1])

4.4   Validation Results

The scatter charts in Fig. 5 show the degree to which the load balancer was able to spread the load across the four engines in our test, for each of the load distribution capability levels, when running simulated cases of the *Rental Equipment* specification. Each chart is represented as a time series graph, where the x-axis specifies a series of equal time slices and the y-axis specifies the number of standard deviations between the set of busyness factors reported for each engine at

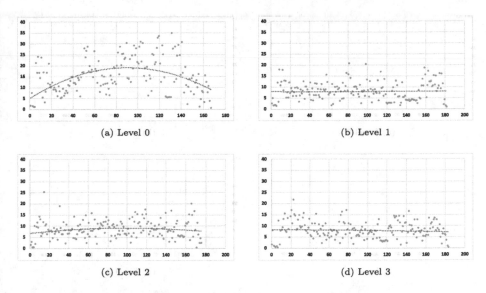

(a) Level 0

(b) Level 1

(c) Level 2

(d) Level 3

Fig. 5. Scatter charts showing simulation results using *Rental Equipment* model at each load distribution capability level

each time slice. The number of standard deviations is a sound measure of the uniformity of load distribution, where a lower value represents a better balancing of engine loads across all engines.

At distribution capability level 0 (i.e. random choice), as expected there is a wide variance in the busyness values of each engine at any particular time slice (Fig. 5(a)). Almost half of the values exceed 15 std. devs. and 14% exceed 25 std. devs. These results illustrate the inappropriateness of typical stateless load balancing algorithms for supporting stateful, possibly long lived process executions.

Setting the load balancer to distribution capability level 1, which calculates a busyness value for each engine at the moment a new case start is requested, can be seen in Fig. 5(b). It is evident that there is an immediate strong improvement over the level 0 random distribution when each engine's current busyness is taken into account, with a narrowing of the range of engine busyness factors at each time slice. Almost all values are under 20 std. devs., and over 92% are less than 15 std. devs. However, since snapshot values are used for level 1 distribution, there is still some variance between engines at times, as should be expected.

When the load balancer is set to distribution capability level 2, which uses a backward sliding window to smooth each engine's busyness factor readings, the narrowing of busyness distributions per time unit becomes even more marked, as can be seen in Fig. 5(c). Capability level 3, which uses a prediction algorithm to envisage future busyness movements, shows a further, but less marked, refinement (Fig. 5(d)).

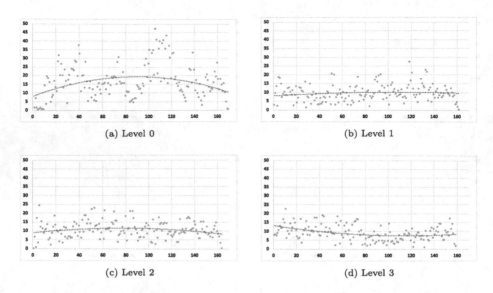

(a) Level 0   (b) Level 1

(c) Level 2   (d) Level 3

Fig. 6. Scatter charts showing simulation results using *Travel Booking* model at each load distribution capability level

A similar set of outcomes for the *Travel Booking* specification can be seen in Fig. 6, which shows that the load balancing component is able to ensure a reasonably even load distribution for models with differing complexity metrics, especially at the higher capability levels.

Finally, Fig. 7 shows the results of a simulation at each level when a random combination of the two test specifications were used, that is, when engines executed a number of each kind of specification simultaneously. While the overall shape of each scatter chart is similar to those when only one specification was used, there is some widening of data points, especially at the lower capability levels, which may indicate more work needs to be done to fine tune the busyness measures and weights to better manage concurrent specifications of different complexities.

It can be seen that the algorithms at each subsequent capability level refine the ability of the load balancing component to maintain a relatively even load across all available engines, with levels 1–3 strongly outperforming the typical stateless load balancing strategy and level 3 with look-ahead prediction performing best of all. By tuning how the busyness of a stateful process engine can be measured, real gains can be achieved in ensuring engines avoid being overloaded to the point where client response times start to suffer. This in turn can help to guarantee Quality of Service thresholds. From our testing, it is clear that implementing an intelligent load algorithm for stateful process management provides improved load balancing outcomes over stateless or random-choice approaches.

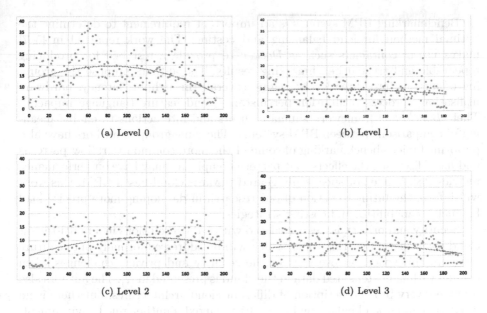

(a) Level 0

(b) Level 1

(c) Level 2

(d) Level 3

Fig. 7. Scatter charts showing simulation results when both test specifications are combined

5   Related Work

An overview of the related literature shows that most of the existing studies focus on the provision of cloud-based SaaS that can be executed for multiple tenants on demand, while the overall stateful process execution is still deployed, operated and managed locally. For example, in [15] the authors discuss a multi-tenant BPM architecture, but utilising only a single-instance Apache ODE system in the cloud and allowing multiple tenants to run their process instances on it. There is no discussion on performance management, or running more than a single instance.

There are a few approaches that require a new BPMS to be developed from the ground up, rather than leveraging existing systems in their entirety in the cloud as discussed in this paper. For example, in [8] the authors propose a hybrid architecture that offers distributed process executions across both client and server sides, although the cloud-based executions are limited to invoked web services. As another example, the research in [13] focusses on an event-based distributed process execution environment coupled with SLA models.

Some of the relevant research emphasizes Quality-of-Service (QoS), such as the generic QoS framework for cloud-based BPMS presented in [12], which comprises QoS specification, service selection, monitoring and violation handling. The authors also present a QoS framework that provides a hierarchical scheduling strategy for costing purposes [20]. Such frameworks guide the design of new support systems based in the cloud, rather than primarily informing the cloud-based hosting of existing BPMS.

Benchmarking BPM systems is an important requirement to determine the optimal method for load balancing and costing. The work reported in [5–7] makes use of containers such as *Docker* to provide a test bed for producing repeatable and trusted benchmarking results. The authors propose the grouping of metrics into three levels: engine (overall throughput, latency, hardware utilisation), process (utilisation per instance) and feature (language impacts). Their *BenchFlow* framework simulates users and interactions to enable meaningful comparisons between BPM systems. The same group of authors have also performed micro-benchmarking of some of the more common workflow patterns and have discussed the effects that pattern complexity may have on performance metrics [18]. While not specifically geared towards cloud-based BPM, this work will inform the future direction of our research in developing more precise load balancing capabilities and costing strategies.

Security and privacy issues associated with BPMS solutions in a multi-tenant cloud is another interesting topic, which will be addressed formally in our future work. In this space, an overview of the security and privacy challenges when dealing with multi-tenanted data in the cloud is presented in [2]. The authors also provide a very broad description of different cloud architectures, but there is no discussion about load balancing in a multi-tenanted, multi-engined environment. In [19] the authors put forward an approach that separates process execution and data management through the use of so-called 'self-guided artifacts', thereby providing data security in a multi-tenanted environment.

Finally, an approach relevant to that presented in this paper is described in [17], where the authors define an architecture for hosting multiple YAWL engines in the cloud for multiple tenants, but only offer a basic load balancing strategy.

6   Conclusion

In this paper, we have detailed a scalable system architecture, with a focus on the design of load balancing strategies, for deploying BPMS in a multi-tenanted cloud environment. Further, we have developed and validated a prototype implementation of a load balancing component that is able to distribute system load over an array of process engines at different capability levels. We have also demonstrated that such a load balancing component can successfully handle a large number of concurrent process instances while balancing system resources, in particular process engines, in a cloud-based BPMS. Given that our design is generic and independent of any specific system or environment, it can be used to serve as a reference for the implementation of specific BPMS in the cloud.

There are several clear directions for future work. One is to develop more precise load balancing capabilities, by considering more complex and accurate measures for runtime process model complexity metrics, and indeed for process engine busyness measures too. Another direction is to develop a further capability level, one which interrogates process logs to determine actual experiential measures of past specification executions and uses that data as input into forecasting methods. A third future direction is to add support for dynamically increasing and/or reducing the number of active process engines at run-time to achieve better scalability and economy of resources. Finally, consideration of machine learning techniques to gradually tune the load balancing algorithms for maximal load smoothing across multiple workflow engines will be another interesting focus of future work in this area.

Acknowledgments. This work is supported by the Research Foundation of Science and Technology Plan Project in Guangdong Province (2016B050502006) and the Research Foundation of Science and Technology Plan Project in Guang-zhou City (2016201604030001).

References

1. van der Aalst, W.M.P., ter Hofstede, A.H.M.: YAWL: yet another workflow language. Inf. Syst. **30**(4), 245–275 (2005)
2. Anstett, T., Leymann, F., Mietzner, R., Strauch, S.: Towards BPEL in the cloud: exploiting different delivery models for the execution of business processes. In: Proceedings of the International Workshop on Cloud Services (IWCS 2009), pp. 670–677. IEEE Computer Society (2009)
3. Cardoso, J.: Control-flow complexity measurement of processes and Weyuker's properties. In: 6th International Enformatika Conference, vol. 8, pp. 213–218 (2005)
4. Chong, F., Carraro, G.: Architecture strategies for catching the long tail. MSDN Library, April 2006
5. Ferme, V., Ivanckikj, A., Pautasso, C., Skouradaki, M., Leymann, F.: A container-centric methodology for benchmarking workflow management systems. In: Proceedings of the 6th International Conference on Cloud Computing and Service Science, pp. 74–84. SciTePress (2016)
6. Ferme, V., Ivanchikj, A., Pautasso, C.: A framework for benchmarking BPMN 2.0 workflow management systems. In: Motahari-Nezhad, H.R., Recker, J., Weidlich, M. (eds.) BPM 2015. LNCS, vol. 9253, pp. 251–259. Springer, Cham (2015). https://doi.org/10.1007/978-3-319-23063-4_18
7. Ferme, V., Ivanchikj, A., Pautasso, C.: Estimating the cost for executing business processes in the cloud. In: La Rosa, M., Loos, P., Pastor, O. (eds.) BPM 2016. LNBIP, vol. 260, pp. 72–88. Springer, Cham (2016). https://doi.org/10.1007/978-3-319-45468-9_5
8. Han, Y.-B., Sun, J.-Y., Wang, G.-L., Li, H.-F.: A cloud-based BPM architecture with user-end distribution of non-compute-intensive activities and sensitive data. J. Comput. Sci. Technol. **25**(6), 1157–1167 (2010)
9. ter Hofstede, A.H.M., van der Alast, W.M.P., Adams, M., Russell, N. (eds.): Modern Business Process Automation: YAWL and its Support Environment, 1st edn. Springer, Heidelberg (2010). https://doi.org/10.1007/978-3-642-03121-2
10. Hollingsworth, D.: Workflow management coalition: the workflow reference model. Technical report, TC00-1003, January 1995
11. Lassen, K.B., van der Alast, W.M.: Complexity metrics for workflow nets. Inf. Softw. Technol. **51**(3), 610–626 (2009)
12. Liu, X., Yang, Y., Yuan, D., Zhang, G., Li, W., Cao, D.: A generic QoS framework for cloud workflow systems. In: 2011 IEEE Ninth International Conference on Dependable, Autonomic and Secure Computing, pp. 713–720. IEEE (2011)

13. Muthusamy, V., Jacobsen, H.-A.: BPM in cloud architectures: business process management with SLAs and events. In: Hull, R., Mendling, J., Tai, S. (eds.) BPM 2010. LNCS, vol. 6336, pp. 5–10. Springer, Heidelberg (2010). https://doi.org/10.1007/978-3-642-15618-2_2

14. Ouyang, C., Adams, M., ter Hofstede, A.H., Yu, Y.: Towards the design of a scalable business process management system architecture in the cloud. In: Trujillo, J. (ed.) ER 2018. LNCS, vol. 11157. Springer, Cham (2018). https://doi.org/10.1007/978-3-030-00847-5_24

15. Pathirage, M., Perera, S., Kumara, I., Weerawarana, S.: A multi-tenant architecture for business process executions. In: 2011 IEEE International Conference on Web services, pp. 121–128. IEEE (2011)

16. Polančič, G., Cegnar, B.: Complexity metrics for process models-a systematic literature review. Comput. Stand. Interfaces **51**, 104–117 (2017)

17. Schunselaar, D.M.M., Verbeek, H.M.W., Reijers, H.A., van der Aalst, W.M.P.: YAWL in the cloud: supporting process sharing and variability. In: Fournier, F., Mendling, J. (eds.) BPM 2014. LNBIP, vol. 202, pp. 367–379. Springer, Cham (2015). https://doi.org/10.1007/978-3-319-15895-2_31

18. Skouradaki, M., Ferme, V., Pautasso, C., Leymann, F., van Hoorn, A.: Microbenchmarking BPMN 2.0 workflow management systems with workflow patterns. In: Nurcan, S., Soffer, P., Bajec, M., Eder, J. (eds.) CAiSE 2016. LNCS, vol. 9694, pp. 67–82. Springer, Cham (2016). https://doi.org/10.1007/978-3-319-39696-5_5

19. Sun, Y., Su, J., Yang, J.: Separating execution and data management: a key to business-process-as-a-service (BPaaS). In: Sadiq, S., Soffer, P., Völzer, H. (eds.) BPM 2014. LNCS, vol. 8659, pp. 374–382. Springer, Cham (2014). https://doi.org/10.1007/978-3-319-10172-9_25

20. Wu, Z., Liu, X., Ni, Z., Yuan, D., Yang, Y.: A market-oriented hierarchical scheduling strategy in cloud workflow systems. J. Supercomput. **63**(1), 256–293 (2013)

Towards Cooperative Semantic Computing: A Distributed Reasoning Approach for Fog-Enabled SWoT

Nicolas Seydoux[1,2]([⊠]) [iD], Khalil Drira[1], Nathalie Hernandez[2],
and Thierry Monteil[1]

[1] LAAS-CNRS, Universit de Toulouse, CNRS, INSA, Toulouse, France
{nicolas.seydoux,khalil.drira,thierry.monteil}@laas.fr
[2] IRIT, Maison de la Recherche, Univ. Toulouse Jean Jaurs,
5 alles Antonio Machado, 31000 Toulouse, France
nathalie.hernandez@irit.fr

Abstract. The development of the Semantic Web of Things (SWoT) is challenged by the nature of IoT architectures where constrained devices are connected to powerful cloud servers in charge of processing remotely collected data. Such an architectural pattern introduces multiple bottlenecks constituting a hurdle for scalability, and degrades the QoS parameters such as response time. This hinders the development of a number of critical and time-sensitive applications. As an alternative to this Cloud-centric architecture, Fog-enabled architectures can be considered to take advantage of the myriad of devices that can be used for partially processing data circulating between the local sensors and the remote Cloud servers. The approach developed in this paper is a contribution in this direction: it aims to enable rule-based processing to be deployed closer to data sources, in order to foster the implementation of semantic-enabled applications. For this purpose, we define a dynamic deployment technique for rule-based semantic reasoning on Fog nodes. This technique has been evaluated according to a strategy improving information delivery delay to applications. The implementation in Java based on SHACL rules has been executed on a platform containing a server, a laptop and a Raspberry Pi, and is evaluated on a smart building use case where both distribution and scalability have been considered.

Keywords: Distributed reasoning · Cloud-Fog processing
Semantic Web of Things · Cooperative semantic computing
SHACL rules

1 Introduction

The maturity of Internet of Things (IoT) communication technologies is fostering a wide variety of industrial and societal applications with responsiveness and privacy requirements, especially useful for home automation or industry

© Springer Nature Switzerland AG 2018
H. Panetto et al. (Eds.): OTM 2018 Conferences, LNCS 11229, pp. 407–425, 2018.
https://doi.org/10.1007/978-3-030-02610-3_23

4.0 scenarios. However, the development of IoT service platforms and related applications faces new challenges emerging from data heterogeneity on the one hand and real-time decision management on the other. Such challenges can be addressed by investigating **semantic interoperability** and **automated reasoning** techniques that respectively allow IoT systems to exchange data with a shared understanding, and to infer new information from existing data and formalized knowledge. These requirements, crucial to IoT applications, are fostering the adoption of the Semantic Web (SW) principles and technologies as interoperability enablers [9]. A new domain has emerged from the interaction between the IOT and the SW, the Semantic Web of Things (SWoT) [15]. The number of connected devices [21] and the volume of generated data to be processed by SWoT systems are growing substantially, requiring **scalability** to be addressed as an additional requirement. Moreover, SW technologies are resource-consuming, and go beyond the capacities of constrained IoT devices, leading to centralizing processing on remote powerful nodes. The SW stack components are often executed on the Cloud, where servers collect and process IoT data before feeding applications with curated data, analytics results and decision instructions [16]. In such a centralized architecture, the Cloud becomes a single point of failure, a bottleneck for communication and a threat for privacy. Storing and processing a large data volume in a central place induces delay [21] and degrades quality of service for IoT applications. It may hamper the development of a variety of applications and inhibit the implementation of time-critical applications. Our objective is to face these issues by applying the Fog computing paradigm to the SWoT infrastructure by considering devices located between constrained sensors and the Cloud as initiated in [13]. Even if constrained, Fog nodes also provide computation capabilities that are starting to be used to process data closer to the sources it is generated from [11,13]. These limited computation capabilities are also leveraged by the development of dedicated SW libraries, such as μJena[1]. The purpose here is not to replace the Cloud by the Fog, but to **complement their capabilities** to overcome their respective limits, thus achieving **Cloud-Fog cooperative semantic computing**.

IoT applications, such as an automatic light manager or a smart city traffic monitor, exploit data collected by device networks for ad-hoc purposes. These applications usually consume IoT data from Cloud nodes, processed to provide high-level information relevant to the application end user [2]. Rules can be used to capture the specific needs of applications by representing and sharing dedicated deduction intent [19]. Existing IoT architectures mainly process rules on the Cloud [4]. In this paper we propose an approach in which Fog nodes **produce information of concern for applications directly** by applying rules on sensor observations, instead of simply forwarding raw data to remote Cloud nodes. Moreover, the diversity of IoT applications implies **multiple strategies for rule deployment**: applications may favor response time, privacy, or energy consumption, and rules should be deployed in the Fog accordingly. For instance, rules applied on nodes closer to sensors can yield deductions faster compared to

---

[1] http://poseidon.ws.dei.polimi.it/ca/?page_id=59.

a centralized approach, both because they receive data earlier and because they process a smaller amount of observations, reducing the end-to-end reasoning time [11, 22]. Identifying such Fog nodes is a possible rule deployment strategy. However, identifying which node should process which rule in order to provide applications with the required results while minimizing resource consumption is challenging, especially in a dynamic setting which characterizes IoT networks.

Our contribution aims at **distributing semantic data processing** in the Fog, according to **adaptable rule and data deployment strategies** defining an **optimal rule deployment** among nodes. We propose Emergent Distributed Reasoning (EDR), a **scalable**, strategy-agnostic approach to semantic processing based on **dynamic** rule deployment in a device network. The core approach is based on a deployment technique enabled by dedicated knowledge representations, and the various strategies are implemented based on modular rules expressed in SHACL, a recent W3C formalism. EDR relies on knowledge local to nodes: each node propagates reasoning rules toward the device network edge. In this paper, EDR is instantiated with a strategy aiming at **increasing application responsiveness**, $EDR_{\mathcal{PT}}$, based on **property types** of observations produced by sensors (e.g., temperature or luminosity).

The remainder of this paper is organized as such: Sect. 2 describes a smart factory use case. In Sect. 3, existing approaches are studied and compared to identify the innovations in EDR. The core contribution is detailed in Sect. 4, with the associated hypothesis and knowledge representations. EDR and $EDR_{\mathcal{PT}}$ are evaluated in Sect. 5, based on the smart factory use case. Finally, the paper is concluded in Sect. 6.

2   Motivating Use Case

Let us consider a production plant divided into two floors, processing different kind of products as illustrated in Fig. 1. These floors are modular: the structure described thereafter is subject to change to adapt to new productions. Each floor is equipped with conveyor belts carrying products from machine to machine for transformation. Devices are organized hierarchically: machines are connected to conveyors that are connected to the floor gateway, that collects and delivers data to the factory datacenter. The factory is equipped with sensors to ensure the safety of workers: each floor is equipped with presence, luminosity particle and temperature sensors, and the workers are equipped with wearables that automatically communicate in BLE[2] with nearby conveyors. Observations from the different sensors are used to identify potentially harmful situations, and then notify the control center, where actions can be taken remotely. Unsafe situations are described with deduction rules, based on the semantic description of observations and of the environment. Examples of rules include "the activation of a machine creating sparks in an atmosphere loaded with particles creates a detonation hazard", or "The presence of a worker near an operating machine in a low luminosity environment is a personal security hazard". Some rules are also

---

[2] https://en.wikipedia.org/wiki/Bluetooth_Low_Energy.

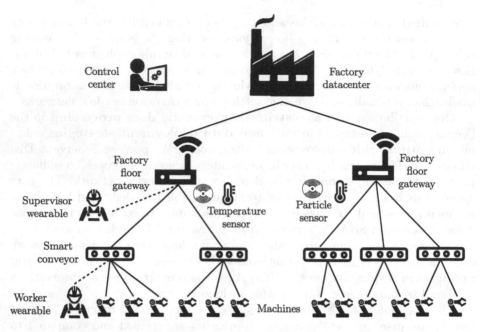

Fig. 1. Fog-enabled smart factory

dedicated to quality insurance: sensors available in the factory, such as temperature sensors, or sensors integrated to machines and to the conveyor, enable the continuous control of production quality. Some operations are temperature-sensitive, and a quality insurance rule is "The detection of a temperature above a certain threshold is a break in the cold chain". Adapting the speed of conveyors to the speed of machines is also part of quality enforcement. All the rules are summarized in Table 1, and their SHACL representation is available online[3].

Safety and quality insurance are time-sensitive applications, which is why the processing of the rules should be as fast as possible. Moreover, the mobility of some sensors (workers wearable), combined to the modularity of the factory floors, are suitable for a dynamic solution adaptative to their evolution over time.

3   Related Work

As the concern of the proposed approach is to deploy reasoning rules among Fog nodes to enable the deduction of application-dedicated information from IoT data, state-of-the-art work dealing with logical rules for the SW, logical rules for the IoT, distributed reasoning and processing on constrained nodes is presented.

Complementary to domain knowledge representation through ontologies, logical rules can be seen as a paradigm for knowledge modeling dedicated to specific

---

[3] https://w3id.org/laas-iot/edr/iiot/iiot.tar.gz.

Table 1. Safety and quality rules

Rule ID	Rule core
R1: Low machine visibility	$Location(l) \wedge Presence(l, o_1) \wedge o_1 = True \wedge Luminosity(l, o_2)$ $\wedge o_2 < 300L \wedge Machine(m) \wedge Activity(m, o_3)$ $\wedge o_3 = True \wedge locatedIn(m, l) \rightarrow LowMachineVisibility(m)$
R2: Low conveyor visibility	$Location(l) \wedge Presence(l, o_1) \wedge o_1 = True \wedge Luminosity(l, o_2)$ $\wedge o_2 < 300L \wedge Conveyor(c) \wedge Activity(c, o_3)$ $\wedge o_3 = True \wedge locatedIn(c, l) \rightarrow LowConveyorVisibility(c)$
R3: No supervision	$Location(l) \wedge Presence(l, o_1) \wedge o_1 = False$ $\wedge Conveyor(c) \wedge Activity(c, o_3) \wedge o_3 = True \wedge locatedIn(c, x)$ $\wedge SupervisorPost(s) \wedge supervises(s, c) \rightarrow NoSupervision(c)$
R4: Fire hazard	$Location(l) \wedge ParticleLevel(l, o_1) \wedge o_1 > 25\%$ $\wedge SparkMachine(m) \wedge Activity(m, o_3) \wedge o_3 = True$ $\wedge locatedIn(m, l) \rightarrow Firehazard(m)$
R5: Cold chain broken	$Location(l) \wedge Temperature(l, o_1) \wedge o_1 > 6^{\circ}C$ $\wedge TemperatureSensitiveMachine(m) \wedge Activity(m, o_3)$ $\wedge o_3 = True \wedge locatedIn(l, m) \rightarrow ColdChainBroken(m)$
R6: Conveyor too fast	$Conveyor(c) \wedge Machine(m) \wedge onConveyor(m, c)$ $\wedge MachineSpeed(m, s_m) \wedge ConveyorSpeed(c, s_c) \wedge s_c > s_m$ $\rightarrow ConveyorTooFast(c)$
R7: Low quality product	$Machine(m) \wedge ProductQuality(m, o_1) \wedge o_1 < 98.5$ $\rightarrow LowQualityProduct(m)$

usages. Logical rules are purely used for deduction: if their preconditions are true, the engine deduces their postconditions. With the goal of facilitating rule reuse, Linked Rules principles have been proposed [8]. They apply to rules the basic principles of Linked Open Data and Linked Open Vocabularies: rules are designated by dereferencable URIs, expressed in W3C-compliant standards, and they can be linked to each other. Different formalisms are available to represent logical rules, such as SWRL[4] and SPIN[5]. SHACL[6] and its extension[7] are the latest W3C standard for rules representation. SHACL aims to represent constraints on an RDF graph, called "shapes", as well as deduction rules. SHACL rules, similarly to SPIN, can be based on SPARQL: it is possible to express a production rule in SHACL as a SPARQL CONSTRUCT query. Rules expressed in the SHACL formalism are used in the EDR approach we propose to provide an interoperable rule representation.

Logical rules being explicit deduction representations, they have been considered in IoT networks to express and share the correlation between sensor

---

[4] https://www.w3.org/Submission/SWRL/.
[5] https://www.w3.org/Submission/spin-modeling/.
[6] https://www.w3.org/TR/shacl/.
[7] https://www.w3.org/TR/shacl-af/.

observations and high-level symptoms since early work on the SWoT [20]. [19] lists numerous works using rules for context-awareness in the IoT. Inspired from the Linked Rules, the Sensor-based Linked Open Rules (S-LOR)[4] is dedicated to rules re-usability for deductions based on sensor observations. Deduction rules are a mechanism similar to Complex Event Processing (CEP) approaches such as [10], but the rule representation shifts from an ad-hoc rule format in CEP to a unified format in the SWoT.

In most existing approaches [4,26], rules are handled by Cloud nodes. An example of Industrial IoT (IIoT) use case enabled by Cloud-based semantic rules processing is presented in [25]. This paper proposes a self-configuring smart factory in which conveyors and machines produce data which is processed in a Cloud where user rules are used to make reconfiguration decisions. Rules are expressed in SWRL. Such architecture raises multiple issues, such as the cost of semantic reasoning that increases rapidly with the size of the Knowledge Base (KB) [11], and the impact on resiliency, since the Cloud node constitutes a single point of failure. A way of overcoming these issues is to consider Fog computing, defined by the open Fog consortium[8] as a "system-level horizontal architecture that distributes resources and services [...] anywhere along the continuum from Cloud to Things". The applicability of such a paradigm to the IoT, compared to pure Cloud computing, is particularly studied, e.g., by [13]. This work identifies key IoT requirements tackled by the Fog computing paradigm, namely **low latency**, **network topology dynamism**, and **scalability**. The constrained nature of Fog nodes (compared to Cloud nodes) must be taken into account: processing power or bandwidth are critical resources.

Most approaches for processing on constrained nodes focus on optimizations enabling such processing for a single node without considering the other, or in distributed cases processing placement is not dynamic: all nodes execute the same rules, or each a predefined static rule set. [3] shows how gateways are Fog nodes capable of enriching data: observations are initially produced by legacy devices in ad-hoc formats. It is the gateway, communicating with devices using protocols adapted to constrained environments, such as CoAP, that enriches the data before forwarding it towards the Cloud. Therefore, observations are enriched on the edge of the network, and only the Fog nodes in direct contact with legacy devices have to perform data enrichment. [5] proposes to execute a different type of rules in the Fog: Event Condition Action (ECA) rules associate a deduction with an action, which is used to automate the response of the system to a stimulus. However, the authors only consider one gateway executing the rules, and the ad-hoc rule format is not suited for rule exchange. Regarding processing distribution in existing work, the dynamic nature of IoT networks should be considered. The topology of a network evolves as devices connect, disconnect, or move geographically. Therefore, a viable distribution of rules at a given moment is not guaranteed to remain optimal in the future, and **the distribution strategy should be adapted to the evolution of the network topology**. [11] does not detail the mobility strategy used for its mobile nodes,

---

[8] http://openfogconsortium.org/.

and each node applies all the rules regardless of their relevance to the messages it aggregates. In [22], rule placement is static, in either Cloud or Fog nodes. [24] focuses on resource placement in a Fog-enabled IoT. The authors compute optimal deployment of application modules based on the representation of available resources on the Fog compared to requirements expressed by applications. Module positions are static, and computed at the time of deployment. Rules are deployed on gateways in an IIoT context in [7]. The rules themselves are not expressed using SW formalisms, and they are not dynamically assigned to Fog devices. However, ad-hoc mechanisms enable rule update at runtime. Rules are combined to a semantic engine proposed in [6] so that rules are expressed based on enriched data.

EDR differs from previous proposals in its focus on the dynamic rule deployment in the SWoT system at runtime, involving all the system's nodes. EDR is only based on either common knowledge, or **knowledge local to the node making the decision** to delegate processing to another node.

4   Enabling Rule Deployment Based on Their Representation

EDR is based on rule deployment in the Fog, in order to enable their placement on nodes according to a deployment optimization strategy, such as the proximity to the sensors producing the data they consume. The EDR core deployment technique is agnostic to the chosen strategy, and includes rule and data propagation. Rules are deployed **neighbor-to-neighbor** between Fog nodes, assuming the nodes are organized in a hierarchical topology as described in Sect. 4.1. The deployment of rules and the propagation of data are based on node local knowledge described with a vocabulary on which node functionalities rely. Both functionalities and the vocabulary are presented in Sect. 4.2. This vocabulary is embedded in rules as described in Sect. 4.3, where the modularity of rules is examined.

4.1   Assumptions, Underlying Architecture and Approach Overview

EDR is based on the hypothesis of a **hierarchical network topology**: nodes are organized in a tree-like structure, and only communicate with neighboring nodes. This assumption is made because such topologies are frequent in IoT networks, represented in studies such as [27], [1] (based on the oneM2M standard[9]), [23], or [22]. The tree root is the Cloud, leaves are devices, and nodes in between are Fog nodes. Applications are not part of the Cloud integrated to the IoT topology: they are executed remotely on personal devices such as smartphones or laptops. **Rules represent applicative needs**: when deductions from sensor observations are required by an application, it injects the rule in the network in order to be provided directly with the deductions, instead of being forwarded

---

[9] http://onem2m.org/.

raw data by the network and applying the rules itself. It is assumed therefore that Fog nodes can communicate with applications directly. Rules are initially submitted by applications to the Cloud node, so it is the only node they know a priori. The Cloud provides a unique permanent interface to the network, and the dynamic Fog topology underneath is therefore transparent for applications. Finally, we assume that since IoT data is strongly bound to a spacio-temporal context [14], the distribution of Fog nodes reflects the distribution of features observed by sensors. EDR is suitable for rules exploiting this spacio-temporal locality by correlating data sharing an identical context, for instance the activity of a machine and the level of particles in the same room needed in rule R4 of Table 1.

To ensure decentralization, the EDR approach is executed in parallel on each node able to perform reasoning in the topology, i.e. Cloud nodes and semantic-computing-enabled Fog nodes. EDR is a neighbor-to-neighbor approach: each node only communicates with its direct neighbors within the IoT network, i.e. its parent and children in the hierarchical topology. A parent node propagates a rule to its child if the parent considers that the child is empowered to apply the rule, based on the associated strategy. In the same way, a node n sends back a rule r to its parent if n considers it can no longer apply r. To enable the deployment of rules, nodes exchange messages describing their capabilities, e.g., their location or the type of data they observe. When a node makes a new deduction based on a rule, it sends the result to all the nodes interested, including the application that submitted the rule. A functional representation of an EDR node is provided in Fig. 2: each node has a local KB, where the information it has about its neighbors in the network topology is stored as well as the observations of the environment. The KB also contains the rules that have been sent to the node by either applications or other nodes, and these rules are used by the inference engine to update the KB. How the node functionalities are related to the KB in the core EDR mechanism to enable the propagation of observations and rules is first described in Sect. 4.2. The modular rule representation embedding the deployment strategy, and the updates of the KB they trigger, are detailed in Sect. 4.3.

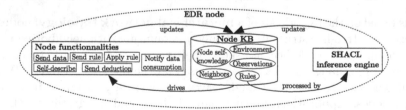

Fig. 2. EDR node functional overview

4.2   A Core Deployment Mechanism Based on a Dedicated Vocabulary

The hierarchical nature of the topology is captured in the EDR vocabulary[10]. The relation between a node n_p and its child n_c is expressed with the triplet $<n_p,lmu{:}hasDownstreamNode,n_c>$[11], based on a nomenclature presented in earlier work [18]. The inverse relation exists, to express the connection between a node n_c and its parent n_p: $<n_c,lmu{:}hasUpstreamNode,n_p>$.

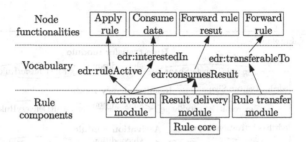

Fig. 3. Relation between node functions and rules modules

Node behavior is made quite simple on purpose, to decorrelate the rule application-specific deployment strategy, e.g., response time reduction or privacy enforcement, from the core deployment technique on which EDR is based, which is application-agnostic. The node functionalities are represented in Fig. 2, and their relation with the EDR vocabulary and the rule modules is shown in Fig. 3. Each functionality relies on dedicated triplets, and the node implements its behavior based on the description held in its KB. The first functionality of a node is to **apply rules**. When a node n receives a new observation, either from its own sensors or lower nodes, n executes the rules r stored in its KB if the description of rule r contains $<r,edr{:}isRuleActive,$true$>$. If processing the new observation with rule r by node n leads to a deduction δ, δ is sent to the nodes n' when $<n',edr{:}consumesResult,r>$ in the KB of n: this is the **rule result delivery** functionality. Especially, the application that submitted the rule to the network, known as the rule originator, is a consumer of rule results, in order to enable deduction delivery to applications. Moreover, if the upper node of node n (denoted u_n) has declared its interest for the type of the new observation α_t, the observation if forwarded toward u_n. Observations are exchanged lazily: if a node n receives an observation of type α_t, and knows no other node interest in such type, the observation is not forwarded. Such interest is represented in node n KB with the triplet $<u_n,edr{:}isInterestedIn,\alpha_t>$. Similarly, node n has to explicitly notify its children that it consumes the data type α_t to receive observations of

---

[10] Used namespaces: **edr**:<https://w3id.org/laas-iot/edr>, **lmu**:<https://w3id.org/laas-iot/lmu>, **sh**:<http://www.w3.org/ns/shacl>, **ex**:<http://example.org/ns>.
[11] Individuals such as $n-p$ and n_c are identified with a URI in the triplets.

this type. This active advertisement is denoted as the **data consumption** functionality: when node n has a triple in its KB such that $<n, edr:interestedIn, \alpha_t>$, it forwards this interest to its children. Finally, when a rule r is present in the KB of node n in a triplet $<r, edr:transferrableTo, n'>$, r if forwarded to node n' by n. This is the **rule transfer functionality**.

These core node functionalities are agnostic of the rule deployment strategy: they are solely based on the representation of themselves and their neighbors by nodes. How the triplets triggering node behaviors are injected in the knowledge base, and their relationship with the rules is detailed in Sect. 4.3.

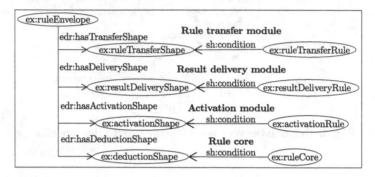

Fig. 4. Rule modules

4.3   Rule Representation and Deployment

EDR rules are composed of several modules, as represented on Fig. 3. Each of these modules enables some node functionalities: therefore, the intelligence is located in the rules, and not in the nodes. The deployment technique can therefore be parametrized at a fine granularity, for each rule (Fig. 4).

Rules are represented in SHACL, and the modules are based on SHACL advanced features named "SHACL rules". Each module is composed of two parts: a SHACL rule, that inserts deductions into the KB, and a SHACL shape that determines whether the rule is applied or not. An example rule is provided online[12]. In the remainder of this section, EDR is instantiated with a rule deployment and data propagation strategy enabled by rule modules referred to as EDR$_{\mathcal{PT}}$. EDR$_{\mathcal{PT}}$ extends the work initially proposed in [17], that does not consider the genericity of EDR. Therefore, in this previous work, no distinction is made between the generic EDR approach and its refinement with EDR$_{\mathcal{PT}}$. EDR$_{\mathcal{PT}}$ is based on the **property types** of data produced by nodes in order to reduce deduction delay for applications. EDR$_{\mathcal{PT}}$, dedicated to a specific strategy, is an instantiation of EDR. These properties can be either environmental properties captured by sensor observations (e.g., luminosity) or higher level properties deduced by other rules (e.g., safety).

---

[12] https://w3id.org/laas-iot/edr/iiot/r1.ttl.

The self-representation of nodes (stored in their KB) includes, but is not limited to, the types of property it observes. A node directly produces observations on a property if one of its children is a sensor dedicated to this property. Moreover, indirect observation of a property is enabled by a proxying mechanism. $EDR_{\mathcal{PT}}$ relies on a proxying mechanism such as in [12], where reasoning nodes act as proxy for the capabilities of legacy nodes unable to process enriched data. In $EDR_{\mathcal{PT}}$, each reasoning-enabled node n has a similar role, with respect to both its sensors and lower nodes that are proxied toward its parent. This mechanism makes a node aware of the types of properties produced by any node below its lower nodes while communicating only with its lower nodes, therefore ensuring the locality of its decisions. The production of observations by node n for a property type ρ_t is denoted $<n,edr:producesDataOn,\rho_t>$. When a child node connects to its parent, it includes such triplets in its self-description. If the parent node n was not a producer of the property type ρ_t, it includes a new triplet in its KB $<n,edr:producesDataOn,\rho_t>$, and forwards this triplet to its own parent. If node n was already a producer for rho_t, its capabilities remain unchanged, and the information propagation stops. A similar proxying mechanism is used to propagate interests: when a node n receives a message from its parent n' containing a triple $<n',edr:isInterestedIn,\rho_t>$, n sends a message to its children producing rho_t (if any) containing a triple $<n,edr:isInterestedIn,\rho_t>$. The knowledge of nodes about their environment is therefore limited to their neighborhood, enabling purely local decisions.

The knowledge of property types observed by remote nodes is used for rule deployment in $EDR_{\mathcal{PT}}$. The operational part of the rule, containing the application-dedicated inference, is referred to as the **rule core** module. It is in this part of the rule that the properties observed by the node are exploited to make deductions. The rule core has a structure similar to other rule modules, with a conditional shape and a deduction SHACL rule. The condition of the rule core only tests that the triple $<r,edr:isRuleActive,true>$ is in the node's KB. To materialize the property types featured in rule core in the following, we use the notations $body\_t(r_x) = \{\gamma_1, ..., \gamma_{n'}\}$ and $head\_t(r_x) = \{\delta_1, ..., \delta_{m'}\}$ where γ_i designates the property type of Γ_i, and δ_j the property type of the deduction Δ_j. It should be noted that not all Γ_i or Δ_j used in the rule are relevant to the $EDR_{\mathcal{PT}}$ approach. For R1, an illustrative rule introduced in the use case Sect. 2, $body\_t(R1) = \{luminosity, presence, machine\ state\}$, and $head\_t(R1) = \{low\ machine\ visibility\}$. When a node receives a rule, or when the knowledge it has on its neighbors is updated, it triggers a reasoning step to process rule modules.

The **rule transfer module** determines on which remote nodes the rule may be deployed. A child node n' is considered for the forwarding of a rule r from node n if, $\forall \rho_t \in body_t(r)$, $<n',edr:producesDataOn,rho_t>$. This condition is expressed as a SPARQL query as part of the SHACL rule being the conditional part of the rule transfer module. The deduction part of the module infers $<r,edr:transferrableTo,n'>$, enabling the rule forwarding mechanism of the node.

The **activation module** detects if the current node is suitable to apply the rule itself. To apply a rule r, a node n must be the lowest common ancestor to the producers of property types in the rule body. Such node has a set $\mathcal{P}$ of children partially producing the rule head. Individually, none of the children produce all the elements of the rule head, but combined, their productions enable the processing of the rule. It is characterized as such: $\exists \mathcal{P}$, such as $\forall n' \in \mathcal{P}$, $<n, lmu{:}hasDownstreamNode, n'>$ and $\exists \{\rho_t, \rho'_t\} \subseteq body(r)$, $<n', edr{:}producesDataOn, \rho_t>$ and $\neg \exists <n', edr{:}producesDataOn, \rho'_t>$, and $\forall \rho_t \in body(r), \exists n' \in \mathcal{P}$, $<n', edr{:}producesDataOn, \rho_t>$. If the conditional part of rule r activation module determines that the current node is suitable to apply r, some deductions are inferred. The activity of rule r is made explicit by the triplet $<r, edr{:}isRuleActive, true>$, and the nodes $n' \in \mathcal{P}$ are identified by $<n', edr{:}partialDataProvider, \rho_t>$. The interest of the rule originator o is also denoted with $<o, edr{:}consumesResult, r>$. These inferences enable both the **rule application** and the **rule result delivery mechanisms** as described in Sect. 4.2.

The **result delivery module** enables the forwarding of deductions to other nodes that are not the originator of the rule, such as the parent n' of a node n if n' applies a rule r' that consumes the deductions made by a rule r applied by n. In EDR$_{\mathcal{PT}}$, the condition of the result transfer module checks if a node expressed interest for the type of deductions yielded by the rule. If there exists a triple $<n', edr{:}interestedIn, \rho_t>$, with n' a remote node and ρ_t an element of the rule r's head $head(r)$, then the result transfer module infers that $<n', edr{:}consumesResult, r>$.

It is worth noting that rule modules only need to be evaluated when the rule is received, or when the topology evolves, e.g., with new productions by children or new consumptions by parents. On the other hand, the rule core must be computed each time a new observation is received by the node. To reduce the computation load, and to only process rule modules when needed, a SHACL functionality is used: the reasoner does not consider shapes or rules r such that $<r, sh{:}deactivated, true>$. The modules of a rule r are therefore only activated for a reasoning step when r is received, or when the topology evolves.

5   Experiments

To evaluate EDR$_{\mathcal{PT}}$, the smart factory use case described in Sect. 2 is simulated on a setup detailed in Sect. 5.2. The purpose of the EDR (instantiated by EDR$_{\mathcal{PT}}$) approach we propose is to directly feed rule deductions to applications. However, for comparison purposes, multiple deduction delivery mechanisms are proposed and compared in Sect. 5.1.

5.1   Deduction Delivery Mechanisms

Unlike rule deployment strategies, deductions delivery mechanisms are decorrelated from the rules: they are variations of the "Forward rule result" functionality described in Sect. 4.2. Therefore, the propagation of rules, the deductions they yielded and data is described as intended according to ad-hoc strategies (here,

EDR$_{\mathcal{PT}}$) through the EDR vocabulary. However, for experimental purpose, this propagation can be altered at the node level, preventing rule deployment or rerouting deduction delivery. Five deduction delivery mechanisms are compared in our experiments:

- **Cloud-Indirect-Raw (CIR)** is the baseline approach: the rules are only kept in the Cloud, and raw observations are forwarded neighbor-to-neighbor from the nodes that collect them toward the central node. The Cloud then delivers deductions to applications. Applications are notified by the Cloud, and not by Fog nodes, in all the following strategies except the last one.
- **Cloud-Direct-Raw (CDR)** is also an approach where rules are not deployed, and only processed in the central Cloud. However, the interest proxying mechanism presented in Sect. 4.3 is altered: node that are not the upper node in the hierarchy propagate the interests they receive without proxying them. Therefore, the observation producers directly send raw observations to the Cloud node, where they are used for rule-based deductions.
- **Cloud-Direct-Processed (CDP)** is a hybrid approach, where rules are deployed in the Fog to be processed as far as possible in the hierarchy, but where deduction are delivered directly to the Cloud instead of sending them to applications. To do so, when forwarding a rule it has received, the Cloud node declares itself as the originator instead of the application. Deductions can also be propagated in the Fog if a node explicitly expressed its interest. Processing rules in the Fog means that the propagation of observations is limited to the Fog nodes applying rules consuming such observations.
- **Cloud-Indirect-Processed (CIP)** is another hybrid strategy for EDR$_{\mathcal{PT}}$: rules are processed in the Fog, and the results are propagated neighbor-to-neighbor towards the Cloud before being delivered to applications. To modify the result delivery behavior, whenever a node propagates a rule, it declares itself as the originator of the said rule.
- **Application-Direct-Processed (ADP)** is the purely decentralized strategy that we propose for EDR$_{\mathcal{PT}}$, where rules are processed in the Fog and deductions are delivered directly to applications that submitted the rules. In this case only, a deduction that has been inferred in the network will not be hosted by the Cloud before being delivered.

5.2 Experimental Setup and Implementation

Variations of a reference building's device network topology, described in the use case in Sect. 2 and displayed in Fig. 1, are simulated. The root node at depth 0 is the Cloud server while other nodes are Fog nodes. Sensors, not represented on the figure, are arbitrarily connected under nodes. Each sensor pushes a random observation to its parent every two seconds. To assess the distributed nature of EDR$_{\mathcal{PT}}$, and its suitability for constrained Fog nodes, the experimental setup includes a Raspberry Pi 3, a laptop and a server, described in Table 2. Each physical machine hosts multiple virtual nodes, composed of an HTTP server, a KB, a SHACL-enabled SPARQL engine, and a code base[13].

---

[13] The code is available at https://framagit.org/nseydoux/edr.

To measure the responsiveness of applications enabled by EDR, the **response time between the reception** of the piece of data **and the deductions** they triggered is measured. Precisely, the response time for the processing of a rule is characterized as the time difference between the moment when the most recent data used in the body of the rule is produced, and the moment when the rule head is received by the application. A dedicated timestamp is associated to each observation once it has been enriched, to avoid any impact of the enrichment process on the measure. For instance, if a conveyor speed observation measured at t_1 and a machine speed observation measured at t_2 trigger a deduction by R7 received by the originating application at t_3, the response time for this particular deduction will be $t_3 - max(t_1, t_2)$.

Table 2. Experimental setup

	RAM	Cores	CPU
Server	32 GB	32	3.0 GHz
Laptop	16 GB	8	2.6 GHz
RPi 3	1 GB	4	1.4 GHz

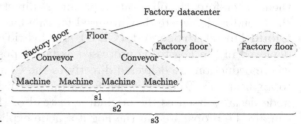

Fig. 5. Simulation topology s*

Experimental measures showed that, for each simulation, the number of deductions is consistant between centralized and distributed approaches: there is no knowledge loss when applying EDR$_{\mathcal{PT}}$ under our assumptions of bound between the Fog topology and the correlation between data.

5.3 Scalability of EDR$_{\mathcal{PT}}$

To assess the scalability of the proposed strategy for EDR, performances have been measured on three topologies, denoted s1, s2 and s3[14], and collectively as s*, as represented on Fig. 5. All s* topologies mimic the use case architecture presented in Fig. 1, with variations in the number of floors. A floor is constituted of two conveyors, each of which supports two machines, with sensors distributed as shown on a JSON blueprint provided online[15], leading to a total of 30 nodes (including both reasoning nodes and sensors). The rules described in Sect. 2 are used. The number of nodes is increased by duplicating floors: s1 has one, s2 two, and s3 three floors, for a total number of respectively 31, 61 and 91 nodes. After presenting deduction delay measures, a deeper analysis of the causes for these delays is provided.

---

[14] Topology representations are available at https://w3id.org/laas-iot/edr/iiot/ scala_syndream/clone_f_⟨0,1,2⟩.ttl respectively.

[15] https://w3id.org/laas-iot/edr/iiot/clone_f_0_blueprint.json.

Due to scaling issues, results are separated in two figures: on Fig. 6, results are shown on the left for deduction propagation strategies CIR and CDR, where observations are sent to the Cloud where they are processed according to rules, and on the right for deduction propagation strategies CIP, CDP and ADP, where $EDR_{\mathcal{PT}}$ is implemented and rules are distributed in the network.

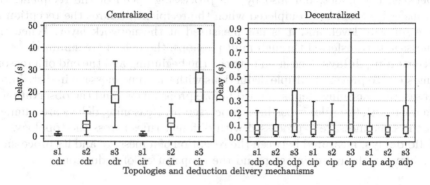

Fig. 6. Scalability measures

The gain in scalability provided by the decentralized approaches appears in the results. In topology s1, the discrepancy between response time for distributed and centralized approaches is reduced, with a median around 0.65 s for CIR and CDR, and 0.065 s for CDP, CIP and ADP. However, in topologies s2 and s3, the gap between centralized and distributed approaches increases dramatically. An increase is also observed for distributed deduction strategies, but it is smaller. Approaches promoting direct communication, i.e. CDR and CDP, perform better that their indirect counterparts, respectively CIR, CIP. This is an expected result, as direct communication reduces the number of hops required for a message (be it an observation or a deduction) to reach its target.

To analyze closely the cause for the increased delay, the journey of a message has been broken down in discrete timestamped events. The first event related to a message is its construction, either by enrichment of an observation or by achieving a deduction. In order to be propagated in the network, a message might be sent from a node n to another node n', which is identified as two events: the sending from node n, and the reception by node n'. Multiple hops are registered, from the first node responsible for the message creation toward any node that is interested in the message content for deduction. When a message is received by a node n, n starts a reasoning step where it tries to make new deductions based on the rules in its knowledge base. Events are logged at the beginning and at the end of reasoning. To detail the delay for each deduction, the journey of the most recent observation leading to the deduction is reconstructed. This journey is built by identifying all consecutive events related to the piece of data leading to the deduction, from its initial enrichment to its processing leading to the deduction, and the delivery of said deduction to the application. The average composition of delays is broken down in Fig. 7, after normalization: each bar represents the share of the delay type in the global delay. Three types of delay have been characterized:

- **Transfer delays**, measured between the emission and the reception of a message. This delay is both impacted by the quality of the network link between two nodes, but also by the processing speed of the recipient: the transfer is considered completed when the recipient declares the reception at the software level, and it is not measured at the network layer. When the message is transferred through multiple hops, the delays are summed.
- **Reasoning delays**, measured between the beginning and the end of a reasoning step. To prevent multiple deductions with the same message in distributed cases, the reasoning step is applied before propagation, and the observation is marked with the rules it has been processed by. Reasoning delays are summed if the same message is processed with different rules across the topology.
- **Idle delays**, measured between the reception of a message and its processing, or between the reasoning step and the propagation of deductions.

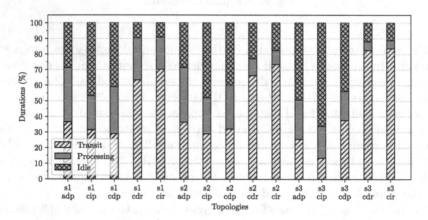

Fig. 7. Breakout of delays (normalized)

A trend that can be observed in the breakout is the increase of the share of transfer time in centralized strategies compared to decentralized ones. An explanation for this phenomenon is the saturation of the network link, combined to an overhead on the central node induced by the necessity to perform all the reasoning. The central node has less CPU time available to declare reception of messages, and therefore the time between the emission event and the reception event is increased. Overall, the limited increase of delays and the balance of the delays breakdown in the distributed settings support our claim that $EDR_{\mathcal{PT}}$ is a scalable approach to rule-base reasoning in the Fog.

6   Conclusion and Future Work

In this paper, we presented EDR, a distributed rule deployment technique for cooperative semantic computing associating Cloud and Fog resources for smart heterogeneous IoT networks. The elaborated solution leverages the Cloud and Fog complementarity to implement an efficient deployment and propagation approach for data, rules and deductions respectively. Such a solution can facilitate the development of SWoT systems and the associated semantically-enabled IoT applications with non-functional requirements related to performance or privacy. The approach we propose is more general than any specific rule deployment strategy chosen by applications according to their requirements: in our contribution, rule deployment strategies are embedded in the rules themselves. We instantiate EDR by $EDR_{\mathcal{PT}}$ with a deployment strategy dedicated to scalability and response time. $EDR_{\mathcal{PT}}$ allows rules to be executed on Fog nodes close to the sensors producing the type of data these rules consume.

The scalability of $EDR_{\mathcal{PT}}$ is studied based on a smart factory use case, where seven rules are propagated across three instances of a reference hierarchical tree-like topology according to five different deduction delivery mechanisms. The results, when measuring responsiveness from an application standpoint, support the performance superiority of the partially or totally distributed deduction delivery mechanism, where rule computation is balanced between Cloud and Fog nodes, compared to the centralized ones, where the whole computation is achieved on Cloud resources.

In future work, we will consider situations where direct communication between Fog nodes and the Cloud or applications is not possible. It is especially interesting in deployments where nodes communicate over ad-hoc networks where border gateways connecting networks can be used as relays to forward observations and deductions. In such a topology, the Application-Direct-Processed and Cloud-Direct-Processed delivery mechanisms cannot be applied, and proposing a new approach derived from Cloud-Indirect-Processed may allow the relaxation of our hypothesis requiring direct communication between the Fog and application. The deployment of rules in the Fog can also be driven by privacy requirements, a predominant concern in the IoT community[16]. Shifting the paradigm from the concentration of data in remote centralized nodes to the propagation of processing close to data producers and consumers enforces the locality of data processing. We intend to elaborate a privacy-aware deployment strategy, to measure how the overhead of traffic, compared to the solution proposed in the present paper, impacts performances, and how to find a trade-off between privacy and performances.

---

[16] https://www.slideshare.net/kartben/iot-developer-survey-2018.

References

1. Alaya, M.B., Medjiah, S., Monteil, T., Drira, K.: Toward semantic interoperability in oneM2M architecture. IEEE Commun. Mag. **53**(12), 35–41 (2015). https://doi.org/10.1109/MCOM.2015.7355582
2. Ali, M.I., et al.: Real-time data analytics and event detection for IoT-enabled communication systems. Web Semant.: Sci. Serv. Agents World Wide Web **42**, 19–37 (2017). https://doi.org/10.1016/j.websem.2016.07.001
3. Desai, P., Sheth, A., Anantharam, P.: Semantic Gateway as a Service architecture for IoT Interoperability. Netw. Internet Arch., 16 (2014). https://doi.org/10.1109/MobServ.2015.51
4. Gyrard, A., Serrano, M., Jares, J.B., Datta, S.K., Ali, M.I.: Sensor-based Linked Open Rules (S-LOR): An Automated Rule Discovery Approach for IoT Applications and its use in Smart Cities. In: International Conference on World Wide Web Companion, pp. 1153–1159 (2017). https://doi.org/10.1145/3041021.3054716
5. Kaed, C.E., Khan, I., Berg, A.V.D., Hossayni, H., Saint-Marcel, C.: SRE: semantic rules engine for the industrial Internet-of-Things gateways. IEEE Trans. Ind. Inform. **14**(2), 715–724 (2018). https://doi.org/10.1109/TII.2017.2769001
6. Kaed, C.E., Khan, I., Hossayni, H., Nappey, P.: SQenloT: semantic query engine for industrial Internet-of-Things gateways. In: 2016 IEEE 3rd World Forum on Internet of Things, WF-IoT 2016, pp. 204–209 (2016). https://doi.org/10.1109/WF-IoT.2016.7845468
7. Kaed, C.E., Khan, I., Van Den Berg, A., Hossayni, H., Saint-Marcel, C.: SRE: semantic rules engine for the industrial Internet- of-Things gateways. IEEE Trans. Ind. Inform. **14**(2), 715–724 (2018). https://doi.org/10.1109/TII.2017.2769001
8. Khandelwal, A., Jacobi, I., Kagal, L.: Linked rules: principles for rule reuse on the web. In: Rudolph, S., Gutierrez, C. (eds.) RR 2011. LNCS, vol. 6902, pp. 108–123. Springer, Heidelberg (2011). https://doi.org/10.1007/978-3-642-23580-1_9
9. Kiljander, J., et al.: Semantic interoperability architecture for pervasive computing and Internet of Things. IEEE Access **2**, 856–873 (2014). https://doi.org/10.1109/ACCESS.2014.2347992
10. Li, Z., Chu, C.H., Yao, W., Behr, R.A.: Ontology-driven event detection and indexing in smart spaces. In: IEEE International Conference on Semantic Computing, pp. 285–292 (2010). https://doi.org/10.1109/ICSC.2010.63
11. Maarala, A.I., Su, X., Riekki, J.: Semantic reasoning for context-aware internet of things applications. IEEE Internet Things J. **4**(2), 461–473 (2017)
12. Nikoli, S., Penca, V., Konjovi, Z.: Semantic web based architecture for managing hardware heterogeneity in wireless sensor network. Int. J. Comput. Sci. Appl. **8**(2), 38–58 (2011)
13. Patel, P., Intizar Ali, M., Sheth, A.: On using the intelligent edge for IoT analytics. IEEE Intell. Syst. **32**(5), 64–69 (2017). https://doi.org/10.1109/MIS.2017.3711653
14. Perera, C., Zaslavsky, A., Christen, P., Georgakopoulos, D.: Context aware computing for the internet of things: a survey. IEEE Commun. Surv. Tutor. **16**(1), 414–454 (2014). https://doi.org/10.1109/SURV.2013.042313.00197
15. Pfisterer, D., et al.: SPITFIRE: toward a semantic web of things. IEEE Commun. Mag. **49**(11), 40–48 (2011). https://doi.org/10.1109/MCOM.2011.6069708
16. Poslad, S., Middleton, S.E., Chaves, F., Tao, R., Necmioglu, O., Bugel, U.: A semantic IoT early warning system for natural environment crisis management. IEEE Trans. Emerg. Top. Comput. **3**(2), 246–257 (2015). https://doi.org/10.1109/TETC.2015.2432742

17. Seydoux, N., Drira, K., Hernandez, N., Monteil, T.: Reasoning on the edge or in the cloud? Internet Technol. Lett., e51

18. Seydoux, N., Drira, K., Hernandez, N., Monteil, T.: Capturing the contributions of the semantic web to the IoT: a unifying vision. In: Maleshkova, M., Verborgh, R., Gyrard, A. (eds.) Proceedings of the Second SWIT Workshop Co-located with ISWC. CEUR Workshop Proceedings, vol. 1930 (2017)

19. Sezer, O.B., Dogdu, E., Ozbayoglu, A.M.: Context-aware computing, learning, and big data in internet of things: a survey. IEEE Internet Things J. **5**, 1–27 (2018)

20. Sheth, A., Henson, C., Sahoo, S.S.: Semantic sensor web. IEEE Internet Comput. **12**(4), 78–83 (2008). https://doi.org/10.1109/MIC.2008.87

21. Shi, W., Dustdar, S.: The promise of edge computing. Computer **49**(5), 78–81 (2016)

22. Su, X., et al.: Distribution of semantic reasoning on the edge of Internet of Things. In: IEEE UbiComp, p. 79, November 2018

23. Szilagyi, I., Wira, P.: Ontologies and semantic web for the Internet of Things - a survey. In: IECON. IEEE (2016)

24. Taneja, M., Davy, A.: Resource aware placement of IoT application modules in Fog-Cloud Computing Paradigm. In: 2017 IFIP/IEEE Symposium on Integrated Network and Service Management, pp. 1222–1228. IEEE, May 2017. https://doi.org/10.23919/INM.2017.7987464

25. Wang, S., Wan, J., Li, D., Liu, C.: Knowledge reasoning with semantic data for real-time data processing in smart factory. Sensors (Switzerland) **18**(2), 1–10 (2018). https://doi.org/10.3390/s18020471

26. Xu, G., Cao, Y., Ren, Y., Li, X., Feng, Z.: Network security situation awareness based on semantic ontology and user-defined rules for internet of things. IEEE Access **5**, 21046–21056 (2017)

27. Zanella, A., Bui, N., Castellani, A., Vangelista, L., Zorzi, M.: Internet of Things for smart cities. IEEE Internet Things J. **1**(1), 22–32 (2014). https://doi.org/10.1109/JIOT.2014.2306328

Enhancing Business Process Flexibility by Flexible Batch Processing

Luise Pufahl[1] and Dimka Karastoyanova[2](✉)

[1] Hasso Plattner Institut, University of Potsdam, Potsdam, Germany
luise.pufahl@hpi.de
[2] University of Groningen, Gronigen, The Netherlands
d.karastoyanova@rug.nl

Abstract. Business Process Management is a powerful approach for the automation of collaborative business processes. Recently concepts have been introduced to allow batch processing in business processes addressing the needs of different industries. The existing batch activity concepts are limited in their flexibility. In this paper we contribute different strategies for modeling and executing processes including batch work to improve the flexibility (1) of business processes in general and (2) of the batch activity concept. The strategies support different flexibility aspects (i.e., variability, looseness, adaptation, and evolution) of batch activities. The strategies provide a systematic approach to categorize existing and future batch-enabled BPM systems. Furthermore, the paper provides a system architecture independent from existing BPM systems, which allows for the support of all the strategies. The architecture can be used with different process languages and existing execution environments in a non-intrusive manner.

Keywords: Batch activities · Business processes · Flexibility
Flexibility strategies · Separation of concerns · Modular architecture

1 Introduction

Business process management (BPM) and the automation of business processes allows organizations to execute their own as well as collaborative business processes more effectively and efficiently. The main artifact of BPM are process models which help to document, analyze, improve, and automate/execute business processes [6]. A process model represents a blueprint for a set of process instances whereby each process instance represents the execution of a single business case [35]. Each process instance has "an independent existence and typically executes without reference to each other" [26]. However, a common phenomenon in operational business processes is batch processing [29]. Batch

The research leading to these results has been partly funded by the BMWi under grant agreement 01MD18012C, Project SMile.

H. Panetto et al. (Eds.): OTM 2018 Conferences, LNCS 11229, pp. 426–444, 2018.
https://doi.org/10.1007/978-3-030-02610-3_24

processing implies that several cases are collected at specific activities to process them as a group to reduce costs or processing time [12]. For example, in the administration, several invoices are collected first before they are approved and checked [1], or in logistics, several individually ordered goods are processed and packaged as one batch by a package delivery service.

Recent research efforts have introduced batch activities [12,14,19] to model and automate batch work in business processes. A batch activity as presented by Pufahl et al. [19] provides several parameters, such as an activation rule, a maximum batch size etc. to allow for configuration of a batch activity individually, which then can be executed automatically by a BPM system. A prototypical implementation of the concept for batch activities in such an existing system is described in [17]. Existing implementation concepts of batch activities are static and limited in their flexibility. Once a process model with a batch activity and its batch configuration is deployed on a BPM system, it can only be adapted by changing the process model and re-deploying it. However, in practice, more flexible approaches are needed. For instance, in the SMile project[1], innovating the last mile logistics of parcels, batch processing is used for the consolidation of parcels during the last mile delivery. Parcels are provided in batches to different carriers whereas different variations of batching are required depending on the individual carrier (e.g., batch size). Also the adaptation of batches is important in case of exceptions, such as the sudden unavailability of a carrier.

Flexibility is besides time, cost, and quality, one important performance dimension [6] and multiple research efforts focus on enabling the flexibility of business processes. These efforts fall into one of the following groups according to the flexibility needs they address: variability, looseness, adaptation, and evolution, which were proposed by Reichert and Weber [23]. In order to address the need for flexibility, in this paper we introduce the so called *flexibility strategies for batch activities* – the modeling, deployment, and execution strategy. Inspired by the principle of separation of concerns [5], the flexibility strategies for batch activities presented prescribe different levels of modularity for the design of the business processes that comprise batch configuration.

Our contribution regarding flexibility is twofold: on the one hand we improve the *flexibility of batch activities* in business processes by specifying different strategies for their possible adaptation during modeling and execution. On the other hand, by introducing the flexible batch activities we contribute another mechanism for process adaptation and therefore *enhance process flexibility in general*.

The presented flexibility strategies for batch activities contribute a *systematic approach for evaluating and categorizing batch capabilities of existing BPM systems* in terms of both architecture and implementation. For each of these strategies, we identify the degree of flexibility supported, the requirements they impose on a BPM system for enacting individual strategies, the corresponding advantages, and challenges. Furthermore, each strategy is compared to existing research works. Finally, we contribute a generic architecture of a supporting sys-

---

[1] http://smile-project.de/.

tem, which can guide the realization of all three or only some of the strategies for a concrete implementation depending on the individual requirements.

In the remainder of the paper, first related work in Sect. 2 regarding batch processing in business processes and existing support for process flexibility in service-oriented environments is discussed. After the batch activity concept is revisited in Sect. 3, the three flexibility strategies are introduced and discussed in detail in Sect. 4. Based on the BPM system requirements described in each flexibility strategy, a generic BPM system architecture for flexible batch processing is presented and discussed in Sect. 5. Finally, the paper concludes in Sect. 6 and highlights directions for future work.

2   Related Work

The related work overview will focus on several aspects. We will review related works in batch processing, works focusing on addressing flexibility needs, and we will present some of the approaches investigating how flexibility, modularity and reusability can be enhanced by using different strategies.

Batch Processing in Business Processes. The need for batch processing in business processes was discussed by several works, such as by Reijers and Mansar [24] as redesign heuristic, by Aalst et al. [2] as an escalation strategy to avoid deadline violations, by Fdhila et al. [7] as an instance-spanning constraint, or in process mining by Martin et al. [13]. A first proposal to integrate activities with a batching behavior is the *Compound activity* by Sadiq et al. [27], which is mainly manually driven by the task performers. Concepts to model batch activities in business processes and execute them automatically can be found in the works of Liu et al. [12] and Natschläger et al. [14] having certain limitations, such as the focus on single resources, the merge of instances during the batch execution, and no proof-of-concept implementation. The batch activity concept by Pufahl [17] is designed based on a requirement analysis, and also provides a prototypical implementation in Camunda², an open-source, java-based BPM system. This concept will provide the basis for the work of this paper.

Although flexibility of batch activities is mentioned in the requirements analysis on integrating batch processing in business processes [20], none of previous mentioned batch activity concepts provide sufficient means to support different flexibility needs. Pflug and Rinderle-Ma [16] provide a sequential batch processing approach for activities with long waiting queues with a dynamic classification of instances at runtime. Still, the critical activity and the clustering algorithm has to be selected at design time. The work by Pufahl et al. [18] provides a concept for flexible batch activity configuration. This approach allows adapting the configuration of a batch activity during runtime if certain events occur. This work focuses only on the adaptation of batch configurations; variability, looseness, and evolution aspects are not targeted.

---

² www.camunda.org.

Flexibility of Business Processes. Multiple research efforts have focused on enhancing the flexibility of business processes. Reichert and Weber [23] distinguish four major flexibility needs, namely (a) variability, i.e., the need to have different variants of a process model to handle different customer groups, product groups etc., (b) looseness, which is the need to leave some aspects unspecified during modeling because cases are unpredictable etc., (c) adaptation as the need to react to exceptions or special cases by adapting the process, and (d) evolution being the need to change process models over time due to changed business processes and which also involves the migration of instances. Regarding the flexibility need variability introduced by [23] the survey presented by [25] gives an overview of available approaches and recommendations as about which approach can be applied for specific context. The looseness need has been addressed in works like [4] using the so called node activities, [28] using the concept of pockets of flexibility, [9] using parameterized service compositions, and [10] where Semantic Web Services are dynamically discovered during process deployment or execution and others. The adaptation need has been the subject of abundant literature; some examples are for choreographies [22,23,33]. As for the evolution need, interesting results have been published in works such as [3,30,32] and others.

In related work dedicated to *adaptation* of processes in service-oriented environments, several approaches have been introduced that specify different strategies for modularization of process models and as a result enhancements in flexibility in the different process life cycle phases. For example, in [31] four possible service binding strategies (i.e., the way of selecting services in workflows have been identified and mapped to the process life cycle), and their impact on the modular structure of the workflow-based applications, the service middleware and its architecture have been discussed. The strategies allow for static service binding and three flavors of dynamic binding: (1) traditional binding where the service middleware/bus selects the endpoint, (2) dynamic binding with service deployment and (3) with software stack provisioning. The concept of Semantic Service Bus (SSB) [10] was introduced with the purpose of enabling service selection at deployment or runtime of processes based on semantic information. Both approaches enable *flexibility on the functional dimension* of processes and the improved modularity of the process model facilitates the enhancement of the flexibility of the processes. The *adaptation of the control flow* of processes has been in the focus of the BPEL'n'Aspects approach by Karastoyanova and Leymann [8]. This work introduces a non-intrusive approach for adapting the control flow of processes during their execution and on per-instance basis. The adaptation operations supported are: *insert*, *delete*, and *skip* activity or activities. Inspired by the separation of concerns principle, the adaptation steps are enabled by modular process models that specify adaptation steps separately from the control flow logic. This allows for dynamically switching on and off the adaptations that need to be performed on process instances, even after they have been started and as long as the change happens (in some cases) at the wavefront of the process instance or after it, i.e. elements in the process that have not been executed yet. Note that the approach can be extended with these adaptation operations so that they can be performed on part of the control logic that has already been executed, i.e. before the wavefront, by using the so called rewinding of processes [33]. The advantages of BPEL'n'Aspects over other approaches sum up to (1) runtime adaptation of processes and specification of the adaptation is enabled on a per-instance basis (2) in a generic manner since adaptation operations are mapped to WS operation calls rather than any other language-specific implementation; (3) a general purpose mechanism to specify when and how a process and an adaptation operation are connected; (4) the approach and the infrastructure can be used with legacy process models and process execution environments. The approach we introduce in this work is inspired by these works on process adaptation and are applied and transfered to the problem of insufficient flexibility of batch processing.

3   Background: The Batch Activity Concept

In this section, we revisit briefly the batch activity concept presented in [17,19] as it is the basis for the rest of the paper. Batch activities can be used to collect several process instances and to process them as a group in order to reduce costs or processing time.

For example, Fig. 1a shows a health care process with two batch activities as a BPMN process diagram [15]. In this process, a blood sample is taken from a patient, if a blood test is needed. Then, the sample is brought to the laboratory where the blood sample is prepared for testing. The actual test is conducted by a blood analysis machine. After the test, the results are published in the central hospital information system where they are accessible by the physicians for evaluation in the respective ward. There are two batch activities in this process. The activity *Transport sample and order to lab* models the fact that a ward nurse does not bring each blood sample separately to the lab but instead she delivers several samples together to save transportation time. The second batch activity is the sub-process which consists of two activities: it collects multiple blood samples before a test on a blood analysis machine is started to save some machine costs.

As the health care process presented above, a process model consists in general of nodes that can be activities, events, or gateways and edges [35]. Each process activity can be a task (i.e., non-decomposable activity) or sub-process (i.e., an activity with internal behavior). Different types of tasks exist, such as *user task* where a task performer gets a work item via worklist manager provided, or a *service task* where a service is called. For integrating batch activities, conceptually activities are extended by a so-called batch model [17] as shown in Fig. 1b comprising different batch configuration parameters. A process designer can specify which instances are grouped in a batch (*groupedBy*), when a batch is started (*activationRule*), what is the maximum allowed size of a batch in number of instances (*maxBatchSize*), and how the batch is executed: in parallel or sequential (*executionOrder*). In the case of the batch activity *Transport sample*

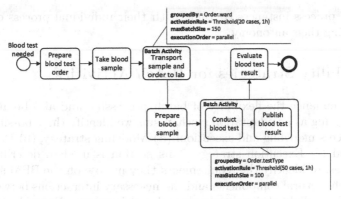

(a) Blood testing process with two batch activities.

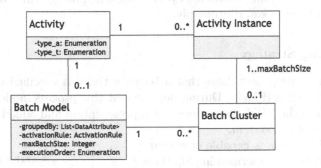

(b) Batch activity concept.

Fig. 1. Batch activity concept and an example process with batch activities from the health care domain.

and order to lab, the blood samples are grouped into batches based on their hospital ward where they were requested (i.e., `Order.ward`). A batch is initialized if at least 20 blood samples are collected or at most after 1 hour waiting time. At maximum 150 blood samples can be processed in one batch and the batch is executed in `parallel` which means that all blood samples of a batch are transported at the same time to the laboratory. In contrast, in `sequential` batch execution the activity instances are still executed one after another but it is used to save setup costs or time.

If an activity has a batch model assigned, its enabled activity instances are *disabled* at runtime, i.e., their execution is paused in order to collect/assign them in/to batches clusters based on the *groupedBy*-parameter. Batch clusters (see Fig. 1b) are responsible for the batch execution. Each batch cluster has at least one activity instance assigned or at most the *maxBatchSize*. A batch cluster is enabled as soon as the activation rule is fulfilled, after which it is provided to the responsible task performer or service. As long as a batch cluster is not started, activity instances can be added to it. After the termination of a batch cluster, the process instances continue with their individual process execution, thus regaining their autonomy.

4   Flexibility Strategies for Batch Activities

In order to enhance the flexibility of batch processing and also be able to use batch processing as an adaptation mechanisms we identify three possible strategies for process modeling and execution: (a) Modeling strategy, (b) Deployment strategy and (c) Execution strategy. This section is used to describe each of the strategies, to identify the requirements they impose on the BPM systems in terms of architectural components and the necessary interactions between them, as well as to highlight their advantages and disadvantages. For each strategy we consider only the relevant process life cycle phases, i.e., design time, deployment, and execution time (a.k.a. runtime phase).

4.1   Modeling Strategy

The modeling strategy postulates that a batch activity is specified in a process model at design time (Fig. 2). During deployment this process model is parsed, transformed into the BPM system internal representation and added to the process repository of the system.

If the BPM system is capable of executing batch activities as described in [20], then anytime a batch activity is reached in a process instance, it will be executed following the execution semantics and under consideration of the batch model that specifies the batch activity. The *requirements*, this strategy poses *on a BPM system* are that (a) it has to be able to parse, transform and store process models containing batch activities and their corresponding configuration information into an internal representation, (b) the execution engine has to implement the execution semantics of a batch activity and extend the life cycle management functionality to be able to deal with instances that belong to a batch cluster: pause the instances

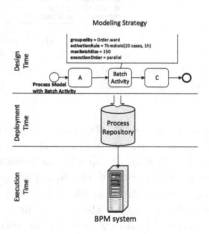

Fig. 2. Modeling Strategy

in their execution, assign them to a cluster, and execute the specific batch behavior as soon as the batch cluster is enabled.

The main *advantage* of the modeling strategy is its simplicity. It only requires the implementation of the parsing and execution of batch activities. However, if there is any change needed in the batch configuration, then the complete process model has to be redesigned and re-deployed. Using this strategy neither improves modularity of the processes nor their flexibility. Despite these disadvantages, existing works on adding a new modeling element to business processes to model and execute batch activities, such as [12,14,19] support only this strategy because of its simplicity. In the case of commercial BPM systems, which do not provide batching support yet, the deployment strategy is considered the most feasible and easy to implement.

4.2   Deployment Strategy

The deployment strategy calls for a modular design and requires keeping the process model and the batch model (batch activity configuration) separate. For the process designer, the modeling of batch activities (i.e., at design time) would not change. What changes is the way the process model and the batch model are stored by the business process modeler. If a batch activity is included in a process model, then its specification, the batch model, is stored in an extra deployment artifact as shown in Fig. 3 as well as a connector describing to which process activity a batch model applies.

This means that the batch activity concept is extended by a connector which realizes an m-to-n relationship between a process activity and a batch model as shown in Fig. 4. Since one batch model can be assigned to multiple activities, and one activity can have multiple batch models, the connector specifies the process context to which a batch configuration applies.

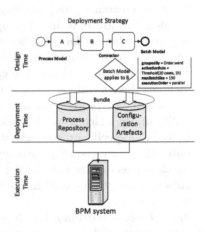

Fig. 3. Deployment Strategy

During deployment, the process model as well as its batch models and connectors are deployed on the BPM system, all as a bundle. As a result, the process model is stored in the process repository, whereas the batch model and connector are stored as configuration artifacts of the BPM system. If an update on the batch model or process model is necessary, all three artifacts – the process model, the batch model, and the connector – have to be deployed again together as bundle.

The *BPM system requirements* we identify are: (a) the process modeling tool should be able to produce separate files/artifacts for the batch model, the process model, and the connector showing which process and which batch configuration go together, as well as the process bundle and a deployment descriptor; (b) the deployment component of the execution environment has to be able to accept the bundle as input and parse, transform and store it accordingly; (c) the execution environment has to support the execution behavior of batch processing. Depending on the implementation approach, the actual connection between the process model and the batch model can be realized either already at deployment time or during the instantiation of processes. The actual redeployment is done by the user. If there is a need to keep track of versions of process and batch models, then the BPM system has to enable this. Similarly, it is an implementation and domain specific concern if the redeployment would lead to process instance migration to the new versions or not, and what would happen with the currently running instances. Existing work on this topic like [3,30,32] present different ways of addressing this issue. It is out of the scope of the strategy definitions to prescribe which approach to use.

Through the concept of separation of concerns (i.e., separating the batch model from the process model), the *advantages* of the deployment strategy are that it improves the modularity, reusability, and also the flexibility of process artifacts. It addressed the *variability* need [23], and allows the *batching over several process models*. In the following, these two aspects are presented in more detail:

- *Batch activity variability*: Several variants of a batch configuration can be defined for a batch activity which is valid for different resources or case types. For instance, a premium customer group can require that its members are only handled optionally as batch, whereas for all other customer groups the batch processing mechanism is always applied [21]. Hence, the connector includes the conditions, a set of data values which the process instances have to fulfill, under which a batch model is connected to a specific process activity. Please note, the other shown parameters of the batch connector are relevant for the execution strategy presented in the following sub-section.

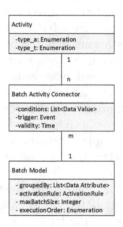

Fig. 4. Batch activity concept with connector

- *Multi-Process Batching*: Batch processing across multiple business processes is enabled by this strategy if one batch model is assigned to several, similar activities in different process models.

As stated, the deployment strategy improves the modularity, maintainability, and reusability which are crucial characteristics from software engineering perspective. However, quick adaptations on the batch model to react on changes in the business environment *on per-instance basis* are not possible with this strategy, since a redeployment is required.

Currently, this strategy is partly supported in the work of [20] in which the batch model is centrally stored in an object life cycle[3] independently from a process model. Independent of the batch processing, a deployment strategy is supported by most BPM systems supporting service orchestrations with or

---

[3] Object lifecycles complement process models and describe allowed actions of business processes on data artifacts across the process-model boundaries.

without flexibility support, like e.g. [8,10,31,33] that are based on open source versions of ActiveBPEL[4] and Apache ODE[5].

4.3   Execution Strategy

The execution strategy allows for assigning a batch model to one or more (or all) running instances of a process model dynamically at execution time. Similar to the deployment strategy, at design time, the process modeling tool has to be able to generate three artifacts for a business process involving batch activities: the process model, the batch model, and the connector. However, the execution strategy postulates that at deployment time, these artifacts are not deployed together as a bundle, but rather independently from each other. It allows for deploying a process model to which only at a later point in time of its life cycle, deployment or execution, a batch model can be designed and/or an existing one assigned by using a separate artifact specifying the connector between them.

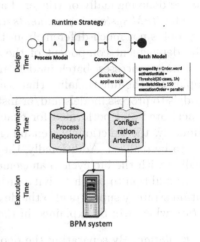

Fig. 5. Execution Strategy

The connector is used to define the exact process instances that will be affected by batch processing. For this, several parameters in the connector shown in Fig. 4 can be defined. The *conditions* consisting of a list of data values defines for which type of process instances a connector between an activity and a batch model is valid. Furthermore, a *trigger* might be given to define under which circumstances a batch model is activated. Therefore, an event is specified that has to occur to activate the corresponding batch model; for instance, the scheduling of a maintenance on the blood testing machine used in the above presented blood testing process. If no trigger is specified, the batch model is immediately activated upon its deployment until its validity ends. The *validity* might be used to specify a point in time or a duration when the assignment of a batch model to an activity or a set of activities ends.

Deleting the assignment of a batch model to instances can also be done at any time during the execution phase. This will affect all process instances which are not yet assigned to a batch cluster or in a batch cluster which is not yet being executed. The clusters that are already in state *running* will still be executed as described by the batch model. Additionally, an update of the process model is possible independently of the batch model and the connector, and updates of a batch model or connector are possible at any time independently of the process model since all these are separate artifacts. When updating the process model,

---

[4] http://www.activebpel.org/.
[5] http://ode.apache.org/.

all possible adaptation mechanisms are allowed at any stage for which abundant literature exists. If a change on the batch activity configuration is necessary, then only the batch model has to be adapted and re-deployed. Same as in the case of deleting an assignment of a batch model; this affects all batch clusters which have not started with their execution yet. The change on the batch configuration can be done manually or triggered automatically.

The *BPM system requirements* are more extensive. Conceptually, the modeling tool is not much different from the one needed for the deployment strategy. The deployment component of the system has to be able to parse, transform and store processes, batch models and connectors. The execution engine has to implement the functionality that will support the dynamic assignment of batch models to process models and/or instances and ensure that the batch processing is activated and performed for exactly these process models and/or instances. Similarly, the functionality enabling the dynamic undeployment of connectors has to be realized. Additionally, the execution environment needs a component dealing with the life cycle management of the batch models and connectors.

In addition to batch activity variability and multi-process batching, the execution strategy supports also the flexibility needs *evolution, looseness* and *adaptation* which will be explained in the following:

- *Evolution*: By separating the deployment of a batch model from the process models, the ability to update the process model or a batch configuration due to changes in the business is further facilitated.
- *Looseness and Process Adaptation*: Batch work does not need to be specified before process execution. A batch configuration can be flexibly assigned and deleted during the execution phase of a business process. Thus, looseness where aspects can be unspecified during modeling is supported. Furthermore, the adaptability of processes is improved since we introduce and allow the use of batch processing as a new runtime adaptation mechanism.
- *Adaptation of Batch Activity*: The batch activity concept itself is also made flexible since batch configurations can be adapted during execution time in case of exception or special cases.

The execution strategy is a non-intrusive approach improving flexibility, modularity, re-usability, and maintainability of process models and also their batch activities due to the separation of concerns. It will allow batch processing independently of the functionality of a BPM system by providing it as an extension functionality. More details on this are discussed in the following section on the architecture. The disadvantage of the execution strategy is that it is much more complex to implement and requires higher effort in terms of investment in technologies and skills of the software implementation team.

To the best of our knowledge this strategy is only partly supported by the work of Pufahl et al. [18] in which adaptation of running batch clusters can be applied by defining batch adaptation rules explicitly which are triggered by external events. In the same work, the BPM system includes explicit functionalities for the batch execution and the adaptation of batch clusters. The approach is less modular and does not support the evolution of batch activities or process models. The execution strategy is realized for control flow adaptation of processes by the BPEL'n'Aspects approach [8]; however it has not been applied for batch processing and does not consider human users as participants in the activity execution but rather only conventional Web Services.

5   Architecture

Based on the requirements on a BPM system identified in the previous section, this section presents the conceptual architecture of a BPM system that supports the introduced flexibility strategies for batch activities. We start with a description of the main components of the architecture in Sect. 5.1 and present their interactions enabling the flexibility strategies in Sect. 5.2. Advantages, disadvantages, and the feasibility of the realization of the architecture are discussed in Sect. 5.3.

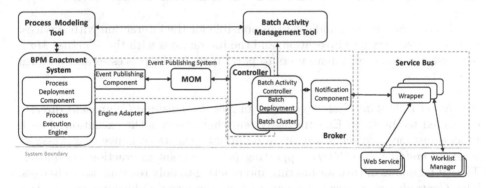

Fig. 6. Architecture for a BPM system realizing the flexibility strategies for batch activities

5.1   Components and Their Functionality

The components of the BPM System architecture supporting the flexibility strategies are presented in Fig. 6. In this figure some of the components are grouped in components on a higher abstraction level (surrounded by red dashed lines) in order to bring structure to the discussion and to denote their higher-level functionality. The system supports the complete life cycle of business processes, namely the modeling, deployment, execution, monitoring and analysis phases; components used for monitoring and analysis are neither discussed nor depicted. Next, the components are presented:

The Process Modeling Tool is used to create process models, batch models, and connectors. It has to be capable of producing the deployment artifacts necessary to support the different strategies (see Sect. 4 for specific details), like deployment descriptor, which typically contains configuration information necessary during deployment, and deployment bundles, or process models separately from the batch models and connectors.

The BPM Enactment System is the execution environment that accepts as input the artifacts produced in the modeling phase. The *Process Deployment Component* parses and stores the models into the system database and prepares the models for instantiation, typically by generating additional artifacts and deploying endpoints at which the model can be instantiated and used. How these steps are performed by the deployment component is strategy- and implementation-specific. The *Process Execution Engine* executed the process behavior and tracks the execution status of all instances of all process model elements. It also delegates the invocation of activity implementations to a corresponding middleware, called the *Service Bus*; we do not show this connection here for brevity.

The Service Bus is the middleware responsible for the interaction with services, including discovery and invocation, and the interactions with the *Worklist Manager*, which allows for human users to participate in the processes and perform certain tasks.

The Event Publishing System contains an *Event Publishing Component* directly connected to the BPM Enactment System that observes the execution of each process instance and sends a notification about each state change to a *message-oriented middleware (MOM)* supporting point-to-point interaction via queues. The MOM allows in turn for filtering and reacting to only relevant state changes. The *Controller* component participating in the event publishing system is a generic component that defines an interface allowing to implement plug-ins for different process extension functionalities that should not be hard-coded in the process language/engine itself [11]. It helps to improve system modularity. An example of such an extension is the runtime adaptation functionality presented in the BPEL'n'Aspects approach [8] that supports inserting new activities in the control flow of a process or deleting activities while process instances are running. For enabling batch processing, a new controller is introduced, called *Batch Activity Controller* which keeps the batching functionality separate from the Process Execution Engine. For enabling the interaction between the BPM Enactment System and the batch activity implementation, the Batch Activity Controller is part of the *Broker* component.

The Broker is a complex component responsible for: (1) enabling the publishing of events of interest from and to the BPM Enactment System so that a reaction upon such events can be defined, (2) notifying the batch implementations (either services or a Worklist Manager) to execute a batch and return the results via the *Notification Component* to the Service Bus, and (3) implementing a reaction to events through the Batch Activity Controller. The Batch Activity Controller implements the behavior as specified in [19], which reacts to the *ready* state of activity instances of batch activities. This means that the Batch Activity Controller will subscribe itself only to those state changes that are pertinent to activities that have a batch model assigned. To enable this, the Batch Activity Controller stores the batch model and the connectors between the batch model and process model, i.e., it possesses the information about the batch configuration. For the *modeling* and *deployment strategy*, the configuration information is provided by the Process Deployment Component of the BPM Enactment System, because the batch configuration is either in-line in the process model, or a part of the process deployment bundle. In the case of the *execution strategy*, the configuration information is provided by the *Batch Activity Management Tool*.

The Batch Activity Management Tool is used to interact with the process modeling tool to create and edit the batch activity models and connectors, and with the Batch Activity Controller to delete, deploy, and undeploy the batch configurations also during process execution. The operations that these architectural components support are described in the following: *Deploying a batch configuration* means that a connector is created that assigns a batch model to process activities; the *undeployment* operation deletes the connector. Assigning batch models to activities using the connectors results in interpreting the batch model together with the connector, i.e., the batch configuration, and creating or deleting a subscription to process instance events (produced by the execution engine). In addition, the controller checks the parameters of the connector definition, like condition, validity and the trigger, and performs an activation step in the case the trigger is evaluated to true. Since the trigger identifies an external event the controller has to be able to subscribe to such events.

5.2   Component Interactions

Regarding the interaction of the components for realizing the flexibility strategies, in Fig. 6, all interactions are depicted by a bi-directional arrows to denote the bi-directional nature.

After a batch configuration is deployed, the *Batch Activity Controller* initiates a subscription to events of activity instances whose activity is assigned to a batch model, more precisely the events indicating that an activity instance is *ready* for execution. If such an event is received, the Batch Activity Controller informs the engine via the *Engine Adapter* to pause the respective activity instance of the batch activity and assigns it to an existing or new batch cluster according to the *batch configuration*. As described in the Sect. 3, the batch cluster itself checks regularly whether its activation rule is fulfilled and in case a batch cluster is enabled the batch execution is started. This functionality is implemented by the Batch Activity Controller for all three strategies.

For the batch execution, the *Broker* deals with the bi-directional communication between the Process Execution Engine and the Service Bus. Therefore, it interacts with the *Service Bus* using notifications and delegates to it the execution of the batch clusters through service invocations, or to humans via the interface provided by *Worklist Manager*. Additionally, it takes care of returning the results of batch executions via the batch activity controller to the concrete process instances.

The communication between the Batch Activity Controller and the Web Services or the Worklist managers is enabled in terms of a pub/sub infrastructure. The actual service calls are done by the so-called wrappers. Wrappers are (1) forwarding batch clusters to the right service invocation or the worklist manager and (2) returning the results of a batch execution back to the Batch Activity Controller. For each of the deployed batch configurations the Batch Activity Controller creates a topic and there is one wrapper that consumes the messages. Upon publication of a message to the topic the wrapper executes the service invocation [11] or delegates the batch cluster to a worklist manager. Upon a batch implementation response, the wrapper publishes the message on a single topic which collects the responses of all batch implementations. The Batch Activity Controller is a subscriber to this topic and processes the responses of the batch implementations as described above.

5.3   Discussion

Recalling the related work, batch processing functionalities are currently designed and implemented as inseparable part of the BPM system. In the modular BPM system architecture we propose here, this functionality is outsourced to support all three proposed flexibility strategies for batch activities. Even though the *execution strategy* caters for highest degree of flexibility, modularity and maintainability, still in some cases the requirements on a solution do not call for as much flexibility, or the effort and budget requirements to implement such a system would not allow its development. Specifically, the effort related extending the Process Modeling Tool and Process Deployment Component of a BPM System are increasing for each strategy: while for the *modeling strategy* only a new modeling element has to be added to the modeler, which needs to be parsed by the deployment component, for the *deployment strategy* different modeling artifacts have to be handled by both components, and for the *execution strategy* the Batch Activity Management Tool is additionally necessary to (un)deploy batch models during process execution. Still, even in the case of the simple *modeling strategy*, the advantage of the modular batch activity component is that it allows for improved maintainability and flexibility at a later point in time if requirements change.

For enhancing an existing BPM system with the Batch Activity Controller functionality two things have to be available: (1) the implementation of the Batch Activity Controller and (2) a mapping between the event model of the process engine and the event model of the presented architecture. The architecture we presented here is designed from its onset with the objective to allow for and encourages the reuse of the Batch Activity Controller implementation as described in [17]. The minimum remaining prerequisite for being able to use the batch processing functionality is the mapping between the event model of the process engine with the event model used in the Controller component. Even though this brings certain implementation complexity, such an approach has been successfully applied for example in the work [11] on a pluggable, generic architecture for enabling extending behavior of BPEL engines and in works like [8] and [34] that separate the control flow adaptation functionality from a process engine for service compositions and choreographies.

6   Conclusion

In this paper, we presented an approach that has the goal to improve the flexibility of batch activities in business processes, while also improving the flexibility of processes in general and increasing the degree of modularity and the maintainability of process artifacts. To this end, we introduced the concept of flexibility strategies for batch activities, namely the *modeling*, *deployment* and *execution* strategies. These strategies help to realize in a systematic way batch processing in business processes with different flexibility degree: from simply allowing batch activities, through batch activity variants and multi-process batching, to flexible batch activities as a dynamic process adaptation mechanism. The individual strategies address one or more flexibility needs for batch work in business processes. Each of the three strategies imposes different requirements on the BPM system and exhibit different advantages and disadvantages. Based on the identified requirements, a conceptual architecture for a BPM system was contributed to support all introduced flexibility strategies.

While we recommend to use the execution strategy for highest degree of flexibility and modularity, we also advise to select the strategy that fits best the business context and the existing BPM system in an enterprise. Our approach can support such a decision since it provides the systematic framework to compare the strategies based on the requirements they impose and the advantages they give.

Our plan for future work is to focus on several aspects. On the one hand, we will work on finalizing the proof-of-concept prototype supporting the flexibility strategies introduced in this paper. In addition to that, we will work towards enabling the monitoring of processes and batch processing, and will consider the different strategies. So far our focus was on the runtime adaptation of a batch configuration targeting batches which have not been started yet. In future, runtime adaptation of batches even after their execution has been started is of particular interest. Since this topic is closely related to the type and properties of participating resources, we will investigate how a resource management can be incorporated into the presented approach in terms of concepts, realization and optimization mechanisms. In this respect, the life cycles of resources and batch activities have to be identified, and the way they intertwine with the life cycle of processes and their instances researched.

References

1. Bizagi Forum. http://feedback.bizagi.com/suite/en/topic/add-existing-entities-to-a-collection. Accessed 12 May 2018
2. van der Aalst, W.M., Rosemann, M., Dumas, M.: Deadline-based escalation in process-aware information systems. Decis. Support Syst. **43**(2), 492–511 (2007)
3. Casati, F., Ceri, S., Pernici, B., Pozzi, G.: Workflow evolution. In: Thalheim, B. (ed.) ER 1996. LNCS, vol. 1157, pp. 438–455. Springer, Heidelberg (1996). https://doi.org/10.1007/BFb0019939
4. Casati, F., Ilnicki, S., Jin, L.J., Krishnamoorthy, V., Shan, M.-C.: Adaptive and dynamic service composition in *eFlow*. In: Wangler, B., Bergman, L. (eds.) CAiSE 2000. LNCS, vol. 1789, pp. 13–31. Springer, Heidelberg (2000). https://doi.org/10.1007/3-540-45140-4_3
5. Czarnecki, K., Eisenecker, U.: Generative Programming: Methods, Tools, and Applications. Addison Wesley, Boston (2000)
6. Dumas, M., La Rosa, M., Mendling, J., Reijers, H.A.: Fundamentals of Business Process Management, vol. 1. Springer, Heidelberg (2013). https://doi.org/10.1007/978-3-642-33143-5
7. Fdhila, W., Gall, M., Rinderle-Ma, S., Mangler, J., Indiono, C.: Classification and formalization of instance-spanning constraints in process-driven applications. In: La Rosa, M., Loos, P., Pastor, O. (eds.) BPM 2016. LNCS, vol. 9850, pp. 348–364. Springer, Cham (2016). https://doi.org/10.1007/978-3-319-45348-4_20
8. Karastoyanova, D., Leymann, F.: BPEL'n'Aspects: adapting service orchestration logic. In: Proceedings of ICWS 2009. IEEE, Los Angeles (2009)
9. Karastoyanova, D., Leymann, F., Nitzsche, J., Wetzstein, B., Wutke, D.: Parameterized BPEL processes: concepts and implementation. In: Dustdar, S., Fiadeiro, J.L., Sheth, A.P. (eds.) BPM 2006. LNCS, vol. 4102, pp. 471–476. Springer, Heidelberg (2006). https://doi.org/10.1007/11841760_41
10. Karastoyanova, D., Wetzstein, B., van Lessen, T., Wutke, D., Nitzsche, J., Leymann, F.: Semantic service bus: architecture and implementation of a next generation middleware. In: Proceedings of ICDE Workshop on Service Engineering (SEIW 2007), pp. 347–354. IEEE Computer Society, April 2007
11. Khalaf, R., Karastoyanova, D., Leymann, F.: Pluggable framework for enabling the execution of extended bpel behavior. In: Di Nitto, E., Ripeanu, M. (eds.) ICSOC 2007. LNCS, vol. 4907, pp. 376–387. Springer, Heidelberg (2009). https://doi.org/10.1007/978-3-540-93851-4_37
12. Liu, J., Hu, J.: Dynamic batch processing in workflows: model and implementation. Futur. Gener. Comput. Syst. **23**(3), 338–347 (2007)
13. Martin, N., Swennen, M., Depaire, B., Jans, M., Caris, A., Vanhoof, K.: Retrieving batch organisation of work insights from event logs. Decis. Support Syst. **100**, 119–128 (2017)
14. Natschläger, C., Bögl, A., Geist, V., Biró, M.: Optimizing resource utilization by combining activities across process instances. Systems, Software and Services Process Improvement. CCIS, vol. 543, pp. 155–167. Springer, Cham (2015). https://doi.org/10.1007/978-3-319-24647-5_13
15. OMG: Business Process Model and Notation (BPMN), V. 2.0 (2011)
16. Pflug, J., Rinderle-Ma, S.: Application of dynamic instance queuing to activity sequences in cooperative business process scenarios. Int. J. Coop. Inf. Syst. **25**, 1650002 (2016)

17. Pufahl, L.: Modeling and enacting batch activities in business processes. Ph.D. thesis, University of Potsdam (2018)
18. Pufahl, L., Herzberg, N., Meyer, A., Weske, M.: Flexible batch configuration in business processes based on events. In: Franch, X., Ghose, A.K., Lewis, G.A., Bhiri, S. (eds.) ICSOC 2014. LNCS, vol. 8831, pp. 63–78. Springer, Heidelberg (2014). https://doi.org/10.1007/978-3-662-45391-9_5
19. Pufahl, L., Meyer, A., Weske, M.: Batch regions: process instance synchronization based on data. In: Enterprise Distributed Object Computing Conference (EDOC), pp. 150–159. IEEE (2014)
20. Pufahl, L., Weske, M.: Batch processing across multiple business processes based on object life cycles. In: Abramowicz, W., Alt, R., Franczyk, B. (eds.) BIS 2016. LNBIP, vol. 255, pp. 195–208. Springer, Cham (2016). https://doi.org/10.1007/978-3-319-39426-8_16
21. Pufahl, L., Weske, M.: Requirements framework for batch processing in business processes. In: Reinhartz-Berger, I., Gulden, J., Nurcan, S., Guédria, W., Bera, P. (eds.) BPMDS/EMMSAD -2017. LNBIP, vol. 287, pp. 85–100. Springer, Cham (2017). https://doi.org/10.1007/978-3-319-59466-8_6
22. Reichert, M., Rinderle, S., Kreher, U., Dadam, P.: Adaptive process management with ADEPT2. In: ICDE 2005, pp. 1113–1114 (2005)
23. Reichert, M., Weber, B.: Enabling Flexibility in Process-aware Information Systems: Challenges, Methods, Technologies. Springer, Heidelberg (2012). https://doi.org/10.1007/978-3-642-30409-5
24. Reijers, H.A., Mansar, S.L.: Best practices in business process redesign: an overview and qualitative evaluation of successful redesign heuristics. Omega **33**(4), 283–306 (2005)
25. Rosa, M.L., Van Der Aalst, W.M.P., Dumas, M., Milani, F.P.: Business process variability modeling: a survey. ACM Comput. Surv. **50**(1), 2:1–2:45 (2017)
26. Russell, N., van der Aalst, W.M.P., ter Hofstede, A.H.M., Edmond, D.: Workflow resource patterns: identification, representation and tool support. In: Pastor, O., Falcão e Cunha, J. (eds.) CAiSE 2005. LNCS, vol. 3520, pp. 216–232. Springer, Heidelberg (2005). https://doi.org/10.1007/11431855_16
27. Sadiq, S., Orlowska, M., Sadiq, W., Schulz, K.: When workflows will not deliver: the case of contradicting work practice. In: International Conference on Business Information Systems (BIS), vol. 5, pp. 69–84 (2005)
28. Sadiq, S., Sadiq, W., Orlowska, M.: Pockets of flexibility in workflow specification. In: S.Kunii, H., Jajodia, S., Sølvberg, A. (eds.) ER 2001. LNCS, vol. 2224, pp. 513–526. Springer, Heidelberg (2001). https://doi.org/10.1007/3-540-45581-7_38
29. Slack, N., Chambers, S., Johnston, R.: Operations and Process Management: Principles and Practice for Strategic Impact. Pearson Education, London (2009)
30. Sonntag, M., Karastoyanova, D.: Concurrent workflow evolution. In: Electronic Communications of the EASST, vol. 37, pp. 1–12. Gesellschaft für Informatik e.V. (GI) (2011). ISSN 1863-2122
31. Vukojevic-Haupt, K., Haupt, F., Karastoyanova, D., Leymann, F.: Service selection for on-demand provisioned services. In: Proceedings of EDOC 2014, pp. 120–127. IEEE, September 2014
32. Weber, B., Rinderle, S., Reichert, M.: Change patterns and change support features in process-aware information systems. In: Krogstie, J., Opdahl, A., Sindre, G. (eds.) CAiSE 2007. LNCS, vol. 4495, pp. 574–588. Springer, Heidelberg (2007). https://doi.org/10.1007/978-3-540-72988-4_40
33. Weiss, A.: Model-as-you-go for choreographies: rewinding and repeating scientific choreographies. IEEE Trans. Serv. Comput. **PP**(99), 1 (2017)

34. Weiß, A., Andrikopoulos, V., Sáez, S.G., Hahn, M., Karastoyanova, D.: ChorSystem: a message-based system for the life cycle management of choreographies. In: Debruyne, C., et al. (eds.) OTM 2016. LNCS, vol. 10033, pp. 503–521. Springer, Cham (2016). https://doi.org/10.1007/978-3-319-48472-3_30
35. Weske, M.: Business Process Management: Concepts, Languages, Architectures, 2nd edn. Springer, Heidelberg (2012). https://doi.org/10.1007/978-3-642-28616-2

Scheduling Business Process Activities for Time-Aware Cloud Resource Allocation

Rania Ben Halima[1,2](✉), Slim Kallel[2], Walid Gaaloul[1], and Mohamed Jmaiel[2,3]

[1] Telecom SudParis, UMR 5157 Samovar, Univ. of Paris Saclay, Paris, France
{rania.ben_halima,walid.gaaloul}@telecom-sudparis.eu
[2] ReDCAD Laboratory, ENIS, University of Sfax, B.P. 1173, 3038 Sfax, Tunisia
{slim.kallel,mohamed.jmaiel}@redcad.tn
[3] Digital Research Center of Sfax, B.P. 275, Sakiet Ezzit, 3021 Sfax, Tunisia

Abstract. Cloud Computing is gaining more and more attention among enterprises thanks to its high performance and low operating cost. Particularly, Cloud resources are used to deploy enterprises' business processes which are constrained by hard timing requirements. Similarly, Cloud providers propose resources in various pricing strategies based on temporal perspective. Taking into consideration both the time constraints and the variety of Cloud pricing strategies helps enterprises to achieve cost-effective process execution plans. Basically, to minimize process costs, stakeholders need to decide the execution time of process activities that overlaps with the temporal interval of the cheapest pricing strategy. In this paper, we present an approach to optimally schedule activities without violating their temporal constraints and capacity requirements. To do so, we use a mixed integer programming model with an objective function under a set of constraints. Our approach has been implemented and the experimental results highlight its performance and effectiveness.

Keywords: Business process · Cloud resource · Pricing
Optimization · Scheduling · Time

1 Introduction

Cloud Computing builds upon advances on virtualization and distributed computing to support cost-efficient usage of computing resources, emphasizing on resource scalability and on demand services [1]. The adoption rate of Cloud Computing is currently driving a significant increase in both the supply and the demand side of this new market for IT utilities [2]. For this reason, Cloud providers are diversifying their pricing strategies (PSs for short) to attract more clients and reduce their unused resources' expenditure. For example, Amazon, one of the largest players in this field, has currently three PSs in place: on-demand, reserved and spot instance charged variously. The cost of each PS depends on some parameters such as time, region, etc. Consequently, enterprises will look to minimize their spending based on less expensive PSs.

© Springer Nature Switzerland AG 2018
H. Panetto et al. (Eds.): OTM 2018 Conferences, LNCS 11229, pp. 445–462, 2018.
https://doi.org/10.1007/978-3-030-02610-3_25

This technological trend has attracted organizations that seek cost-effective ways to deploy and execute their business processes (BPs for short). The latter are influenced by a wide range of temporal constraints, which rise from legal, regulatory, and managerial rules [3]. Time perspective is a critical dimension to consider as it is closely related to customer satisfaction and cost reduction. The temporal constraints need to be viewed from multiple perspectives namely, the relative and absolute temporal constraints [4]. Absolute constraints can be inflexible (i.e. tied to a specific time point) or flexible [5].

To optimize their expenses, enterprises look to allocate cheaper Cloud resources (CRs for short) to run their BPs. Since activities' temporal constraints can be flexible, one can define the process activities start and finish times to overlap with the less expensive CRs temporal availabilities while respecting process requirements and resource constraints. In other words, one needs to define a scheduling execution plan to minimize the process deployment cost.

Many research works have been made in the context of resource allocation and management in the BPM field [6–9]. Whereas, few works deal with task scheduling in Cloud [10,11]. To the best of our knowledge, optimally scheduling BP activities with a wide range of temporal constraints while combining different PSs is not yet handled.

The variety of Cloud providers PSs and the activity requirements (temporal, RAM, vCPU, penalty violation, etc.) makes the optimization problem more complex. Therefore, we define an approach to derive a scheduling execution plan for activities that can be optimally executed using CRs. Concretely, given: (i) a BP with its set of activities' requirements (time, RAM, vCPU) and (ii) a set of CRs delivered from a set of Cloud providers having different PSs properties, our approach provides the optimal scheduling of BP activities that results in the minimal CR allocation cost.

The remainder of this paper is organized as follows: in Sect. 2, some preliminary concepts related to BP and CRs are presented. Section 3 motivates the problem with a real case from France Telecom Orange labs. Next, Sect. 4 details our proposed mixed integer program. Section 5 evaluates the experimental results. An overview about related work is given in Sect. 6. Finally, we present our conclusions and future works in Sect. 7.

2   Preliminaries

This section introduces the main concepts and definitions related to BPs, CR and Cloud provider PSs. Besides, we present the temporal constraints of both: CRs and process activities.

2.1   Cloud Resource

The Cloud Computing delivers three important types of resources which are: computing, networking, and storage [7]. For the sake of simplicity, in this paper, we consider just computing resources. The latter have specific capacities expressed in terms of memory amount RAM and a virtual core number $vCPU$.

Definition 1 *(Cloud resource). A CR is a tuple (id, Cap) where:*

- *id is its unique identifier;*
- *Cap is a tuple that defines the resource capacities, $Cap = (RAM, vCPU) \in \mathbb{N} \times \mathbb{N}$*

2.2 Cloud Providers Pricing Strategies

Each Cloud provider defines a set of PSs to sell its resources. For example, Microsoft proposes prepaid subscription PS. Google proposes a per-minute billing strategy. While, Amazon has three PSs: *on-demand*, *reserved* and *spot*. *On-demand* instances are purchased at a fixed cost per hour. With *reserved* strategy a customer can save up to 30% in resource cost compared to on-demand strategy cost when he reserves a resource capacity for 1 or 3 years [12]. However, bidding for spare unused Amazon instances offers to consumers *spot* instances with a significant discount up to 90% compared to on-demand instances. The bid price is the price to pay per instance hour based on the spot price history available via the Amazon EC2 API and the AWS Management Console [12]. If the spot price overcomes the bid price, the spot instances will be interrupted. Therefore, Amazon proposes spot instances with predefined durations in hourly increments up to six hours in length [8] without an interruption risk.

In the following, we consider Amazon PSs as they are the more expressive (other Cloud provider's PS can expressed using Amazon's PSs) and Amazon is the most popular Cloud provider (see Definition 2).

Definition 2 *(CR pricing strategy). A CR PS St is defined as a triplet $St = (type, TC, c)$:*

- *type is St strategy type;*
- *TC defines the temporal constraints imposed by the strategy St;*
- *c is the unit hour cost proposed by strategy St, $c \in \mathbb{R}$*

We denote by:

- $\mathcal{R} = \{R_i \mid 1 \leq i \leq n\}$ the set of CRs. For a resource $R_i \in \mathcal{R}$, we write $R_i = (id_i, t_i, Cap_i)$;
- $Pr_i = \{Pr_{ij}, \forall j \in \{1, \cdots, p\}\}$ the set of CR providers (where $\forall i \in \{1, \cdots, n\}$);
- $\mathcal{S} = \{St_{ijk} \mid 1 \leq i \leq n, 1 \leq j \leq p, 1 \leq k \leq s\}$ the set of Cloud PSs of each Pr_{ij}. For a specific strategy $St_{ijk} \in St_{ij}$, we write $St_{ijk} = (T_{ijk}, TC_{ijk}, c_{ijk})$.

The temporal constraints of Cloud PSs are:

- **Relative temporal constraints:** specify the time duration allowed for using an instance expressed in terms of a time interval $[MinAvR_i, MaxAvR_i]$ with $1 \leq MinAvR_i \leq MaxAvR_i$.

– **Absolute temporal constraints:** specify the start and finish times of resource temporal availability. For that, we propose the following temporal constraints :

- Start Using No Earlier Than $(SUNET(R_i))$, Finish Using No Earlier Than $(FUNET(R_i))$
- Start Using No Later Than $(SUNLT(R_i))$, Finish Using No Later Than $(FUNLT(R_i))$

We refer the reader to [8] for more details about the pricing temporal constraints.

We present in Tables 1 and 2 a set of CRs $\mathscr{R}=\{R_1, R_2, R_3, R_4, R_5\}$ proposed by Amazon and Microsoft whose instance operating system is Linux and the availability zone of the instances is us-east-1a. Each resource R_i is characterized by a RAM_i and a $vCPU_i$ and different PS costs. For example, $R_2 = $ r3.large is an Amazon instance, its memory $RAM_2 = 15\,\mathrm{GB}$, and $vCPU_2 = 2$. The cost of R_2 defined by $Pr_{21} = $ "Amazon" in various PSs are c_{21k} (where $k \in \{1, \cdots, 4\}$). For instance, the PS $St_{213} = $ (spot, [1h,6h], 0.142\$) defines that the unit hour cost of the spot predefined strategy (T_{213}) is equal to c_{213}=0.142\$ and the instance is temporally available for $MinAvR_2 = 1$ hour and $MaxAvR_2 = 6\,\mathrm{h}$.

Table 1. Virtual Machine Instance Properties by $Pr_{i1} = $ Amazon EC2

Instances	RAM	vCPU	On-demand c_{i11}	Reserved (no upfront)c_{i12}	Spot pred dur c_{i13}	Spot non-pred dur c_{i14}
$R_1 = $ m4.xlarge	16 GB	4	0.215\$/h	0.147\$/h	0.129\$/h [0h,1h]	0.0491\$/h [06 pm,01 am$^{(+1)}$]
					0.142\$/h [1h,6h]	0.0386\$/h [01 am,06 pm]
$R_2 = $ r3.large	15 GB	2	0.166\$/h	0.105\$/h	0.096\$/h [0h,1h]	0.0225\$/h [03 am,10 pm]
					0.102\$/h [1h,6h]	0.0381\$/h [10 pm,03 am$^{(+1)}$]
$R_3 = $ m3.2xlarge	30 GB	8	0.532\$/h	0.380\$/h	0.293\$/h [0h,1h]	0.0787\$/h [10 am,9 pm]
					0.372\$/h [1h,6h]	0.0863\$/h [09 pm,10 am$^{(+1)}$]

Table 2. Virtual Machine Instance Properties by $Pr_{i2} = $ Microsoft

Instances	RAM	vCPU	On-demand c_{i21}
$R_4 = $ A2m v2	16 GB	2	0.128\$/h
$R_5 = $ D4 v2	28 GB	8	0.387\$/h

2.3 Business Process Model

We formally define a process model as a directed graph where nodes are activities gateways or events and edges are control dependencies. BPs can be constrained with a wide range of temporal requirements [3,5]. Those temporal constraints can be relative or/and absolute:

– **Relative temporal constraints:** are used to specify requirements such as activity duration and temporal dependency.

- *Activity duration:* specifies the turnaround time of a BP activity. We assume that each activity a_q (where $q \in \{1, \cdots, r\}$) has a duration expressed in terms of a time interval $[MinD_{a_q}, MaxD_{a_q}]$ where $MinD_{a_q}$ (respectively $MaxD_{a_q}$) specifies a_q minimum (respectively maximum) duration with $1 \leq MinD_{a_q} \leq MaxD_{a_q}$ For instance, a1 and a5 duration is $MinD_{a_q} = 1$h and $MaxD_{a_q} = 2$h.

- *Temporal dependency:* is a relationship between two activities, say a_l and a_q (where $l \neq q$), in which one activity depends on the start or finish of another activity in order to begin or end [13]. Four temporal dependencies $(TD(q, l, du))$ can be specified such as: Start-to-Finish (SF), Start-to-Start (SS), Finish-to-Start (FS) and Finish-To-Finish (FF) [5].

  For example, a temporal dependency FS[0h,4h] between activities a1 and a5 is defined expressing that the time lag between the finish time of a1 and the start time of a5 shall be between 0 h and 4 h.

- **Absolute temporal constraints:** can be specified to define the start and finish times of process activities. These temporal constraint can be inflexible (i.e. tied to specific time point) or flexible [5]. The inflexible temporal constraints $(AInT)$ are:

  - Must Start On (MSO_{a_q})
  - Must Finish On (MFO_{a_q})

  A flexible temporal constraint does not specify a specific time point for a process or an activity a_q, but rather imposes scheduling upper and/or lower bounds [5]. The flexible temporal constraints (AFT) considered are:

  - Start No Earlier Than $(SNET_{a_q})$, Finish No Earlier Than $(FNET_{a_q})$
  - Start No Later Than $(SNLT_{a_q})$, Finish No Later Than $(FNLT_{a_q})$

    In the business process Fig. 1, we can specify absolute temporal constraints for activities. For example, a_3 should start at 11 am (MSO_{a_3}).

  For further details about activities temporal constraints, we refer interested reader to [5].

Definition 3 *(Business Process Model). A BP model is a tuple (N,E,F,Req$_A$):*

- *N is the set of nodes. It includes the set of activities $\mathcal{A}$, gateways G and events Ev;*
- *$F : \mathcal{A} \longrightarrow T$ assigns temporal constraints to activities;*
- *Req$_A$ is the set of activities requested capacities;*
- *$E \subseteq N \times N$ is the set of edges;*

We present in Fig. 1 a real BP "the service supervision process" from France Telecom/Orange labs. Table 3 shows its set of activities $\mathcal{A}$, temporal constraints T and activities requested capacities Req_A.

While activities and some PSs have temporal constraints, it is necessary to ensure the matching between their temporal constraints. For that, in work [8], we proposed an approach to verify formally that CR allocations in a given BP are temporally consistent before finding the best scheduling execution plan.

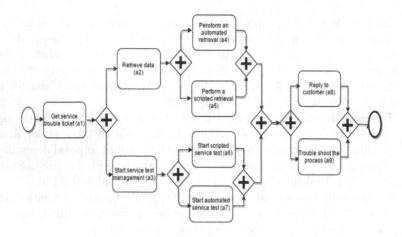

Fig. 1. A service supervision process

Table 3. Process activities requirements

Activities	a1	a2	a3	a4	a5	a6	a7	a8	a9
Durations	[1h,2h]	[2h,3h]	[1h,1h]	[1h,4h]	[1h,2h]	[2h,3h]	[1h,1h]	[1h,1h]	[1h,2h]
Penalties	0.7$	0$	0$	0.2$	0$	0$	0$	0$	0$
RAM	16 GB	15 GB	16 GB	28 GB	16 GB	28 GB	30 GB	16 GB	15 GB
vCPU	4	2	4	8	4	8	8	2	2

Activities need a set of resources capacities Req_A, expressed in terms of RAM_{a_q} and $vCPU_{a_q}$. Table 3 shows the activities requested capacities (see Sect. 2.3). For instance, activities a_1 and a_3 require a virtual machine having a capacity of 16 GB memory and 4 cores. Furthermore, some process activities are critical. So, a penalty cost [8] p_q (where $q \in \{1, \cdots, r\}$) should be added when the activity will not be executed properly. For example, in Table 3, activity a_4 has a penalty cost $p_4 = 0.2\$$. The latter is added if a_4 is performed by a resource bought using a spot instance strategy with an interruption risk.

3   Optimal Resource Allocations in BPs

Enterprise's big challenge is to minimize its expenditure. Thus, it should use the suitable CR to run the activities at an optimal cost while respecting all the process constraints. For the process presented in Fig. 1, we note that more than one resource, in Tables 1 and 2, can satisfy activity requested capacities, shown in Table 3. For example, R_1 and R_5 have sufficient amount of RAM and vCPU to perform a_1, a_2, a_3, a_5, a_8 and a_9. But, it is difficult to decide which is the best CR to allocate for each activity.

Besides, due to the variety of PSs the same resource has different cost values and cheapest one may have a limited temporal availability [8]. Moreover, there are critical activities which have penalty costs if they are interrupted. So, in some cases it may be better to select more expensive PS than a cheaper one with a risk of interruption.

In Table 4, we present different process scheduling execution plan computed based on the process start time and activities' durations. Then, we apply our previous work [6] (see Sect. 6) to find from the set of PSs, the ones that guarantee to the enterprise to have the minimal cost without violating temporal constraints of both: activities and Cloud provider PSs. For instance, for the second execution choice, R_2 is assigned to a_2 as a spot instance with non predefined duration. Whereas, a spot instance with a predefined duration is allocated for a_2 in the third execution choice. This is due to CR temporal availability specified for each PS. Consequently, activity start and finish times can lead to select more expensive PSs to respect activities temporal constraints. Thus, we need to match activities execution time with the temporal interval of cheapest instances.

Table 4. Activities Start and Finish Times and Assigned Instance

Activities	First execution choice			Second execution choice			Third execution choice		
	Start	Finish	Instance	Start	Finish	Instance	Start	Finish	Instance
a_1	02 am	04 am	$R_1 Spot$	02 pm	04 pm	$R_1 Res$	00 am	02 am	$R_1 Spot$
a_2	04 am	07 am	$R_2 SpotNo$	04 pm	07 pm	$R_2 SpotNo$	02 am	05 am	$R_2 Spot$
a_3	04 am	05 am	$R_1 Res$	04 pm	05 pm	$R_1 SpotNo$	02 am	03 am	$R_1 Res$
a_4	07 am	11 am	$R_3 Spot$	07 pm	11 pm	$R_3 Spot$	05 am	09 am	$R_3 Spot$
a_5	08 am	10 am	$R_1 SpotNo$	08 pm	10 pm	$R_1 Spot$	06 am	08 am	$R_1 SpotNo$
a_6	05 am	08 am	R_5	05 pm	08 pm	$R_3 SpotNo$	03 am	06 am	R_5
a_7	05 am	06 am	$R_3 Res$	05 pm	06 pm	$R_3 Res$	03 am	04 am	$R_3 Res$
a_8	11 am	12 am	$R_3 SpotNo$	11 pm	$00^{(+1)}$ am	R_4	09 am	10 am	R_4
a_9	11 am	01 pm	$R_2 Spot$	11 pm	$01^{(+1)}$ am	$R_2 Spot$	09 am	11 am	$R_2 SpotNo$
Process cost	3.887\$			3.12\$			4.016\$		

As a consequence, the challenge is to find for each activity: (i) the CRs that fulfill its requested capacities and (ii) its start and finish times to be able to use the best PS without violating temporal constraints, and with a minimization of penalties costs. The optimal scheduling should ensure: (i) minimization of CR expensive cost while guaranteeing activities requirements in terms of time constraints and requested capacities and (ii) minimization of critical activities penalties costs. To solve this problem, we present in the next section an approach based on MIP.

4   Mathematical Formulation of the Linear Program

We formulate the activity assignment and scheduling problem for resources proposed by Cloud providers in various PSs as a Mixed Integer Programming (MIP)

model. Indeed, mathematical programming, especially MIP, because of its rigorousness, flexibility and extensive modeling capability, has become one of the most widely explored methods for process scheduling problems [14].

We introduce first the necessary inputs and the decision variables. Then, we present our MIP (objective function and constraints). In this manner, a scheduling execution plan and an optimal allocation will be provided.
The inputs include:

- The set of activities $\mathcal{A}$, the set of the activities' requirements, $Req_{\mathcal{A}} = \{Req_{a_q}, a_q \in \mathcal{A}\}$ and the set of activities temporal constraints $T_{\mathcal{A}} = \{T_{a_q}, a_q \in \mathcal{A}\} = AFT \cup AInT$;
- The set of CRs $\mathcal{R} = \{R_i, \forall i \in \{1, \cdots, n\}\}$ needed by activities $\mathcal{A}$
- The set of CR providers $(\forall i \in \{1, \cdots, n\})$ $Pr_i = \{Pr_{ij}, \forall j \in \{1, \cdots, p\}\}$ and the set of Cloud PSs St_{ij} of each Pr_{ij}.

In the following, we present the decision variables of our mathematical problem:

- $X_{ijkq} \in \{0, 1\}$ is a binary decision variable for the assignment of a resource i from provider j in strategy k to activity q.
- $V_q \in \{0, 1\}$ is a binary decision variable that indicates if activity a_q uses a spot instance with non-predefined duration and its penalty cost is not null.
- S_{a_q} (respectively F_{a_q}) is an integer decision variable used to define the start (respectively finish) time of each activity $a_q \in \mathcal{A}$. S_{a_q} and F_{a_q} represent a_q flexible absolute temporal constraints (Sect. 2.3).

The objective function (Eq. (1)) of the model minimizes the CRs cost to run the BP. It seeks to: (i) select for each activity the CR that has the required capacities (ii) select the suitable Cloud PSs for each CR and (iii) define the start and end times of each BP activity to overlap with the cheaper PSs temporal constraints.

$$MinC = \sum_{i=1}^{|\mathcal{R}|} \sum_{q=1}^{|\mathcal{A}|} \sum_{j=1}^{p} \sum_{k=1}^{s} d_q c_{ijk} X_{ijkq} + \sum_{q=1}^{|\mathcal{A}|} p_q V_q \tag{1}$$

The total execution cost includes two terms. The first minimizes the sum of resources costs. Minimizing this quantity is done by deciding the start and end times of activities to be able to allocate the cheapest Cloud instances satisfying all the constraints. This quantity is given by the multiplication of the activity duration d_q (where $d_q \in [MinD_q, MaxD_q]$) by the resource R_i hourly unit strategy cost c_{ijk} defined by the provider Pr_{ij}. The second term minimizes the penalty cost p_q values added when critical activities are performed by spot instances with an interruption risk.

In our work, we assume that the transferred data size between instances is small. Therefore, the data transfer time is about some seconds which is negligible compared to activities duration expressed in hours.
The objective function is subject to the following sets of constraints:

1. **Process Temporal constraints:** The following four constraints are used to ensure the respect of BP temporal constraints. For instance, Eq. (2) guarantees that the start time (S_{a_l}) of each activity is after the finish time (F_{a_q})

of all its predecessors. Equation (3) ensures the respect of inflexible temporal constraints. Equation (4) ensures that the time lag between the end and start times discovered by our MIP for each activity should respect activities temporal durations. Equation (5) limits the process execution temporal duration t defined by the enterprise (deadline).

$$max(F_{a_q}) + TD(q, l, du) \leq S_{a_l}, \forall a_l \in \mathcal{A} \text{ and } q \neq l \tag{2}$$

$$S_{a_q} == MSO_{a_q} \ \& \ F_{a_q} == MFO_{a_q}, \forall a_q \in \mathcal{A} \tag{3}$$

$$0 \leq F_{a_q} - S_{a_q} \leq d_q, \forall a_q \in \mathcal{A} \tag{4}$$

$$S_{a_q} \geq 0 \ \& \ S_{a_q} \leq t \ \& \ F_{a_q} \geq 0 \ \& \ F_{a_q} \leq t, \forall a_q \in \mathcal{A} \tag{5}$$

In our example (Fig. 1), a_2 and a_3 are the successors of a_1 then, they can start after the finish time of a_1 (Eq. (2)), i.e., $F_{a_1} \leq S_{a_2}$ and $F_{a_1} \leq S_{a_3}$. In addition, if we assume that a_1 must start at $MSO_{a_1} = 08am$ and a_4 must finish at $MFO_{a_4} = 05\,pm$ then we have inflexible temporal constraints for those two activities. That is why, in the MIP solution, we should have: $S_{a_1} = MSO_{a_1} = 08am$ and $F_{a_4} = MFO_{a_4} = 05\,pm$ (Eq. (3)). Furthermore, a_2 has a maximum duration equal to $d_q = 3\,h$. Then, the time lag between the start and finish times of a_2 shall be between a_2 temporal duration value (Eq. (4)). The process has a temporal duration that should be respected. Therefore, the start and finish times of activities should be limited. For instance, if the process should finish in one day, then $t = 24$ and all the start and finish times of activities should be included in the temporal interval [0,24] (Eq. (5)).

2. **Resource constraints:** Equations (6) and (7) ensure that the resource's capacities in processing and memory satisfy the activity requirements.

$$\sum_{i=1}^{|\mathcal{R}|} \sum_{j=1}^{p} \sum_{k=1}^{s} min(RAM_i)X_{ijkq} \geq RAM_{a_q}, \forall q \in \{1, \cdots, r\} \tag{6}$$

$$\sum_{i=1}^{|\mathcal{R}|} \sum_{j=1}^{p} \sum_{k=1}^{s} min(vCPU_i)X_{ijkq} \geq vCPU_{a_q}, \forall q \in \{1, \cdots, r\} \tag{7}$$

For instance, activity a_1 should be assigned to a resource R_i that can satisfy its requirements. Indeed, the minimal RAM_i and $vCPU_i$ of the selected R_i should be equal or grater to RAM_{a_1} and $vCPU_{a_1}$ (Eqs. (6) and (7)).

3. **Pricing strategies constraints:** To restrict the time span allowed to use a resource R_i ($[MinAvR_i, MaxAvR_i]$), Eqs. (8) and (9) are used. Whereas, Eqs. (10) and (11) guarantee that an activity a_q should be performed when its required resource R_i is available.

$$\sum_{i=1}^{|\mathcal{R}|} \sum_{j=1}^{p} \sum_{k=1}^{s} MinAvR_i X_{ijkq} \geq d_q, \forall q \in \{1, \cdots, r\} \tag{8}$$

$$\sum_{i=1}^{|\mathscr{R}|}\sum_{j=1}^{p}\sum_{k=1}^{s} MaxAvR_i X_{ijkq} \geq d_q, \forall q \in \{1,\cdots,r\} \tag{9}$$

$$\sum_{i=1}^{|\mathscr{R}|}\sum_{j=1}^{p}\sum_{k=1}^{s} SUNET(R_i) X_{ijkq} \leq S_{a_q} V_q, \forall q \in \{1,\cdots,r\} \tag{10}$$

$$\sum_{i=1}^{|\mathscr{R}|}\sum_{j=1}^{p}\sum_{k=1}^{s} FUNET(R_i) X_{ijkq} \geq F_{a_q} V_q, \forall q \in \{1,\cdots,r\} \tag{11}$$

Each activity has temporal constraints that should be respected by the temporal constraints PS of its allocated resource (Eqs. (8) and (9)). For example, a_1 has a minimum duration equal to 1 h and maximum duration equal to 2 h. R_i is a spot instance with a predefined duration can be assigned to a_1. Thus, its temporal duration values should be grater than d_1 i.e., $MinAvR_i \geq d_1$ and $MaxAvR_i \geq d_1$. When defining activities start and end times, one has to select the resource type, the provider and the PS having the cheapest cost while ensuring the respect of PSs constraints. For example, R_1 from Pr_{11} in strategy $St_{114} = $ (spot non predefined, [01am, 06 pm], 0.0386\$) will be allocated for a_1, then $S_{a_1} \geq SUNET(R_1) = 01am$ and $F_{a_1} \leq FUNET(R_1) = 06\,pm$ (Eqs. (10) and (11)).

4. **Interruption constraint:** When the instance has an interruption risk ($str_k = 1$), and the activity has a penalty cost so p_q should be added (Eq. (12)).

$$\sum_{i=1}^{|\mathscr{R}|}\sum_{j=1}^{p}\sum_{k=1}^{s} X_{ijkq} str_k = V_q, \forall a_q \in \mathcal{A}, p_q > 0 \text{ and } str_k = 1 \tag{12}$$

In the process Fig. 1, some activities have penalties costs. Thus, the latter will be added when the activity execution is interrupted. For instance, a_1 penalty cost is not null, therefore, if the resource assigned to a_1 is a spot instance with non predefined duration, a_1 penalty cost will be considered when computing the optimal cost (Eq. (12)).

5. **Assignment Constraint:** Eq. (13) ensures that each instance type, one Cloud provider and one PS is used by one activity.

$$\sum_{q=1}^{|\mathcal{A}|} X_{ijkq} = 1, \forall i \in \{1,\cdots,n\}, \forall j \in \{1,\cdots,p\}, \forall k \in \{1,\cdots,s\} \tag{13}$$

6. **Placement constraint:** This constraint is used to guarantee that each task uses an only one instance type, Cloud provider and PS.

$$\sum_{i=1}^{|\mathscr{R}|}\sum_{j=1}^{p}\sum_{k=1}^{s} X_{ijkq} = 1, \forall a_q \in \mathcal{A} \tag{14}$$

For instance, a_1 can use only one Cloud instance R_1 from $Pr_{11} = $ Amazon from $St_{113} = $ spot strategy and this R_1 performs only a_1 then $X_{1131} = 1$ (Eqs. (14) and (13)).

7. **Binary constraints:** Eqs. (16) and (15) impose that the decision variables X_{ijkq} and V_q should be either 0 or 1 (binary variables).

$$X_{ijkq} \in \{0,1\}, \quad \forall q \in \{1, \cdots, r\}, i \in \{1, \cdots, n\}, j \in \{1, \cdots, p\}, k \in \{1, \cdots, s\} \tag{15}$$

$$V_q \in \{0,1\}, \quad q \in \{1, \cdots, r\} \tag{16}$$

5   Evaluation

To evaluate our approach, we implemented our Mixed Integer Program (MIP) using IBM-ILOG Cplex Optimization Studio V12.6.3 on a laptop with a 64-bit Intel Core 2.3 GHz CPU, 8 Go RAM and Windows 10 as OS. We start by demonstrating the feasibility of our proposed MIP for scheduling process activities (Fig. 2) (Sect. 5.1). Then, we study the impact of varying the process structure on the cost (Sect. 5.2.1). Next, we analyze the scheduler performance while varying the number of process activities, the maximum number of CRs proposed by each Cloud provider in different PSs (Sect. 5.2.2). Moreover, we study the effect of inflexible temporal constraints (Sect. 5.2.2) and evaluate the impact of deadline constraint (Eq. (5)) (Sect. 5.2.3). Finally, we compare the performance of our approach against two resource allocation methods (Sect. 5.3). Due to space limitation, more experimental results and implementation details can be found at {http://www-inf.it-sudparis.eu/SIMBAD/tools/BPpricing}.

5.1   Approach Feasibility

We present in Fig. 2 the scheduling plan of the BP (Fig. 1) presented as motivating example. The inputs are: (i) the set of activities described in Table 3 and (ii) the set of Cloud resources presented in Tables 1 and 2. Indeed, the execution periods are depicted as green rectangles with a tag on it defining activity name and the allocated resource is mentioned in red colour.

5.2   Performance Analysis

For our experimental evaluation, we generated a set of data defined randomly from the ranges presented in Table 5.

5.2.1   And Split/Join Constraints

The process structure is one of the factors that can have impacts on the scheduling execution plan and then on the process deployment cost. In our work, we consider that BP models have only the AND split/join branching. So, we vary the number of this branching on the process models. The results are shown in Fig. 3. We note that usually the process cost is less expensive when the number of AND is higher: most of activities are executed in parallel (AND). We deduce that the parallelism usually helps to reduce the objective function.

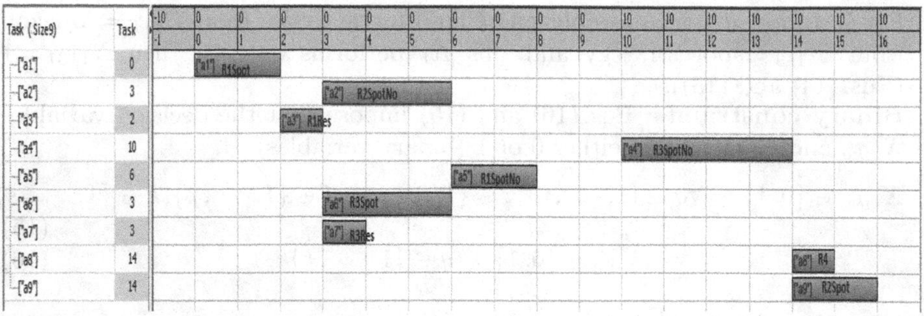

Fig. 2. Gantt chart of the service supervision process

Table 5. Data input ranges

Information	Type	Range
Providers' number	integer	$[1, \cdots, 2]$
Amazon strategies' number	integer	$[1, \cdots, 4]$
Microsoft strategies' number	integer	$[1, \cdots, 1]$
vCPU number	integer	$[2, \cdots, 10]$
RAM amount	double	$[15, \cdots, 30]$
Compute price	double	$[0.01\$, \cdots, 0.532\$]$
Requirement in vCPU	integer	$[2, \cdots, 10]$
Requirement in RAM	double	$[15, \cdots, 30]$
Activities' number	integer	$[2, \cdots, 20]$
Activities' durations	integer	$[1, \cdots, 5]$
Penalty cost	double	$[0\$, \cdots, 1\$]$

5.2.2 Temporal Flexibility Constraint

The search space of our MIP can be limited if the process have some inflexible temporal constraints. For that, in a first experiment we evaluate the Cplex time and objective function when all the temporal constraints are flexible, and when the rate of temporal constraints is about 50% while varying the number of process activities and the maximum number of CRs proposed by each Cloud provider in different PSs. In Table 6, we present the experimental results. We note that the objective function and the response time values are not very high and they have lower values when the temporal constraints are all flexible. In fact, the convergence to optimal solutions is faster and the cost value is cheaper while the research space is restricted.

In the second experiment, we vary the rate of inflexible temporal constraints. Figure 3b shows the total process cost values. The cost increases when the number of inflexible temporal constraints increase. Indeed, we obtain the highest cost is when the rate is 75% because the start and finish times of activities are defined before the resolution of our MIP. Therefore, our MIP does not have a large flexibility to define activities' scheduling in order to choose the cheapest CRs. This led to raise the process cost while those temporal constraints should be respected.

In both cases, we conclude that our method performs better when the inflexible temporal constraints rate is low.

5.2.3   Deadline Constraint

In our MIP, we use a deadline constraint that limits the end times of activities (in Eq. (5)). Thus, we investigate the Cplex time and objective function values when this constraint is removed. As depicted in Table 6, without deadline constraint, we can notice that the objective function is higher and the CPLEXs computational time is considered as a good indicator of the importance of Eq. (5). In fact, the latter limits the research domain of our MIP and gives faster and better solutions. Finally, we conclude that the deadline constraint helps to converge rapidly to optimal solutions.

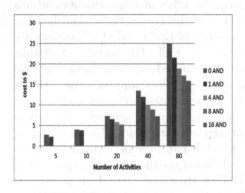

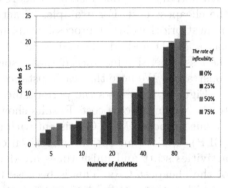

(a) AND Split/Join Variation   (b) Flexibility Evaluation

Fig. 3. AND Split/Join Variation

Table 6. Experimental Results

Nb Activities	Nb Providers	Nb Strategies	Nb VM Types	100% Flexibility	50% Flexibility	MIP without deadline constraint
5	2	14	5	$ob = 2.24\$$	$ob = 3.4\$$	$ob = 3.83\$$
				$t_{cp} = 0.42$s	$t_{cp} = 0.58$s	$t_{cp} > 1$ h
10	2	14	5	$ob = 3.85\$$	$ob = 5.12\$$	$ob = 6.38\$$
				$t_{cp} = 2.4$s	$t_{cp} = 4.9$s	$t_{cp} > 1$ h
20	2	40	17	$ob = 5.63\$$	$ob = 6.95\$$	$ob = 7.59\$$
				$t_{cp} = 22.6$s	$t_{cp} = 29.8$s	$t_{cp} > 1$ h
40	2	60	20	$ob = 10.05\$$	$ob = 11.8\$$	$ob = 13.02\$$
				$t_{cp} = 35.1$s	$t_{cp} = 42$s	$t_{cp} > 1$ h
80	2	170	50	$ob = 18.9\$$	$ob = 20.5\$$	$ob = 23.8\$$
				$t_{cp} = 51.03$s	$t_{cp} = 57$s	$t_{cp} > 1$ h

5.3   Comparison with Other Approaches

In this experiment, our approach is compared to two other approaches. The first one is our previous work [6]. Using a Binary Linear Program (BLP) we derive the optimal allocation cost based on the fixed and predefined activities start and finish times. The selection of the best CR allocation and the best PS is done while respecting the set of temporal constraints and resources properties (described in Sects. 3 and 6). The second is a priority based scheduling algorithm ("Priority+FCFS" [15]). The priority scheduling algorithm is a non-preemptive algorithm and one of the most common scheduling algorithms. Its basic idea is straightforward: a priority is assigned for each activity. Activity with highest priority is to be executed first and so on. Equal-Priority activities are scheduled in FCFS (First Come First Served) order [15]. Priority can be decided based on different factors such as memory requirements, time requirements or any other resource requirement. In our case, priority is based on the process control flow. In other words, the highest priority is given to the first activity and the lowest one is given to the last activity. We assign, in the order of decreasing priority, for each activity only one instance that respects its different requirements and has the cheapest cost. For example, an activity a is the first activity (i.e., has the highest priority) in the process, needs an amount of RAM and vCPU capacities, has a MinD and MaxD and a penalty cost not null. This activity will take the less expensive instance R that respects its required capacities and temporal constraints. We note that each instance is assigned to one and only one activity.

Figure 4 represents the performance of the evaluated methods with respect to the whole process cost. Results show that the cheapest process cost is given by our proposed MIP. Consistent with our expectation, Binary Linear Program (BLP) [6] gives an optimal allocation cost but it can be more minimized when activities start and finish times are defined based on the temporal constraints of the cheapest CRs. This is because the PSs are chosen based on predefined activities start and finish times. In fact, if the instance is temporally available it will be assigned to the activity.

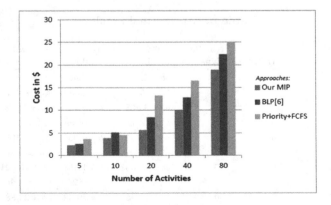

Fig. 4. Approaches Comparison

While "Priority+FCFS" scheduling algorithm performs worse than our scheduling approach because it tends to assign less expensive instances based on the arrival time of each activity while respecting its required capacities and temporal requirements. This is because CR allocation is done after the scheduling of process activities. However, using our MIP we define the scheduling plan at the same time of CRs and PSs selection. Finally, we note that our MIP saves up to 40% in process cost in comparison with both methods.

6 Related Work

There is a wealth of prior work in time-based CR optimization. Wang et al. [16] propose an approach to optimize the data-center net profit with deadline-dependent scheduling by jointly maximizing revenues and minimizing electricity costs. Other approaches are proposed to optimize not only the CRs cost but also the processing time [10,11,17]. Indeed, based on the number of available VMs and their properties, the set of tasks and their requirements, etc an optimal resource assignment and scheduling is provided. However, authors do not deal with CRs allocated from different PSs to run processes where the temporal constraint are more complex than the simple deadline. While in our approach, we schedule process activities in order to overlap with temporal constraints of cheaper PSs. In this way, we optimize the CR costs under a set of constraints.

Mastelic et al. [18] define an approach that predicts the BP execution path, estimates and optimizes the allocated CRs cost based on process metrics and resources properties. In [19], authors propose an algorithm, that takes as inputs: the process structure, the amount of resources, the activities requirements and VM cost, to allocate CRs for a dynamic workflow. In [20], authors present an algorithm based on the meta-heuristic optimization technique, particle swarm optimization (PSO), to minimize the total workflow execution cost. However, authors do not take into account the different PSs of the CRs.

Employee scheduling is concerned with the allocation of employees to specific shifts (time-slots) in order to satisfy certain types of work demand based on employees' qualifications [21]. Indeed, different approaches based on mixed integer programming (MIP) model are developed. For instance, Al-Yakoob et al. [21] and Afilal et al. [22] propose MIP models to allocate human resources under a set of constraints such as employee availability. Havur et al. [23] offer an approach for automated resource allocation in BP that relies on Answer Set Programming (ASP) to find an optimal solution. While our purpose is to define activities start and finish times to use the cheapest computing resources having different PSs.

There has been various types of scheduling algorithm in distributed computing system such as First Come First Served (FCFS), Round Robin scheduling algorithm (RR), Min-Min algorithm, Max-Min algorithm and priority scheduling algorithm. The main advantage of job scheduling algorithm is to achieve a high performance computing and the best system throughput [15]. However, our purpose is to derive the best scheduling execution plan for process activities to optimize the resource allocation cost.

Compared to [6] where the execution plan is given as input, our objective is to schedule BP activities to overlap with the temporal constraints of the less expensive PSs in order to optimize more deployment cost. In fact, the outputs of our approach are: an optimal allocation cost and an optimal scheduling execution plan while respecting a set of temporal constraints and requirements. Furthermore, in the current work we enrich the set of constraints considered in [6] with a deadline constraint (Eq. 5). Then, we evaluate the influence of this constraint on the objective function value and the Cplex time.

Huang et al. [9] propose a mechanism in which the resource allocation optimization problem is modeled as Markov decision processes and solved using reinforcement learning. Li et al. [10] propose a method to solve the issue by applying stochastic integer programming for optimal resource scheduling in Cloud computing. In [11], authors propose a MIP to optimize the Cloud resources cost and the total processing time.

We argue that serverless computing has emerged as a new compelling paradigm for the deployment of applications services. This paradigm has advantages for both consumers and providers. The consumer pays only for what resources he uses without servers or resources' provisioning and management needs. The Cloud providers are able to control the entire development stack, reduce operational costs by efficient optimization and management of CRs [24]. However, Cloud providers impose certain limits on the amount of resources that can be consumed. For example, AWS Lambda offers just 512 MB as a temporary disk space. Therefore, it will be better to allocate standard VMs and use our approach to optimize the process cost deployment in CRs than using serverless computing.

7   Conclusion

In this paper, we propose an optimal scheduling of BP activities performed in CRs, using a Mixed Integer Program (MIP). Indeed, we mapped the problem to deriving the suitable CR, its cheaper pricing cost and a process activity execution timing. Our MIP is defined through an objective function which minimizes the process cost under a set of constraints. The latter includes activities requirements, CRs capacities and PSs temporal constraints. The experimental results show the effectiveness and performance of our approach.

Some potential threats to validity exist in our work. First, since we have assumed that a BP model has only the type of branching ANDsplit, we need to consider other branching types such as ORsplit and XORsplit. Second, we evaluated our work using solutions provided by IBM-ILOG Cplex Optimization Studio. In future work, we intend to test and analyze our approach using a simulation environment, i.e., CloudSim [25]. Moreover, in this work we assumed that the spot price is not full dynamic and is known at the process design time. Thus, we aim to study deeply the dynamic fluctuation of the spot instance cost and to analyze the impact of this aspect on process cost deployment at runtime. Finally, in the evaluation we consider that the process activities number is about

100. So, we aim in the future to deal with processes composed of huge number of activities such as scientific workflows.

References

1. Papagianni, C., Leivadeas, A., Papavassiliou, S., Maglaris, V., Cervello-Pastor, C., Monje, A.: On the optimal allocation of virtual resources in cloud computing networks. IEEE Trans. Comput. **62**(6), 1060–1071 (2013)
2. Van den Bossche, R., Vanmechelen, K., Broeckhove, J.: Cost-optimal scheduling in hybrid IaaS clouds for deadline constrained workloads. In: Cloud Computing (CLOUD), pp. 228–235. IEEE (2010)
3. Cheikhrouhou, S., Kallel, S., Guermouche, N., Jmaiel, M.: Enhancing formal specification and verification of temporal constraints in business processes. In: IEEE International Conference on Services Computing, pp. 701–708 (2014)
4. Bohnenkamp, H., Belinfante, A.: Timed testing with TorX. In: Fitzgerald, J., Hayes, I.J., Tarlecki, A. (eds.) FM 2005. LNCS, vol. 3582, pp. 173–188. Springer, Heidelberg (2005). https://doi.org/10.1007/11526841_13
5. Gagne, D., Trudel, A.: Time-BPMN. In: IEEE Conference on Commerce and Enterprise Computing, CEC 2009, pp. 361–367. IEEE (2009)
6. Halima, R.B., Kallel, S., Gaaloul, W., Jmaiel, M.: Optimal cost for time-aware cloud resource allocation in business process. In: 2017 IEEE International Conference on Services Computing, SCC 2017, pp. pp. 361–367 (2017)
7. Boubaker, S., Gaaloul, W., Graiet, M., Hadj-Alouane, N.B.: Event-B based approach for verifying cloud resource allocation in business process. In: International Conference on Services Computing, SCC 2015, pp. 538–545 (2015)
8. Ben Halima, R., Kallel, S., Klai, K., Gaaloul, W., Jmaiel, M.: Formal verification of time-aware cloud resource allocation in business process. In: Debruyne, C. (ed.) OTM 2016. LNCS, vol. 10033, pp. 400–417. Springer, Heidelberg (2016). https://doi.org/10.1007/978-3-319-48472-3_23
9. Huang, Z., van der Aalst, W.M., Lu, X., Duan, H.: Reinforcement learning based resource allocation in business process management. Data Knowl. Eng. **70**(1), 127–145 (2011)
10. Li, Q., Guo, Y.: Optimization of resource scheduling in cloud computing. In: International Symposium on Symbolic and Numeric Algorithms for Scientific Computing (2010)
11. Hu, M., Luo, J., Veeravalli, B.: Optimal provisioning for scheduling divisible loads with reserved cloud resources. In: 18th IEEE International Conference on Networks, pp. 204–209. IEEE (2012)
12. Amazon ec2 (2012). https://aws.amazon.com/ec2/. Accessed 20 May 2017
13. Cheikhrouhou, S., Kallel, S., Guermouche, N., Jmaiel, M.: Toward a time-centric modeling of business processes in BPMN 2.0. In: The 15th International Conference on Information Integration and Web-based Applications & Services, p. 154 (2013)
14. Floudas, C.A., Lin, X.: Mixed integer linear programming in process scheduling: modeling, algorithms, and applications. Ann. Oper. Res. **139**(1), 131–162 (2005)
15. Salot, P.: A survey of various scheduling algorithm in cloud computing environment. Int. J. Res. Eng. Technol. **2**(2), 131–135 (2013)
16. Wang, W., Zhang, P., Lan, T., Aggarwal, V.: Datacenter net profit optimization with individual job deadlines. In: Proceedings of Conference on Information Sciences and Systems (2012)

17. Chaisiri, S., Lee, B.S., Niyato, D.: Optimization of resource provisioning cost in cloud computing. IEEE Trans. Serv. Comput. **5**, 164–177 (2012)
18. Mastelic, T., Fdhila, W., Brandic, I., Rinderle-Ma, S.: Predicting resource allocation and costs for business processes in the cloud. In: SERVICES, pp. 47–54 (2015)
19. Fakhfakh, F., Kacem, H.H., Kacem, A.H.: A provisioning approach of cloud resources for dynamic workflows. In: CLOUD 2015, pp. 469-476 (2015)
20. Rodriguez, M.A., Buyya, R.: Deadline based resource provisioningand scheduling algorithm for scientific workflows on clouds. IEEE Trans. Cloud Comput. **2**(2), 222–235 (2014)
21. Al-Yakoob, S.M., Sherali, H.D.: Mixed-integer programming models for an employee scheduling problem with multiple shifts and work locations. Ann. Oper. Res. **155**(1), 119–142 (2007)
22. Afilal, M., Chehade, H., Yalaoui, F.: The human resources assignment with multiple sites problem. Int. J. Model. Optim. **5**(2), 155 (2015)
23. Havur, G., Cabanillas, C., Mendling, J., Polleres, A.: Automated resource allocation in business processes with answer set programming. In: Reichert, M., Reijers, H.A. (eds.) BPM 2015. LNBIP, vol. 256, pp. 191–203. Springer, Cham (2016). https://doi.org/10.1007/978-3-319-42887-1_16
24. Baldini, I., et al.: Serverless computing: current trends and open problems. In: Chaudhary, S., Somani, G., Buyya, R. (eds.) Research Advances in Cloud Computing, pp. 1–20. Springer, Singapore (2017). https://doi.org/10.1007/978-981-10-5026-8_1
25. Calheiros, R.N., Ranjan, R., Beloglazov, A., De Rose, C.A., Buyya, R.: Cloudsim: a toolkit for modeling and simulation of cloud computing environments and evaluation of resource provisioning algorithms. Softw.:Pract. Exp. **41**(1), 23–50 (2011)

Designing Process Diagrams – A Framework for Making Design Choices When Visualizing Process Mining Outputs

Marit Sirgmets[1], Fredrik Milani[1(✉)], Alexander Nolte[1], and Taivo Pungas[2]

[1] University of Tartu, Tartu, Estonia
{marit.sirgmets,milani,alexander.nolte}@ut.ee
[2] Datamob, Tallinn, Estonia
taivo@pungas.ee

Abstract. Modern information systems can log the executions of the business processes it supports. Such event logs contain useful information on the performance and health of business processes. Event logs can be used in process analysis with the aid of process mining tools. Process mining tools use various diagrams to visualize the output of analysis made. Such diagrams support the visual exploration of the event logs, facilitating process analysis, and usefulness of process mining tools. However, designing such diagrams is not an easy task. Oftentimes neither the developer nor the end-user know how to visualize the outputs created by process mining algorithms, nor do they know where the interesting information is hidden. Designing diagrams for process mining tools require taking design decisions that, on the one hand allow flexible exploration, and on the other hand, are simple and intuitive. In this paper, we investigate how existing process mining outputs are visualized and their underlying design rationale. Our analysis show that process diagrams, the most common type of diagrams used, are designed with next to no guidance from data visualization principles. Based on our findings, we propose a framework to support developers when designing visualization for process mining outputs. The framework is based on data visualization theory and practices within process mining visualization. The effectiveness and usability of the framework is tested in a case study.

Keywords: Process mining · Process visualization · Process diagram design Framework

1 Introduction

Data is an essential resource for organizations. The ability of a business to analyze and interpret data and make informed business decisions based on analysis results is crucial for a company to survive. The availability of vast amounts of data from various resources makes it hard for analysts to identify interesting or problematic aspects. Visualizations are crucial for process analysts and decision makers to explore and analyze complex data and make informed decisions. Such visualizations are however

© Springer Nature Switzerland AG 2018
H. Panetto et al. (Eds.): OTM 2018 Conferences, LNCS 11229, pp. 463–480, 2018.
https://doi.org/10.1007/978-3-030-02610-3_26

oftentimes created by developers with little knowledge about how to design useful and easy to use visualizations.

One area which is facing this challenge is process mining [1]. In process mining, knowledge is retrieved from execution logs and analyzed from different perspectives such as control flow, resources, and data [2]. Designing visualization of process mining outputs is a complex task [1]. Developers are faced with numerous design questions when composing process diagrams where little or no visualization standards exist. They are also oftentimes left alone when making critical decisions about how to design such visualizations since there is a lack of proper support and guidance. Useful tips are scattered in various resources that process mining developer might or might not be aware about. Moreover, most available resources are presented in a generalized way covering aspects of holistic diagram design [3–5]. This makes such approaches hard to use for developers in the context of process mining due to the necessity to adjust them to the specifics of this domain. Such adjustments are also time-consuming and difficult for developers, who are not professional designers.

For this work we followed a design science approach [6]. We address a real-life problem creating a framework for process mining developers, who are tasked with designing process diagrams. The framework identifies common design issues when visualizing process maps and proposes ideas for solutions. It includes topics related to visual encoding as well as interaction. The framework is based on process mining visualization practices and data visualization theory that is adjusted to process mining. We then conduct a formative evaluation in a real-life project to assess the feasibility of our proposed framework and to identify means for improvement. For the evaluation we recruited developers, who were designing a process diagram as a part of developing a process mining tool. Results of this evaluation point towards the usefulness of this framework for developers of process maps and provide hints for its improvement.

This paper is organized as follows. Section 2 presents the state of the art of visualization within the field of process mining. Section 3 presents our proposed framework. Next, Sect. 4 presents the evaluation of the framework while Sect. 5 concludes the paper.

2 State of the Art

We conducted a review to answer three research questions. The first is "which process mining techniques use visualization". The second is "how are current process mining techniques visualized" and finally, "how do developers decide on how to design visualizations". We searched for related articles on Scopus and Web of Science using the keywords "visual", "process", and "mining". These electronic libraries were chosen as they constitute the main venues for publication within the field of process mining.

The search yielded more than 2000 results which was filtered in three rounds. In the first round, duplicates were removed. In the second round, papers clearly out of scope, such as those on coal mining or data mining, were excluded. The remaining papers were examined and filtered based on the following criteria. Papers less than 3 pages, not accessible, not in English, or older than 10 years were excluded (exclusion criteria). The remaining list of papers further examined and included if they fell within the domain of process mining, introduced a visualization technique, and mentioned design choices for visualization (inclusion criteria). The final list consisted of 28 papers[1]. Data about the paper (meta data), process mining technique, proposed visualization, platform or tool where visualization is implemented, design process of the visualization, and evaluation of the visualization were extracted from each paper.

2.1 Process Mining Techniques Using Visualization

To answer the first research question, we examined the extent to which visualization is used to communicate the output of different process mining techniques. Our review showed that visualization was mostly used for process discovery (generating process models from event logs), process performance (measuring cost, time, and quality aspects of process executions), and process comparison (comparing several processes or checking a model against its event log). Visualization was also used for predictive monitoring (predicting future outcome or upcoming execution paths of a process instance), organizational mining (discovery of organizational structures and communication between units), model repair (improving discovered models based on event logs), deviance mining (uncovering causes of deviant executions of a process), compliance monitoring (surveillance of compliance or violations against regulations in the process execution), and concept drift (changes in the process execution over time). We did not identify any studies visualizing process optimization (identifying improvement opportunities) or process decomposition (clustering models into high-level functions).

A total of 13 papers used visualizations for only one process mining output (single purpose visualization). A few notable examples are process comparison [7, 8], organizational mining [9, 10], performance analysis [11], predictive monitoring [12], and deviance mining [13]. Several studies (a total of 10) concurrently visualized several outputs. An example is the InterPretA tool [14] that visualizes outputs from deviance mining and performance analysis. Another example is "Event Streamer" [15] that visualizes both discovery of declarative processes and concept drift. The remaining studies (5 papers) did not mention any specific process mining technique. Instead, they proposed methods for general exploration of process logs. For example, the tool Event Explorer [1] can be used when an analyst does not know a priori, which specific analysis technique to select.

2.2 Visualization of Process Mining Techniques

The second research question aimed at identifying how process mining outputs are visualized. Our review revealed that node-link diagrams is the most commonly used method for visualizing process mining outputs. For instance, node-link diagrams were used to show the relationship between activities (process diagram) [7, 16] or connections between resources (social network diagrams) [9, 10]. The second most common type of diagrams used were bar-, pie-, and line charts. Such diagrams were often used for visualizing process performance [17, 18]. Performance was also visualized using

---

[1] The list of papers is available at https://babook.cs.ut.ee/pmviz_framework/.

box plots for value distribution [14] and gauge charts [19]. Hierarchical process relationships, such as medical treatment processes and their sub-processes, were commonly
visualized using tree maps [1]. Scatterplots were used to visualize correlations of care-
process parameters, such as correlation of number of treatment activities and patient's
length of hospital stay [18]. Other chart types used were stream graphs [2] for visualizing live process instance flows and turtle graphic trace map [20] for detecting flow
differences amongst process variations.

The charts used followed prevalent and conventional styles. For instance, when
using stacked bar charts, the length represented the value and the color hue distinguished the sub-groups [14]. In the case of node-link diagrams, we noted a greater
variability in how outputs were visualized. The variability was expressed by unique
combinations of visual and interactive elements. In addition to portrayal of the base
topology, other visual channels such as shapes, colors, and sizes were utilized to
represent additional data elements. The extend of the variability seems to indicate a
lack of visualization standards. In summary, we identified eleven different types of
diagrams (not necessarily complete) used in the 28 studies reviewed (see Table 1).

Table 1. Diagrams used for process mining visualization

Diagram type	Number of studies	*References*
Node-link diagram	24	[1, 7–11, 14–31]
Bar/triangle chart	7	[12, 16, 18, 19, 22, 25, 30]
Pie chart	4	[11–13, 17]
Line/area chart	3	[1, 14, 15]
Tree-map	2	[1, 18]
Scatterplot	2	[13, 18]
Parallel coordinate plot	1	[32]
Box plot	1	[14]
Gauge chart	1	[19]
Instance stream graph	1	[15]
Turtle graphics trace map	1	[33]

2.3   Methods for Visualizing Process Mining Outputs

The noted variety of visualization designs prompted us to identify how design choices
were reached. We noted that most papers focused on the proposed algorithm and as
such, presented the algorithm outputs without presenting a rationale for design choices
taken. Although the identified studies did not explicitly follow a systematic method, they
drew inspiration and used input from mainly four sources. These sources are (1) existing
practices, (2) domain expert input, (3) visualization theory, and (4) argumentation.

The first input source refers to critical analysis based on a reviewing process mining
related literature and tools. An example is Bachhofner et al. [11] who noted that existing
solutions only visualize one performance metric on process diagrams. To address this
limitation, they proposed a tool that allows for concurrently representing several

performance metrics. Domain expert input refers to cases where design choices were based on real-life task requirements or user feedback. A notable example is a tool specifically built for users in a hospital by Basole et al. [18]. The involved domain experts provided feedback to the proposed visualizations. This iterative process resulted in the first versions of visualization being discarded. One study stood out as it employed a systematic design framework based on visualization theory. Wynn et al. [16] modified design science methodology by using process mining knowledge, visualization principles, and evaluation of visualization as input for design choices. For instance, Wynn et al. [16], in using size of diagram elements to express continuous variables, grounded this decision in research conducted by Moody [34]. The most common rationale for design choices however, is argumentation. Arguments behind design decisions were generally along the lines of "by watching the displays' content and simultaneously performing selection on the business process model, ...differences in the selected sets of data become intuitively visible...". [19] or "we chose this representation because it makes comparisons more natural for the user" [14]. The argumentation was not grounded in common practices, supportive theory, or the result of comparing alternative choices. One could deduce that the arguments were somewhat arbitrarily chosen.

2.4 Summary

Our literature review has shown that most process mining techniques use visualization to present their output. The process mining use cases not using visualization are decomposition and optimization. Decomposition relies on algorithmically restructuring of processes and thus do does not require visualization. Optimization is commonly based on metrics where weaknesses in existing process executions are identified. Such weaknesses might not require specific visualization. Nevertheless, it thus appears that visualization is an integral part of most process mining techniques.

Our review also revealed that a variety of diagram types are used to visualize process mining outputs. The most common is by means of node-link diagrams. The listing of various visual and interactive elements overlaid with node-link diagrams however, seem to indicate a lack of standard or structured way of making design choices. Our review also showed that developers of process mining techniques did not employ a systematic or structured method when making design choices. Design choices are rather oftentimes reached arbitrarily. This is somewhat surprising considering the crucial role visualization plays when exploring and analyzing complex data. Taken together, these results reveal a gap in the visualization design practices within the process mining field. There is thus a need for a specifically tailored visualization framework that supports developers to design useful visualizations for the output of process mining techniques.

3 Framework

This chapter describes our proposed framework for guiding developers of process mining techniques in composing process diagrams. The first part sets the foundation of the framework. The second part moves on to describe its development process and the final part presents the structure and content of the framework.

3.1 Foundation

We propose a framework that is specifically tailored for process mining techniques. The framework serves to support a developer when designing diagrams, it is not a tool that offers suggestions when given requirements as an input. The primary audience is thus developers of process mining algorithms, who do not have professional experience in the design field. The framework aims to aid the aforementioned developers in making informed design decisions. Hence, the output of the framework is a set of decisions options that a developer can decide on to compose a visualization, not a ready-made composition or a mock-up. It should be noted that in this context, design refers to the structure (requirements) of the visualization of process mining output and not for instance its appearance.

The framework focuses on process diagrams because our review showed that they are the most prolific type of diagram. Most process mining techniques require an understanding of the topology of processes, which is usually supported by node-link diagrams, i.e. process diagrams. Moreover, far too little attention has been given to the design of the visualization of process mining tools, resulting in limited guidance in designing process diagrams [1].

Our framework is based on the literature analysis describe in Sect. 2. While the identified studies provided a plethora of aspects to consider, they did not provide a sufficient foundation to shape a framework. There is, therefore, a need for a foundational data visualization theory to build upon. To this end, we chose Munzner's visualization theory [3] for two main reasons. First, Munzner's work [3] proposes an overarching framework for designing and analyzing data visualization. The framework considers all aspects of the visualization process, from domain and data analysis to validation. Furthermore, the core of Munzner's work, how to visualize data, is well aligned with our purpose. Secondly, Munzner's work is based on well-accepted academic work on data visualization theory (c.f. [35–37]). Other frameworks such as those by Few [38], Ware [39], Cairo [40], Wilkinson [35], and Tufte [36] were considered but found not appropriate for our purpose. They either mostly focused on dashboards [38], considered presentational rather than explorative data visualization [40], focused on how visualization is perceived [39], or discussed theoretical foundations rather than practical implementation of visualization [35, 36]. Munzner's work in contrast is user-centric, considers representation and interaction, is systematically categorizied and organized, and addresses specifics of network data and node-link diagrams.

Munzner presents the data visualization process as a nested model where the output of one layer serves as an input to the next. Munzner considers four layers, domain situation, data/task abstraction, visual encoding and interaction idiom, and algorithm [3]. The question of "how to visualize", which is the focus of our framework, lays in the third layer – visual encoding and interaction idiom. Munzner breaks this part into several questions which are then decomposed further into additional sub-questions. Together, the questions form a hierarchical design tree for design choices [3].

The structure of Muzner's framework is generic. It can therefore be used for a wide range of data visualization cases. This characteristic of the framework enables designers, who aim to expand their awareness of different visualization possibilities, to explore data visualization for a multitude of contexts. However, its generic nature makes it unsuitable for developers who face design choices when visualizing process diagrams. The wide spectrum and the vast materials to consult when designing process diagrams, will most likely be more confusing than constructive for a developer. To address this limitation, we propose a framework that is adjusted and specialized for the context of designing process diagrams within the domain of process mining techniques.

3.2 Development

The framework was developed in three steps. The first step was to identify questions that should be considered when designing process diagrams and set them into a logical sequence. During the second step, the questions were enrichened with alternative answers. The third step addressed understandability aspects of the framework. For this step, we developed illustrations to improve the understandability of the used concepts and terminology.

The aforementioned design questions were extracted from Munzner's theory [3] and mapped against process mining visualization practices. We only included questions that were applicable to process mining techniques. Complementary questions were included where Munzner's theory failed to cover design aspects essential to process mining techniques. For instance, most process flow diagrams within process mining are directed whereas Munzner's theory does not cover directed node-link diagrams sufficiently. Therefore, we added for instance, the question of "how is the sequence of the process shown?" This question is derived from process mining practice and not from Munzner's visualization theory. The final selection contained 62 questions which were considered relevant for our framework.

We then structured the questions using a top-down approach. We identified two main areas – encoding and interaction. Each of these areas was divided into two subcategories. Encoding was divided into *arrange* and *map* and interaction was divided into *reduce* and *change* (see Fig. 1). The remainder of the questions were structured along these four subcategories. After dividing the questions, we identified the dependencies between the questions i.e., one question cannot be answered before some other decisions have already been made. For example, the decision about the basic elements of a diagram must be taken before designing the details of that same diagram. These dependencies defined the sequence and hierarchy of the questions. In cases where there did not exist dependencies, the questions were ordered according to Munzner's visualization theory [3].

In the second step, the questions were enriched with alternative options. When a developer has to take design decisions, they are not served with alternative options. To address this limitation, our framework proposes alternative solutions for each question. The alternative solutions were extracted from the visualization theory we selected as foundation [3]. For instance, the first question in the framework is "what is the base diagram?". Munzner lists three alternative solutions to this question, which are all included in our framework – node-link diagram, adjacency matrix or enclosure [3]. If Munzner's theory did not provide suitable options for the process mining context or if options were missing, we drew examples from process mining diagrams to identify suitable options. An example is the answer to the question of "where does the embedded data appear?". The options added are pop-up window or pane that appears on the diagram itself, covering parts of it [17], or in a separate area next to the diagram [14]. When required, we searched for additional supporting theoretical material. For instance, the options to the question of "how are the basic elements ordered" were taken from Colligan et al. [41] who conducted a comparative study on the effectiveness of hierarchical versus sequential visualization of care-processes. The strengths and weaknesses were extracted together with the specific answers from the visualization theory or inspired by general principles from the theory [3]. In cases, where dualistic pros and cons were irrelevant, common practices with brief reasoning extracted from the literature study were listed instead of theoretical trade-offs.

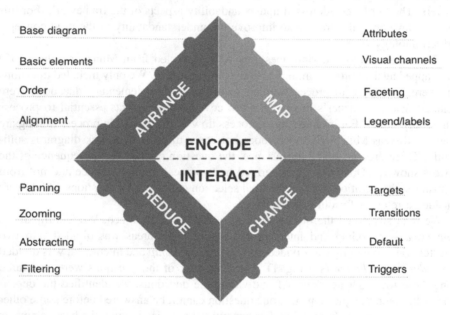

Fig. 1. Reference model of the framework. (Color figure online)

The last step in the development of our framework aimed at improving the comprehension of the framework. This was achieved by adding visual illustrations. All visual illustrations were inspired by examples drawn from state of art studies or data visualization theory. For example, the illustration next to the question of how to solve occlusion in animations was inspired by the work of de Leoni et.al. [17], who propose a process animation tool called Log On Map Replayer (see Fig. 2). Examples were also inspired by modelling languages, such as using the pool and lane concepts commonly known from BPMN [42] to give an example of the use of spatial region in process mining (see Fig. 3).

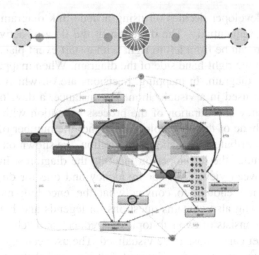

Fig. 2. Illustrations in the framework were inspired by existing tools and visualizations. An illustration for a question about solving occlusion in an animation (upper image) was inspired by Log On Map Replayer solution (lower image) [17].

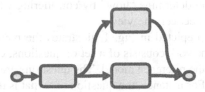

Fig. 3. Illustrations used in the framework were also inspired by modeling languages, such as illustrating the use of spatial region as employed in BPMN [42].

3.3 Overview

In this section, we introduce the structure and main contents of the framework. The main contents of the framework are illustrated in Fig. 1[2]. The model should be read from inside out as the topics are in a hierarchical order. The topics are divided into two building blocks – "encoding" and "interaction". Encoding contains questions about visual aspects of the diagram while interaction covers questions about how to manipulate the diagram. In encoding, the developer chooses to visualize the frequency and duration of process activities as two separate layers of a process diagram. In interaction, the developer decides how the user of the diagram can switch between these two layers.

Encoding is further divided into two sub-topics – "arrange" and "map". Arrange covers questions on the basic structure of the diagram, such as which type of base diagram to use as well as the ordering and alignment of diagram elements. For instance,

---

[2] The framework and detailed instructions on how to use it are available at
  https://owncloud.ut.ee/owncloud/index.php/s/lQxL4P1Iq2Z9ofN .

when encoding, a developer decides on using a node-link diagram where nodes (activities) are ordered sequentially. The direction of the flow of activities is also determined. The direction can be from left to right with a start event placed on the left-hand and the end event on the right-hand side of the diagram. When mapping, the focus is on the aesthetics of the diagram. In mapping, decisions are on which attributes (such as color and shape) are used in a visualization. For instance, a developer might wish to visualize the frequency and duration of the process execution with color saturation – e.g. the darker the shade of the node color, the higher the number of process instances or duration. As two attributes, frequency and duration, are shown on the same diagram, the mapping also guides the developer on faceting the diagrams. In this example, the user can switch between views, one for frequency and one for duration. Using only color saturation is not enough to convey what the encoding means. To facilitate understanding, mapping also contains questions on legends and labels.

Interaction also consists of two sub-topics – "reduce" and "change". Reduce refers to which data the user can choose to be visualized. The user can e.g. use zoom and pan (scroll) to highlight specific aspects of interest. Filtering and abstraction allow for select subsets of the dataset to be visualized. Change on the other hand, refers to changing the diagram. Change considers what the user can change, how the changes transition from one image to another, and which user actions trigger changes. In mapping, developers chose to facet duration and frequency as two separate versions of the same diagram. In "change", a developer can determine "how" by considering which is to be the default view and how to switch between the views.

The reference model depicted in Fig. 1 illustrates the main topics covered in the framework. The full framework consists of a set of questions, categorized according to the main components of the reference model. The questions are structured according to the topic they belong to. For instance, the question of "what is the base diagram" is part of "arrange" which in turn, is under "encode". As such, the framework systematically guides the developer when visualizing process mining outputs through a set of questions. Table 2 provides an example for questions and their structure. For example, on the reference model, encoding is divided to two parts – "arrange" and "map", which corresponds to the question (level 1) of "how to encode data". This question is then further divided to two questions (level 2), namely "how to arrange data" and "how to map data". At the third level, the questions are broken down into detailed sub-questions. The framework also provides further considerations for each alternative answer of the detailed sub-questions questions.

In Fig. 4, we illustrate an example of alternative answers to a sub-question at the fourth level, namely for the question of "where does the embedded data appear". The available alternatives are "on the diagram" and "off the diagram". The framework also indicates further consideration (strengths and weaknesses) of each alternative. Embedding the data in the diagram makes it easier to track but requires space which might occlude other relevant parts of the diagram. Also, if it is important to see the diagram while drilling into detailed level of data, "off the diagram" might be a better alternative. "Off the diagram" refers to presenting detailed information on a separate pane which does not cover the diagram. The downside is the space required for the pane will come at the expense of the space allocated for the diagram. Thus, the framework provides the questions and relevant considerations for each of alternative solutions (Fig. 4).

Table 2. Hierarchy of the questions in process visualization framework

Level 1	Level 2	Level 3
How to encode data?	How to arrange data?	What is the base diagram?
		What are the basic elements of the diagram?
		How are the basic elements ordered?
		How is the diagram aligned?
	How to map data?	Which attributes are shown on the diagram?
		Which channels express the attributes?
		How is the data faceted on the diagram?
		How does the user know the meaning of the channels?
How to design interaction?	How can the user change the visualization?	What can be changed on the diagram?
		How do the changes appear?
		What is the default appearance?
		How can the changes be triggered?
	How can the user reduce data?	Does the diagram need panning?
		Does the diagram need zooming?
		Does the diagram need abstracting?
		Does the diagram need filtering?

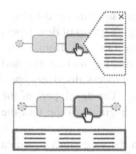

1.2.3.7 Where does the embedded data appear?

O On the diagram

+ element and embedded data are close – easy for eyes to track.	- pop-up windows occlude parts of the base diagram.

O Off the diagram

+ the full process is in the view when the details are shown.	- space-consuming.

Fig. 4. An example of lowest level of question and its considerations of the framework.

4   Evaluation

In order to evaluate the usefulness of the framework we conducted a case study it in a real-life project. The aim of the evaluation was to identify strengths and weaknesses of the framework as well as means for further improvement. In the following we will describe the design of the evaluation (Sect. 4.1) before discussing our findings.

4.1   Design

The evaluation was designed as a case study since we aimed to explore its usefulness in a real-life context [43]. Case studies are suitable for answering "why" and "how" questions, particularly in cases where context can provide insightful information and the research requires an observational approach [44].

The aim of the framework is to support developers to create visualizations of process mining diagrams, specifically process maps. It thus has to fulfil the following four main criteria: (1) it has to be understandable by developers of such visualizations; (2) it has to be relevant in the given context; (3) it has to be complete in order to support developers to create visualizations that are useful to and useable by process analysts and (4) it has to be easy to use by the target audience to ensure a balance between the time and effort developers spend using the framework when designing visualizations. The evaluation thus focused on perceived understandability, relevance, completeness, and usefulness of the framework.

The unit of the analysis was defined as follows:

- The effect of the framework on data visualization design tasks executed by developers of process mining tools.

The effect was observed through the lens of the following questions that address the understandability, relevance, completeness and usefulness:

- How is the framework understandable/unclear for developers?
- How is the framework relevant/irrelevant for the process of designing visualizations for process mining diagrams?
- Which aspects are potentially missing from the framework?
- How easy is it to use the framework?

The case study took place in the context of a project that aims to visualize data from a queuing management system used to manage border crossing. A group of developers were building a process mining tool that would help to translate data from aforementioned queuing management system into insightful information to improve and innovate the queuing process. The focus of development was process discovery, performance analysis, predictive monitoring, and deviance mining. The developers of the tool used the framework to design visualizations for process diagrams of the tool the team was developing.

We chose three members (See Table 3) as participants in the case study – a data scientist and two researchers. All participants have had experience in developing process mining techniques. The data scientist (P1) was currently working on a PhD thesis in process mining field and has worked on industry projects related to process mining. Both of the researchers have about 10 years of experience in developing tools in the context of process mining. In addition, all the participants have had previous experience in data visualization. The first participant had been using data visualization mostly for presentation purposes. The second participant (P2) had become acquainted with data visualization concepts through practice as well as theory. The third participant (P3) had developed process mining tools that include visual presentations – some lectures s/he holds require familiarity with the data visualization literature. None of the participants were professional visualization designers.

Table 3. Case study participants

ID	Experience	Field	Project focus
P1	4 years	Industry projects Academic research	Predictive monitoring
P2	10+ years	Academic research	Process discovery and performance analysis
P3	10 years	Academic research Teaching	Deviance mining

The procedure of the study was conducted as follows. Each participant was invited to an individual session, which was divided to three parts. The session started with a semi-structured interview during which the participants were asked to explain the project they were working on and their role in the project in more detail. They were also asked to explain potential issues they were facing during the project. After the interview the participants were asked to explain their initial visualization ideas before they were introduced to the framework. The participants were then asked to use the framework – which they received in a printed form – for their respective visualization task. The final part of each session was a semi-structured interview during which we asked the participants for their opinions about the framework and their perspective on the impact it had on their visualization task. Each session was audio recorded and the researcher conducting the studies took additional field notes.

4.2 Discussion

The following results are structured along the aforementioned evaluation dimensions of understandability, relevance, completeness, and usefulness.

Understandability. All participants found the framework to be understandable (*"Definitely it was easy"*, P2). This was evident by the participants not struggling with aspects, such as which questions to answer or the meaning of illustrations or tables. Also, all participants were able to identify the purpose of the framework (*"to get a better understanding, to formulate a visualization task better"*, P1). The participants reported the framework to be useful for tool improvement, making vague visualization ideas more concrete and using it as an inspiration point for designing new visualizations. One participant referred to it as a catalogue of tested ideas (*"sort of a catalogue with some already tested practices"*, P2), which can be revisited several times during the design process. Another participant saw its use in user surveys to identify the solution that the target users would prefer. Two participants pointed out the potential to develop the framework into a mock-up tool, which would turn answers to questions in the framework into sample visualization.

Even though, the basic understandability of the framework was good, all the participants highlighted aspects that could be improved. Two participants found parts of the terminology to be confusing. One participant found the terms easy but added that this is due to his familiarity with literature on visualization theory. One participant suggested a glossary (*"maybe it would be helpful if somewhere were those [definitions], so that they can be immediately looked up"*, P1), where the terms could be easily looked up. Another issue mentioned by two participants was targeting of the. Both mentioned two potential reasons for their issues related to the understandability of the questions: The wording of the questions – it was not clear if the questions are about existing solutions or prospective preferences. For example, "How is the diagram aligned?" refers to something that already exists, while wording such as "How would you like the diagram to be aligned?" is aimed for the designer to think about how s/he would design a future diagram. The second potential reason for aforementioned confusion was the sequence of topics and transitions (*"sometimes it is difficult to follow the sequence of questions"*, P3). The questions move from one topic to another with abrupt transitions and the user may miss that the target of the question has switched.

Relevance. The participants found the framework relevant for their work in particular related to developing new ideas or improving and clarifying existing ideas of process mining visualizations. The purpose of the framework was easy to understand for the participants and it was found relevant for all process mining visualization tasks which include tool improvement as well as inspiration and a guidance for making ideas more concrete. All participants mentioned that the framework helped them to develop new ideas for their respective visualization task and they would recommend it to their colleagues who struggle with similar tasks (*"yes, I think it helps to put ideas together, especially in the initial stage of development"*, P1).

Completeness. Most of suggestions for adding and changing the framework stem from aforementioned understandability issues (*"I don't know how complete the idea gets, maybe you can add even more alternatives"*, P3). For instance, examples of real tools and a glossary of definitions were suggested to improve the clarity of the framework. Also, the transitions between topics were brought out as a potential place to improve the comprehensibility of the framework. One participant suggested a solution for clarifying the targets of the questions by reducing the topics in the framework. For example, focusing on one of the main topics – representation of data or interactivity – and allowing users to explore the selected topic in more depth, while discarding the other. One participant saw a possibility to include more questions specifically about embedded data – how to visualize data that is shown in the pop-up windows.

All participants also saw a potential to digitalize the framework (*"if the framework was digital, then it would be very comfortable to see the final result"*, P1). Two participants mentioned the potential to develop the questions in the framework into a mock-up tool that could show an example diagram based on the selected alternatives. One participant suggested to hide the positive and negative aspects in the default view and provide the user with a respective option that allows him to reveal them if necessary.

Usefulness. One participant estimated the level of required focus high, while two others thought it required little effort to use it. The participants also mentioned that terms and targets of questions were the most difficult to understand. The time required for going through the framework on paper varied from 25 to 45 min. However, two participants mentioned that it should be used repeatedly during the visualization development process. All participants thought that time and effort they put into using the framework was worthwhile. The framework helped them to develop new visualization ideas and make existing ones more concrete in a relatively short time.

4.3   Limitations

The evaluation of the framework had a number of limitations. First of all, the framework was evaluated on experienced developers with familiarity of visualized outputs from process mining techniques. As such, the framework might require further instructions for novice developers. Furthermore, the evaluation did not include representatives of intended end-users of the visualization such as process analysts. Also, the evaluation focused on developing new node-linked diagrams. Although visualization within this field predominantly uses node-linked diagrams, the suitability of the framework for other types of base diagrams was not covered in the evaluation.

5   Conclusion

During the course of this paper we presented the development and evaluation of a framework that supports developers in designing process mining diagrams. Our work showed the importance of visualizations in process mining field and revealed the complexity of the design tasks developers are facing. Regardless of the importance and complexity of the visualizations, most of diagrams are currently designed by developers with little to no training in developing visualizations and with no systematic support. Design decisions are instead often based on a combination of logical argumentation, existing practices and domain input.

The proposed framework is based on two cornerstones – existing process mining visualizations and data visualization theory. Majority of the topics covered in the framework have their foundation in the visualization theory forwarded by Munzner [27]. However, adjustments were made to the theory to make it relevant to process mining. The framework consists of questions and alternative answers with strengths and weaknesses. In addition, illustrations that are specific to process mining, were designed and added to the framework to increase the comprehensibility through visual examples.

We evaluated the framework in a case study with three developers. The evaluation revealed that the developers found the framework relevant and balanced in terms of how much effort it requires and how beneficial it is to the task at hand. The main value of the framework was found in making vague ideas concrete, coming up with new ideas, and improving existing ones. The evaluation also revealed potential means for improvement such as clarification of terms.

In the future, we aim to extend the framework to other types of visualizations such as dashboards used in process mining. Another possible venue is improving the format of the framework by developing an online tool.

References

1. Gschwandtner, T.: Visual analytics meets process mining: challenges and opportunities. In: Ceravolo, P., Rinderle-Ma, S. (eds.) SIMPDA 2015. LNBIP, vol. 244, pp. 142–154. Springer, Cham (2017). https://doi.org/10.1007/978-3-319-53435-0_7
2. van der Aalst, W.M.P.: Process Mining. Springer, Heidelberg (2016). https://doi.org/10.1007/978-3-662-49851-4
3. Munzner, T.: Visualization Analysis and Design. AK Peters/CRC Press, Boca Raton (2014)
4. Rogers, Y., Sharp, H., Preece, J.: Interaction Design: Beyond Human - Computer Interaction. Wiley, West Sussex (2011)
5. Unger, R., Chandler, C.: A Project Guide to UX Design: For User Experience Designers in the Field or in the Making. New Riders, Berkeley (2012)
6. Bichler, M.: Design science in information systems research. Wirtschaftsinformatik. **48**, 133–135 (2006)
7. Gall, M., Wallner, G., Kriglstein, S., Rinderle-Ma, S.: Differencegraph - a ProM plugin for calculating and visualizing differences between processes. In: CEUR Workshop Proceedings, vol. 1418, pp. 65–69 (2015)
8. Bolt, A., de Leoni, M., van der Aalst, W.M.P.: A visual approach to spot statistically-significant differences in event logs based on process metrics. In: Nurcan, S., Soffer, P., Bajec, M., Eder, J. (eds.) CAiSE 2016. LNCS, vol. 9694, pp. 151–166. Springer, Cham (2016). https://doi.org/10.1007/978-3-319-39696-5_10
9. Slaninova, K., Martinovic, J., Drazdilova, P., Snashel, V.: From Moodle log file to the students network. In: Herrero, Á., et al. (eds.) International Joint Conference SOCO'13-CISIS'13-ICEUTE'13. Advances in Intelligent Systems and Computing, vol. 239, pp. 641–650. Springer, Cham (2014). https://doi.org/10.1007/978-3-319-01854-6_65
10. Jalali, A.: Supporting social network analysis using chord diagram in process mining. In: Řepa, V., Bruckner, T. (eds.) BIR 2016. LNBIP, vol. 261, pp. 16–32. Springer, Cham (2016). https://doi.org/10.1007/978-3-319-45321-7_2
11. Bachhofner, S., Kis, I., Di Ciccio, C., Mendling, J.: Towards a multi-parametric visualisation approach for business process analytics. In: Metzger, A., Persson, A. (eds.) CAiSE 2017. LNBIP, vol. 286, pp. 85–91. Springer, Cham (2017). https://doi.org/10.1007/978-3-319-60048-2_8
12. Jorbina, K., et al.: A web-based tool for predictive process monitoring. In: CEUR Workshop Proceedings, vol. 1920 (2017)
13. Bose, R.P.J.C., van der Aalst, W.M.P.: Discovering signature patterns from event logs. In: 2013 IEEE Symposium on Computational Intelligence and Data Mining (CIDM). CIDM 2013. IEEE Symposium Series of Computational Intelligence (IEEE SSCI 2017). SCI 2013, pp. 111–118 (2013)
14. Dixit, P.M., Caballero, H.S.G., Corvo, A., Hompes, B.F.A., Buijs, J.C.A.M., van der Aalst, W.M.P.: Enabling interactive process analysis with process mining and visual analytics. In: Proceedings of 20th International Joint Conference on Biomedical Engineering Systems and Technology, vol. 5, pp. 573–584 (2017)

15. Burattin, A., Cimitile, M., Maggi, F.M.: Lights, camera, action! business process movies for online process discovery. In: Fournier, F., Mendling, J. (eds.) BPM 2014. LNBIP, vol. 202, pp. 408–419. Springer, Cham (2015). https://doi.org/10.1007/978-3-319-15895-2_34
16. Wynn, M.T., et al.: ProcessProfiler3D: a visualization framework for log-based process performance comparison. Decis. Support Syst. **100**, 93–108 (2017)
17. de Leoni, M., Suriadi, S., ter Hofstede, A.H.M., van der Aalst, W.M.P.: Turning event logs into process movies: animating what has really happened. Softw. Syst. Model. **15**, 707–732 (2016)
18. Basole, R.C., Park, H., Gupta, M., Braunstein, M.L., Chau, D.H., Thompson, M.: A visual analytics approach to understanding care process variation and conformance. In: ACM International Conference Proceeding Series, p. 6 (2015)
19. Gulden, J., Attfield, S.: Business process models for visually navigating process execution data. In: Reichert, M., Reijers, H.A. (eds.) BPM 2015. LNBIP, vol. 256, pp. 583–594. Springer, Cham (2016). https://doi.org/10.1007/978-3-319-42887-1_47
20. Cordes, C., Vogelgesang, T., Appelrath, H.-J.: A generic approach for calculating and visualizing differences between process models in multidimensional process mining. In: Fournier, F., Mendling, J. (eds.) BPM 2014. LNBIP, vol. 202, pp. 383–394. Springer, Cham (2015). https://doi.org/10.1007/978-3-319-15895-2_32
21. Lucas, W., Xu, J., Babaian, T.: Visualizing ERP usage logs in real time. ICEIS Proceedings of 15th International Conference on Enterprise Information Systems, vol. 3, pp. 83–90 (2013)
22. Pini, A., Brown, R., Wynn, M.T.: Process visualization techniques for multi-perspective process comparisons. In: Bae, J., Suriadi, S., Wen, L. (eds.) AP-BPM 2015. LNBIP, vol. 219, pp. 183–197. Springer, Cham (2015). https://doi.org/10.1007/978-3-319-19509-4_14
23. Mannhardt, F., de Leoni, M., Reijers, H.A.: The Multi-perspective Process Explorer. BPM (Demos) **1418**, 130–134 (2015). CEUR Workshop Proceedings
24. Leemans, S.J.J., Fahland, D., van der Aalst, W.M.P.: Exploring processes and deviations. In: Fournier, F., Mendling, J. (eds.) BPM 2014. LNBIP, vol. 202, pp. 304–316. Springer, Cham (2015). https://doi.org/10.1007/978-3-319-15895-2_26
25. Vogelsang, T., Appelrath, H.-J.: Multidimensional process mining with PMCube explorer. In: CEUR Workshop Proceedings, vol. 1418, pp. 90–94 (2015)
26. Hipp, M., Strauss, A., Michelberger, B., Mutschler, B., Reichert, M.: Enabling a user-friendly visualization of business process models. In: Fournier, F., Mendling, J. (eds.) BPM 2014. LNBIP, vol. 202, pp. 395–407. Springer, Cham (2015). https://doi.org/10.1007/978-3-319-15895-2_33
27. Slaninová, K., Vymětal, D., Martinovič, J.: Analysis of event logs: behavioral graphs. In: Benatallah, B., et al. (eds.) WISE 2014. LNCS, vol. 9051, pp. 42–56. Springer, Cham (2015). https://doi.org/10.1007/978-3-319-20370-6_4
28. Leemans, S.J.J., Fahland, D., van der Aalst, W.M.P.: Process and deviation exploration with inductive visual miner. In: CEUR Workshop Proceedings, vol. 1295, pp. 46–50 (2014)
29. Kecman, P., Goverde, R.M.P.: Process mining of train describer event data and automatic conflict identification. WIT Trans. Built Environ. **127**, 227–238 (2012)
30. van Dongen, B.F., Adriansyah, A.: Process mining: fuzzy clustering and performance visualization. In: Rinderle-Ma, S., Sadiq, S., Leymann, F. (eds.) BPM 2009. LNBIP, vol. 43, pp. 158–169. Springer, Heidelberg (2010). https://doi.org/10.1007/978-3-642-12186-9_15
31. Gall, M., Wallner, G., Kriglstein, S., Rinderle-Ma, S.: A study of different visualizations for visualizing differences in process models. In: Jeusfeld, M.A., Karlapalem, K. (eds.) ER 2015. LNCS, vol. 9382, pp. 99–108. Springer, Cham (2015). https://doi.org/10.1007/978-3-319-25747-1_10

32. Gupta, N., Anand, K., Sureka, A.: Pariket: mining business process logs for root cause analysis of anomalous incidents. In: Chu, W., Kikuchi, S., Bhalla, S. (eds.) DNIS 2015. LNCS, vol. 8999, pp. 244–263. Springer, Cham (2015). https://doi.org/10.1007/978-3-319-16313-0_19

33. Štolfa, J., Štolfa, S., Kopka, M., Snśšel, V.: Adaptation of turtle graphics method for visualization of the process execution. In: Abraham, A., Krömer, P., Snasel, V. (eds.) Afro-European Conference for Industrial Advancement. AISC, vol. 334, pp. 327–334. Springer, Cham (2015). https://doi.org/10.1007/978-3-319-13572-4_27

34. Moody, D.: The "physics" of notations: toward a scientific basis for constructing visual notations in software engineering. IEEE Trans. Softw. Eng. 35, 756–779 (2009)

35. Wilkinson, L.: Grammar of Graphics. Springer, New York (1999). https://doi.org/10.1007/978-1-4757-3100-2

36. Tufte, E.: The Visual Display of Quantitative Information. Graphics Press, Cheshire (1983)

37. Ware, C.: Information Visualization: Perception for Design. Morgan Kaufman, Hampshire (1999)

38. Few, S.: Information Dashboard Design (2006)

39. Ware, C.: Visual thinking: for design. Ergonomics 53, 138–139 (2008)

40. Cairo, A.: The Functional Art: An Introduction to Information Graphics and Visualization. New Riders, Berkeley (2013)

41. Colligan, L., Anderson, J.E., Potts, H.W.W., Berman, J.: Does the process map influence the outcome of quality improvement work? A comparison of a sequential flow diagram and a hierarchical task analysis diagram. BMC Health Serv. Res. 10, 7 (2010)

42. Object Management Group: Business process model and notation (BPMN) (2011)

43. Baxter, P., Jack, S.: Qualitative case study methodology: study design and implementation for novice researchers. Qual. Rep. 13, 544–559 (2008)

44. Yin, R.K.: Case Study Research and Applications: Design and Methods. SAGE Publications, Los Angeles (2018)

A Viewpoint for Integrating Costs in Enterprise Architecture

João Miguens[1]($\boxtimes$), Miguel Mira da Silva[1,2]($\boxtimes$), and Sérgio Guerreiro[1,3]($\boxtimes$)

[1] Instituto Superior Técnico, University of Lisbon, Av. Rovisco Pais 1, 1049-001 Lisbon, Portugal
{joao.miguens,mms,sergio.guerreiro}@tecnico.ulisboa.pt
[2] INOV - Inesc Inovação, Av. Duque de Ávila 23, 1000-005 Lisbon, Portugal
[3] INESC-ID, Rua Alves Redol 9, 1000-029 Lisbon, Portugal

Abstract. Managing and controlling costs is a major concern in every organization. To cope with costs complexity, enterprise models are referred in the literature as solutions to understand that partial view of the reality. Enterprise Architecture (EA) is one of the solutions that is widely used to cover many of the stakeholder's concerns and viewpoints. However, EA lacks a viewpoint for representing costs. In this paper we present an ArchiMate viewpoint as an extension of any EA model, allowing costs representation. The proposal is grounded in the Time-Driven Activity Based Costing (TDABC) method. Our proposed viewpoint includes the ArchiMate concepts, their relationships and properties enabling the calculation and representation of costs in EA according to TDABC. To validate the usefulness and applicability of the proposal a well-known Harvard case study related with Heart Bypass Surgery is used to formulate a questionnaire. Then, it was applied to EA experts and the results are analysed using a quantitative approach. The paper concludes that representing costs in the architecture is considered as a good option for stakeholders to analyse and draw conclusions about the costs of their business processes and also to enables a better display of cost information when compared to unstructured representations.

Keywords: ArchiMate · Costing · Enterprise Architecture · TDABC Viewpoint

1 Introduction

Managing and controlling costs is a major concern for organizations, since it plays a crucial role for business management [1]. This concern is a priority because costs are fundamental in any organization [2–4] and according stakeholders' opinion, by addressing this concern it could help them to decrease the costs related to the organisation' business [5]. This is relevant because in order to reduce costs, managers need to know the principal determinants of cost. This is not always obvious due to the complex interplay of the set of cost drivers

© Springer Nature Switzerland AG 2018
H. Panetto et al. (Eds.): OTM 2018 Conferences, LNCS 11229, pp. 481–497, 2018.
https://doi.org/10.1007/978-3-030-02610-3_27

[1], that if poorly understood can compromise the total cost estimation, affecting the quality of the organization' cost information. According to Kaplan and Cooper [6] poor cost information about the organization' cost objects, increases the likelihood of leading to a bad competitive strategy, which may affect the organisation' performance, having direct impact on its results.

EA is considered as a valuable organization's information basis, that initially was only used for having a blueprint of organizations in their several domains, but that over the years evolved into a powerful way to actively managing an organisation, helping it to cope with a several number of its stakeholder's concerns and viewpoints [7,8].

However this evolution was mainly focused on EA functional properties leaving aside EA quantitative aspects like cost measures, which are also important. Although EA is recognised as a viable solution to represent organization' costs [5,11], it lacks a viewpoint for representing costs, not allowing to fulfil one of its stakeholder's main concern, like the EA costs concern [5].

ArchiMate is an EA modelling language that covers a broad range of aspects (from technology to business layer), allowing to represent EAs, costs and also the relation between those costs and their sources (EA elements).

This paper presents an approach for estimate, manage and represent the cost analysis of EAs. This approach is based on a mapping between ArchiMate and Time-Driven Activity-Based Costing [9], which allowed us to create an ArchiMate viewpoint framing a set of concerns, for supporting EA cost analysis.

Our proposal has one major objective: a specification of the concepts, relationships and attributes that capture the fundamental information for representing and reasoning upon costs of an EA description.

The remainder of this paper is structured as follows. Research Methodology describes the method used to guide our research. Then, Research Problem presents the problem that this proposal intends to address, as well the motivation to solve it. Section 4 presents the theoretical background, followed by the related work where it is introduced the existing literature regarding the problem domain. In Sect. 5, we detail our proposal, as well the main objectives that we want to achieve with its use. To demonstrate our work, in Sect. 6 it is presented the application of our proposal to a case study. The evaluation process and criteria are defined in Sect. 7. Lastly, some concluding remarks are given in Sect. 8.

2   Research Methodology

Design Science Research Methodology (DSRM) [10] was chosen as the research methodology to be used throughout this research.

This methodology aims at developing solutions (artifacts) to important business problems overcoming traditional research methods whose final result was most of the times merely explanatory and theoretical, not being applicable to the problem found. Such artifacts include constructs (vocabulary and symbols), models (abstractions and representations), methods (algorithms and practices) and instantiations (implemented and prototype systems) [10]. This IS methodology is widely-adopted by researchers due to its appropriateness to researches that intend to create, evaluate and improve new and innovative IT artifacts, which is the case of our proposal.

DSRM is an iterative cycle composed of six steps [10]:

- **Problem Identification:** present and explain the research problem together with its importance to the relevant community;
- **Objectives Definition:** describe the solution' objectives inferred from the problem definition, in order to solve the identified problem;
- **Design and Development:** describe the functionality of the artifact that will be developed as well the models and methods applied to create it;
- **Demonstration:** demonstrates both the feasibility and the utility of the artifact to solve one or more instances of the problem. Examples of demonstrations are experimentations, simulations, cases studies, proofs, or other appropriate activity;
- **Evaluation:** measure and classify the usefulness, quality and the effectiveness of the artifact. It can be done according to several methods (observational, analytical, experimental, testing or descriptive);
- **Communication:** communicates the problem and its relevance, the research proposal (artifact developed), and the utility of the artifact to researchers and other relevant audiences.

3   Research Problem

EA is a powerful tool, being one of its main purposes to assist managers in the process of decision making by offering an insight into cross-domain architecture relations, typically through projections and intersections of underlying models [7]. In order to provide such support, EA models should be amenable to quantitative analyses of various properties (e.g. performance, cost, etc.), thus providing a basis for fact-based EA-related decision-making. According to the results of a survey [5], where several Chief Information Officers (CIOs) (which is the EA primary stakeholder) were interviewed, the main concern about business, that they were expecting EA would help them to decide about, is the EA cost related concern. However, even though costs are fundamental for an organization to guide their decision making process [2–4] and being EA a crucial instrument to help stakeholders in this process, there is little or no explicit support to include cost information on EA, thereby allowing better decisions.

Although ArchiMate, which is a modelling language (used by scholars and practitioners worldwide) and open standard that can be used for modelling EA descriptions, allows to model costs through the motivation element "Value" [11] it is not possible to fully address the EA costs concern. The main reason for this limitation is because it does not exist a selection of a relevant subset of the ArchiMate concepts and their relationships that allow to model this concern with the necessary level of detail regarding all the necessary cost information (how much cost a resource or what is the capacity of a resource), so that is possible to apply a cost model while modelling an EA, adorning the relevant entities, like resources or activities with pertinent cost information. So there is a challenge that has to be overcome, which is: **the lack of solution to integrate the knowledge from EA models with cost models**. Otherwise EA will continue to lack an important aspect regarding decision support, that would allow to have an higher degree of transparency and would help stakeholders, like managers or CIOs, to perform EA-related decision based on costs more easily.

4   Theoretical Background

This section intends to give a brief overview of the context where this work is inserted. To do this we start by explaining the main concepts related to identified problem and also the ones that are going to be used to address it.

4.1   Enterprise Architecture

Enterprise Architecture (EA) allows to describe, design and control several parts of an organisation's structure (business processes, applications, technology) by applying architecture principles and practices to guide organisations to fulfil their organisational objectives [7]. This way it provides an holistic view of the organisations, based on views that allows to see the relationship between artifacts and architectures, as also the existent dependencies between the various layers of the enterprise [7]. The alignment between the organisation' layers allow to have a blueprint of the organization, which is used to manage and align assets, people, operations and projects to support business goals and strategies, improving the governance of its processes and systems [12,13].

4.2   ArchiMate

ArchiMate graphical modelling language is a standard from The Open Group used by scholars and practitioners worldwide [14] to describe, analyse and visualize an architecture representation in its several domains [7]. Thus allowing EA' stakeholders to access and communicating the consequences of decisions and changes within and between the several organization domains by providing a graphical notation to represent EA over time [7].

The language core defines the elements (concepts and relationships) that are necessary to model an EA [7]. ArchiMate core has a layered look due to its division in layers: business, application and infrastructure. Regarding the aspects ArchiMate distinguishes in each layer the several elements in three groups according to its characteristics: Active, behaviour or passive structure.

Respecting visualization and according with the Zachman Framework [15], ArchiMate allows to have a separation of concerns through the use of views. These views are defined by viewpoints, which is a sub-set of concepts and relations that allows to model and focus on a certain aspect of the EA. The main purpose of viewpoints is to serve as a mean of communication about certain aspects of an architecture. The aspects and the viewpoint content are based on the stakeholders' concerns. So, even though there is only a single EA for an organization, all of its stakeholders can benefit from it, since through viewpoints, each stakeholder can view the EA according to its needs.

Extending ArchiMate. ArchiMate is one of the most expressive modelling notations in the EA domain. Therefore the language' core, embedded in the ArchiMate metamodel, contains only the basic concepts and relationships that serve general EA modelling purposes. However, the language should also be able to facilitate, through extension mechanisms, domain-specific purposes.

For this purpose ArchiMate allows to extend its core by adding attributes to ArchiMate concepts and relationships. For example when interested in performing detailed quantitative analysis using key performance indicators [16].

Since one of the goals of the ArchiMate language [7] is the integration of detailed design models, together with the characteristics presented before, it is perfectly suited to address the problem previously identified.

4.3   Costing Method

Generally, there are two possible approaches of estimating costs: bottom-up or top-down [17]. Top-down approaches are based on resource's use estimates, being their results dependent on the quality of the estimate used. Bottom-up approaches such as activity-based costing (ABC), quantify the amount of each resource that is used to produce a service and apportion costs accordingly to estimate unit costs [17]. Therefore, the results of these approaches are more feasible, however, comparing with the top-down, they require more effort to implement [17]. In spite of its advantages, it has been difficult for many organizations to implement ABC due to problems related with its implementation and data collection [9].

Time-Driven Activity Based Costing. Due to several criticisms of the ABC, its authors without fully abandoning the concept (after all it helped many companies identifying important costs [18]) came up with a newer and refined ABC. This method was the Time-Driven Activity Based Costing (TDABC), that was simpler, more flexible and easier to implement.

TDABC only requires two estimates: (1) capacity cost rate (CCR) of a resource (calculated by dividing the cost of a resource by its practical capacity (amount of time that a resource is available to work)), and (2) time needed to execute each activity. With these two measures it is possible to determine the cost of an activity, by multiplying the time taken by its unit cost.

To solve the ABC low flexibility problem, due to fact that a small variation in an activity originated a new one, TDABC introduced the concept of time-equation, that is a linear equation that represents costs in function of the time and resources it consumes, allowing to represent conditions and cycles.

An example of this, is the case of the activity "Process Order" in a logistics department, that for special needs, extra time is needed to finish. In ABC this would result in two activities, "process normal order" and "process special order". In TDABC its simpler because this variation can be expressed in the same equation (Eq. 1), reducing the number of existing activities in comparison with ABC. This equation means that processing an order consumes 5 time units of the logistics department for every order, plus 3 additional time units for each special order. This way it is possible to consider a greater complexity of the reality avoiding an exponential growth of the number of activities by grouping all the activity' variations in a single time equation.

$$\text{Order processing} = 5 + 3 * [\text{ If special }] \tag{1}$$

Another advantage of the TDABC is the possibility of measuring the unused capacity that can give the ability to carry out performance analysis [9].

We decided to focus on Activity-Based methods, like TDABC, since our objective is to calculate the cost of a service or product, and some of the other existent costing methods are outdated (e.g. direct costing) or are focused on value (e.g. Lean) or on quality (e.g. Just-In-Time), and so on. Also, there is a clear parallelism between the activities defined in Activity-Based methods and Business Processes in ArchiMate, which will make the connection between the diagrams and the costing template easier.

5 Related Work

The related work consists of contributions dealing with EA cost analysis. One of the first initiatives where this non-functional EA aspect was addressed was [19], where associated to a possible EA scenario was its probable cost. This way stakeholders would have a criteria to decide about the best scenario to choose. However this cost may not be entirely representative, since it was estimated based on unfounded assumptions instead of its absolute value.

In the case of [14] even if it is presented an ArchiMate costs viewpoint to address the EA costs concern, the way costs are calculated is based on simplistic traditional cost methods, which may distort the real value of a cost object.

Similar to [14], in [20] ArchiMate is used to model EA' costs, however even being a viable solution to represent costs, the cost analysis method used in this work in some cases may lead to incorrectly estimates due to the way how the resources' cost is allocated. For example the cost of a behaviour element (e.g. business process) is the sum of the cost of all resources used by that element, however in case we are considering a resource that is used by more than one business process at the same time, its cost cannot be considered twice, instead it as to be divided between them according an adequate criteria (e.g. if two business processes are being performed at the same time, in the same facility, but in different rooms, the facility cost has to be divided by the two business processes, using as criteria the space occupied by each one of them). Furthermore, this proposal does not consider all the relevant kind of resources, not allowing to address all the possible cost analysis scenarios. For example, in case of a human resource, it can be modelled either by a business actor (the only option considered in [20], which is the most common) or by a business role (in some organizations an employee can have different costs according the role performed).

In [11] ArchiMate is used to represent the business models of an organization using Business Model Canvas (BMC), serving a bigger purpose than only to address the EA costs concern. The cost analysis method used in this work is an updated version of [20], overcoming one of its predecessor limitations by considering all the resources concepts necessary to model all the possible scenarios. However in BMC only the key resources are considered, leaving aside indirect resources, which even if not directly linked to a cost object are necessary to run the business and may have a considerable impact in its final cost.

All approaches previously mentioned, although being a valid option to represent costs have some issues regarding the way costs are calculated. This is mainly due to the absence of the use of an appropriate costing method. However according our literature review there are two approaches [21, 22], where EA' costs are calculated according an appropriate costing method. In these proposal the authors present a method to apply TDABC in a business process (modelled in Business Process Model and Notation). Both proposals are valid options to estimate organisation' costs, however, even if considering all the different cost sources, they allocate all those costs to human resources, making difficult to know what are the resources necessary to realize a certain product or service and also to distinguish their impact on its the final cost.

All the existing solutions although related to the identified problem do not solve it entirely, because none of them considers all the relevant aspects necessary to solve the problem. Some of the key aspects that a solution for the problem identified in this paper should take into account are: allow cost analysis according an appropriate cost model (like TDABC, which we believe, that in addition to its advantages is the costing method most appropriate to use in cases like this), based on the organisation' EA (in order to benefit of its advantages) and easy to analyse and understand (facilitating its use and allowing to communicate effectively and clearly - with transparency - its results to relevant stakeholders).

6   Proposal

This section presents an ArchiMate viewpoint that allows to apply a costing method (TDABC) while modelling an EA, making possible to verify the costs associated to an EA. The aim is to support communication and decision-making amongst stakeholders during the cost analysis process. The viewpoint specifies the concepts, relationships and attributes that must be taken into account when considering the cost analyses of a certain cost object.

6.1   Objectives

The main objectives that we intend to fulfil with the use of this proposal by **improving cost representation in the EA** are:

- Express and reason upon the underlying elements regarding EA costs concern;
- Enable stakeholders to know their EA real costs;
- Aid in the communication and decision-making towards EA costs;

6.2   The Viewpoint

Our proposal is based on two artifacts: a mapping (between TDABC and Archi-Mate) and an ArchiMate viewpoint (based on the previous mapping, with the purpose of explaining how to use ArchiMate in order to make possible to apply TDABC while modelling an EA).

6.3   TDABC-ArchiMate Mapping

So, after selecting the methods and approaches that will be the basis of our proposal, the initial step to reach our first artifact was to identify the TDABC main concepts and only then map them in the possible ArchiMate elements (first column of Table 1).

The ArchiMate concepts suitable to be related with the TDABC concepts, are not in a single layer, since a cost method in order to be useful needs several types of information from the different layers (business, technological, etc.) [9].

To define a valid correspondence between the two we started by comparing the concepts existing in TDABC with the ones defined by ArchiMate. Table 1 shows and motivates the proposed correspondence resulting from this comparison. Remind that this is a unilateral mapping relation and not bilateral one.

As can be seen, not all the TDABC concepts have a correspondent ArchiMate element (these being modelled as other ArchiMate elements attributes), even being ArchiMate a very rich language. This is due to ArchiMate not consider the time, which is a vital concept for TDABC (time is used to describe the resources' capacity and the unit time estimate of an activity). Thus, the only way to model time is extending ArchiMate using for that purpose the attributes of an element. In our case, those elements will be the elements that represent a resource and the ones that represent an activity.

"Final cost" was the only cost related concept that was modelled implicitly because there is only one cost for a cost object (what does not have a big impact on the diagram complexity) and because it is the main result that we aim to check with the use of this viewpoint.

Regarding TDABC-ArchiMate resources mapping, we based on Technology Business Management (TBM) framework[1] to name the different types of resources (as can be observed on Fig. 1). The main reason why we did this was that the TBM framework provides a standard taxonomy to describe cost sources and also because one of its main stakeholders be the CIO (same as ours). Thus, by using this taxonomy we are assuring that the terms used are understood and mean the same thing for everyone, allowing to create a basis for costs transparency and providing a standard taxonomy for reporting costs.

---

[1] https://www.tbmcouncil.org/learn-tbm/tbm-taxonomy.

Table 1. Defining the correspondence between TDABC and ArchiMate

TDABC	ArchiMate	Justification
Activity	Business Process	An activity [in the TDABC] is defined as any event, unit of work or task with a specific goal. This definition conforms to the ArchiMate definition of "Business Process", since both define it as something with an objective, which this case is to realize a cost object
Resource Pool (Labour, hardware, software...)	(Resource)-Business Actor - Business Role - Application Component - Node - Business Service	A resource [in the TDABC] is defined as a factor required to complete an activity. This definition reproduces almost literally the definition of "Resource" ArchiMate element. Due to this and according with [7], the particular cases of a Resource are: Business Actor; Business Role; Application Component; Node. Besides these we considered also the "Business Service" to represent outside resources (e.g. cleaning, security, maintenance, etc.), which are not directly involved in the production of a cost object, but without which it would not be possible obtain one. These types of resource are usually used to serve other resources, assuring their normal usage. ArchiMate [7] does not consider a "Business Service" it as a resource, but according TDABC [9] and other authors [23,24], it clearly is one
Practical capacity (min)	Resource Attribute	Practical capacity [in the TDABC] refers to how much a resource can really do within a given period, being less than its theoretical capacity. ArchiMate cannot model or express time, remaining as the only solution to use its extension mechanism and representing it through an attribute, in this case a resource attribute
Cost of capacity supplied ($)	Resource Attribute	This equivalence was already presented and established in another work. "The only ArchiMate concept that can be used to model cost is value. Another option is to specify the costs as an attribute of the architectural elements generating them (e.g., a human, technical or informational resource). However, in such case (as opposed to modelling cost as value), the modelling of cost sources is explicit, while that of costs themselves is implicit" [11]
Capacity Cost Rate ($/min)	Resource Attribute	The resource's capacity cost rate [in the TDABC] is calculated by dividing its cost by its capacity, usually expressed as a cost per hour. So, this measure is a cost per unit of time, thus is still a cost. So it can be explained by the same reason presented in the row above
Unity Time Estimate (min)	Business Process Attribute	Same reason as "Practical capacity"
Cost Object	- Business Process - Business Service - Business Product - Stakeholder -Business Object	A cost object [in the TDABC] is an item for which costs are separately measured. There are several types of cost objects, of which we highlight the following three: Output (which is the most common within an organization, allowing to estimate the cost of their outputs (e.g. products, services)); Operational (internal to the organization, allowing to know the cost of performing a certain behaviour (e.g. process)), or External (extrinsic to the organisation, allowing to determine the cost of dealing with a certain entity (e.g. customer, order)). The first two types have a direct match to ArchiMate concepts. Regarding the last type, its first particular case (customer) is a specialization of the ArchiMate concept "Stakeholder" [7], whereas the second corresponds to a "Business Object" (see Time Driver row of this table for more information)
Time Driver	Business Object Attribute	A time driver [in the TDABC] is a factor that influences the duration of activities. In ArchiMate the element which can be associated to an activity (represented by a "Business Process" according our mapping) and that can represent an information asset that are relevant for a business process, in this case influencing its duration due to its information content, is the "Business Object". For example the time required to process an order is proportional to the number of items ordered, which is described in the order, which in ArchiMate is represented as a Business Object)
Time equation	Business Process Attribute	TDABC estimates the resources demand by a time-equation [9]. Just like "Practical Capacity" this concept is also related with time, intending to model the time that an activities took to be performed. By the same reason given for "Practical Capacity" we choose an attribute to model this concept
Final Cost ($)	Value	Same reason as "Cost of capacity supplied"
Activity Cost ($)	Business Process Attribute	Same reason as "Cost of capacity supplied" and "Final Cost"

Table 2. Relationship matching

TDABC relationship	ArchiMate relationship
is a	Specialization
associate	Association
is realized	Realization
assigned	Assignment
serve	Serving
has	Element attribute

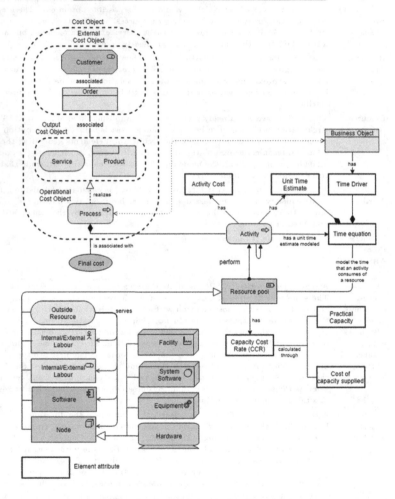

Fig. 1. ArchiMate costs viewpoint

After the most suitable matches for TDABC were found in ArchiMate, we analysed the relationships between them in order to match TDABC relationships with ArchiMate relationships. The result of this step is presented in Table 2.

After completing all the necessary mapping both for concepts and relations we were able to create the second artifact mentioned in the beginning of this section, the costs viewpoint (Fig. 1).

6.4 Viewpoint Classification

This viewpoint, like described in Fig. 2, represents the costs of an organisation's cost object, according to the EA that supports it, allowing to depict the resources necessary to create it and the cost of each one of them, and also the cost and unit time estimate of the activities that are executed to create it. This viewpoint could benefit many management processes where cost analyses are needed, like price setting, budgeting or planning, facilitating processes and helping managers to inform relevant stakeholders easily.

Costs Viewpoint		
Stakeholders	Managers, CEO, CIO, others	
Concerns	Cost information, resources, activities and costs objects of an organization	
Purpose	Deciding, informing	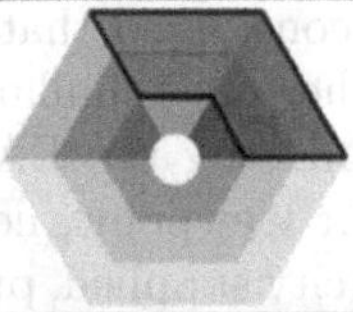
Abstraction level	Coherence, overview	
Layer	Business, application, technology, physical	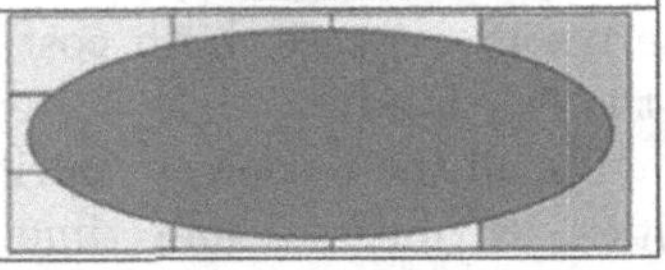
Aspects	Passive structure, behavior, active structure, motivational	

Fig. 2. Costs viewpoint description. Adapted from [26]

7 Demonstration

This section presents an application of our viewpoint towards representing the costs of an EA. To demonstrate the utility of our proposal we focus on the healthcare sector due to the cost crisis that is affecting this sector [25]. One of the main reasons for this crisis is the lack of understanding about how much costs to deliver patient care [24,25]. This has brought a new necessity to this sector to gain a deeper understanding of their business costs through real-world evidence instead of assumptions and self-reported data [2,24,25].

Regarding this worrisome situation, we applied our proposal to this problem in order to aid managers to have a better view of their costs. For that purpose a Harvard case study [24] was used to model the EA that was supporting a healthcare service. In this case study [24], the authors describe the several phases of coronary artery bypass graft surgery (CABG) and their cost. In this demonstration we have only modelled the pre-operative phase (cost object under analysis),

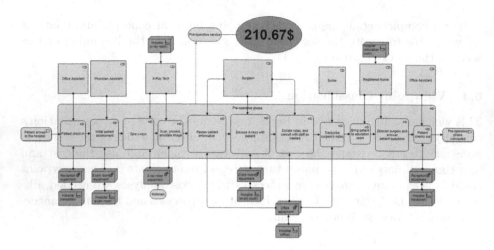

Fig. 3. Pre-operative phase of CABG surgery service supporting EA

because we considered that this phase was rich enough to apply our proposal, being the other ones similar, not adding an extra value to our demonstration.

After modelling the EA (Fig. 3), based on the case study description and according our viewpoint, next step was to add the attributes, first to resources (cost of capacity supplied, practical capacity and CCR) and then to the activities. In the resources' case, both cost and time-related attributes were withdrawal from the case study.

Having all the resources' attributes filled, the following step was to add the activities' attributes - unit time estimate and activity cost - to each one of the activities required to perform the CABG pre-operative service. The first attribute was obtained directly from the case study, while the second one was calculated for each activity according Eq. 2. In the activities' case, only the time equation attribute was not considered, because the case study description lacks this information.

$$\text{Activity Cost} = (CCR_{resource_1} + \ldots + CCR_{resource_n}) * UnitTimeEstimate \quad (2)$$

With the EA modelled and both the resources and activities' attributes filled we were able to calculate the average cost for this service (using Eq. 3), obtaining an average value of 210.67$.

$$\text{Service Cost} = ActivityCost_1 + \ldots + ActivityCost_1 1 \quad (3)$$

As it can be seen in Fig. 3, not all the elements modelled in this view have the same size. This is due to resource's size be proportional to their CCR and the activities' size be proportional to their cost. This way we can easily verify which are the most expensive activities and resources, facilitating the perception of the main cost determinants of a cost object.

8   Evaluation

To evaluate our proposal a generic evaluation model for IS artifacts and evaluation criteria was used [27]. Several evaluation criteria were chosen, taking into account different evaluation contexts. As an evaluation of an instantiation of our viewpoint taking into account its goals achievement (efficacy), environment consistency with people (utility, ease of use and understandability) and also a structural assessment (simplicity and consistency). To evaluate these criteria, a quantitative assessment was made by evaluating the answers collected in a questionnaire about our proposal and the demonstration shown before.

8.1   Questionnaire

The questionnaires were answered by 31 EA experts, with more than six months experience. Before answering to the questionnaire subjects were introduced about the context of our proposal in an introductory session.

The object under analysis in the questionnaire was the viewpoint we created. For that purpose subjects were asked to evaluate an instance of our viewpoint (view presented in Sect. 7). The questionnaire was pre-tested with EA experts with between 10 and 15 years of experience in the field. The pre-test resulted in minor adjustments of the wording.

The questionnaire comprises three sections, including a total of 29 questions. In order to characterize the respondents' profile, first section has asked them to classify their experience regarding our proposal theme (costs and EA). The second section intended to assess how well respondents comprehended our work and if the objectives that we defined for our proposal were met. For that purpose respondents had to answer several questions about specifics aspects of the demonstration used in the questionnaire (Fig. 4). In the last section respondents have been asked to assess the model' quality regarding the criteria previously chosen, on a 5-point Likert scale [28], where 1 is the lowest point on the scale, and 5 is the highest point on the scale.

Regarding section two, it also helped us performing data cleaning, by ignoring answers whose results were negative in this section (showing that the respondent did not understood our proposal) and positive in the last one.

Due to space limitations, only questionnaire' Sect. 2 is presented (Fig. 4).

8.2   Questionnaires Data Analysis and Results Discussion

In order to draw significant conclusions from the questionnaire results, the confidence intervals statistical method was used based on the given result sample data. The data used in the analysis correspond to the answers to the questionnaires. All data analysis was conducted in Excel.

According to the answers in the questionnaire' first section we concluded that subjects had all some considerable experience in the EA field (45% know well and 55% know very well this theme). However, as it was already expected, none of the subjects was familiarized with the cost subject of our proposal, so to

2.1. Can you identify the final cost of the pre-operative phase modelled above?

☐ Yes   ☐ No

2.2. If you answered yes to the question 2.1. State the final cost you identified.

_____

2.3. From the following list, where several type of resources are present, mark with X the ones that were used/modeled in the above view.

☐ Personnel   ☐ Software ☐ Hardware   ☐ Machine/Equipment

☐ Physicall Space ☐ Eletricity ☐ Raw materials ☐ Material/Consumables

2.4. From the several resources needed to perform the pre-operative phase modelled in the view above, can you identify the one with the highest cost?

☐ Yes   ☐ No

2.5. If you answered yes to the question 2.4. Identify the name of the resource with the highest cost.

_____

2.6. If you answered yes to the question 2.5. Identify the capacity cost rate (CCR) ($/min) of the resource, which you identified in the previous question.

_____

2.7. In the diagram presented previously, what is the activity with the lowest cost?

_____

Fig. 4. Questionnaire Sect. 2

evaluate the cost information relevance and correctness we will have to repeat the evaluation with a different sample (e.g. costs experts).

Section two' results revealed that almost 80% of the respondents answered correctly to most of the questions (see Fig. 4), being the main difficulty question 2.3., related with resource identification (Fig. 4). When asked to identify from a list what were the different type of resources necessary to perform the service presented in the questionnaire (which the correct answer was: personnel, equipment, materials, physical space), some of the subjects forget to mark materials (due to its minimal size, consequence of its low cost) or marked besides the correct resources, hardware (considering that it was included in equipment).

Regarding last part, the results of the quantitative assessment of our the proposal' demonstration based on the aforementioned criteria can be seen in the diagram 5. The diagram for each question takes the form of a box that spans the range of values where possible target users will answer to the question, according a confidence level of 95%. Besides this, it also allows to visualize the median (line splitting the box) and the maximum and minimum value obtained for each answer (extremes).

By analysing the results obtained it is verified that an average above of 3 was obtained for every question, which associated to a confidence level of 95%, makes acceptable to conclude that all target users will answer these questions in the range of scores, presented by the confidence intervals on Fig. 5.

Based on the previous findings we can assert that our proposal had a positive assessment from the questionnaire respondents, even without costs experience.

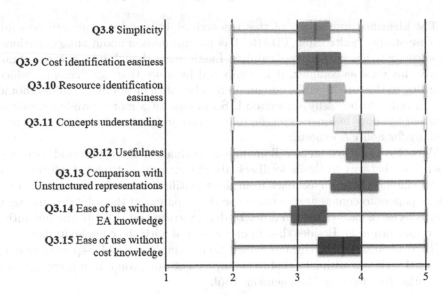

Fig. 5. Questionnaire results

This provides an initial validation regarding the achievement of the solution' objectives. This result lead us to believe that our proposal is a valid option to represent and reason about the costs of an EA, adding an extra aspect to EA, which is relevant to its stakeholders, but that until now was not being addressed.

9   Conclusions

Costs are fundamental in any organization and according EA stakeholders, estimating and managing costs is one of the major concerns they pretend to address with the use of EA, since in their opinion, it can help them to decrease the costs related with business organization. However, so far EA does not support it.

The EA costs viewpoint presented in this paper intends to address this concern by improving cost representation in the EA, allowing stakeholders to know the costs of their activity, offering them a way to trace the costs of the organisation outputs to each one of the organisation' resources.

To validate the usefulness and applicability of the proposal a Harvard case study related with Heart Bypass Surgery was used to formulate a questionnaire.

Based on our findings, we know that our viewpoint is compliant with Archi-Mate, is a good option to address the problem identified (even for someone with few experience in the costs domain) and also that is a valid and alternative option to represent costs when compared to unstructured representations. However when asked about if someone without experience in the EA domain would be able to fully understand the information shown on a view like the one used in the demonstration, results were lower.

The identified limitations of this research at its current stage are twofold: this case study despite using TDABC has no information about time equations, not allowing to assess their applicability. Lastly, even though EA experts did not qualify this view as complex, it is composed by more than 30 elements, which according to Horton [29] is the maximum number of elements that a model should have in order to be easily understood. So in case of a more complex scenario, eventually it may become unreasonable to comprehend and analyse the view, specially for non EA experts.

We believe the next steps will include an evaluation of our proposal with cost experts, so that we can obtain feedback about the view' cost related information. We also intend to obtain feedback from the scientific community through submission of papers to conferences, resume the development of the solution, and then carry a series of case-studies in order to demonstrate and validate the instantiation of our proposal. Besides this, in order to deal with the complexity issue, we will take advantage of ArchiMate views, facilitating stakeholder-specific visualisations. Lastly our ultimate goal is to implement our proposal in a commercial Enterprise Architecture Management Tool.

References

1. Shank, J.K.: Strategic cost management: new wine, or just new bottles. J. Manag. Account. Res. 1(1), 47–65 (1989)
2. Demeere, N., Stouthuysen, K., Roodhooft, F.: Time-driven activity-based costing in an outpatient clinic environment: development, relevance and managerial impact. Health Policy 92(2–3), 296–304 (2009)
3. Everaert, P., Bruggeman, W., Sarens, G., Anderson, S.R., Levant, Y.: Cost modeling in logistics using time-driven ABC. Int. J. Phys. Distrib. Logist. Manag. 38(3), 172–191 (2008)
4. Niazi, A., Dai, J.S., Balabani, S., Seneviratne, L.: Product cost estimation: technique classification and methodology review. J. Manuf. Sci. Eng. 128(2), 563–575 (2006)
5. Lindström, Å., Johnson, P., Johansson, E., Ekstedt, M., Simonsson, M.: A survey on CIO concerns-do enterprise architecture frameworks support them? Inf. Syst. Front. 8(2), 81–90 (2006)
6. Cooper, R., Kaplan, R.S.: Measure costs right: make the right decision. Harv. Bus. Rev. 66(5), 96–103 (1988)
7. Lankhorst, M.: Enterprise Architecture at Work - Enterprise Modelling, Communication and Analysis, vol. 36, 2nd edn. Springer, Heidelberg (2013). https://doi.org/10.1007/978-3-642-29651-2
8. Ahlemann, F.: Strategic Enterprise Architecture Management: Challenges, Best Practices, and Future Developments. Springer, Heidelberg (2012). https://doi.org/10.1007/978-3-642-24223-6
9. Kaplan, R.S., Anderson, S.R.: Time-driven Activity-based Costing: A Simpler and More Powerful Path to Higher Profits, p. 266. Harvard Business School Press Books, Boston (2007)
10. Peffers, K., Tuunanen, T., Rothenberger, M., Chatterjee, S.: A design science research methodology for information systems research. J. Manag. Inf. Syst. 24(3), 45–77 (2008)

11. Iacob, M.E., Meertens, L.O., Jonkers, H., Quartel, D.A.C., Nieuwenhuis, L.J.M., van Sinderen, M.J.: From enterprise architecture to business models and back. Softw. Syst. Model. **13**, 1059–1083 (2012)
12. Gama, N.: Building sustainable information systems (2013)
13. Pereira, C. M., Sousa, P.: A method to define an enterprise architecture using the Zachman framework. In: Proceedings of the 2004 ACM Symposium on Applied Computing, pp. 1366–1371 (2004)
14. Lagerström, R., Johnson, P., Ekstedt, M., Franke, U., Shahzad, K.: Automated probabilistic system architecture analysis in the multi-attribute prediction language (MAPL): iteratively developed using multiple case studies. Complex Syst. Inform. Model. Q. (11), 38–68 (2017)
15. Zachman, J.A.: A framework for information systems architecture. IBM Syst. J. **26**(3), 276–292 (1987)
16. Miguel, N.: Modeling a process assessment framework in ArchiMate (2014)
17. Cunnama, L., Sinanovic, E., Ramma, L., Foster, N.: Using topdown and bottomup costing approaches in LMICs: the case for using both to assess the incremental costs of new technologies at scale. Health Econ. **25**(S1), 53–66 (2016)
18. Kaplan, R.S., Anderson, S.R.: Time-driven activity-based costing. Harv. Bus. Rev. **82**(11), 131 (2004)
19. Kazman, R., Asundi, J., Klein, M.: Quantifying the costs and benefits of architectural decisions. In: Proceedings of International Conference on Software Engineering, pp. 297–306 (2001)
20. Gunasekaran, A. (ed.): Global Implications of Modern Enterprise Information Systems: Technologies and Applications. IGI Global (2008)
21. Lourenço, A.G.: Analyzing cost and profitability using process-based ABC (2013)
22. de Andrade, J.E.R.: Templates for calculating IT services costs (2014)
23. Gregório, J., Russo, G., Lapão, L.V.: Pharmaceutical services cost analysis using time-driven activity-based costing: a contribution to improve community pharmacies' management. Res. Soc. Adm. Pharm. **12**(3), 475–485 (2016)
24. Erhun, F.: Time-driven activity-based costing of multivessel coronary artery bypass grafting across national boundaries to identify improvement opportunities: study protocol. BMJ Open **5**(8), e008765 (2015)
25. Kaplan, R.S., Porter, M.E.: How to solve the cost crisis in health care. Harv. Bus. Rev. **89**(9), 46–52 (2011)
26. ter Doest, H., Iacob, M., Lankhorst, M., van Leeuwen, D., Slagter, R.: Viewpoints functionality and examples, pp. 1–92 (2004)
27. Prat, N., Comyn-Wattiau, I., Akoka, J.: Artifact evaluation in information systems design-science research-a holistic view. In: PACIS (2014)
28. Likert, R.: A technique for the measurement of attitudes. In: Archives of Psychology (1932)
29. Horton, W.: Illustrating Computer Documentation. Wiley, New York (1991)

Coop-DAAB: Cooperative Attribute Based Data Aggregation for Internet of Things Applications

Sana Belguith[1]($\boxtimes$), Nesrine Kaaniche[2], Mohamed Mohamed[3], and Giovanni Russello[4]

[1] School of Computing, Science and Engineering, University of Salford, Manchester, UK
belguith.sana@gmail.com
[2] SAMOVAR, Telecom SudParis, University Paris-Saclay, Paris, France
nesrine.kaaniche@telecom-sudparis.eu
[3] IBM Research, Almaden Research Center, San Jose, CA, USA
mmohamed@us.ibm.com
[4] The Cyber Security Foundry, The University of Auckland, Auckland, New Zealand
g.russello@auckland.ac.nz

Abstract. The deployment of IoT devices is gaining an expanding interest in our daily life. Indeed, IoT networks consist in interconnecting several smart and resource constrained devices to enable advanced services. Security management in IoT is a big challenge as personal data are shared by a huge number of distributed services and devices. In this paper, we propose a **Coop**erative **D**ata **A**ggregation solution based on a novel use of **A**ttribute **B**ased signcryption scheme (Coop-DAAB). Coop-DAAB consists in distributing data signcryption operation between different participating entities (i.e., IoT devices). Indeed, each IoT device encrypts and signs in only one step the collected data with respect to a selected sub-predicate of a general access predicate before forwarding to an aggregating entity. This latter is able to aggregate and decrypt collected data if a sufficient number of IoT devices cooperates without learning any personal information about each participating device. Thanks to the use of an attribute based signcryption scheme, authenticity of data collected by IoT devices is proved while protecting them from any unauthorized access.

Keywords: IoT data aggregation · IoT applications
Resource-constrained devices
Constant size attribute based signcryption

1 Introduction

The Internet of Things (IoT) applications are deployed in several fields such as health care, smart cities, smart monitoring [1,5]. IoT systems connect loosely

© Springer Nature Switzerland AG 2018
H. Panetto et al. (Eds.): OTM 2018 Conferences, LNCS 11229, pp. 498–515, 2018.
https://doi.org/10.1007/978-3-030-02610-3_28

defined objects, gateways and services that may exchange data about peoples' body state, life events, habits, location or professional information. In most of the cases, this data is sensitive and should be processed and managed with high security measures. Security management in IoT is a big challenge as sensitive data are shared by a huge number of distributed services and devices. To face the exponential growth of data being generated by IoT devices and to efficiently ensure their collection, aggregation and sharing, fog and cloud computing are usually used to assist IoT devices since these latters are resource constrained. Although the usage of these environments has clear benefits, it brought new security and privacy threats as the data might be outsourced to untrusted environments. To countermeasure these threats, the data is usually obfuscated before being outsourced. In addition, the collected data from different IoT devices is generally aggregated to considerably save the energy resources and extend the lifetime of the IoT devices. However, the aggregation affects the security properties that might be provided by the protection schemes. For instance, the data needs to be authenticated at the aggregation phase to ensure that it is outsourced from benign devices while preserving their privacy.

Several works have been proposed to ensure data authentication based on signature schemes. However, these techniques are usually combined with encryption to provide data contents' secrecy. This combination incurs heavy computation and communication overhead due to the cumulative costs of encryption and signature. Signcryption [18] has been proposed by Zheng et al. as a cryptographic mechanism combining signature and encryption in only one phase. Signcryption allows an entity to encrypt and sign the ciphertext while incurring reduced computation costs compared to executing the encryption and the signature algorithms separately.

Attribute based signcryption (ABSC) is a signcryption primitive that ensures fine grained access control, data origin authenticity and data confidentiality thanks to the combination of attribute based encryption and signature in one logic step. Similar to other attribute based techniques, ABSC introduces one main drawback related to high computation and storage costs which depends on the size of used access policies. To mitigate this limitation, thus, attribute based signcryption schemes with constant computation costs and ciphertext sizes have been proposed [2].

Contributions. This paper extends our previous work [4] and it introduces Coop-DAAB, a cooperative privacy preserving attribute based signcryption mechanism based on the constant-size attribute based signcryption (ABSC) technique [2]. As such, our Coop-DAAB construction consists on performing the combined signing and encrypting processes of a set of data devices' inputs in a secure collaborative manner. The main idea behind Coop-DAAB relies on the distribution of the signcrypting operation among different devices, with respect to selected sub-sets of a general access predicate. That is, each device signcrypts its input data and sends the partial signcrypted information to an untrusted aggregator. This latter is capable of decrypting the received data only if a sufficient number of IoT devices cooperates.

Paper Organization. Section 2 introduces potential applications while identifying major security requirements. Section 3 introduces the network model and details Coop-DAAB concrete construction. The security analysis of Coop-DAAB is discussed in Sect. 4. Finally, performances analysis is detailed in Sect. 5 and related work is discussed in Sect. 6 before concluding in Sect. 7.

2 Motivating Applications and Requirements

The deployment of IoT devices is gaining an expanding interest in our daily life. Indeed, IoT networks consist in interconnecting several smart and resource constrained devices to enable advanced services. Security management in IoT is a big challenge as personal data are shared by a huge number of distributed services and devices. As reducing the amount of transmitted data can effectively save energy, aggregation services are generally applied to derive succinct contents. This technique can be an alternative that aims at providing security enhancement while reducing processing and communication overheads in several applications, namely *vehicular networks, mobile crowd sensing* and *service level agreement monitoring.*

Intelligent Transport Systems (ITSs). In ITSs, connected cars are responsible for continuously publishing data related to their location and traffic status among the network. These broadcasted data need to be genuine in order to avoid injecting false data in the network. However, ensuring data authentication should not lead to a privacy leakage of personal data. To fulfill tis trade-off, collected data should be authenticated using privacy preserving techniques. The most prominent C-ITS solutions today allows the detection of traffic jam based on received data from the connected vehicles. Each connected car sends a set of collected data information, based on embedded sensors, to a central node. This latter needs to make sure that the received data are authentic. Then, to decrypt data, the aggregator merges received data from different devices and decrypts the contents to monitor the state of the road and supply the centralized infrastructure with necessary information.

Mobile Crowd Sensing. The popularity of increasingly capable human carried mobile devices such as smartphones and smart-watches involved several embedded sensors. This led to the appearance of the Mobile Crowd Sensing (MCS) paradigm. MCS is a sensing paradigm able to outsource collected data by sensors to a group of participating users, namely (crowd) workers. MCS is mainly based on the use of mobiles devices and their resources. Thus, data aggregation can be an efficient technique to ensure data collection while saving costs at the crowd worker side. Obviously, privacy preservation of the participating entities and data authentication should be considered in such environment to ensure the collection of benign data.

SLA Monitoring. Service Level Agreements (SLAs) are widely used to describe the agreements between customers and service providers regarding the quality of the provided services [11]. To measure the providers compliance to the SLA, monitoring tools should be setup to collect specific metrics that can quantify the Quality of Service (QoS). These monitoring mechanisms should run continuously or periodically but in either cases can generate huge amounts of data. Eventually, this data is shared with other components responsible for performing analysis, reporting and executing SLA enforcement measures. The amounts of monitoring data can be huge and sending it through the wire can be an overkill for any system. Moreover, in most of the cases, this data is confidential and should be processed in a secure manner. Consequently, monitoring modules can act as an aggregator to reduce the amount of generated data. And to protect data from any tampering, they should be collected, authenticated and aggregated before being used by SLA services to estimate the QoS.

As described above, it is clear that designing a secure, privacy preserving and efficient data aggregation solution, for IoT applications has a high importance. The proposed scheme must fulfill the following properties:

- **data confidentiality** – the proposed aggregation mechanism should protect data contents from being accessed by unauthorized users.
- **data origin authenticity** – the aggregation scheme needs to ensure that data are created by authorized entities.
- **privacy** – the secrecy of devices' access pattern must be protected. Indeed, the aggregator must be able to verify data origin authenticity without leaking extra information about signing devices.
- **low computation and storage costs** – low processing complexities and storage costs need to be provided, by the proposed mechanism mainly for resource-constrained devices.

3   Coop-DAAB: Cooperative Attribute based Data Aggregation Scheme

In this section, we expose our network model then we present a general overview of Coop-DAAB. Finally, we detail the different phases of the proposed solution.

3.1   Network Model

Coop-DAAB network model relies on five different actors: an attribute authority, an administrator, a set of IoT devices $\{d_i\}_{\{i=1,\cdots,M\}}$, a set of gateway aggregating entities $\{Gw_j\}_{\{j=1,\cdots,L\}}$ and a selected IoT trusted node. These different entities are shown in Fig. 1 and defined as follows:

- **attribute authority** AA – is responsible for bootstrapping the whole system in the initialization phase. We assume that AA is a trusted entity. It also issues certified attributes and related secret keys for the aggregating gateways and IoT devices.
- **administrator** $admin$ – is the system administrator responsible for generating the general access predicates used to encrypt data contents. In addition, $admin$ generates the signing access predicate submitted to the IoT devices to sign the collected data.

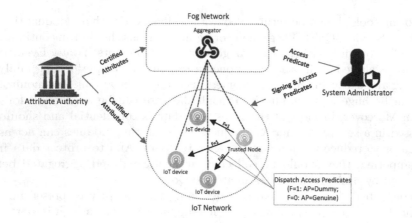

Fig. 1. Network model

- **aggregating entity** (Gw) – is considered as a local network gateway. It is responsible of collecting, deciphering and verifying the authenticity of data contents.
- **IoT device** (d) – collects, encrypts data then signs ciphertexts before forwarding to the aggregating entity.
- **trusted IoT node** (d_s) – is a trusted selected IoT device, periodically assigns to each involved IoT device d_i a sub-access predicate to be used for encrypting data.

3.2   Overview

In this paper, we design a new cooperative privacy preserving encryption scheme, for IoT signed data contents, denoted by Coop-DAAB with constant ciphertext size. Our proposal relies on the constant size attribute based signcryption proposed by Belguith et al. [2], which has been extended to support collaborative encryption of a set of data inputs, collected by IoT devices and gathered by an aggregator with respect to his granted privileges.

Our proposed Coop-DAAB construction involves three phases, namely, SYS_INIT, DATA_SIGNCRYPT and DATA_AGG.

During the first SYS_INIT phase, three randomized algorithms are executed. First, the attribute authority AA performs the stp and keygen algorithms to generate the global public parameters and derive secret keys associated with each involved entity's attributes (i.e., IoT device, gateway). In addition, the system administrator executes accgen algorithm to generate the general access signing and enciphering predicates, denoted by Γ_s and Γ_e.

The second phase occurs periodically, such that each involved IoT device d_i has to signcrypt its collected data content and independently sends the resulting signcrypted information to the aggregating gateway. Note that the time period T is specified by the system administrator, during the SYS_INIT phase. For the second DATA_SIGNCRYPT phase, two algorithms are performed, namely enc and sign, by each involved IoT device. For this purpose, a trusted selected IoT device (d_s) periodically assigns to each involved IoT node a sub-access predicate. That is, the general enciphering access predicate is split by d_s into a set of dummy and genuine sub-access predicates. Note that a dummy access predicate refers to an access predicate such that the aggregating gateway Gw does not satisfy the required threshold (i.e., generally, the aggregator does not have any required attributes), while for a genuine access predicate, the gateway may have some of the required attributes. As such, when an IoT device d_i is assigned a dummy access predicate, denoted by γ_{d,d_i}, it has to encipher and sign its collected data content. And, when assigned a genuine access predicate, denoted by γ_{g,d_i}, it enciphers and signs a neutral group element 1_G. Afterwards, each IoT device d_i sends the signcrypted result to the aggregating node.

During the DATA_AGG phase, the aggregating gateway Gw gathers signcrypted data contents from involved IoT devices. Note that the aggregating gateway ignores assigned sub-access predicates, and is only aware of the general enciphering access predicate Γ_e, published by the system administrator. During this phase, four different algorithms are run by the aggregating gateway Gw, namely vrfchunk, agg, dec and vrfd algorithms. As such, Gw first verifies signed data chunks based on vrfchunk algorithm. Then, it merges the received signcrypted results, in order to generate a global ciphertext, based on the agg algorithm, deciphers the resulting global ciphertext, relying on dec algorithm and checks the authenticity of the received global data content, based on vrfd algorithm.

3.3   Coop-DAAB Phases

This section details the Coop-DAAB main phases and their algorithms.

3.3.1   Sys_Init Phase

This first phase includes three randomized algorithms defined as follows:

- stp—the setup algorithm is executed by the attribute authority AA. The stp is responsible for generating the public parameters pp and the master secret key msk while taking as input security parameter ξ. For this purpose, the trusted authority defines a bilinear setting $(\hat{e}, \mathbb{G}_1, \mathbb{G}_2, \mathbb{G}, g, h)$ of prime order p such that $\hat{e} : \mathbb{G}_1 \times \mathbb{G}_2 \to \mathbb{G}$, $g \in \mathbb{G}_1$ and $h \in \mathbb{G}_2$. In addition, It specifies an encoding function τ such that $\tau : \mathbb{U} \to (\mathbb{Z}/p\mathbb{Z})^*$, where $\mathbb{U}$ is the attribute universe of cardinal n. The function τ is chosen such that for each encoded attribute values $\tau(a) = x$ are pairwise different.

  Then, the stp algorithm selects a set $\mathcal{D} = \{d_1, ..., d_{n-1}\}$ consisting of $n-1$ pairwise different elements of $(\mathbb{Z}/p\mathbb{Z})^*$ (i.e.; dummy users), which must also be different to the values $\tau(a_i)$, for all $a_i \in \mathbb{U}$. Note that for any integer i lower or equal to $n-1$, we denote as $\mathcal{D}_i$ the set $\{d_1, ..., d_i\}$. Finally, the stp algorithm computes u defined as $u = g^{\alpha \cdot \gamma}$ and outputs the global public parameters pp as follows:

$$pp = \{\mathbb{G}_1, \mathbb{G}_2, \mathbb{G}, \hat{e}, h, u, \{h^{\alpha \gamma^i}\}_{\{i=0,\cdots,2n-1\}}, \mathcal{D}, \tau, \hat{e}(g^\alpha, h)\}$$

The master key is set to be $\mathtt{msk} = (g, \alpha, \gamma)$ where α, γ are two values randomly selected from $(\mathbb{Z}/p\mathbb{Z})^*$.

- keygen—the attribute authority executes the keygen algorithm once it receives a key generation request from any participating entity (i.e.; IoT device, aggregating gateway). We denote by E, any participating entity, such that $E \in \{d, Gw\}$. The keygen algorithm takes as input the participating entity's set of attributes, denoted by A_E and the master key of the attributes authority $\mathtt{msk}$ and generates the corresponding secret key sk_E.
  For any subset $A_E \subset \mathbb{U}$ of attributes associated with E, keygen picks a random value $r_E \in (\mathbb{Z}/p\mathbb{Z})^*$ and derives the secret key as follows:

$$sk_E = (\{g^{\frac{r_E}{\gamma + \tau(a)}}\}_{a \in A_E}, \{h^{r_E \gamma^i}\}_{i=0,\cdots,m-2}, h^{\frac{r_E-1}{\gamma}})$$
$$= (sk_{E_1}, sk_{E_2}, sk_{E_3})$$

Remark 1. Communication overhead optimization for IoT devices' key distribution—Considering resource constraints of IoT devices, in terms of storage, processing and communications, our Coop-DAAB scheme assumes that IoT devices are pre-configured with corresponding secret keys. That is, the IoT device's manufacturer is responsible for contacting the attribute authority to physically embed the secret keys into the IoT device.

- accgen—the system administrator runs the accgen algorithm. It takes as input the attributes universe $\mathbb{U}$ and outputs the time period T, the access signing predicate Γ_s and the general access enciphering predicate Γ_e. Recall that T permits to regulate processing and transmitting signcrypted contents by IoT devices, periodically. Each access predicate is represented by a set of attributes $S \in \mathbb{U}$ and a threshold value t, such that at least t attributes need to be satisfied by an IoT device to sign or encrypt a data message. In the following, we denote by $\Gamma_s = (S_s, t_s)$ the access signing predicate, mainly used by IoT devices to prove the authenticity of their data contents, and $\Gamma_e = (S_e, t_e)$ the access enciphering predicate, mainly used by the aggregating gateway to decipher the resulting data contents.

3.3.2 Data_SignCrypt Phase

The second phase occurs periodically, such that each involved IoT device d_i has to signcrypt its collected data content and independently sends the resulting signcrypted information to the aggregating gateway. Recall that a trusted selected IoT device d_s periodically assigns to each involved IoT node a sub-access predicate. That is, the general enciphering access predicate Γ_e is divided into a set of dummy and genuine sub-access predicates such that:

$$\Gamma_e = \bigcup \{\{\gamma_{g,d_i}\}_{i=1,\cdots,k}, \{\gamma_{d,d_i}\}_{i=k+1,\cdots,l}\}$$

l represents the number of all participating IoT devices, k is randomly selected by d_s and represents the number of IoT devices that are assigned genuine access predicates such that they have to encipher and sign a neutral group element 1_G, while the remaining $l - k$ devices are assigned dummy access predicates such that they encipher and sign their collected data contents[1].

This phase consists of two randomized algorithms, performed by each involved IoT device, detailed hereafter:

- enc—this algorithm takes as input a message m_i and the assigned sub-access tree γ_{F,d_i}, where $F = 0$ presents a genuine sub-access predicate and $F = 1$ denotes a dummy sub-access predicate. It outputs the encrypted message C_{d_i}. Note that each sub-access predicate is presented as $\gamma_{F,d_i} = (S_F, t_F)$, where S_F is the set of required deciphering attributes by γ_{F,d_i}, $|S_F| = s_F$ is the number of attributes of S_F and t_F is the threshold value w.r.t. γ_{F,d_i}. First, the device d_i picks a random $\kappa_i \in (\mathbb{Z}/p\mathbb{Z})^*$ and computes the enciphered message C_{d_i} defined as $C_{d_i} = (C_{1,i}, C_{2,i}, C_{3,i})$, defined as follows:

$$
\begin{cases}
C_{1,i} = (g^{\alpha \cdot \gamma})^{-\kappa_i} \\
C_{2,i} = h^{\kappa_i \alpha \cdot \prod_{a \in S_F} (\gamma + \tau(a)) \prod_{d \in \mathcal{D}_{n+t_F-1-s_F}} (\gamma + d)} \\
C_{3,i} = \hat{e}(g,h)^{\alpha \cdot (\kappa_i + m_i)} \cdot f(m_i) = K_i \cdot f(m_i)
\end{cases}
$$

where f is a bijective and semi-homomorphic function that is specified by *admin* (i.e.; f depends on the use case), supporting the following property: $\prod_i f(m_i) = f(\sum_i m_i)$. For example, f may be the exponential function exp, where $\prod_i exp(m_i) = exp(\sum_i(m_i))$.

- sign—this algorithm is performed by the IoT device d_i to sign his encrypted data input, with respect to $\Gamma_s = (S_s, t_s)$, defined by the system administrator, where $S_s \subset \mathbb{U}$ is an attribute set of size $s = |S_s|$ such that $1 \leq t_s \leq |S_s|$. Let A_{d_i} be the subset of attribute set related to the signing IoT device where $|A_{d_i} \cap S_s| = t_s$.

For this purpose, each device d_i uses his secret key sk_{d_i} and the aggregate algorithm `aggreg` [6][2] to output a signature σ_{d_i}. Indeed, d_i first computes T_{1,d_i} such as:

$$
T_{1,d_i} = \texttt{aggreg}(\{g^{\frac{r_{d_i}}{\gamma + \tau(a)}}, \tau(a)\}_{a \in A_{d_i}}) = g^{\frac{r_{d_i}}{\prod_{a \in A_{d_i}} (\gamma + \tau(a))}}
$$

Then, d_i defines the polynomial $P_{(A_{d_i}, S_s)}(\gamma)$ such as:

$$
P_{(A_{d_i}, S_s)}(\gamma) = \frac{1}{\gamma} \left(\prod_{a \in S_s \cup \mathcal{D}_{n+t_s-1-s} \setminus A_{d_i}} (\gamma + \tau(a)) - B_{1,d_i} \right)
$$

Where $B_{1,d_i} = \prod_{a \in S_s \cup \mathcal{D}_{n+t_s-1-s} \setminus A_{d_i}} \tau(a)$

---

[1] Assigning sub-access dummy and genuine predicates may be set via activating a flag F, where $F = 0$ presents a genuine sub-access predicate and $F = 1$ denotes a dummy sub-access predicate.

[2] Coop-DAAB relies on the aggregate algorithm `aggreg` introduced by Delerablee et al. [9].

Afterwards, using the element $sk_{d_{i_2}}$, the signcrypting device d_i derives B_{2,d_i} as follows:

$$B_{2,d_i} = h^{r_{d_i} P_{(A_{d_i}, S_s)}(\gamma)/B_{1,d_i}}$$

In the sequel, d_i generates the signature $\sigma_{d_i} = (\sigma_{1,i}, \sigma_{2,i}, \sigma_{3,i})$ defined as:

$$\begin{cases} \sigma_{1,i} = T_{1,d_i} \cdot g^{\frac{m_i}{\prod_{a \in A_{d_i}}(\gamma + \tau(a))}} \\ \sigma_{2,i} = sk_{d_{i_3}} \cdot B_{2,d_i} \cdot h^{m_i P_{(A_{d_i}, S_s)}(\gamma)/B_{1,d_i}} \\ \sigma_{3,i} = \hat{e}(g^\alpha, h)^{m_i} \end{cases}$$

Finally, the signcrypting IoT device d_i outputs the signcryption of the message m_i as follows:

$$\Sigma_i = (C_{d_i}, \sigma_{d_i}, B_{1,i})$$

Remark 2. Processing cost optimization for IoT devices' encryption and signature algorithms—Considering resource constraints of IoT devices, in terms of storage and processing, our Coop-DAAB scheme assumes that several elementary functions (i.e., computation of signcrypted message elements based on public parameters, such as $P_{(A_{d_i}, S_s)}(\gamma)$, $C_{2,i}^{\kappa_i^{-1}}, \cdots$) are outsourced to a semi-trusted device, with sufficient computation capacities [3]. As such, only a few number of exponentiations and multiplication is required by each single IoT device.

3.3.3   Data_Agg Phase

The DATA_AGG phase involves four algorithms, executed by the aggregating gateway Gw, defined as follows:

- vrfchunk—before starting the aggregation process, Gw has to verify that each received signcrypted message m_i has been correctly signed by a related device d_i. The vrfchunk algorithm takes as input the set of received signcrypted messages $\{\Sigma_i\}_{\{i=1,\cdots,l\}}$ and the public parameters pp. It outputs a boolean value $b \in \{0, 1\}$, where 0 means *reject* and 1 means *accept*. For this purpose, the aggregating gateway Gw has to check the following equality:

$$\sigma_{3,i} \overset{?}{=} \hat{e}(u^{-1}, \sigma_{2,i}) \cdot \hat{e}(g^\alpha, h)^{-1} \cdot \hat{e}(\sigma_{1,i}^{\frac{1}{B_{1,i}}}, h^{\alpha \cdot \prod_{a \in S_s \cup \mathcal{D}_{n+t_s-1-s}}(\gamma + \tau(a))}) \quad (1)$$

Note that the aggregating gateway Gw performs the following algorithms of DATA_AGG phase relying on correctly signed data contents. That is, Gw withdraws inaccurate signatures.

- agg—this algorithm takes as input the set of signcrypted data chunks $\{\Sigma_i\}_{\{i \in [1,l]\}}$, where l is the number of participating IoT devices[3]. It outputs an aggregated signcrypted data message Σ w.r.t. a global message $M = \sum_{i=1}^{l}(m_i)$, as follows:

---

[3] For ease of presentation, we consider that all received signcrypted contents are correctly verified.

$$\begin{cases} C_1 = \prod_{j=1}^{l} C_{1,i} \\ C_2 = \prod_{j=1}^{l} C_{2,i} \\ C_3 = \prod_{i=1}^{l} C_{3,i} = \prod_{i=1}^{l} K_i \cdot f(m_i) \\ \sigma_3 = \prod_{i=1}^{l} \sigma_{3,i} \end{cases}$$

- dec—This algorithm is executed by the aggregating gateway. It takes as input the aggregated signcrypted message Σ, the set of attributes A_{Gw}, the secret key sk_{Gw} of the aggregating gateway and the enciphering access predicate Γ_e. It outputs the message M, such that $M = \sum_{i=1}^{l} m_i$. Indeed, any aggregating gateway Gw having a set of attributes A_{Gw} where $|A_{Gw} \cap S_e| = t_e$ can verify and decrypt the signcrypted message under the access policy $\Gamma_e = (S_e, t_e)$. Then, for all $a \in A_{Gw}$, Gw has to aggregate its secret keys related to the required attributes such as:

$$A_2 = \mathbf{aggreg}(\{g^{\frac{r_{Gw}}{\gamma + \tau(a)}}, \tau(a)\}_{a \in A_{Gw}}) = g^{\frac{r_{Gw}}{\prod_{a \in A_{Gw}}(\gamma + \tau(a))}} \tag{2}$$

Afterwards, Gw defines the polynomial $P_{A_{Gw}}(\gamma)$ such as:

$$P_{A_{Gw}}(\gamma) = \frac{1}{\gamma}\left(\prod_{a \in S_e \cup \mathcal{D}_{n+t_e-1-s_e} \backslash A_{Gw}} (\gamma + \tau(a)) - B_{Gw} \right)$$

where $B_{Gw} = \prod_{a \in S_e \cup \mathcal{D}_{n+t_e-1-s_e} \backslash A_{Gw}} \tau(a)$

Afterwards, Gw uses the aggregated secret key A_2 and the $sk_{E_{Gw}}$ key element to compute:

$$[\hat{e}(C_1, h^{r_{Gw} P_{A_{Gw}}(\gamma)}) \cdot \hat{e}(A_2, C_2)]^{\frac{1}{B_{Gw}}} = e(g, h)^{(\sum_{i=1}^{l} \kappa_i \cdot a) \cdot r_{Gw}}$$

Then, the aggregating gateway Gw deduces the deciphering key $K = \prod_{i=1}^{l} K_i$ such as:

$$K = \hat{e}(C_1, sk_{Gw_3}) \cdot \sigma_3 \cdot \hat{e}(g, h)^{\sum_{i=1}^{l} \kappa_i \cdot r_{Gw} \cdot \alpha}$$
$$= \hat{e}(g, h)^{\alpha \cdot (\sum_{i=1}^{l} m_i + \kappa_i)}$$

Finally, the aggregating gateway Gw recovers the message $M = \sum_{i=1}^{l} m_i$ as follows:

$$M = f^{-1}\left(\frac{C_3}{K}\right) = f^{-1}\left(f\left(\sum_{i=1}^{l} m_i\right)\right) = \sum_{i=1}^{l} m_i$$

- vrfd—to verify the authenticity of the signature of the resulting message M, the gateway Gw uses the retrieved message M and the aggregated σ_3. That is, Gw verifies the correctness of following equality:

$$\sigma_3 \stackrel{?}{=} \hat{e}(g^\alpha, h)^M \tag{3}$$

4 Security Discussion

In this section, we analyse the security resistance of the proposed cooperative aggregation scheme Coop-DAAB with respect to the threat model presented in Sect. 4.1 while proving the fulfillment of the security challenges defined in Sect. 2.

4.1 Threat Model

In order to design a relevant aggregation scheme to secure IoT assisted applications, we define two potential attackers: *malicious IoT external device* and *honest but curious aggregating gateway*, defined as follows:

- *honest but curious aggregating gateway* – this aggregating gateway is honest in term of executing the protocols included in our proposed Coop-DAAB scheme. However, it may be curious about the participating entities sensitive data.
- *malicious IoT device* – this IoT device may be an external device trying to access aggregated data, to forge the signed data contents or to transmit false inputs. This adversary aims at persuading the gateway that he is a genuine IoT user.

4.2 Confidentiality

In our proposed Coop-DAAB scheme, data are encrypted before being stored using an attribute based signcryption scheme. Therefore, the secrecy of data is inherited from the used ABSC scheme.

Theorem 1. Coop-DAAB *ensures the secrecy of both encrypted data chunks and aggregated data contents.*

Sketch of Proof. The proof of Theorem 1 is twofold. First, the secrecy of sign-crypted data contents depends on the security of the signcryption algorithm used to encrypt data chunks provided by IoT devices. Thus, Coop-DAAB inherits the indistinguishability property from [2], such that if a malicious attacker knows some data about the plaintext, it can not leak information about the ciphertext. Note that in ABSC schemes, the adversary may try to overcome the indistinguishability property using his own attributes or by colluding with other compromised users. Indeed, similar to [2], in Coop-DAAB the users secret keys are randomised using the r_E value which is unique for each participating entity. This stops the collusion attacks as users can not put their secret keys together and override their access rights. In addition, sub-access encrypting predicates used to encrypt data chunks are not communicated to the aggregator Gw. Furthermore, Gw's attributes do not satisfy sub-access policies used for encrypting genuine data contents phase thanks to the use of dummy sub-access polices. Consequently, an aggregator cannot deduce data chunk content.

Second, the secrecy of resulting aggregated contents depends on the consistency of the aggregating algorithm agg, such that aggregated data contents are only accessed by the authorized aggregator. Indeed, the general enciphering access predicate is published by the system administrator to the involved aggregating entities. As such, thanks to the use of the ABSC scheme, data are only accessed by users whose attributes match the defined access policy [2].

4.3 Privacy

Theorem 2. Coop-DAAB *ensures the privacy property, such that signing attributes are indistinguishable against a curious aggregating entity.*

Sketch of Proof. The proof of Theorem 2 states that an aggregating entity cannot guess which attributes have been used to signcrypt the data chunk. That is, let us consider a signing IoT device d, holding two different sets of attributes $A_{d,1}$ and $A_{d,2}$, that both satisfy a fixed access predicate $\Gamma^* = (S^*, t^*)$. Thus, d randomly selects a set $A_{d,b}$, where $b \in \{1, 2\}$ to sign a data message m, chosen by a malicious entity. As $|A_{d,b} \cap S^*| = |A_{d,1} \cap S^*| = |A_{d,2} \cap S^*| = t^*$, we deduct that the two signatures computed with respect to the two set of attributes $A_{d,1}$ and $A_{d,2}$ have similar distribution. Moreover, the used signcryption scheme [2] is demonstrated to be privacy preserving in the standard model. That is, while signing data, the identity of the signcrypting entity and its set of attributes are kept hidden from any verifying entity. Thus, Coop-DAAB guarantee that the applied signature does not leak any extra information about the signing entity neither its signing attributes except what can be already inferred from the used signing access policy.

4.4 Access Control to Data

Theorem 3. Coop-DAAB *provides an access to authenticated data contents.*

Sketch of Proof. The resistance of Coop-DAAB against unauthorized access to data relies on the correctness of the aggregation agg and decryption dec algorithms, as detailed in Lemmas 1 and 2. In addition, the support of data origin authentication is provided by the correctness of the signing algorithm sign (c.f., Lemma 3) and its resistance against forgery attacks (c.f., Lemma 4).

Lemma 1. *Correctness of the aggregation of data chunks algorithm* agg—*The correctness of the aggregation of received signcrypted data chunks is detailed hereafter.*

$$\begin{cases} C_1 = \prod_{i=1}^l (g^{\alpha \cdot \gamma})^{-\kappa_i} \\ C_2 = \prod_{j=1}^l h^{\kappa_i \alpha \cdot \prod_{a \in S_F} (\gamma + \tau(a)) \prod_{d \in \mathcal{D}_{n+t_F-1-s_F}} (\gamma + d)} \\ C_3 = \prod_{i=1}^l \hat{e}(g, h)^{\alpha \cdot (\kappa_i + m_i)} \cdot f(m_i) \\ \sigma_3 = \prod_{i=1}^l \hat{e}(g^\alpha, h)^{m_i} \end{cases}$$

$$\begin{cases} C_1 = (g^{\alpha \cdot \gamma})^{-\sum_{i=1}^l \kappa_i} \\ C_2 = h^{\sum_{i=1}^l \kappa_i \alpha \cdot \prod_{a \in S_e}(\gamma + \tau(a)) \prod_{d \in \mathcal{D}_{n+t_e-1-s_e}}(\gamma + d)} \\ C_3 = \hat{e}(g,h)^{\alpha \cdot \sum_{i=1}^l(\kappa_i + m_i)} \cdot f(\sum_{i=1}^l m_i) \\ \sigma_3 = \hat{e}(g^\alpha, h)^{\sum_{i=1}^l(m_i)} \end{cases}$$

Lemma 2. *Correctness of the decryption algorithm* dec—*After aggregating its secret key relying on Eq. 2, the aggregator calculates the decryption key K by executing the following equations:*

$$\hat{e}(g,h)^{\sum_{i=1}^l \kappa_i \cdot r_{Gw} \cdot \alpha} = (\hat{e}(C_1, h^{r_{Gw} P_{A_{Gw}}(\gamma)})) \cdot \hat{e}(A_2, C_2))^{\prod_{a \in S_e \cup \mathcal{D}_{n+t_e-1-s_e} \backslash A_{Gw}} \tau(a)}$$

$$\hat{e}(g,h)^{\sum_{i=1}^l \kappa_i \cdot \alpha} = \hat{e}(C_1, h^{\frac{r_{Gw}-1}{\gamma}}) \hat{e}(g,h)^{\sum_{i=1}^l \kappa_i \cdot r_{Gw} \cdot \alpha}$$

$$\hat{e}(g, \sigma_3) = \hat{e}(g, h^{\alpha \cdot \sum_{i=1}^l m_i}) = \hat{e}(g,h)^{\alpha \cdot \sum_{i=1}^l m_i}$$

Finally, the aggregator may decrypt the message as follows:

$$M = f^{-1}\left(\frac{C_3}{\hat{e}(g,h)^{\alpha \cdot \sum_{i=1}^l m_i} \cdot \hat{e}(g,h)^{\alpha \cdot \sum_{i=1}^l \kappa_i}}\right)$$

$$= f^{-1}\left(\frac{C_3}{K}\right) = f^{-1}\left(f\left(\sum_{i=1}^l m_i\right)\right) = \sum_{i=1}^l m_i$$

Lemma 3. *Correctness of the signature verification algorithms* vrfchunk *and* vrfd—*First, the correctness of the algorithm* vrfchunk *relies on the correctness of Eq. 4. In the following, we set the quantities* $\circledS = \hat{e}(u^{-1}, \sigma_{2,i}) \cdot \hat{e}(g^\alpha, h)^{-1} \cdot \hat{e}(\sigma_{1,i}^{\frac{1}{B_{1,i}}}, h^{\alpha \cdot \prod_{a \in S_s \cup \mathcal{D}_{n+t_s-1-s}}(\gamma + \tau(a))})$ *and* $\tau_\gamma(a) = \gamma + \tau(a)$. *The aggregating entity has to check if* $\sigma_{3,i}$ *is equal to* $\circledS$ *such as:*

$$\circledS = \hat{e}(u^{-1}, \sigma_{2,i}) \cdot \hat{e}(g^\alpha, h)^{-1} \tag{4}$$

$$\cdot \hat{e}(\sigma_{1,i}^{\frac{1}{B_{1,i}}}, h^{\alpha \cdot \prod_{a \in S_s \cup \mathcal{D}_{n+t_s-1-s}}(\tau_\gamma(a))})$$

$$= \hat{e}(g^{-\alpha\gamma}, h^{\frac{r_{d_i}-1}{\gamma}} \cdot h^{(r_{d_i}+m_i) P_{(A_{d_i}, S_s)}(\gamma)/B_{1,d_i}})$$

$$\cdot \hat{e}(g^\alpha, h)^{-1} \cdot \hat{e}(\sigma_{1,i}^{\frac{1}{B_{1,i}}}, h^{\alpha \cdot \prod_{a \in S_s \cup \mathcal{D}_{n+t_s-1-s}}(\tau_\gamma(a))})$$

$$= \hat{e}(g^{-\alpha}, h^{(r_{d_i}+m_i) \prod_{a \in S_s \cup \mathcal{D}_{n+t_s-1-s} \backslash A_{d_i}}(\tau_\gamma(a))})$$

$$\cdot \hat{e}(g,h)^{\alpha(1-r_{d_i})} \cdot \hat{e}(g^\alpha, h^{(r_{d_i}+m_i)}) \cdot \hat{e}(g^\alpha, h)^{-1}$$

$$\cdot \hat{e}(\sigma_{1,i}^{\frac{1}{B_{1,i}}}, h^{\alpha \cdot \prod_{a \in S_s \cup \mathcal{D}_{n+t_s-1-s}}(\tau_\gamma(a))})$$

$$= \hat{e}(g^{-\alpha}, h^{\frac{r_{d_i}+m_i}{B_{1,d_i}} \prod_{a \in S_s \cup \mathcal{D}_{n+t_s-1-s} \backslash A_{d_i}} \tau_\gamma(a)}) \hat{e}(g^\alpha, h)^{m_i}$$

$$\cdot \hat{e}(g^{\frac{r_{d_i}+m_i}{B_{1,d_i} \prod_{a \in A_{d_i}}(\tau_\gamma(a))}}, h^{\alpha \cdot \prod_{a \in S_s \cup \mathcal{D}_{n+t_s-1-s}}(\tau_\gamma(a))})$$

$$= \sigma_{3,i}$$

Second, the correctness of the signature verification vrfd *of the resulting aggregated data is based on the correctness of Eq. 3. Such that, the aggre-*

gating entity relies on the received message $M = \sum_{i=1}^{l}(m_i)$, to check that $\sigma_3 = \hat{e}(g^\alpha, h)^{\sum_{i=1}^{l}(m_i)} = \hat{e}(g^\alpha, h)^M$.

Lemma 4. *Unforgeability of the signing algorithm* sign—*As our* Coop-DAAB *is based on [2], it supports the unforgeability property of the signing algorithm, such that a malicious external device cannot provide a valid signcrypted data message as it does not satisfy the signing predicate Γ_s. Indeed, thanks to the randomization of the secret keys, unauthorized entities can not pool their attributes together to sign the data chunks. Thus, only authorized devices can generate genuine signcrypted data chunks.*

5 Performance Analysis

In this analysis we compare the computation and the storage overheads of the closely related attribute based signature schemes. In most ABSC schemes, the size of a signcrypted data grows along with the size of the encryption access policies [7,13]. As Coop-DAAB relies on the constant ciphertext size ABSC scheme introduced in [2], it presents an efficient aggregation scheme in terms of processing and storage overheads. Indeed, as shown by Table 1, our contribution introduces a ciphertext size which is independent from the size of the used access policy.

Table 1. Features, computation and storage costs comparison of ABSC schemes

Scheme	Type	Access policy	Key size	Signcryption size	Signcrypt time	Unsigncrypt time
[13]	CP-ABE	Monotone	$l_s + 2, l_e + 2$	$\mathcal{O}(l_s) + \mathcal{O}(l_e)$	$\tau_p + E_T + E_1(3 + l_e + 3l_s)$	$\tau_p(3l_s + 5) + 2l_s E_T + 2l_s E_1$
[7]	CP-ABE	Threshold	$3l_e, 2l_s$	$\mathcal{O}(M) + \mathcal{O}(l_e) + \mathcal{O}(l_s)$	$E_1(2\mathcal{O}(M) + ml_e + 4l_e + 3 + l_s)$	$2E_1 + \tau_p(2t + 2 + l_s) + tE_T$
[16]	KP-ABE	Monotone	$m + l_s, m + l_e$	8	$E_1(10 + l_e + 4l_s)$	$6\tau_p + E_1(l_e + 3l_e)$
Coop-DAAB	CP-ABE	Aggregate-Threshold	$n_E + m + 1$	8	$E_1(t_s+2)+2E_2+2E_T$	$E_1 t_e + 7\tau_p + 2E_T$

s denote both the size of the signing policy and the encryption policy. t_e and t_s, $\mathcal{O}(M)$ and m respectively define the encryption threshold, the signing threshold value, the size of the plaintext message M and the cardinal of the attributes' universe $\mathcal{U}$. n_E presents the cardinal of the set of attributes of the user. E_1, E_2, E_T represent exponentiation cost in $\mathbb{G}_1$, $\mathbb{G}_2$, $\mathbb{G}_T$, while τ_p is the cost of a pairing operation.

Attribute based techniques have been implemented in resources-constrained devices in several research works [3,8]. Based on our on going-implementation, we deduce that in DATA_SIGNCRYPT phase, the computation overhead raises linearly with the size of the access signing predicate (t_s) (c.f., Fig. 3), due to the execution of aggreg [6] during the sign algorithm. Similarly, dec computation cost grows linearly with the size of the access enciphering predicate t_e. The algorithms vrfchunk, agg, enc and vrfd is independent of the sizes of the encryption and signing access policy. Recall that agg algorithm only involves multiplications which requires a negligible computation cost compared to pairing and exponentiation times.

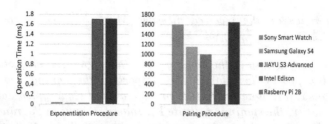

Fig. 2. Elementary functions computation costs

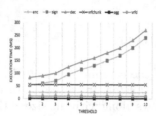

Fig. 3. Estimation of Coop-DAAB computation costs

The performances of attribute based techniques have been studied in several research works especially in IoT applications [3, 8]. Our ongoing implementation of the Coop-DAAB's Proof of Concept (PoC) consists in evaluating the execution costs of the most known elementary cryptographic functions while performed in different IoT devices as presented in [3]. To evaluate the performances of Coop-DAAB, we executed three main cryptographic operations in different types of IoT devices, namely bilinear maps and exponentiation functions (cf. Fig. 2) (Table 2).

Table 2. Selected devices [3]

Device	Type	Processor
Sony SmartWatch 3 SWR50	Smart watch	520 MHz Single-core Cortex-A7
Samsung I9500 Galaxy S4	Smartphone	1.6 GHz Dual-Core Cortex-A15
Jiayu S3 advanced	Smartphone	1.7 GHz Octa-Core 64bit Cortex A53
Intel edison	IoT development Board	500 MHz Dual-Core Intel AtomTM CPU, 100 Mhz MCU
Raspberry Pi 2 model B	IoT development Board	900 MHz Quad-Core ARM Cortex-A7

6 Related Work

IoT networks consist in interconnecting several smart and resource constrained devices to enable advanced services. Data aggregation has been widely explored, yet, few research works have focused on data aggregation in IoT networks while considering data secrecy and privacy preserving requirements.

Shi et al. [17] have proposed a privacy preserving aggregation technique. In this scheme, an aggregator is able to collect participants' data and run statistics over them without learning private information about each participant.

In [14], the authors design a data aggregation scheme adapted for fog computing systems. This solution relies on the use of the Chinese Remainder Theorem to aggregate received data from different parties. This scheme applies one-way hash chain to ensure data origin authentication. In the same vein, Lyu et al. proposed PPFA, a Privacy Preserving Fog-enabled Aggregation for smart grid environments [15]. Their construction relies on fog nodes computation resources to perform heavy computation-consuming functions. Later, Hu et al. introduced a privacy preserving data aggregation scheme for IoT applications [10]. The proposed scheme relies on Secure Multiparty Computation (SMC) techniques, such that each device has to first divide sensory data, locally keeps one piece, and sends the remaining pieces to other group devices. Then each IoT device adds the received shares and the held piece together to get immediate result. The [10] construction provides heavy computation and communication costs. Hence, it makes it unsuitable for resource-constrained devices.

Recently, in 2018, a lightweight aggregation signature scheme for IoT environments have been proposed in [12]. This scheme is based on the use of a set homomorphic signature scheme to aggregate received signed data from IoT devices without learning secret keys of each participant. However, data are transmitted in clear text, between the different involved devices.

Signcryption schemes are generally considered as a logic combination of encryption and signature schemes that enables a data owner to encrypt and sign data in one step. This cryptographic technique allows data origin authentication as the receiver verifies the data owner signature before decrypting. Attribute based signcryption schemes have been introduced by Gagné et al. [7], in 2010. This first proposed construction combines attribute based encryption and signature schemes, based on the same access structure. As such, data secrecy and data origin authentication and flexible fine-grained access control features are provided. Nevertheless, Gagné et al. construction suffers from an important communication overhead that increases dependently with the number of attributes involved in the access structure. Recently, Belguith et al. introduced a constant-size threshold attribute based signcryption, referred to as t-ABSC, for cloud applications [2]. The proposed construction ensures both fine-grained access control and data origin authentication, thanks to the usage of two different access policies, assigned respectively to the enciphering and signing process. As such, users' privacy and outsourced data confidentiality are ensured. In addition, the size of signcrypted messages does not depend on the number of attributes involved in the threshold access structure, which makes t-ABSC scheme suitable for bandwidth-limited applications and resource-constrained devices.

7 Conclusion

Several security and privacy concerns have raised, due to the emergence of Cloud assisted IoT applications, considered as highly dynamic and distributed environments. This lead us to design a new cryptographic mechanism to ensure cooperative data aggregation for IoT applications while preserving devices' privacy, thanks to the attractive properties of attribute based cryptographic techniques.

The proposed Coop-DAAB scheme enables an edge device, i.e., aggregator to collect sensory data from different IoT devices and verify their authenticity using an attribute based signcryption scheme. The privacy of involved IoT devices is ensured, thanks to the intrinsic properties of the signing procedure, as it does not reveal more information other than the accuracy of data integrity verification.

Furthermore, compared to most closely related work, Coop-DAAB is suitable for resource-constrained devices based on an ongoing implementation of the proposed construction and a detailed theoretical performance analysis w.r.t. computational, communication and storage costs. The analysis clearly shows that the size of the signcrypted data does not depend on the number of attributes involved in the threshold access structure.

References

1. Atwady, Y., Hammoudeh, M.: A survey on authentication techniques for the internet of things. In: Proceedings of the International Conference on Future Networks and Distributed Systems, p. 8. ACM (2017)
2. Belguith, S., Kaaniche, N., Laurent, M., Jemai, A., Attia, R.: Constant-size threshold attribute based signcryption for cloud applications. In: SECRYPT 2017: 14th International Conference on Security and Cryptography, vol. 6, pp. 212–225. Scitepress (2017)
3. Belguith, S., Kaaniche, N., Laurent, M., Jemai, A., Attia, R.: PHOABE: securely outsourcing multi-authority attribute based encryption with policy hidden for cloud assisted IOT. Comput. Netw. **133**, 141–156 (2018)
4. Belguith, S., Kaaniche, N., Mohamed, M., Russello, G.: C-ABSC: cooperative attribute based signcryption scheme for internet of things applications. In: Proceedings of the International Conference on Services Computing IEEE SCC, p. 6. IEEE (2018)
5. Coates, A., Hammoudeh, M., Holmes, K.G.: Internet of things for buildings monitoring: experiences and challenges. In: Proceedings of the International Conference on Future Networks and Distributed Systems, p. 38. ACM (2017)
6. Delerablée, C., Pointcheval, D.: Dynamic threshold public-key encryption. In: Wagner, D. (ed.) CRYPTO 2008. LNCS, vol. 5157, pp. 317–334. Springer, Heidelberg (2008). https://doi.org/10.1007/978-3-540-85174-5_18
7. Gagné, M., Narayan, S., Safavi-Naini, R.: Threshold attribute-based signcryption. In: Garay, J.A., De Prisco, R. (eds.) SCN 2010. LNCS, vol. 6280, pp. 154–171. Springer, Heidelberg (2010). https://doi.org/10.1007/978-3-642-15317-4_11

8. Guo, L., Zhang, C., Yue, H., Fang, Y.: PSaD: a privacy-preserving social-assisted content dissemination scheme in DTNs. IEEE Trans. Mob. Comput. **13**(12), 2903–2918 (2014)
9. Herranz, J., Laguillaumie, F., Ràfols, C.: Constant size ciphertexts in threshold attribute-based encryption. In: Nguyen, P.Q., Pointcheval, D. (eds.) PKC 2010. LNCS, vol. 6056, pp. 19–34. Springer, Heidelberg (2010). https://doi.org/10.1007/978-3-642-13013-7_2
10. Hu, C., et al.: An efficient privacy-preserving data aggregation scheme for IoT. In: Chellappan, S., Cheng, W., Li, W. (eds.) WASA 2018. LNCS, vol. 10874, pp. 164–176. Springer, Cham (2018). https://doi.org/10.1007/978-3-319-94268-1_14
11. Kaaniche, N., Mohamed, M., Laurent, M., Ludwig, H.: Security SLA based monitoring in clouds. In: 2017 IEEE International Conference on Edge Computing (EDGE), pp. 90–97, June 2017
12. Kaâniche, N., Jung, E.E., Gehani, A.: Efficiently validating aggregated IoT data integrity (2018)
13. Liu, J., Huang, X., Liu, J.K.: Secure sharing of personal health records in cloud computing: ciphertext-policy attribute-based signcryption. Future Gen. Comput. Syst. **52**, 67–76 (2015)
14. Lu, R., Heung, K., Lashkari, A.H., Ghorbani, A.A.: A lightweight privacy-preserving data aggregation scheme for fog computing-enhanced IoT. IEEE Access **5**, 3302–3312 (2017)
15. Lyu, L., Nandakumar, K., Rubinstein, B., Jin, J., Bedo, J., Palaniswami, M.: PPFA: privacy preserving fog-enabled aggregation in smart grid. IEEE Trans. Ind. Inf. (2018)
16. Rao, Y.S., Dutta, R.: Efficient attribute-based signature and signcryption realizing expressive access structures. Int. J. Inf. Secur. **15**(1), 81–109 (2016)
17. Shi, E., Chan, H., Rieffel, E., Chow, R., Song, D.: Privacy-preserving aggregation of time-series data. In: Annual Network & Distributed System Security Symposium (NDSS). Internet Society (2011)
18. Zheng, Y.: Digital signcryption or how to achieve cost(signature & encryption) $\ll$ cost(signature) + cost(encryption). In: Kaliski, B.S. (ed.) CRYPTO 1997. LNCS, vol. 1294, pp. 165–179. Springer, Heidelberg (1997). https://doi.org/10.1007/BFb0052234

On Cancellation of Transactions in Bitcoin-Like Blockchains

Önder Gürcan[1](✉), Alejandro Ranchal Pedrosa[1,2,3], and Sara Tucci-Piergiovanni[1]

[1] CEA LIST, Point Courrier 174, 91191 Gif-sur-Yvette, France
{Onder.Gurcan,Alejandro.Ranchal-Pedrosa,Sara.Tucci}@cea.fr
[2] LIP6 – Sorbonne Université, Paris, France
[3] EIT Digital, Stockholm, Sweden

Abstract. Bitcoin-like blockchains do not envisage any specific mechanism to avoid unfairness for the users. Hence, unfair situations, like impossibility of cancellation of transactions explicitly or having unconfirmed transactions, reduce the satisfaction of users dramatically, and, as a result, they may leave the system entirely. Such a consequence would impact significantly the security and the sustainability of the blockchain. Based on this observation, in this paper, we focus on *explicit* cancellation of transactions to improve the fairness for users. We propose a novel scheme with which it is possible to cancel a transaction, whether it is confirmed in a block or not, under certain conditions. We show that the proposed scheme is superior to the existing workarounds and is implementable for Bitcoin-like blockchains.

1 Introduction

Bitcoin, introduced by Nakamoto [13], is the core of decentralized cryptocurrency systems. Participants following this protocol can create together a distributed, economical, social and technical system where anyone can join and leave. User participants create and broadcast transactions across the network for being confirmed. Miner participants try to confirm them as a block by solving a computational puzzle (mining). Successful miners broadcast their block to the network to be chained to the blockchain, being rewarded for their success. In addition, all participants validate all data (transactions and blocks) broadcast across the network. It is a very attractive technology, since it maintains a *public, immutable* and *ordered* log of transactions which guarantees an *auditable* ledger, accessible by anyone.

The security and sustainability of blockchains, however, are not trivial and require increased participation, since each participant validates the diffused data, and keeps a replica of the entire blockchain. Participants consider worthwhile to join and stay in the system over time only if they find it *fair* [7]. Miner participants find the system fair if they are able to create blocks as they expected, and user participants find the system fair if they manage to cancel their transactions

© Springer Nature Switzerland AG 2018
H. Panetto et al. (Eds.): OTM 2018 Conferences, LNCS 11229, pp. 516–533, 2018.
https://doi.org/10.1007/978-3-030-02610-3_29

and/or their transactions are confirmed as they expected. Considering miners, several formal studies have been conducted so far [5,6,15,18], concluding that Bitcoin-like blockchains are not promoting participation of miners.

In [7], it has been for the first time shown that Bitcoin-like blockchains are unfair for user participants. It has also been discussed that for the time being it is not possible to explicitly cancel a transaction. There are only some workarounds that a user can try in order to cancel its transaction. One of these workarounds consist of trying to replace an old transaction with the new one with higher fees (Replace-By-Fee (RBF)[1]). This way, it is possible to create another transaction attempting to spend the same inputs but sending the money to the issuer itself. This is a workaround for cancellation of unconfirmed transactions by *implicitly* marking them to prevent their further use. However, once a transaction is confirmed, there is no workaround to reverse the situation. Since double spending is not allowed, the miners will not put both transactions in their blocks. However, there is no guarantee that the double spending transaction will arrive to the miners before the confirmation of the first transaction inside a block. Since it is like that, once a user decides to issue a transaction, s/he can never abandon this decision. Considering that a user is a rational agent that aims to maximize its utility by choosing to perform the actions with the optimal expected outcomes [17], this implies the utility is going to be minus infinity [7]. In the decision theory terms, this would mean assuming a user having an infinite interest on a transaction, which is hard to assume in realistic settings.

To this end, we propose a novel scheme where it is possible to cancel a transaction by rolling back its state whether it is confirmed in a block or not under certain conditions. This way, the utility of user agents can be maximized and, consequently, their willingness to leave the system decreases.

The contributions of this paper are as follows:

- A novel scheme based on a novel type of transaction that enables *explicit* cancellation of transactions.
- The implementation of such mechanism in Bitcoin-like blockchains.

The remainder of this paper is organized as follows. Section 2 gives the related work. Section 3 provides a formalization of the existing Bitcoin-like blockchain data structure considered in this paper. Section 4 provides a high-level Bitcoin-like blockchain protocol description as a rational multi-agent model where agents are using the aforementioned data structure. The proposed scheme for cancellation of transactions is presented in Sect. 5. Section 6 presents an analysis of the proposed scheme and, finally, the discussion and conclusions are provided in Sects. 7 and 8, respectively.

2   Related Work

This study is based on our previous study on user (nodes that do not participate to the mining) fairness in blockchain systems [7]. The closest works in blockchain systems to our study are [1,4,8,12].

---

[1] https://en.bitcoin.it/wiki/Replace_by_fee, last access on 16 July 2018.

Herlihy and Moir in [8] study the user fairness and consider as an example the original Tendermint[2] [3,9]. The authors discussed how processes with malicious behaviour can violate fairness by choosing transactions [4] then they propose modifications to the original Tendermint to make those violations detectable and accountable. Helix [1] and HoneyBadgerBFT [12], on the other hand, attempt to design consensus protocols focusing on assuring a degree of fairness among the users by being resilient to transaction censorship where initially encrypted transactions are included in blocks, and only after their order is finalized, the transactions are revealed.

Another notion of fairness applied to the user side concerns the fair exchange in the e-commerce context [2] which is extended to the Bitcoin-Like scenario in [10] more in the sense that if there are two players performing an exchange then either both of them get what they want or none of them.

3   Blockchain Model

We model a blockchain ledger as a dynamic, append-only tree $B = \{b_0 \xleftarrow{\#_0} b_1 \xleftarrow{\#_1} ... \xleftarrow{\#_{l-1}} b_l\}$ where each block b_i ($0 < i \leq l$) contains a cryptographic reference $\#_{i-1}$ to its previous block b_{i-1}, $l = |B|$ is the length of B, b_0 is the root block which is also called the *genesis block*, b_l is the furthest block from the genesis block which is referred to as the *blockchain head*, $h = |b_i|$ is the height of b_i (the length of the path from b_i block to b_0)[3] and $d = |B| - |b_i|$ is the depth of b_i (the length of the path from b_i block to b_l).

A block b_{i-1} can have multiple children blocks, which causes the situation called a *fork*. The *main branch* is then defined as the longest path l from any block to b_0 and is denoted as $B^\star$ where $|B^\star| = l$ and $B^\star \subseteq B$ such that $|B^x| < |B^\star|$ for all branches $B^x \subset B$ where $B^x \neq B^\star$. All branches other than the main branch are called *side branches*. If at any time, there exists more than one longest path with a length l (i.e. there are multiple heads), the blockchain ledger B is said to be *inconsistent* and thus $B^\star = \emptyset$. This situation disappears when a new block extends one of these side branches and creates $B^\star$. The blocks on the other branches are discarded and referred as *stale blocks*.

3.1   Block Model

We denote a block as $b_i = \langle \mathbf{h}_i, \Psi_i \rangle$ where $\mathbf{h}_i$ is the block header and Ψ_i is the block data. The block data Ψ_i contains a set of transactions organized as a Merkle tree [11]. The set of transactions θ_m are selected by the miner from its memory pool.

---

[2] Jae Kwon and Ethan Buchman. Tendermint. https://tendermint.readthedocs.io/en/master/specification.html, last access on 25 July 2018.

[3] For each transaction tx inside b_i, the block height $|tx|$ is the equal to h also.

Here it is important to note that, blocks have limited sizes[4] and thus the total size of the selected transactions can not exceed this limit[5].

3.2   Transaction Model

We model a transaction as $tx = \langle I, O \rangle$ where I is a list of inputs ($I \neq \emptyset$) and O is a list of outputs ($O \neq \emptyset$). Each input $i \in I$ references to a previous unspent output for spending it. Each output then stays as an Unspent Transaction Output (UTXO) until an input spends it. If an output has already been spent by an input, it cannot be spent again by another input (no double spending). We model the outputs as $o_i = \langle s_i, \mathcal{c}_{o_i} \rangle$ where s_i is a set of tuples $s_i = \{(n_i, conds_{n_i})\}$ that define the conditions $conds_{n_i}$ for the receiver $n_i \in N$ to become owner of the coin $\mathcal{c}_{o_i}$ ($\mathcal{c}_{o_i} \geq 0$). The input that wants to spend the particular coin must satisfy at least one list of conditions $conds_{n_i}$, since an output (a coin) might be spendable by two different receivers $n_i, n_j \in N$ independently, although it will ultimately be spent by only one, on a first-come, first-served basis. All inputs of a transaction have to be spent in that transaction and the total input coins $\mathcal{c}_I$ has to be greater than or equal to the total output coins $\mathcal{c}_O$. The fee f_{tx} of a transaction tx is then modeled as $f_{tx} = \mathcal{c}_I - \mathcal{c}_O$. Depending on the fee to be paid, if there are still some coins left to be spent, the sender can add an output that pays this remainder to itself.

4   Network Model

In this section we provide a high-level Bitcoin protocol description as a rational multi-agent model[6] where rational agents chooses their actions/behavior with respect to their perceptions in order to maximize their utility.

We model the blockchain network as a dynamic directed graph $G = (N, E)$ where N denotes the dynamic rational agent (vertex) set, E denotes dynamic directed link (edge) set. A link $\langle n, m \rangle \in E$ represents a directed link $n \rightarrow m$ where $n, m \in N$, n is the owner of the link and n is the neighbor of m.

4.1   Agent Model

Each agent $n \in N$ has a list of its neighbors N_n where $N_n \subseteq N$ and $\forall m \in N_n | \langle n, m \rangle \in E$. An agent n can communicate one or more of its neighbors by exchanging messages of the form $\langle n, msg, d \rangle$ where n is the sender, msg is the type and d is the data contained. Using such messages, the (user) agents issue transactions (by creating transactions messages and diffusing them to the network) to send coins to each other.

---

[4] The current maximum block size in Bitcoin is 1 MB. See https://bitcoin.org/en/glossary/block-size-limit, last access on 13 July 2018.

[5] Average block size for Bitcoin is given in https://blockchain.info/charts/avg-block-size, last access on 13 July 2018.

[6] This description is based on the system and rational models given in [7].

Each agent n has a memory pool Θ_n in which it keeps *unconfirmed transactions* that have input transactions, an orphan pool $\bar{\Theta}_n$ in which they keep *unconfirmed transactions* that have one or more missing input transactions (orphan transactions) and a blockchain ledger B_n in which they keep confirmed transactions where $\Theta_n \cap \bar{\Theta}_n = \emptyset$, $\Theta_n \cap B_n = \emptyset$ and $\bar{\Theta}_n \cap B_n = \emptyset$ always hold.

A (user) agent n can turn to be a miner agent if it chooses to create blocks for confirming the transactions (mining) in its memory pool Θ_m, and n is said to be a miner node if it started mining but has not stopped yet. The set of miner agents is then denoted by M where $M \subseteq N$. In order to mine, $n \in M$ has to solve a cryptographic puzzle (i.e. Proof of Work) using its hashing power. The successful miners are awarded by a fix amount of reward plus the totality of the transaction fees.

4.2 Behavior Model

A rational agent behaves according to its local perceptions and local knowledge, models uncertainty via expected values of variables or actions, and always chooses to perform the actions with the optimal expected outcome (among all feasible actions) for maximizing its utility [17]. Each rational agent $n \in N$ has a set of actions A_n and a utility function $\mathcal{U}_n$. Using A_n and $\mathcal{U}_n$, n uses a decision process where it identifies the possible sequences of actions to execute. We call these sequences as rational behaviors of n and denote as β. The objective of n is to choose the behaviors that selfishly keep $\mathcal{U}_n$ as high as possible.

We model the utility function of a rational agent $n \in N$ as

$$\mathcal{U}_n = u_0 + \sum_{i=1}^{k} \mathcal{U}(\beta_i) \tag{1}$$

where u_0 is the initial utility value, $k \geq 0$ is the number of behaviors executed so far and $\mathcal{U}(\beta_i)$ is the utility value of the behavior β_i. An agent $n \in N$ finds a system (i.e. the blockchain network) G *fair*, if the total satisfaction of its expectations $\mathcal{U}_n$ is above a certain degree τ_n where $\tau_n < u_0$ [7].

A utility value $\mathcal{U}(\beta_i)$ can also be interpreted as the *degree of satisfaction* experienced by the realization of β_i. The utility value $\mathcal{U}(\beta_i)$ is calculated as

$$\mathcal{U}(\beta_i) = \mathcal{R}(\beta_i) - \mathcal{C}(\beta_i) \tag{2}$$

where $\mathcal{R}(\beta_i)$ is the overall reward gained and $\mathcal{C}(\beta_i)$ is the overall cost spent for the execution of β_i.

When an agent needs to choose a behavior for execution, it needs to calculate its expected utility value. The expected value $\mathcal{E}(\beta_i)$ depends on the probabilities of the possible outcomes of the execution of β_i. We model the expected value as

$$\mathcal{E}(\beta_i) = \sum_{j=1}^{m}(p_j \cdot \mathcal{U}(\beta_i^j))$$

$$= \sum_{j=1}^{m}(p_j \cdot (\mathcal{R}(\beta_i^j) - \mathcal{C}(\beta_i^j))) \qquad (3)$$

$$= \sum_{j=1}^{m}(p_j \cdot \mathcal{R}(\beta_i^j)) - \sum_{j=1}^{m}(p_j \cdot \mathcal{C}(\beta_i^j))$$

$$= \mathcal{R}^{\mathcal{E}}(\beta_i) - \mathcal{C}^{\mathcal{E}}(\beta_i)$$

where $m > 0$ is the number of possible outcomes, $\mathcal{U}(\beta_i^j)$ is the utility value of the possible jth outcome β_i^j, p_j is the probability of this outcome such that $\sum_{j=1}^{m} p_j = 1$, and $\mathcal{R}^{\mathcal{E}}(\beta_i)$ and $\mathcal{C}^{\mathcal{E}}(\beta_i)$ are the expected gain and the expected cost of β_i respectively.

In the following, we list the user and miner agents behaviors important for this study[7] conforming to the above description. To formalize the behaviors, we model a round based approach (like Garay et al. [6]) in which miner agents start creating a new block with at the beginning of the round and a round ends when a new block is successfully created by one of the miners. Both user and miner agents make their decisions on a roundly basis. This round-based model implicitly assumes that the block sent at the end of the round is immediately delivered by all participants, i.e. communication delay is negligible with respect to block generation time.

Miner Agent Behaviors are as follows:

– *Selecting transactions.* When creating the next block, there is no required selection strategy and no known way to make any particular strategy required, but there are two transaction selection strategies popular among miners to include them into their blocks: (1) Selecting the transactions with the highest fees to attempt to maximize the amount of fee income they can collect[8]. Since the size of blocks is limited, *miners could decide to deliberately exclude an unconfirmed transaction that has already been received with a lower fee.* This behavior would obviously delay the confirmation time of that transaction and affects its confirmation probability. (2) Selecting the transactions with highest amount of coins moved to attempt to maximize the market value of the cryptocurrency [14]. Since the size of blocks is limited, *miners could decide to deliberately exclude an unconfirmed transaction that has already been received with a lower amount.* In this study, it is assumed that miners are using the 1st selection strategy and this behavior is modeled as $P(f)$ probability function.

---

[7] A detailed list of behaviors, along with their pseudo-codes, can be found in [7].
[8] https://en.bitcoin.it/wiki/Transaction_fees, last access on 20 July 2018.

User Agent Behaviors are as follows:

- *Issuing transactions.* We model issuing a transaction with a specific fee f as with the action of the form $issueTransaction(b, ¢)$ where $b \in N$ is the receiver and $¢$ is the amount of coins (Algorithm 1). For simplicity, it is assumed that a users agent has an ordered set of fees $\{f_1, f_2, \ldots, f_k\}$ to use. It is also assumed that $f_i < f_{i+1}$ where $0 < i < k$ and $0 \leq P(f_i) < P(f_{i+1}) \leq 1$ where $P(f)$ is the probability of a transaction with a fee f to be confirmed.
- *Leaving because of unfairness.* If at any time, an agent a finds G *unfair* ($\mathcal{U}_a \leq \tau_a$), it may decide to leave G if from its points of view it will not be possible to increase its overall utility above τ_a by calculating the expected values of its possible future behaviors. In other words, a may decide to leave G if $\mathcal{U}_a + \sum_{j=k}^{m} \mathcal{E}(\beta_j) \leq \tau_a$ where $\beta_k, \ldots, \beta_m$ are sufficiently enough desired future behaviors of a.

Algorithm 1. The $issueTransaction(b,¢)$ action of a user agent a where a wants to send $¢$ to b using an input set I by paying a fee f and returning the remaining $¢_r$ to itself. Note that each output o_i is required to be signed by the public key pk of the receiver. After creation, the transaction tx is diffused to the neighbors N_a. For more details see [7].

```
1: action issueTransaction(b, ¢)
2:     I ← selectUnspentTransactionOutputs(B_a, ¢)
3:     o_1 ← ⟨(b, pk_b),¢⟩
4:     f ← estimateFee(I, {o_1})
5:     ¢_r ← ¢_I−¢−f
6:     o_2 ← ⟨{a, pk_a},¢_r⟩
7:     tx = ⟨I, {o_1, o_2}⟩
8:     Θ_a ← Θ_a ⋃{tx}
9:     sendMessage(⟨a,"inv",ℋ(tx)⟩,N_a)
```

In the next section, we present our proposed improvements to allow cancellation of transactions.

5 Cancellation of Transactions

In this section, we improve the models given in Sects. 3 and 4 to enable cancellation of transactions. We first make necessary definitions (Sect. 5.1), then provide a new model for canceling transactions (Sect. 5.2) and finally provide user agent behaviors for canceling transactions (Sect. 5.3).

5.1   Definition

We define cancellation of transactions as *intentionally marking a transaction by its issuer to prevent its further use*, i.e. negating the effect of a transaction by its issuer and thus creating a new state as if the issuer of the transaction has never sent money to the recipients[9]. Here it should be emphasized that such cancellation can only be performed by the issuer of the transaction, since the receiver can always create another transaction that returns the money back to the original account and sign it. However, such a case is not considered as *cancellation* but as *refunding*.

5.2   Cancellation-Enabled Transaction Model

In this subsection, we improve the transaction model given in Sect. 3.2 to allow cancellation of transactions, as defined in Sect. 5.1.

We consider that a transaction can be canceled by its issuer before a predefined *cancellation timeout* r_c. Such a condition impacts directly in the required amount of time to consider that a transaction is final[10]. We define a *final transaction* as a transaction that is not cancellable any more. However, it should be noted that this definition is relative to a branch of the blockchain. If a *side branch* that does not contain a transaction becomes longer than the current *main branch* that contains that transaction, technically that transaction is said to be rolled back, however this is different from canceling it.

Being $tx = \langle I, O \rangle$ a valid transaction issued by $n \in N$, a *cancellation transaction* txc is a transaction issued by n that gives ownership of the coins back to the user $n \in N$ that issued tx, although perhaps in a different address. In reality, our proposed solution gives more freedom than this definition, as we discuss further on.

5.3   Cancellation Behaviors of User Agents

Consider a user agent $a \in N$ that issues a transaction tx at round r_0 for sending ¢ coins to user node $b \in N$, with a fee of f_{tx}, has an interest I on tx and a waiting cost $C(¢)$.

Assuming that tx is confirmed at most at round n, the cumulative expected value $\mathcal{E}$ is:

$$\mathcal{E}(\beta_0) = \sum_{r=1}^{n} \overline{P(f)}^{r-1} \cdot P(f) \cdot (I - f) - \sum_{r=1}^{n} \overline{P(f)}^{r-1} \cdot C(¢)^{r-1} \qquad (4)$$

where $\mathcal{R}(\beta_0) = (I - f)$ and $\mathcal{C}(\beta_0) = C(¢)$ (based on Eq. 1 in [7]).

Now suppose user agent a decides to cancel tx at round r_1 ($r_0 \leq r_1 < r_c$). There are two possible user agent behaviors to cancel tx:

---

[9] It is important to note that the fee paid to the miner is not considered.

[10] This complies with previous usage of transactions that are 'non-final': https://bitcoin.stackexchange.com/questions/9165/whats-are-non-final-transactions, last access on 13 July 2018.

- β_1: *Canceling with cancellation transaction,*
    - *Explicit* cancellation mechanism for transactions that are even confirmed, proposed by us in this paper (see Algorithm 2).
- β_2: *Canceling with a Replace-By-Fee (RBF) transaction*
    - *Implicit* cancellation mechanism for only unconfirmed transactions, proposed by Bitcoin.

where user agents calculate the expected values considering the gain and the cost of each behavior as in [7]. For simplicity, and without loss of generality, we consider that fees are not changing per round.

β_1: Canceling with Cancellation Transaction. User creates a transaction txc at round r_1 with a fee f_{txc}, where f_{txc} has a value such that txc is supposed to be appended to the blockchain in less than round $r_c + r$, being r the round in which tx hits the blockchain.

The expected gain $\mathcal{R}^{\mathcal{E}}(\beta_1)$ of behavior β_1 is then as follows:

$$
\begin{aligned}
\mathcal{R}^{\mathcal{E}}(\beta_1) &= \left(\sum_{r=r_0}^{n} \overline{P(f_{tx})}^{r-1} P(f_{tx}) \cdot \left(\sum_{s=r+1}^{r+r_c} \overline{P(f_{txc})}^{s-1-r} P(f_{txc}) \right) \right) \\
&\quad \cdot (I - f_{tx} - f_{txc}) \\
&= \left(\sum_{r=1}^{n} \overline{P(f_{tx})}^{r-1} P(f_{tx}) \cdot \left(\sum_{s=1}^{r_c} \overline{P(f_{txc})}^{s-1} P(f_{txc}) \right) \right) \\
&\quad \cdot (I - f_{tx} - f_{txc})
\end{aligned}
\tag{5}
$$

where $\mathcal{R}(\beta_1) = (I - f_{tx} - f_{ftxc})$, and r and s ($r_0 < r \le s$) are the rounds in which tx and txc are appended to the blockchain, respectively. Note that, for $r_1 < r$, the transaction txc is orphan and thus it can be appended in the blockchain only at $r = s$ or $r < s$. Furthermore, if $r_1 > s$, then the best option is to use this behavior (since RBF transactions cannot be used anymore). Therefore, we compare for cases where $r_0 \le r_1 \le r$. For further simplification, without loss of generality, we use $r_0 = 1$.

The series given in Eq. 5 follow two nested geometric distributions $X \curvearrowleft (p_{tx}), Y \curvearrowleft (p_{txc})$, with probabilities of success $p_{tx} = P(f_{tx})$ and $p_{txc} = P(f_{txc})$. Therefore, considering $x = q_{tx} = 1 - p_{tx}$ and $y = q_{txc} = (1 - p_{txc})$, we have the following solution to the series:

$$
\mathcal{R}^{\mathcal{E}}(\beta_1) = \left(\sum_{r=1}^{n} x^{r-1} p_{tx} \cdot \left(\sum_{s=1}^{r_c} y^{s-1} p_{txc} \right) \right) \cdot (I - f_{tx} - f_{txc})
\tag{6}
$$

We can extract all constant values into $c = p_{tx} \cdot p_{txc} \cdot (I - f_{tx} - f_{txc})$, and consider the solution to

$$
g_{\beta_1}(n) = \sum_{r=1}^{n} x^{r-1} \cdot \sum_{s=1}^{r_c} y^{s-1} \cdot c = \sum_{r=0}^{n-1} x^{r} \cdot \sum_{s=0}^{r_c-1} y^{s} \cdot c
\tag{7}
$$

As such, notice that, applying basics of geometric convergent series, the inner series $\sum_{s=0}^{r_c-1} y^s$ converges to $\frac{1-y^{r_c}}{(1-y)}$ and thus the outer one:

$$g_{\beta_1}(n) = \frac{(1-x^n)(1-y^{r_c})}{(1-x)(1-y)} \cdot (1-x)(1-y)(\mathcal{I} - f_{tx} - f_{txc})$$
$$= (1-x^n)(1-y^{r_c})(\mathcal{I} - f_{tx} - f_{txc}) \tag{8}$$

And thus, being the expected gain when $g_{\beta_1}(n)$ tends to infinity:

$$\lim_{n\to\infty} g_{\beta_1}(n) = (1-y^{r_c})(\mathcal{I} - f_{tx} - f_{txc}) \tag{9}$$

As for the cost resulted from the Time Value of Money (TVM)[11], both transactions use the same amount money, but we consider that the TVM remains for as long as the transaction is not fully canceled. As such, the expected cost of behavior $\mathcal{C}^{\mathcal{E}}(\beta_1)$ of β_1 is modeled as follows:

$$\mathcal{C}^{\mathcal{E}}(\beta_1) = \sum_{r=1}^{\infty} \overline{P(f_{tx})}^{r-1} \cdot \left(\sum_{s=r_1}^{r+r_c} \overline{P(f_{txc})}^{s-1-r_1} \cdot C(\mathcal{C}_{tx})^{s-1} \right) \tag{10}$$

Where $C(\mathcal{C})$ represents the waiting cost derived of the time value of money, per round. These series are identical to $g_{\beta_1}(n)$ applying $x = q_{tx}$, $y = q_{txc} \cdot C(\mathcal{C}_{tx})$ and the constant c by $d = C(\mathcal{C}_{tx})^{r_1}$, therefore, we define $h_{\beta_1}(n)$:

$$h_{\beta_1}(n) = \sum_{r=1}^{n} x^{r-1} \cdot \sum_{s=1}^{r+r_c-r_1} y^{s-1} \cdot d = \sum_{r=0}^{n-1} x^r \cdot \sum_{s=0}^{r+r_c-r_1-1} y^s \cdot d \tag{11}$$

For which the result is analogously obtained as for $g_{\beta_1}(n)$ (note that $|xy| < 1$ to guarantee convergence):

$$h_{\beta_1}(n) = \left(\frac{1-x^n}{(1-y)(1-x)} - y^{r_c-r_1} \cdot \frac{1-(xy)^n}{(1-y)(1-xy)} \right) \cdot d \tag{12}$$

And thus, again, being the cost when $h_{\beta_1}(n)$ tends to infinity:

$$\lim_{n\to\infty} h_{\beta_1}(n) = \left(\frac{1}{(1-y)(1-x)} - y^{r_c-r_1} \cdot \frac{1}{(1-y)(1-xy)} \right) \cdot d$$
$$= \left(\frac{1}{(1-q_{txc}C(\mathcal{C}_{tx}))(1-q_{tx})} - (q_{txc}C(\mathcal{C}_{tx}))^{r_c-r_1} \right. \tag{13}$$
$$\left. \cdot \frac{1}{(1-q_{txc}C(\mathcal{C}_{tx}))(1-q_{tx}q_{txc}C(\mathcal{C}_{tx}))} \right) \cdot C(\mathcal{C}_{tx})^{r_1}$$

β_2: Canceling with a Replace-By-Fee (RBF) Transaction. User agent issues a transaction txr with a greater fee than the fee of tx, hoping to replace

---

[11] The concept that indicates that money available at the present time worths more than the identical sum in the future due to its potential earning capacity.

Algorithm 2. The *issueCancellableTransaction(b, ¢, r_c)* and *issue CancellationTransaction(tx)* actions of a user agent *a*. The former contains the necessary constraints (e.g., *txc* is only valid if current block-height $|B_a|$ is less or equal than $|tx| + r_c$) for the latter action to work.

```
 1: action issueCancellableTransaction(b, ¢, r_c)
 2:     I ← selectUnspentTransactionOutputs(B_a, ¢)
 3:     o_1 ← ⟨{(b, sk_b), (a, sk_a and |B_a| ≤ |tx| + r_c)},¢⟩
 4:     f ← estimateFee(I, f_tx)
 5:     ¢_r ←¢_I−¢−f
 6:     o_2 ← ⟨{a, pk_a},¢_r⟩
 7:     tx = ⟨I, {o_1, o_2}⟩
 8:     Θ_a ← Θ_a ∪{tx}
 9:     sendMessage(⟨a,"inv",H(tx)⟩,N_a)
10:
11: action issueCancellationTransaction(tx)
12:     i_1 ← tx.o_1
13:     o_1 ← ⟨(a, pk_a),¢⟩
14:     f ← estimateFee(i_1, f_tx)
15:     ¢_r ←¢_I−¢−f
16:     txc = ⟨i_1, o_1⟩
17:     Θ_a ← Θ_a ∪{txc}
18:     sendMessage(⟨a,"inv",H(txc)⟩,N_a)
```

it before *tx* hits the blockchain (i.e. $r_c = 0$). Such *implicitly* tells miner agents to ignore *tx*. In this case, we model the reward as follows:

$$\mathcal{R}^{\mathcal{E}}(\beta_2) = \left(\sum_{r=r_0}^{\infty} \overline{P(f_{tx})}^{r-1} P(f_{tx}) \cdot \left(\sum_{s=r_1}^{r} \overline{P(f_{txr})}^{s-1-r_1} P(f_{txr}) \right) \right) \cdot (\mathcal{I} - f_{txr})$$

$$(14)$$

and its expected cost:

$$\mathcal{C}^{\mathcal{E}}(\beta_2) = \sum_{r=r_0}^{\infty} \overline{P(f_{tx})}^{r-1} \cdot \left(\sum_{s=r_1}^{r} \overline{P(f_{txr})}^{s-1-r_1} \cdot C(¢_{tx})^{s-1} \right) \qquad (15)$$

For both, the process is analogous to extract $g_{\beta_2}(n)$ and $h_{\beta_2}(n)$, obtaining:

$$g_{\beta_2}(n) = \left(\frac{1-x^n}{1-x} - y^{r_1} \cdot \frac{1-(xy)^n}{1-xy} \right) \cdot (1-x)(\mathcal{I} - f_{txr}),$$

$$\text{with } x = q_{tx}, y = q_{txr} \qquad (16)$$

$$\lim_{n \to \infty} g_{\beta_2}(n) = \left(\frac{1}{1-q_{tx}} - q_{txr}^{r_1} \cdot \frac{1}{1-q_{tx}q_{txr}} \right) \cdot (1-q_{tx})(\mathcal{I} - f_{txr}) \qquad (17)$$

and

$$h_{\beta_2}(n) = \left(\frac{1-x^n}{(1-y)(1-x)} - y^{r_1} \cdot \frac{1-(xy)^n}{(1-y)(1-xy)} \right) \cdot d,$$

$$\text{with } x = q_{tx}, y = q_{txr} \cdot C(¢_{tx}), d = C(¢_{tx})^{r_1}, |xy| < 1 \qquad (18)$$

$$lim_{n\to\infty}h_{\beta_2}(n) = \left(\frac{1}{(1-y)(1-x)} - y^{r_1} \cdot \frac{1}{(1-y)(1-xy)}\right) \cdot d$$

$$= \left(\frac{1}{(1-q_{txr} \cdot C(\mathbb{C}_{tx}))(1-q_{tx})} - (q_{txr} \cdot C(\mathbb{C}_{tx}))^{r_1}\right. \tag{19}$$

$$\left. \cdot \frac{1}{(1-q_{txr} \cdot C(\mathbb{C}_{tx}))(1-q_{tx}q_{txr} \cdot C(\mathbb{C}_{tx}))}\right) \cdot C(\mathbb{C}_{tx})^{r_1}$$

In the next section, we will compare β_1 and β_2 by quantitatively analysing the equations formulated in this section.

6   Analyses and Results

We analyzed the rational behaviors of user agents proposed in Sect. 5.3 using gnuplot 5.2.4[12]. It is assumed that the initial utility values u_0 of all agents are the same and high enough from the threshold τ. In the following, we provide results of these analyses employing synthetic data.

Recalling Eq. (4), it is clear that, depending on the values of $\overline{P(f)}$ and $C(\mathbb{C})$, the expected value $\mathcal{E}$ may or may not converge to $-\infty$. For the geometric distributions of $\mathcal{R}^{\mathcal{E}}(\beta_1)$ and $\mathcal{R}^{\mathcal{E}}(\beta_2)$, we can see that they always converge, since all three $\overline{P(f_{tx})}$, $P(f_{txc}) P(f_{txr})$ are positive, and strictly less than 1. For the expected costs, $C^{\mathcal{E}}(\beta_1)$ converges to the value listed in Eq. (13) if and only if $|q_{txc} \cdot C(\mathbb{C}_{tx})| < 1$. Analogously, $C^{\mathcal{E}}(\beta_2)$ converges to the value of Eq. (19) if and only if $|q_{txr} \cdot C(\mathbb{C}_{tx})| < 1$.

Obviously, many values will have a strong impact in the results. Specifically, the correlation between the fees of a transaction f_{tx} and how much they increase the probability of such transaction to be included in each round $P(f_{tx})$. There are some online results on the average fees in Bitcoin and the average amount of blocks (rounds) a transaction takes with each fee[13]. Again, it is easy to see that the average amount of blocks is the expected value of a geometric distribution of which the probability p is the one we are looking for, and, therefore, one can obtain this value solving the series. However, this is out of the scope of this document. Furthermore, we show further on that the cancellation transaction behavior can be better even under optimistic values for this probability, in which a small increase of the fee f_{tx} incurs in a big increase in the probability $P(f_{tx})$.

Figure 1 left shows the gain, cost, and expected values $\mathcal{E}$ of both behaviors, as functions on the number of rounds, and fixing the rest of the variables. Notice that here we consider that 20 Satoshis give a probability of $P(20) = 0.5$, while $P(50) = 0.6$. In this case, we can see how our approach is better, regardless of the round. To the right, we compare our approach with different assumptions on the increase required in the fee to increase the probability of a transaction being included. We can see how our approach is not always the best, and for example when only 10 Satoshis are required to increase the probability to $P(30) = 0.6$, using a Replace-By-Fee transaction is a better behavior.

---

[12] http://www.gnuplot.info/, last access on 24 July 2018.
[13] https://bitcoinfees.earn.com/, last access on 24 July 2018.

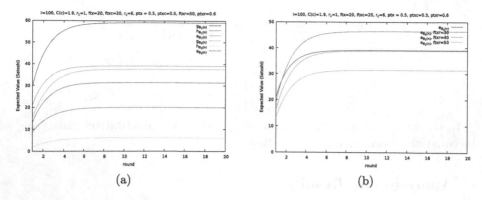

Fig. 1. Expected reward g, cost h, and value functions e as functions on the number of rounds, comparing both behaviors (a). Also, comparison of different assumptions on the relationship between f_{tx} and $P(f_{tx})$ for canceling with cancellation transaction behavior β_1 and canceling with a RBF transaction behavior β_2 (b).

One can note that another important variable fixed in Fig. 1 is r_1, that is, the round at which the user decides he wants to cancel the transaction tx, issued at round 1. In Fig. 1, we assume $r_1 = 1$. However, it is important to consider that the user may want to cancel the transaction later than when they issued it. Finally, r_c can also have an impact in the results, since it increases the time in which a transaction can be cancellable by the cancellation transaction behavior β_1, but it also increases the cost of such behavior, since it can lead to a greater waiting time. In the following, we study how the results vary depending on r_1 and r_c.

For such cases where the series converge, we consider the values in rounds in the infinite (that is Eqs. (9, 13, 17, 19)), and tweak other values. Firstly, it is easy to note that behavior β_1 also converges for $r_c \to \infty$:

$$
\lim_{n \to \infty, r_c \to \infty} \mathcal{E}(\beta_1)(n, r_c) = \lim_{n \to \infty, r_c \to \infty} g_{\beta_1}(n, r_c) - h_{\beta_1}(n, r_c)
$$

$$
= (\mathcal{I} - f_{tx} - f_{txc}) - \frac{C(\mathbb{c}_{tx})^{r_1}}{(1 - q_{txc}C(\mathbb{c}_{tx}))(1 - q_{tx})} \tag{20}
$$

While $\lim_{n \to \infty, r_c \to \infty} \mathcal{E}(\beta_2)(n)$ does not depend on r_c by construction. It is possible to see, however, how the gain decreases for β_2 when r_1 increases, while in β_1 this is irrelevant. Nevertheless, the cost increases with r_c for β_1, since this leads to higher waiting cost.

Figure 2 left compares several values depending on r_1, when the amount of rounds n tends to infinity. One can see how, for $r_c = 6$, it is a better approach to use β_1 for $r_1 > 4$. Furthermore, even if r_c tends to infinity, and being optimistic in terms on the correlation between fees and probability of hitting the blockchain, β_1 seems to be a better behavior for $r_1 > 5$. Nevertheless, if one decides within the first rounds to cancel the transaction, then it is better to use behavior β_2 and issue a Replace-By-Fee transaction. Recall, however, that Fig. 2 plots values for

$n \to \infty$. For constant number of rounds, with $r_1 = 1$, Fig. 1 already showed that β_2 is only a better approach when one can be optimistic about the probability of a transaction with a slightly higher fee hitting the blockchain.

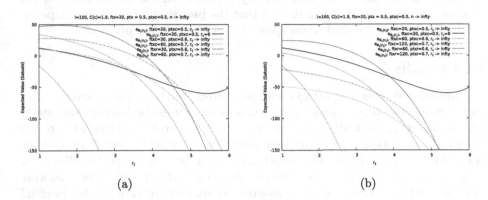

Fig. 2. Expected values for canceling with cancellation transaction behavior β_1 and canceling with a RBF transaction behavior β_2 as a function on r_1, with different values for r_c, f_{tx} and $P(f_{tx})$ where $\mathcal{I} = 100$. To the left, optimistic values for $P(f_{tx})$ (better for β_2).

As a result, we suggest always creating the initial transaction tx as if it can be canceled by a cancellation transaction txc. However, depending on the conditions, the user may choose to issue a Replace-by-Fee transaction txr as well as trying to cancel tx. Nevertheless, a transaction that can be canceled in r_c rounds leaves this transaction as non-final for the first r_c, increasing the block-depth required for a receiver of coins to consider full ownership of such coins. The receiver can however move the coins to a new UTXO. Besides, txr is more flexible and may work even if tx is prepared for being canceled by txc.

In general, it seems as a good approach to firstly issue an RBF transaction, and after round r_1 issue a cancellable transaction such that $f_{txc} + f_{tx} = f_{txr}$. Otherwise, miners will always choose the one that gives higher reward. It is possible that the left side should be greater, $f_{txc} + f_{tx} > f_{txr}$, to account for more space used by two transactions, in such a way that the reward for the miner equalizes.

7   Discussion

In this section, we discuss the proposed mechanism, and the results obtained from several perspectives: *fairness, security* and *implementability*.

7.1   Fairness

To the best of our knowledge, this is the first study focusing on providing an *explicit mechanism for cancellation of transactions* to improve the fairness for *users.* In other words, the proposed mechanism is a first step towards guaranteeing fairness for users, given that fairness is the overall satisfaction of rational agents. Moreover, we showed that the users as rational agents have no incentive to choose the Replace-by-Fee behavior (β_2) since its *cancellation timeout* is much less shorter. As a result, it can be said that the proposed mechanism is superior to the Replace-by-Fee workaround provided by the existing protocol.

7.2 Security

To avoid double-spending attacks and inconsistencies, blockchains need a selection strategy based on a predefined criteria. Bitcoin uses the *longest chain* strategy for selecting the main branch. Moreover, there can be situations where the network is partitioned for some time and then reconnects, with[14] or without any malicious participant. The question is: which branch should be followed in case the *cancellation transaction* exists only in one branch, and there is another transaction that is spending the transaction we wish to cancel in another branch? Should we throw away one or more branches that creates such inconsistencies? Which branch should be followed?

In our opinion, pruning any chain (i.e. reducing the length of the chain) is dangerous and weakens its security level, which is directly proportional to the work that must be done to *replace* such a chain and allow an attacker to target the chain more easily. Thus, we think that the blockchain should remain *append-only*. Furthermore, we claim that the existing longest chain rule should remain the same. Even though with this setting the *cancellation* might be at times ignored, the security of the blockchain is more important than what the user desires. Furthermore, an upgrade to support cancellation of transactions should be backward compatible, not to require a hard-fork for existing Bitcoin-like blockchains.

7.3 Implementability

As shown in Sect. 6, the proposed cancellation mechanism is feasible. This section shows its implementability in Bitcoin-like blockchains. Many such blockchains (Bitcoin, Bitcoin Cash, Litecoin, etc.) already give support to a similar feature as the one described in this document, motivated by the implementation of 2nd-layers in their network, such as the Lightning Network [16].

Bitcoin, for instance, provides support for lightning, thanks to the implementation of Bitcoin Improvement Proposal (BIP) 112[15] in the system. BIP 112 implements the opcode[16] OP_CHECKSEQUENCEVERIFY (typically referred to as

---

[14] If there is a malicious participant that is partitioning the network, this is called a man-in-the-middle-attack.

[15] https://github.com/bitcoin/bips/blob/master/bip-0112.mediawiki, last access on 28 August 2018.

[16] Operation codes from the Bitcoin Script language which push data or perform functions within a pubkey script or signature script.

OP_CSV), that *prevents a non-final transaction from being selected for inclusion in a block until the corresponding input has reached the specified age, as measured in block-height or block-time.* In this case, a non-final transaction refers to the fact that it has not reached the specified age. As such, BIP 112 already offers functionality of giving preference to some specific node to spend an UTXO.

Algorithm 3. Node a sends coins to node b in a transaction.

```
1: IF
2:    <pubkey of node b> CHECKSIG
3: ELSE
4:    "30d" CHECKSEQUENCEVERIFY DROP
5:    <pubkey of node a> CHECKSIG
6: ENDIF
```

For example, node a can pay node b in a transaction with the redeem script given in Algorithm 3. In this case, node b can spend the output at any time, while node a needs to wait 30 days. After such time, any of the two can independently spend the output. However, we would like the inverse functionality where node a can still cancel the issued transaction in the first 30 days, if node b has not spent this output (Algorithm 4). Before 30 days, both node a and node b can independently spend the output. After 30 days, only node b can spend it.

Algorithm 4. Node a cancels the transaction.

```
1: IF
2:    <pubkey of node b> CHECKSIG
3: ELSE
4:    "-30d" CHECKSEQUENCEVERIFY DROP
5:    <pubkey of node a> CHECKSIG
6: ENDIF
```

As illustrated in BIP 68[17], and by James Prestwich[18], OP_CSV compares the top stack item to the input's sequence_no field. Thus, the top stack item is parsed just as the sequence_no field for nSequence (the input-level relative time-lock). That is, it interprets 18 of the 32 bits (the remaining 14 bits are still undefined). There are two special flags: the disable and the type flag. The disable flag (bit 31, the 32nd least significant bit) specifies that logs are disabled. The type flag (bit 22, the 23rd least significant bit) specifies the type of information: if set, the remaining 16 least significant bits are interpreted in units of 512 s granularity, if

---

[17] https://github.com/bitcoin/bips/blob/master/bip-0068.mediawiki, last access on 28 August 2018.

[18] https://prestwi.ch/bitcoin-time-locks/, last access on 28 August 2018.

not, they are interpreted as block-height. The flag $(1 \leq\leq 22)$ is the highest order bit in a 3-byte signed integer for use in bitcoin scripts as a 3-byte PUSHDATA with OP_CHECKSEQUENCEVERIFY (BIP 112), as detailed in the specification of BIP 68.

At the time of writing, using OP_CSV with value 0x00400001 and 0x00C00001 applies the same timelock: a relative locktime of 512 s. Moreover, the specification says OP_CSV "errors if the top stack item is less than 0". We propose, however, to consider the sign bit as the sign flag, and interpret instead this negative value exactly as detailed in the aforementioned example.

OP_CSV is useful for considering a relative time from the inclusion of the transaction in the blockchain. If, instead, one would like to specify an absolute time, or a block-height, this is possible using OP_CHECKLOCKTIMEVERIFY (OP_CLTV) (see BIP 65[19]). Analogously, we propose the same consideration for OP_CLTV. These two simple backward compatible features can be implemented after proposing them in a new BIP. Also, for simplicity, and to comply with the current definition of OP_CSV and OP_CLTV, we propose referring to each upgrade as OP_CHECKSEQUENCEVERIFYINVERSE (OP_CSVI) and OP_CHECKLOCKTIMEVERIFYINVERSE (OP_CLTVI).

As such, we show that it is possible, and convenient, to implement the required opcodes to give support for cancellation transactions in Bitcoin, making use of the sign bit, that was useless until now, although it was respected as the sign bit. We do this in a backward compatible way.

8   Conclusions

In this paper, we proposed a novel *explicit* transaction cancellation mechanism that cancels issued transactions under certain conditions. Such a mechanism increases the fairness for the users and thus increases the security and sustainability of the blockchain system. To avoid security issues and related complex analyses, the proposed mechanism sticks as much as possible to the original Bitcoin protocol, introducing mechanisms to improve the degree of fairness of the system. To this end, we showed the implementation of our approach for Bitcoin, and consider also its implementability for Bitcoin-like blockchains.

References

1. Asayag, A., et al.: Helix: a scalable and fair consensus algorithm. Technical report, Orbs Research (2018). https://orbs.com/wp-content/uploads/2018/07/Helix-Consensus-Paper-V1.2-1.pdf
2. Asokan, N.: Fairness in Electronic Commerce (1998)
3. Buchman, E.: Tendermint: byzantine fault tolerance in the age of blockchains. Ph.D. thesis, University of Guelph, June 2016

---

[19] https://github.com/bitcoin/bips/blob/master/bip-0065.mediawiki, last access on 28 August 2018.

4. Carlsten, M., Kalodner, H., Weinberg, S.M., Narayanan, A.: On the instability of bitcoin without the block reward. In: Proceedings of the 2016 ACM SIGSAC Conference on Computer and Communications Security, pp. 154–167. ACM (2016)
5. Eyal, I., Sirer, E.G.: Majority is not enough: bitcoin mining is vulnerable. In: Böhme, R., Brenner, M., Moore, T., Smith, M. (eds.) International Conference on Financial Cryptography and Data Security, pp. 436–454. Springer, Heidelberg (2014)
6. Garay, J., Kiayias, A., Leonardos, N.: The bitcoin backbone protocol: analysis and applications. In: Oswald, E., Fischlin, M. (eds.) EUROCRYPT 2015. LNCS, vol. 9057, pp. 281–310. Springer, Heidelberg (2015). https://doi.org/10.1007/978-3-662-46803-6_10
7. Gürcan, Ö., Del Pozzo, A., Tucci-Piergiovanni, S.: On the bitcoin limitations to deliver fairness to users. In: Panetto, H. (ed.) OTM 2017. LNCS, vol. 10573, pp. 589–606. Springer, Cham (2017). https://doi.org/10.1007/978-3-319-69462-7_37
8. Herlihy, M., Moir, M.: Enhancing accountability and trust in distributed ledgers. CoRR abs/1606.07490 (2016). http://arxiv.org/abs/1606.07490
9. Kwon, J.: Tendermint: consensus without mining. Technical report, Tendermint (2014). https://tendermint.com/static/docs/tendermint.pdf
10. Liu, J., Li, W., Karame, G.O., Asokan, N.: Towards fairness of cryptocurrency payments. arXiv preprint arXiv:1609.07256 (2016)
11. Merkle, R.C.: A digital signature based on a conventional encryption function. In: Pomerance, C. (ed.) CRYPTO 1987. LNCS, vol. 293, pp. 369–378. Springer, Heidelberg (1988). https://doi.org/10.1007/3-540-48184-2_32
12. Miller, A., Xia, Y., Croman, K., Shi, E., Song, D.: The honey badger of BFT protocols. In: Proceedings of the 2016 ACM SIGSAC Conference on Computer and Communications Security, CCS 2016, pp. 31–42. ACM, New York (2016). http://doi.acm.org/10.1145/2976749.2978399
13. Nakamoto, S.: Bitcoin: a peer-to-peer electronic cash system (2008). https://bitcoin.org/bitcoin.pdf
14. Pappalardo, G., di Matteo, T., Caldarelli, G., Aste, T.: Blockchain inefficiency in the bitcoin peers network. CoRR abs/1704.01414 (2017). http://arxiv.org/abs/1704.01414
15. Pass, R., Seeman, L., Shelat, A.: Analysis of the blockchain protocol in asynchronous networks. IACR Cryptology ePrint Archive 2016, 454 (2016)
16. Poon, J., Dryja, T.: The bitcoin lightning network, pp. 1–22 (2015)
17. Russell, S.J., Norvig, P.: Artificial Intelligence - A Modern Approach, 3rd edn. Pearson Education, London (2010)
18. Sapirshtein, A., Sompolinsky, Y., Zohar, A.: Optimal selfish mining strategies in bitcoin. In: Grossklags, J., Preneel, B. (eds.) FC 2016. LNCS, vol. 9603, pp. 515–532. Springer, Heidelberg (2017). https://doi.org/10.1007/978-3-662-54970-4_30

Spam Detection Approach for Cloud Service Reviews Based on Probabilistic Ontology

Emna Ben-Abdallah[✉], Khouloud Boukadi[✉], and Mohamed Hammami

Mir@cl Laboratory, Sfax University, Sfax, Tunisia
emnabenabdallah@ymail.com, khouloud.boukadi@gmail.com

Abstract. Online reviews provide a vision on the strengths and weakness of products/services, influencing potential customers' purchasing decisions. The fact that anybody can leave a review provides the opportunity for spammers to write spam reviews about products and services for different intents. To counter this problem, a number of approaches for detecting spam reviews have been proposed. However, to date, most of these approaches depend on rich/complete information about items/reviewers, which is not the case of Social Media Platforms (SMPs). In this paper, we consider well known spam features taken from the literature to them we add two new ones: the user profile authenticity to allow the detection of spam review from any SMP and opinion deviation to verify the opinion truthfulness. To define a common model for different SMPs and to cope with the incompleteness of information and uncertainty in spam judgment, we propose a Review Spam Probabilistic Ontology (RSPO) based approach. Probabilistic Ontology is defined using Probabilistic Web Ontology Language (PR-OWL) and the probability distributions of the review spamicity is defined automatically using a learning approach. The herein reported experimental results proved the effectiveness and the performance of the approach.

Keywords: Spam review detection · Probabilistic ontology
Social media

1 Introduction

Nowadays, online reviews are an important source of information for consumers to evaluate online services and products before deciding which product and which provider to choose. In fact, they have a significant power to influence consumers' purchasing decisions. Through social network sites (SNS) such as Facebook, which are considered as the most used one according to the statistics presented in Pew 2018 [7], consumers can freely give feedback, exhibit their reactions to a post or product, share their opinion with their peers and also share their grievances with the companies. However, SNSs cannot yet detect spam reviews and even fake profiles in-time, and hence discriminating between real

© Springer Nature Switzerland AG 2018
H. Panetto et al. (Eds.): OTM 2018 Conferences, LNCS 11229, pp. 534–551, 2018.
https://doi.org/10.1007/978-3-030-02610-3_30

and fake profiles is difficult for non-technically savvy users. Being aware of this, an increasing number of companies have organized spammer review campaigns, in order to promote their products and gain an advantage over their competitors by manipulating and misleading consumers. Hence, this makes trust arise as a crucial factor on the web.

Research on this topic has cast the problem of spam review and spammer user detection into a binary classification: a review is either credible or spam and a user is either honest or spammer. To this end, spam feature clues (behavioral and linguistic features) are defined to identify the spam reviews and spammer users. These features are determined from meta-information (date of review, rate, history of the user, etc.) and from review text. Behavioral features are mostly geared from platform review sites such as Yielp and Amazon where the meta-information about the user's history are almost available. Contrariwise, this is not always the case of SNSs like Facebook. Several existing studies [15, 27] consider the review text for tackling spam reviews by using linguistic features such as, the average content and maximum content similarity; however such features are not considered to analyze the spamicity of the opinion. We believe that it is important to analyze the opinion for the spam review detection. In other words, spammer generally does not give the right opinion to defame or to promote a product/service.

To present the features, many approaches relied on graph/network based methods [26, 27]. However, they do not pay attention to the concepts heterogeneity, for example the "profile" concept in Facebook is the same as "account" concept in review sites, also in review sites the "review" concept is similar to the "feedback" concept in Facebook. Since the social media environment is open, distributed, and semantically enabled, it is not only necessary to have spam detection techniques but also to empower these techniques with semantics to facilitate the quality access and the retrieval of credible reviews from any social media plateform. Besides, the spam judgments are subjective and uncertain in nature. In fact, we cannot affirm the clue of spamicity, or we cannot affirm that the review is spam or the reviewer is a spammer if it/he has spam features. For instance, one reviewer may use a fake profile to hide his identity but he writes a credible review, and vice versa. Moreover, if a review has some features depicting that it is a spam review, while others indicate that it is a credible one; thus leading to a confusing situation. Therefore, an approach that aims at resolving the heterogeneity problem of reviews description and reviewers of social media platforms and supporting the uncertainty of the spam review assessment is of paramount importance. This paper focus on how to reveal spammers and spam reviews from any social media platforms. Moreover, it sheds light on how inferring spam firstly from incomplete and ambiguous information related to spam feature clues and secondly by supporting the uncertainty of the spam judgment.

To cope with the problems mentioned above, we propose to rely on ontology to resolve the heterogeneity problem of social media platforms. However, traditional ontology does not support the uncertainty reasoning [9]. The probabilistic ontology has the merit of supporting the uncertainty, which could be used to asses the spamicity of reviews in SMPs. Besides, we rely on learning based method to generate the probability distribution of the review spamicity. The choice of a learning based method to predict the review spamicity can be explained by two reasons: First, if the probability distributions are defined manually by domain experts, this can decrease the spam review detection performance. In fact, experts can not predict all spammers and spam review behaviors. Second, spammers may take advantage of the design and update their review to deceive the detection process.

Our proposed approach introduces Review Spam Probabilistic Ontology (RSPO) which describes relevant concepts for the detection of spammer users and spam reviews from social media platforms with the aim of facilitating the retrieval of credible reviews and the detection of spam information. This probabilistic ontology is defined using PR-OWL [11] and infers the degree of spamicity of reviews based on MEBN-learning (Multi Entity Bayesian Network learning) method [25].

The rest of the paper is organized as follows. Section 2 aims to define the probabilistic ontology and the PR-OWL language. Section 3 presents spam features used in this paper as clues of spamicity. The proposed RSPO ontology is depicted in Sect. 4. Experimental evaluations are presented in Sect. 5. Section 6 discusses the related works before drawing some conclusions and discussing some future work in Sect. 7.

2 Background

This section presents a brief overview of the probabilistic ontology and the Uncertainty Modelling Process for the Semantic Web (UMP-SW) methodology which form the basis of our work. A probabilistic ontology is an explicit, formal knowledge representation that expresses knowledge about a domain of application. This encompasses: types of entities, properties, relationships, processes and events that happen with the entities, statistical regularities that characterize the domain, inconclusive, ambiguous, incomplete, unreliable, and dissonant knowledge, and uncertainty about all the above forms of knowledge. Probabilistic ontologies are used for the purpose of comprehensively describing knowledge about a domain and the uncertainty associated with that knowledge in a principled, structured, and sharable way [9]. This has given birth to a number of new languages such as: PR-OWL [11], OntoBayes [30] and BayesOWL [12]. In this paper, we rely on PR-OWL to represent the RSPO. Actually, PR-OWL not only provides a consistent representation of uncertain knowledge that can be reused by different probabilistic systems, but also allows applications to perform plausible reasoning with that knowledge, in an efficient way [9]. This can be explained by the fact that PR-OWL is based on Multi-Entity Bayesian Network (MEBN) logic. MEBN extends Bayesian Networks (BN) to achieve first-order expressive power. MEBN represents knowledge as a collection of MEBN Fragments (MFrags), which are organized into MEBN Theories (MTheories). An MFrag (see Fig. 4) contains random variables (RVs) and a fragment graph representing dependencies among these RVs. An MFrag represents a repeatable pattern of knowledge that can be instantiated as many times as needed to form a BN addressing a specific situation called situation-specific Bayesian Networks (SSBN), and thus can be seen as a template for building and combining fragments of a Bayesian network. An MFrag can contain three kinds of nodes: context nodes which represent conditions under which the distribution defined in the MFrag is valid, input nodes which have their distributions defined elsewhere and condition the distributions defined in the MFrag, and resident nodes with their distributions defined in the MFrag. Each resident node has an associated class local distribution which defines its distribution as a function of the values of its parents, namely Local Probability Distribution (LPD). The RVs in an MFrag can depend on ordinary variables. We can substitute different domain entities for the ordinary variables to make instances of the RVs in the MFrag.

In order to model and implement PR-OWL ontologies, Carvalho et al. proposed the Uncertainty Modelling Process for the Semantic Web (UMP-SW) methodology [10]. This methodology is consistent with the Bayesian network modelling methodology [18] and includes three main steps: model the domain, populate its Knowledge Base (KB), and perform reasoning based on both the model and the KB. The modelling step consists of three major stages: requirements, analysis and design, and implementation. These stages are borrowed from the Unified Process (UP) with some modifications to fit the ontology modelling domain.

3   Spam Feature Description

To infer the degree of spamicity/credibility of a review and reviewer, this paper relies on spam features [15, 23] that fall into the categories as follows:

1. Review-Behavioral (RB) based features: This type of feature is based on the review meta-information and not on the review text itself. The RB category encompasses two features:
   - Early Time Frame (ETF): Spammers often review early to inflict spam as the early reviews can greatly impact people's sentiment on a product/service [22].

$$v_{etf} = \begin{cases} 0 & (T_i - F_i) \notin [0, \delta] \\ 1 - \frac{T_i - F_i}{\delta} & (T_i - F_i) \in [0, \delta] \end{cases} \tag{1}$$

   Where $T_i - F_i$ denote the period between the r_i (review i) date and the first review date. $\delta = 7$ months is a threshold for denoting earliness. $etf(r_i)$ takes value 1 if v_{etf} is greater than 0.5 otherwise it takes value 0.
   - Rate Deviation (RD) [22]: Spammers attempt to promote or demote products/services, their ratings can deviate from the average ratings given by other reviewers. Rating deviation is thus a possible behavior demonstrated by a spammer. This feature attains the value of 1 if the rating deviation of a review exceeds some threshold β ($\beta = 0.63$).

$$rd_i = \begin{cases} 1 & \frac{rt_{ij} - avg_{e \in E_{*j}} r(e)}{4} > \beta \\ 0 & otherwise \end{cases} \tag{2}$$

Where rt_{ij} refers to the rating given by the reviewer i towards an item j.

2. Review-Linguistic (RL) based features: Features in this category are based on the review text. In this work, we use two main features in RL category:

   - Ratio of Exclamation Sentence containing '!' (RES) [19]: Spammer put '!' in their sentences as much as they can to increase impression on users and highlight their reviews among other ones.

$$res(r_i) = \begin{cases} 1 & contain'!' \\ 0 & otherwise \end{cases} \tag{3}$$

   - Number of the first Personal Pronoun (NPP) [19]: Studies show that spammers use second personal pronouns much more than first personal pronouns.

$$npp(r_i) = \begin{cases} 1 & true \\ 0 & false \end{cases} \tag{4}$$

3. User-Behavioral (UB) based features: Relate to each user and encapasses two main features:

   - Reviewing Burstiness (BST) [21]: Spammers, always write their spam reviews in short period of time for two reasons: first, because they want to impact readers and other users, and second because they are temporal users, they have to write as much as reviews they can in short time.

$$v_{bst} = \begin{cases} 0 & (L_i - F_i) \notin [0, \tau] \\ 1 - \frac{L_i - F_i}{\tau} & (L_i - F_i) \in [0, \tau] \end{cases} \tag{5}$$

Where τ is the time window parameter representing a burst ($\tau = 28$ days). $L_i - F_i$ present the time interval between the first and the last reviews written by the user i (u_i). $bst(u_i)$ takes value 1 if v_{bst} is greater than 0.5 otherwise it takes value 0.

   - Negative Ratio (NR) [21]: Spammers tend to write reviews which defame businesses which are competitor with the ones they have contact with, this can be done with destructive reviews, or with rating those businesses with low score. Hence, ratio of their scores tend to be low.

$$nr(u_i) = \begin{cases} 1 & average\_rate\_of\_user_{u_i} \leq 2 \\ 0 & otherwise \end{cases} \tag{6}$$

4. User-Linguistic (UL) based features: These features, which are extracted from the users' language, show how the users are describing their feelings or opinions about what they have experienced as a customer of a business. We use this type of features to understand how a spammer communicates in terms of wording. The Average Content Similarity (ACS) is considered in this work since it is largely adopted in the litterature.

   - ACS [14]: As crafting a new review every time is time consuming, spammers are likely to copy reviews across similar products. It is thus useful to capture the content similarity of reviews (using cosine similarity) of the same author. We choose the maximum similarity to capture the worst spamming behavior.

$$acs(u_i) = \begin{cases} 1 & u_i \quad has\_similar\_reviews \\ 0 & otherwise \end{cases} \qquad (7)$$

5. Profile Authenticity (PA) feature: Besides the features depicted above, we propose in this work a new feature, namely profile authenticity, to detect spammers. It is more likely that people who write spam reviews hide their identities, especially in social network sites where it is easy to create a fake account. To differentiate between fake profiles and authentic ones, we choose the four most famous profile elements: the profile picture (exist or not), the number of friends, his location and professional information. Considering professional information, it can be explained by the fact that the proposed approach will be applied to detect spam reviews of cloud services, which are generally used by enterprises and not by individual users. Hence, a spammer may hide his enterprise, his workplace as well as his job.

6. Opinion Deviation (OD) feature: The use of opinion deviation feature aims at detecting, first, the unusual reviews (for example, $< 3 < 3 < 3 < 3 < 3 < 3$; Great!!!!!!!!!!!!!!!!!!); second the without-feature-reviews (for example, in the field of cloud service, reviews that do not contain any service property, such as World's Best Service!!! Just < 3 You!!!); and third the divergent opinions compared to the majority of reviews.

   In fact, many approaches have been used to detect deviations among which we can mention, the clustering based approach which is the most commonly developed [17]. For this reason, we use the clustering technique to identify divergent opinions (see Fig. 1) by calculating the outliers for each object. This factor depends on the distance from the object to the centroid of the cluster to which the object belongs. The algorithm starts iteratively by first finding the object with the maximum distance d_{max} to the cluster centroid thus:

$$d_{max} = max_i\{||x_i - C_i||\}, \quad i = 1, 2, ..., N \qquad (8)$$

Outlier factors o_i, for each object are then calculated. An outlier factor (deviation) value for each object x_i is calculated using the Eq. 9.

$$o_i = \frac{||x_i - C_i||}{d_{max}} \qquad (9)$$

Where $||x_i - C_i||$ is the distance between each object x_i and its allocated cluster centroid C_i. d_{max} is the maximum distance of a certain object to the cluster centroid/center. After all iterations, each object will have an outlier factor value that represents the object's deviation degree. All outlier factor values of the dataset are normalized to the range $[0, 1]$. The outlier factor value is compared with a predefined threshold value T that lies between 0 and 1. An outlier factor with a greater value is more likely to be a deviation. The object for which $o_i > T$ is considered a deviation. In order to annotate data for the clustering, we adopt our previous work [8]. In particular, aligned with the cloud service domain, each review is presented as a set of service properties sp_j associated with their sentiment scores (as depicted in Fig. 2). The sentiment score is computed using [8] which presents a normalized average of the reviewer's sentiments scores about a service property in each review (the score of each sentiment is extracted from SentiwordNet [13]).

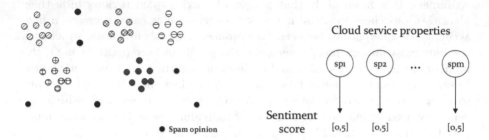

Fig. 1. Example of outlier objects   **Fig. 2.** Review form

4   Review Spam Probabilistic Ontology Modelling

After defining the features that will be used as clues to detect spammers and spam reviews, we should deal with the problem of modelization of these features and how to infer if the review is spam or not from the latter ones. The main challenges that hamper the review spam detection are: first, the subjectivity expectation of spamicity judgment, which makes the review spam inference uncertain and second, the incompleteness of spam features. This is can be explained by the fact that information about the user's history, review and profile is not always available on SMP. For this purpose, Review Spam Probabilistic Ontology (RSPO) is proposed in this work. The details of the RSPO modelling are presented in this section. The RSPO is created using the Uncertainty Model for the Semantic Web (UMP-SW) presented in Sect. 2. In particular, we deal with the RSPO modelling through three stages: Requirements, Analysis and Design, and Implementation.

4.1   Requirements

The main goal is to identify the likelihood of a particular review being spam. Requirement discipline draws out the goals, queries, and evidence for a particular system. To ensure the traceability of requirements, a specification tree is used. Each of the requirements is linked to its 'parent' requirement and every evidence is linked to its parent query, which in turn is linked to its higher-level goal. This arrangement helps trace the requirements.

Overall Goal of the RSPO is to determine either a review is credible or spam.

(1) Query: Does the reviewer have an authentic profile or a fake one?
  - Evidence: Look at the location information if it is available (on the reviewer's profile);
  - Evidence: Look at the enterprise information if it is available (on the reviewer's profile);
  - Evidence: Look at the job information if it is available (on the reviewer's profile);
  - Evidence: Look at the picture if it is available (on the reviewer's profile);
  - Evidence: Look at the friendship network number if it is greater than 50 (on the reviewer's profile);
(2) Query: Did the user write reviews to describe his experiences as a customer of a certain business?
  - Evidence: Look if the reviewer has similar reviews or not;
(3) Query: Has the reviewer a normal behaviour or suspicious one?
  - Evidence: Look at the Early Time Frame feature if it is greater than 0.5 or not;
  - Evidence: Look at the Rating Deviation feature if it is equal to 1;
(4) Query: Has the review a normal content or suspicious one?
  - Evidence: Look at the review text if it contains '!';
  - Evidence: Look if the reviewer uses second personal pronouns or not;
(5) Query: Did the reviewer describe in the review text his feeling or opinion about a real experience with a product/service?
  - Evidence: Look at the reviewer's feature-based opinion if it is deviated from the majority of reviewing feature-based opinions.

4.2 Analysis and Design

Analysis and Design is the second broad step of the UMP-SW methodology. Once goals and evidences to achieve them are identified, modelling the entities, attributes, relationships, and applicable rules can be started. This step also specifies the semantics of the model. We rely on the UML diagram to present the semantic model of RSPO. The UML diagram in Fig. 3 depicts the entities, attributes and relations and describes the objects, attributes, and relationships necessary to represent the RSPO. As depicted in Fig. 3, two main categories of spam feature are defined, Behavior Feature and Linguistic Feature. Behavior Feature has in turn three sub-categories such as *Profile Behavior Feature*, *User Behavior Feature* and *Review Behavior Feature*. The *Linguistic Feature* has also three sub-categories, *User Linguistic Feature*, *Review Linguistic Feature*, and *Opinion Deviation Feature*. Two possible linguistic values of spamicity level are defined: low and high. The *SpamicityLevel* class has four *has-Type* relations since each profile has a spamicity level, each user has a spamicity level and each review has two spamicity levels, the first one is based on review-features and the second presents the overall spamicity level which is based on the aggregation of the other spamicity levels. This provides a starting point to actually define entities/concepts of the probabilistic ontology. Since UML has a poor support to complex rule definitions required for uncertainty, the probabilistic rules are

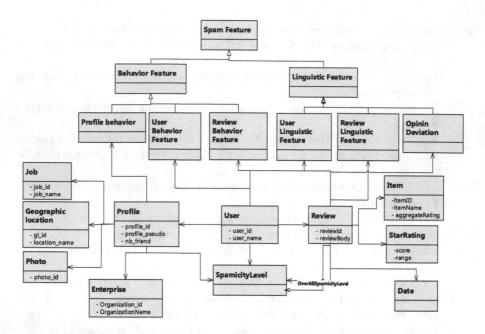

Fig. 3. UML diagram for review spam probabilistic ontology

specified separately. These rules are very useful when implementing the model in PR-OWL to specify the LPDs. Examples of the probabilistic rules required for the RSPO are presented as follows:

- If the majority of reviewer' star rates is between 1 and 2 then it is more likely that he tends to defame businesses which are competitor. Indeed, we can consider him as a spammer reviewer. At the same time, if he has an authentic profile then it is more likely to be a credible reviewer.
- If a review opinion agrees with the majority of reviews' opinions reviewing the same item, then it is more likely to be a credible review. Meanwhile, if its rating deviates from the average ratings then it is more likely to be a spam review.

Such probabilistic rules model the uncertain knowledge. These rules help in establishing causal relation between random variables.

4.3   Implementation

This phase starts by choosing the modelling language for the probabilistic ontology. In this work we use PR-OWL 2, which is supported by the UnBBayes PR-OWL 2 Plugin [20]. The entities, their attributes, and relations identified earlier are mapped to PR-OWL/MEBN constructs. The first step to go through is to map the entities, their attributes, and relations to PR-OWL, which uses essentially MEBN terms. Once the entities are defined, the uncertain characteristics

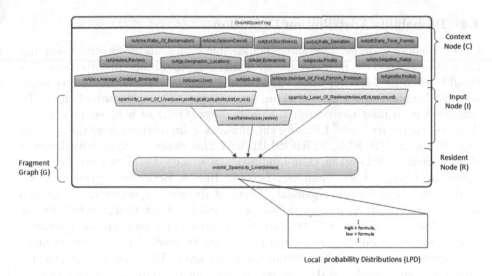

Fig. 4. MFrag for identifying the overall degree of review spamicity

should be identified. Uncertainty is represented in MEBN as random variables (RVs). In UnBBayes, an RV is first defined in its Home MFrag. Grouping RVs into MFrags closely follows the grouping observed in the Analysis and Design stage. Typically, an RV represents an attribute or a relation in the designed model. For instance, the RV *Spamicity_Level_Of_User* maps to the attribute *spamicityLevel* of the class *User* and the RV *hasProfile* maps to the relation hasProfile(Profile,User) (see Fig. 3). As a predicate relation, *hasProfile* relates a User to one Profile, the same way the class *Profile* is related to one User. Hence, the possible values (or states) of this RV are True or False. Each RV is represented as a resident node in its home MFrag. Once all resident RVs are created, their relations are defined by analyzing dependencies. This is achieved by looking at the rules defined in the semantic model of the RSPO. Rules consist in defining probability distribution of the resident node over its random variable instances. In our work, we define 21 MFrags including 21 resident nodes associated with their probability distributions. Figure 4 presents an MFrag example dealing with the overall spamicity level of a review. The resident node *overall_Spamicity_Level(review)* depends on three input nodes: the spamicity level of the reviewer, the spamicity level of the review and the relationship between the review and reviewer. The ordinary variables such as, user, review, profile, etc. can be filled in with different entities of type User, Review and Profile, etc. to make different instances of this MFrag as needed and reason about a specific situation. The local probability distributions of the defined MFrags are depicted in the next section.

4.4 Probability Distribution Definition

Once a random variable, its arguments, possible values, and respective mappings have been defined, it is necessary to define its probability distributions. We should define an LPD for each resident node (these local distributions apply only if all context nodes in the MFrag are satisfied). The main aim of the LPDs definition is to infer the overall spamicity level (OSL) of a given review. For doing so, Spamicity levels' LPDs for the different spam feature categories, namely SLP, SLUL, SLUB, SLU, SLRB, SLRL and SLR denoting Spamicity Level of Profile, Spamicity Level of User Linguistic, Spamicity Level of User Behavior, Spamicity Level of User, Spamicity Level of Review Behavior, Spamicity Level of Review Linguisitc and Spamicity Level of Review respectively, are defined according to the spam feature values (see Table 1). After that, we rely on the MEBN learning method [25] to learn the relationships between the spamicity level of spam feature categories and the spamicity level of the review in order to generate automatically the overall spamicity level LPD. The MEBN learning uses a relational model (RM) as a data schema for the dataset. The annotation of the review dataset is conducted in conjunction with cloud instructors from the IT department of the University of Sfax (considered as experts). The goal is to annotate the collected cloud service reviews from different SMPs (more details about the collected reviews are depicted in Sect. 5) with spam or credible reviews by relying on cloud service benchmarking tools, such as cloudHarmony [1] and Cloudlook [2]. To this end, the instructors organized themselves into four groups, where each group examined around 1000 reviews. Afterwards, they conducted a cross-validation process among the different groups.

4.5 RSPO Knowledge Base Population

The population of the RSPO is mainly based on three steps:

1. Data collection and pre-processing: this step consists in collecting and pre-processing information about user and review from SMPs (for more details the reader can refer to [8]).
2. Spam features detection: this step aims to detect and compute spam features of both users and collected reviews.
3. RSPO instantiation: this step instantiates automatically classes and relations using KARMA[1] tool, which is an information integration tool that enables users to quickly and easily integrate data from a variety of data sources. It maps structured sources to RSPO in order to build semantic descriptions.

4.6 Review Spam Probabilistic Ontology Reasoning

Once the probabilistic ontology is implemented and populated, it is possible to realize plausible reasoning through the process of creating a Situation-Specific

---

[1] http://usc-isi-i2.github.io/karma/.

Table 1. An excerpt of LPDs' definition.

MFrag name	LPD
SLUL	if any user have (hasAverage_Content_Similarity = 1) [high = 0.9, low = 0.1] else [high = 0.1, low = 0.9]
SLUB	if any user have (hasBustiness = 1 & hasNegativeRatio = 1) [high = 0.9, low = 0.1] else [if any user have ((hasBustiness = 1 & hasNegativeRatio = 0) | (hasBustiness = 0 & hasNegativeRatio = 1)) [high = 0.5, low = 0.5] else [high = 0.1, low = 0.9]]
SLU	if any user have (SLUP = high & SLUB = high & SLUL = high) [high = 0.9, low = 0.1] else [if any user have((SLUP = low & SLUB = high & SLUL = high) | (SLUP = high & SLUB = low & SLUL = high) | (SLUP = high & SLUB = high & SLUL = low)) [high = 0.7, low = 0.3] else [if any user have((SLUP = low & SLUB = low & SLUL = high) | (SLUP = high & SLUB = low & SLUL = low) | (SLUP = low & SLUB = high & SLUL = low)) [high = 0.3, low = 0.7] else [high = 0.1, low = 0.9]]]

Bayesian Network (SSBN). UnBBayes has implemented an algorithm that creates an SSBN for a particular query. An example of reasoning is shown in Fig. 5. Information about review and reviewer are extracted from the provider DigitalOcean official Facebook page. As depicted in Table 2, the user's history information is missing. Consequently, we cannot compute neither the User Linguistic Features nor the User Behavior Features. In this case our approach does not consider these two categories in the spam judgment by given the same probability of the two values of the degree of being spam (high and low). Figure 5 presents the generated SSBN of the review. Given spam feature values, the inference system generates the probabilities of the two values of the degree of spamicity of the review: the probabilities of high and low. When returning to the example, the overall spamicity of the review is considered as high by 65.6% and as low by 34.4%.

Table 2. Examples of reviews. A: Available; NA: Not Available

Reviewer	Service	Rating	Socia media platform category	Content	Review information	User information	Profile information
mimi	DigitalOcean	5	SNS	Good!!	A	NA	A

5 Experiments and Results

This section presents the experimental evaluation part of this study including the datasets, the defined metrics as well as the obtained results.

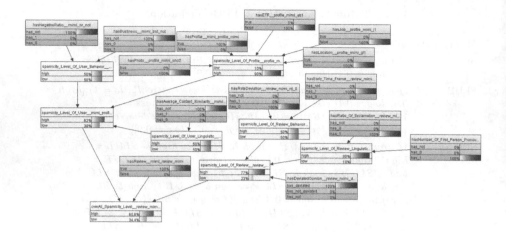

Fig. 5. SSBN for the query overall spamicity level of mimi's review

Table 3. Review datasets

	SNS	RLI	RF	All
#Reviews	1000	1000	1000	3000
% Spam reviews	20%	5%	13%	13%

5.1 Datasets and Evaluation Metrics

Datasets: Table 3 includes a summary of the used datasets and their characteristics. These datasets include the reviewers' impressions and comments about the quality of cloud services. As per this table, the datasets are categorized into three categories according to the review source:

- SNS dataset, includes reviews collected from Facebook pages as a SNS.
- Review platforms with LinkedIn authenticate dataset (RLI), includes reviews collected from online review platforms that obligate the access with LinkedIn account such as TrsutRadius [6] and G2Crowd [4].
- Review platforms Free dataset (RF), includes reviews collected from online review platforms when anyone can put a review without any authentication, such as hostadvice [5] and cloudReview [3].

We take 80% of the annotated reviews from each dataset category as a training dataset in order to learn the MEBN model and 20% are taken as test dataset to evaluate the effectiveness and the performance of the approach.

Evaluation Metrics: To evaluate the performance of our approach, four metrics are used: Precision (P), Recall (R), F1-score (F1) and Accuracy (A).

$$P = \frac{TP}{TP + FP} \tag{10}$$

$$R = \frac{TP}{TP + FN} \tag{11}$$

$$F1 = \frac{2 \times p \times R}{P + R} \tag{12}$$

$$A = \frac{TP + TN}{TP + FP + FN + TN} \tag{13}$$

In the context of this paper, the false positive (FP) refers to the number of credible reviews that are misidentified as spam ones, while the true positive (TP) refers to the number of correctly identified spam reviews. Similarly, the false negative (FN) refers to the number of spam reviews that are misidentified as credible ones, while the true negative (TN) refers to the number of correctly identified credible reviews.

5.2 Experimental Results

This section demonstrates the RSPO based approach effectiveness and performance.

(1) Overall effectiveness and performance analysis: Table 3 demonstrates that the RSPO based approach detects spam reviews from the three datasets (SNS, RLI and RF). In addition, Fig. 6 illustrates the RSPO performance in terms of precision, recall, F1-score and accuracy. As for this figure, the RSPO based approach has a high performance over the three datasets (around 90% for all metrics).

(2) Dataset impression on spam detection: Our experiments revealed a number of spam reviews in RLI dataset, but this number is much more important in SNS and RF. In fact, the RLI platforms, such as trustRadius, mainly verify the credibility of users (they mention "verified user"), who are obliged to use their LinkedIn identities prior to posting their reviews. Contrariwise, in Social Network, such as Facebook any person can create a fake profile and write a review. Table 3 shows the huge difference between the number of detected spam reviews and spammers found in review platforms and those of Facebook pages.

(3) Spam feature Analysis: The combination of Spam features can be a good hint for achieving better performance. The PA achieves better performance with RF dataset (around 90% of accuracy). Moreover, even in SNS dataset, the PA realizes a good result. In fact during our experiments, we noticed that when a spammer wants to promote a service (by giving a 5 star rate), he uses an authentic profile so that the profile authenticity feature cannot reflect the real state of the review (we found 24% of accuracy in this case). However, when he wants to defame a service, he hides his identity. Therefore, the PA in this case represents an important feature for the detection of spam (83% of accuracy). Besides, the opinion deviation feature (OD) achieves a greater influence on the performance of the spam review detection result in most datasets (especially for SNS and RLI datasets).

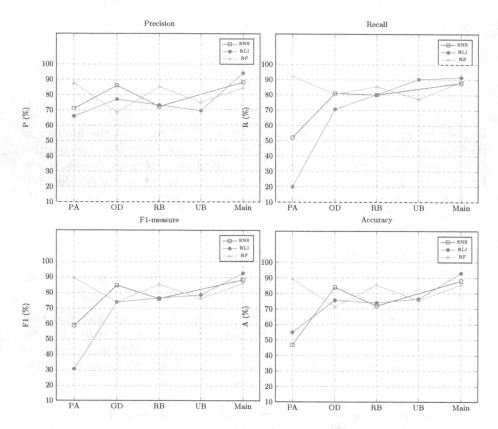

Fig. 6. Spam detection performance from different SMPs with different datasets. **RB:** Review Based features; **UB:** User Based features; **Main:** All spam features.

6 Related Work

In the last decade, a great number of research studies focus on the problem of spotting spammers and spam reviews. However, since the problem is non-trivial and challenging, it remains far from fully solved. To detect spammers, there are four categories of features in the literature, including review-behavioral, user-behavioral, review-linguistic, and user-linguistic.

Fei et al. in [15] consider the burstiness of each review to find spammers and spam reviews on Amazon. They build a network of reviewers appearing in different bursts. Then, they model reviewers and their co-occurrence in bursts as a Markov Random Field (MRF), and employ the Loopy Belief Propagation (LBP) method to infer whether a reviewer is a spammer or not. Shehnepoor et al. [27] propose a spam detection framework, namely NetSpam. This framework is established based on metapath concept and graph-based method to label reviews. The authors also introduce the importance of spam features to obtain better results on Yelp and Amazon Web sites. Xue et al. in [29] use the rate

Table 4. Analysis of the spam review detection approaches

Study	Review information	User history	SMP	Support of incomplete information
[15]		√	Amazon	
[27]	√	√	Yielp	
[26, 29]		√	Yielp	
[16]	√	√	Amazon	
[24]	√	√	TripAdvisor, Yielp	√

deviation of a specific user and employs a trust-aware model to find the relationship between users to compute the final spamicity score. Savage et al. in [26], in turn, use the rate deviation to identify opinion spammers. They focus on the differences between user rating and the majority of honest users using a binomial model. Xie et al. in [28] use a temporal pattern (time window) to find singleton reviews (reviews written just once) on Amazon. Further, Heydari [16] proposes a spam detection system which investigates rate deviation, content based factors and activeness of reviewers in suspicious time intervals captured from time series of reviews by a pattern recognition technique. Mukherjee et al. [24] present a consistency model using limited information for detecting non-credible reviews. To do so, they rely on latent topic models leveraging review texts, item ratings, and timestamps. The above spam review detection approaches are summarized in Table 4. Compared to the existing works, the RSPO based approach covers the almost SMP including SNS and review platforms which was obviously not the case for the other approaches. This is by adding profile authenticity feature. Moreover, unlike the proposed works, our approach deals with missing information and uncertainty about spam judgment using a probabilistic reasoning.

7   Conclusion

Spam review is a continuing problem for consumers looking to be guided by online reviews in making their purchasing decisions. In the current study, we have introduced a Review Spam Probabilistic ontology (RSPO) based approach, which describes relevant concepts for the detection of spammer users and spam reviews from any social media platform. In addition, the RSPO can infer the degree of spam given incomplete information about spam features thanks to the probabilistic reasoning. In order to outperform the spam review detection, we extended the spam features with two new ones, *profile authenticity* and *opinion deviation* features. The experiments showed the improvement achieved by these two features. Moreover, they demonstrated the performance and the effectiveness of the RSPO based approach for the spam review detection from real data extracted from different categories of SMPs. As a future endeavor, we plan to investigate the presented spam review detection approach by proposing a credible cloud service recommendation approach through online reviews.

References

1. Cloudharmony. cloudharmony.com
2. Cloudlook. www.cloudlook.com
3. Cloudreviews. cloudreviews.com
4. G2crowd. g2crowd.com
5. Hostadvice. hostadvice.com
6. Trustradius (2013). trustradius.com
7. Social media use in 2018 (2018). http://www.pewinternet.org/2018/03/01/social-media-use-in-2018/
8. Ben-Abdallah, E., Boukadi, K., Hammami, M.: SMI-based opinion analysis of cloud services from online reviews. In: Abraham, A., Muhuri, P.K., Muda, A.K., Gandhi, N. (eds.) ISDA 2017. AISC, vol. 736, pp. 683–692. Springer, Cham (2018). https://doi.org/10.1007/978-3-319-76348-4_66
9. Carvalho, R.: Probabilistic ontology: representation and modeling methodology, January 2011
10. Carvalho, R., Laskey, K.B., Costa, P., Ladeira, M., Santos, L.L., Matsumoto, S.: Unbbayes: modeling uncertainty for plausible reasoning in the semantic web (2012)
11. Carvalho, R.N., Laskey, K.B., Costa, P.C.: PR-OWL: a language for defining probabilistic ontologies. Int. J. Approx. Reason. **91**, 56–79 (2017). https://doi.org/10.1016/j.ijar.2017.08.011, http://www.sciencedirect.com/science/article/pii/S0888613X17301044
12. Ding, Z., Peng, Y., Pan, R.: BayesOWL: uncertainty modeling in semantic web ontologies. In: Ma, Z. (ed.) Soft Computing in Ontologies and Semantic Web, pp. 3–29. Springer, Heidelberg (2006). https://doi.org/10.1007/978-3-540-33473-6_1
13. Esuli, A., Sebastiani, F.: Sentiwordnet: a publicly available lexical resource for opinion mining. In: Proceedings of the 5th Conference on Language Resources and Evaluation (LREC 2006), pp. 417–422 (2006)
14. Fei, G., Mukherjee, A., Liu, B., Hsu, M., Castellanos, M., Ghosh, R.: Exploiting burstiness in reviews for review spammer detection, pp. 175–184, January 2013
15. Fei, G., Mukherjee, A., Liu, B., Hsu, M., Castellanos, M., Ghosh, R.: Exploiting burstiness in reviews for review spammer detection. In: ICWSM (2013)
16. Heydari, A., Tavakoli, M., Salim, N.: Detection of fake opinions using time series. Expert Syst. Appl. **58**(C), 83–92 (2016). https://doi.org/10.1016/j.eswa.2016.03.020, https://doi.org/10.1016/j.eswa.2016.03.020
17. Jiang, S.Y., Yang, A.M.: Framework of clustering-based outlier detection. In: 2009 Sixth International Conference on Fuzzy Systems and Knowledge Discovery, vol. 1, pp. 475–479, August 2009. https://doi.org/10.1109/FSKD.2009.94
18. Laskey, K.B., Mahoney, S.M.: Network engineering for agile belief network models. IEEE Trans. Knowl. Data Eng. **12**(4), 487–498 (2000). https://doi.org/10.1109/69.868902
19. Li, F., Huang, M., Yang, Y., Zhu, X.: Learning to identify review spam. In: Proceedings of the Twenty-Second International Joint Conference on Artificial Intelligence, IJCAI 2011, vol. 3, pp. 2488–2493, AAAI Press (2011). https://doi.org/10.5591/978-1-57735-516-8/IJCAI11-414

20. Matsumoto, S., et al.: UnBBayes: a Java framework for probabilistic models in AI (2011)
21. Mukherjee, A., Venkataraman, V., Liu, B., Glance, N.: What yelp fake review filter might be doing?, pp. 409–418, January 2013
22. Mukherjee, A., et al.: Spotting opinion spammers using behavioral footprints. In: Proceedings of the 19th ACM SIGKDD International Conference on Knowledge Discovery and Data Mining, KDD 2013, pp. 632–640. ACM, New York (2013). https://doi.org/10.1145/2487575.2487580
23. Mukherjee, A., Venkataraman, V., Liu, B., Glance, N.S.: What yelp fake review filter might be doing? In: ICWSM (2013)
24. Mukherjee, A., Dutta, S., Weikum, G.: Credible review detection with limited information using consistency features. In: Frasconi, P., Landwehr, N., Manco, G., Vreeken, J. (eds.) ECML PKDD 2016. LNCS (LNAI), vol. 9852, pp. 195–213. Springer, Cham (2016). https://doi.org/10.1007/978-3-319-46227-1_13
25. Park, C.Y., Laskey, K.B., Costa, P.C.G., Matsumoto, S.: Multi-entity Bayesian networks learning for hybrid variables in situation awareness. In: Proceedings of the 16th International Conference on Information Fusion, pp. 1894–1901, July 2013
26. Savage, D., Zhang, X., Yu, X., Chou, P., Wang, Q.: Detection of opinion spam based on anomalous rating deviation. Expert Syst. Appl. **42**(22), 8650–8657 (2015). https://doi.org/10.1016/j.eswa.2015.07.019, http://www.sciencedirect.com/science/article/pii/S0957417415004790
27. Shehnepoor, S., Salehi, M., Farahbakhsh, R., Crespi, N.: NetSpam: a network-based spam detection framework for reviews in online social media. IEEE Trans. Inf. Forensics Secur. **12**(7), 1585–1595 (2017). https://doi.org/10.1109/TIFS.2017.2675361
28. Xie, S., Wang, G., Lin, S., Yu, P.S.: Review spam detection via temporal pattern discovery. In: Proceedings of the 18th ACM SIGKDD International Conference on Knowledge Discovery and Data Mining, KDD 2012, pp. 823–831. ACM, New York (2012). https://doi.org/10.1145/2339530.2339662
29. Xue, H., Li, F., Seo, H., Pluretti, R.: Trust-aware review spam detection. In: 2015 IEEE Trustcom/BigDataSE/ISPA, vol. 1, pp. 726–733, August 2015. https://doi.org/10.1109/Trustcom.2015.440
30. Yang, Y., Calmet, J.: OntoBayes: an ontology-driven uncertainty model. In: International Conference on Computational Intelligence for Modelling, Control and Automation and International Conference on Intelligent Agents, Web Technologies and Internet Commerce (CIMCA-IAWTIC 2006), vol. 1, pp. 457–463, November 2005. https://doi.org/10.1109/CIMCA.2005.1631307

Formal Modelling and Verification of Cloud Resource Allocation in Business Processes

Ikram Garfatta[1]([⊠]), Kais Klai[2], Mohamed Graiet[3], and Walid Gaaloul[4]

[1] Faculty of Sciences, University of Monastir, Monastir, Tunisia
ikram.garfatta@gmail.com
[2] LIPN, CNRS UMR 7030, University of Paris 13, Villetaneuse, France
kais.klai@lipn.univ-paris13.fr
[3] ISIMM, Universiy of Monastir, Monastir, Tunisia
mohamed.graiet@imag.fr
[4] Telecom SudParis, UMR 5157 Samovar, University of Paris-Saclay, Paris, France
walid.gaaloul@telecom-sudparis.eu

Abstract. Cloud environments have been increasingly used by companies for deploying and executing business processes to enhance their performance while lowering the operating cost. Nevertheless, the combination of business processes and Cloud environments is a field that needs to be further studied since it lacks an explicit and formal description of the resource perspective in the existing business processes and especially of Cloud-related properties, namely vertical/horizontal elasticity. Therefore, this field cannot yet fully benefit of what Cloud environments can offer. Besides the lack in formalization, there is also a need for a verification method to check the correctness of allocations. In fact, without formal verification, the designer can easily model erroneous allocations which lead to runtime errors if left untreated at design-time. In this work, we address the above shortcomings by proposing a formal model for the Cloud resource perspective in business processes using the Coloured Petri net formalism, which can be used to check the correctness of Cloud resource allocation at design-time.

Keywords: Business Process Models · Formal verification
Cloud resources · Elasticity · Coloured Petri net

1 Introduction

Finding the right compromise between the best performance and the lowest cost has always been sought by enterprises that deal with Business Process Management Systems. Combining BPM with Cloud environments has proved to be of great benefit for such enterprises in their quest, especially considering that properties such as elasticity and shareability are at the very essence of the definition of Cloud Computing. Researchers have been investigating the possibilities

© Springer Nature Switzerland AG 2018
H. Panetto et al. (Eds.): OTM 2018 Conferences, LNCS 11229, pp. 552–567, 2018.
https://doi.org/10.1007/978-3-030-02610-3_31

of hosting entire Business Processes in the Cloud, going to the extent of proposing the Business Process as a Service (BPaaS) as a new service model that takes part in the Cloud computing paradigm [23].

On the other hand, the resource perspective in BPM has caught the interest of researchers since many activities may require certain resources for their proper execution. Many studies have dealt with the human resources management [6,7], but non-human resources, especially Cloud resources, were seldom considered.

Taking advantage of Cloud resources, however, is not only achieved by implementing the whole Business Process on the Cloud. A Business Process (BP) can use resources managed by a Cloud provider without having to be part of that Cloud environment. The communication between the BP and the Cloud provider has to be well defined for the interaction to be flawless. In an effort to outline this relationship between the two domains, a first attempt was made in [10]. The idea was to extend the Business Process Model to include configurable Cloud resources allocation. The extension comprises three operators essentially, which are designed to define the way the Cloud provider assigns the resources to the process activities taking into account both the elasticity and shareability concepts, and emphasizing on the type and capacity of the required resources. These proposed operators, though designed to deal with configurable allocations, can also be used for non-configurable ones if we consider them after configuration. Despite the visual representation it offers, this proposition remains unverified and lacking formalism. In fact, the interaction between the Business Process and the Cloud resources provider, however intuitive it may seem, is a very intricate task. Thus, using the operators in [10] can easily result into faulty allocations. For instance, two activities may request the same resource to consume at the same time with a total required capacity that exceeds the capacity the resource can provide. This allocation, though representable using the operators in [10], is a situation that ought to be avoided.

To be able to avoid unsound allocations, a verification method needs to be applied at design time to insure a correct behaviour at runtime. To achieve this goal, we propose a formal model to represent the Cloud resource allocation mechanism in BP models, using Coloured Petri Nets [12]. Having formalized the allocation aspect, we are then able to verify essential structural and behavioural properties by formally analysing and verifying the model.

The remainder of this paper is organized as follows: the related work is presented in Sect. 2. Section 3 introduces basic concepts used in the other sections and our proposed model is detailed in Sect. 4. Results for a case study are presented in Sect. 5. Finally, we conclude and provide insights for future work in Sect. 6.

2   Related Work

Business processes usually include activities that need resources to carry out their expected tasks. The required resources can be classified into two categories: human resources which are basically the workforce needed for the execution of the activities, and non-human resources. In literature, more attention has been drawn to the first category, whereas only a few researches have dealt with the second one. Our work focuses on this disregarded category, and more precisely on Cloud resources.

The representation of the resource perspective in the context of business processes has been the subject of some works in literature. For example, in [18], the authors have extended the BPMN 2.0 metamodel in order to model and visualize the resource perspective requirements. This extension was however proved to be against the workflow resource patterns [17]. This work was then followed by the proposition of an approach [8,19] that enables the implementation of the requirements of the resource perspective in extended BPMN models and resource structure models into BPEL definitions in an attempt to provide a support to the lifecycle (i.e., definition, implementation, verification and validation) of the resource perspective requirements in the development of PAISs based on WfMSs.

Petri net has been prominently present in research dealing with the verification of the structural aspect in business processes. This formalism was used in [1] to define a number of workflow patterns that formalize the control flow in business processes. Other works used Petri net to verify workflow specifications [20] with focus on checking their soundness [21]. In [17], series of workflow resource patterns were proposed to capture the various ways in which human resources are represented and used in business processes. Using ordinary Petri net to formalize Cloud resource allocation would result in massive impractical models. We use Coloured Petri net to get a concise and more elegant representation without losing the benefit of formality that Petri net offers. Another asset of our CPN model is its support for multi-tenancy. Not only does it allow the allocation of multiple resources to one activity, it also covers the sharing of the same Cloud resources by multiple process instances that may pertain to different processes.

Despite the research effort put into the modelling of human resource, works that deal with non-human resources, especially Cloud resources, are still scarce and no formal patterns have been proposed for their modelling. For instance, an extension to BPMN was proposed in [10]. This work aimed at representing the Cloud perspective but lacked expressivity and formality. In our work, we present a formal model that offers a more rigorous expressiveness. In a previous work [5], an Event-B formal specification of the Cloud resource perspective was proposed as a step towards its validation. This work, however, does not treat the resource management as minutely as we do in our present work by detailing the interaction between the Cloud provider and business activities. This is probably due to the fact that Event-B is not the most suitable language to formally describe behavioural properties.

Cloud properties have recently been addressed in some researches in an attempt at their modelling and analysis. A temporal logic called CLTLt(D) (Timed Constraint LTL) was used by [4] to formalize elasticity in Cloud-based systems. In [2], authors define a formal framework for the description and evaluation of service-based business processes elasticity. Elasticity mechanisms and strategies for service-based business processes were described in [14] using Petri net. An analytical model based on Markov chain was used in [24] to evaluate elasticity strategies and help Cloud providers decide on which strategy to implement. In [13] authors propose an elasticity model description language for StratModel on which they based their framework for the evaluation of elasticity strategies.

Although the elasticity property has been relatively recurrent in literature when business processes are discussed in a Cloud computing context, these works do not approach elasticity from the same point that we consider in this present work. In fact, they all regard elasticity as a behaviour of the whole system where this property manifests as adding/removing copies of a service or resizing its capacity depending on the considered type of elasticity. In this paper however, we are interested in elasticity at a lower lever, that is to say, we focus on elasticity as the behaviour of each service that requires some Cloud resources to finish its execution. This property concerns the made requests as they may vary from one execution to another of the same business process, as well as it concerns the required resources which can exhibit an elastic behaviour to satisfy the requests.

3   Preliminaries

In this section, we present essential concepts for the comprehension of this paper.

3.1   Cloud Resource Allocation in Business Processes

The allocation mechanism was treated in [11] where a resource perspective extension to BPMN was proposed. This extension allows a description of the resources as well as their allocation management. This work was then followed by [10] where the focus was directed on the configurable allocation of Cloud resources. The authors considered three types of Cloud resources, namely network, storage and compute resources and focused on two key properties, namely elasticity and shareability of resources. They adopted a pattern-based modelling approach and hence their proposition consisted of three operators mainly: the assignment, elasticity and sharing/batching operators. The application of the said operators allows for a customizable selection of the required resources by a business process while taking into account their properties. Using this approach results in models that are better in terms of expressivity than the models in which the resource perspective had to be hard-coded, while being lower in terms of complexity. These operators, however designed to express a configurable resource allocation, can be used in their configured form to express a basic allocation. Since we are not interested in our work in the configuration aspect of the allocation, we present these operators stripped of the configuration-related notions.

The resource assignment operator can be considered as the main operator since it is the one that expresses which resources are actually allocated to which activity. Two parameters are used in this operator, namely a *type* and a *range*. The former basically expresses whether all of the connected resources are assigned to the activity (type AND), only one of them is assigned to the activity (type XOR) or an unfixed number of the connected resources are assigned to the activity (type OR). On the other hand, the *range* parameter is but a configuration guideline specified by the Cloud provider to additionally constraint the minimal and maximal number of resources of each type that should be assigned to each activity. The second operator is used to express the elasticity aspect of a set of resources. It is used to specify the way the resources scale up and down (vertical, horizontal or hybrid elasticity). The third and last operator deals with the shareability of a resource. It specifies the activities allowed to share a specific resource.

The Considered Allocation Mechanism. Although the model proposed in [10] presents a considerable improvement compared to having the resource allocations hard-coded in an ad-hoc manner, it still presents a number of issues. In fact, the proposed modelling approach is informal and the operators should be described formally. Furthermore, even though the elasticity operator is used to set an elasticity type for each of the provided resources, it does not actually allow the modelling of the desired behaviour rigorously. As a matter of fact, the interaction between the Cloud resources provider and the different business processes' activities cannot be depicted using such operators.

In our work, we approach Cloud resource allocation in a way that is inspired from the work in [10] and yet treats it formally and more elaborately. We propose a model that separates the Cloud provider from the requesting activities and therefore is able to represent the interaction between the two parties. This model allows the modelling of parallel requests as well as exclusive requests which correspond to an AND- and XOR-typed assignment operator respectively. Basically, a Cloud provider offers a number of resources. Each resource is characterized by a type, a provided capacity and a number of provided instances. An activity makes a request for a number of instances of some resource of a certain type indicating the requested capacity. Elasticity in our work is not a mere description of a resource property. It is considered at three different levels:

1. Elasticity at the level of the request: an elastic request is a request for a non fixed number of resource instances that have a possibly varying capacity. In other words, both the number of instances and their requested capacity can vary at each execution within the range of a fixed interval.
2. Elasticity at the level of the provider: a resources administered by a Cloud provider can be elastic, i.e., it may have the capability of scaling-up and down its provided capacity. It can be either horizontally or vertically elastic. A horizontally elastic resource scales-up/down by creating/destroying resource instances. A vertically elastic resource scales-up/down by increasing/decreasing the capacity of its instances.

   Two strategies can be considered when dealing with elasticity. The first is a *proactive* strategy: the provider predicts the need to scale-up/down a resource before the scaling becomes a necessity. This can be triggered by reaching capacity thresholds or a certain usage percentage. The second is a *reactive* strategy: the provider proceeds to scale-up when an elastic resource is no longer able to satisfy an incoming request.

3. Elasticity at the level of the requester: an activity's needs may vary during the execution of the process, and consequently, it can choose to ask for more or less resources at runtime.

3.2 Coloured Petri Net

Petri nets have proven to be one of the best formalisms to model and analyse concurrent systems. Nevertheless, the basic Petri net model is not suitable for the modelling of many systems encountered in IT. In fact, trying to describe a real system using Petri nets usually results in a very large and complex model that may even be inaccurate. Moreover, the tokens in a Petri net are often mapped into objects or resources in the modelled system. However, a simple Petri net token does not make a suitable representation for an object with attributes. To solve these problems many authors propose extensions of the basic Petri net model. Several authors have extended the basic Petri net model with *coloured* or *typed tokens* [22] which have values. A large Petri net model can therefore be represented in a much more compact and manageable manner using a *Coloured Petri net.*

Coloured Petri Nets (CP-nets or CPNs) [12] is a graphical language designed to construct models of concurrent systems and analyse their properties. It combines the capabilities of Petri nets, which serve as basis for the graphical notation, with the capabilities of the programming language CPN ML, which is based on the high-level functional programming language Standard ML [15], to define data types. CP-nets are generally used to model systems where concurrency and communication are key characteristics, such as business processes and workflows.

We refer the reader to [12] for a formal definition of Coloured Petri nets.

4 A Coloured Petri Net Formal Modelling of Cloud Resource Allocation in BP

In this section, we detail the provider's and requester's sub-models separately while highlighting the way we represent elasticity at its different levels. Then we explain the way our model can be integrated with the BP's control flow model.

4.1 Overview of the Model

In this work, we propose a coloured Petri net model to represent the cloud resources perspective in BPs. As shown in Fig. 1, our model separates the requester from the provider, resulting in two sub-models communicating asynchronously through three buffers: buffer_ request, buffer_ response and buffer_ release.

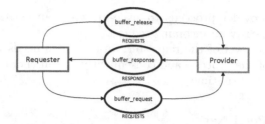

Fig. 1. Global view on the model

Colour Sets. To be able to represent the specific aspects of Cloud resource allocation, we create the following main colour sets which we use in our model:

- id_p: an integer representing the process identifier.
- id_i: an integer representing the process's instance identifier.
- id: a couple (p: id_p, i: id_i) to identify a unique process instance.
- id_a: an integer representing the activity's instance identifier.
- id_r: an integer representing the resource identifier.
- $restype$: an enumeration to specify the type of a resource (a compute, storage or network resource).
- $capacity$: an integer representing the capacity of a resource.
- $instnumber$: an integer representing the number of instances of a resource.
- $elasticity$: an enumeration to specify the type of resource elasticity (none, vertical or horizontal).
- $instance$: a couple (cap: $capacity$, inb: $instnumber$) to represent the number of resource instances with a capacity cap.
- $resource$: an octuple (r: id_r, t: $restype$, $init\_cap$: $capacity$, $init\_inst$: $instnumber$, $total\_cap$: $capacity$, $total\_inst$: $instnumber$, $inst\_list$: $list$ ($instance$), el: $elasticity$) where $init\_cap$ is the resource instances initial capacity, $init\_inst$ is the initial number of provided resource instances, $total\_cap$ is the total provided capacity, $total\_inst$ is the total number of provided instances, $inst\_list$ is a list of the resource's instances and el is the resource's elasticity type.
- $demand$: an octuple (r: id_r, t: $restype$, $inst\_min$: $instnumber$, $inst\_max$: $instnumber$, $inst$: $instnumber$, cap_min: $capacity$, cap_max: $capacity$, cap: $capacity$) where $inst\_min$ and $inst\_max$ are the lower and upper bounds of the possible number of requested instances, $inst$ is the actual number of requested instances, cap_min and cap_max are the lower and upper bounds of the possible requested capacity and cap is the actual requested capacity.
- $request$: a couple (a: id_a, ld: $list\ demand$) where id_a identifies the requesting activity and $list\ demand$ specifies a set of requested resources.
- $requests$: a couple (i: id, req: $request$) where id identifies the process as well as the instance and $request$ is a list of requested resources by a specific activity.
- $response$: a couple ($reqs$: $requests$, b: $bool$) indicating whether a certain request has been satisfied.

For the sake of simplifying the expressions used in our model later on, we denote by l_R the list of resource identifiers of the elements of a *list* of *resources l*.

4.2   The Provider's Sub-model

Upon the reception of a request, and depending on the availability of resources, the provider either allocates the requested resources to the requesting activity or sends back a response indicating the rejection of the request. Upon the release of resources, the provider takes them back into consideration as available resources.

We propose the model in Fig. 2 to represent the behaviour of the provider.

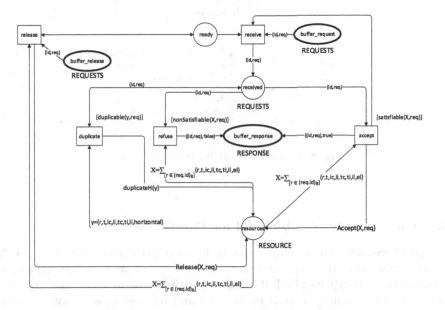

Fig. 2. The provider's model with reactive elasticity

The presence of a request in the buffer_ request enables the *receive* transition. If that request is satisfiable (enough resources are available in the place *resource*) a positive response is deposited in the buffer_ response and the provided resources are updated accordingly through the $Accept(X, req)$ function (the requested number of instances is subtracted from the corresponding resource's provided instances), otherwise if the request is unsatisfiable (not enough instances available of some non elastic requested resource) a negative response is sent through the buffer_ response. The case where resources need to be duplicated for the request to become satisfiable (i.e., reactive elasticity) is treated through the *duplicate* transition.

Released resources are communicated via the buffer_ release. The firing of the *release* transition updates the values of the the corresponding resource from the *resources* place by adding the released resource's instances to the provided resource's instances list.

Proactive Elasticity for the Provider's Sub-model. As previously mentioned in Sect. 3.1, a Cloud provider may consider a proactive strategy to manage the elasticity of its resources. To do so, the model in Fig. 3 needs to be composed with the one in Fig. 2 by merging the *resources* place of the two models. Conditions 1 to 4 are used to trigger the duplication/consolidation actions. Transitions *duplicateH* and *consolidateH* are responsible for the management of horizontally elastic resources. Transitions *duplicateV* and *consolidateV* are responsible for the management of vertically elastic resources.

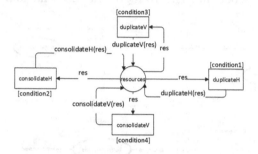

Fig. 3. Proactive elasticity for the provider

4.3   The Requester's Sub-model for Parallel Requests

In need of resources to accomplish its task, an activity sends a request to the provider indicating the type of resources it needs as well as the number of instances and capacity required. If it gets a positive response for all the requested resources, the activity proceeds to execute and releases those resources upon completion. Otherwise, the activity fails to execute. This behaviour corresponds to the use of the assignment operator with its *type* parameter set to *AND*.

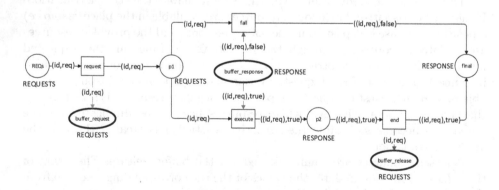

Fig. 4. The requester's model for parallel requests

We propose the model in Fig. 4 to represent the behaviour of the requester when making a demand for multiple resources. In order to request some resources, the task has to go through a three-step procedure:

1. announce its request to the provider (the blue arc) via buffer_ request and wait for its response;
2. start its execution and consume the acquired resources (the green arc) as soon as a positive response to the request is available in buffer_ response;
3. and lastly release those resources (the purple arc) through buffer_ release and finish its execution successfully.

In case the request had not been met (a corresponding negative response is present in buffer_ response), the task fails (the red arc), and skips to a final state that indicates its failure. We note that we use lists to represent a request for multiple resources. It is worth mentioning that to model the assignment of a single resource, we simply use lists containing one element.

4.4   The Requester's Sub-model for an Exclusive Choice of Requests

An activity may need to make an exclusive choice between a number of requests and leave the decision for runtime. This behaviour corresponds to the one expressed by the assignment operator with an *XOR type*.

To model this using CPN, we propose the model in Fig. 5. This model is similar to the previous one, with the few following changes: the place *REQs* and transition *request* are duplicated as many times as the number of exclusive requests. A place *pXOR* is added to make sure that only one *request* transition is fired, and therefore only one request is made at runtime.

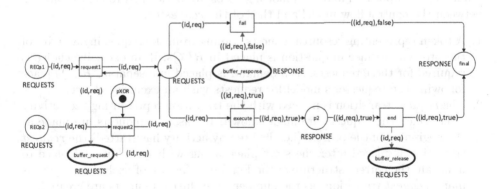

Fig. 5. The requester's model for requests with an exclusive choice

The Requester's Sub-model with Support for Elasticity. As previously mentioned in Sect. 3.1, a requester may need more or less resources than it had already requested. To be able to represent this behaviour, an elasticity mechanism that allows a requester to update its request just before its execution needs to be implemented. Figure 6 shows the model we propose to support such a property. To avoid a more complex model, we use $req3$ to refer to a request smaller than the initial request req and $(req + req2)$ to refer to a request larger than the initial request req.

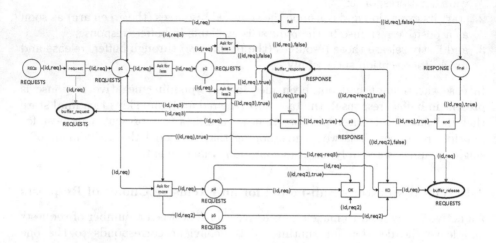

Fig. 6. The requester's model for requests with support for elasticity

4.5   Composing the Control Flow and the Resource Perspective

As soon as knowledge of resource requests is provided, the pattern for the *requester* is duplicated and the following steps are performed to create the link between the control flow model and that of the *requester*:

1. A token representing resource requests coming from the requesting activity of the process instance in question is placed in *REQs*. If an exclusive choice is required for these requests, each demand is placed in a separate *REQs* place following the requester's model for requests with an exclusive choice.
2. The *request* transition is merged with the transition representing the activity after which the requests are made known. In case the requests are made at the beginning of the process (i.e., before any activity has started) the *request* transition is placed after the start place along with an additional place to maintain the correct structure of the Petri net. In case of exclusive choice, as many *request* transitions as the number of exclusive requests are created.
3. A place typed *request* is added between the newly created transition in step 2 and the transition of the requesting activity with the corresponding request inscribed on both its in and out arcs. In the case of an exclusive choice, this is done for each *request* transition.
4. The *execute* transition of the requester's model is merged with the transition representing the requesting activity in the control flow model.

It is worth mentioning that our proposed model supports multi-tenancy at multiple levels. As a matter of fact, it allows the modelling of multiple resources used by a single activity. Besides, it allows the modelling of, not only multiple processes but also multiple instances, sharing the same pool of Cloud resources. This is achieved by using the same provider model composed with multiple requester models pertaining to different activities of different processes' instances.

5 Evaluation

In order to evaluate our work, we established the corresponding model for a case study from France Telecom/Orange labs [3] which, due to the lack of space, we thoroughly present on our web page[1].

5.1 Case Study

The model of our case study has been implemented and validated using CPN Tools which offer a palette to analyse the net's state space. In this section we discuss some of the results indicated by the resulting state space analysis report. The full CPN model along with the generated report file can be found on our web page[1] where we present our case study with different provided resources to showcase errors detection when the allocation requests are unsatisfiable and lead to a deadlock. In fact, the allocations' correctness amounts to having a corresponding CPN model whose dead markings correspond to final markings only. The presence of a dead marking that does not correspond to a final marking translates the fact that some made request is unsatisfiable and prevents a complete execution of the business process. This can be deducted by inspecting the report file of the CPN model in question.

5.2 Checking Behavioural Properties

A major strength of Petri nets is their support for the analysis of many properties and problems associated with the modelled systems [16]. Those properties can be classified into two categories: those which depend on the initial marking, and those which are independent of the initial marking. The former type of properties is referred to as marking-dependent or behavioural properties, whereas the latter type of properties is called structural properties.

In this section, we are interested in basic behavioural properties. The properties *boundedness*, *reversibility* and *liveness* are independent of each other.

Boundedness. The state space report shows that all of the places of our illustrating example's model are bounded. This translates the fact that the number of tokens in every place does not exceed some finite number for any marking

---

[1] http://www-inf.it-sudparis.eu/SIMBAD/tools/CLoudResourceBP.

reachable from the initial marking. We are particularly interested in the bound-edness of the three buffer places which are shown to be bounded by the number of initially made requests. By verifying this property, it is guaranteed that there is no overflows in the buffers, no matter what firing sequence is taken.

Reversibility. To be able to derive this property from the model of our proposed example, we use a short-circuited version of the model where we added a transition *repeat* that retrieves answers from the *final* place and restores them as requests in the *REQs* place in order to simulate a continuous flow of requests.

The state space report indicates that all of the markings are home markings which translates the fact that for each marking M reachable from the initial marking M_i, M_i is reachable from M. Thus, we can always get back to M_i.

Reachability. Reachability is a fundamental basis for studying the dynamic properties of any system [16]. The *reachability problem* for Petri nets is the problem of finding if a given marking M in a net (P, M_i) is reachable from the initial marking M_i. This property can be relaxed if we are only interested in the markings of a subset of places. In this case, we talk about a *submarking reachability problem*. This problem can be defined as the problem of finding whether M' is reachable from M_i, where M' is any marking whose restriction to a given subset of places agrees with that of a given marking M.

In our case, it is interesting to make sure that the initial marking of the resource perspective's model is reachable. This is proved by the fact that all of the markings (including the initial marking) are home markings (see Sect. 5.2).

Liveness/Deadlock-Freedom. The concept of liveness is closely related to the complete absence of deadlocks in operating systems. In fact, a live Petri net guarantees a deadlock-free operation, regardless of the firing sequence.

The resulting state space analysis report for our illustrating example's model (the short-circuited version) indicates that all of our model's transitions are live, which means that every transition remains fireable for any firing sequence. Therefore we can say that the whole CPN model is live and can deduct that it is deadlock-free. In other words, a final marking (a marked "final" place) is always reachable. In case of a faulty allocation, this property cannot be satisfied and a deadlock is indeed detected.

If we limit this property to the requester's and provider's sub-models only, without taking into account the control flow's sub-model, we find that it is always satisfied which means that our model makes sure that, even if a request cannot be satisfied and the requesting activity fails, which may result in the failure of the whole BP, the model can still manage incoming requests from other BPs. In other words, our proposed model supports multi-tenancy at multiple levels.

6   Conclusion

Business process deployment in the Cloud has been a hot topic over the last years. The Cloud resource perspective however, has rarely been broached in the context of BPM and the need to formalize this perspective has yet to be studied. To address this issue, we proposed in this paper a Coloured Petri Net formal model for Cloud resource allocation in Business Processes.

The CPN model that we propose can be manipulated to map the BPMN extension proposed in [10]. However, we chose to represent an allocation mechanism that is richer and more detailed. Therefore, BPMN cannot be used to represent the Cloud resource allocation and elasticity aspect as we treat it using coloured Petri net in this work. From a designer's point of view, it would be easier to use a more business-related language than to model a business process using Petri net. Consequently, we will be proposing in future work a BPMN extension that includes all the relevant information needed to perform an automatic transformation into our CPN model which can then be used to verify the initial model.

This formalization that we propose is a first step towards the verification of Cloud resource allocation in business processes. We have so far worked on the verification of generic properties such as liveness and deadlock-freedom of the model. In the future work, we will detail how to use our model to verify other properties which are more specific to resource management and therefore improve the correctness verification of the allocations at design-time. To do so, we consider proving domain-specific properties like the shareability of Cloud resources. Such properties and others can be expressed in LTL (Linear Temporal Logic) and verified using tools as Helena [9] and CPN Tools.

References

1. van der Aalst, W.M.P., ter Hofstede, A.H.M., Kiepuszewski, B., Barros, A.P.: Workflow patterns. Distrib. Parallel Databases **14**(1), 5–51 (2003)
2. Amziani, M., Melliti, T., Tata, S.: Formal modeling and evaluation of service-based business process elasticity in the cloud. In: 2013 IEEE 22nd International Workshop on Enabling Technologies: Infrastructure for Collaborative Enterprises (WETICE), pp. 284–291. IEEE (2013)
3. Assy, N., Yongsiriwit, K., Gaaloul, W., Yahia, I.G.B.: A framework for semantic telco process management - an industrial case study. In: 14th International Conference on Intelligent Systems Design and Applications, ISDA 2014, 28–30 November 2014, Okinawa, Japan, pp. 44–49 (2014). https://doi.org/10.1109/ISDA.2014.7066276
4. Bersani, M.M., Bianculli, D., Dustdar, S., Gambi, A., Ghezzi, C., Krstic, S.: Towards the formalization of properties of cloud-based elastic systems. In: Proceedings of the 6th International Workshop on Principles of Engineering Service-Oriented and Cloud Systems, pp. 38–47. ACM (2014)
5. Boubaker, S., Mammar, A., Graiet, M., Gaaloul, W.: An event-B based approach for ensuring correct configurable business processes. In: IEEE International Conference on Web Services, ICWS 2016, 27 June – 2 July 2 2016, San Francisco, CA, USA, pp. 460–467 (2016). https://doi.org/10.1109/ICWS.2016.66

6. Cabanillas, C., Knuplesch, D., Resinas, M., Reichert, M., Mendling, J., Ruiz-Cortés, A.: RALph: a graphical notation for resource assignments in business processes. In: Zdravkovic, J., Kirikova, M., Johannesson, P. (eds.) CAiSE 2015. LNCS, vol. 9097, pp. 53–68. Springer, Cham (2015). https://doi.org/10.1007/978-3-319-19069-3_4

7. Cabanillas, C., Norta, A., Resinas, M., Mendling, J., Ruiz-Cortés, A.: Towards process-aware cross-organizational human resource management. In: Bider, I., et al. (eds.) BPMDS/EMMSAD -2014. LNBIP, vol. 175, pp. 79–93. Springer, Heidelberg (2014). https://doi.org/10.1007/978-3-662-43745-2_6

8. Chiotti, O., Stroppi, L.J.R., Villarreal, P.: Extending the WS-humantask architecture to support the resource perspective of BPEL processes (2014)

9. Evangelista, S.: The Helena Petri net tool (2013). http://www.lipn.univ-paris13.fr/~evangelista/helena/

10. Hachicha, E., Assy, N., Gaaloul, W., Mendling, J.: A configurable resource allocation for multi-tenant process development in the cloud. In: Nurcan, S., Soffer, P., Bajec, M., Eder, J. (eds.) CAiSE 2016. LNCS, vol. 9694, pp. 558–574. Springer, Cham (2016). https://doi.org/10.1007/978-3-319-39696-5_34

11. Hachicha, E., Gaaloul, W.: Towards resource-aware business process development in the cloud. In: 29th IEEE International Conference on Advanced Information Networking and Applications, AINA 2015, 24–27 March 2015, Gwangju, South Korea, pp. 761–768 (2015)

12. Jensen, K., Kristensen, L.M.: Coloured Petri Nets: Modelling and Validation of Concurrent Systems, 1st edn. Springer, Heidelberg (2009). https://doi.org/10.1007/b95112

13. Jrad, A.B., Bhiri, S., Tata, S.: STRATModel: elasticity model description language for evaluating elasticity strategies for business processes. In: Panetto, H. (ed.) OTM 2017. LNCS, vol. 10573, pp. 448–466. Springer, Cham (2017). https://doi.org/10.1007/978-3-319-69462-7_29

14. Klai, K., Tata, S.: Formal modeling of elastic service-based business processes. In: 2013 IEEE International Conference on Services Computing (SCC), pp. 424–431. IEEE (2013)

15. Milner, R., Tofte, M., Harper, R.: Definition of Standard ML. MIT Press, Cambridge (1990)

16. Murata, T.: Petri nets: properties, analysis and applications. Proc. IEEE **77**(4), 541–580 (1989)

17. Russell, N., van der Aalst, W.M.P., ter Hofstede, A.H.M., Edmond, D.: Workflow resource patterns: identification, representation and tool support. In: Pastor, O., Falcão e Cunha, J. (eds.) CAiSE 2005. LNCS, vol. 3520, pp. 216–232. Springer, Heidelberg (2005). https://doi.org/10.1007/11431855_16

18. Stroppi, L.J.R., Chiotti, O., Villarreal, P.D.: Extending BPMN 2.0: method and tool support. In: Dijkman, R., Hofstetter, J., Koehler, J. (eds.) BPMN 2011. LNBIP, vol. 95, pp. 59–73. Springer, Heidelberg (2011). https://doi.org/10.1007/978-3-642-25160-3_5

19. Stroppi, L.J.R., Chiotti, O., Villarreal, P.D.: Extended resource perspective support for BPMN and BPEL. In: CIbSE, pp. 56–69 (2012)

20. Van Der Aalst, W.M.: Workflow verification: finding control-flow errors using Petri-net-based techniques. Bus. Process. Manag. **1806**, 161–183 (2000)

21. Van Der Aalst, W.M., et al.: Soundness of workflow nets: classification, decidability, and analysis. Form. Asp. Comput. **23**(3), 333–363 (2011)

22. Van Hee, K., Verkoulen, P.: Integration of a data model and high-level Petri nets. In: Proceedings of the 12th International Conference on Applications and Theory of Petri Nets, Gjern, pp. 410–431 (1991)
23. Woitsch, R., Utz, W.: Business process as a service (BPaaS). In: Janssen, M., et al. (eds.) I3E 2015. LNCS, vol. 9373, pp. 435–440. Springer, Cham (2015). https://doi.org/10.1007/978-3-319-25013-7_35
24. Yataghene, L., Ioualalen, M., Amziani, M., Tata, S.: Using formal model for evaluation of business processes elasticity in the cloud. In: Drira, K., et al. (eds.) ICSOC 2016. LNCS, vol. 10380, pp. 33–44. Springer, Cham (2017). https://doi.org/10.1007/978-3-319-68136-8_3

Integrating Digital Identity and Blockchain

Francesco Buccafurri[1], Gianluca Lax[1(✉)], Antonia Russo[1],
and Guillaume Zunino[2]

[1] University of Reggio Calabria, Reggio Calabria, Italy
{bucca,lax,antonia.russo}@unirc.it
[2] ENSICAEN, Caen, France
guillaume.zunino@ecole.ensicaen.fr

Abstract. Blockchain is a recent technology whose importance is rapidly growing. One of its native features is pseudo-anonymity, since users are referred by (blockchain) addresses, which are hashed public keys with no link to real identities. However, when moving from the use of blockchain as simple platform for cryptocurrencies to applications in which we want to automatize trust and transparency, in general, there is not the need of anonymity. Indeed, there are situations in which secure accountability, trust and transparency should coexist (e.g., in supply-chain management) to accomplish the goal of the application to design. Blockchain may appear little suitable for these cases, due to its pseudo-anonymity feature, so that an important research problem is to understand how to overcome this drawback. In this paper, we address this problem by proposing a solution that mixes the mechanism of public digital identity with blockchain via Identity-Based-Encryption. We define the solution and show its application to a real-life case study.

Keywords: Digital identity · Blockchain · IBE

1 Introduction

Blockchain [34] is a recent technology used in many application contexts, such as financial services, industry 4.0, smart city, share trading. It was defined in [34] and allows us to replace a single centralized party managing a service with a distributed ledger of replicated, shared, and synchronized digital data spread across different servers. Data are saved in a growing list of records, called blocks, and each block contains a cryptographic hash of the previous block, a timestamp, and transaction data. Blockchain can record transactions between two parties efficiently and in a verifiable and permanent way [27]: it is managed by a peer-to-peer network of nodes running a common protocol for validating blocks. Once saved, the data in a block cannot be modified without alteration of all previous blocks, which requires a too high power computation.

Blockchain has several features: it is completely decentralized, since there is no central authority regulating data; it guarantees irreversible transactions,

© Springer Nature Switzerland AG 2018
H. Panetto et al. (Eds.): OTM 2018 Conferences, LNCS 11229, pp. 568–585, 2018.
https://doi.org/10.1007/978-3-030-02610-3_32

because once a transaction is generated, there is no way to delete or modify it; it is a trustless system, since it allows the transfer of sensitive information on a non-trust network by trusting the system on the whole not the system participant; it shows a pseudo-anonymous nature, since anybody can create a blockchain address to be used for transactions and it is no way to trace back it to his/her identity if appropriate precautions are taken [33]. It is worth noting that anonymity, in the original notion of blockchain, is a fundamental feature, as blockchain is born with the cryptocurrencies in mind and, for many years, cryptocurrencies were the sole applications for blockchain.

However, in the last years, also thanks to the advent of new blockchains and smart contracts, we are witnessing the shift from the use of blockchain as simple platform for cryptocurrencies to complex applications in which we want to automatize trust and transparency, and to take advantage from the other features of blockchain. In these cases, in general, we do not need anonymity anymore. Indeed, there are situations in which accountability, trust and transparency should strictly coexist, and accountability should be implemented by allowing a secure association with real-life identities. This requirement may derive from many different needs: it might be just an opportune measure to prevent unresolvable disputes, or it could derive from compliance with the law. Observe that, an approach in which users simply auto-declare their identity by a Blockchain transaction is not enough if there is no certainly of such an identity, because a user could declare a fake identity.

For these cases, blockchain appears little suitable, especially when the domain of the involved actors is open and not confined inside a single organization, which is a prerequisite for the suitability of blockchain itself. Consider, for instance, the management of the flow of goods and services (supply chain) [25]: it involves the movement and storage of raw materials, of work-in-process inventory, and of finished goods from a point of origin to a point of consumption. Typically, a supply chain is managed by a platform, a sets of technologies and processes promoting information sharing and coordination. There exist platforms for same day e-commerce home delivery in which consumers use a smart phone to browse and shop a broad range of products aggregated from nearby retail stores. Then, customer orders are handled by nearby independent couriers for pick-up and delivery to the customer. However, the platform acts as a trusted third party, thus it has to be always online and trusted by all participants. If at least one of the two conditions does not hold, using a blockchain makes sense. In this case, it should be necessary that anybody generating a transaction can be identified, but the current version of blockchain allowing pseudo-anonymous transactions does not help us. For all the above reasons, an important research problem is to understand how to overcome the native pseudo-anonymity of blockchain in order to support identity-aware applications.

In this paper, we address this problem by proposing a solution that mixes the mechanism of public digital identity with blockchain via Identity-Based-Encryption. We found this way the most suitable and not explored (so far) approach, because it accomplishes all the aimed requirements. Identity-Based-

Encryption (IBE) gives a direct role to the notion of identity, so allowing a direct link between the pair of cryptographic keys used to sign and verify a transaction and the identity of the transaction signer. On the other hand, public digital identity allows us to give a concrete definition of the identity to be used in IBEs by solving one of the problems of the concrete solutions based on IBEs, which is the proof of identity to the party issuing private keys (i.e., the Private Keys Generator).

As public digital identity, we use the notion compliant with eIDAS [7], a recent European Union regulation on electronic identification fully effective from 2016. It establishes the principle of mutual recognition and reciprocal acceptance of interoperable electronic identification schemes among Member States, and we chose it because (1) it is expected that, in the next years, eIDAS will be used by the most of EU citizens, (2) it is based on robust cryptographic primitives so that it can be considered secure, and (3) it has full legal effect.

We observe that an attempt of direct integration of public digital identity with a blockchain-based application would not provide a good result in terms of trust. Indeed, we should require that some entity of the application (even a smart contract if we adopt a blockchain like Ethereum) should play as a Service Provider of the public digital identity system (like in [19]). This implicitly requires the trust in this node for the assessment of identity, and this does not reach the goal in a satisfactory way from the security point of view, because it requires that the service providers (internal to the application domain) are trusted third parties. In contrast, the use of IBEs requires that only Identity Providers (and this is an assumption accepted also in eIDAS) and the Private Keys Generator of IBEs are trusted parties, that are parties external to the application. Clearly, Identity Provider and Private Keys Generator might also coincide.

It is worth noting that the approach proposed in this paper has the ambition to mix state-of-the-art techniques and methodologies to meet concrete needs. As a matter of fact, this paper is developed within the project called "Id Service: Digital Identity and Service Accountability" [6] funded by the Ministry of Economic Development (MISE), whose aim is studying innovative methods and techniques for designing and developing infrastructures for the accountability of cooperative services, also based on blockchain infrastructures, and their validation in virtual environments.

The paper is structured as follow. In Sect. 2, we introduce the notion of digital identity and the related technologies. In Sect. 3, we present Identity-Based Encryption, which is used to binding a digital identity and a public key. In Sect. 4, we present the idea underlying our solution. In Sect. 5, we describe the concrete proposal aiming at associating a public digital identity with a blockchain transaction. In Sect. 6, we instantiate our proposal to a specific scenario and we provide the technical details about how our solution works. The related work is discussed in Sect. 7. Finally, in Sect. 8, we draw our conclusions.

2   Digital Identity

A digital identity is defined as information on an entity used by computer systems to represent an external agent that may be a person, organization, application, or device [5]. Another similar definition given by ISO/IEC 24760-1 reports digital identity as a set of attributes related to an entity [1]. In this section, we briefly survey the main technologies related to digital identity and describe that used in our proposal.

Open Authorization (OAuth) [10] is an open access delegation protocol used by users to provide a third party (typically a site or an application) with the ability to access their personal information registered on a site without providing them with credentials to access this site. This protocol is widely used, especially in social networks, by many big companies (examples are Facebook, Twitter, Google) to allow their users to share profile information with third parties. OAuth is designed to use the HTTPs protocol for communication and exploits the release to the third party of tokens by an authorization server, once the user approves the proxy. These tokens are used as credentials to access shared information.

OpenID is another decentralized authentication protocol promoted by the OpenID non-profit foundation. By this protocol, a site administrator is supported in managing the users' authentication procedure, because no credential for user's login has to be stored. By OpenID, user access different sites with the same digital identity and password. In this protocol, the third party that handles authentication is the OpenID identity provider, while a site compatible with OpenID is called a relying party. The protocol is distributed among the identity providers and there is no central entity that manages authentication or decides who can act as a provider or identity provider. The first version of OpenID was published in 2005 by Brad Fitzpatrick, creator of the LiveJournal community and with the name Yadis (yet another distributed identity system). In 2007, Symantec included OpenID as a supported standard. In 2008, the OpenID 2.0 release was published and carried out by several major providers (Yahoo, Google, IBM, Microsoft, VeriSign, MySpace). The third and latest version, called OpenID Connect, was released in 2014.

Windows CardSpace [16] is a Microsoft software for digital identity management released in 2007. Born with the purpose of providing an environment robust against phishing attacks, CardSpace stores digital identities and provides a graphical interface for their management. When an application or a site needs to obtain information about the user, it generates a request for that information. The request is intercepted by CardSpace, which starts a graphical interface that shows the information stored and associated with that application or site. At this point, CardSpace contacts the digital identity provider to obtain the information to be shared, which is returned as a signed XML file, to guarantee its authenticity and integrity. In 2011, Microsoft registered a development of CardSpace, due to the technological changes and feedback received from partners and users. At the same time, Microsoft has shifted interest towards the U-Prove project. U-Prove is an advanced cryptographic technology, combined with identity solutions on existing standards, aimed to find a compromise to the eternal dilemma between identity and privacy guarantee with two important privacy-preserving features: (1) unlinkability and (2) selective disclosure of attributes.

In this paper we refer to a specific notion of digital identity, which is *public digital identity*, which means that it is recognized by law in a Country or at international level making the basis for non-repudiable accountable applications. There is a concrete instantiation of this notion in the European Union. It is based on the Regulation (EU) N. 910/2014 [7] on electronic identification and trust services for electronic transactions in the internal market (eIDAS Regulation), issued on 23 July 2014 and fully effective from 1 July 2016. It has the purpose of providing a normative basis at EU level for fiduciary services and providing the means of Member States' electronic identification to increase the security and effectiveness of e-business services and e-business and e-commerce transactions in the European Union. Thanks to the principle of mutual recognition and reciprocal acceptance of interoperable electronic identification schemes, eIDAS wants to simplify the use of electronic authentication against public administrations, both by companies and by citizens. Each Member State maintains it own electronic identification systems, which have to be accepted by all other member states. For example, Italy has notified to the EU Commission the institution of *SPID*, the Italian public system for the management of the digital identity of citizens and businesses [15]. Thanks to the eIDAS regulation, it is possible for Italian citizens to access the online services of other EU countries (university services, banking, public administration services, other online services) using SPID credentials, and at the same time, European citizens in possession of recognized national digital identities within the eIDAS framework will have access to the services of Italian public administrations. It is expected that in the next years, eIDAS will involve the most of EU people. This consideration, as well as the high security of this identification mechanism, suggested us to exploit eIDAs-compliant identification schemes as solution for the management of digital identity in our proposal.

3   Identity-Based Encryption

Asymmetric cryptography is based on the use of a public and private key for each user. Public keys are typically arranged by a Public Key Infrastructure, which binds public keys with the respective identities of entities (like people and organizations) through a process of registration and issuance of certificates by a certificate authority (CA). However, there are cases in which pre-distribution of keys is inconvenient or infeasible due to technical restraints: in these situation, Identity-based Encryption is a solution.

Identity-based Encryption (IBE) [9] allows any party to generate a public key from a known identity value (for example, an e-mail address). A trusted third party, called the Private Key Generator (PKG), generates the corresponding private key. To operate, the PKG first publishes a master public key, and retains the corresponding master private key (referred to as master key). Given the master public key, any party can compute a public key corresponding to an identity by suitably combining the master public key with the identity value. To obtain a corresponding private key, the party authorized to use the identity ID contacts the PKG, which uses the master private key to generate the private key for the identity ID. The operations carried out in an IBE scheme are summarized in Fig. 1.

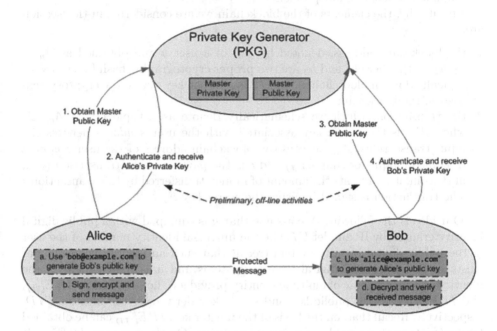

Fig. 1. Operations carried out in an IBE scheme.

As a result, parties may encrypt messages (or verify signatures) with no prior distribution of keys between individual participants, once their identity is known and well-defined. However, to decrypt or sign messages, the authorized user must obtain the appropriate private key from the PKG, by proving the possession of the proper identity. The most used IBE systems have been proposed by Boneh-Franklin [22] and by Sakai-Kasahara [36].

4   The Ideal Solution

We recall that the basic goal of this paper is to integrate blockchain and public digital identity. In this section, we sketch what we identify as the ideal solution of the problem above, in the sense that it implements the above integration in the most direct and strong way.

Suppose we have an IBE system with Private Key Generator PKG (see Sect. 3) and a public identity digital system with identity provider IP (assumed unique, w.l.o.g.). For simplicity, we assume we are not considering blockchains allowing smart contracts (i.e., Blockchain 2.0), even though the generalization to every kind of blockchain is straightforward. Therefore, we focus our attention just on the elements related to our problem, which are the blockchain addresses and, consequently, the form of transactions. Obviously, the organization of blocks, the consensus protocol, the mining process, and the other aspects of the blockchain are outside the scope of our problem.

Specifically, the elements of the blockchain we are considering in this section are:

1. the blockchain address, denoted by A_u, of a user u and obtained as $A_u = h_1(h_2(P_u))$, where h_1 and h_2 are two proper cryptographic hash functions (as typically done in blockchains), and P_u is a public key of u in the cryptosystem used in the blockchain;
2. the transaction, which we schematically denote as a tuple $\langle P_{u_s}, i, A_{u_r}, c \rangle$, where P_{u_s} is the public key associated with the user *sender*, i denotes the input transactions, A_{u_r} denotes the blockchain address of the user *recipient* (assumed unique for simplicity) and c is the payload of the transaction (e.g., in Bitcoin, it represents the amount of money transferred by this transaction). The transaction is signed by using the secret key S_{u_s}.

Our idea is the following. We assume that u is equipped with a public digital identity granted by IP and let UID be the universal identity number of the user in the public digital identity system (recall that such an identification number exists in real-life public digital identity systems and it is independent of the identity provider, in case of multiple identity providers). Let denoted by IBE_{UID}^P and IBE_{UID}^S the IBE public key and secret key derived by the identity UID, respectively. Recall that, on the basis of the master key, IBE_{UID}^P can be obtained by any party with no need of further information. On the contrary, IBE_{UID}^S is released by PKG through a secure channel to any party able to demonstrate to be the owner of the identity UID. What we require is that PKG becomes a service provider in the public digital identity system, which means that it recognizes in a secure way the identity of people by leveraging the federated authentication protocol involving IP and a (strong) authentication session of the user at IP. Therefore, in order to release secret keys, PKG will require a secure authentication session done according to the protocol of the public digital identity system.

This allows us to design a blockchain in which the address of the user u, recognized in the public digital identity system by the identifier UID, is obtained as: $A_u = h_1(h_2(IBE_{UID}^P))$ (we recall that h_1 and h_2 are two cryptographic hash functions). Therefore, the sources and the recipients of a transaction are derived directly from UIDs, thus from public digital identities, and impersonation is not possible provided that it is not possible in the public digital identity system. Specifically, a transaction $\langle P_{u_s}, i, A_{u_r}, c \rangle$ done by the user u_s with identity UID_s and having as recipient the user u_r with identity UID_r, is signed by the IBE secret key $IBE_{UID_s}^S$ and verified by the IBE public key $IBE_{UID_s}^P$, which everyone can compute on the basis of the IBE public master key, once the identity UID_s is known. This allows us also to represent the transaction as: $\langle UID_s, i, UID_r, c \rangle$.

This representation reflects a nice feature of our solution, in which blockchain addresses are intensionally always existing in the blockchain domain, even though they are not materialized, provided that the corresponding identities exist in the public digital identity system. As a consequence, a given transaction moving a token (or money) to a user u may exist in the blockchain without requiring any action from u on the blockchain (the creation of a key-pair), as identities are implicitly blockchain addresses.

One could argue that a similar solution makes us lose the full decentralization of the blockchain paradigm. This is necessarily true if we want to rely on the current notion of public digital identity system, which is inherently centralized. However, a different notion of digital identity could be applied, also fully decentralized and based on blockchain itself like [8] or [30].

It is worth noting that the ideal solution here presented implicitly requires that blockchain (public and private) keys are compliant with the adopted IBE scheme (for example, RSA [35]). Unfortunately, this is not the case of existing blockchains: for an instance, Bitcoin blockchain adopts the elliptic curve secp256k1 [32], which is not compliant with any IBE scheme and a definition of an IBE scheme on this cryptographic scheme is not feasible.

For this reason, to give a more practical value (also for the industrial nature of the research project in which this paper is located) to this paper, we implement in the next section a workaround that allows us to basically obtain the same result by leveraging any existing blockchain. Specifically, we chosen Bitcoin blockchain because it is one of the most used, but any other blockchain could be considered, also by extending the approach toward smart-contract-supporting blockchains like Ethereum. Consider that, in this case, any solution (like [19]) that implements the integration between the public digital identity system and the blockchain by directly giving the role of service provider to smart contracts, does not reach the goal in a satisfactory way from the security point of view, because it requires that the service providers (internal to the application domain) are trusted third parties (TTPs). Conversely, in our solution, TTPs are only TTPs of the external systems (i.e., the identity provider of the public digital identity system and the Private Key Generator of the IBE system).

5   A Practical Solution

Starting from the considerations done in the previous section, in this section we provide a practical solution that does not relax any security feature w.r.t. the ideal one. It is practical in the sense that it does not require changes of blockchain formats and protocol, thus operating on the exiting ones. For the sake of presentation, we describe the solution on the Bitcoin blockchain, which is widely used.

The actors in our scenario are:

- *Users*, physical or legal people using a public digital identity for authentication.
- *Identity Providers*, which create and manage public digital identities.

- *IBE Services*, public or private organizations providing the mapping between a public digital identity and a pair of asymmetric encryption keys (called IBE keys).
- a *Blockchain*, a Distributed Ledger.

In our proposal, we can identify the following operations.

1. *Digital Identity Issuing*. First, a user creates his/her public digital identity. To do this, he/she must be registered to one of the *Identity Providers*, which is responsible for the verification of the user identity before issuing the public digital identity and the security credentials.
   A public digital identity is identified by the pair $\langle username, IP \rangle$, where IP is the identifier of the identity provider that issued the public digital identity and *username* is a string. Moreover, there exists a string *UID* (Universal ID), which identifies a public digital identity. For example, the user X registered by the Identity Provider Y is identified by the UID X@Y. It is worth noting that UIDs are supported by the Public Digital Identity Systems.
2. *IBE private key gathering*. To obtain the IBE private key, a user contacts the Private Key Generator (PKG) of the IBE service to receive the master public key, if it is not already known. Then, the Private Key Generator, by acting as a service provider of the public digital identity system, authenticates the user by an eIDAS-compliant scheme, as illustrated in Fig. 2.
   First, the user using a browser (`User Agent`) sends to PKG a request for gathering the IBE private key (Step 1). Then, PKG replies with an authentication request to be forwarded to *Identity Provider* (Step 2). If the received request is valid, *Identity Provider* performs a challenge-response authentication with the user (Steps 3 and 4). In case of successful user authentication, *Identity Provider* prepares the statement of user authentication, which is forwarded to PKG (Step 6). Finally, PKG provides the user with the IBE private key (Step 7).
3. *Blockchain Registration*. First, the user generates a pair of private and public blockchain keys, and, starting from the public one, the blockchain address A is computed. Then, the user generates on the blockchain a transaction from A to A, having as payload $\langle UID, E(A) \rangle$, where UID is the universal ID of the public digital identity of the user, and $E(A)$ is the encryption of the user's blockchain address by the user's IBE secret key. By this transaction, the user links her/his public digital identity to the blockchain address A: indeed, by computing $E(A)$, the user proves the knowledge of the IBE secret key associated with this UID.
4. *Transaction*. When a user S (sender) wants to carry out a transaction with a user R (receiver), the following operations are done:
   (a) S obtains the universal ID of R, say IUD_r.
   (b) S searches for the transaction having IUD_r in the payload: this is the transaction done by R in the blockchain Registration step.
   (c) S extracts from this transaction the blockchain address of R, say A_r.

(d) S generates a blockchain transaction from her/his blockchain address A_s to A_r (the value of the payload depends on the application).

Now, it should be easy to understand how to know the public digital identity of a user involved in a blockchain transaction. Consider a blockchain transaction from the (blockchain) address A_s to the (blockchain) address A_r, and assume we are interested in knowing the identity of the user associated with A_r[1].

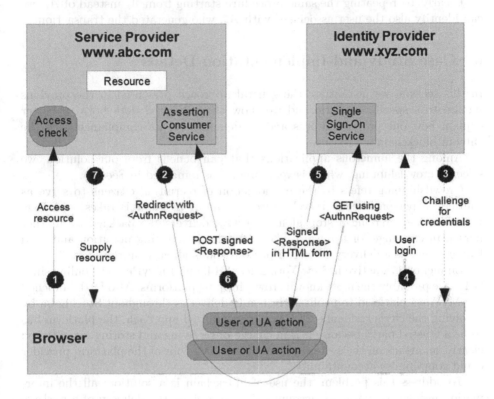

Fig. 2. Data flow in an authentication process.

The first operation to do is to search for the transaction having A_r as sender and receiver (i.e., the transaction done in the Blockchain Registration step). If it is not found, this means that A_s did not execute the protocol correctly, because she/he generated a transaction to an unregistered user (clearly, it is not possible that the registration transaction of A_r has been deleted because blockchain transactions are immutable). Thus, we assume that this registration transaction, say T, is found.

---

[1] For the sake of presentation and to avoid to introduce new notations, in the following, with a little abuse of notation, we use the address A_u also to refer to the user u, thus meaning "the user associated with the address A_u".

Now, after verifying the authenticity and integrity of T (i.e., that it has been signed by the blockchain public key associated with the address A_r), the payload $\langle p_1, p_2 \rangle$ is extracted.

Next, the IBE public key $IBE_{p_1}^K$ derived from the string p_1 is computed, as described in Sect. 3 and used as public key to decipher p_2. If the decryption of p_2 corresponds to A_r, then we are sure that the receiver (i.e., A_r) of the transaction T is associated with the public digital identity p_1.

Clearly, by repeating the same procedure starting from A_s instead of A_r, we can identify also the user associated with A_s, who generated the transaction.

6   Case Study and Implementation Details

In this section, we instantiate the general approach presented in the previous section to a specific scenario and we show the generated data both to better explain how our proposal works and to demonstrate its compliance with the Bitcoin blockchain.

Among the numerous applications that can benefit from our solution, we selected crowdshipping, which is very timely (as remarked in Sect. 7).

Crowdshipping refers to the phenomenon of recruiting citizens to serve as couriers: a person already traveling from point A to point B takes a package with him and, making a stop along the way, delivers the package to another person in exchange for a reward. The objective is reducing pollution and road traffic using, as a delivery carrier, a person who is already on the move.

Zipments [17], active in New York since 2014, and PiggyBee [11], online since 2012, are probably the most known crowdshipping platforms. Also DHL launched the MyWays platform to facilitate last-mile deliveries throughout Stockholm by involving the city's residents [2]. Being a centralized approach, the platform has to be a trusted party because it is in charge of receiving and storing log activity: clearly, an attack on the system or a malicious behavior of the platform provider could compromise accountability.

To address this problem, the use of blockchain is a solution: all the information needed to guarantee accountability, especially the delivery of a package between two users, is stored in the blockchain. In particular, we considered the basic step of a crowdshipping system, which occurs when a user, say Alice, delivers a package to another user, say Bob. Alice needs both: (1) to be sure that the person receiving the package is Bob and (2) to have a proof of delivery. Our solution guarantees both the goals without using a centralized crowdshipping platform.

We implemented a Java prototype to test our solution in a crowdshipping scenario: it is composed of a module implementing the IBE system and a module implementing the access to the blockchain. We did not need to implement the identification scheme compliant with eIDAS, because it is a service used by our prototype. We show all the operations carried out by the two users and the generated data.

1. *Digital Identity Issuing.* Both Alice and Bob have a public digital identity: thus, they have been identified by an identity provider, say `example.com`, which gave each of them a public digital identity and a credential for authentication (typically, a password). Now, assume that the username of Alice is `alice` and the username of Bob is `bob`. Thus, the UIDs of Alice and Bob are `alice@example.com` and `bob@example.com`, respectively. Observe that, for the sake of presentation, we used the same identity provider (i.e., `example.com`) for both the users: however, no problem arises in case the public digital identities are issued by different identity providers, because the solution does not depend on the particular UID of the user.

2. *IBE private key gathering.* To obtain the IBE private key, a user connects to the site of the IBE system by the browser (i.e., the user agent) and sends a request for accessing the service. Observe that the IBE system acts as a service provider in this step, because it needs to authenticate the user before issuing the private key. Then, the IBE system replies to the user agent with an authentication request to be forwarded to the identity provider. The identity provider is selected according to the user's UID.

   If the received request is valid, the identity provider performs a challenge-response authentication with the user. In case of successful user authentication, the identity provider prepares the *assertion* containing the statement of the user authentication for the IBE service provider. The assertion contains the reference to the request message, the authenticated user, the identity provider, the personal information about the authenticated user, the temporal range of validity, and the description of the authentications context. The assertion is signed by the identity provider to guarantee integrity and authenticity.

   Now, the assertion returned to the user agent is forwarded via `http POST` Binding to the IBE service provider. The IBE system verifies the assertion and provides the user with her/his IBE private key. We denote by IBE_U^S the IBE private key of the user U.

   Concerning the user's IBE public key, they are computed starting from the master public key and the user's UID. We denote by IBE_U^P the IBE public key of the user U.

   In Table 1, the IBE public and private keys of Alice and Bob are reported: they are represented by Base58Check encoding [3], which is used for blockchain addresses (see later).

3. *Blockchain Registration.* Each user needs to have a private and a public blockchain key. The private key is a randomly generated 256-bit string. The public key is generated by the private one by means of a cryptographic function named *elliptic curve point multiplication*. In particular, the used algorithm is Curve Digital Signature Algorithm (ECDSA) and the elliptic curve is `secp256k1` [32]. The use of these functions is necessary to guarantee the compatibility of our solution with blockchain.

   We denote by BKC_U^S and BKC_U^P the blockchain private key and public key of the user U. In Table 1, the blockchain public and private keys of Alice and Bob are reported.

The blockchain address A of a user is computed from the public key K as $A = \text{RIPEMD160}(\text{SHA256}(K))$, where SHA256 [14] is a cryptographic hash developed by National Security Agency (NSA) and returns a 256-bit digest, whereas RIPEMD160 [13] is a cryptographic hash designed in the open academic community and returns a 160-bit digest.

We denoted by A_U the blockchain address of the user U. In Table 1, the blockchain addresses of Alice and Bob are reported. Observe that blockchain addresses are usually represented by Base58Check, an encoding similar to Base64 but modified to remove non-alphanumeric characters and letters which might look ambiguous when printed. It is therefore designed for human users who manually enter the data by copying from some visual source.

Finally, each user generates on the blockchain a transaction with her/his address as both sender and receiver, having as payload $\langle UID, E(A) \rangle$, where UID is the universal ID of the public digital identity of the user, and $E(A)$ is the encryption of the user's blockchain address done by the user's IBE private key.

4. *Transaction.* Now, both Alice and Bob have their public digital identity associated with a blockchain address. Suppose that Alice has to deliver a package with $ID = AB123$ to Bob and, consequently, she needs a proof of delivery from Bob. In a real-life situation, we can image that carriers run a mobile app on their smartphones to manage transaction generations. We can suppose also that the package ID is a QRcode [12] printed on the box, so it can be easily read by the mobile app running on carrier's smartphone. Moreover, the same mobile app can show another QRcode reporting the UID of the owner, in such a way that when Alice has to deliver the package to Bob, Bob can show his UID by his mobile smartphone and vice versa.

Once the package ID and the UID of Alice have been collected, Bob's mobile app generates a transaction to A_{Alice} (i.e., the Alice's blockchain address) including in the payload the type of operation carried out (i.e., package receiving) and the id of the product. This transaction is signed by Bob with the blockchain private key and stored on the blockchain.

Alice can read on the blockchain this transaction and checks its correctness: clearly, this is done by the app mobile. This transaction represents the proof of delivery of the package from Alice to Bob.

Observe that in some context it could be necessary also an additional proof: in this case, Alice can generate a transaction to Bob having, in the payload, "package sending" as the type of operation and the id of the product, in such a way that Bob can proof the reception of the package from the correct user (i.e., Alice).

7   Related Work

In this section, we survey the most important proposals of the state of the art related to our approach.

Table 1. Value of the data generated in our running example.

Symbol	Value
IBE_{Alice}^{S}	2UBUArXzNjLYArGyk46pn6yJVrik5x5sFRne1H2ACznMeBeAfvJyMdbdY 5aDofJ2NxjTQHwCUdpRiPMfKo4Y7CNhrwVq4KDFTiA3KavkyX7b7dE U1CVB1SwEZkMQL2KoD5erHSVsCwcKiqm6yPESsZMPWhEUkWio47D SETVP672AYrnw4E8zMH18gvrnPaERiLd9KL8Z9ZhYui7NL7NWA4oJ8kq GXSsJX85gTheFyswfSXya4HrwfXtQYotrpq7uuS7rKgGGsAhazE7Ceg6mY cMUSdbPg9drR21EUx3LrD3z3sp8QhFvDBpLkdhEZMGsngjDgdu24rZZh M5beQUCC56WVWefRE
IBE_{Alice}^{P}	C9vPE,185xNAn8quf9s4vrechMHXj6PDZakHJ532JYvGbQt7obcLqyyeLub 7VTXcY8qPg3FJj3TvPgEV3CAEx4K9U8diTkGj1xe2dZFicQLWk68KnGy eDZimALtNHm2hEvjFJSDinMtBEQAZrZUXGBrqN7QFT5imVLV821kEJ b1sMh3HMfnhEmucXsn1MDKgCymhkrXHKsFfFRyta9wzQvQKRmdNhT 5FruGXTV8VbA6XG1VaszpUco7EiUzLfayZdA8YWw3umeZDr5bi3AenxM WHXw6ptZFPg7rkf2YgmbHKqFnBCEyDT2gsF8XF9S1rkWjiqsKxef1ZC63 czWiRvfAZ3A5xfo7R9QAsk
IBE_{Bob}^{S}	18NxP33Ub9UwGTA1SRVT5ih4Ji7oq5mFmowxqUHAxQKYgzcUtR5aECK c2sgf6Xmwxo1jzQnPkq1D4LfZp5egDFBCQ2mhUNtL5mwqrkGUaR3qgkDf Dgpu57WntwmyXo8KVoNvHRgcwRSEUzBdMgnGA1JDTaxbCeyf98ebh4D tDxCoygjumt3khenAwnkytiwxiXXnjgXtBxaTPwhBYr1ZVtCmJmR9yoLTn msCQevuL57DtpQc492jwymgMi957FQwe1bDB1WDENgogBX87bTQjqRk mYyK74EFuN1bFPFjbZJ3WAMPWnRCVBawic9eumHWPfwmSfg38jFgP 9f2tJrNKWVX7msGFWQ
IBE_{Bob}^{P}	Meip8,185xNAn8quf9s4vrechMHXj6PDZakHJ532JYvGbQt7obcLqyyeLub7 VTXcY8qPg3FJj3TvPgEV3CAEx4K9U8diTkGj1xe2dZFicQLWk68KnGye DZimALtNHm2hEvjFJSDinMtBEQAZrZUXGBrqN7QFT5imVLV821kEJb 1sMh3HMfnhEmucXsn1MDKgCymhkrXHKsFfFRyta9wzQvQKRmdNhT5 FruGXTV8VbA6XG1VaszpUco7EiUzLfayZdA8YWw3umeZDr5bi3AenxM WHXw6ptZFPg7rkf2YgmbHKqFnBCEyDT2gsF8XF9S1rkWjiqsKxef1ZC6 3czWiRvfAZ3A5xfo7R9QAsk
BKC_{Alice}^{S}	z4KNrFydhCHUt15g9N3MDX4Di2WfuE1JzMMZHRFtVCWkpx6DTH1H VTqBCtdwCL1ERwVfeto3A5pU8G8Fgkv8V2G
BKC_{Alice}^{P}	2f6eQs8MtzDWd1dj95LncxAfEseGNvn7LhbUYrg8kPSUNp5gYKN9zwkvwm ZtPFoFpjPrqdEgeYj3jzbvzKvComRQ7iF9JSE3JGY6UfNBYshkZzXG8qkJ WM93MDDSz6rJzYfAZ8rVU6n9xLLH2CeSSfhv5QZW9MqjcT3v7Mpsahh HtYHUK7
BKC_{Bob}^{S}	22PMoQ4VMGhGwep4KxxqHLB4JtPZFJ5AvLWar5ndP3fvGHaRbsEW2H yw55qAZR9TMyG9Z49P3tApibixFZo2SwXV
BKC_{Bob}^{P}	2f6e6iHZ4yg6h5oTaNj5tvTpgbhbjGJ63DUrA7Zi9n8aut1wtWQ9ELNt3iMeZ 7taavtwv55bZUY2MQdNFmAhoozstxUt6km8j791bWwDtaZGmkxv7vb5bzy sLtnDre8fgQ8GfLuv3F5yEmzweGv69S
A_{Alice}	1QCw797xQbc7UjbpeMeCcdxyj9SpbVnvN4
A_{Bob}	1JCn8rVTLTKf8H1hJQc9Jx2s9FTEwPNUDk

In [18], the authors review applications relying on blockchain. They high-light the potential benefit of such technology in manufacturing supply chain and a vision for the future blockchain ready manufacturing supply chain is proposed. The paper [20] provides a high level understanding of how blockchain technology will be a powerful tool to improve supply chain operations. It illustrates theoretical and conceptual models for use of open and permissioned blockchain in different supply chain applications with real life practical use cases as is being developed and deployed in various industries and business functions. The paper [29] states that digital supply chain integration is becoming increasingly dynamic. Access to customer demand needs to be shared effectively, and product and service deliveries must be tracked to provide visibility in the supply chain. Business process integration is based on standards and reference architectures, which should offer end-to-end integration of product data. The authors of this study investigate the requirements and functionalities of supply chain integration, concluding that cloud integration can be expected to offer a cost-effective business model for interoperable digital supply chains. Moreover, they explain how supply chain integration through the blockchain technology can achieve disruptive transformation in digital supply chains and networks. In [28], the authors highlight that the need for blockchain-based identity management is particularly noticeable in the Internet age, as we have faced identity management challenges since the dawn of the Internet. They observe that blockchain technology may offer a way to circumvent this problem by delivering a secure solution without the need for a trusted, central authority. It can be used for creating an identity on the blockchain, making it easier to manage for individuals, giving them greater control over who has their personal information and how they access it. The proposed solution stores users' encrypted identity, allowing them to share their data with companies and manage it on their own terms.

Bitnation [4] is the world's first Decentralised Borderless Voluntary Nation (DBVN). Bitnation started in July 2014 and hosted the first blockchain for refugee emergency ID, marriage, birth certificate, World Citizenship and more. The website proof-of-concept, including the blockchain ID and Public Notary, is used by tens of thousands of Bitnation Citizens and Embassies around the world. In [24], the authors focus on Public Digital Identity System (SPID), the Italian government framework compliant with the eIDAS regulatory environment. They observe that a drawback limiting the real diffusion of this framework is that, despite the fact that identity and service providers might be competitor private companies, SPID authentication results in the information leakage about the customers of identity providers. To overcome this potential limitation, they propose a modification of SPID to allow user authentication by preserving the anonymity of the identity provider that grants the authentication credentials. This way, information leakage about the customers of identity providers is fully prevented. The paper [37] focuses on pseudonymisation, a concept that was only recently formally introduced in the EU regulatory landscape. In particular, it attempts to derive the effects of the introduction of pseudonyms (or pseudonymous credentials) as part of the eIDAS Regulation on electronic identification and trust services and, ultimately, to compare them with the effects of pseudonymisation within the meaning of the General Data Protection Regulation (the GDPR). The paper examines how eIDAS conceives pseudonymisation and explains how this interpretation would translate in practical uses in the context of a pan-European interoperability framework. In [23], an advanced electronic signature protocol that relies on a public system for the management of the digital identity is proposed. This proposal aims at implementing an effective synergy to provide the citizen with a unique, uniform, portable, and effective tool applicable to both authentication and document signature. In [21], the authors propose a security framework that integrates the blockchain technology with smart devices to provide a secure communication platform in a smart city. The authors observe that, despite a number of potential benefits, digital disruption poses many challenges related to information security and privacy. In [26], the authors explore an environment in which in-store customers supplement company drivers can take on the task of delivering online orders on their way home. The results of their computational study provide insights into the benefits for same-day delivery of this form of crowdshipping, and demonstrate the value of incorporating and exploiting probabilistic information about the future.

The study carried out in [31] highlights that passengers and freight mobility in urban areas represents an increasingly relevant component of modern city life. On one side, it fosters economic growth, but, on the other, it also generates high social costs. Congestion and pollution are two problems policy-makers want to curb adopting appropriate measures. In this context, this paper analyses the feasibility and behavioral levers that might facilitate the diffusion of crowdshipping in urban areas. Two are the main objectives the paper. The first is to investigate under which conditions passengers would be willing to act as crowdshippers. The second is to find out under which conditions people would be willing to receive their goods via a crowdshipping service. Crowdshipping can generate positive impacts, such as the reduction of total and ad-hoc trips, by optimizing, through sharing, the use of resources and infrastructures.

From the brief review of the state of the art here reported it clearly emerges both the importance of securely identifying the entities operating in real-life applications that can also benefit from blockchain, and the originality of our proposal that, to the best of our knowledge, is the first combining IBEs and blockchain.

8 Conclusion

In this paper, we discussed about the benefits deriving from the possibility of binding the sender or the receiver of a blockchain transaction to a public digital identity. We proposed an architecture to do this, which exploits eIDAS-compliant identification schemes for handling public digital identities and Identity-based Encryption for associating a digital identity with a public key. This architecture has been implemented by a Java prototype and used to validate the proposal in a crowdshipping scenario. To the best of our knowledge, this is the first attempt to create a non-anonymous blockchain, which can be used in all cases in which the author of a transaction has to be identified with certainty and legal effect.

As future work, we plan to investigate the possibility to use blockchain 2.0 to solve the accountability problem by a smart contract, for example, to allow the inclusion of new rules and conditions in the product delivery process. Moreover, we need to evaluate the dependence of our solution on the regulation and technological changes or advances in the use of available mechanisms for a more explicit and transparent digital identification.

Acknowledgment. This paper has been partially supported by the project "Id Service - Digital Identity and Service Accountability" funded by the Ministry of Economic Development (MISE), project code number F/050238/03/X32 and by INdAM – GNCS Project 2018 "Processing and analysis of Big Data modeled as graphs in different application contexts".

References

1. ISO/IEC 24760–1:2011: Information technology – Security techniques – A framework for identity management – Part 1: Terminology and concepts (2011). http://standards.iso.org/ittf/PubliclyAvailableStandards/index.html
2. DHL crowd sources deliveries in Stockholm with MyWays (2013). http://www.dhl.com/en/press/releases/releases_2013/logistics/dhl_crowd_sources_deliveries_in_stockholm_with_myways.html
3. Base58Check (2018). https://en.bitcoin.it/wiki/Base58Check_encoding
4. Bitnation Pangea—Your Blockchain Jurisdiction (2018). https://tse.bitnation.co/
5. Digital identity (2018). https://en.wikipedia.org/wiki/Digital_identity/
6. Digital Identity and Service Accountability (2018). http://www.okt-srl.com/ricerca-pon-mise-idservice.html
7. eIDAS - Interoperability Architecture (2018). https://ec.europa.eu/futurium/en/content/eidas-regulation-regulation-eu-ndeg9102014
8. IBM Blockchain (2018). https://www.ibm.com/blockchain/solutions/identity
9. ID-based Encryption (2018). https://en.wikipedia.org/wiki/ID-based_encryption
10. OAuth Community Site (2018). https://oauth.net/
11. PiggyBee: CrowdShipping - Crowdsourced delivery (2018). https://www.piggybee.com/
12. QR code (2018). https://en.wikipedia.org/wiki/QR_code
13. RIPEMD (2018). https://en.wikipedia.org/wiki/RIPEMD
14. SHA-2 (2018). https://en.wikipedia.org/wiki/SHA-2
15. SPID Sistema Pubblico di Identità Digitale (2018). https://www.spid.gov.it/
16. Windows CardSpace (2018). https://en.wikipedia.org/wiki/Windows_CardSpace
17. Zipments: Same Day Delivery Service (2018). https://zipments.com/
18. Abeyratne, S.A., Monfared, R.P.: Blockchain ready manufacturing supply chain using distributed ledger (2016)
19. Angiulli, F., Fassetti, F., Furfaro, A., Piccolo, A., Saccà, D.: Achieving service accountability through blockchain and digital identity. In: Mendling, J., Mouratidis, H. (eds.) CAiSE 2018. LNBIP, vol. 317, pp. 16–23. Springer, Cham (2018). https://doi.org/10.1007/978-3-319-92901-9_2
20. Banerjee, A.: Blockchain technology: supply chain insights from ERP. Adv. Comput. **111**, 69–98 (2018)
21. Biswas, K., Muthukkumarasamy, V.: Securing smart cities using blockchain technology. In: 2016 IEEE 18th International Conference on High Performance Computing and Communications, IEEE 14th International Conference on Smart City, IEEE 2nd International Conference on Data Science and Systems (HPCC/SmartCity/DSS), pp. 1392–1393. IEEE (2016)

22. Boneh, D., Franklin, M.: Identity-based encryption from the Weil pairing. In: Kilian, J. (ed.) CRYPTO 2001. LNCS, vol. 2139, pp. 213–229. Springer, Heidelberg (2001). https://doi.org/10.1007/3-540-44647-8_13

23. Buccafurri, F., Fotia, L., Lax, G.: Implementing advanced electronic signature by public digital identity system (SPID). In: Kő, A., Francesconi, E. (eds.) EGOVIS 2016. LNCS, vol. 9831, pp. 289–303. Springer, Cham (2016). https://doi.org/10.1007/978-3-319-44159-7_21

24. Buccafurri, F., Fotia, L., Lax, G., Mammoliti, R.: Enhancing public digital identity system (SPID) to prevent information leakage. In: Kő, A., Francesconi, E. (eds.) EGOVIS 2015. LNCS, vol. 9265, pp. 57–70. Springer, Cham (2015). https://doi.org/10.1007/978-3-319-22389-6_5

25. Christopher, M.: Logistics & Supply Chain Management. Pearson, London (2016)

26. Dayarian, I., Savelsbergh, M.: Crowdshipping and same-day delivery: employing in-store customers to deliver online orders. Optimization Online (2017)

27. Iansiti, M., Lakhani, K.R.: The truth about blockchain. Harv. Bus. Rev. **95**(1), 118–127 (2017)

28. Jacobovitz, O.: Blockchain for identity management (2016)

29. Korpela, K., Hallikas, J., Dahlberg, T.: Digital supply chain transformation toward blockchain integration. In: Proceedings of the 50th Hawaii International Conference on System Sciences (2017)

30. Lewko, A., Waters, B.: Decentralizing attribute-based encryption. In: Paterson, K.G. (ed.) EUROCRYPT 2011. LNCS, vol. 6632, pp. 568–588. Springer, Heidelberg (2011). https://doi.org/10.1007/978-3-642-20465-4_31

31. Marcucci, E., Le Pira, M., Carrocci, C.S., Gatta, V., Pieralice, E.: Connected shared mobility for passengers and freight: investigating the potential of crowdshipping in urban areas. In: 2017 5th IEEE International Conference on Models and Technologies for Intelligent Transportation Systems (MT-ITS), pp. 839–843. IEEE (2017)

32. Mayer, H.: ECDSA security in bitcoin and ethereum: a research survey. CoinFaabrik, 28 June 2016

33. Moreno, F., Trivedi, S.: Staying Anonymous on the Blockchain: Concerns and Techniques (2017). https://securingtomorrow.mcafee.com/mcafee-labs/staying-anonymous-on-the-blockchain-concerns-and-techniques/

34. Nakamoto, S.: Bitcoin: a peer-to-peer electronic cash system (2008)

35. Rivest, R.L., Shamir, A., Adleman, L.: A method for obtaining digital signatures and public-key cryptosystems. Commun. ACM **21**(2), 120–126 (1978)

36. Sakai, R., Kasahara, M.: ID based cryptosystems with pairing on elliptic curve. IACR Cryptology ePrint Archive, p. 54 (2003)

37. Tsakalakis, N., Stalla-Bourdillon, S., O'hara, K.: What's in a name: the conflicting views of pseudonymisation under eIDAS and the general data protection regulation (2016)

Context-Aware Predictive Process Monitoring: The Impact of News Sentiment

Anton Yeshchenko[1]([⊠]), Fernando Durier[2], Kate Revoredo[2], Jan Mendling[1], and Flavia Santoro[2]

[1] Vienna University of Economics and Business (WU), Vienna, Austria
{anton.yeshchenko,jan.mendling}@wu.ac.at
[2] Federal University of the State of Rio de Janeiro (UNIRIO),
Rio de Janeiro, Brazil
{fernando.durier,katerevoredo,flavia.santoro}@uniriotec.br

Abstract. Predictive business process monitoring is concerned with forecasting how a process is likely to proceed, covering questions such as what is the next activity to expect and what is the remaining time until case completion. Process prediction typically builds on machine learning techniques that leverage past process execution data. A fundamental problem of a process prediction methods is the data acquisition. So far, research on predictive monitoring utilize data, which is internal to the process. In this paper, we present a novel approach of integrating the external context of the business processes into prediction methods. More specifically, we develop a technique that leverages the sentiments of online news for the task of remaining time prediction. Using our prototypical implementation, we carried out experiments that demonstrate the usefulness of this approach and allowing us to draw conclusions about circumstances in which it works best.

Keywords: Predictive process monitoring · External context
Sentiment analysis of news

1 Introduction

Business Process Monitoring is the phase of the Business Process Management lifecycle [7] that relates to the identification and assessment of issues and opportunities for runtime process improvement. Predictive monitoring [14] is concerned with predicting how a process is likely to proceed, covering questions such as what will be remaining time until process instance finishes, the next activity to expect or the outcome. These methods typically build on statistical methods or machine learning techniques and execution data of singular or multiple cases. It is a fundamental problem in process prediction that process execution only partially depends on internal factors.

On the other hand, contextual information has brought benefits in many scenarios of Business Process Management [1,23], since it provides knowledge

© Springer Nature Switzerland AG 2018
H. Panetto et al. (Eds.): OTM 2018 Conferences, LNCS 11229, pp. 586–603, 2018.
https://doi.org/10.1007/978-3-030-02610-3_33

related to the goal, organization, and environment of the business process [17]. In real life, business processes are complex and involve collaboration of the participants who interact outside of the system and with events of the world. Processes can be dependent on external factors, such as weather, a population density of the place where the process is running, or even the psychological climate under which participants interact are facts that are not reflected in the event logs. The lack of techniques to encode these external factors and enrich event logs prevent a more accurate process analysis.

In this paper, we address the challenge of enriching the process event log with an external context in order to improve the analysis of the process performance. Furthermore, due to the broad dissemination of information in the current days, we argue that a relevant contextual information is a sentiment about media content published while the process instance is running consequently influencing the group of process participants involved in this instance. Our contribution is a technique that enriches the event log with the context that affects activity execution in an encoding for process prediction. A comparison of predictors of the remaining time of an instance learned based on a pure event log and based on an event log enriched with external context was performed. Sentiment about web news was considered a source of context. The potential of the approach was shown through an evaluation using a real dataset considering news from a relevant newspaper.

This paper is structured as follows. Section 2 describes the research problem along with its requirements and summarizes prior research. Section 3 details our proposal. Section 4 evaluates its applicability on real-world event logs and news collected from a newspaper with broad impact. Section 5 concludes the paper and provides future directions.

2 Background

2.1 Problem Statement

In this paper, we focus on a specific class of processes which heavily rely on manual (human executed) tasks. In this type of processes external events may interfere with the workload, i.e. they change the way the ta are performed. This change may impact business process performance indicators which are quantifiable metrics focused on measuring the progress towards a goal or strategic objective aimed to control and improve the business process performance [22]. Some examples of these metrics are the remaining execution time of a process or the outcome including the likelihood of a fault in the system or abnormal termination of a running instance. As an example, consider the health insurance claim process depicted in Fig. 1.

This process starts with a claim on behalf of a patient done by a medical center attendant. The patient is asked to hand over some documents required by the health insurance office. Then, these documents are assessed for conformance. Once the claim is approved, the patient is allowed to receive health treatment

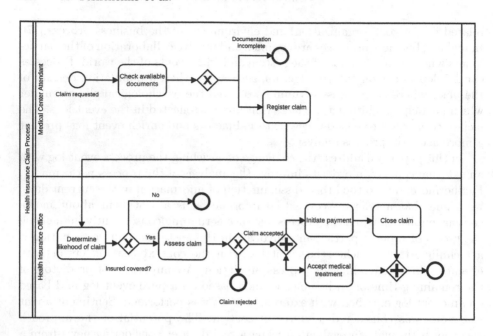

Fig. 1. Health insurance claim process

and the process finishes. Patients are usually in a state of anxiety and they urge for fast treatment, thus a plausible indicator is the remaining execution time.

Human behavior driven by specific events may affect such kind of processes. For example, recently in Brazil, the occurrence of an unknown disease but with similar symptoms to a typical harmless disease brought a state of alert in the population. People observing these symptoms immediately run to a medical center. As a consequence, the medical center became crowded and the health insurance claim process was delayed, sometimes even collapsed. In this regards, monitoring these news and evaluating the impact on the process would allow the process managers to anticipate a possible collapse and mitigate it: e.g., by speeding up process execution by reducing some document checking or by using additional resources.

On the other hand, *predictive business process monitoring* is an emerging research area [14] aimed at giving runtime insights about ongoing process executions. Process information contained in the event logs is used in order to make predictions about the process indicators such as remaining time [31] until completion of the case, next activities to be performed in the process [4], or the outcome [28]. Process monitoring proved itself to be a valuable asset when planning and managing resources and making customer-oriented decisions. In this regards, process prediction may benefit from considering external events. In our scenario prediction considering the sentiment about the online mention reporting the new disease could allow the health insurance office to prepare itself for the higher demand for medical treatment in the next weeks.

The research question addressed by this paper is:

RQ: How to enrich event log with sentiments from media content in order to increase the accuracy of predictive monitoring?

2.2   Related Work

Verenich et al. [31] affirm that diverse predictive process methods have been already proposed for different prediction goals, like an outcome, the next activity of a process instance or performance indicators, e.g., the remaining cycle time of a process instance. The last one is the focus of our paper.

Besides, the analysis of literature shows that process prediction has considered different data input [15]. These data define the *context* [2] in which the process is running and therefore are used to improve its execution. *Process context* is the knowledge potentially relevant to the execution of the process, available at the start of the execution of the process, and not impacted via the execution of the process [25]. This knowledge can be *internal* to the process, as the level of priority of a determined activity, or *external*, such as the weather at the location where the process occurs. The knowledge can be accessible in a *structured* format, such as, the resource who executed some activity; in an *unstructured* format, such as, textual messages exchanged by the resources while executing the process; or in *semistructured* format, such as the seasonal changes for the number of applications, or a priori knowledge for particular process instances about external events, such as staff sick leave.

Table 1 provides a summary of researches classified according to the type of data used. We can observe that the majority of the works on process prediction relies on knowledge internal to the process and stored in a structured format.

Table 1. Related process prediction papers

Approach	Association to the process	Type of data
Senderovich et al. [24]	Internal	Structured
Frey et al. [9]		
Verenich et al. [30]		
Di Francescomarino et al. [3]		
Navarin et al. [18]		
Marquez-Chamorro et al. [16]		
Polato et al. [21]		
Maggi et al. [14]		
Polato et al. [20]		
Teinemaa et al. [27]	Internal	Structured and unstructured
Folino et al. [8]	Internal	Structured and semistructured
Di Francescomarino et al. [4]		

The works [4, 8, 24] use the context of the process, such as the overall process performance at the moment [24], workload and seasonal changes to the process [8], and the unexpected but known a priori onetime process changes [4]. Although these works explore the use of unconventionally available data, they do not explore information from external sources.

We argue that other knowledge may affect process execution, especially those external to the process. For instance, processes heavily dependent on people may be impacted by events provoked by people such as the increase in health care demand. Moreover, this external context is not always available in a structured format. Therefore, in this paper, we address the gap of external unstructured context for process prediction.

3   Proposed Technique

We propose a technique that extracts external unstructured context found in newspapers or other media such as twitter to enrich the business process logs in order to improve the predictive monitoring results. The technique takes as input an event log and produces an enriched log appended with a media content sentiment. The overall process is depicted in Fig. 2 and consists of five main steps. The first one involves the extraction of process metadata to be used for collecting the appropriate media content, which is done in the second step. Once the content is retrieved, their sentiment is extracted and used to enrich the event log in steps three and four respectively. The process finishes with a prediction model being learned from the enriched log.

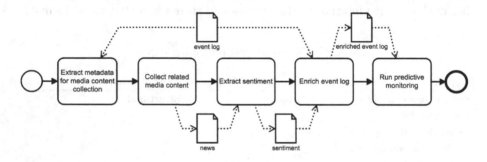

Fig. 2. Our proposed technique to enrich an event log with sentiment.

In the following sections each of the steps of our technique is detailed.

3.1   Extract Metadata for Media Content Collection

An event log is a set of recorded traces, that corresponds to process executions. Each trace consists of a multi-set of time-ordered activities and can be enriched with data defining the context in which the process is executed.

Table 2 depicts an example of a trace for the process described in Fig. 1. This trace, *Case*01, consist of the activities executed alongside their execution time (*timestamp*). Moreover, this trace provides internal context, such as, the location where the process runs and the resource that executed each of the activities.

Table 2. An example of trace indicating the sequence of activities executed in each time and enriched with contextual data (location and resource)

Case ID	Activity	*Timestamp*	*Location*	*Resource*
Case01	Request claim	15.12.2017	Rio De Janeiro	Customer 1
Case01	Check available documents	16.12.2017	Rio De Janeiro	Hospital manager 1
Case01	Register claim	16.12.2017	Rio De Janeiro	Hospital manager 1
Case01	Determine likelihood of claim	25.12.2017	Rio De Janeiro	Insurance staff 1
Case01	Insurance is not covered	25.12.2017	Rio De Janeiro	Insurance staff 1

Our aim at this step of the technique is to extract metadata about the business process in question as to guide the collection of media content that may interfere in the process. This metadata should contain information about the *time interval* of the process to consider, *search keywords* for gathering content and *location* information. Furthermore, *media content sources* are also defined.

The first and the last available timestamps in the event log define the *time interval*. *Location* information is either directly gathered from the event log, if it is provided with a context, or from another source such as a process description. *Media content sources* and *search keywords* are extracted manually, reviewing the event log, descriptions of the process, and consulting the process manager. Table 3 depicts the metadata extracted for the health insurance claim process of the Fig. 1.

Table 3. Metadata of the illustrative process of Fig. 1.

Process log	Time interval	News keywords	Location	Media content sources
Health insurance claim	01.2015 07.2018	Unknown disease medical situation hospitals	Brazil	New York news O Globo Folha de S. Paulo

3.2   Collect Related Media Content

In this step, we collect external contextual data as the information from online media sources. The aim of the technique is to find related online content to the process in question. For that, the use of automated web data extraction methods is suitable, as a method of copying particular information from the web by means of a web crawler or a bot [29].

For each collected news our approach saves its unique *timestamp, location* the content refers to, and the *body* of the text of the web post itself. A media content information item then is represented as a 3-tuple *MediaContent* = *<timestamp, location, body>*. Table 4 depicts examples of news collected considering the metadata showed in Table 3.

Table 4. Sample of media content collected based on the metadata from Table 3

Timestamp	Location	Text body
31.12.2015	Brazil	"Alarm Spreads in Brazil Over a Virus....Brazil a little-known virus.." [a]
20.03.2018	Brazil	"Fearing New Outbreaks, Brazil Will Vaccinate....Hoping to stave off another deadly.." [b]

[a] https://www.nytimes.com/2015/12/31/world/americas/alarm-spreads-in-brazil-over-a-virus-and-a-surge-in-malformed-infants.html.
[b] https://www.nytimes.com/2018/03/20/world/americas/yellow-fever-brazil-vaccinate.html.

3.3   Extract Sentiment

The media content as gathered in Table 4 describe an event, that for a given time, in a given location, may have an influence on how people behave or feel towards a particular phenomenon. That, in turn, might have an influence on business processes that depend on people. For the task of identifying how the news might influence people, we use techniques from *sentiment analysis*.

Sentiment analysis [13], or also called *opinion mining* is the field of computer science that analyzes people's opinions, sentiments, attitudes and emotions towards subjects from the text. Opinions are very important to organizations and businesses, for analysis and understanding of public opinions. While businesses could improve a product by knowing the opinion about their offerings, the governments could monitor public opinions about new policies.

In our work, we use sentiment analysis techniques to extract and associate quantifiable scores to the collected textual information. For the purposes of our technique, we define sentiment scores that also hold information about a place and the time when the emotion happened. Thus, for each news uniquely identified by its timestamp and location, we associate sentiment scores based on the analysis of the news body. These scores are defined by different techniques that based on their learning methods are able to retrieve different sentiments, henceforth called sentiment types [13].

Usually, sentiment analysis techniques use supervised machine learning models for building models. As input, a pre-tagged word-based corpus with sentiment is used. Most algorithms use word counting techniques to extract features for building sentiment extraction models. Usually, Naive Bayes, Decision trees or Neural Networks are the choices for the models [13]. Due to the complexity of training data collection, and since sentiment analysis techniques have been broadly used the news as input, we decided against our own sentiment extraction algorithm implementation and favored third-party techniques. There are many open-source and free-to-use solutions available[1].

We formally define media content sentiment as a n-tuple ζ = $<timestamp, location, s_1, s_2, ..., s_N>$ where $s_i \in [-1, 1]$ is a real valued number that represents the sentiment as a scale from negative to positive for N different sentiment types.

Table 5 depicts sentiment scores for the data illustrated on Table 4. In this example the sentiment for the first item is highly negative while for the second one is more close to neutral.

Table 5. Sample sentiment scores for media content in Table 4.

News sent.	Timestamp	Location	s_1	s_2
ζ_1	31.12.2015	Brazil	−0.865	−0.954
ζ_2	20.03.2018	Brazil	−0.322	−0.199

3.4   Enrich Event Log

In this step we describe three algorithms that we propose towards enriching the event log with sentiments from media content, i.e. with external contextual information. The input for the *enrichment algorithms* are an event log L with the set of traces $\sigma \in L$, and a set of media content sentiment $Z = \{\zeta_i \mid i \in \overline{1, M}\}$, where M is the number of items collected. The difference among the three algorithms concerns the time window used for aggregating sentiment: (i) media content in the same day, (ii) collected information happening since previous event occured and (iii) from k days before, independently of previous events. As output for the algorithms we have the enriched event log enL.

The first technique to enrich the log with media content sentiment is considering the articles and mentions aired on the same day as the business event ocurred. As an example, consider Fig. 3 which shows 18 days of a month on the top row, and the events happening on the 5th, 10th and 13th of the month. The *same day* technique enriches these three events with the corresponding sentiment from news happening in these three days. Algorithm 1 details how the technique works. For each activity of each trace the sentiment score of the new information published in the same day is included in the log. If sentiment score for the corresponding day is not found, we search for the content in previous days from the same location. The first day to have the sentiment is the one considered and we search at most k days before. If no sentiment score is found than the log is enriched with zeros meaning a neutral sentiment. Note that the function $date()$ present in the algorithm extracts a day out of the activity's or sentiment's timestamp.

---

[1] https://www.npmjs.com/package/wink-sentiment
     http://text-processing.com/demo/sentiment/.

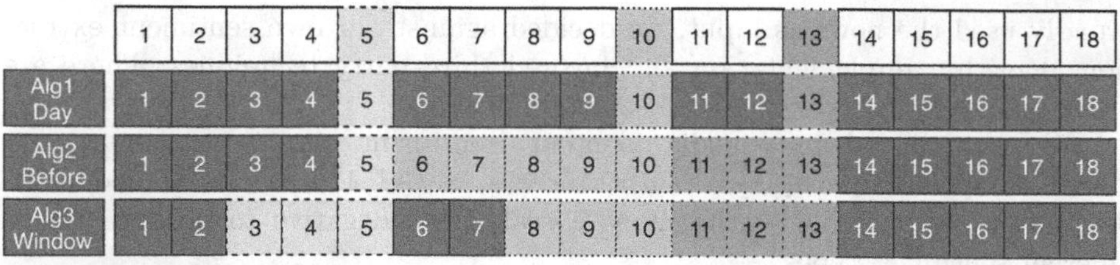

Fig. 3. Enrichment methods comparison

Input: L, Z
Output: SAME DAY enriched log enL

1 $enL = L$;
2 **for** $\sigma \in enL$ **do**
3   **for** $a_i \in \sigma$ **do**
4   **if** $\exists \zeta_i \in Z$ *such that* $(date(a_i), location(a_i)) = (date(\zeta_i), location(\zeta_i))$
            then
5              $\mid$ enrich σ with sentiment scores of ζ_i;
6   **else**
7              $Z_{aux} = \{\zeta_j | date(\zeta_j) \in [date(a_i) - k_{days}, date(a_i)] \wedge location(a_i) = location(\zeta_j)\}$;
8   enrich σ with sentiment scores of ζ_j with max timestamp from Z_{aux};

9 **return** enL

Algorithm 1. SAME DAY enrichment

The second technique to enrich the event log is based on the assumption that all events that happen before the event, have an influence on it. Third line in Fig. 3 illustrates how this technique works. All days after the first event on the 5th and before 10th have an influence on the event on the 10th.

Algorithm 2 details how the second technique works. The main difference with respect to the Algorithm 1 is the aggregation of sentiment scores from the days before the event until the previous event (lines 7–9). For the enrichement considering the first event Algorithm 1 is called (line 5).

The third technique proposed assumes that an event in a particular day is influenced by online media content from few days before. The fourth line at Fig. 3 illustrates the WINDOW cumulative enrichment with the windows size of 3. The algorithm for this technique is similar to the Algorithm 2. The only difference is the interval of days to be considered, thus in line 7 the interval is changed to $[date(a_i) - window\_size, date(a_i)]$, where the $window\_size$ is a parameter.

Table 6 illustrates each of the three techniques.

3.5 Run Predictive Monitoring

Predictive monitoring is separated in two phases: training machine learning prediction model, that relies on the enriched event log, and inference on the ongoing trace, when the prediction model is used to anticipate a process indicator.

Input: L, Z
Output: BEFORE enriched log enL
1 $enL = L$;
2 **for** $\sigma \in L$ **do**
3 | **for** $a_i \in \sigma$ **do**
4 | | **if** a_i *is the first event in the* σ **then**
5 | | | call Algorithm 1;
6 | | **else**
7 | | | **for** $day \in [date(a_{i-1}), date(a_i)]$ **do**
8 | | | | s_{day} = sentiment scores of ζ_i, where $(day, location(a_i)) = (date(\zeta_i), location(\zeta_i))$;
9 | | | enrich σ with sentiment scores of average s_{day} ;

10 **return** enL

Algorithm 2. BEFORE enrichment

Table 6. Example of the event log enrichment

(a) Media content sentiment scores

Sent.	Timestamp	Location	s_1	s_2
ζ_1	10.07.2018	Rio De J.	-0.42	-0.65
ζ_2	11.07.2018	Sao Paulo	0.12^5	-0.03^5
ζ_3	11.07.2018	Rio De J.	-0.22	-0.42
ζ_4	12.07.2018	Rio De J.	0.54	0.77
ζ_5	13.07.2018	Sao Paulo	$-0.77^{1/3/5}$	$-0.91^{1/3/5}$
ζ_6	13.07.2018	Sao Paulo	$-0.58^{1/3/5}$	$-0.62^{1/3/5}$
ζ_7	14.07.2018	Sao Paulo	0.01^4	0.46^4
ζ_8	15.07.2018	Rio De J.	-0.38	-0.41
ζ_9	16.07.2018	Sao Paulo	$0.26^{4/6}$	$-0.19^{4/6}$
ζ_{10}	17.07.2018	Rio De J.	0.35	0.28
ζ_{11}	17.07.2018	Sao Paulo	$-0.97^{2/4/6}$	$-0.81^{2/4/6}$

(b) Example of enriched log

Algorithm	Activity	Timestamp	Location	s_1	s_2
None	Req. Claim	13.07.2018	Sao Paulo		
	Check Doc.	17.07.2018	Sao Paulo		
SAME DAY	Req. Claim	13.07.2018	Sao Paulo	-0.68^1	-0.77^1
	Check Doc.	17.07.2018	Sao Paulo	-0.97^2	-0.81^2
BEFORE	Req. Claim	13.07.2018	Sao Paulo	-0.68^3	-0.77^3
	Check Doc.	17.07.2018	Sao Paulo	-0.23^4	-0.18^4
WINDOW	Req. Claim	13.07.2018	Sao Paulo	-0.41^5	-0.52^5
$w\_size = 3$	Check Doc.	17.07.2018	Sao Paulo	-0.36^6	-0.50^6

Predictive monitoring positions itself amongst *supervised* learning problems. For the task of remaining time prediction, for instance, the training data is represented as $D = \{(x_1, y_1), (x_1, y_1), ..., (x_n, y_n), n \in \mathbb{N}\}$ and the task is learning a *regression* function $f(x, \theta) \approx y$, that for each trace based on the feature vector x_i finds the best representation for the remaining time y_i. Usually, multiple feature vectors are extracted from each trace. For remaining time prediction, the feature vectors represent the k-first events of the trace called *prefix* of size k, along with its remaining time y. What is needed as the model should be able to generalize the prediction at any moment after the start of the process.

In [31] different encodings to represent an *event log* as a set of feature vectors x are presented and compared for the task of remaining time prediction. In the Table 7 the example of the index-based encoding is shown. The Request claim and Check documents activities are abbreviated as A and B respectively.

In this step, predictive monitoring algorithms are used to learn a predictor for the enriched event log and evaluated on the testing set of traces.

Table 7. Example of the event log encoding

(a) Example of enriched log

Activity	Timestamp	Location	s_1	s_2
Req. Claim	13.07.2018	Sao Paulo	-0.68	-0.77
Check Doc.	17.07.2018	Sao Paulo	-0.97	-0.81
Assess cl.	29.07.2018	Sao Paulo	0.19	0.03

(b) Enriched event log with index-based encoding

Activ_1	Time_1	Loc_1	s_1_1	s_2_1	Activ_2	Time_2	Loc_2	s_1_2	s_2_2	Remtime
A	13.07.2018	1	-0.68	-0.77	-	-	-	-	-	16 days
A	13.07.2018	1	-0.68	-0.77	B	17.07.2018	1	-0.97	-0.81	12 days

4   Evaluation

In this section we evaluate our technique. Section 4.1 describes the event logs used. In Sect. 4.2 we describe the external data and the enrichment of the event logs and in Sect. 4.3 we empirically evaluate our method.

4.1   Event Logs

Our approach focuses on the business processes that might be influenced by external events and news. Therefore for a thorough evaluation, we selected event logs about processes that depend on people behavior. For that, we analyzed publicly available logs[2]. The selection procedure is to identify if: A business process is directly dependent on customer behavior; a process includes many manual activities; the log has a potential for the use of predictive monitoring techniques. Following these requirements 4 real-life logs were chosen. We briefly describe the characteristics and their suitability below.

BPI2012 [5] and **BPI2017**[6]: event log describing the application process for a personal loan or overdraft from a Dutch Financial Institute. The latter considers a redesigned process and a different time period. We deemed these logs suitable for our evaluation as the process depends on the applying customers for the bank loans. On the other hand, the process also depends on the overall economic situation. Therefore we hypothesize that news about economics, business, taxes, etc. might help in understanding how the market for personal loans behaves.

BPI2013 [26]: real-life log from Volvo IT Belgium VINST system. This event log has data from *Handle incidents process*, which its main purpose is to timely restore normal service operation. This system operates within European countries. We chose this event log, since it contains many manual activities, and depends on the inflow of customers. News about the car industry (such as problems with cars, vehicle recalls, etc.) might influence customers, leading them to panic, and therefore to the increased inflow of incidents reported, or created.

---

[2] https://data.4tu.nl/repository/collection:event_logs_real.

Road Traffic Fine. Management process [12]: event log from Italian road police that describes processes about managing traffic fines. It contains events related to notifications, payments, and appeals. The event log is suitable for our research since it has a customer-oriented nature, i.e. the presence of many people-depended activities. It is also interesting due to its long timespan, around 13 years. Even though the log is specific to Italy, we included the search for general news about traffic fine policies in European Union, since those also influence the process, as Italy adheres to the general European traffic guideline changes.

Table 8 summarizes the features of these logs and the metadata extracted. For the extraction of the metadata, we consulted the descriptions of the log for keyword selection. For other information, the Disco process mining[3] was used.

Table 8. The four event logs considered for evaluation with metadata

Log	BPI2012	BPI2013	BPI2017	Road Traffic Fine
Traces	7554	13087	31508	150370
Mean case duration	12.1 days	8.6 days	21.9 days	48.8 weeks
Number of activities	13	6	26	11
Time interval	10.2011 - 03.2012	01.2010 - 06.2012	01.2016 - 02.2017	01.2000 - 07.2013
News keywords	finances debt personal loan overdraft tax break economics public finance loan economy technology business politics education health	Volvo car industry car technology Vinst car problems Volvo recall economy technology business	as BPI2012	traffic ticket traffic fine illegal parking road traffic fines road fine traffic policies vehicle policy traffic rules
Region	The Netherlands Dutch EU policies EU economics	All European countries	as BPI2012	Italy EU

4.2 Event Logs Enrichment

In this section, we describe implementation details of the proposed technique. For collecting the news we developed a specialized software full source code available on github[4]. We considered the New York Times online news source, due to its

---

[3] https://fluxicon.com/disco.
[4] https://github.com/FernandoDurier/News-Sentiment-Analyzer.

reliable and global coverage. The newspaper has a public News Archive API. A callable RESTful HTTP POST that required a private development key, year and month as an input. This endpoint returns a set of metadata (headline, abstract, author, news desk, publication time etc.). We developed a Node.js script for news collection. Table 9 summarizes information about the media content collected. Noticeably, the biggest set of keywords and amount of news collected per day were found for the processes described by BPI2012 and BPI2017.

Table 9. Collected media content

Log	BPI2012	BPI2013	BPI2017	Road traffic fine
News collected	2092	2395	5292	54289
Number of articles per day	12.53	2.63	13.33	10.95

For different types of sentiment extraction we used supervised methods based on the pre-tagged AFFIN165 [19] word list, implemented in the npm sentiment[5] (later as *npms*) and npm wink[6] (*npmw*) modules. Implementation relies on the text from the articles comparing each word in the text to the ones in the AFFIN list and determining its score. After scoring each word in the text the sentiment is calculated. All data are then grouped according to their locations, and the timestamp while averaging the sentiment scores. At last, the data is saved in the CSV format, following the definition of the news sentiment presented in Sect. 3.3.

Techniques SAME DAY, BEFORE and WINDOW were developed in Python.[7] We used the window size of 5 for the WINDOW enriching technique in order to include the strong sentiment of the news that happen over a weekend and might be otherwise discarded because business processes function during working days (newspapers publish more news over a weekend than weekdays).

Table 10 depicts an example of the enrichment for BPI2017. One instance of the news collected was about *the Netherlands potential of becoming a new financial capital of Europe*[8] (see Table 10a). Table 10b shows the sentiment score extracted, and Table 10c the corresponding enriched event log excerpt using SAME DAY algorithm.

4.3 Evaluation with Predictive Monitoring

In order to thoroughly evaluate the proposed technique of process log enrichment, we use state of the art algorithms of remaining time prediction of business processes. We choose the best-performing methods and parameters according

---

[5] https://www.npmjs.com/package/sentiment.

[6] https://www.npmjs.com/package/wink-sentiment.

[7] https://github.com/yesanton/Context-aware-predictive-process-monitoring-the-impact-of-news-sentiment.

[8] https://www.nytimes.com/2016/07/01/business/after-brexit-finding-a-new-london-for-the-financial-world-to-call-home.html.

Table 10. Example of the enrichment process output

(a) Sentiment

Timestamp	Location	Text of an article
01.07.2016	The Netherlands	*"The race is on to be the new London. Unless Britain finds a way.."*

(b) Media content sentiment

Timestamp	Location	npms	npmw
01.07.2016	The Netherlands	-0.134	0.280

(c) Activity enriched

CaseID	Activity	Resource	Timestamp	npms	npmw
Application_315416169	W_Call incomplete files	User_128	1.07.2016 10:44:34	-0.134	0.280

to [31]. They are summarized in the Table 11. We also considered the availability of implementation of the models and the time of training multiple models on the conventional CPU for the possibility of experiments reproduction. As of [31] we chose the Extreme gradient boosting model (XGboost), which is an ensemble method that combines many decision trees for efficiency and accuracy of the results. In turn, decision tree uses a tree-like structure to build classification or regression models. A tree structure is used to partition datasets into smaller and smaller subsets of decisions based on questions to the input data.

We used XGboost in conjunction with index-based (lossless, considers all information about events and also internal context), and last-state (considers all information of the sequence of events, and only the last event internal context) encodings for event logs. For building models and evaluation we used the open source implementation of Nirdizati[9] predictive process monitoring suite [10].

The model trained on the non-enriched event log is used as baseline for comparison with the models learned from the enriched logs resulted from SAME DAY, BEFORE, WINDOW techniques. Mean absolute error (MAE) was considered as metric for evaluation as it is being a standard for prediction task.

The result of the combination of techniques and logs is depicted in Table 12. As usual, some traces have fewer activities than the others, there are fewer traces for longer prefixes, so the results of MAE presented are weighted averaged for all prefixes. Figure 4 depicts MAE for different prefix size and different algorithms.

The Table 12 shows that proposed techniques outperform the baseline for all logs with *last-state encoding* and most logs with *index-based encoding*. Enriching algorithm BEFORE has an impressive performance, having best MAE for most cases. This technique performs best with *BPI2013* and *Road Traffic Fine* logs. Both of the logs have relatively few distinct activities (see Table 8), and *Road Traffic Fine* log has a very long average time of a case. That can explain BEFORE has the promising performance, as the impact of the news is settling down on the process actors long before activities are to be executed. *Road Traffic Fine* has a small improvement with the WINDOW and SAME DAY because *Road Traffic Fine*

---

[9] https://github.com/nirdizati/nirdizati-training-backend.

Table 11. Evaluation parameters

Parameter	Explanation	Value
Core method	The method used for training	XGboost
Log encoding	Feature representation of the log (x, y)	index-based encoding last-state
n_estimators	Number of decision trees in the ensemble	400
Learning_rate	How the contribution of each tree in ensemble deteriorates over learning time	0.01
Subsample	Fraction of observation to be sampled for each tree	0.7
colsample_bytree	Fraction of features x_i to be sampled for each tree	0.7
max_depth	Maximum tree depth for the xgboost	5

Table 12. Results of the evaluation

Event log	Enrich. technique	Weighed MAE Last-state encoding	Weighed MAE Index-based encoding	Event log	Enrich. technique	Weighed MAE ± std Last-state encoding	Weighed MAE ± std Index-based encoding
BPI2012	None	5.963	6.237	BPI2017	None	9.449	**9.214**
	SAME DAY	**5.281**	5.859		SAME DAY	9.430	9.226
	BEFORE	5.310	**5.850**		BEFORE	**9.429**	9.229
	WINDOW	5.352	5.925		WINDOW	**9.429**	9.239
BPI2013	None	9.945	8.467	Road Traffic Fine	None	217.119	202.655
	SAME DAY	11.168	10.904		SAME DAY	212.889	205.353
	BEFORE	**5.689**	**4.912**		BEFORE	**203.107**	**196.805**
	WINDOW	8.926	8.7821		WINDOW	207.715	198.222

process has few activities per trace while a long duration of the case so one time events (or for a small window size of 5) have ineligible influence. *BPI2012* and *BPI2017* being of the same nature, also behave similarly in terms of improvements with enrichment techniques. These logs show a good performance of the SAME DAY, that is understandable since these processes are executed swiftly (with the average case duration of 12.1 and 21.9 days respectively) while having many events per case. That could identify a fast change of activities and dependence on the news.

As a general result, we can observe that on the examples of *BPI2012* and *BPI2017*, the cases occur very fast, with many events, might be less of a candidate for the proposed techniques, while the cases that take longer time (say from a month per case) and have fewer activities could benefit greatly, as examples of the *Road Traffic Fine* and *BPI2013* show.

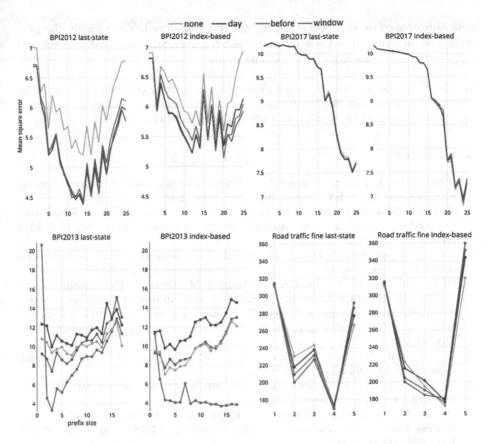

Fig. 4. Prediction error for different prefix lengths (Measured in MAE)

5   Conclusions

In this paper, we presented an approach for automatic event log enrichment for business process predictive monitoring tasks based on sentiment from online media content. The paper includes techniques and guidelines for collecting media content, extracting the sentiment, as well as three novel algorithms for log enrichment. Finally, the experiments based on the remaining time prediction task show the benefits and a potential of the approach. We found that our approach based on the BEFORE algorithm performs consistently better than a baseline, therefore proving that an external context is influential for the process, and its potential for the process mining tasks should be utilized.

Future works include evaluation with other types of acquiring external contextual information (social media, blogs, twitter, etc.). The research on crisis management with microblogging [11] can be explored in the context of business processes. Furthermore, a deep analysis of sensitivity of the approach regarding the set of keywords and source of news chosen. Also, other types of prediction algorithms based on enriched event log are to be explored. We envision the possibility of building an automatic framework for obtaining external context and support business processes with runtime monitoring.

Acknowledgements. This work is partially funded by the EU H2020 program under MSCA-RISE agreement 645751 (RISE_BPM), FFG Austrian Research Promotion Agency (project number: 866270), and UNIRIO (PQ-UNIRIO N01/2018).

References

1. vom Brocke, J., Zelt, S., Schmiedel, T.: On the role of context in business process management. Int. J. Inf. Manag. **36**(3), 486–495 (2016)
2. da Cunha Mattos, T., Santoro, F.M., Revoredo, K., Nunes, V.T.: A formal representation for context-aware business processes. Comput. Ind. **65**(8), 1193–1214 (2014)
3. Di Francescomarino, C., Dumas, M., Maggi, F.M., Teinemaa, I.: Clustering-based predictive process monitoring. IEEE Trans. Serv. Comput. (2016)
4. Di Francescomarino, C., Ghidini, C., Maggi, F.M., Petrucci, G., Yeshchenko, A.: An eye into the future: leveraging a-priori knowledge in predictive business process monitoring. In: Carmona, J., Engels, G., Kumar, A. (eds.) BPM 2017. LNCS, vol. 10445, pp. 252–268. Springer, Cham (2017). https://doi.org/10.1007/978-3-319-65000-5_15
5. van Dongen, B.: BPI challenge 2012 (2012). https://doi.org/10.4121/uuid: 3926db30-f712-4394-aebc-75976070e91f
6. van Dongen, B.: BPI challenge 2017 (2017). https://doi.org/10.4121/uuid: 5f3067df-f10b-45da-b98b-86ae4c7a310b
7. Dumas, M., Rosa, M.L., Mendling, J., Reijers, H.A.: Fundamentals of Business Process Management, 2nd edn. Springer, Heidelberg (2018). https://doi.org/10.1007/978-3-662-56509-4
8. Folino, F., Guarascio, M., Pontieri, L.: Discovering context-aware models for predicting business process performances. In: Meersman, R. (ed.) OTM 2012. LNCS, vol. 7565, pp. 287–304. Springer, Heidelberg (2012). https://doi.org/10.1007/978-3-642-33606-5_18
9. Frey, M., Emrich, A., Fettke, P., Loos, P.: Event entry time prediction in financial business processes using machine learning: a use case from loan applications. In: Proceedings of the 51st Hawaii International Conference on System Sciences 2018, pp. 1386–1394. IEEE Computer Society (2018)
10. Jorbina, K., et al.: Nirdizati: a web-based tool for predictive process monitoring (2017)
11. Kireyev, K., Palen, L., Anderson, K.: Applications of topics models to analysis of disaster-related twitter data. In: NIPS Workshop on Applications for Topic Models: Text and Beyond, Whistler, Canada, vol. 1 (2009)
12. de Leoni, M., Mannhardt, F.: Road traffic fine management process (2015). https://doi.org/10.4121/uuid:270fd440-1057-4fb9-89a9-b699b47990f5
13. Liu, B.: Sentiment Analysis: Mining Opinions, Sentiments, and Emotions. Cambridge University Press, Cambridge (2015)
14. Maggi, F.M., Di Francescomarino, C., Dumas, M., Ghidini, C.: Predictive monitoring of business processes. In: Jarke, M., et al. (eds.) CAiSE 2014. LNCS, vol. 8484, pp. 457–472. Springer, Cham (2014). https://doi.org/10.1007/978-3-319-07881-6_31

15. Marquez-Chamorro, A.E., Resinas, M., Ruiz-Corts, A.: Predictive monitoring of business processes: a survey. IEEE Trans. Serv. Comput. **PP**, 1 (2017)
16. Marquez-Chamorro, A.E., Resinas, M., Ruiz-Corts, A., Toro, M.: Run-time prediction of business process indicators using evolutionary decision rules. Expert Syst. Appl. **87**(C), 1–14 (2017)
17. Metzger, A., et al.: Comparing and combining predictive business process monitoring techniques. IEEE Trans. Syst. Man Cybern.: Syst. **45**(2), 276–290 (2015)
18. Navarin, N., Vincenzi, B., Polato, M., Sperduti, A.: LSTM networks for data-aware remaining time prediction of business process instances. CoRR (2017)
19. Nielsen, F.Å.: AFINN, March 2011. http://www2.imm.dtu.dk/pubdb/p.php?6010
20. Polato, M., Sperduti, A., Burattin, A., de Leoni, M.: Data-aware remaining time prediction of business process instances. In: 2014 International Joint Conference on Neural Networks (IJCNN), pp. 816–823, July 2014
21. Polato, M., Sperduti, A., Burattin, A., de Leoni, M.: Time and activity sequence prediction of business process instances. CoRR (2016)
22. del Río-Ortega, A., Resinas, M., Cabanillas, C., Ruiz-Cortés, A.: On the definition and design-time analysis of process performance indicators. Inf. Syst. **38**(4), 470–490 (2013)
23. Rosemann, M., Recker, J., Flender, C.: Contextualisation of business processes. Int. J. Bus. Process. Integr. Manag. **3**(1), 47–60 (2008)
24. Senderovich, A., Di Francescomarino, C., Ghidini, C., Jorbina, K., Maggi, F.M.: Intra and inter-case features in predictive process monitoring: a tale of two dimensions. In: Carmona, J., Engels, G., Kumar, A. (eds.) BPM 2017. LNCS, vol. 10445, pp. 306–323. Springer, Cham (2017). https://doi.org/10.1007/978-3-319-65000-5_18
25. Sindhgatta, R., Ghose, A., Dam, H.K.: Context-aware analysis of past process executions to aid resource allocation decisions. In: Nurcan, S., Soffer, P., Bajec, M., Eder, J. (eds.) CAiSE 2016. LNCS, vol. 9694, pp. 575–589. Springer, Cham (2016). https://doi.org/10.1007/978-3-319-39696-5_35
26. Steeman, W.: BPI challenge 2013 (2013). https://doi.org/10.4121/uuid:a7ce5c55-03a7-4583-b855-98b86e1a2b07
27. Teinemaa, I., Dumas, M., Maggi, F.M., Di Francescomarino, C.: Predictive business process monitoring with structured and unstructured data. In: La Rosa, M., Loos, P., Pastor, O. (eds.) BPM 2016. LNCS, vol. 9850, pp. 401–417. Springer, Cham (2016). https://doi.org/10.1007/978-3-319-45348-4_23
28. Teinemaa, I., Dumas, M., Rosa, M.L., Maggi, F.M.: Outcome-oriented predictive process monitoring: review and benchmark. CoRR (2017)
29. Vargiu, E., Urru, M.: Exploiting web scraping in a collaborative filtering-based approach to web advertising. Artif. Intell. Res. **2**(1), 44 (2012)
30. Verenich, I., Dumas, M., La Rosa, M., Maggi, F.M., Di Francescomarino, C.: Complex symbolic sequence clustering and multiple classifiers for predictive process monitoring. In: Reichert, M., Reijers, H.A. (eds.) BPM 2015. LNBIP, vol. 256, pp. 218–229. Springer, Cham (2016). https://doi.org/10.1007/978-3-319-42887-1_18
31. Verenich, I., Dumas, M., Rosa, M.L., Maggi, F.M., Teinemaa, I.: Survey and cross-benchmark comparison of remaining time prediction methods in business process monitoring. CoRR (2018)

Deadlock-Freeness Verification of Cloud Composite Services Using Event-B

Aida Lahouij[1($\boxtimes$)], Lazhar Hamel[2], and Mohamed Graiet[3]

[1] ISITCOM, Sousse, Tunisia
aida.lahouij@gmail.com
[2] ISIMM, Monastir, Tunisia
lazhar.hamel@isimm.rnu.tn
[3] ENSAI RENNES, Bruz, France
mohamed.graiet@ensai.fr

Abstract. With the emergence of the Cloud computing paradigm, interests were focused on representing and verifying the Cloud architecture in a formal way in order to prevent eventual failures and deadlocks. Service composition promotes reuse, interoperability, and loosely coupled interaction. However, verifying the correctness of a composite service remains a tedious task. To ensure the correctness of a composition, critical properties such as deadlock freeness must be verified. In this paper, we propose a novel formal approach to verify Cloud composite services correctness based on the Event-B method. We consider that both behavioral incompatibilities and conflicted resource may lead the composite service execution to failure. Event-B provides rigorous mathematical reasoning that helps building trust on developed software. A verification and validation approach combining both model proofs and model checking is finally performed to check the soundness of the proposed model.

Keywords: Cloud · Composite service · Behavior
Resource provisioning · Resource management · Deadlock · Formal
Event-B · Verification

1 Introduction

Nowadays, service composition is a very tempting area of research specially with the emergence of cloud computing [13] and the additional challenges it brings. Service composition is an emerging technique to develop applications by composing existing services in order to build more powerful and complex ones. The result is called *composite service* and its constituting services are called *component services* [23]. Although there have been many contributions related to service composition, ensuring the deadlock freeness of the composite service behavior remains a challenge.

In fact, verifying the deadlock freeness of a composition usually relies on the formal verification of its behavioral properties [29] (e.g., interoperability, Reachability, Liveness, and Persistence). Such verification typically depends on the

© Springer Nature Switzerland AG 2018
H. Panetto et al. (Eds.): OTM 2018 Conferences, LNCS 11229, pp. 604–622, 2018.
https://doi.org/10.1007/978-3-030-02610-3_34

formal modeling of the composition behavior via a modeling language with clear semantics. This task is not trivial for Cloud managers since it involves several operations such as discovery, compatibility checking, selection and deployment. Similarly to a non Cloud environment, service composition raises the need for design-time approaches to check the correct interactions between the different components of a composite service. However, for Cloud-based compositions, specific constraints must be considered such as resources management, elasticity, and multi-tenancy. Several works have been carried out in this area. Some verification approaches are based on formal languages (e.g., Event-B [14], SOG [8,18], and Process Algebra [27], etc.). Although these approaches respect the majority of principles of the Cloud composite service, they mainly consider the resource allocation verification without verifying the composition correct behavior.

In line with our previous work [16,19], in this paper, we propose a formal approach based on the Event-B [7] method to model and verify Cloud service composition. Such verification is necessary, in order to avoid incorrect composition behavior and unnecessary execution for an erroneous composition. With First-order logic and set theory as underlying mathematical notation, the Event-B method allows us to specify and model software systems in a mathematically sound way. The use of formal methods is now a necessity to create reliable software for critical and complex systems.

Our approach is able to ensure the correctness of a composition through the verification of the deadlock freeness of the composite service which implicitly involves the following behavior properties (Interoperability, Reachability, Liveness, and Persistence). In other words, the proposed approach must:

- check if the resources involved into the composition can be linked together. This is related to data type compatibility between the linked resources where the output of a resource should be of the same type of the input of another resource.
- guarantee that the desired final composition state is reachable from the initial state.
- ensure that all component services participating in the composition will be invoked during composition execution.
- when parallel component services are executed simultaneously, ensure that the occurrence of one service will not disable another.
- check the resource allocation requirements.

As we can notice, these properties are complex, so the designed composition behavior at runtime can easily deviate from users' needs. To cope with the aforementioned problems, our approach considers the component services protocols by analyzing the several states they may move into. We adopt the bidirectional compatibility checking approach proposed in [11], to check the behavior compatibility between the N component services of the composite service. The proposed approach verifies also the resource allocation properties of the involved component services. Two levels of refinement have been performed, the first level models the behavior properties of the composition and the second level adds details on the resource properties.

The reminder of the paper is organized as follows; in the following section, we present our motivations. An overview of the Event-B method is performed in Sect. 3. Our formal model is introduced in Sect. 4. The verification approach is performed in Sect. 5. Section 6 presents a comparison with related works. Finally, we conclude and provide insights for future works, in Sect. 7.

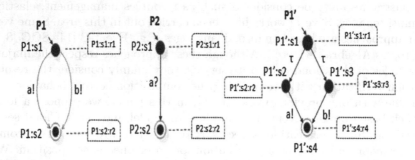

Fig. 1. P1 and P2 interoperate successfully, but P1' and P2 can deadlock

2   Motivations

Analyzing services protocols helps us to find out possible interoperability issues. Indeed, although one service can behave as expected by its partner from an external point of view, interoperability issues may occur because of unexpected internal behaviors that services can execute. For instance, Fig. 1 shows two versions of one service protocol with (P1) and without (P1') its internal behavior. As we can see, P2 and P1 can perfectly interoperate because each service can send (respectively receive) the messages expected (respectively sent) by its partner. However, if we consider P1' that describes what the service actually does. Executing the internal choice action τ, may move the state P1':s1 to state P1':s3 while P2 is still in state P2:s1. At this point, P1' and P2 cannot exchange messages, and the system deadlocks. This issue would not have been detected with P1. In this work, we introduce our approach to check component services behavior based on the bidirectional compatibility notion introduced in [11].

Analyzing allocated resources at each state is also necessary to ensure the correctness of the composition. Indeed, the Cloud environment provides three types of resources: computing, networking and storage. The used resources can be elastic or not. An elastic resource instance has the ability to change its capacity to accommodate the workload and handle concurrent requests. However, non-elastic resources are classical resources with fixed capacity that cannot be changed at runtime. A Cloud resource may also be shareable or non-shareable. A commonly shareable resource means that two or more services can be executed at the same time. An exclusively shareable resource can handle two or more services but not at the same time. For instance, if at the global state GS1, the resource needed by P1 at the local state P1:s1 is the same resource required by P2 at P2:s1 (P1:s1:r1 = P2:s1:r1), and if the resource is non-shareable or is exclusively shareable, the composite service execution deadlocks.

To overcome these problems, we propose an Event-B based formal approach that covers the following requirements: (1) Protocol matching verification: In order to verify the correctness behavior of the composition, the component services protocols are checked according to the bidirectional compatibility notion. (2) Resource allocation verification: deadlock must be avoided while allocating resources to the component service for example s_i is waiting for a resource held by s_j, which in turn, is also waiting for a resource held by s_i.

3 Overview of the Event-B Method

The B-method was developed by Jean-Raymond Abrial [3] and has been used in major safety-critical system applications in Europe such as the Paris Metro Line 14. It has robust, commercially available tool support for specification, design, proof and code generation. Event-B is an evolution of the B-method also called classical B [5]. Event-B [7] reuses the set-theoretical and logical notations of the B method and provides new notations for expressing abstract systems or simply models based on events [9]. Through sequential refinement, this formal method enables incremental development of software step by step from abstract level to more detailed levels and possibly to code level. The complexity of a system is mastered thanks to the refinement concept allowing to gradually introduce the different parts that constitute the system starting from an abstract model to a more concrete one. A stepwise refinement approach produces a correct specification by construction since we prove the different properties of the system at each step. Event-B is supported by the eclipse-based RODIN platform [6] on which different external tools (e.g. provers, animators, model-checkers) can be plugged in order to animate/validate a formal development.

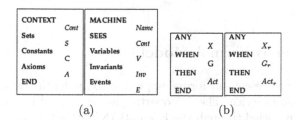

(a)   (b)

Fig. 2. The Event-B specification

An Event-B specification (Fig. 2) is made of two elements: context and machine. A context describes the static part of an Event-B specification. An Event-B context is optional and contains essentially the following clauses: the clause SETS that describes a set of abstract and enumerated types, the clause CONSTANTS that represents the constants of the model and the clause AXIOMS that contains all the properties of the constants and their types. A context can optionally extend another one by adding its name in the clause EXTENDS. A context is referenced by the machine in order to use its sets and constants by adding its name in the clause SEES. An Event-B machine describes the dynamic part of an Event-B specification. It is composed of a set of clauses organized as follows; the clause VARIABLES representing the state variables of the model, the clause INVARIANTS defining the invariant properties of the system that it must allow, at least, the typing of variables declared in the clause VARIABLES and finally the clause EVENTS containing the list of events related to the model. An event is modeled with a guarded substitution and fired when its guards are evaluated to true. The events occurring in an Event-B model affect the state described in the clause VARIABLES. A machine can optionally refine another one by adding its name in the clause REFINES.

An event consists of a guard and a body (Fig. 2b). When the guard is satisfied, the event can be activated. When the guards of several events are satisfied at the same time, the choice of the event to enable is deterministic.

Refinement is a process of enriching or modifying a model in order to augment the functionality being modeled, or/and explain how some purposes are achieved. Both Event-B elements context and machine can be refined. A context can be extended by defining new sets and/or constants together with new axioms. A machine is refined by adding new variables and/or replacing existing variables by new ones that are typed with an additional invariant. New events can also be introduced to implicitly refine a skip event.

Proof-based development methods integrate formal proof techniques in the development of software systems. The main idea is to start with a very abstract model of the system under development. We then gradually add details to this first model by building a sequence of more concrete ones. As such, an Event-B model is controlled by means of a number of proof obligations, which guarantee the correctness of the development. Our previous works have proven the Event-B method efficiency on the modeling and verification of several aspects of composite services [16, 19].

4   The Event-B Formal Model

In this section, we present our formal approach to model the behavioral properties of a composite service. These properties are expressed in terms of requirements that are modeled through the Event-B INVARIANTS and EVENTS.

Figure 3 depicts the formalization architecture of our Event-B model. Our model abstraction is provided in five levels; the machine StrucM0 sees the context StructC0 and models the structural properties of the Cloud service composition. SemM0 refines StrucM0 and introduces the semantic properties of the composition. BehM0 refines the SemM0 by adding the behavioral properties of the composition. At this level, we focus on the behavioral based service selection. The selection is generic and performed at design time. This machine sees BehC0

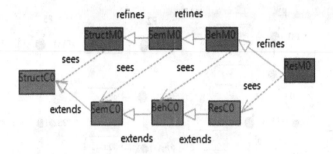

Fig. 3. CloudModel architecture

which extends SemC0. The machine ResM0, refines the BehM0, where details about resource allocation are added. At this level, we aim to avoid deadlocks in resource provisioning by defining adequate events. This machine sees ResC0 which extends BehC0.

In this paper, we content to present the two behavior and resource abstraction levels as follows:

- *BehM0* models the behavior requirements and verifies the behavioral properties of the composite service by checking the protocols matching of the component services.
- The machine *ResM0*, refines the *BehM0*, to model and verify the resource requirements.

4.1   Modeling the Behavior Requirements

At this level, we introduce our formal approach to verify the composite service behavioral requirements. This verification is performed at design time (i.e. the discovery and the selection of component services). We adopt the bidirectional compatibility (BC) notion introduced in [11] to check protocol matching. It is the most intuitive notion of compatibility which requires that when one service can send a message, there is another service which eventually receives that message, and when one service is waiting to receive a message, then there is another service which must eventually send that message. Furthermore, the protocols must be deadlock-free. Figure 4 shows three protocols that are not compatible with respect to BC. The incompatibility is detected at the global state (P1:s2, P2:s2, P3:s1), reachable from the initial state, since P1, P2 and P3 are not able to reach any global state in which message d at state P1:s2 can match. In other words, n services are bidirectional compatible if their protocols global states are compatible on the send and the receive interactions and are deadlock free. Therefore, we start by defining the global state of a composite service (*Beh1*).

Beh1. A global state is a global observation of local states and communication channels.

The sets *Services, Protocols, GSs, LSs,* and *Channels* are created in the context *BehC0* denoting respectively the services set, the protocols set, the global

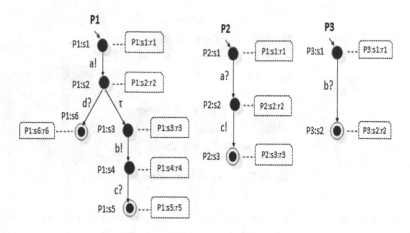

Fig. 4. The bidirectional compatibility

states set, the local states set and the channels set. Each service is represented in our model by its protocol. This relation is expressed by the total function in the axiom *axm1*. A total function is used since each service must have one protocol.

> **CONTEXT** BehC0
> **EXTENDS** SemC0
> **SETS**
>   Services, Protocols, LSs, GSs, Channels
> **CONSTANTS**
>   ProtOf, GP, LSsOf, GSsOf, LSsOfGS
> **AXIOMS**
>   **axm1:** $ProtOf \in Services \rightarrow Protocols$
>   **axm2:** $LSsOf \in Protocols \rightarrow LSs$
>   **axm3:** $GP \in \{ProtOf\} \rightarrow Protocols$
>   **axm4:** $GSsOf \in GP \rightarrow GSs$
>   **axm5:** $LSsOfGS \in GSs \rightarrow LSs$
>   **axm6:** $ChannelsOfGS \in GSs \rightarrow Channels$
> **END**

Each protocol has a set of local states (*axm2*) and global states (*axm4*). The set of local states of a global state are defined in the total function *LSsOfGS* (*axm5*). The channels of a global state are represented by the total function *ChannelsOfGS* (*axm6*). A global protocol is the protocol composed of the n services protocols (*axm3*).

Back to the bidirectional compatibility notion, two protocols are compatibles if, at first, their global states are compatibles on the send interactions (*Beh2*). This requirement is modeled by the *CompatibleStatesSI* invariant.

Beh2. A global state GS_i is said to be Send_compatible ($GSi \in CompatibleSt-atesS$), if for each local state LS_{ik}: (a) (i) when a message m is sent at this state ($send \mapsto m \in ActionsOf(pk)$), it must be received by another service at state

LS_{il} at the same current global state ($\exists LSil, pl \cdot GSi \mapsto LSil \in LSsOfGS \land pl \mapsto LSil \in LSsOf \land receive \mapsto m \in ActionsOf(pl)$). Or, (ii) there exists a global state GS_j reachable from GS_i ($\exists GSj, LSjl, pl \cdot GSi \mapsto GSj \in ReachableGSs$), where m is received at its local state LS_{jl} ($GSj \mapsto LSjl \in LSsOfGS \land pl \mapsto LSjl \in LSsOf \land receive \mapsto m \in ActionsOf(pl)$), (b) the queue q of the receiving protocol must be able to accept m ($EffSizeOfQ(q) \leq SizeOfQ(q)$), and (c) the average rate of receiving m is lower than the delay defined by LSjl ($AvArRateOf(m) \leq DelayOf(LSjl)$).

MACHINE BehM0
REFINES SemM0
SEES BehC0
INVARIANTS
> CompatibleStatesSI: $\forall GSi \cdot GSi \in GSs \land GSi \in CompatibleStatesS \Rightarrow (\forall LSik, pk \cdot GSi \mapsto LSik \in LSsOfGS \land pk \mapsto LSik \in LSsOf \Rightarrow (\forall m \cdot send \mapsto m \in ActionsOf(pk) \Rightarrow ((\exists LSil, pl \cdot GSi \mapsto LSil \in LSsOfGS \land pl \mapsto LSil \in LSsOf \land receive \mapsto m \in ActionsOf(pl) \land (\forall q \cdot pl \mapsto q \in QueuOf \Rightarrow EffSizeOfQ(q) \leq SizeOfQ(q)) \land (AvArRateOf(m) \leq DelayOf(LSil))) \lor (\exists GSj, LSjl, pl \cdot GSi \mapsto GSj \in ReachableGSs \land GSj \mapsto LSjl \in LSsOfGS \land pl \mapsto LSjl \in LSsOf \land receive \mapsto m \in ActionsOf(pl) \land (\forall q \cdot pl \mapsto q \in QueuOf \Rightarrow EffSizeOfQ(q) \leq SizeOfQ(q)) \land (AvArRateOf(m) \leq DelayOf(LSjl))))))$

END

The second constraint that must be preserved by the n protocols, in order to be bidirectional compatibles, is that when a service is waiting to receive a message, then this message must eventually be sent (*Beh3*). *Beh3* is introduced in the invariant *CompatibleStatesRI*.

Beh3. A global state GS_j is said to be Receive_compatible ($GSj \in Compatible-StatesR$), if for each local state LS_{jl}: (a) (i) when it is waiting to receive a message m ($receive \mapsto m \in ActionsOf(pl)$), it must be sent at the local state LS_{jk} at the same current global state GS_j ($\exists LSjk, pk \cdot GSj \mapsto LSjk \in LSsOfGS \land pk \mapsto LSjk \in LSsOf \land send \mapsto m \in ActionsOf(pk)$). (ii) Or, there exists a global state GS_i reachable from GS_j ($GSi \mapsto GSj \in ReachableGSs$), where m is eventually sent ($GSi \mapsto LSik \in LSsOfGS \land pk \mapsto LSik \in LSsOf \land send \mapsto m \in ActionsOf(pk)$), (b) the queue of the pk is able to receive m ($EffSizeOfQ(q) \leq SizeOfQ(q)$), and, (c) the average rate of receiving m is lower than the delay defined by $LSjl$ ($AvArRateOf(m) \leq DelayOf(LSjl)$).

> **CompatibleStatesRI:** $\forall GSj \cdot GSj \in GSs \wedge GSj \in$
> $CompatibleStatesR \Rightarrow (\forall LSjl, pl \cdot GSj \mapsto LSjl \in$
> $LSsOfGS \wedge pl \mapsto LSjl \in LSsOf \Rightarrow (\forall m \cdot receive \mapsto$
> $m \in ActionsOf(pl) \wedge (\forall q \cdot pl \mapsto q \in QueuOf \Rightarrow$
> $EffSizeOfQ(q) \leq SizeOfQ(q)) \wedge (AvArRateOf(m) \leq$
> $DelayOf(LSjl)) \Rightarrow ((\exists LSjk, pk \cdot GSj \mapsto LSjk \in$
> $LSsOfGS \wedge pk \mapsto LSjk \in LSsOf \wedge send \mapsto m \in$
> $ActionsOf(pk)) \vee (\exists GSi, LSik, pk \cdot GSi \mapsto GSj \in$
> $ReachableGSs \wedge GSi \mapsto LSik \in LSsOfGS \wedge pk \mapsto LSik \in$
> $LSsOf \wedge send \mapsto m \in ActionsOf(pk)))))$

The last condition imposed by the bidirectional compatibility notion is deadlock freeness. The n interacting protocols must be deadlock-free i.e. their initial global state is deadlock free ($Beh4$). This requirement is modeled by the invariant $DeadLockfree$.

Beh4. A global state GS_i is deadlock free ($GSi \in DeadLockFree$) if: (a) its local states are finals ($\forall LSik \cdot GSi \mapsto LSik \in LSsOfGS \Rightarrow LSik \in FinalLSs$). Or, (b) there exists GS_j reachable from GS_i and is deadlock free (i.e. its local states are finals) ($\exists GSj \cdot GSi \mapsto GSj \in ReachableGSs \Rightarrow GSj \in DeadLockFree$).

> **DeadLockfree:** $\forall GSi \cdot GSi \in DeadLockFree \Rightarrow$
> $((\forall LSik \cdot GSi \mapsto LSik \in LSsOfGS \Rightarrow LSik \in FinalLSs) \vee$
> $(\exists GSj \cdot GSi \mapsto GSj \in ReachableGSs \Rightarrow GSj \in$
> $DeadLockFree))$

In order to verify the service compatibility, we need to check every global state that can be reached during system execution. Let us consider the service protocols given in Fig. 4. The set of global states which can be reached from (P1:s1, P2:s1, P3:s1) for P1, P2, and P3 is the following: (P1:s2, P2:s1, P3:s1), (P1:s2, P2:s2, P3:s1), (P1:s3, P2:s2, P3:s1), (P1:s3, P2:s2, P3:s2), (P1:s4, P2:s2, P3:s2), (P1:s4, P2:s3, P3:s2), (P1:s5, P2:s3, P3:s2). It provides the set of global states that n interoperating services can reach, in one or more steps, from a current global state $(s1, ..., sn)$ through synchronisations or independent evolutions ($Beh.5$). This requirement is defined in the invariant $ReachableGs$.

Beh5. A global state GS_j is reachable from another global state GS_i ($GSj \mapsto GSi \in ReachableGSs$), if starting from each local state of GS_i we can reach a local state of GS_j ($\forall LSi \cdot GSi \mapsto LSi \in LSsOfGS \Rightarrow (\exists LSj \cdot LSj \mapsto LSi \in ReachableLSs$).

> **ReachableGS:** $\forall GSj, GSi \cdot GSj \in GSs \wedge GSi \in$
> $GSs \wedge GSj \mapsto GSi \in ReachableGSs \Rightarrow (\forall LSi \cdot GSi \mapsto LSi \in$
> $LSsOfGS \Rightarrow (\exists LSj \cdot LSj \mapsto LSi \in ReachableLSs))$

The invariant $ReachableLs$ models the requirement $Beh6$ regarding local states reachability.

Beh6. A local state LS_j can be reached from a local state LS_i ($LSj \mapsto LSi \in ReachableLSs$): (a) if there exists a transition (action) from LS_i to LS_j ($\exists act \cdot$

$act \in Actions \wedge act \in dom(ActionsOf(pi)) \wedge act \in dom(ActionsOf(pj)))$. Or, (b) there exists a Local state LS_k reachable from LS_i where LS_j is reachable from LS_k ($\exists LSk \cdot LSi \mapsto LSk \in ReachableLSs \wedge LSk \mapsto LSj \in ReachableLSs$).

$$
\begin{array}{l}
\textbf{ReachableLS:} \quad \forall LSi, LSj, pi, pj \cdot LSj \mapsto LSi \in \\
ReachableLSs \wedge ProtOfS(LSi) = pi \wedge ProtOfS(LSj) = \\
pj \Rightarrow (\exists act \cdot act \in Actions \wedge act \in dom(ActionsOf(pi)) \wedge \\
act \in dom(ActionsOf(pj))) \vee (\exists LSk \cdot LSi \mapsto LSk \in \\
ReachableLSs \wedge LSk \mapsto LSj \in ReachableLSs))
\end{array}
$$

In the events clause, we have introduced the BC event to model the candidate services selection based on the bidirectional compatibility ($Beh7$).

Beh7. N services protocols are bidirectional compatible if their protocols global states are compatibles on the send and the receive interactions and are deadlock free.

$$
\begin{array}{l}
\textbf{Event BC } \langle\text{ordinary}\rangle \mathrel{\widehat{=}} \\
\textbf{any} \\
\quad c, s \\
\textbf{where} \\
\quad \textbf{grd4:} \quad \forall Gs, Gip, ip \cdot \{ip\} \mapsto Gip \in GP \wedge Ss \subseteq \\
\qquad dom(\{ip\}) \wedge \{ip\} \mapsto Gip \mapsto Gs \in GSsOf \Rightarrow Gs \in \\
\qquad CompatibleStatesR \wedge Gs \in CompatibleStatesS \wedge \\
\qquad Gs \in DeadLockFree \\
\textbf{then} \\
\quad \textbf{act1:} \ SerOf := SerOf \cup \{c \mapsto s\} \\
\textbf{end}
\end{array}
$$

This event ensures a service selection based on the behavioral properties. As introduced previously, only services whose protocols are compatible are selected. The N-compatibility based selection is performed to the set of services outcoming from the functional selection. The n services that their protocols preserves the invariants above ($CompatibleStatesR$, $CompatibleStatesS$ and $DeadLockFree$), are selected ($Beh7, grd4$). The obtained set is added to the set of services of the composition ($act1$).

4.2 Modeling the Resource Requirements

In this section, we define our second level of refinement that verifies the resource allocation properties. Constraints on resource allocation are introduced through Event-B invariants. At this level, we extend the context $BehC0$ by adding a new context $ResC0$ that contains sets and constants related to the resource perspective.

> **CONTEXT** ResC0
> **EXTENDS** BehC0
> **SETS**
>> Resources, RType
> **CONSTANTS**
>> ResOF, ResTOf, IsShareable, IsElastic, IsExcShar,
>> IsComShar, Storage, Network, Compute
> **AXIOMS**
>> axm2: $ResTOf \in Resources \rightarrow RType$
>> axm3: $IsShareable \in Resources \rightarrow BOOL$
>> axm4: $IsElastic \in Resources \rightarrow BOOL$
>> axm5: $IsExcShar \in Resources \rightarrow BOOL$
>> axm6: $IsComShar \in Resources \rightarrow BOOL$
>> axm7: $RType = \{Storage, Network, Compute\}$
> **END**

We introduce two new sets named *Resources* and *RType* to represent, respectively, all available resources and their types. To map each resource to its adequate type, we have defined the *ResTOf* constraint through a total function between the *Resources* set and the *RType* set (*axm2*). The resource properties are defined via Boolean total functions: *IsShareable*, *IsElastic*, *IsExcShar*, *IsComShar* denoting respectively if the resource is shareable, elastic, exclusively shareable or commonly shareable (*axm3, axm4, axm5, axm6*). The *RType* set elements are given in *axm7*.

> **MACHINE** ResM0
> **REFINES** BehM0
> **SEES** ResC0
> **VARIABLES**
>> ResOF, ResCaOf, ReqRCapacity, ReqR
> **INVARIANTS**
>> inv1: $ResOF \in Services \rightarrow Resources$
>> inv2: $ResCaOf \in Resources \rightarrow \mathbb{Z}$
>> inv3: $ReqR \in LSs \rightarrow Resources$
>> inv4: $ReqRCapacity \in LSs \rightarrow \mathbb{Z}$
> **END**

In the machine *ResM0* that refines *BehM0*, we have introduced variables and invariants related to the resource perspective. The variable function *ResOF* is defined to map each service to its resources set (*inv1*). In invariant *inv2*, the function *ResCaOf* determines each resource capacity. The resource requested by the service at each local state (*LSs*) is represented by the variable function *ReqR* (*inv3*). The resource capacity requested by the service at each state is given by a total function between the local states set (*LSs*) and the integer set (*inv4*).

When N services are provisioned with the same resource at the same time, the composite service may deadlock. For instance, when two or more services are provisioned with a resource r that is exclusively shareable, which means that it can handle two or more services but not at the same time. If these services are requiring this resource at the same time, a deadlock situation occurs. In this work, the provisioned resources of candidate services are checked at design time to avoid such situations. The following requirements must be preserved when allocating resources to component services:

Res1. A resource is allocated to a component service only if its capacity is able to handle the component service requirements, for example when a service required 1Go of disk space, the available capacity of the allocated storage resource must be greater than 1Go, or is elastic.

Res2. A shareable and non elastic resource is allocated to more than one component service only if its capacity is able to handle their requirements at once.

Res3. A shareable resource with limited capacity is allocated to more than one component service only if it is elastic.

Res4. A resource is allocated to more than one component service at the same time only if it is commonly shareable.

Res5. A global state is considered as resource deadlock free if it preserves the requirements *Res1* or *Res2* or *Res3* or *Res4*.

In the INVARIANTS clause, we have introduced the invariant *DiffRes* to model the requirement *Res1* where the resources required at the global state GS_i, are different from each other ($ReqR(LSik) \neq ReqR(LSil)$) and each one capacity is higher than the requested capacity ($ReqRCapacity(LSik) \leq ResCaOf(Req-R(LSik))$) or the resource is elastic ($IsElastic(ReqR(LSik)) = TRUE$).

> **DiffRes:** $\forall GSi \cdot GSi \in DiffRes \Rightarrow$
> $(\forall LSik, LSil \cdot GSi \mapsto LSik \in LSsOfGS \wedge GSi \mapsto LSil \in$
> $LSsOfGS \wedge LSik \neq LSil \Rightarrow ((ReqR(LSik) \neq ReqR(LSil)) \wedge$
> $((ReqRCapacity(LSik) \leq ResCaOf(ReqR(LSik)) \vee$
> $(IsElastic(ReqR(LSik)) = TRUE)) \wedge$
> $(ReqRCapacity(LSil) \leq ResCaOf(ReqR(LSil)) \vee$
> $IsElastic(ReqR(LSil)) = TRUE))))$

The requirement *Res2* is modeled through the invariant *ShNotElas*. If at a global state GS_i, N ($n >= 2$) services are requiring the same resource ($ReqR(LSik) = ReqR(LSil)$) at different local states ($LSik \neq LSil$), then the resource must be shareable ($IsShareable(ReqR(LSik)) = TRUE$) and its capacity is higher than the capacity required by the N component services ($(ReqRCapacity(LSik) + ReqRCapacity(LSil)) \leq ResCaOf(ReqR(LSik))$).

> **ShNotElas:** $\forall GSi \cdot GSi \in ShNotElas \Rightarrow$
> $(\forall LSik, LSil \cdot GSi \mapsto LSik \in LSsOfGS \wedge GSi \mapsto$
> $LSil \in LSsOfGS \wedge LSik \neq LSil \Rightarrow ((ReqR(LSik) =$
> $ReqR(LSil)) \wedge (IsShareable(ReqR(LSik)) =$
> $TRUE \wedge IsElastic(ReqR(LSik)) = FALSE) \wedge$
> $((ReqRCapacity(LSik) + ReqRCapacity(LSil)) \leq$
> $ResCaOf(ReqR(LSik)))))$

The requirement *Res3* is modeled through the invariant *ShAndElas*. If at a global state GS_i, N ($n >= 2$) services are requiring the same resource ($ReqR(LSik) = ReqR(LSil)$) at different local states ($LSik \neq LSil$), then the resource must be shareable ($IsShareable(ReqR(LSik)) = TRUE$). If the resource capacity is lower than the capacity needed by the N services, it must be elastic ($IsElastic(ReqR(LSik)) = TRUE$).

> **ShAndElas:** $\forall GSi \cdot GSi \in ShAndElas \Rightarrow$
> $(\forall LSik, LSil \cdot GSi \mapsto LSik \in LSsOfGS \land GSi \mapsto$
> $LSil \in LSsOfGS \land LSik \neq LSil \Rightarrow ((ReqR(LSik) =$
> $ReqR(LSil)) \land (IsShareable(ReqR(LSik)) = TRUE \land$
> $IsElastic(ReqR(LSik)) = TRUE)))$

The requirement *Res4* is modeled through the invariant *CommSh*. If at a global state GS_i, N ($n >= 2$) services are requiring the same resource ($ReqR(LSik) = ReqR(LSil)$) at the same local state ($LSik = LSil$), then the resource must be commonly shareable ($IsShareable(ReqR(LSik)) = TRUE$).

> **CommSh:** $\forall GSi \cdot GSi \in ShAndElas \Rightarrow$
> $(\forall LSik, LSil \cdot GSi \mapsto LSik \in LSsOfGS \land GSi \mapsto$
> $LSil \in LSsOfGS \land LSik = LSil \Rightarrow ((ReqR(LSik) =$
> $ReqR(LSil)) \land (IsComShar(ReqR(LSik)) = TRUE)))$

A global state GS_i is considered as resource deadlock free if it preserves the invariants $DiffRes$ and $ShNotElas$ and $ShAndElas$ (Res5).

> **NoBlockingRes:** $\forall GS \cdot GS \in NBRes \Rightarrow GS \in$
> $DiffRes \lor GS \in ShNotElas \lor GS \in ShAndElas$

Finally, we have performed the candidate service selection, in the Event *NoBlockingResP*. This event refines the *BidirectionalComplementarity* Event by adding constraints on the services provisioned resources. A composite service is considered resource deadlock free if each global state of the global protocol is ($grd2$).

> Event NoBlockingResP ⟨ordinary⟩ $\widehat{=}$
> refines BidirectionalComplementarity
>
> **any**
> c, s, ip
> **where**
>   **grd2:** $\forall Gs, Gip \cdot ip \mapsto Gip \in GPOf \land ip \mapsto Gip \mapsto$
> $Gs \in GSsOf \Rightarrow Gs \in NBRes$
> **then**
>   **act1:** $SelectedServices := SelectedServices \cup \{s\}$
> **end**

5   Model Verification and Validation

In this section, we describe the steps followed in order to verify and validate our model. The verification covers the static and dynamic properties of the model. Static properties are expressed through invariants. Invariants of the model must hold in all states of the model; they hold at the initial state and are preserved by each event. We refer by dynamic properties to the temporal properties of the system. Such properties could not be expressed through invariants. They express the different states of the system at different animation times. The LTL are used to ensure the dynamic properties verification.

The validation consists on observing the behavior of the specification. The Rodin platform [1] provides the plug-in ProB for the animation and the validation of Event-B specifications. It gives us the possibility to play different scenarios by showing at each stage the values of each variable and distinguishing the enabled events from the disabled ones.

5.1   Verification By Proof Obligations

The term proof obligation is mentioned in this section regularly. What is meant by a proof obligation is, a theorem that needs to be proved in order to verify the correctness of the model [26]. Proof obligations (POs) are automatically generated by Proof Obligation Generator tool of the Rodin Platform. These POs ensure that each event preserves the invariants. The name of this PO is evt/inv/INVT where for each event, we have to establish that:

$$\forall S, C, X.(A \wedge G \wedge Inv \Rightarrow [Act]Inv)$$

where the event actions Act must preserve the invariant Inv.

Modeling in Event-B relies entirely on the interplay between editing models and analysing their proof obligations. Proof obligations are generated not only to ensure that each event preserves the invariants, but also to verify that the refinement had been correctly performed.

In our case, 75 proof obligations have been generated: 36% of them are automatically discharged by the automatic prover. It fails to discharge the remaining proofs due to the numerous steps they require and not on account of their difficulty. To finish discharging these proofs, we resorted to the interactive prover and helped it find the right steps and rules to apply. The proof statistics are given in Table 1.

5.2   Validation By ProB Model Checker

ProB is an animator, constraint solver and model checker for the B-Method. It allows fully automatic animation of B specifications, and can be used to systematically check a specification for a wide range of errors. The constraint-solving capabilities of ProB can also be used for model finding, deadlock checking and test-case generation [21]. Thanks to ProB we have played and observed different

Table 1. Proof statistics

M/C	Total POs	Automatic	Interactive
BehC0	0	0	0
ResC0	0	0	0
BehM0	42	15	27
ResM0	33	12	21
Overall	75	27	48

scenarios in order to check the behavior of our model. Before proceeding to animation, we have given values to the carrier sets, constants and variables of the model. The animation consists on the following steps (1) we start by firing the SETUP-CONTEXT event that gives values to the constants and carrier sets in the context, (2) we then fire the INITIALISATION event to set the model in its initial state, for the initialisation, we have used our motivating example (3) we, finally, proceed the steps of the scenario to check. At each step, the animator computes all guards of all events, and enables the ones with true guards, and shows parameters which make these guards true. After event firing, substitutions are computed and the animator checks if the invariants still hold.

For instance, we animated the complete behavior of the composite service during candidate services selection process while verifying the different states in which it may move. We have successfully applied the animation of PROB on our final level of refinement model as follows: (1) We start by the functional selection which is not subject of this paper. However we can't proceed to the behavioral selection without selecting services according to the functions provided by the composite service. (2) The results of the functional selection are refined by the BC event. A behavioral selection is performed in order to obtain the set of services whose protocols match. (3) An other refinement is performed by the resource based selection to avoid inter-blocking situation.

6   Related Works

The deadlock freeness verification of composite service is becoming more challenging specially in the heterogeneous Cloud environment. Actually, few works are handling the behavior verification in the Cloud environment. Not to mention the lack of works addressing the resource provisioning and management. Such problems are difficult to deal with using only simulation methods. Formal methods [4] prevail, in such situations, to capture the semantics and behavior of complex and critical systems. Formal methods use mathematical models for the analysis of computing, communication, and industrial systems in order to establish system correctness with mathematical rigor which makes them highly recommended verification techniques in this field.

For instance, there are several semi-formal languages and supporting tools that have been used for modeling and verification of Cloud concepts. In [22], a semantic-aware model checking (SAMC) approach were proposed to capture the simple semantic information of the target system. This method does not provide any formal semantics. The major inconvenient of semi-formal verification is the lack of informal semantics. Therefore, the formal methods are more suitable to the Cloud context. The benefits of using formal methods to ensure the correctness are well-proven.

Authors in [12], presented an abstract formalization of federated cloud work-flows using the Z notation [28]. The work-flow is modeled as an abstract data type upon which various operations are possible. In [10], the authors used Labeled Transition System Analyser (LTSA) to present a λ calculus model for analyzing and verifying the resources used in web service applications in cloud computing environment. Early research efforts often focus on non-functional behavior in terms of response time and financial cost namely the work in [2], where authors presented a model for web services Event Condition Action (ECA). They used SPIN model checker [17] in order to verify service agreement property. And so on, the mentioned works commonly use a formal method to model and verify the Cloud open issues however none of them have addressed the deadlock freeness of Cloud services nor compositions.

The work in [24] used High-Level Petri Nets (HLPN) to analyze and model the structural and behavioral properties of three open source VM-based cloud management platforms: Open Nebula, Eucalyptus and Nim-bus. Recently, the work in [25] also used Markov Decision Processes (MDP) for the cloud elasticity modeling, and then used PRISM model checker in order to model and verify several elasticity decision policies that aim to maximize user-defined utility functions. Model checking suffers from the state-space explosion problem that makes exhaustive verification very difficult for large and complex systems. In addition, it is computationally expensive to cover all the state space of the system model. In fact, abstraction is a powerful technique that enables fitting big systems into model checkers, yet, it has not been explored well for modeling and verification of cloud systems.

Our aim, in this paper, is to overcome the aforementioned verification limits by introducing a new approach, that takes advantage of the Event-B modeling method which enables the modeling of systems by means of abstraction techniques. The proposed model focuses on the deadlock freeness verification of composite Cloud services.

Otherwise, existing approaches focuses either on the behavior or on the resource allocation verification but not both of them. The works in [8,18] focus on the verification of deadlock freeness of business processes focusing only on the resource perspective. *In this work, we believe that both perspectives must be considered in order to ensure a deadlock free composite service.*

For instance, the works in [14,15] intend to manage resource allocation at run-time. An Event-B model is performed. However managing the resource allocation at run-time is not enough and in some critical systems it is mandatory to perform verification at design-time to avoid disastrous consequences.

7   Conclusion

To sum up, in this work, we have introduced a new formal approach based on the Event-B formal method to verify the Cloud composite service deadlock freeness. We have succeeded to introduce a new approach that combines both behavior and resource properties thanks to the Event-B refinement concept. At the first level of refinement, we have modeled behavioral properties of the composition and defined constraints on the n selected services in order to prevent interacting protocols deadlocks. Then we have specified the second level of refinement to model resource properties and check the resources provisioned for each candidate service. The obtained model was verified and validated by mean of proof obligation and model checking tools of the Rodin platform.

In the near future, we aim to extend this work by considering the resource elasticity problems. We also plan to extend the proposed model by adding QoS properties as presented in [20]. We then aim to deal with dynamic reconfiguration and adaption of Cloud composite service.

References

1. Event-B and the Rodin platform. http://www.event-b.org/
2. Abdelsadiq, A., Molina-Jimenez, C., Shrivastava, S.: A high-level model-checking tool for verifying service agreements. In: Proceedings of 2011 IEEE 6th International Symposium on Service Oriented System (SOSE), pp. 297–304, December 2011
3. Abrial, J.R.: The B tool (abstract). In: Bloomfield, R.E., Marshall, L.S., Jones, R.B. (eds.) VDM 1988. LNCS, vol. 328, pp. 86–87. Springer, Heidelberg (1988). https://doi.org/10.1007/3-540-50214-9_8
4. Abrial, J.R.: Faultless systems: yes we can! Computer 42(9), 30–36 (2009)
5. Abrial, J.: The B-Book - Assigning Programs to Meanings. Cambridge University Press, Cambridge (2005)
6. Abrial, J.-R., Butler, M., Hallerstede, S., Voisin, L.: An open extensible tool environment for Event-B. In: Liu, Z., He, J. (eds.) ICFEM 2006. LNCS, vol. 4260, pp. 588–605. Springer, Heidelberg (2006). https://doi.org/10.1007/11901433_32
7. Abrial, J.-R., Mussat, L.: Introducing dynamic constraints in B. In: Bert, D. (ed.) B 1998. LNCS, vol. 1393, pp. 83–128. Springer, Heidelberg (1998). https://doi.org/10.1007/BFb0053357
8. Boubaker, S., Klai, K., Schmitz, K., Graiet, M., Gaaloul, W.: Deadlock-freeness verification of business process configuration using SOG. In: Maximilien, M., Vallecillo, A., Wang, J., Oriol, M. (eds.) ICSOC 2017. LNCS, vol. 10601, pp. 96–112. Springer, Cham (2017). https://doi.org/10.1007/978-3-319-69035-3_7
9. Cansell, D., Méry, D.: The Event-B modelling method: concepts and case studies. In: BjØrner, D., Henson, M.C. (eds.) Logics of Specification Languages. EATCS, pp. 47–152. Springer, Heidelberg (2008). https://doi.org/10.1007/978-3-540-74107-7_3

10. Chen, J., Huang, L., Huang, H., Yu, C., Li, C.: A formal model for resource protections in web service applications. In: 2012 International Conference on Cloud and Service Computing, pp. 111–118, November 2012
11. Durán, F., Ouederni, M., Salaün, G.: A generic framework for n-protocol compatibility checking. Sci. Comput. Program. **77**(7–8), 870–886 (2012)
12. Freitas, L., Watson, P.: Formalising workflows partitioning over federated clouds: multi-level security and costs. In: 2012 IEEE Eighth World Congress on Services, pp. 219–226, June 2012
13. Furht, B., Escalante, A.: Handbook of Cloud Computing, 1st edn. Springer, Boston (2010). https://doi.org/10.1007/978-1-4419-6524-0
14. Graiet, M., Mammar, A., Boubaker, S., Gaaloul, W.: Towards correct cloud resource allocation in business processes. IEEE Trans. Serv. Comput. **10**(1), 23–36 (2017)
15. Graiet, M., Hamel, L., Mammar, A., Tata, S.: A verification and deployment approach for elastic component-based applications. Formal Asp. Comput. **29**(6), 987–1011 (2017)
16. Graiet, M., Lahouij, A., Abbassi, I., Hamel, L., Kmimech, M.: Formal behavioral modeling for verifying SCA composition with Event-B. In: 2015 IEEE International Conference on Web Services, ICWS 2015, New York, NY, USA, 27 June–2 July 2015, pp. 17–24 (2015)
17. Holzmann, G.J.: The model checker spin. IEEE Trans. Softw. Eng. **23**(5), 279–295 (1997)
18. Klai, K., Tata, S., Desel, J.: Symbolic abstraction and deadlock-freeness verification of inter-enterprise processes. Data Knowl. Eng. **70**(5), 467–482 (2011). Business Process Management 2009
19. Lahouij, A., Hamel, L., Graiet, M.: Formal modeling for verifying SCA dynamic composition with Event-B. In: 24th IEEE International Conference on Enabling Technologies: Infrastructure for Collaborative Enterprises, WETICE 2015, Larnaca, Cyprus, 15–17 June 2015, pp. 29–34 (2015)
20. Lahouij, A., Hamel, L., Graiet, M., Elkhalfa, A., Gaaloul, W.: A global SLA-aware approach for aggregating services in the cloud. In: Debruyne, C. (ed.) OTM 2016. LNCS, vol. 10033, pp. 363–380. Springer, Cham (2016). https://doi.org/10.1007/978-3-319-48472-3_21
21. Laili, Y., Tao, F., Zhang, L., Cheng, Y., Luo, Y., Sarker, B.R.: A ranking chaos algorithm for dual scheduling of cloud service and computing resource in private cloud. Comput. Ind. **64**(4), 448–463 (2013)
22. Leesatapornwongsa, T., Hao, M., Joshi, P., Lukman, J.F., Gunawi, H.S.: SAMC: semantic-aware model checking for fast discovery of deep bugs in cloud systems. In: Proceedings of the 11th USENIX Conference on Operating Systems Design and Implementation, OSDI 2014, pp. 399–414. USENIX Association, Berkeley (2014)
23. Lemos, A.L., Daniel, F., Benatallah, B.: Web service composition: a survey of techniques and tools. ACM Comput. Surv. **48**(3), 33:1–33:41 (2015)
24. Malik, S.U.R., Khan, S.U., Srinivasan, S.K.: Modeling and analysis of state-of-the-art VM-based cloud management platforms. IEEE Trans. Cloud Comput. **1**(1), 1 (2013)
25. Naskos, A., et al.: Cloud elasticity using probabilistic model checking, May 2014
26. Padidar, S.: A study in the use of Event-B for system development from a software engineering viewpoint. http://www.ai4fm.org/papers/MSc-Padidar.pdf
27. Papapanagiotou, P., Fleuriot, J.: Formal verification of web services composition using linear logic and the pi-calculus. In: 2011 IEEE Ninth European Conference on Web Services, pp. 31–38, September 2011

28. Woodcock, J., Davies, J.: Using Z: Specification, Refinement, and Proof. Prentice-Hall Inc., Upper Saddle River (1996)
29. Yang, Y., Tan, Q., Xiao, Y.: Verifying web services composition based on hierarchical colored petri nets. In: Proceedings of the First International Workshop on Interoperability of Heterogeneous Information Systems, IHIS 2005, pp. 47–54. ACM, New York (2005)

A Formal Model for Business Process Configuration Verification Supporting OR-Join Semantics

Souha Boubaker[1,2]([✉]), Kais Klai[2], Hedi Kortas[2], and Walid Gaaloul[1]

[1] Telecom SudParis, UMR 5157 Samovar, Universite Paris-Saclay, Paris, France
souha.boubaker@gmail.com
[2] LIPN, CNRS UMR 7030, University of Paris 13, Paris, France

Abstract. In today's industries, similar process models are typically reused in different application contexts. These models result in a number of process model variants sharing several commonalities and exhibiting some variations. Configurable process models came to represent and group these variants in a generic manner. These processes are configured according to a specific context through configurable elements. Considering the large number of possible variants as well as the potentially complex configurable process, the configuration may be a tedious task and errors may lead to serious behavioral issues. Since achieving configuration in a correct manner has become of paramount importance, the analysts undoubtedly need assistance and guidance in configuring process variants. In this work, we propose a formal behavioral model based on the Symbolic Observation Graph (SOG) allowing to find the set of deadlock-free configuration choices while avoiding the well-known state-space explosion problem and considering loops and OR-join semantics. These choices are used to support business analysts in deriving deadlock-free variants.

Keywords: Business process management
Configurable process model · Formal verification

1 Introduction

When considering today's changing business requirements, a high degree of flexibility in the design and management of business processes is becoming a strong need. Configurable process models [10,18] are introduced as a way to model variability as reference models. Such models represent a large family of related process variants that share several commonalities while exhibiting some variations. The configurable model can be configured by selecting one design option for each configurable element (capturing the variation points) in order to accommodate different contexts requirements while still achieving the business goal of the process. The obtained configured processes are called *variants*. For instance, in Fig. 1, a simplified example of a configurable process model designed by a

H. Panetto et al. (Eds.): OTM 2018 Conferences, LNCS 11229, pp. 623–642, 2018.
https://doi.org/10.1007/978-3-030-02610-3_35

process provider for a hotel booking agency is presented. The process is modeled using the Configurable Business Process Model and Notation (C-BPMN) [5,13], a configurable extension to BPMN[1]. The travel agency has a number of branches in different countries. Depending on the specific needs of each country, each branch performs a different variant of this process model in terms of structure and behavior. For instance, a process tenant may need the exclusive execution of one path among the outputs of *S1* (i.e. configurable connector modeled with a thicker border). This refers to configuring *S1* to an *XOR-split*. Another tenant may choose to execute both paths concurrently by configuring *S1* to an *AND-split*.

As a configurable element represent a design-time *configuration choice* [18], then any design mistake should be avoided in order to ensure a correct execution of the derived variants. Furthermore, configurable processes may be large with complex inter-dependencies between the different possible configurations. Consequently, the configuration can not be done manually and a verification phase is essential. So far, a number of approaches have addressed the verification of the process configuration correctness. Some of them have only discussed the syntactical correctness (e.g. [10,18]), others have attempted to verify behavioral correctness but have faced the exponential number of state-space problems (e.g. [12]). Very few have addressed the configuration behavior verification while trying to reduce state explosion problem (e.g. [2,4]) but still suffer from the exponential complexity of generating their reachability graph.

In our previous work [8], we proposed an approach for the verification of process models configuration that contributes greatly to resolving the state explosion problem. In the current paper, we extend this approach, on the one hand, by considering configurable processes with cycles and synchronizing OR-joins, and on the other, by respecting additional configuration constraints: activity configuration and connectors configuration by restricting output or input branches. We use the Symbolic Observation Graph (SOG) [11,14,15] as a basis to define, for a given configurable process model, the set of all configuration options that satisfy specifically the deadlock-freeness property in a reduced graph. That way, the combinations of options are found prior to configuration time. And then, these combinations will serve to guide the business analyst in deriving deadlock-free variants without computing correctness at each step (as in [2]).

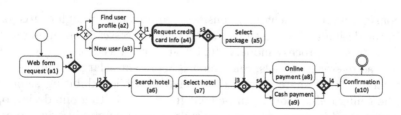

Fig. 1. A configurable hotel booking process model

---

[1] BPMN 2.0 specification http://www.omg.org/spec/BPMN/2.0/.

The SOG is a versatile symbolic representation formalism that allows to build an abstraction of the reachability state graph of a formally modeled system (e.g. using Petri net). In our case, this abstraction is achieved by observing the configurable elements of the process (that label the SOG arcs) and hiding non configurable elements inside the aggregates (the SOG nodes). Moreover, without limiting the generality of our approach, we propose to use Configurable BPMN (C-BPMN) as input notation. BPMN is highly adopted by business analysts since it is considered as the internationally recognized industry standard notation for business process description.

Figure 2 depicts the proposed SOG-based approach allowing to obtain deadlock-free process variants. First, as depicted on the left-hand side of the figure, a C-BPMN configurable process model is introduced. This process model is then mapped into a Petri net-based model; and new semantics is defined to take into account configurable elements (step 1, see Sect. 3). Afterward, we extend the SOG construction algorithm by three main points (step 2): (i) by observing and highlighting configurable connectors and activities in the graph arcs; (ii) by hiding non-configurable elements' states in aggregates (see Sect. 4); and (iii) by restricting the graph nodes to the ones leading to deadlock-free variants (see Sect. 4.2). As a result, we obtain a reduced SOG that groups the behavior of all correct configurations. The set of correct configuration combinations is then extracted. The last three steps are performed on-the-fly during the SOG construction. The set of correct configurations is finally supplied to the business analyst (step 3) in order to derive correct variants, with no need to compute correctness at each intermediate configuration step.

Organizationally, in Sect. 2, some preliminary concepts about Petri nets and Workflow nets are described. New Petri-net-based models for business process models and configurable process models as well as their semantics are defined in Sect. 3. Then, in Sect. 4, we illustrate our approach based on the Symbolic Observation Graph associated with the defined configurable formal model. We discuss the related work in Sect. 6. Finally, we conclude and provide insights for future work.

2 Preliminaries and Notations

In order to represent the C-BPMN process model and its corresponding execution semantics, we propose to use the Petri-nets as a pivot formalism. Actually, Petri nets offer a formal model that the large majority of modeling languages can be mapped into it. In the following, first, we formally define a Petri net and its corresponding notations and semantics. Then, we define a Workflow net and its deadlock-freeness requirements.[2]

Definition 1 (Petri Net). *A Petri net is a tuple* $N = \langle P, T, F, W \rangle$ *s.t.:*

---

[2] Note that our approach does not rely on specific Petri nets properties but can be applied to any formal model as soon as states and transition relations are well defined.

626   S. Boubaker et al.

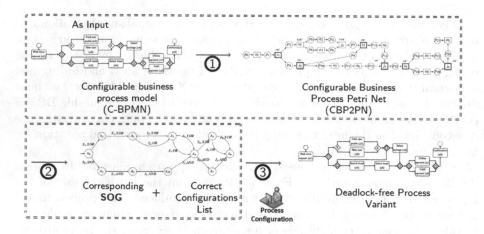

Fig. 2. Our approach overview

- *P is a finite set of places and T a finite set of transitions with $(P \cup T) \neq \emptyset$ and $P \cap T = \emptyset$,*
- *$F \subseteq (P \times T) \cup (T \times P)$ is a flow relation,*
- *$W : F \to \mathbb{N}^+$ is a mapping assigning a positive weight to arcs.*

Each node $x \in P \cup T$ of the net has a pre-set and a post-set defined respectively as follows: $^\bullet x = \{y \in P \cup T \mid (y,x) \in F\}$, and $x^\bullet = \{y \in P \cup T \mid (x,y) \in F\}$. The preset (resp. postset) is extended to sets of nodes X as $^\bullet X = \bigcup_{x \in X} {}^\bullet x$ (resp. $X^\bullet = \bigcup_{x \in X} x^\bullet$). Also, the flow relation F, its inverse F^{-1} and its reflexive transitive closure F^* are extended respectively as follows: $F(X) = \bigcup_{x \in X} F(x)$, $F^{-1}(X) = \bigcup_{x \in X} F^{-1}(x)$ and $F^*(X) = \bigcup_{x \in X} F^*(x)$. For a transition t, $W^-(t) \in \mathbb{N}^{|P|}$ (resp. $W^+(t) \in \mathbb{N}^{|P|}$) denotes the vector where, $\forall p \in P$, $W^-(t)(p) = W(p,t)$ (resp. $W^+(t)(p) = W(t,p)$). A marking of a Petri net N is a function $m : P \to \mathbb{N}$.

Semantics. Let m be a marking of $t \in T$

- a transition t is said to be enabled by m, denoted by $m \xrightarrow{t}$, iff $W^-(t) \leq m$.
- when t is enabled by m, its firing leads to a new marking m', denoted by $m \xrightarrow{t} m'$, s.t. $m' = m - W^-(t) + W^+(t)$.

For a finite sequence $\sigma = t_1 \dots t_n$, $m_i \xrightarrow{\sigma} m_n$ denotes the fact that σ is enabled by m_i i.e., $m_i \xrightarrow{t_1} m_1 \xrightarrow{t_2} m_2 \to \dots \xrightarrow{t_n} m_n$. Given a set of markings S, we denote by $Enable(S)$ the set of transitions enabled by elements of S. The set of markings reachable from a marking m in N is denoted by $R(N,m)$. The reachability graph of a Petri net N, denoted by $G(N,m_i)$ (m_i is the initial marking), is the graph where nodes are elements of $R(N,m_i)$ and an arc from m to m', labeled with t, exists iff $m \xrightarrow{t} m'$. The set of markings reachable from a marking m, by firing the transitions of a subset T' only is denoted by $Sat(m,T')$. By extension, given a set of markings S and a set of transitions T', $Sat(S, T') = \bigcup_{m \in S} Sat(m, T')$. For a marking m, $m \not\rightarrow$ denotes that m is a dead marking (i.e., there is no transition s.t. $m \xrightarrow{t}$ which means $Enable(\{m\}) = \emptyset$).

Definition 2 (WF-Net). *Let $N = \langle P, T, F, W \rangle$ be a Petri net and F^* is the reflexive transitive closure of F. N is a Workflow net (WF-net) iff:*

- *there exists exactly one input place $i \in P$, s.t. $|{}^{\bullet}i| = 0$,*
- *there exists exactly one output place $o \in P$, s.t. $|o^{\bullet}| = 0$,*
- *each node is on a directed path from the input place to the output place, i.e. $\forall n \in P \cup T, (i, n) \in F^* and (n, o) \in F^*$.*

3   Formal Model for Configurable Business Processes

A business process consists of a set of activities that are performed in coordination in an organizational and technical environment. These activities jointly realize a business goal [20]. In order to obtain an abstract formal definition of a BPMN process model, we formally define a new model based on *Petri nets*, called Business Process Petri Nets (*BP2PN*). Then, we extend the *BP2PN* to take into account configurable elements, leading to a new model, namely the Configurable Business Process Petri Nets (*CBP2PN*). In this work, we extend existing mapping work from BPMN models to Petri nets (e.g. [9]) by preserving connectors' blocks as transitions representing configurable transitions.

3.1   Business Process Petri Nets (BP2PN)

Definition 3 (BP2PN). *A BP2PN is a tuple $B = \langle P, T \cup OP, F, W, O, L \rangle$ where:*

- *$\langle P, T \cup OP, F, W \rangle$ is a WF-Net,*
- *$F \subseteq (P \times T \cup OP) \cup (T \cup OP \times P)$ is the flow relation,*
- *$O : OP \rightarrow \{OR^-, OR^+, XOR^-, XOR^+, AND^-, AND^+\}$ is a mapping that assigns a type to each operator,*
- *$L : T \rightarrow \{on, off\}$ is a function that assigns for each transition a label on or off.*

BP2PN is a *Workflow net* such that, the set of places P corresponds to the set of conditions determining the enabling of a task or a connector; and the set of transitions $T \cup OP$ corresponds to the set of tasks and connectors. These nodes are interconnected through a set of arcs using the flow relation F. Each connector must either be a join (depicted using the $-$ exponent on the right in Definition 4) or a split (the $+$ exponent) while having a type: OR, XOR or AND. For simplicity reasons, we enrich the model with transitions labels. A transition in T is labeled by *on* by default, however, it is labeled by *off* if it has been excluded from the model and thus it is considered as a silent transition (further details in the next section).

For simplicity reasons, we define the notion of a syntactically correct BP2PN, and we particularly consider the deadlock-freeness property as follows.

Definition 4 (Syntactically correct BP2PN). *A BP2PN* $B = \langle P, T \cup OP, F, W, O, L \rangle$ *is syntactically correct iff:*

- *For each place* $p \in P$: $|{}^{\bullet}p| = 1$ *and* $|p^{\bullet}| = 1$,
- *For each transition* $t \in T$: $|{}^{\bullet}t| = 1$ *and* $|t^{\bullet}| = 1$,
- *for each* $t \in OP$ *and* $O(t) \in \{OR^-, XOR^-, AND^-\}$: $|{}^{\bullet}t| \geq 2$ *and* $|t^{\bullet}| = 1$.
- *for each* $t \in OP$ *and* $O(t) \in \{OR^+, XOR^+, AND^+\}$: $|{}^{\bullet}t| = 1$ *and* $|t^{\bullet}| \geq 2$.

Definition 5 (Deadlock-free BP2PN). *Let* $B = \langle P, T \cup OP, F, W, O, L \rangle$ *be a BP2PN and* m_i, m_f *be the initial (i.e. only i is marked) and final (i.e. only o is marked) markings respectively. B is said to be deadlock-free iff* $\nexists m \in (R(N, m_i) \setminus \{m_f\})$ *s.t.* $m \nrightarrow$.

According to Definition 5, a BP2PN is deadlock-free if there is no dead marking m reachable from the initial marking m_i.

Notations. Before defining semantics, given a transition t and a marking m, let us define $L_{off}(t)$, $Nearest(t)$, $W^{--}(t)$, ${}^{\bullet}t^-$ and $EP_m(t)$ as follows:

- $L_{off}(t) = \{t' \in T \mid L(t') = off \wedge t'^{\bullet} \cap {}^{\bullet}t \neq \emptyset\}$ is the set of **off** transitions t', i.e. $L(t') = off$, producing tokens in some input places of t;
- $Nearest: T \times P \longrightarrow 2^P$ is a function that assigns to a transition t and its input place p having as predecessor an **off** transition, the nearest place s.t. its preset transition is either **on** transition or a connector; so $Nearest(t, p)$ is defined as follows:

$$Nearest(t, p) = \begin{cases} \bot \quad undefined & \text{if} \quad p \notin {}^{\bullet}t \ \vee \ {}^{\bullet}p \notin L_{off}(t) \\ \{p' \in (F^{-1})^{*\bullet}p \mid ({}^{\bullet}p' \subset T \wedge L({}^{\bullet}p') = on) \vee {}^{\bullet}p' \in OP\} & \text{else.} \end{cases}$$

We apply the backward (i.e. inverse) of the transitive closure of F on the place p until finding the first **on** transition or connector.

- $Nearest(t) = \bigcup_{p \in L_{off}(t)^{\bullet}} Nearest(t, p)$ is the set of nearest places starting from all **off** transitions that have as output, input places of t
- $W^{--}(t)$ is the weight allowing the enabling of t s.t.:

$$W^{--}(t)(p) = \begin{cases} W^-(t)(p) & \text{if } {}^{\bullet}p \notin L_{off}(t) \\ W^-(t)(p') & \text{if } p = Nearest(t, p') \\ 0 & \text{else.} \end{cases}$$

- ${}^{\bullet}t^- = Nearest(t) \cup \{p \in {}^{\bullet}t \mid m(p) \geq W^-(t)(p)\}$ is the set of preset places of t but also nearest places of t having predecessor transitions in $L_{off}(t)$;
- $EP_m(t) = \{p \in {}^{\bullet}t^- \mid m(p) \geq W^{--}(t)(p)\}$ is then the set of marked input places of t at the marking m.

In the following, we use ${}^{\bullet}t^-$ to refer to *presets* of a transition t, and we use $W^{--}(t)$ to refer to the *weight* vector of t presets. These notations allow for the following definition of the fireability of BP2PN transition.

Semantics. We define new semantics of *BP2PN* models by taking into account the new features of the underlying Petri net (i.e. connectors and the *on/off* labeling function).

Definition 6 (Enabled/fireable transition)
  Let $t \in T \cup OP$ be a transition, and let m be a marking of a BP2PN; m enables t **iff** one of the following conditions holds:

1. if $(t \in T \wedge L(t) = on)) \vee (t \in OP \wedge O(t) \notin \{XOR^-, OR^-\})$ then, m enables t **iff**:
   - $\forall p \in {}^\bullet t \setminus L_{off}{}^\bullet, \; m(p) \geq W^-(t)(p) \quad \wedge \quad \forall p' \in Nearest(t), \; m(p') \geq W^-(p'^\bullet)$
2. if $t \in OP \wedge O(t) = XOR^-$ then, m enables t **iff**:
   (a) $|EP_m(t)| = 1$
3. if $t \in OP \wedge O(t) = OR^-$ then, m enables t **iff** (a) **and** (b) hold:
   (a) $EP_m(t) \neq \emptyset$
   (b) $\nexists m' \in R(N, m) \Rightarrow EP_m(t) \subset EP_{m'}(t)$

We denote by $m \xrightarrow{t}$ the fact that m enables t by one of the above mentioned conditions. If m enables t, t is said to be fireable in m except when $t \in T \wedge L(t) = off$.

Using this enabling rule, an XOR^- is enabled if *one* of its *presets* is marked (see Fig. 3). An ordinary transition or an AND^- is enabled if all its *presets* are marked. And, an OR^- is enabled, if (a) at least one of its *presets* is marked, and (b) there is no other token that may mark one of its *presets* in the future.

Definition 7 (Firing of a transition). *Given a fireable transition t, in a marking m, we define $Pre(t)$ and $Post(t)$ as vectors in $\mathbb{N}^{|P|}$ s.t.:*

- $Pre(t) = W^{--}(t)_{EP_m(t)}, \quad \forall t \in T \cup OP$
- $Post(t)$ is defined as follows:
  1. if $t \in OP \wedge O(t) = XOR^+$ (resp. $O(t) = OR^+$), then $Pre(t) = W^{--}(t)$ and $Post(t) = W^+(t)_{\{p\}}$ for some $p \in t^\bullet$ (resp. $Post(t) = W^+(t)_S$ for $S \subseteq t^\bullet$);
  2. Otherwise, $Post(t) = W^+(t)$.

The firing of (a fireable transition) t at a marking m leads to a marking m' defined by $m' = m - Pre(t) + Post(t)$.

In fact, $Pre(t)$ corresponds to the number of tokens to consume in the input places of the transition t or/and the input places of the *off* transitions in $L_{off}(t)$. $Post(t)$ is the number of produced tokens after firing t. As in [8], this semantics leads to non-deterministic firing. For instance, having a split transition OR^+ with two output places p_1 and p_2, its firing leads to 3 possible reachable markings m_1 (only p_1 is marked), m_2 (only p_2 is marked), and m_{1_2} (both places are marked).

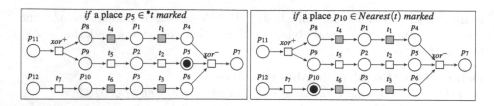

Fig. 3. Enabling rule examples for XOR^- having $L_{off} = \{t_1, t_3\}$

3.2   Configurable Business Process Petri Nets (CBP2PN)

Definition 8 (CBP2PN). *A CBP2PN is a tuple* $CB = \langle P, T \cup OP, F, W, O, L, C \rangle$ *where:*

- $\langle P, T \cup OP, F, W, O, L \rangle$ *is a BP2PN where L applies on non configurable elements only, and assigning on to each of them.*
- $C : T \cup OP \rightarrow \{true, false\}$ *is a function determining the configurable elements (i.e. any* $t \in T \cup OP$ *s.t.* $C(t) = true$).

Back to our running example, our *C-BPMN* process is mapped onto *CBP2PN* in Fig. 4. Configurable transitions are highlighted with a thicker border. This example includes 5 configurable transitions: *a4, s1, s3, j2* and *j3*.

A configurable activity may be needed in a process variant and not in another depending on specific requirements. We denote by T^c the set of configurable activities i.e. $T^c = \{t \in T \mid C(t) = true\}$. The configuration of an activity is reduced to labeling it with *on* or *off*. Moreover, we denote by OP^c the set of configurable operators s.t. $OP^c = \{t \in OP \mid C(t) = true\}$. A configurable operator $c^c \in OP^c$ includes a generic behavior which is restricted during the configuration phase. A connector may be configured by (i) changing its type (e.g. from OR to AND), or/and (ii) restricting its incoming or outgoing branches; w.r.t. the set of configuration constraints [18] defined in Table 1 (having $x \in \{+, -\}$). Each row corresponds to a configurable connector that can be configured to one or more of the connectors in columns. Thus, these constraints allow to specify which regular connector's type may be used in the derived process variant. For example, a configurable *OR* can be configured to any connector's type while a configurable *AND* can only be configured to an *AND*. A connector may also be configured to a sequence *SEQ* by preserving only one branch (cf. last column of Table 1). But, an AND connector should never be configured to a *SEQ*. In addition to the constraints in Table 1, Definition 9 formally defines the possible configurations for each transition type. Note that, when configuring all configurable elements of a *CBP2PN* we obtain a *BP2PN*.

Table 1. Constraints for connectors configuration [18].

OP^c	$Conf$			
	OR^x	XOR^x	AND^x	SEQ
OR^x	√	√	√	√
XOR^x		√		√
AND^x			√	

Definition 9 (CBP2PN configuration). *Let CB be a CBP2PN, then the configuration $Conf_T$ of a configurable transition $t^c \in T^c$ and the configuration $Conf_{OP}$ of a configurable connector $t^c \in OP^c$, are defined as follows:*

- $Conf_T : T^c \rightarrow \{on, off\}$
- $Conf_{OP} : OP^c \rightarrow \{OR^-, OR^+, XOR^-, XOR^+, AND^-, AND^+\} \times 2^P$
  s.t. $Conf_{OP}(t^c) = (op, s)$ where $s \subseteq {}^\bullet t^c$ if t^c is a join connector, and $s \subseteq t^{c\bullet}$ if t^c is a split connector.
  - *If $op \in \{AND^+, AND^-\}$ then $card(s) \geq 2$*
  - *If $card(s) = 1$ then op configuration corresponds to a SEQ*

One possible configuration of the *CBP2PN* in Fig. 4 can be done by selecting the following configuration choices: (i) the credit card verification activity is discarded (i.e., $Conf_T(a_4) = off$), (ii) s_1 is configured to a regular XOR^+ (i.e., $Conf_{OP}(s_1) = (XOR^+, \{p_2, p_{10}\}))$, (iii) j_2 is configured to AND^- (i.e., $Conf_{OP}(j_2) = (AND^-, \{p_{10}, p_9\}))$, (iv) s_3 and s_4 are configured to XOR^+ (i.e., $Conf_{OP}(s_3) = (XOR^+, \{p_{14}, p_9\})$ and $Conf_{OP}(s_4) = (XOR^+, \{p_{17}, p_{19}\}))$; and (v) j_3 and j_4 are configured to XOR^- (i.e., $Conf_{OP}(j_3) = (XOR^-, \{p_{13}, p_{15}\})$ and $Conf_{OP}(j_4) = (XOR^-, \{p_{18}, p_{20}\}))$. Hence, the resulting process is a *BP2PN* since it does not contain any configurable transitions.

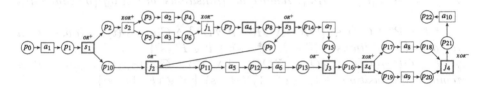

Fig. 4. The CBP2PN of the configurable process in Fig. 1

Semantics. The semantics of a *CBP2PN* is defined such that any reachable marking by any possible instance of a configuration is represented. Thus, we consider a configurable transition as the union of all its possible configurations. That way, we can define its enabling and firing rules as if it is the union of all executable configured transitions. Since a configuration of a transition $t^c \in T^c$ to its corresponding regular transition $t \in T$ is ensured by labeling it as $L(t) = Conf_T(t^c)$, its semantics remains the same as previously defined (see Definitions 6 and 7). Regarding a configurable connector, its fireability and firing rules follow new semantics as follows.

We denote by $AllConf_{OP}(t^c)$ the set of all possible configurations of $t^c \in OP^c$ (w.r.t Table 1). For example, $AllConf_{OP}(s_1) = \{(OR^+, \{p_2, p_{10}\}), (XOR^+, \{p_2, p_{10}\}), (AND^+, \{p_2, p_{10}\}), (OR^+, \{p_2\}), (OR^+, \{p_{10}\})\}$; where each couple represents a possible configuration of the configurable connector s_1. Note that, the last two couples correspond to the configuration of the split connector s_1 to a sequence, then the configured type is not relevant, i.e., it may be any type.

Definition 10 (Enabling and firing of a configurable connector). *Let m be a marking of a CBP2PN and t^c be a transition of OP^c, the enabling and firing rules are as follows:*

- *m enables t^c, denoted by $m \xrightarrow{t^c}$, iff $\exists x \in AllConf_{OP}(t^c)$ s.t. $m \xrightarrow{x}$*
- *when m enables t^c, for some configuration x of t^c, the firing of t^c from m, under configuration x, leads to a marking m', denoted by $m \xrightarrow{t^c,x} m'$ iff $m \xrightarrow{x} m'$*

Using this semantics, the reachability marking graph associated with a *CBP2PN* covers all the possible behavior of all its possible configurations. For instance, having the *CBP2PN* of Fig. 4, the configurable transition s_1 could be configured either to: (i) AND^+ with all of its output places marked, (ii) XOR^+ with only one of the output places marked, or (iii) OR^+ with one or more output places marked.

Definition 11 (Configured BP2PN). *Let $\underline{CB}$ be a CBP2PN and a set of configurations choices leading to resulting BP2PN, P, that is constructed as follows:*

- *$P = P \setminus \{p \in t^\bullet | t \in T \wedge L(t) = off\}$: remove all places that are postsets to transitions labeled off*
- *$T = T \setminus \{t \in T | L(t) = off\}$: remove all transitions labeled off (or configured off);*
- *$\forall t, t \in OP^c \wedge O(t) \in \{OR^+, XOR^+, AND^+\} \wedge Conf_{OP}(t) = (op, s)$ then remove all transitions $\in \{t' \in F^*(t^\bullet \setminus s) \mid t' \notin F^*(s)\}$*
- *$\forall t, t \in OP^c \wedge O(t) \in \{OR^-, XOR^-, AND^-\} \wedge Conf_{OP}(t) = (op, s)$ then remove all transitions $\in \{t' \in (F^-1)^*(^\bullet t \setminus s) \mid t' \notin (F^-1)^*(s)\}$*

A BP2PN is said to be configured if none of its transitions is configurable: $T^c = OP^c = \emptyset$.

Definition 12 (Correct CBP2PN). *Let CB be a CBP2PN. CB is said to be correct if at least one deadlock-free BP2PN could be configured from it.*

Using Definition 12, the *CBP2PN* of Fig. 4 is considered correct since one can configure at least one correct variant by choosing OR type as configuration choice for all its configurable connectors (the correctness of such a variant is proven in Sect. 4.2). However, incorrect variants could be derived from this process as well. For instance, one can choose the alternatives presented earlier that lead to a deadlock caused by an exclusive choice XOR^+ (i.e. $s1$) followed by a synchronizing join AND^- (i.e. j_2). In this situation, in order to be enabled, the transition AND^- will be waiting for both places p_9 and p_{10} to be marked, however only one could be marked. So, the resulting derived variant could never terminate properly. Here, in the reachability graph, a dead marking from which no transition could be enabled, is detected. In the following section, we propose to use the SOG in order to abstract the reachability graph of a CBP2PN, and to extract all the deadlock-free possible configurations.

4 Symbolic Observation Graph for Process Configuration

In this paper, we check the behavior correctness of all possible configurations of a configurable model. This refers to verifying the reachability graph that covers them all, which leads to the state space explosion problem. In order to reduce this issue, we propose to use the Symbolic Observation Graph (SOG) [11,15].

4.1 Symbolic Observation Graph (SOG)

We propose to specifically observe the configurable transitions and highlight all possible configurations leading to a deadlock-free $BP2PN$. Hence, given a $CBP2PN$, the set of observed transitions, denoted by Obs, is the set of configurable elements, i.e. $Obs = OP^c \cup T^c$. Any other transition belongs to the set of unobserved transitions, denoted by $UnObs$, i.e., $UnObs = (T \cup OP) \setminus Obs$. In such a way, we construct the SOG as a graph where each node is a set of states linked by unobserved transitions and each arc is labeled by an observed transition. Nodes of the SOG are called aggregates and may be represented and managed efficiently using decision diagram techniques (e.g., Binary Decision Diagrams BDDs). As a result, by highlighting observed transitions, the SOG represents the behavior of all process configurations in only one reduced graph.

In the following, the SOG definition and construction algorithm are adapted to the defined $CBP2PN$ semantics. Before defining the SOG associated with a $CBP2PN$, we formally present an aggregate as follows.

Definition 13 (Aggregate).   Let $N = \langle P, T \cup OP, F, W, O, L, C \rangle$ be a $CBP2PN$. Let m_i and m_f be the corresponding initial and final markings respectively. An aggregate a of N w.r.t. Obs is a couple $\langle S, f \rangle$ satisfying:

- $S \subseteq R(N, m_i)$ is a set of reachable markings, where $\forall s \in S$:
   1. $(\exists (s', u) \in R(N, m_i) \times UnObs \mid s \xrightarrow{u} s') \Leftrightarrow s' \in S$;
   2. $(\exists (s', o) \in R(N, m_i) \times Obs \mid s \xrightarrow{o} s') \wedge (/\exists (s'', u) \in S \times UnObs \mid s'' \xrightarrow{u} s') \Leftrightarrow s' \notin S$.
- $d \in \{true, false\}$; $d = true$ iff S contains a dead state.
- $f \in \{true, false\}$; $f = true$ iff S contains a final state.

In addition to the d and f attributes of an aggregate, the above definition specifies the states that must belong to an aggregate (the aggregation criterium) and those that must be excluded: (1) For any state s in the aggregate, any state s' being reachable from s by the occurrence of an unobserved transition, belongs necessarily to the same aggregate. (2) For any state s in the aggregate, any state s' which is reachable from s by the occurrence of an observed transition is necessarily outside the aggregate, unless s' is reachable from a state s' in the aggregate by an unobserved transition.

In order to define the SOG, we first introduce the operation $Out(a, t)$. It returns, for an aggregate a and an observed transition t, the set of states outside a that are reachable from some state in a by firing t, i.e., $Out(a, t) = \{s' \in R(N, m_i) \mid \exists s \in a.S, s \xrightarrow{t} s'\}$.

Definition 14 (Deterministic SOG). *Let* $N = \langle P, T \cup OP, F, W, O, L, C \rangle$
be a CBP2PN. Let m_i *and* m_f *be the corresponding initial and final markings
respectively. The Deterministic Symbolic Observation Graph (SOG) associated
with* N, *over Obs, is a graph* $\mathcal{G} = \langle A, Obs, \rightarrow, A_0, \Omega \rangle$ *where:*

1. *A is a non empty finite set of aggregates satisfying:*
    - $\forall a \in A, \ \forall t \in Obs, \ Out(a,t) \neq \emptyset \implies \exists a' \in A \ s.t. \ a' = Sat(Out(a,t), UnObs)$
2. $\rightarrow \subseteq A \times Obs \times A$ *is the transition relation where:*
    - $((a,t,a') \in \rightarrow) \Leftrightarrow ((t \in Obs) \wedge Out(a,t) \neq \emptyset \wedge a' = Sat(Out(a,t), UnObs))$
3. A_0 *is the initial aggregate s.t.* $A_0.S = Sat(m_i, UnObs)$.
4. $\Omega = \{a \in A \mid m_f \in a.S\}$.

The nodes of the symbolic observation graph are aggregates (1). The finite set
of aggregates A of a SOG is defined in a complete manner so that the necessary
aggregates are represented. Point (2) defines the transitions relation: there exists
an arc, labeled with an observed transition t, from a to a' iff a' is obtained by
saturation on the set of reached states $(Out(a,t))$ by the firing of t from $a.S$.
The last two points of Definition 14 characterize the initial aggregate and the
set of final aggregates respectively.

Starting from the initial marking, the original SOG construction algorithm
introduced in [11] follows a classical depth first search based traversal of the
built aggregates. Each aggregate is built by a transitive closure application on
unobserved transitions. The successor a' of an aggregate a is built by, first, firing
an observed transition from states of a, then by adding all the reachable states by
unobserved transition. The correctness of the *SOG* associated with a *CBP2PN*
was characterized and proved in our previous work [8].

In the following, we propose to adapt the original SOG construction algorithm [11], associated with a *CBP2PN*, in three ways. First, by adopting the
new semantics. Second, the deadlock-freeness property is checked on the fly,
such that any aggregate containing a deadlock state is not inserted in the graph
and so are all the underlying paths. Finally, the set of correct configurations is
extracted on-the-fly.

4.2 Extracting Correct Configurations Using the SOG

In this section, we present the core contribution of this paper: A construction
algorithm of the SOG associated with a *CBP2PN*. Regarding to the original
SOG construction algorithm [11], Algorithm 1 allows to reduce the SOG, by
removing, on-the-fly, the paths involved in incorrect configurations, and by saving, within the initial aggregate the correct configurations. To reach this goal,
two new attributes are added to an aggregate: (1) c, which is the set of correct
(possibly partial) configurations, starting from this aggregate (and leading to a
final aggregate). (2) nc, which is the set of incorrect (possibly partial) configurations, starting from this aggregate (leading to a dead one). Once the *SOG* is
built, the set of correct configurations will be saved within the initial aggregate.

In the following, we go through Algorithm 1 to explain the main steps while using a SOG example in Fig. 5 for illustration. Note that the main novelties of this algorithm w.r.t. the original one in [11], are underlined.

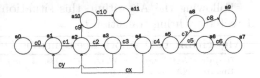

Fig. 5. An example of a SOG

Two main data are used: The SOG graph $\mathcal{G}$, containing aggregates and edges, and a stack containing the to-be-treated aggregates associated with the set of fireable observed transitions F_{obs}.

The first step of Algorithm 1 (lines 5–10) allows to build the initial aggregate and to push it onto the stack. Then, the main loop (lines 11–56) processes the set of to-be-treated aggregates as follows: a stack item (line 12) and the corresponding current observed transition in F_{obs} (line 14) are picked, and the successor of the current aggregate by that transition, if any, is calculated using the semantics of Subsect. 3.2 (lines 15–20). This includes the computation of the dead (line 19) and final (line 20) attributes of the obtained successor aggregate.

If the latter aggregate is deadlock-free (line 21), has not already been explored (line 22), and if it is not in the stack (line 23), then it is pushed onto the stack with its set if fireable observed transitions (lines 24–25). For instance, considering the example in Fig. 5, we start the construction of this SOG by pushing the aggregate a_0 onto the stack then consecutively push aggregates until a final one (i.e. does not enable any observable transition). As depicted by step (1) in Fig. 6, a_7 is found as a final aggregate, then it is popped from the stack (line 45) and we start the loop again by picking a_6 to consider its remaining observed transitions (step (2)). In this case, there are no other remaining observed transitions, so we go back again to a_5 (step (3)), and so on. Since this first path leads to a final aggregate, the corresponding configuration is considered correct. Then before popping an aggregate, its set of correct configuration is updated (green list in aggregates of Fig. 6). However, if the obtained aggregate holds a dead state, e.g., the aggregate a_9 in step (5), then the corresponding fired observed transition, e.g., $c8$, is concatenated to the incorrect configurations of its predecessor a_8 (line 39 in Algorithm 1), depicted by the red list in aggregates of Fig. 6. Obviously, a' is not pushed onto the stack and no edge is created. Then, we recursively verify its predecessors starting from a using the function $recRemoveAggregate(a, t)$ (line 40). Using this function, each predecessor aggregate is removed only if the states enabling the current one becomes dead (i.e. there is no other enabled transition from that state). In this case, its successors are also recursively eliminated in case they do not have other predecessors. If the newly built successor aggregate a' is still in the stack (line 26), then it is not entirely treated and may still have other observed transitions to fire (that is why it is not yet popped from the stack). In fact, this means that a' is actually both successor and predecessor of the aggregate a. In step (9) of Fig. 6, after popping a_5, a new observed transition cx is fired from a_4. However, the obtained aggregate a_2 is actually in the stack because it is not entirely treated. So, a_2 is both successor and predecessor of a_4 which means that the corresponding *CBP2PN* contains a loop.

Following the Algorithm, this situation leads to a loop between a and a' detected by firing transition t.

Algorithm 1. Deadlock-free Symbolic Observation Graph

Require: $N\langle P, T \cup OP, F, W, O, L, C\rangle$, Obs, m_i, m_f
Ensure: $\mathcal{G}\langle A, Obs, \rightarrow, A_0, \Omega\rangle$, C
1: Vertices $A=\emptyset$; vertex a, a'; {Aggregates}
2: Vertices $C=\emptyset$; {Correct configurations}
3: set $S, S', UnObs = (T \cup OP) \setminus Obs$, F_{obs}, F'_{obs};
4: stack st; Edges E$= \emptyset$;
5: $S = Sat(\{m_i\}, UnObs)$; {first Aggregate}
6: $a.S = S$;
7: $a.d = DetectDead(a.S)$;
8: $a.f = IsFinal(a)$;
9: $F_{obs} = fireableObs(a)$; {fireable observed transitions of a}
10: $st.Push(\langle a, F_{obs}\rangle)$;
11: **while** $st \neq \emptyset$ **do**
12:     $\langle a, F_{obs}\rangle = st.Top()$;
13:     **if** $(F_{obs} \neq \emptyset)$ **then**
14:         $t = F_{obs}.next()$;
15:         $S' = Out(a.S, t)$
16:         **if** $(S' \neq \emptyset)$ **then**
17:             $S' = Sat(S', UnObs)$;
18:             $a'.S = S'$;
19:             $a'.d = DetectDead(a'.S)$;
20:             $a'.f = IsFinal(a')$;
21:             **if** $(\neg a'.d)$ **then** {there is no dead state in a'}
22:                 **if** $(\nexists x \in A$ s.t. $x == a')$ **then** {a' found for the first time}
23:                     **if** $(\nexists y \in st$ s.t. $y.key() == a')$ **then** {a' is not in the stack}
24:                         $F'_{obs} = fireableObs(a')$;
25:                         $st.Push(\langle a', F'_{obs}\rangle)$;
26:                     **else** {a' not yet entirely treated and popped from the stack}
27:                         a' exists in the stack $\Rightarrow$ it is both a successor and a predecessor of a;
28:                         $UpdateIncomplete(a, a', t)$;
29:                         $\rightarrow = \rightarrow \cup \{a, \langle t, Conf(t)\rangle, a'\}$;
30:                     **end if**
31:                 **else** {a' is an existing aggregate}
32:                     free a';
33:                     Let a' be the already existing aggregate;
34:                     $\rightarrow = \rightarrow \cup \{a, \langle t, Conf(t)\rangle, a'\}$;
35:                     $UpdateC(a, a', t)$;
36:                     $UpdateNC(a, a', t)$;
37:                 **end if**
38:             **else** {there is a dead state in a'}
39:                 $a.nc = a.nc \cup \{\langle t, Conf(t)\rangle\}$;
40:                 $recRemoveAggregate(a, t)$
41:             **end if**
42:         **end if**
43:         $a.c = a.c \cup \{\langle t, Conf(t)\rangle\}$;
44:         $CompareCorrect(a)$;
45:         $st.Pop()$;
46:         $A = A \cup \{a\}$;
47:         **if** $(m_i \in a.S)$ **then**
48:             $C = a.c$;
49:             **while** a is incomplete **do**
50:                 $UpdateCLoop(Succ(a))$;
51:                 $UpdateNCLoop(Succ(a))$;
52:                 $a \leftarrow Succ(a)$;
53:             **end while**
54:         **end if**
55:     **end if**
56: **end while**

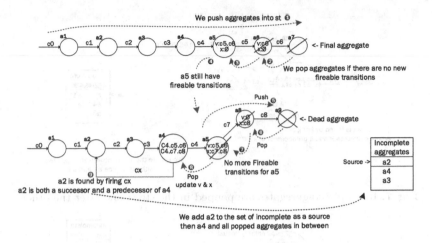

Fig. 6. An example of SOG construction in case of a CBP2PN having a loop

To deal with it, we add a new aggregate's attribute: *incomplete* that represent a structure to save the set of incomplete aggregates starting from a'. Updating this attribute is ensured by the function *UpdateIncomplete* (line 28). Then, we add the arc from a to a' (line 29). This allows to keep an information about this loop in order to consider it in a further step while updating the set of correct and incorrect configurations c and nc. As can be seen in Fig. 6, the aggregate a_2 has been saved as a source in the *incomplete* list, and below, the aggregate a_4 is firstly added. Then, once this later is popped, a_3 is added. Afterwards, once the source aggregate a_2 is entirely treated and popped from the stack (cf. Fig. 7), we use the functions *UpdateCLoop* and *UpdateNCLoop* (lines 49–53) to update the lists of correct and incorrect configurations. Thus, we update these lists by concatenating the configurations of the different other paths with the one containing the loop. This is done consecutively for all incomplete successors (cf. Fig. 8).

If the newly built successor aggregate a' has already been treated (lines 31–36), then the current aggregate a inherits from a' its correct and incorrect configuration (to which the transition linking a to a' is added). This is ensured by functions *UpdateC* and *UpdateNC* (lines 35–36). The function *UpdateC* also verifies that, starting from the same aggregate a, a correct configuration do not include an existing (or to-be-treated) incorrect one, as in this case it leads to a deadlock in a different transitions' firing order. This way, correct and incorrect configurations are computed backwards starting from the final aggregate to the initial one.

It is worth noting that before popping an aggregate from the stack and storing it in the graph (lines 45–46), a final check is carried out on its correct configurations by the function *CompareCorrect* (line 44). Actually, many observed transitions may be fired from the same aggregate, so some of the corresponding correct configurations may refer to the same one. Hence, a correct sequence

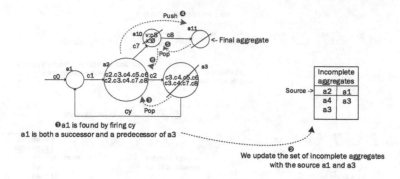

Fig. 7. Incomplete aggregates are popped in the stack one after the other

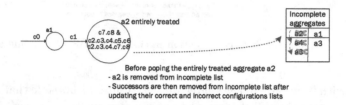

Fig. 8. Updating incomplete aggregates in order to consider the loop configuration

is preserved if, for every first fired observed transition op, (i) it is fireable by the states that have fired another sequence starting by op (i.e. different configurations), or (ii) if their common operators have the same configured type (i.e. the same configurations but in a different order). Otherwise, the sequence is considered as incorrect and is eliminated.

Finally, the set of correct configurations is obtained from the initial aggregate, the last one popped from the stack. As a result, each path of the obtained SOG starting from the initial aggregate and leading to a final aggregate, represents one possible configuration and belongs to the set of configurations C. In this case, this configuration leads to a deadlock-free *BP2PN*. Note that, different paths could represent a configuration (e.g. two concurrent configurable connectors). The obtained reduced SOG of our example contains 29 nodes and 76 arcs, and 135 correct configurations. Which is very reduced in comparison with the size of the original graphs of configurations (162 possible configurations of our example). For instance, the configuration of s_1, j_2, s_3 and j_3 to OR, s_4 and j_4 to XOR, and a_4 to on leads to a *BP2PN* whose reachability graph contains 655 states and 1674 arcs, while in our SOG this configuration is ensured by 8 nodes and 7 arcs. Due to the lack of space, we could not illustrate the entire SOG (available at[3]).

---

[3] http://www-lipn.univ-paris13.fr/~klai/SignavioExtensionSOG.

5   Usage and Proof of Concept

As a proof of concept, we have developed an extension to the Signavio Process Editor. Signavio is an open source web application for modeling business processes in BPMN that, initially, does not support neither process configuration nor the correctness of designed variants. To fill this gap, first, we developed a plugin that allows to start the configuration procedure by charging an input file including the list of correct configurations. This list is provided by our SOG-based application that computes the corresponding SOG of the designed process and returns the set of its correct possible configurations[4]. Afterwards, we have particularly extended a function that initially allows to load a list of possible connectors types (default) that may replace a designed connector. Using this extension, only the correct configurations provided by the SOG are charged in this list. Thus, the analyst may choose only one type that is allowed by this list. For example, in Fig. 9, after configuring s_1 to XOR, the only allowed configurations for j_2 are OR and XOR (see footnote 3 for more details about this implementation).

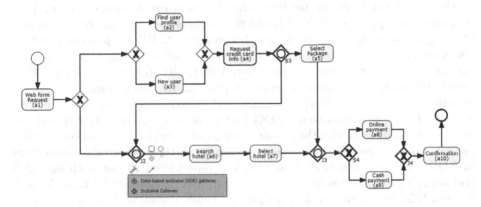

Fig. 9. An example of the configuration extension in Signavio

6   Discussion

In order to facilitate the design of configurable process models, a range of process modeling languages have been recently extended with variable elements such as Event-driven Process Chain (EPC) (e.g. [1,18]), Business Process Model and Notation (BPMN) (e.g. [5,13]), Unified Modeling Language (UML) (e.g. [17]), and Yet Another Workflow Language (YAWL) (e.g. [10]). Based on some of them, a number of approaches have attempted to reach correct process configuration either syntactically [10,18] or behaviorally. Traditionally, behavioral correctness

---

[4] For more details see http://www-inf.it-sudparis.eu/SIMBAD/tools/SOGImplementation.

related to process configuration can be handled by verifying every single possible configuration using existing work on verification of business processes and work-flows [3] and some existing tools such as Woflan [19]. However, these methods are too time-consuming and lead to the state space explosion problem. Authors in [12] discuss the Provop approach [13] for ensuring soundness of process vari-ants derived by options. However, this approach is not feasible in large processes and runs into the state space problem. In [2], Petri net was used to formalize and verify correctness and soundness properties of Configurable EPC (C-EPC) processes. They derive propositional logic constraints that guarantee the behav-ioral correctness of the configured model. However, in these approaches authors achieve correctness by checking constraints at each configuration step. Also, authors impose that the C-EPC process model should be syntactically correct. In our work, we propose a model that finds all possible correct configurations at design time instead of configuration time without any restriction on the input C-BPMN process. This allows the process analyst to derive correct processes without intermediate computing.

In [4], they focused on the behavioral correctness of the configured model and moved the checking up to design time. Their approach is based on partner synthesis and considers the weak termination as a correctness criterion. This technique was applied on C-YAWL and the configuration is build by hiding and blocking transitions, while our approach configures C-BPMN processes (which offers an increasing business support and an easier modeling experience) by changing configurable elements behavior. Practically, the synthesis algorithm used in [4] constructs an automaton that represents the configuration guidelines. In order to evaluate our work, we conducted an exhaustive study in which we compare their automaton to our SOG and show how significantly reduced the latter is in comparison with the former in terms of states number and complexity. For example, in our SOG, the configuration of all the configurable connectors to XOR and the configurable task to ON is represented by one path having 8 nodes and 7 arcs, whereas in their automaton, it is represented by 16 nodes and 33 arcs. It is also worth mentioning that this configuration is the simplest one to represent in their case. In fact, the configuration of connectors to OR gets much higher numbers while it is the same in the SOG. Due to the limited space, further details about the evaluation aforementioned are available in our web page[5].

[16], which is applied on C-EPC using questionnaire models, and [6], which is applied on C-BPMN using configuration guidelines, have attempted to provide guidance to analysts for process configuration, however, these approaches espe-cially ensure domain compliant variant and they do not consider any correctness criterion.

In our previous work [7], a formal approach for deriving correct process variants from a C-BPMN was proposed. It models the process using Event-B language and verifies the different constraints and properties using predicates. These predicates must be satisfied by each configuration step. This work con-tributes essentially to prevent structural correctness issues in process models

---

[5] http://www-inf.it-sudparis.eu/SIMBAD/tools/SOGEvaluation.

configuration using a systematic design. However, structural correctness may be not sufficient. In [8], we especially propose a behavioral approach to achieve the correctness of the process configuration, and in this current work, we propose to extend this approach by considering the new semantics of the synchronizing OR-join and processes with loops.

7   Conclusion

In this work, we propose an approach for ensuring deadlock-free process configuration. We use a SOG-based behavioral model to find all deadlock-free configurations while considering synchronizing OR-joins and loops. Behavioral anomalies are excluded on-the-fly while constructing the SOG of the configurable process model. As a result, we obtain a reduced graph in which only correct instances of process variants are preserved. Then, each path from the initial aggregate to the final one represent a possible correct configuration. The list of these configurations serves as an input of a plugin that was developed to assist the designers in the configuration of their processes.

As future work, we plan to first take into account other types of process configurations such as resource configuration. Then, we aim to adapt the SOG construction algorithm in order to integrate other specific properties such as domain constraints.

References

1. van der Aalst, W.M.P., Dreiling, A., Gottschalk, F., Rosemann, M., Jansen-Vullers, M.H.: Configurable process models as a basis for reference modeling. In: Bussler, C.J., Haller, A. (eds.) BPM 2005. LNCS, vol. 3812, pp. 512–518. Springer, Heidelberg (2006). https://doi.org/10.1007/11678564_47
2. Aalst, W.V.D., et al.: Preserving correctness during business process model configuration. Form. Asp. Comput. **22**(3–4), 459–482 (2008)
3. Aalst, W.V.D., et al.: Soundness of workflow nets: classification, decidability, and analysis. Formal Asp. Comput. **23**(3), 333–363 (2011)
4. Aalst, W.V.D., Lohmann, N., Rosa, M.L.: Ensuring correctness during process configuration via partner synthesis. Inf. Syst. **37**(6), 574–592 (2012)
5. Assy, N.: Automated support of the variability in configurable process models. Ph.D. thesis, University of Paris-Saclay, France (2015)
6. Assy, N., Gaaloul, W.: Extracting configuration guidance models from business process repositories. In: Motahari-Nezhad, H.R., Recker, J., Weidlich, M. (eds.) BPM 2015. LNCS, vol. 9253, pp. 198–206. Springer, Cham (2015). https://doi.org/10.1007/978-3-319-23063-4_14
7. Boubaker, S., Mammar, A., Graiet, M., Gaaloul, W.: A formal guidance approach for correct process configuration. In: Sheng, Q.Z., Stroulia, E., Tata, S., Bhiri, S. (eds.) ICSOC 2016. LNCS, vol. 9936, pp. 483–498. Springer, Cham (2016). https://doi.org/10.1007/978-3-319-46295-0_30
8. Boubaker, S., Klai, K., Schmitz, K., Graiet, M., Gaaloul, W.: Deadlock-freeness verification of business process configuration using SOG. In: Maximilien, M., Vallecillo, A., Wang, J., Oriol, M. (eds.) ICSOC 2017. LNCS, vol. 10601, pp. 96–112. Springer, Cham (2017). https://doi.org/10.1007/978-3-319-69035-3_7

642   S. Boubaker et al.

9. Dijkman, R., Dumas, M., Ouyang, C.: Semantics and analysis of business process models in BPMN. Inf. Softw. Technol. **50**(12), 1281–1294 (2008)
10. Gottschalk, F., Aalst, W.V.D., Jansen-Vullers, M., Rosa, M.L.: Configurable workflow models. Int. J. Coop. Inf. Syst. **17**, 177–221 (2008)
11. Haddad, S., Ilié, J.-M., Klai, K.: Design and evaluation of a symbolic and abstraction-based model checker. In: Wang, F. (ed.) ATVA 2004. LNCS, vol. 3299, pp. 196–210. Springer, Heidelberg (2004). https://doi.org/10.1007/978-3-540-30476-0_19
12. Hallerbach, A., Bauer, T., Reichert, M.: Guaranteeing soundness of configurable process variants in Provop. In: IEEE Conference on Commerce and Enterprise Computing, CEC, pp. 98–105 (2009)
13. Hallerbach, A., Bauer, T., Reichert, M.: Capturing variability in business process models: the Provop approach. J. Softw. Maint. **22**(6–7), 519–546 (2010)
14. Klai, K., Tata, S., Desel, J.: Symbolic abstraction and deadlock-freeness verification of inter-enterprise processes. In: Dayal, U., Eder, J., Koehler, J., Reijers, H.A. (eds.) BPM 2009. LNCS, vol. 5701, pp. 294–309. Springer, Heidelberg (2009). https://doi.org/10.1007/978-3-642-03848-8_20
15. Klai, K., Tata, S., Desel, J.: Symbolic abstraction and deadlock-freeness verification of inter-enterprise processes. Data Knowl. Eng. **70**(5), 467–482 (2011)
16. La Rosa, M., et al.: Questionnaire-based variability modeling for system configuration. Softw. Syst. Model. **8**(2), 251–274 (2008)
17. Razavian, M., Khosravi, R.: Modeling variability in business process models using UML. In: Fifth International Conference on Information Technology: New Generations. ITNG 2008, pp. 82–87, April 2008
18. Rosemann, M., Aalst, W.V.D.: A configurable reference modelling language. Inf. Syst. **32**(1), 1–23 (2007)
19. Verbeek, H., Basten, T., Aalst, W.V.D.: Diagnosing workflow processes using Woflan. Comput. J. **44**(4), 246–279 (2001)
20. Weske, M.: Business Process Management: Concepts, Languages, Architectures. Springer, Heidelberg (2007)

Change Detection in Event Logs by Clustering

Agnes Koschmider[✉] and Daniel Siqueira Vidal Moreira

Institute AIFB, Karlsruhe Institute of Technology, Karlsruhe, Germany
agnes.koschmider@kit.edu, daniel.moreira@student.kit.edu

Abstract. The detection of changes in event logs recording the behavior of flexible processes is especially challenging and process mining algorithms generate useless "spaghetti" models out of them. Due to this, existing approaches for change detection in event logs recording the behavior of flexible processes can only localize a change point, which is of no avail when it comes to explain *when*, *why* and *how* a process model changed and *will* change. The aim of this paper is to present a novel clustering technique laying the foundation to determine a variety of changes and to foresee changes. In order to do this, four algorithms have been developed. We report the results of evaluations on synthetic as well as real-life data demonstrating the efficiency of the approach and also its broad scope of application for event and sensor data.

1 Introduction

Event log files are used as input to any process mining algorithm. Often, the aim of these algorithms is to derive an as-is model of the process that created these logs that can be used to further analyze the actual process execution. Process mining algorithms can be applied to historical data or real-time data where log files of low-level sensor data is processed. For such data, change detection is especially challenging. Changes or drifts might occur while the process is running or under analysis [1]. Many of the different kinds of event, e.g., a sudden change in sensor data, do not comply with the strict assumptions of process mining methods. The changed log is, therefore, not considered by process mining methods. The challenge of change detection increases when event logs record the behavior of unstructured processes. Unstructured or flexible processes are processes with little structure and process mining algorithms have problems dealing with such log files. They create spaghetti-like process models, which are harmful with respect to analysis and the comparison of spaghetti-like process models is almost not feasible. An automated detection of changes in event logs and also in those that record the behavior of unstructured processes would be of great avail for process mining and event log analysis in general. This paper presents an approach for change detection in sensor and event data that might

© Springer Nature Switzerland AG 2018
H. Panetto et al. (Eds.): OTM 2018 Conferences, LNCS 11229, pp. 643–660, 2018.
https://doi.org/10.1007/978-3-030-02610-3_36

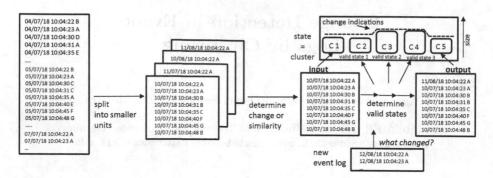

Fig. 1. Overview of the approach for change detection: changes are detected by the analysis of clusters' size. A change point means that the cluster size is higher than that of neighbor clusters as found for C2 and C3 and C4 and C5.

also be recorded from unstructured processes. For this, we define a novel clustering approach in order to structure event logs. Changes are detected and foreseen by analysis of the clusters over time. Firstly, an event log is split into clusters of size 1 for which a new notion of behavioral morphing is defined. The number of clusters must not be specified beforehand, but it will not change after the application of the behavioral morphing approach, only the size of clusters increases. Figure 1 broadly summarizes our approach for change detection. An event log is first split into smaller event logs according to, for example, time or location and then transformed into a particular data structure (not shown in this figure and explained in Sect. 2). The event log must record at least a date, time-stamp and event or activity and might track the behavior of any sensor connected environments (e.g., CRM, medical processes, smart production). Also (graphical) process models might be used as source, which are transformed into the appropriate required data format. Then, an input and output event log are selected and their behavioral similarity is iteratively calculated where each iteration represents one cluster. If new event data emerges then similarities to existing clusters are calculated and the size of those clusters that are most similar to the new event log increases. Through the analysis of the change evolution and the clusters' size, different kinds of changes can be detected with this approach, an insight which was impossible with previous process mining methods. Besides change detection, the approach presented in this paper also addresses similarity calculation between event logs as use case (see Fig. 2).

The structure of this paper is as follows. Section 2 presents the approach of behavioral morphing to structure event logs. Section 3 describes change detection and prediction based on the analysis of the clusters' size. Evaluation results are summarized in Sect. 4. The paper ends with a summary and an outlook on future work.

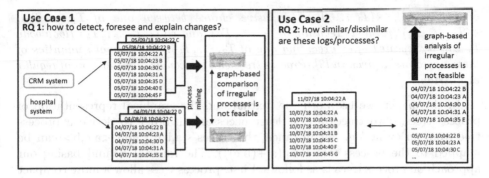

Fig. 2. This paper addresses research questions related to detection and prediction of changes and to the similarity of event logs by means of changes.

2 Approach to Structure Event Logs

To provide solutions for the research questions **RQ1** and **RQ2** from Fig. 2, this approach works as follows: First, the approach requires an input and an output, which can be event logs or graphical process models such as BPMN diagrams. An event log L (over a set of all events Σ) is a multiset of traces. A trace t is a sequence of events (possibly empty) $t \in \Sigma^*$. In Fig. 2 the trace <A,B,C,A,D,E,F> can be identified for the event log from 05/07/18 where A can be the event "registration off". Events can also refer to sensor IDs. Then a trace with the events <M45,M44,M37,M38> refers to e.g., motion sensors in a storehouse. The output event log (see Fig. 1 "output") is called a *target*. The rationale behind a *target* is to have a reference point to be reached. Given an event log with events of the past 30 days where changes should be detected. For this, the log is divided e.g., according to phases of days (i.e., morning, afternoon, evening, night). The target can then be the trace of the last two days of the 30 days. The target could also be the activities of a particular customer and then our approach is used to indicate changes in the consumption behavior. After the selection of the target, the input event log (or graphical process models) has to be selected, which is called the *origin*. With an input and output event log, first a transformation into process trees or behavior-oriented trace is conducted and then a behavioral morphing between the trees is applied aiming to determine clusters and finally changes. All terms and the corresponding algorithms are explained subsequently.

2.1 Preliminaries

We define a process tree and behavior-oriented trace as follows.

Definition 1. Process trees [2,3] *allow for defining block-structured process models in a recursive way. Let Σ be a set of activities. Each activity $a \in \Sigma$ is a process tree over Σ. Let $T_1, \ldots, T_k$ be process trees over Σ, then the following expressions are also process trees over Σ: $\rightarrow (T_1, \ldots, T_k)$ (sequence* of $T_1, \ldots, T_k$), $\times(T_1, \ldots, T_k)$ (exclusive choice between one of $T_1, \ldots, T_k$), $+(T_1, \ldots, T_k)$ (parallel execution of $T_1, \ldots, T_k$), and $xl(T_1, T_2, \ldots, T_k)$ (loop body T_1 can be repeated after executing one of $T_2, \ldots, T_k$). The formal semantics of these operators is given in [3]. Process trees can be seen as an extension of regular expressions.

In this paper, we use both the textual and the graphical representation of process trees. Two traces are at least necessary in order to derive a process tree from an event log. For instance, the traces <a,b,c> and <a,c,b> can be mapped to the process tree (seq,(a,+(b,c))). The rational behind basing our approach on process trees is as follows. First, process trees allow a more compact representation of traces (see Fig. 3) and efficient parsing techniques for trees exist. Second, with our approach process trees can be derived for unstructured processes. While process mining approaches aim to derive a (sound) as-is model of the process that created these logs and thus they aim to derive (sound) process trees, this assumption of soundness is relaxed in our approach. Our approach aims to identify *valid process trees* (and not sound) and thus event logs can be structured even for unstructured processes.

Definition 2. Behavior-oriented trace *A behavior-oriented trace bt is the textual representation of a process tree.*

Definition 3. Valid process tree *A process tree, represented either as behavior-oriented trace or graphically as process tree is valid if each branch is syntactically correct (i.e. complete).*

Figure 3 shows an event log and their traces. A process tree can be derived from these traces. In this example, the process has the following semantics. The events a and b are executed in sequence, followed by events c,d and e in a loop and f and g in a loop. This graphical process tree corresponds to the following behavior-oriented trace: (seq,(a,b,(xl,(xl,c,d,e))),f,g)). Figure 4 shows the identification of valid and invalid process trees for the input (seq,(a,b,(xl,(xl,c,d,e))),f,g)) and the output (seq,(a,(xl,(xl,(x,c,d,e)),tau),tau),b))). For this input and output three valid process trees are determined with the morphing, deletion and insertion algorithms as explained below. In this example, not valid process trees exist when missing child nodes for branching nodes are found. Thus, they cannot be mapped to a syntactically correct process model. For instance, the first process tree is invalid due to a missing silent activity for the redo loop operator. A redo loop has to be defined with at least two children. "The first child is the *do* part and the other children are *redo* parts" [1]. The black nodes in Fig. 4 within the redo loop operator are placeholder for one of these children. This means that at least two subnodes are required for redo loop operators. Placeholder nodes might be replaced later by concrete events. If one of these subnodes is missing, then the process tree cannot be mapped to a syntactical correct graphical process model.

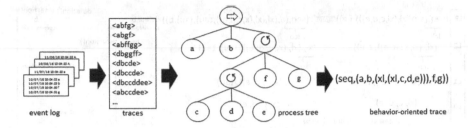

Fig. 3. This figure shows how a graphical process tree and a behavior-oriented tree are determined from an event log.

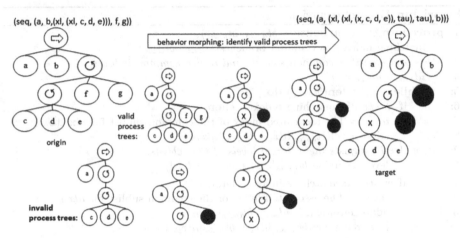

Fig. 4. Identification of valid process trees for an input and output process trees. Exemplary, invalid process trees are shown.

2.2   Algorithms to Determine Valid Process Trees

To derive valid process trees the concepts of recursion, longest common branches, insertion and deletion operations are used. Algorithms 1, 2 and 3 determine all valid process trees between an input and an output. For this, Algorithm 1 (level recursion) traverses a process tree or behavior-oriented trace as array list and uses longest common branches as search indices. The advantages of array lists (compared to hash keys) as data format is an ordered storage of objects, possibility to use duplicate items and objects can be accessed by an index. Duplicates occur for redo loops and for placeholders (black nodes). Also with lists, the identification of longest common branches can be carried out extremely efficiently. The decomposition of lists and the identification of valid process trees is conducted recursively as exemplary explained in Fig. 5. After decomposition into lists, the tree is first traversed from left to right. As soon as a branching is found, the algorithm traverses bottom-up and intends to find *longest common branches* (lcb). A lcb is defined by at least a 2-gram item. The benefit of lcbs is their use as search indices for elements and thus similar elements can be

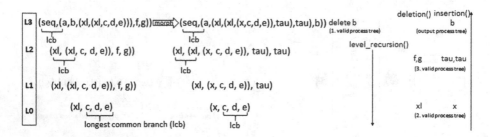

Fig. 5. Three valid process trees exist for the input and output from Fig. 4.

Algorithm 1. Level recursion

1: **procedure** LEVEL_RECURSION(bt_{as_is}, bt_{to_be})
2:     matching_indexes = common_branches(bt_{as_is}, bt_{to_be})
3:   *This part identifies common sublists and return common indexes*
4:     $idx_{tb} = idx_{is} = 0$
5:     **while** $idx_{is} < $ **len**(bt_{as_is}) **do**
6:         **if** *a:* idx_{is} in matching_indexes: **then**
7:             $idx_{tb} = $ The associated index of the common sublist for idx _tb
8:             adapt_level($bt_{as\_is}[idx_{is}]$, $bt_{to\_be}[idx_{tb}]$)
9:   *Adapt _level() adjusts the different levels of the sublists.*
10:             level_recursion($bt_{as\_is}[idx_{is}]$, $bt_{to\_be}[idx_{tb}]$)
11:         **if** *b:* idx_{tb} in matching_indexes: **then**
12:             $idx_{is} = $ The associated index of the common sublist for idx_{tb}
13:             adapt_level($bt_{as\_is}[idx_{is}]$, $bt_{to\_be}[idx_{tb}]$)
14:             level_recursion($bt_{as\_is}[idx_{is}]$, $bt_{to\_be}[idx_{tb}]$)
15:         **if** *c:* $bt_{as\_is}[idx_{is}]$ is **instance**(list)
16:         **and** $bt_{to\_be}[idx_{tb}]$ is **instance**(list): **then**
17:             adapt_level($bt_{as\_is}[idx_{is}]$, $bt_{to\_be}[idx_{tb}]$)
18:             level_recursion($bt_{as\_is}[idx_{is}]$, $bt_{to\_be}[idx_{tb}]$)
19:             matching_indexes.append(idx_{is},idx_{tb})
20:   *Matching indexes is extended with sublists without common events*
21:         **if** *d:* $bt_{as\_is}[idx_{is}]$ is **instance**(list): **then**
22:             adapt_level($bt_{as\_is}[idx_{is}]$, $bt_{to\_be}[idx_{tb}]$)
23:             level_recursion($bt_{as\_is}[idx_{is}]$, bt_{to_be})
24:         idx_{tb} += 1 **if not lastIndex**(bt_{to_be})
25:         idx_{is} += 1
26:     **if** idx_{is} is lastIndex(bt_{as_is}) **then**
27:         $bt_{as_is} = $ delete(bt_{as_is},bt_{to_be},matching_indexes)
28:     **return** bt_{as_is}

---

identified faster than comparing each element with each other. Figure 5 shows the application of the morphing, deletion and insertion algorithms for the input (seq,(a,b,(xl,(xl,c,d,e))),f,g)) and the output (seq,a,(xl,(xl,(x,c,d,e)),tau),tau),b)).

First, all elements until the first branching are compared resulting in the identification of (seq,a) as the first lcb and the deletion of b in the input tree. Then, the first valid process tree is (scq,a,(xl,(xl,c,d,e)),f,g)). Next, the morphing algorithm identifies a branching and navigates to the deepest list. It identifies the next lcb resulting in the deletion of xl and the insertion of x. Due to the second lcb, the second valid process tree is (seq,(a,(xl,(x,c,d,e))),f,g)). The third lcb is found at level L2 resulting in the deletion of the events f and g and the insertion of tau,tau. Finally, the third process tree is (seq,(a,(xl,(xl,(x,c,d,e)),tau),tau)))). Eventually, the event b has to be inserted in order to obtain the output process tree (seq,(a,(xl,(xl,(x,c,d,e)),tau),tau),b))).

Algorithm 1 works as follows. **Line 2-5**: the method common_branches() compares all pairs of sublists of the current level of two process trees, sorts them descending by the number of common sequences. Common indexes is the return value. **Line 5-25**: the while loop iterates over each level using the information from matching_indexes. The cases a to d analyze all possible occurrences (lists with lcb or without any lcs or single events) of the input to the output. The method adapt_level() adjusts the different levels of the sublists, if this can reduce the number of transformation steps. Finally, level_recursion() is called again in order to analyze the reduced subtrees according to the principle divide and conquer. **Line 25-28**: the deletion() method is called and removes elements that cannot be found on the identical level.

The deletion algorithm works as follows: **Line 5-14**: the matching_blocks() method is used to determine the longest common branches on identical levels. The return values either contain lcbs or single events. **Line 5-19**: within the loop all events that do not exist at same levels (between the input and output) are deleted. Insertion of elements is inverse to deletion. Insertion of

---

Algorithm 2. Deletion

1: **procedure** DELETION(bt_{as_is}, bt_{to_be}, matching_indexes)
2:     j = lastmatch = 0
3:     $blocked_{as\_is}$,$blocked_{to\_be}$ = matching_blocks(bt_{as_is},bt_{to_be},matching_indexes)
4:     $blocked_{as\_is}$ and $blocked_{to\_be}$ contain the indexes of the common sequences.
5:     **for** i, is_val in **enumerate**(bt_{as_is}) **do**
6:         **if** is_val is operator **then continue**
7:         **if** i in blocked_is **then**
8:             j = lastmatch = $blocked_{to\_be}$[0] + 1
9:             **del** $blocked_{as\_is}$[0] **and** $blocked_{to\_be}$[0]
10:            **continue**
11:        **else:**
12:            **del** bt_{as_is}[i]
13:            output(bt_{as_is})
14: *In the output method, the validity of the process tree is checked.*
15:        **if** is_val is lastElement(bt_{as_is})
16: **and** as_is operator **is not** to_be operator **then**
17:            **del** bt_{as_is}[0]
18:            output(bt_{as_is})
19:     **return** bt_{as_is}

---

elements (see Algorithm3) is required if the target has elements that are not included in the input process tree. `Line 2-3`: Missing_elements() determines the elements of the iput, which do not exist in the output. The return value is stored in to_insert. to_insert contains the information which element or list to insert and at which index. `Line 4-10`: A loop iterates the list to_insert. An if statement checks whether the run variable points to a list or an element. If not to a list: The value (element) of to_insert[i] is inserted into the particular index in the input process tree. If to a list: recursive call of insertion() with the sublists. The last return value of insertion() is the transformed input process tree, which equals the output. After the application of the level recursion and deletion or insertion algorithms, valid process trees are identified with a cluster size of 1. This means that the behavioral morphing splits an event log into k clusters of size 1.

Algorithm 3. Insertion

1: **procedure** INSERTION(bt_{as_is}, bt_{to_be})
2:     to_insert = missing_elements(bt_{as_is}, bt_{to_be})
3:     Comment: *returns a list with values and indexes to insert into*
4:     **for** i, val in **enumerate**(to_insert) **do**
5:         **if** val **is not instance**(list) **then**
6:             bt_{as_is} = insert(bt_{as_is}, val)
7:             Comment: *elementwise insertion of value which is not in bt_{as_is}*
8:             output(bt_{as_is})
9:         **else**
10:             bt_{as_is} = insertion(bt_{as_is}[index of to_insert], to_insert[i])
11:     **return** bt_{as_is}

---

3   Application of Behavior Morphing

All algorithms are fully implemented in Python. Several plug-ins were integrated into our software prototype. To transform CSV files into XES format (XML-based event log format) we use the .csv to .xes plugin. The inductive visual miner is used to graphically visualize process trees from XES files. Then, graphical process trees are transformed into behavior-oriented traces, which are parsed as array lists.

3.1   Application on Synthetic Data

The morphing approach has been tested on a different number of events. The evaluation results on synthetic event logs are summarized in Table 1.The evaluation indicates a correlation between the number of clusters and the similarity between the input and output. With respect to the performance, models with little overlap still requires a low transformation time (morphing) despite the recursion implementation, which demonstrates the efficiency of the approach.

Table 1. Analysis of performance

No. of events	Nes. depth (bt_{as})	Common events	Nes. depth (bt_{tb})	Transformation time	No. of clusters
15	5	4	3	4.6 ms	5
26	7	20	8	11.4 ms	9
41	9	26	6	24.6 ms	8
100	4	88	4	111.5 ms	15
225	4	75	4	536.2 ms	131
1000	3	970	3	12.3 s	24

3.2 Application on Graphical Process Models

To demonstrate the practicability of our approach, we applied the morphing approach on BPMN diagrams as shown in Figs. 6 and 7. Figure 6 shows an as-is CRM
process model. A new customer has three options how to register: online communication via conventional channels (a), contact via third party (affiliates) (c)
or customer recommendations (d). Then, two registration steps follows (e) and
(f) and a customer service (g) is requested. The registration is repeated if errors
occur (h) during the registration or if the customer is denied (j). After successful
registration, the customer can decide whether (s)he wants to complete the registration (l) or abort the onboarding (m). The company now wants to implement
a new registration path in order to make the process more agile. Figure 7 shows
the company's new target (to-be) model. An on-line registration option is implemented (i) in order to replace a registration in the store. Furthermore, the second
registration step in the store (f) is removed. Additionally, the company decides
to end offline campaigns (a) and instead to target customers via the Internet (b).
Assumption supporting the process change (i.e., giving hints to accept the to-be
process) will be given in Sect. 4. The morphing approach is applied as preprocess-
ing step aiming to identify the customer acceptance after implementing the new
processes. Both process models (see Figs. 6 and 7) are transformed into behavior-
oriented traces and the morphing algorithm is applied with the $input_{asis}$
(seq,(x,a,c,d),(xl,(seq,(xl,(and,(seq,e,f),g),h),k),j),(x,l,m)) and the $ouput_{tobe}$
(seq,(x,b,c,d),(xl,(x,(seq,(xl,(and,e,g),h),k)i),j),(x,l,m)). Three valid states are
determined in 11.2 ms with $common\_branches_{asis}$: (seq,(xl,(and,(seq,e,f)
,g),h),k) and $common\_branches_{tobe}$: (seq,(xl,(and,e,g),h),k). Each valid state is
considered as a cluster. Thus, 5 clusters result. Next, we explain how the size of
clusters can increase.

3.3 Change of Cluster Size: Assign New Event Logs to Clusters

The size of clusters in our approach changes if additional event data emerges and
has been identified as another path of one of the existing clusters. This means
that a new event log is compared with each cluster (see Fig. 8, steps 1 and 2) and
then the lowest similarity value of the comparison decides to which cluster this
new event log is most similar. The similarity between event logs is calculated with

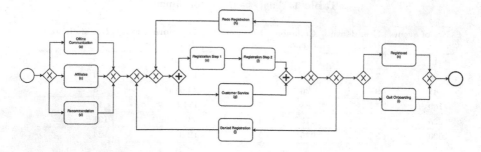

Fig. 6. Input process model: as-is process of registration.

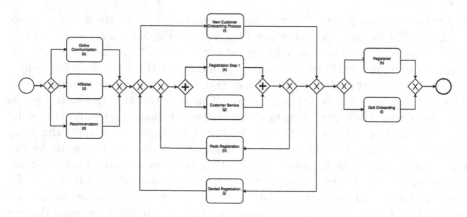

Fig. 7. Output process model: to-be process model of registration.

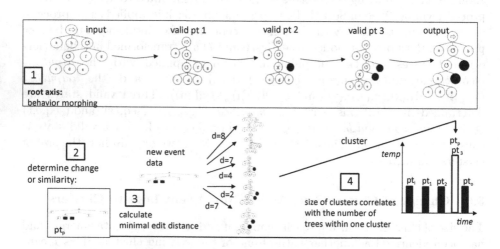

Fig. 8. After behavioral morphing each process tree on the root axis correspond to a cluster of size 1. The size increases through new similar event data.

the Levenshtein distance in combination with the level recursion and valid process trees. Levenshtein's edit distance aims to find the shortest sequence of edits (deletion, insertion, replacement) required to transform one string into another. In our context, Algorithms 4 intends to find the minimal distance between the new event data (represented as process tree) and the input, valid process trees and the output, see Fig. 8, step 3. In Fig. 8 the new event data is most similar to cluster 4 and thus the size of this cluster increases to 2. The advantage of our similarity approach in combination with morphing compared to a pure Levenshtein distance is a more appropriate consideration of the structure of process models. To demonstrate this, the following input and output behavior-oriented traces are given: $bt_i = (\text{seq},(a,(x,(xl,b,c),d)),e)$ and $bt_o = (\text{seq},(a,b,c,d,e))$. The Levenshtein distance determines two edits, which means that the traces are quite similar. This, however, is not appropriate since both traces are two widely different process models, see Fig. 9. The morphing algorithm determines four valid process trees, which is reasonable. In the case of identical minimal edits (between a new event log and two clusters), then the current implementation also considers the order of elements, the number of lcb and the common elements between lcb in one list in the similarity measurement. In the future we intend to include quantitative values such as cost functions (see [4]) in the similarity calculation.

Algorithm 4. Assignment

1: **procedure** ASSIGNMENT(new_trace, is_to_be_states)
2:     H = float('inf')
3:     **for** j, morphing_state in **enumerate**(is_to_be_states) **do**
4:         d = levenshtein_distance(new_trace, morphing_state)
5:         **if** d <= H **then**
6:             H = d
7:             idx = j
8:     **return** idx

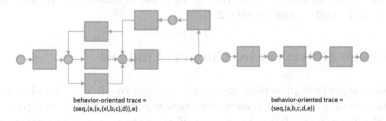

behavior-oriented trace =
(seq,(a,(x,(xl,b,c),d)),e)

behavior-oriented trace =
(seq,(a,b,c,d,e))

Fig. 9. The Levenshtein distance determines 2 edits required to transform the left trace into the right trace. The morphing algorithm determines 4 valid states between the traces, which is more reasonable.

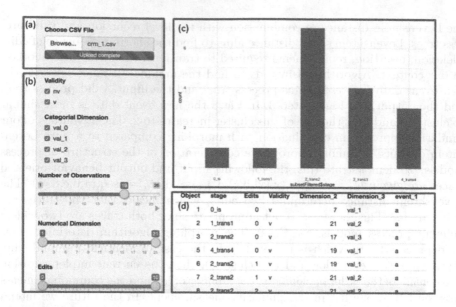

Fig. 10. Visualization of clusters: (a) upload file, (b) filtering of process trees, (c) visualization of the cluster centroids as bar charts, (d).

Giving the use case from Sect. 3.2, a new behavior-oriented trace ($newInput$) with (seq,b,(xl,(seq,(xl,(and,e,g),h),k),j),l)) exists and should be compared with the input ($input_{asis}$) and the output ($output_{tobe}$). The objective of the comparison is to answer the research question (**RQ 1**) (see Fig. 1). For this, a morphing between $newInput$ and $input_{asis}$ was performed in 11.8 ms resulting in 8 valid states. A morphing between $newInput$ and $output_{tobe}$ required 14.8 ms and found 5 valid states. This comparison shows a higher similarity between $newInput$ and $output_{tobe}$. Then, a morphing to all valid states between $input_{asis}$ and $output_{tobe}$ was conducted. This analysis shows that $newInput$ is a valid state of $output_{tobe}$. This means that the current running process instance is very close to the to-be process model from Fig. 7.

The next section presents how the size of clusters matters for change analysis and thus we intend to answer **RQ 2**.

4 Change Detection by Cluster Analysis

The previous section presented the morphing approach used to cluster event logs. The size of a cluster increases when new event data is assigned to a cluster due to the largest similarity by means of changes required to transform one tree into another. The simulation of the clusters over a specified period allows changes to be detected and predicted as explained below. The change analysis is demonstrated with the visualization component of the process tree clustering.

To visualize the clusters, we use the shiny package from RStudio[1] and visualize clusters as bar charts and not bubbles (which is rather common). Bar charts allow to present categorical data such as outside temperature, phase of day or interacting entities (e.g., human-beings, machines) proportional to the number of their occurrences. Additionally, the number of the clusters must not be specified before clustering. Figure 10 shows the visualization component.

First, a XES-file must be uploaded including the input behavior-oriented trace, valid states and the output behavior-oriented trace (Fig. 10(a)). Optionally, the number of edits or further meta data can be attached. Then, clusters can be filtered according to categorical data (Categorial Dimension), see Fig. 10(b) or the number of process trees to be considered (Number of Observations). One observation is an additional event log. Edits and Numerical Dimension can be used to identify process trees that require a certain number of edit operations and that have particular numerical properties. Changes are localized by the simulation of the bar charts.

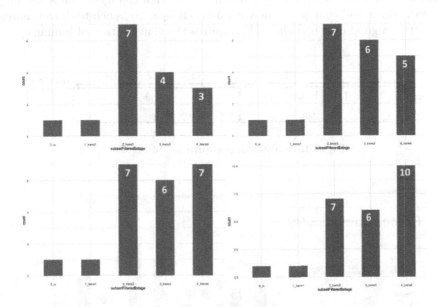

Fig. 11. Indications for changes based upon the analysis of the clusters' size.

We assume that statement on changes in the CRM use case should be given. For this, the clusters of the CRM use case are visualized in Fig. 11 by the bar charts 0_{is} to 4_{trans4}, where 0_{is} corresponds to the input CRM process model from Fig. 6. New observations are considered and the clustering is simulated. The simulation shows that new traces are assigned to cluster t_{trans2}, i.e., they are most similar to this cluster. By further simulation, it can be observed that

---

[1] https://www.r-project.org/nosvn/pandoc/shiny.html.

new traces are becoming more and more similar to the target process (see Fig. 7). The increase of the cluster size on the right side of the diagram indicates a higher similarity to the output. An increase of the clusters' size on the left hand side, would indicate a higher similarity to the output. This cluster development can be interpreted as an acceptance of the new target process by customers (see Fig. 7).

More precise assumptions on changes can be made by the comparison of behavior-oriented traces within one cluster (see Fig. 12). The comparison in Fig. 12 gives hints to an increased duration of similar events. In this figure we use duration as comparison factor. One might use any other sensor value (e.g., temperature, interacting entities). The Fig. 12 explains changes due to an increased duration time of events. Figure 13 visualizes the forecast for the next day of one event log. To predict the value, a trend line was drawn and the correlation coefficient was determined. A value close to 1 increases the prediction quality. In this example, the prediction would be that the value increases than decreases. Eventually, our approach can only give hints as to what changed, that something changed, why it had changed or how it might change. To benchmark the change (i.e., is this a good or a bad change?) requires the consultation of humans.

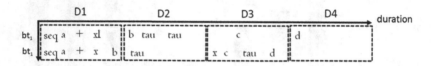

Fig. 12. Analysis of the duration of events.

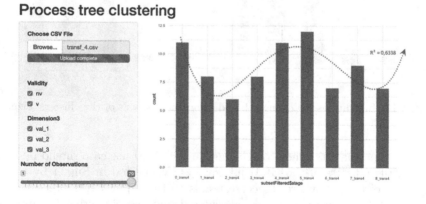

Fig. 13. This diagram shows the prediction for an event log.

5   Evaluation

We applied the algorithms on a smart home data set[2]. The data set tracks events with motion sensors of 28 days. Each day had around 5300 events. We removed the duplicates and then split the data set according to days and then according to the phases of the day (morning, afternoon, evening and night) and ended with around 1100 events for each phase. The first of the 28 days was selected as input and the last day as the output. The application of the morphing approach for the afternoon phase of the input and the afternoon phase of the output determined 106 valid process trees in 14ms. The Levenshtein distance would require more than 140 edits to transform one tree into another (note again that Levenshtein distance disregards the structural similarity of process models, as discussed in Sect. 2.3). Then, we selected the event log of the afternoon of the following days and run Algorithm 4. The change history for this use case is shown in Fig. 14. In contrast to the approach for event log detection of [5], the application of our approach shows several change points.

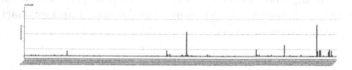

Fig. 14. Detection of Changes for a Smart Home Data Set.

6   Comparison of Related Works

Numerous approaches to cluster event logs exist in the literature [4,6–9]. These approaches have in common that they intent to improve process discovery either determined by partition of the event logs [4,6,7,9,10] or based on a balanced density between the clusters [8]. As a consequence, they do not consider the quality of the underlying process model and are not suitable to detect changes to the full extent. The clustering technique of [11] takes a process model as input for clustering and structures traces in order to mine (sound) process models. Due to this assumption traces might remain unclustered and thus changes might be disregarded. Our approach inspects the complete event log and cluster all traces. It also considers the quality of the underlying process model and thus event log structuring is determined more appropriately. Our cluster technique is complementary to trace clustering [12], which might be used to preprocess an event log and to reduce the complexity needed to determine process trees. In contrast to approaches that calculate the similarity of traces or process models

---

[2] http://casas.wsu.edu/datasets/.

based on vector-based approaches [13,14], our approach can capture the structure of process models and also recognize similarities and changes within subprocess models. This advances our clustering technique to existing approaches. For instance, [14] determines the similarities between process models based on a cosine similarity. Two process models are translated into two vectors: an activity vector to indicate common activities and a transition vector to compare the similarity of the control-flow. The similarity of the process models is determined by the included angle. The process models are then grouped according to an agglomerative clustering. Our clustering technique uses process trees as input and can therefore use the editing operations as an explanation for changes and the simulation allows to explain changes for quite similar process models.

Approaches for change detection in process mining (e.g., [5,15,16]) consider drift detection by means of missing activities or deviations of common activities. Our approach can detect more changes than existing approaches and can also be applied to event logs of unstructured processes.

The approach presented in this paper is a revised version of the idea on process tree clustering presented in [17]. This paper updates and extends the concept of morphing and cluster analysis presented in [17]. The formalization of the approach and change analysis were also developed and evaluations were conducted, which is missing in [17].

7   Conclusion and Outlook

This paper presented a process tree-based approach for clustering of event logs. This clustering technique was primary developed with the intention of change detection and prediction, which are possible through the analysis of the clusters' size and the movements of clusters between the origin and the target. The cluster visualization with bar charts allows substantial indications of changes. In contrast to a pure machine-learning solution to clustering and change detection, no large input sample is required. In data mining settings, clusters are evaluated based upon the idea of "maximizing intra-cluster similarity and minimizing intercluster similarity" [9]. This means that a small distance between all elements within a cluster and a large distance between the clusters should be produced. The clusters derived from event logs with the morphing approach fulfill this requirement and thus make our clustering approach efficient.

In the future, we plan to combine our approach with machine learning techniques in order to allow change period estimation and also to consider quantitative values in the deletion and insertion algorithms in order to refine our notion of longest common branches. This would mean that changes could be detected even earlier than through a time-based simulation. Additionally, we intend to give recommendation on the best selection of origin and target event logs as discussed in Sect. 3. This requires a lot of evaluations with real data sets.

References

1. van der Aalst, W.M.P.: Process Mining - Data Science in Action, 2nd edn. Springer, Heidelberg (2016). https://doi.org/10.1007/978-3-662-49851-4
2. Buijs, J.C.A.M., van Dongen, B.F., van der Aalst, W.M.P.: A genetic algorithm for discovering process trees. In: IEEE Congress on Evolutionary Computation, pp. 1–8. IEEE (2012)
3. Leemans, S.J.J., Fahland, D., van der Aalst, W.M.P.: Discovering block-structured process models from event logs - a constructive approach. In: Colom, J.-M., Desel, J. (eds.) PETRI NETS 2013. LNCS, vol. 7927, pp. 311–329. Springer, Heidelberg (2013). https://doi.org/10.1007/978-3-642-38697-8_17
4. Bose, R.J.C., Van der Aalst, W.M.: Context aware trace clustering: towards improving process mining results. In: Proceedings of the 2009 SIAM International Conference on Data Mining, pp. 401–412. SIAM (2009)
5. Ostovar, A., Maaradji, A., La Rosa, M., ter Hofstede, A.H.M.: Characterizing drift from event streams of business processes. In: Dubois, E., Pohl, K. (eds.) CAiSE 2017. LNCS, vol. 10253, pp. 210–228. Springer, Cham (2017). https://doi.org/10.1007/978-3-319-59536-8_14
6. Sun, Y., Bauer, B., Weidlich, M.: Compound trace clustering to generate accurate and simple sub-process models. In: Maximilien, M., Vallecillo, A., Wang, J., Oriol, M. (eds.) ICSOC 2017. LNCS, vol. 10601, pp. 175–190. Springer, Cham (2017). https://doi.org/10.1007/978-3-319-69035-3_12
7. Song, M., Günther, C.W., van der Aalst, W.M.P.: Trace clustering in process mining. In: Ardagna, D., Mecella, M., Yang, J. (eds.) BPM 2008. LNBIP, vol. 17, pp. 109–120. Springer, Heidelberg (2009). https://doi.org/10.1007/978-3-642-00328-8_11
8. Hompes, B.F.A., Buijs, J.C.A.M., van der Aalst, W.M.P., Dixit, P.M., Buurman, J.: Detecting changes in process behavior using comparative case clustering. In: Ceravolo, P., Rinderle-Ma, S. (eds.) SIMPDA 2015. LNBIP, vol. 244, pp. 54–75. Springer, Cham (2017). https://doi.org/10.1007/978-3-319-53435-0_3
9. Weerdt, J.D., vanden Broucke, S., Vanthienen, J., Baesens, B.: Active trace clustering for improved process discovery. IEEE Trans. Knowl. Data Eng. 25(12), 2708–2720 (2013)
10. Evermann, J., Thaler, T., Fettke, P.: Clustering traces using sequence alignment. In: Reichert, M., Reijers, H.A. (eds.) BPM 2015. LNBIP, vol. 256, pp. 179–190. Springer, Cham (2016). https://doi.org/10.1007/978-3-319-42887-1_15
11. Chatain, T., Carmona, J., van Dongen, B.: Alignment-based trace clustering. In: Mayr, H.C., Guizzardi, G., Ma, H., Pastor, O. (eds.) ER 2017. LNCS, vol. 10650, pp. 295–308. Springer, Cham (2017). https://doi.org/10.1007/978-3-319-69904-2_24
12. Bose, R.P.J.C., van der Aalst, W.M.P.: Trace clustering based on conserved patterns: towards achieving better process models. In: Rinderle-Ma, S., Sadiq, S., Leymann, F. (eds.) BPM 2009. LNBIP, vol. 43, pp. 170–181. Springer, Heidelberg (2010). https://doi.org/10.1007/978-3-642-12186-9_16
13. Greco, G., Guzzo, A., Pontieri, L., Sacca, D.: Discovering expressive process models by clustering log traces. IEEE Trans. Knowl. Data Eng. 18(8), 1010–1027 (2006)
14. Jung, J.-Y., Bae, J., Liu, L.: Hierarchical clustering of business process models. Int. J. Innov. Comput. Inf. Control 5, 4501–4511 (2009)
15. Tax, N., Genga, L., Zannone, N.: On the use of hierarchical subtrace mining for efficient local process model mining. In: SIMPDA, Volume 2016 of CEUR Workshop Proceedings, pp. 8–22. CEUR-WS.org (2017)

16. Maaradji, A., Dumas, M., Rosa, M.L., Ostovar, A.: Detecting sudden and gradual drifts in business processes from execution traces. IEEE Trans. Knowl. Data Eng. **29**(10), 2140–2154 (2017)
17. Koschmider, A.: Clustering event traces by behavioral similarity. In: de Cesare, S., Frank, U. (eds.) ER 2017. LNCS, vol. 10651, pp. 36–42. Springer, Cham (2017). https://doi.org/10.1007/978-3-319-70625-2_4

Dimensions for Scoping e-Government Enterprise Architecture Development Efforts

Agnes Nakakawa[1](✉), Flavia Namagembe[1], and Erik H. A. Proper[2,3]

[1] Makerere University, P. O. BOX 7062, Kampala, Uganda
{anakakawa, fnamagembe}@cis.mak.ac.ug
[2] Public Research Centre Henri Tudor,
1855 Luxembourg-Kirchberg, Luxembourg
e.proper@acm.org
[3] Radboud University Nijmegen, P.O. BOX 9010,
6500 GL Nijmegen, The Netherlands

Abstract. Inspired by developed economies, many developing economies are adopting an enterprise architecture approach to e-government implementation in order to overcome challenges of e-government interoperability. However, when developing an enterprise architecture for a complex enterprise such as the e-government enterprise, there is need to rationally specify scope dimensions. Addressing this requires guidance from e-government maturity models that provide insights into phasing e-government implementations; and enterprise architecture approaches that provide general insight into key dimensions for scoping enterprise architecture efforts. Although such insights exist, there is hardly detailed guidance on scoping initiatives associated with developing an e-government enterprise architecture. Yet the success of such business-IT alignment initiatives is often affected by scope issues. Thus, this paper presents an intertwined procedure that draws insights from e-government maturity models and enterprise architecture frameworks to specify critical aspects in scoping e-government enterprise architecture development efforts. The procedure was validated using a field demo conducted in a Uganda public entity.

Keywords: e-Government maturity · e-Government enterprise architectures

1 Introduction

Integration and interoperability are major drawbacks of e-government growth in developing economies [1, 2]. Literature (e.g. [3–5, 29]) indicates that adopting an enterprise architecture approach to e-government implementation helps to holistically address these issues. Since enterprise architecture steers change and integration in business-IT alignment contexts [6], adopting e-government – a mechanism of business-IT alignment in government service delivery – can be holistically guided by an *e-government enterprise architecture*. Although the success of enterprise architecture development depends on several factors, scope is among the critical ones [7–13]. If scope dimensions are not rationally specified, the success of an enterprise architecture effort is affected [11]. Yet if the enterprise is complex such as the e-government enterprise, defining the scope of its architecture becomes an intertwined problem.

© Springer Nature Switzerland AG 2018
H. Panetto et al. (Eds.): OTM 2018 Conferences, LNCS 11229, pp. 661–679, 2018.
https://doi.org/10.1007/978-3-030-02610-3_37

From literature (see Sect. 2), addressing such a problem requires insights from e-government maturity models (because they guide phasing of e-government implementations) and enterprise architecture approaches (because they articulate key aspects that inform the scoping of architecture development efforts). This implies the need to investigate: *how e-government maturity models can supplement enterprise architecture approaches to provide detailed guidance on scoping e-government enterprise architecture development efforts.* Thus, this paper presents an intertwined procedure that can guide stakeholders on key dimensions to consider when Scoping e-Government Enterprise Architecture development efforts (SGEA). Section 2 presents related work and research approach used, Sect. 3 presents the design of SGEA and its instantiation using a field demo, Sect. 4 highlights key findings from the demo, and Sect. 5 concludes the paper and highlights future research efforts.

2 Related Work and Research Approach

Section 2.1 motivates design of SGEA and Sect. 2.2 discusses the research method.

2.1 Related Work on Scoping e-Government and Architecture Development

e-Government maturity models [14–19] provide a phased approach that enables governments to measure the progress in e-government development and to produce a robust citizen-centric and responsive government [20]. e-Government maturity models play a central role by offering generic concepts that inform, shape, and direct e-government deliberations, investments, and research. However, they hardly provide guidance on how to scope efforts towards attaining the specific stages and features of e-government growth that they articulate (as indicated in Table 1).

Table 1. Harmonization of stages in e-government maturity models

Baum & Maio [19]	UN & ASPA [14]	Deloitte & Touché [18]	Layne & Lee [17]	Hiller & Bélanger [16]; Moon [15]
(1) Web presence	(1) Emerging (2) Enhanced presence	(1) Information publishing/dissemination	(1) Catalogue	(1) Simple information dissemination
(2) Interaction	(3) Interactive presence	(2) Official two-way communication	-	(2) Two-way communication
(3) Transaction	(4) Transaction presence	(3) Multipurpose portals	(2) Transaction	(3) Service and final transaction
(4) Transformation	(5) Seamless or fully integrated presence	(4) Portal personalization (5) Clustering of common services (6) full integration and enterprise transaction	(3) Vertical Integration (4) Horizontal integration	(4) Political participation

Since e Government maturity models use somewhat different terms to refer to specific stages, Table 1 extends the taxonomy in [21] by using specific features to synthesize names of stages with a bias of the naming used in [19] as indicated in column 1. Indicators for achieving each stage are as follows.

Web Presence: websites are used to increase access to formal and catalogued information [14–18] such as agency contacts, frequently asked questions, publications, trends/news updates [19] or dynamic information feeds from active databases [14].

Interaction: portals enable two-way communication [14–16, 18], posting online comments or inquiries, sending and receiving email, downloading documents, submitting forms; and authenticating users to access particular services [19].

Transaction: portals provide services across departments [18], e-forms and e-payment capabilities to allow online completion of transactions [14, 17, 19] on applying and renewing of verification documents and procurement of services [14–17, 19].

Transformation: A universal portal [14] that clusters common services; allows users to customize their views [18]; and enables vertical and horizontal service integration in order to deliver proactive service in a seamless and personalized way, that accommodates needs of specific groups of customers [15–19].

Although indicators for each stage are explicit, detailed guidelines are hardly available on how efforts should be scoped and aligned so as to achieve each stage. Inspired by [3–5, 25], such details can be obtained by designing architecture views as blueprints of capabilities for achieving a desired stage of e-government growth. However, prior to designing such blueprints, there is need to rationally specify their scope. Hence the need to review the extent to which architecture approaches inform scoping.

Scoping Enterprise Architecture Creation. Lack of a clear scope is a major managerial pitfall in architecture development [7, 8, 10]. Yet, on the one hand, architecture maturity models provide indicators for accomplishing specific stages in architecture development [23] without providing guidelines for scoping architecture creation. On the other hand, majority of enterprise architecture frameworks do not adequately address issues on scoping architecture initiatives [9] but emphasize the relevance of scoping and aspects to consider as shown in Table 2.

Overall Research Gap. Existing efforts in Tables 1 and 2 hardly provide detailed guidance on explicitly determining scope of an e-government enterprise architecture. Moreover, instantiating insights in Table 2 when developing an e-government enterprise architecture is not trivial because the e-government enterprise is contextually heterogeneous and multidimensional (Sect. 3.1 elaborates this). Thus, this research is motivated to devise SGEA by blending insights from Tables 1 and 2.

Table 2. Available insights on scoping an enterprise architecture effort

Zachman Framework [13]	**Scope/contextual/perspective phase** involves establishing: • Internal and external limits of an enterprise; goals/scope, strategies, and performance measures of the enterprise; • Elements that need to be considered in other phases; and • Major enterprise timelines
Integrated Architecture Framework – IAF [12]	**Manage complexity by using abstraction levels i.e. the *"why-what-how-with what"* reasoning pattern,** but use the *"why"* and *"what"* to address scope because: • *"Why"* defines business context/aspects that vary across enterprises (e.g. drivers/strategy/vision, objectives, principles, and scope of a problem), and • *"What"* specifies services to be supported
The Open Group Architecture Framework – TOGAF [11]	**Consider four dimensions when scoping architectures:** • *Enterprise* coverage – define all internal and external units of the enterprise and specify those to be considered; • *Architecture* domains – specify domains relevant in achieving target state; • *Level* of detail – specify level of abstraction of concepts in architecture views; • *Resources* (i.e. time, finances, and existing architectural assets of the enterprise) for architecture development

2.2 Research Approach

Design Science supports the development of feasible artifacts that help an enterprise to solve a significant problem or embrace a given opportunity towards achieving effective and efficient operations [22]. Instantiating design science to guide the development of SGEA implied undergoing three cycles. *Relevance cycle* involves highlighting the research gap and potential solutions and verifying applicability of an artifact [22]. Section 2.1 highlights the research gap motivating SGEA design, and Sect. 3 demonstrates its applicability as elaborated below. *Rigor cycle* emphasizes the need to draw from existing foundational approaches when creating an artifact [22]. As indicated in Sects. 2.1 and 3, two knowledge fields indicated in Tables 1 and 2 inform the design of SGEA. *Design cycle* involves building and continuously refining the artifact with respect to the problem context [22]. Section 3 discusses the design of SGEA and Sect. 4 highlights how insights from SGEA validation informed its refinement.

Validation of SGEA in a Developing Country. Design Science artifacts need to be exposed to the problem environment so as to investigate their applicability and use findings to refine them accordingly [22, 33]. Design science artifacts can be validated using lab demos, lab experiments, field demos, field experiments [26], case study, action research [33]. The choice of approach depends on the purpose of the artifact and the nature of problem it addresses. The purpose of SGEA is to provide systematic guidance on aspects to consider when scoping efforts towards developing an e-government enterprise architecture. Validation of such an artifact requires that it is used at the start of a real-life e-government development project, to specify and document scope dimensions of that project using a scope specification matrix. To achieve this, a field demo is used.

A field demo allows researchers to demonstrate usability of an artifact in practice by instantiating it with respect to problems in the business environment [26]. Thus, to demonstrate the usability of SGEA, a field demo was conducted using one of Uganda's public entities (as detailed in Sect. 3). The field demo involved three bilateral structured walkthrough sessions, each lasting for at most one-hour. Each session involved the ICT/Systems Manager of the entity and one of the researchers. Subjects of discussion during the sessions and respective instantiations are presented in Sect. 3.

3   Design of SGEA and Its Applicability in a Developing Country

In design science, a feasible artifact is obtained through exploring how available and proven means can be used to achieve a desired solution [22]. Thus, this section demonstrates how existing foundational insights in two fields are adopted and synthesized to design SGEA. First, from e-government maturity models in Table 1, the model in [19] is adopted as justified in Sect. 2.1, while insights from other models are used to elaborate descriptions of specific stages of maturity. Second, from enterprise architecture frameworks, TOGAF [11] is adopted because it provides clearer insights

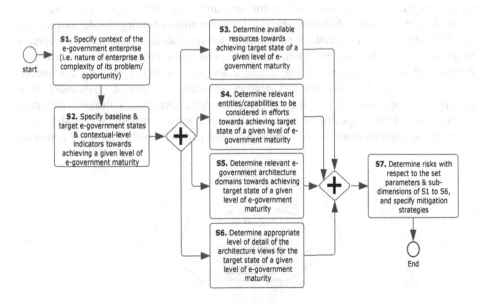

Fig. 1. Dimensions for SGEA

on scoping, while insights from other architecture frameworks or studies are used to elaborate particular aspects of scope dimensions for the e-Government Enterprise (**GE**) architecture. Thus, SGEA adapts TOGAF's guidelines on scoping architecture efforts to support e-government realization with respect to insights on e-government maturity. SGEA, shown in Fig. 1, is a high level process for guiding deliberations on scoping GE architectures as blueprints for implementing interoperable e-government solutions. Each step in Fig. 1 is a pointer to detailed aspects presented in Sects. 3.1 to 3.7, with interludes of text boxes representing instantiations from the field demo. Appendix 1 shows a template of the scope specification matrix that can be used to document scope dimensions derived using SGEA. Sections 3.1 to 3.7 discuss SGEA components or dimensions that are coded as S1 to S7 for cross-referencing purposes.

3.1   Specify Context of the GE (S1)

An enterprise architecture development effort should be tailored to the enterprise context, planned scope of architecture engagement, and enterprise goals [9]. Thus, S1 indicates the need to define contextual features of the GE using two sub-dimensions (i.e. S1.1 and S1.2) that shape deliberations on other dimensions as shown in Fig. 1.

S1.1. Nature of the GE. Understanding the nature of an enterprise implies specifying the "gross size" of that enterprise [13] or defining entities that constitute the full breadth and depth of its operations [11]. The *gross size* of the GE comprises three parameters (i.e. S1.1.1 to S1.1.3) below.

S1.1.1. Structural Nature of the GE. Government is a large and highly complex enterprise characterized by conventionally interconnected sectors [25], which fulfill their specific mandates by establishing agencies, departments, and units. Thus, aligning ICTs with capabilities of such an enterprise yields the **GE** [30], which structurally comprises three tiers (i.e. national, sector, and unit tiers) with each tier having seven scope dimensions that stakeholders need to consider when planning the development of the GE architecture to enable e-government interoperability. In Fig. 2, a heptagon shape is used to illustrate these tiers and the seven scope dimensions of each tier are represented at the vertexes of Fig. 2 and elaborated in Fig. 1 and its discussions. *National level* in Fig. 2 comprises entities mandated as overall regulators of e-government implementation in a country. *Sector level* comprises entities mandated to regulate and deliver specific services towards economic growth. Sector-specific e-government implementations are regulated by the national level. *Unit level* comprises

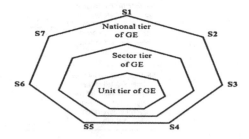

Fig. 2. Three levels/tiers that constitute the GE

agencies/departments/units within a particular sector, and directly interact with each other to deliver a wide range of e-government services.

Specific questions that prompt actors to specify values for S1.1.1 are provided in the text box below, and clarified using an instantiation from the field demo.

Question S1.1.1: What is the composition of the national/sector/unit tier of the GE in a given country? How many entities are at any/each GE tier and what are their key mandates? Which of these entities should be focused on in a given e-government enterprise architecture development effort?
Instantiation: *The national tier of Uganda's GE comprises Ministry of ICT and the sector tier comprises 30 entities [27], each comprising several agencies/ directorates/departments that are perceived as unit tier entities of the GE. However, in the field demo validation of SGEA, an entity at unit level was purposively selected due to resource limitations. Thus, the Directorate of Revenue Collection (DRC) of Kampala City Council Authority was selected due to the swiftness it exhibited in responding to the call for participation that was sent out to a sample of unit level entities in Uganda's GE.*

S1.1.2. Service Portfolio of the GE. The full scope of the e-government landscape comprises three categories of e-government solutions, i.e.: e-administration suite with solutions that support internal processes of a public entity; e-citizen or e-service suite with solutions that support engagements between a public entity and its customers; and e-society suite with solutions that support engagements between a public entity and other public, private, not for profit, or civil society entities [32]. Accordingly, these solutions support electronic transactions and interactions that yield Government to: Government (G2G) services, Citizen/consumer (G2C) services, Business and non-profit agency (G2B) services [31], and Employees (G2E) services [2]. These services are accessible by international, national, regional, provincial, and local communities [31]. These classifications imply that each tier in Fig. 2 delivers three categories of services that need to be considered when deliberating on the scope of the GE architecture for a specific entity at any tier. Questions for deliberation under S1.1.2 and corresponding instantiations from the field demo are provided in the text box below.

Question S1.1.2: What is the composition of the service portfolio for the chosen entity/entities at any/each GE tier in terms of e-administration (e.g. G2E), e-customer or e-service (e.g. G2C), and e-society (e.g. G2G, G2B)? Which of the three categories of e-government services should be focused on in a given e-government architecture development effort?
Instantiation: *DRC service portfolio comprises all services that involve sensitization of taxpayers and establishment of strategic and operational policies to guide mobilization and collection of revenue from all types of businesses in Uganda's capital city, so as to promote economic growth [28]. Thus, the GE service portfolio of DRC includes three categories of e-government services, i.e.: e-administration, e-services, and e-society. The field demo focused on the e-services category.*
Question S1.1.3: What are the socio-cultural and political concerns or priorities of the chosen entity/entities at any/each GE tier, and what are the implications of these on choices in S1.1.1 and S1.1.2?
Instantiation: *Specifications on this are intentionally excluded for confidentiality reasons.*

S1.1.3. Socio-Cultural and Political Environment of the GE. The operational environment of a specific entity at any/each GE tier in Fig. 2 comprises social, cultural, and political factors that have a bearing on the success of e-government implementation and enterprise architecture development. Thus, inter and intra environmental scanning of the enterprise helps to understand the organizational structure, culture, level of commitment, and nature of politics in the enterprise as well as the implied or possible obstacles that may arise [8]. For instance, the level of management commitment and support has a direct bearing on the assurance of: resources for enterprise architecture development, and guidance on enterprise-specific protocols for successful collaboration of all key stakeholders [10]. Thus, S1.1.3 not only involves understanding socio-cultural and political factors and their implications on scoping GE architecture development efforts, but also helps to acquire indicators or implicit information for gauging resource availability levels in scope dimension S3 (Sect. 3.3). The text above shows specific questions for eliciting specify values for S1.1.3.

S1.2. Complexity of Problem or Opportunity of the GE. Complexity of a problem/opportunity is specified using two parameters (i.e. S1.2.1 and S1.2.2).

S1.2.1. Type of Problem/Opportunity in the GE. This parameter indicates the need to: (a) classify a problem/opportunity of an entity at any/each tier of the GE with respect to the three categories of e-government services offered by the GE as indicated in S1.1.2; and (b) determine which category of e-government service should be actualized in order to address the problem or seize the opportunity. Understanding S1.2.1 when scoping the GE architecture effort helps to obtain information relevant for: determining target state of e-government maturity in scope dimension S2 (in Sect. 3.2); and selecting reference models/architectures that can be adapted in scope dimension S5 (in Sect. 3.5). The text box below shows specific questions for eliciting specify values for S1.2.1, and corresponding instantiation.

Question S1.2.1: What is the problem/opportunity of the entity at any/each tier of the GE, and which category of e-government service(s) should be implemented by the GE so as to overcome or seize the opportunity?
Instantiation: *DRC faced a problem of overwhelming client volumes and operational complexity, that called for the need to streamline management of revenue collection through re-engineering existing business processes and developing a supporting e-government solution. Categorically, based on [6,23], this is a problem of establishing IT-reliant business processes and implementing quality management systems respectively. Thus, the categories of e-government services required to address the problem are e-administration and e-services.*

S1.2.2. Magnitude of Problem/Opportunity in the GE. Any enterprise transformation prompts stakeholders to seek information on the part of the enterprise that is to be impacted, corresponding boundaries of the impact, and other related projects that are to be indirectly affected [23]. Therefore, S1.2.2 indicates the need to identify internal and external entities at any/each GE tier that are directly and indirectly affected by the specified problem or opportunity; or that support the achievement of the category of e-government service specified in S1.2.1. Understanding S1.2.2 when scoping the GE architecture effort helps to obtain information relevant for decision making in scope dimensions S4 and S6 (in Sects. 3.4 and 3.6 respectively). The following text box shows specific questions for eliciting specify values for S1.2.2.

Question S1.2.2: Which internal capabilities within the chosen entity/entities in S1.1 and external entities or capabilities are (or are to be) directly and indirectly affected by the issues or specified category of e-government service(s) in S1.2.1?
Instantiation: *DRC comprises 3 business capabilities (i.e. Valuation and Revenue Collection; Compliance and Inspectorate; Research and Business Analysis), and all were directly affected by the problem. Also, out of the 10 sister directorates to DRC, 8 were indirectly affected by the problem in DRC.*

3.2   Determine Baseline and Target State of e-Government Maturity (S2)

This involves sub-dimensions S2.1 to S2.3 below. Specific questions for eliciting values for S2.1 to S2.3, and field demo instances are provided in the text boxes below.

S2.1. Specify baseline e-government maturity level. This indicates the need to assess maturity level of e-government establishment of the focus entity/entities specified in dimension S1. This can be done by using the e-government maturity models in Table 1 (Sect. 2.1) to derive an e-government maturity assessment checklist. However, while deriving such a checklist, there is need to ensure that maturity assessment features accommodate the three categories of services that constitute the service portfolio of the GE (as indicated under S1.1.2). This is because e-government maturity models summarized in Table 1 emphasize maturity assessment features for initiatives under e-citizen/e-service category and some instances under e-society category. They are silent about maturity assessment features for initiatives under e-administration category and some instances under e-society category. However, designing the assessment checklist that addresses this gap is beyond the scope of this paper.

> **Question S2.1:** Using e-government maturity models in [14 - 19], what is the maturity level of the e-services/e-citizen category of e-government services in the entity/entities chosen in S1? What is the maturity level of the e-administration and e-society categories of e-government services?
> **Instantiation:** *Baseline target state of DRC was one-way interaction stage because the DRC website had forms for business people to download and apply for various services.*

S2.2. Specify legally acceptable level and affordable target level of e-government maturity. The target state of e-government maturity in a given entity may depend on the laws and unique constraints associated with exercising the mandate of that entity. In other words, complexities and governing laws in delivering a particular service may not permit implementation of e-government solutions beyond a given stage of maturity. Thus, this sub-dimension indicates the need to: (a) assess complexities and risks associated with implementing each stage of e-government maturity in the focus entity/entities specified in dimension S1; (b) determine the legally acceptable level of e-government maturity for the focus entity/entities specified in dimension S1; and (c) determine the affordable or appropriate target state of a given level of e-government maturity in the focus entity/entities specified in dimension S1. Managerial issues in enterprise architecture efforts such as restricted rules in enterprise operations, ambiguity of goals and strategies [8], and rapidly changing operational conditions [10] are accommodated by specifying both the legally acceptable level of e-government maturity and the affordable target state of e-government maturity. This is because an entity may be legally cleared to achieve a given level of e-government maturity, but resources and situational context may not allow in a given period. Such an entity can then determine and pursue the affordable target state. The affordable or appropriate target state is specified by considering the urgency required in addressing the GE problem in S1.2 and existing legal framework in S3.1 or resources in S3.

> **Question S2.2:** What are the risks associated with implementing each stage of e-government maturity in the entity/entities chosen in S1? What is the legally acceptable level of e-government maturity for the entity/entities chosen, and why? Given existing resources, what is the affordable/ appropriate target state, and why?
> **Instantiation:** *Although the legally acceptable level of e-government maturity for DRC is "transformation" stage, the affordable/appropriate target state was "two-way interaction" level due to two reasons: urgency in streamlining processes towards effective delivery of its mandate and limited resource envelop to support attainment of target state.*

S2.3. Specify context-specific indicators for the e-government target state. This sub-dimension indicates the need to specify explicit indicators/milestones towards realizing the target state of e-government maturity for the focus entity/entities specified in dimension S1. These milestones inform/guide the selection of features associated with other dimensions in scoping the GE architecture (that are discussed in Sects. 3.3 to 3.7). The contextual indicators are derived by instantiating a given stage of an e-government maturity model with respect to the service portfolio (specified in S1.1.2) of the focus entity/entities specified in dimension S1. In addition, these context-specific indicators are not only for the e-government solutions as end products, but also for the operational and governance framework associated with delivering a given level of e-government maturity in the GE (as elaborated below).

If target state is web presence stage across any/all GE tiers, should the operational and governance framework have centralized or decentralized policies and procedures for web content management aspects such as: type of content, language and cultural constraints for content generation, update frequency, feedback and risk control/mitigation strategies, reliability/continuity strategies)?

If target state is interaction stage across any/all GE tiers, which categories of services should be partially completed online and which ones should not as per the operational and governance framework, what is the appropriate extent of online engagement for either categories, and the appropriate feedback and risk control/mitigation strategies?

If target state is transaction stage across any/all GE tiers, which categories of services should be fully handled online and which ones should not as per the operational and governance framework, what are the corresponding feedback and risk control/mitigation strategies, are the legal capabilities established to support full transaction handling for all entities at each GE tier?

If target state is transformation stage across any/all GE tiers, which category of services quality for vertical integration in each GE tier and for horizontal integration across all tiers as per the operational and governance framework, and what are the corresponding legal capabilities?

> **Question S2.3:** What are the entity-specific indicators/milestones towards realizing the target state of e-government maturity in the entity/entities chosen in S1?
> **Instantiation:** *To achieve two-way interaction state in DRC, indicators include: an e-government capability that allows business owners to register businesses online, appear at DRC offices for identity verification, receive a token number for making online payments of licenses and other fees.*

3.3 Determine Available Resources for e-Government Implementation (S3)

Resource-related problems in enterprise architecture development include: outdated legal/regulatory documents and infrastructure; limited capabilities/expertise in management/leadership, change management, enterprise architecture, human resource establishment; and limited budget to provide mitigations for the resource gaps [8]. Therefore, this dimension helps to obtain relevant information for planning the development of the GE architecture with respect to contents in the *"resource envelop"*. To achieve this, two perspectives apply. From an *e-government perspective:* OECD [24] articulates that the success of e-government implementation requires resources such as a comprehensive legislative and regulatory framework, financial sustainability framework, and a strategic-operational framework for reducing digital divide and establishing shared technology infrastructure. From an *architecture perspective:* TOGAF (2009) emphasizes that when securing finances to facilitate architecture development and when determining the appropriate time required to deliver architecture products (to respond to a situation), it is vital to consider all possible ways of reusing existing enterprise resources (such as preliminary/earlier architecture products, reference models, available human resource skillset, and enterprise-wide awareness levels among key stakeholders on enterprise architecture development).

Accordingly, five sub-dimensions (i.e. S3.1 to S3.5) that shape the GE resource envelop are derived from the above two perspectives. For each sub-dimension, there is need to specify the strengths and gaps with respect to achieving the target stage of a given level of e-government maturity (as specified in scope dimension S2) and the magnitude of the identified gaps. The text boxes below show questions that prompt entity-specific values for S3.1 to S3.5 and related instantiations from the field demo.

S3.1. Information Resources. Baseline information resources (such as management frameworks, policies/regulations, principles, tools) need to be identified, updated, and re-used during enterprise architecture development [11]. In addition, the constraints that existing information resources imply on efforts towards achieving the target state specified in S2 need to be determined.

> **Question S3.1:** What are the existing information resources in the entity/entities chosen in S1, and what constraints do they imply on efforts towards achieving the target state specified in S2?
> **Instantiation:** *Existing information resources in DRC included: process flow models of the DRC operational framework, website with service portfolio information, data requirements for the DRC operational framework, downloadable application forms that request for DRC services*
> **Question S3.2:** What are the existing skills among all key stakeholders in the entity/entities chosen in S1 with respect to undertaking activities required to achieve the target state specified in S2?
> **Instantiation:** *Although there was high level of awareness on the use of system engineering and project management approaches in e-government adoption in DRC, there was limited expertise in enterprise architecture adoption in e-government implementation.*

S3.2. Human Resources. There is need to determine existing technical and other skills or capacity (among all internal and external stakeholders of the GE) with respect to designing, implementing, adopting, and maintaining capabilities relevant in achieving the target state specified in S2.

S3.3. ICT Infrastructure. Existing e-government solutions and ICT infrastructure and their functionality status need to be determined in order to devise mechanisms and strategies of realizing the required synergy and interoperability towards achieving the target state specified in S2. For example, it is vital to specify the: existing technology infrastructure, legacy systems with respect to rapidly changing technologies; the extent to which IT infrastructure is shared; the extent of cohesion of existing e-government solutions; and the extent of the digital divide problem [24]. Therefore, S3.3 enables entities at any GE tier to implement e-government solutions (for achieving the target state specified in S2) that are consistent with existing, ongoing, or planned e-government solutions.

Question S3.3: What are the existing e-government solutions and ICT infrastructure, what is their functionality status, and what is their role in efforts achieving the target state specified in S2?
Instantiation: *Existing e-government solutions and infrastructure were being used by sister directorates to DRC*

S3.4. Finances. Finances available to facilitate the achievement of the target state of a given level of e-government maturity depend on the country-specific programmatic planning and partnership funding programme [24]. S3.4 indicates the need to specify the mode of funding and peculiarities that shape its availability to facilitate efforts towards achieving the target state specified in S2.

Question S3.4: What are the funding sources to facilitate efforts towards achieving the target state specified in S2, and what are the availability constraints?
Question S3.5: What is the timeframe for achieving the target state specified in S2 with respect to constraints implied by S3.1 to S3.4?
Instantiation: *S3.4: Since this was a field demo, the finances dimension was not considered comprehensively. S3.5: The duration to conduct the scoping demo was one week because the duration of the larger project that had to use scope specifications from SGEA was two months.*

S3.5. Time. Time available to achieve the target state of a given level of e-government maturity depends on: (a) the urgency required in resolving the challenge or embracing an opportunity of the GE; or (b) country-specific funding mechanisms and donor conditions in terms of financial planning periods or mode of sponsoring/funding as indicated in S3.4. Thus, S3.5 indicates the need to specify the timeframe for achieving the target state specified in S2 with respect to with respect to constraints implied by S3.1 to S3.4.

3.4   Determine Suitable Extent of Engagement in the GE (S4)

Defining the scope of enterprise architecture development involves determining all internal and extended units of an enterprise, units that are to be impacted by the architecture initiative, and units that are within and those that are outside the scope of the initiative [11]. SGEA adapts this principle using two aspects that are explored at different levels/steps due to contextual and understandability issues. The first aspect is discussed in Sect. 3.1, where all entities that are to be directly and indirectly affected by the problem and solution/opportunity experienced by an entity at any/each GE tier in Fig. 2 are specified under parameter S1.2.2. However, due to resource constraints, it may be difficult to afford accommodating all entities listed under S1.2.2.

Thus, the second aspect is accommodated in this dimension, which indicates the need to specify a fraction of those entities that can be engaged in efforts towards achieving particular milestones associated with the target state of a given level of e-government maturity (as specified in Sect. 3.2), with respect to contents in the resource envelop (as specified in Sect. 3.3). Also, given the complex contextual nature of the GE, it is imperative that the extent of intervention in developing the GE architecture is specified at each tier with respect to target state and resource envelop. This can take a top-down approach (from national-sector-unit levels) or bottom-up approach depending on output from dimensions in Sects. 3.1, 3.2 and 3.3. Sub-dimensions S4.1 to S4.3 apply as discussed below. Specific questions and field demo instances for S4.1 to S4.3 are provided in the text box below.

S4.1. National extent of engagement. This indicates the need to determine a fraction of existing sectors in a country and a fraction of partners at national, regional, and international levels that are relevant in achieving the target state of a given level of e-government maturity at national level.

> **Question S4.1:** Which sectors in a country and which partners at national, regional, and international levels are relevant in achieving the target state specified in S2?
> **Question S4.2:** Which existing and planned units/departments/agencies in a given sector, partner sectors/entities, and specific partner entities at national, regional, and international levels are relevant in achieving the target state specified in S2?
> **Question S4.3:** Which internal and external/partner entities of a given unit, and existing/planned business capabilities are relevant in achieving the target state specified in S2?
> **Instantiation:** *S4.1 & S4.2 do not apply due to the selection of DRC as indicated in S1.1.1.*
> *S4.3: All the three business capabilities (listed in S1.2) were considered in designing the e-government business architecture for DRC. However, the 8 sister/affiliate entities/directorates (listed in S1.2) could not be considered due to scope values chosen under S1 to S3*

S4.2. Sector level extent of engagement. This indicates the need to determine: a fraction of existing and planned units/departments/agencies that constitute a given sector, a fraction of partner sectors/entities, and a fraction of specific partner entities at national, regional, and international levels that are relevant in achieving the target state of a given level of e-government maturity at sector level.

S4.3. Unit level extent of engagement. This indicates the need to determine a fraction of internal and external/partner entities that constitute a given unit and a fraction of their existing and planned business capabilities that are relevant in achieving the target state of a given level of e-government maturity at unit level.

Specifying governance structures in the enterprise architecture process helps to: (a) exhaustively identify all relevant stakeholders because missing some key stakeholders may cause questioning of architecture deliverables; and (b) appropriately articulate responsibilities of key stakeholders and implied measures of engagement [10]. Thus, specifying entities to be considered at any/each GE tier in a given period of developing GE architecture helps to specify the roles and responsibilities of specific key stakeholder groups and the GE architecture governance team.

3.5   Determine Relevant Domains of the e-Government Architecture (S5)

Managerial architecture development challenges such as fuzzy strategies for actualizing enterprise goals [8] are addressed through specifying architecture domains that are relevant in delineating the implementation of each e-government implementation strategy/goal. An enterprise architecture comprises five architecture domains (i.e. business, data, application, technology, and security), but limitations in enterprise resources may not permit all the domains to be developed at once [11]. Hence the need to specify relevant architecture domains for guiding e-government implementations in a specific entity at any GE tier. S5 comprises sub-dimensions S5.1 and S5.2 as discussed below. Specific questions and field demo instances for S5.1 and S5.2 are provided in the text boxes below.

S5.1. Determine overall purpose of the GE architecture. This indicates the need to specify the purpose of the GE architecture in realizing the target state of a given level of e-government maturity (that was specified in Sect. 3.2). An enterprise architecture is an instrument for: (1) assessing impact of a strategy before actual implementation of a strategy; (2) specifying business and ICT requirements for realizing a transformation; (3) informing and contracting service providers of specific capabilities towards realizing the desired state; and (4) guiding decision making during a transformation [23]. Thus, specifying any/all of these purposes as the overall purpose of the GE architecture in achieving the target state of e-government helps to guide decision making on SGEA dimensions that are discussed in Sects. 3.3, 3.4, 3.6, and 3.7.

Question S5.1: What is specific purpose/role of the GE architecture in achieving the target state specified in S2?
Instantiation: *General purpose of the GE architecture is to specify business and ICT requirements for the e-government capability that enables features specified in S2.3.*

S5.2. Determine relevant domains at any/each GE tier that should be considered in designing the e-government architecture for achieving the target state of a given level of e-government maturity. Relevant domains are selected based on: the overall purpose of the GE architecture (specified in S5.1), the specified extent of engagement (in Sect. 3.4), resource envelop (in Sect. 3.3), desired target state of e-government maturity (in Sect. 3.2), the GE nature (in Sect. 3.1), and the focus area of each architecture domain. In determining the focus area of each architecture domain, the following definitions are derived based on TOGAF [11] and coded as follows.

S5.2.1. e-Government Business Architecture: specifies business capabilities (and their interrelationships) that are needed in a specific entity at any/each GE tier, so as to establish a responsive operational framework for delivering the full landscape of e-administration, e-citizen/e-service, and e-society services and to realize the e-government implementation and governance strategy.

S5.2.2. e-Government Data Architecture: specifies the logical and physical data capabilities that are needed in a specific entity at any/each GE tier to support S5.2.1.

S5.2.3. e-Government Application Architecture: specifies a suite of specific electronic solutions that need to be deployed/realized in a specific entity at any/each GE tier in order to provide agile e-administration, e-citizen/e-service, and e-society services services; and the interoperability implications of these solutions towards reliably supporting S5.2.1 and S5.2.2.

S5.2.4. e-Government Technology Architecture: specifies the suite of software and hardware capabilities that are needed in a specific entity at any/each GE tier to support S5.2.1 to S5.2.3 and S5.2.5.

S5.2.5. e-Government Security Architecture: specifies the range of all security-related mechanisms and provisions that need to be established in a specific entity at any/each GE tier to protect all resources in S5.2.1 to S5.2.4, so as to increase reliability and agility of e-government services.

> **Question S5.2:** How relevant is the GE business architecture, GE data architecture, GE application architecture, GE technology architecture, and GE security architecture in fulfilling the specific purpose of the GE architecture as specified in S5.1?
> **Instantiation:** *The e-Government Business Architecture domain was selected to show the cohesion of revenue with other departments, in order to determine the scope of access or use of the e-government solution and implied information exchanges.*

3.6 Determine Fitting Level of Detail of the e-Government Architecture (S6)

Understanding architecture descriptions is usually difficult because builders and users/consumers/implementers thereof are often different people [10]. To improve understandability of architecture descriptions, this dimension indicates the need to specify the appropriate level of detail that should be considered in designing the selected domains for the GE architecture (as specified in Sect. 3.5). The level of detail in enterprise architecture views can be at *high/vision level* – showing only major capabilities; *medium/moderately detailed level,* or *extremely detailed level* – showing capabilities in a fine granularity mode [11, 13]. Accordingly, the appropriate level of detail for specific domains of the GE architecture may vary across entities and GE tiers because it is determined basing on: the purpose of the domain (in Sect. 3.5), entities selected at each GE tier (in Sect. 3.4), size of resource envelop (in Sect. 3.3), target state of e-government maturity (in Sect. 3.2), and nature of a given GE tier (in Sect. 3.1). For example: at national level, level of detail for selected domains of the GE architecture could be vision level; at sector level it could be intermediate detailed level; and at unit level it could be extremely detailed level. The text box below shows questions that prompt deliberations and responses on this dimension.

> **Questions S6:** What is the appropriate level of detail that should be considered in designing the selected domains for the GE architecture with respect to choices arising from dimensions S1 to S5?
> **Instantiation:** *The appropriate level of detail for the e-government business architecture for DRC was vision level due to resource limitations and other aspects arising from choices in S1 to S5.*

3.7 Assess Risks in the Specified Scope of the e-Government Architecture (S7)

The scope of an architecture effort and the existence of several enterprise architecture approaches with limited procedural and consistent guidance, have a bearing on the complexity involved in creating and maintaining architecture models for heterogeneous and dynamic social systems [10]. Thus, risk analysis is not only a vital step in the enterprise architecture development process [11], but also crucial in specifying the scope of efforts on developing enterprise architectures. Sections 3.1 to 3.6 attempt to curb the complexity by providing procedural insights into scope dimensions S1 to S6 that require deliberation when planning architecture development efforts for the GE. Thus, basing on output or specified features for dimensions S1 to S6, this dimension indicates the need for two sub-dimensions S7.1 and S7.2 as discussed below.

S7.1. Evaluate possible risks in adopting the proposed concatenations. It is vital to assess risks of specific contextual values/features for each scope dimension and its corresponding parameters in Sects. 3.1 to 3.6. Thus, this sub-dimension indicates the need to: (a) compare possible risks of proposed scope dimensions with envisioned risks in adopting the concatenations of alternative contextual values/features for each scope dimension and corresponding parameters; and (b) determine corresponding risk mitigation strategies for the proposed scope and alternative scope and the implications of these on the resource envelop dimension in S3 (discussed in Sect. 3.3). This may result into feedback loops in dimensions S1 to S6 to address concerns raised from risk analysis.

S7.2. Choose and document the appropriate concatenation of values for all scope dimensions. This sub-dimension indicates the need to: (a) specify selected features/values for each scope dimension of the GE architecture from Sects. 3.1 to 3.6; (b) provide justification for each selected value/feature with respect to a given concatenation of scope values; and (c) document alternative concatenations of scope values along with reasons why they have been deemed inappropriate. The chosen scope values are represented using a scope specification matrix for the GE architecture. A template for this is provided in Appendix 1.

Questions S7.1: What are the possible risks of the proposed features of scope dimensions S1 to S6? What are the envisioned risks in adopting the concatenations of alternative contextual values/features for scope dimensions S1 to S6? What are the risk mitigation strategies for the proposed scope and alternative scope dimensions? What are the implications of the mitigation strategies on the resource envelop in dimension S3?

Questions S7.2: Based on findings in S7.1, what is the justification for each selected value/feature of a given scope dimension and the concatenation thereof? What is the reason why the alternative concatenations of scope values are inappropriate?

Instantiation: *Scoping started at unit tier, thus chances of not engaging key stakeholders in S4.1 and S4.2 were high, and this would affect the quality of the GE architecture in terms of interoperability.*

---

4   Key Insights from the Field Demo on SGEA

From Sect. 2.2 and text boxes in Sects. 3.1 to 3.7, this section highlights how the field demo helped to refine SGEA. SGEA was used at the start of a larger e-government project to specify and document scope dimensions of the project using the template in Appendix 1. This yielded the italicized instantiation phrases presented in text boxes in Sects. 3.1 to 3.7. The actual e-government architecture views that were obtained after the scope specifications can not be included herein because they are beyond the scope of SGEA and the focus of this paper. The field demo yielded three major aspects.

First, parameter S1.2.2 was originally part of dimension S4 (i.e. level of detail for the GE architecture views), but during the field demo it was noted that it has to be shifted to be part of dimension S1 to allow proper reasoning on parameters and values associated with dimensions S1 to S3. *Second,* parameter S7.2 had to be amended in order to ensure that values chosen under each parameter of a given scope dimension are justified. This amendment has been addressed as indicated in Sect. 3.7. *Third,* further research needs to be done to: (a) derive a comprehensive checklist that assesses the baseline maturity level of an entity with respect to all the three categories of e-government services as specified in Sects. 3.1 and 3.2; (b) derive a comprehensive documentation framework of findings resulting from the risk-mitigation analysis of scope dimensions as discussed in Sect. 3.7; and (c) derive additional standardized formats/templates (that serve as feeder templates to the scope specification matrix/template in Appendix 1) for documenting the specific features/values of the scope dimensions, sub-dimensions, and parameters across e-government projects. Aspects (a) to (c) arose as mechanisms to address use-related challenges that were faced when instantiating SGEA scope dimensions during the field demo.

It was noted that if each e-government project in a given entity at any/each GE tier has such a scope specification matrix, it helps to provide early insight into possible counts of integration and interoperability that should be accommodated between and among e-government projects.

5   Conclusion and Future Work

This paper demonstrates that scoping an e-government enterprise architecture development initiative is not a trivial task, but a multi-layered iterative procedure that considers a number of intertwined aspects in order to obtain a well thought out scope specification. This multi-layered synthesis, coined herein as SGEA, draws from the field of e-government maturity models and scoping insights from the field of enterprise architecture development. SGEA not only guides scoping, but allows a specific entity at any GE tier to implement e-government solutions that are interoperable with e-government solutions that are existing, ongoing, or planned within that entity or in other GE entities/tiers. A field demo validation of SGEA revealed the need to underpin some SGEA dimensions with supporting tools or frameworks to allow systematic assessments to be done. Accordingly, future developments of SGEA include development of a comprehensive context-specific checklist for assessing all categories of e-government maturity, standard templates for documenting values for scope dimensions and parameters, and a framework for supporting risk assessment of selected features or values of SGEA scope dimensions.

Acknowledgments. Authors appreciate the Systems manager at KCCA for participating in the field demo and anonymous reviewers of this paper.

Appendix 1. Template for the Scope Specification Matrix of SGEA

Dimension	Sub dimension	Parameters	Scope dimensions of the GE architecture
S1	S1.1	S1.1.1	

References

1. Bwalya, K.J., Mutula, S.: A conceptual framework for e-government development in resource-constrained countries. Inf. Dev. J. **32**(4), 1183–1198 (2016)
2. Alshehri, M., Drew, S.: Challenges of e-government services adoption in Saudi Arabia from an e-ready citizen perspective. World Acad. Sci. Eng. Technol. **66**, 1053–1059 (2010)
3. Her Majesty's UK Government.: UK government reference architecture Government ICT Strategy, version 1.0 (2012)
4. Ask, A.: The Role of Enterprise Architecture in Local e-Government Adoption. (Ph.D. thesis), Örebro University, Sweden. (2012)
5. Janssen, M., Kuk, G.: A complex adaptive system perspective of enterprise architecture in electronic government. In: 39th HICSS, 4–7 January, Kauai, Hawaii (2006)
6. Lankhorst, M., et al.: Enterprise Architecture at Work: Modelling, Communication, and Analysis. Springer, Heidelberg (2005). https://doi.org/10.1007/3-540-27505-3
7. Lauvrak, S., Michaelsen, V.M. Olsen, D.H.: Benefits and challenges with enterprise architecture: a case study of the Norwegian labour and welfare administration. In: NOKOBIT, vol. 25, no. 1 (2017). Bibsys Open Journal Systems
8. Banaeianjahromi, N., Smolander, K.: Understanding obstacles in enterprise architecture development. In: ECIS 2016 Proceedings at Association of Information Systems Electronic Library (AISeL). Research Papers 7 (2016)
9. Buckl, S., Schweda, C.M.: On the State-of-the-Art in Enterprise Architecture Management Literature. Technical report. Technische Unversität München (2011). https://mediatum.ub.tum.de/1120938
10. Lucke, C., Krell, S., Lechner, U.: Critical issues in enterprise architecting – a literature review. In: 16th AMCIS, Lima, Peru, 12–15 August (2010)
11. The Open Group Architecture Forum.: The Open Group Architecture Framework Version 9. Van Haren Publishing, Zaltbommel (2009)
12. Van't Wout, J., Waage, M., Hartman, H., Stahlecker, M., Hofman, A.: The Integrated Architecture Framework Explained: Why, What, How. Springer, Heidelberg (2010). https://doi.org/10.1007/978-3-642-11518-9. ISBN 978-3-642-11517-2
13. Zachman, J.A.: Excerpted from the zachman framework: a primer for enterprise engineering and manufacturing (2003). http://www.zachmaninternational.com
14. United Nations Division for Public Economics and Public Administration, American Society for Public Administration.: Benchmarking e-Government: A Global Perspective – Assessing the Progress of the UN Member States (2002). https://publicadministration.un.org/egovkb/portals/egovkb/documents/un/english.pdf. Accessed 15 Jan 2018
15. Moon, M.: The evolution of e-government among municipalities: rhetoric or reality. Public Adm. Rev. **62**, 424–433 (2002). https://doi.org/10.1111/0033-3352.00196
16. Hiller, J.S., Belanger, F.: Privacy Strategies for Electronic Government. PricewaterhouseCoopers, Arlington (2001)
17. Layne, K., Lee, J.: Developing fully functional e-government: a four stage model. Gov. Inf. Q. **18**, 122–136 (2001)
18. Deloitte & Touche: The citizen as customer, CMA Management, Electronic Government: Third International Conference, vol. 74, no. 10, pp. 58 (2001)
19. Baum, C., Maio, D.: Gartner's Four phases of e-Government model, Gartner's group (2000). http://aln.hha.dk/IFI/Hdi/2001/ITstrat/Download/Gartner_eGovernment.pdf
20. Coursey, D., Norris, F.D.: Models of e-Government: are they correct? An empirical assessment. Public Adm. Rev. **68**(3), 523–536 (2008)

21. Siau, K., Long, Y.: Synthesizing e-government stage models – a meta-synthesis based on meta-ethnography approach. Ind. Manag. Data Syst. **105**(4), 443–458 (2005)
22. Hevner, A.R.: A three cycle view of design science research. Scand. J. Inf. Syst. **19**(2), 87–92 (2007)
23. Op't Land, M., Proper, E., Waage, M., Cloo, J., Steghuis, C.: Enterprise Architecture: Creating Value by informed Governance. Springer, Berlin (2008). https://doi.org/10.1007/978-3-540-85232-2
24. OECD: Challenges for E-Government Development, 5th Global Forum on Reinventing Government, Mexico City (2003). http://unpan1.un.org/intradoc/groups/public/documents/un/unpan012241.pdf
25. Saha, P.: Advances in Government Enterprise Architecture. IGI Global Information Science Reference, Hershey (2008)
26. Wieringa, R.: Design Science Research Methodology: Principles and Practice. Tutorial/Masterclass on Design Science methodology. SIKS, Netherlands (2010)
27. Government of Uganda: Uganda Ministries (2016). http://www.statehouse.go.ug. Accessed 24 Feb 2018
28. Kampala City Council Authority. https://www.kcca.go.ug. Accessed 25 Feb 2018
29. Nakakawa, A., Namagembe, F.: Requirements for developing interoperable e-government systems in developing countries – a case of Uganda. Electron. Gov. Int. J. (2018, in press)
30. Heeks, R.B.: Success and failure in egovernment projects page, egovernment for development project. University of Manchester, UK (2008). http://www.egov4dev.org/success/definitions.shtml. Accessed 20 June 2018
31. Heeks, R.B.: Understanding and measuring e-government: international benchmarking studies. In: UNDESA Workshop on Understanding the Present and Creating the Future for E-Participation and E-Government, Budapest, Hungary (2006)
32. Heeks, R.B.: E-Government in Africa: Promise and Practice, I-Government Working Paper Series Paper No. 13, University of Manchester, Manchester, UK. (2002)
33. Hevner, A.R., March, S.T., Park, J., Ram, S.: Design science in information systems research. MIS Q. **28**(1), 75–105 (2004)

Author Index

Printed in the United States
By Bookmasters